Contributions to Finance and Accounting

The book series 'Contributions to Finance and Accounting' features the latest research from research areas like financial management, investment, capital markets, financial institutions, FinTech and financial innovation, accounting methods and standards, reporting, and corporate governance, among others. Books published in this series are primarily monographs and edited volumes that present new research results, both theoretical and empirical, on a clearly defined topic. All books are published in print and digital formats and disseminated globally. This book series is indexed in Scopus.

Peter Brusov • Tatiana Filatova

Efficiency of Investment Projects

Modern Investment Models and Their Applications

Springer

Peter Brusov
Financial University under the
Government of the Russian Federation
Moscow, Russia

Tatiana Filatova
Financial University under the
Government of the Russian Federation
Moscow, Russia

ISSN 2730-6038 ISSN 2730-6046 (electronic)
Contributions to Finance and Accounting
ISBN 978-3-032-01027-8 ISBN 978-3-032-01028-5 (eBook)
https://doi.org/10.1007/978-3-032-01028-5

This Springer imprint is published by the registered company Springer Nature Switzerland AG
The registered company address is: Gewerbestrasse 11, 6330 Cham, Switzerland

Dedicated to the eternal love
that touched us with its wing
and made us happy,
and without which life is meaningless.

Preface

Existing investment models are unable to correctly assess the efficiency of investments. This is due to both the lack of self-consistent models based on first principles and the inability of correctly assessing the discount rate. The monograph is devoted to the description of innovative investment models created by the authors recently and well tested in practice, and their application. Investment models with debt repayment at the end of the project and with uniform debt repayment have been developed. The effectiveness of the investment project is considered from two perspectives: the owners of equity and debt and the equity holders only. These models were modified for the conditions of real implementation of investment models (with frequent payments of income tax and interest on debt both at the end of periods and as advance payments, variable payments, etc.). The influence of these types of payments on the efficiency of investment projects is studied. The efficiency of an investment project is considered from two points of view: the owners of equity and debt, as well as only the equity. The latest versions of two main theories of capital structure (Brusov–Filatova–Orekhova (BFO) and Modigliani–Miller (MM)) are described, allowing to correctly assess the discount rate when assessing the efficiency of investment projects (long term as well arbitrary duration). The practical application of the created investment models is described, as well as ratings of investment models.

Krondurg, Moscow, Russia — Peter Brusov
Krondurg, Moscow, Russia — Tatiana Filatova
9 May 2025

Competing Interests The authors have no competing interests to declare that are relevant to the content of this manuscript.

Contents

Part II Capital Structure Theories

About the Authors

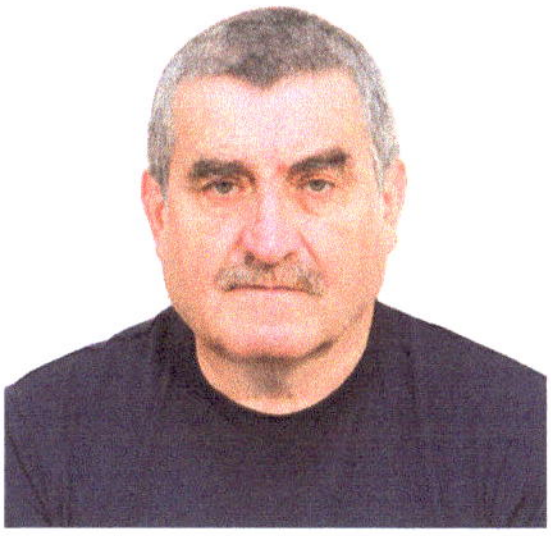

Peter Brusov is professor at the Financial University under the Government of the Russian Federation in Moscow (Russia). He is an expert in financial management, corporate finance, investments, ratings, taxation, and business valuation. He is a co-author of the famous Brusov–Filatova–Orekhova theory, which replaced the well-known Modigliani–Miller theory. As well he is a co-author of the CAPM 2.0 model, which generalized the famous CAPM model for accounting the financial risk along with business risk. He is the author of innovative investment models that have found application in real investment practice.As well he developed a new approach to business valuation and company value.

He has been a visiting Professor of Northwestern University (USA), Cornell University (USA), Osaka City University (Japan), and National Chung Cheng University (Taiwan) among other places. He is the author of over 550 research publications including 9 monographs, numerous textbooks, and articles.

Tatiana Filatova is professor at the Financial University under the Government of the Russian Federation in Moscow (Russia). She is an expert in financial management, corporate finance, investments, ratings, taxation, and business valuation. Tatiana Filatova is a co-author of the famous Brusov–Filatova–Orekhova theory, which replaced the well-known Modigliani–Miller theory. As well she is a co-author of the CAPM 2.0 model, which generalized the famous CAPM model for accounting the financial risk along with business risk. She is the author of innovative investment models that have found application in real investment practice. As well she developed a new approach to business valuation and company value.

In the past 20 years, she has been Dean of the faculties of financial management, management, state and municipal government at the Financial University. Tatiana Filatova is the author of over 300 research publications including 8 monographs, numerous textbooks, and articles.

Part I
Efficiency of Investment Projects

Chapter 1
Introduction

Existing investment models are unable to correctly assess the efficiency of investments. This is due to both the lack of self–consistent models based on first principles and the inability of correctly assessing the discount rate. The innovative investment models created by the authors recently were modified for the conditions of real implementation of investment models (with frequent payments of income tax and interest on debt both at the end of periods and as advance payments, variable payments, etc.) (Brusov & Filatova 2021, 2022b, 2022c, 2023, 2025a, 2025b; Brusov et al. 2020, 2021a, 2021b, 2022a 2022b, 2022c, 2023a, 2023b, 2023c, 2024a, 2024b, 2024c). The influence of these types of payments on the efficiency of investment projects is studied. The efficiency of an investment project is considered from two points of view: the owners of equity and debt, as well as only the equity. The latest versions of two main theories of capital structure (Brusov–Filatova–Orekhova (BFO) and Modigliani–Miller (MM)) are described, allowing to correctly assess the discount rate when assessing the efficiency of investment projects (long–term as well arbitrary duration). The practical application of the created investment models is described, as well as ratings of investment models.

The structure of the book is as follows. **The first part** is Introduction, where the main performance indicators of the investment project are described and compared. **The second part** is devoted to description of the two main theories of capital structure (the Brusov–Filatova–Orekhova (BFO) theory and its perpetual limit—the Modigliani–Miller (MM) theory), which allow calculating all financial indicators of a company of any age (BFO Theory) or perpetual (MM Theory). The recent modifications of these theories, taking into account the following conditions of the real functioning of the company: variable income; inflation; frequency and method of paying income tax (advance payments or payments at the end of periods) are described and discussed in detail. This allows to correctly calculate one of the most important parameter of valuation of the efficiency of investment projects–discount rate.

In **the third part** we describe the modern investment models, created by the authors and well tested in the real economy. These models used by us in subsequent

P. Brusov, T. Filatova, *Efficiency of Investment Projects*, Contributions to Finance and Accounting, https://doi.org/10.1007/978-3-032-01028-5_1

parts for investigation of different problems of investments, such as influence of debt financing, leverage level, taxing, project duration, method of financing, and some other parameters on efficiency of investments and other problems. In the Chap. 3 the investment models with debt repayment at the end of the project and their application are considered while in the Chap. 4 the investment models with uniform debt repayment and their application are considered.

In **the fourth part** the modern investment models, created by the authors, have been modified for frequent payments of tax on profit and of interest on debt payments at the ends of periods as well as by advance payments.

The **fifth part** considers portfolio investing. It consists of two chapters. Chapter 7 describes in detail the general theory of the portfolio, including the Markowitz and Tobin portfolios. Chapter 8 is devoted to describing portfolio investing in bonds.

The **sixth part** describes the applications of two main theories of capital structure (Brusov–Filatova–Orekhova (BFO) and Modigliani–Miller (MM)) for modification of methodology of project ratings.

In the **seventh part** the new effects in investment, which describe the unusual features, which could be important.

The **eighth part considers the** applications of developed investment models in financial and nonfinancial spheres.

The conclusions are described in **ninth part**.

References

Brusov, P., & Filatova, T. (2021). The Modigliani–Miller theory with arbitrary frequency of payment of tax on profit. *Mathematics, 9*(11), 1198. https://doi.org/10.3390/math9111198

Brusov, P., & Filatova, T. (2022a). Influence of method and frequency of profit tax payments on company financial indicators. *Mathematics, 10*(14), 2479. https://doi.org/10.3390/math10142479

Brusov, P., & Filatova, T. (2022b). Generalization of the Brusov–Filatova–Orekhova theory for the case of variable income. *Mathematics, 10*(19), 3661. https://doi.org/10.3390/math10193661

Brusov, P., & Filatova, T. (2023). Capital structure theory: Past, present, future. *Mathematics, 11*(3), 616. https://doi.org/10.3390/math11030616

Brusov, P., & Filatova, T. (2025a). *Business valuation and company value: New qualitative approaches considering business and financial risks* (850 pp). Springer Nature.

Brusov, P., & Filatova, T. (2025b). Qualitatively new approach to business valuation and company value. *Mathematics, 13*(1), 80. https://doi.org/10.3390/math13010080

Brusov, P. N., Filatova, T. V., Orekhova, N. P., Kulik, V. L., Chang, S. I., & Lin, Y. C. G. (2020). Modification of the Modigliani–Miller theory for the case of advance payments of tax on profit. *Journal of Reviews on Global Economics, 9*, 257–267.

Brusov, P., Filatova, T., Chang, S. I., & Lin, G. (2021a). Innovative investment models with frequent payments of tax on income and of interest on debt. *Mathematics, 9*, 1491.

Brusov, P., Filatova, T., Orekhova, N., Kulik, V., Chang, S. I., & Lin, G. (2021b). Generalization of the Modigliani–Miller theory for the case of variable profit. *Mathematics, 9*, 1286.

Brusov, P., Filatova, T., & Orekhova, N. (2022a). *Generalized Modigliani–Miller theory: Applications in corporate finance, investments, taxation and ratings* (362 pp). Springer Nature.

Brusov, P., Filatova, T., & Kulik, V. (2022b). Benefits of advance payments of tax on profit: Consideration within the Brusov–Filatova–Orekhova (BFO) theory. *Mathematics, 10*, 2013.

Brusov, P., Filatova, T., Orekhova, N., Kulik, V., Chang, S. I., & Lin, G. (2022c). The generalization of the Brusov–Filatova–Orekhova theory for the case of payments of tax on profit with arbitrary frequency. *Mathematics, 10*, 1343.

Brusov, P., Filatova, T., & Orekhova N. (2023a). *The Brusov–Filatova–Orekhova theory of capital structure: Applications in corporate finance, investments, taxation and ratings* (769 pp). Springer Nature.

Brusov, P., Filatova, T., Veniamin, K, Chang, S. I., Lin, G., Chang, L. M. (2023b). *Precision finance: Capital structure theories approach reality.* In Proceedings of the 23rd international conference on electronic business (ICEB–2023).

Brusov, P., Filatova, T., Veniamin, K, Chang, S. I., Lin, G., Chang, L. M. (2023c). *Can CAPM (Capital Asset Pricing Model) accurately value assets?* In Proceedings of the 23rd international conference on electronic business (ICEB–2023).

Brusov, P. N., Filatova, T. V., & Kulik, V. L. (2024a). Capital asset pricing model (CAPM) 2.0: Account of business and financial risk. *Finance: Theory and Practice, 28*(2), 128–142. https://doi.org/10.26794/2587-5671-2024-28-2-128-142

Brusov, P. N., Filatova, T. V., & Kulik, V. L. (2024b). Application of the company's "Golden age" effect in the economic practice. *Finance: Theory and Practice, 28*(3), 61–83. https://doi.org/10.26794/2587-5671-2024-28-3-61-83

Brusov, P. N., Filatova, T. V., & Kulik, V. L. (2024c). Incorporating CAPM into capital structure theories: Accounting for business and financial risks. *Finance: Theory and Practice, 28*(5), 83–108. https://doi.org/10.26794/2587-5671-2024-28-5-83-108

Chapter 2
Evaluation of the Efficiency of Investment Projects

Investments play a crucial role in economy and finance. Investments in tangible and intangible assets are a necessary condition for structural adjustment and economic growth. They provide the enhancement of existing basic funds and industries and the creation of new ones. The role of investment is increased many times at the current stage. In this respect, the role of the evaluation of the efficiency of investment projects, which allows for the realization of the most effective projects in the context of scarcity and limited investment resources, increases. Since virtually most investment projects use debt financing, the study of the influence of capital structure and debt financing on the efficiency of investment projects and determining the optimal capital structure is especially important at the present time. This is why in spite of the fact that a lot of different types of investment models have been developed—stochastic, dynamics, investment banking valuation models, etc.—one of the main problem, which has been discussed during the last few decades, is the impact of debt financing on the efficiency of investment projects and on the investment decisions of companies Brusov & Filatova 2025; Brusov et al. 2023a, 2023b, 2024a, 2024b, 2024c).

2.1 Types and Kinds of Investments

Investments can be classified according to various criteria:

- by investment objects (real and financial);
- by investor participation in the investment process (direct and indirect);
- by investment period (long-term, medium-term, short-term);
- by regional criteria (internal and external);
- by forms of ownership of investment resources (private, public, foreign, joint).

P. Brusov, T. Filatova, *Efficiency of Investment Projects*, Contributions to Finance and Accounting, https://doi.org/10.1007/978-3-032-01028-5_2

2.2 Evaluation of the Efficiency of Investment Projects

The basis of the methods of efficiency analysis is the comparison of investment projects with alternative investment opportunities in terms of maximizing income and company value with a comparable degree of risk.

There are two groups of parameters characterizing the efficiency of investment projects: those based on discounted estimates and those based on accounting estimates. The first group of parameters includes: net present value, return on investment index, internal rate of return, modified internal rate of return, discounted payback period. The second group of parameters includes: payback period and investment efficiency coefficient. In this case, when calculating cash flows, one should not take into account those that do not change due to the adoption of a given investment decision (cash flows related to the past period; costs that would have been incurred regardless of the implementation of the investment project).

2.2.1 *Payback Period*

The payback period (PP) is the minimum period during which the projected total income will exceed the initial investment. The faster a project can recoup its initial costs, the more preferable it is. PP analyzes the project's income and the rate of its receipt, rather than the project's profitability and its total profit. The calculation of the payback period depends on the uniformity of the expected income.

If the income is distributed uniformly, then

$$PP = \frac{I}{D}, \tag{2.1}$$

where I is the investment costs, D is the annual income of the project.

In the case of nonuniform distribution of income by years, the payback period is found as follows:

$$PP = \min n, \text{at which} \sum_{i=1}^{n} P_i = I \tag{2.2}$$

For a more accurate calculation of the payback period PP, it is necessary to take into account the fractional part of the year, assuming uniform distribution of income also during the year.

The main advantages of the PP method are the simplicity of calculations and taking into account the liquidity of the project. Due to this, by cutting off the most dubious and risky projects, in which the main cash flows occur at the end of the period, the PP method is used as a simple method for assessing investment risk. It is used for small firms with insignificant cash turnover, as well as for express analysis of projects in conditions of resource shortage.

The disadvantages of the PP method include the following:

1. the choice of the payback period can be subjective;
2. the method does not take into account the profitability of the project beyond the payback period and cannot be used when comparing options with the same payback periods but different lifespans; the method does not take into account the time value of money;
3. the method is not suitable for evaluating projects related to fundamentally new products;
4. the accuracy of calculations using this method largely depends on the frequency of dividing the project lifespan into planning intervals.

2.2.2 *Discounted Payback Period (DPP)*

If the time factor is taken into account when calculating the payback period of an investment project, i.e. discounting of cash flows is performed (usually at the WACC rate), such a method of evaluating the effectiveness of an investment project is called the method of evaluating using the discounted payback period of an investment project. It is based on the application of the following formulas, in which,

$$DPP = \min n, \text{ at which} \sum_{i=1}^{n} \frac{P_i}{(1 + WACC)^i} = I \tag{2.3}$$

and, as in the case of the method of assessment using the payback period of an investment project PP, DPP means the minimum period during which the projected total income of the project will exceed the initial investment.

The DPP method is a more advanced method for evaluating investment projects compared to the PP method. It takes into account a larger number of cash flows from capital investments, has a clear criterion for the acceptability of projects (when using DPP, a project is accepted if it pays for itself during its life), and determines the liquidity of the project.

The main disadvantage of the DPP method is that it does not take into account all cash flows after the completion of the project.

2.2.3 *Investment Efficiency Ratio*

Accounting rate of return (Accounting Rate of Return—ARR) (investment efficiency ratio). The ratio calculated using this method is similar to the return on capital ratio. It shows what part of the investment costs is reimbursed in the form of profit within 1 year. By comparing the calculated rate of return with the minimum or average level of profitability, the investor can come to a conclusion about the advisability of further analysis of this investment project.

$$ARR = \frac{PN}{NR}, \tag{2.4}$$

where PN is net profit (profit minus deductions to the budget), NR is investment costs.

There are several ways to define the concepts of "income" and "invested capital". Different estimates of income may or may not include financial costs, depreciation and taxes. In ARR analysis, income is defined as "earnings before interest and taxes", which includes depreciation. ARR is usually used in two versions depending on the definition of invested capital, which can include either the initial invested capital or the average invested capital over the life of the investment.

The accounting rate of profit is calculated using the formula

$$ARR = \frac{PB}{C}, \tag{2.5}$$

where PB is the average balance sheet profit per year; C is the initial invested capital.

The advantage of the ARR method is the simplicity of calculations and assessment of the profitability of the project.

The disadvantages of the ARR method are as follows:

1. the time value of money is not taken into account. The return on investment is calculated as the average reported profit, although income is earned over different periods of time and may change;
2. book profit, rather than cash flows, is used to assess the profitability of projects. There are many ways to calculate book profit, which makes it possible to manipulate the ARR indicator. Inconsistencies in the calculation of profit lead to significant differences in ARR values. Often, these differences are the result of accounting;
3. income from the disposal of old assets replaced by new ones is not taken into account;
4. the possibility of reinvesting the income received is ignored.

When the calculation with average invested capital is used, the initial costs and the residual value of the investment are averaged to reflect the value of assets related to each other over the entire useful life of the investment. The higher the residual value of the investment, the larger the denominator in the ARR formula becomes and the smaller the calculated rate of return itself.

2.2.4 Net Present Value, NPV

Net Present Value (NPV) is the difference between the discounted amount of revenue from a project and the amount of costs incurred during the project.

The decision-making criterion in the NPV method is the same for all types of investments and companies: if the NPV value is positive, the investment project should be accepted, and vice versa.

NPV is calculated using the formula

$$NPV = \sum_{i=1}^{n} \frac{P_i}{(1+r)^i} - I, \tag{2.6}$$

where P_i is the income for the i-th year of the investment project; r is the cost of capital attracted for the investment project; I is the amount of the initial investment.

If the project involves not a one-time investment, but a consistent investment of financial resources over t years, then the formula for calculating NPV will look like this:

$$NPV = \sum_{i=1}^{n} \frac{P_i}{(1+r)^i} - \sum_{i=0}^{m} \frac{I_i}{(1+r)^i}.. \tag{2.7}$$

A positive NPV value means that the current value of income exceeds the current value of costs, therefore, an increase in the welfare of investors should be expected.

The advantages of the NPV method include the fact that this method:

1. demonstrates whether, and if so, by how much, the welfare of investors increases;
2. takes into account the time value of money.

The disadvantages of the NPV method include the fact that when using it:

1. it is difficult to objectively estimate the rate of return (discount rate). Its choice is a crucial moment in NPV analysis, as it determines the relative value of cash inflows and outflows occurring at different periods of time. The rate used in assessing NPV should reflect the rate of return taking into account risk;
2. it is difficult to estimate such uncertain parameters as moral and physical depreciation of fixed assets, changes in the company's activities. This can lead to an incorrect assessment of the service life of fixed assets;
3. the NPV value does not adequately reflect the result when comparing projects with different initial costs with the same value of net present cash flows; projects with a higher net present value and a long payback period and projects with a lower net present value and a short payback period.

2.2.5 Profitability Index (Profitability) of Investments

Profitability Index (Profitability Index—PI) is a modification of the NPV method, since it also uses information on discounted cash flows. However, instead of finding the difference between the initial costs and the present value of future income, PI determines the ratio of these indicators and is calculated using the following formula:

$$PI = \frac{\sum_{i=1}^{n} \frac{P_i}{(1+r)^i}}{I}. \tag{2.8}$$

The PI index shows the relative profitability of the project, or the discounted value of cash flows from the project per unit of investment. If PI > 1, then the project's profitability is higher than that required by investors, and the project is attractive. Thus, this method can be used to rank projects with different costs and expected lifespans. This is an advantage of this method compared to the NPV method.

The choice of the discount rate when calculating NPV and PI has a significant impact on the final result of the calculation, and therefore on its interpretation. The discount rate depends on the rate of inflation, the minimum real rate of return and the degree of investment risk. The minimum rate of return is considered to be the lowest guaranteed level of return on the capital market, i.e. the lower limit of the cost of capital.

2.2.6 Internal Rate of Return of the Project

The internal rate of return of the project (IRR) (internal rate of return, internal rate of return, internal payback) is the discount rate at which the net present value of the project will be zero, i.e. the present value of expected income is equal to the present value of costs (taking into account the reimbursement of investments):

$$IRR = r, \text{at which } NPV(r) = 0. \tag{2.9}$$

IRR is calculated by iteratively selecting the discount rate values when calculating the NPV of the project. IRR is found from the equation:

$$NPV = \sum_{i=1}^{n} \frac{P_i}{(1 + IRR)^i} - I = 0, \tag{2.10}$$

or

$$\sum_{i=1}^{n} \frac{P_i}{(1 + IRR)^i} = I. \tag{2.11}$$

The linear interpolation method and the chord and tangent method are used to solve this equation. Two values of the discount factor $r_1 < r_2$are selected in such a way that the function $NPV = f(r)$changes its value in the interval $[r_1, r_2]$. Then apply the formulas:

$$IRR = r_1 + \frac{f(r_1)}{f(r_1) - f(r_2)}(r_2 - r_1), IRR' = r_1 + \frac{f(r_1)}{f'(r_1)}, \tag{2.12}$$

where IRR is the value of the internal rate of return that minimizes the positive value of the NPV indicator; IRR′ is the value of the internal rate of return that maximizes the negative value of the NPV indicator.

The IRR indicator characterizes:

1. the lower guaranteed level of profitability of an investment project. Taking into account the investment risk of the project, if the IRR exceeds the average cost of capital (for example, the rate on long-term bank loans) in a given industry, this project can be considered attractive;
2. the maximum rate of payment for attracted sources of financing for the project, at which the latter remains break-even (for example, this can be the maximum rate on loans when assessing the effectiveness of total investment costs);
3. the maximum level of return on investment, which can be a criterion for the feasibility of additional investments in the project.

The advantages of the IRR method include its objectivity, information content, independence from the absolute size of investments, as well as the ability to:

1. assess the relative profitability of the project;
2. compare projects with different levels of risk (projects with a higher level of risk should have a higher internal rate of return).

However, the IRR method also has its disadvantages, namely:

1. possible subjectivity in choosing the standard return;
2. high dependence on the accuracy of the assessment of future cash flows.

2.2.7 MIRR, Modified Internal Rate of Return

MIRR is a modified version of IRR to remove possible uncertainties in the calculations.

Like IRR, MIRR characterizes the discount rate at which the total present value of income from investments made is equal to the cost of these investments. Since the method used to calculate IRR sometimes leads to uncertainties, the modified IRR value is calculated somewhat differently.

MIRR is the internal rate of return adjusted for the reinvestment rate.

From a practical point of view, the most significant drawback of the internal rate of return is the assumption of a constant interest rate when discounting the cash flows generated by the investment. For projects providing rates of return close to the firm's hurdle rate, problems with reinvestment do not arise, since it is quite reasonable to assume that there are many investment options that generate a return whose rate is close to the cost of capital. However, for investments that provide very high or very low rates of return, the assumption that new cash inflows must be reinvested can distort the true return on the project. The concept of reinvestment-adjusted internal rate of return was introduced to counter this distortion inherent in the traditional IRR. Despite its cumbersome name, the reinvestment-adjusted internal rate of return, or MIRR, also known as the modified internal rate of return, is actually much easier to calculate manually than the IRR. And this is precisely because of the assumption made about reinvestment.

Let us give the procedure for calculating the modified internal rate of return MIRR:

1. Calculate the total discounted value of all cash outflows and the total accrued value of all cash inflows.

   Discounting is performed at the price of the project financing source (cost of attracted capital, financing rate or required rate of return on investment (Capital Cost, CC or WACC), i.e. at the hurdle rate. Accrual is performed at an interest rate equal to the level of reinvestment.

   In other words:

   (a) All values of income generated by investments are reduced to the end of the project. For reduction, a rate equal to (WACC) is used.
   (b) All investments and reinvestments are reduced to the beginning of the project. For reduction, the discount rate is used.

   The accrued value of inflows is called the net terminal value (NTV).
2. Set the discount rate, taking into account the total present value of outflows and the terminal value of inflows. The discount rate that aligns the present value of investments (PV) with their terminal value is called MIRR, i.e. MIRR is defined as the rate of return at which all expected revenues, discounted to the end of the project, have a present value equal to the value of all required costs.

   Formula for calculating the modified internal rate of return (MIRR):

$$\sum_{k=0}^{n} \frac{I_k}{(1+i)^k} = \frac{\sum_{k=1}^{n} CF_k (1+d)^{n-k}}{(1+MIRR)^n} \tag{2.13}$$

   where CF_k is the cash inflow in period k = 1, 2, … n; I_k is the cash outflow in period k = 0, 1, 2, … n (in absolute value); i is the hurdle rate (discount rate); d is the reinvestment rate (interest rate based on the possible income from reinvestment of the received positive cash flows or the rate of return on reinvestment), WACC can be used as d; n is the number of periods.

On the left side of the formula is the value of investments (capital investments) discounted at the price of capital, and on the right side is the accrued value of cash receipts from the investment at a rate equal to the reinvestment rate.

Note that the MIRR formula makes sense if the terminal value of inflows exceeds the sum of discounted cash outflows (the inflow of money is greater than its outflow).

$$MIRR = \left(\frac{\sum_{k=1}^{n} CF_k (1+d)^{n-k}}{\sum_{k=0}^{n} \frac{I_k}{(1+i)^k}} \right)^{1/n} - 1. \quad (2.14)$$

The MIRR criterion always has a single value and can be used instead of the IRR indicator to evaluate projects with unusual cash flows. The project is acceptable to the initiator if the MIRR is greater than the hurdle rate (the price of the source of financing).

Using MIRR, compound interest on annual cash flows is calculated at a more appropriate interest rate. Each cash flow and its interest are carried forward to the end of the investment. Then the resulting future values are summed up and the result is compared with the original investment. Instead of determining the internal rate of return by considering several cash inflows, we calculate the IRR for one expense and one cash inflow.

Using MIRR instead of IRR always muffles the effect of investments. Not very profitable investments for which the rates of return are below the hurdle rate or the reinvestment rate will always look better using MIRR than IRR, since in the first case the cash flows will bring higher returns than in the second. On the other hand, particularly profitable investments, for which the rate of return is higher than the hurdle rate, will have a lower MIRR for the same reason.

The MIRR method does not have the problem of multiple determinations of the internal rate of return as the IRR method.

In practice, the MIRR indicator is rarely used, which cannot be considered justified.

2.3 Internal Rate of Return

Internal rate of return is one of the most important parameters for assessing the effectiveness of investment projects. In this paragraph, we will consider this parameter in as much detail as possible. This is important for understanding its role in assessing the effectiveness of investment projects.

2.3.1 The Concept of Internal Rate of Return

Internal rate of return is one of the most important parameters for assessing the effectiveness of investment projects. In this paragraph, we will consider this parameter in as much detail as possible. This is important for understanding its role in assessing the effectiveness of investment projects.

Let us consider in detail one of the most important concepts in investment theory—the internal rate of return, and also investigate the dependence of net present income (*NPV*) on the discount rate.

The investment process described by the financial flow has the following form:

$$CF = \{(t_0, C_0), (t_1, C_1), (t_2, C_2), \dots, (t_n, C_n)\}. \tag{2.15}$$

The value of C_k (K = 0, 1, ..., n) represents the balance of investment costs and net income for the *k*–th period, updated at the end of this period. A negative payment in (2.16) means that investment costs exceeded net income, a positive payment means that net income exceeded investment costs.

By means of the reduction rate *i*, we calculate the current value of the flow (2.16), called in this context net present value (NPV):

$$NPV(i) = \frac{C_0}{(1+i)^{t_0}} + \frac{C_1}{(1+i)^{t_1}} + \frac{C_2}{(1+i)} + \dots + \frac{C_n}{(1+i)^{t_n}} \tag{2.16}$$

If *NPV* has a negative value, it means that the income does not recoup the costs at the accepted rate of return *i*.

So, let's consider the dependence of *NPV* on the conversion rate (the accepted rate of return) *i*. First, let's pay attention to the simplest typical case when all costs are carried out at the initial moment, and then the investor begins to receive income. Let the financial flow have the form:

$$CF = \{(0, -K), (1, C_1), (2, C_2), \dots, (n, C_n)\}, \tag{2.17}$$

where $K > 0$—initial investment, all payments C_k, $k = 1, 2, \dots, n$ are non–negative, and there is at least one positive among them.

Then

$$NPV(i) = -k + \sum_{k=1}^{n} \frac{C_k}{(1+i)^k} \tag{2.18}$$

For $i > -1$, the net reduced income *NPV(i)* is a decreasing function of the reduction rate *i* (Fig. 2.1).

On the one hand, with i tending to -1 (on the right), each summand $\frac{C_k}{(1+i)^k}$ with a positive C_k tends to infinity, which means that the sum (2.19) tends to infinity.

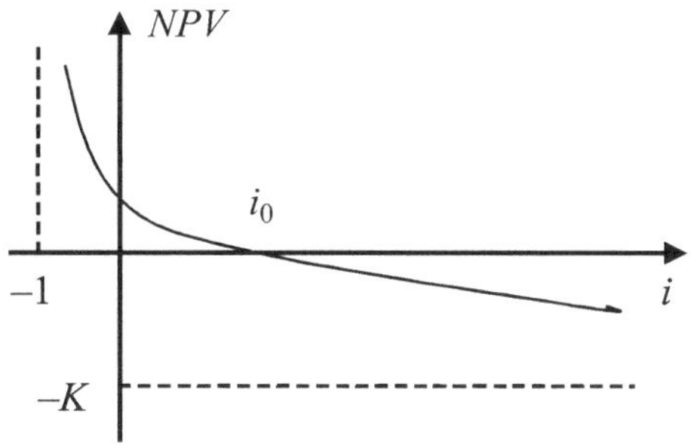

Fig. 2.1 Dependence of net present income on the discount rate

Hence, $PV(i) > 0$ for i is sufficiently close to -1. On the other hand, with unlimited growth of i, all summands $\frac{C_k}{(1+i)^k}$ tend to zero, and the sum (2.19) tends to $-K$. Hence, $PV(i) < 0$ for sufficiently large i. Thus, for $i > -1$, the continuous function $PV(i)$, decreasing, changes the sign from plus to minus. Therefore, $PV(i)$ turns to zero for some $i = i_O$. The value of i_O is called the internal rate of return of the payment flow (2.19).

The internal rate of return serves as the boundary of the interest rates for which the project has a positive and negative applied value: if $i >$ i0, then $PV(i) < 0$, if $i <$ i0, then $PV(i) > 0$.

Since $PV(0) = \sum_{k=1}^{n} C_k - K$, $i_O > 0$ if and only if $\sum_{k=1}^{n} C_k > K$, that is the net amount of income exceeding the initial investment.

Example 2.1 Find the internal rate of return of the flow:

$$CF = \{(0, -8000), (1, 6000), (2, 5000)\}. \tag{2.19}$$

Let's make the equation:

$$-8000 + 6000(1+i)^{-1} + 5000(1+i)^{-2} = 0.$$

By replacing x = (1 + i)$^{-1}$, we get the quadratic equation:

$$5000 \cdot x^2 + 6000 \cdot x – 8000 = 0, \text{ or } 5 \cdot x^2 + 6 \cdot x – 8 = 0.$$

Solving it, we receive

$$(1+i)^{-1} = \frac{-3 \pm \sqrt{9+40}}{5} = \frac{-3 \pm 7}{5}.$$

Of the two roots -2 and 0.8, the second one suits us. Solving the equation

$$(1+i)^{-1} = 0.8,$$

we receive $i \approx 0.25$.

Consequently, the internal rate of return of the flow (1.124) is 25%.

In general, it is said that the financial flow (1.120) has an internal rate of return if the equation

$$NPV(i) = 0 \tag{2.20}$$

has a unique solution $i_0 > -1$; the solution of equation (2.17) is called the internal rate of return of the flow (2.16).

Consider an arbitrary financial flow

$$CF = \{(t_0, C_0), (t_1, C_1), (t_2, C_2), \ldots, (t_n, C_n)\}, \tag{2.21}$$

in which negative payments can alternate with positive ones, assuming, however, that it has an internal rate of return. We will additionally assume that $C_0 < 0$ and $C_n > 0$. We will represent the CF stream as the difference of two non–negative streams A and B:

$$CF = A - B, \tag{2.22}$$

where

$$\begin{aligned} A &= \{(t_0, A_0), (t_1, A_1), (t_2, A_2), \ldots, (t_n, A_n)\}, \\ B &= \{(t_0, B_0), (t_1, B_1), (t_2, B_2), \ldots, (t_n, B_n)\}, \end{aligned} \tag{2.23}$$

Supposing $A_k = \max(0, C_k)$, $B_k = \max(0, -C_k)$. Let us consider τ_A and τ_B as average terms of flows A and B.

$$\frac{A_0}{(1+i)^{t_0}} + \frac{A_1}{(1+i)^{t_1}} + \ldots + \frac{A_2}{(1+i)^{t_n}} = \frac{A_0 + A_1 + \ldots + A_n}{(1+i)^{\tau_A}} \tag{2.24}$$

$$\frac{B_0}{(1+i)^{t_0}} + \frac{B_1}{(1+i)^{t_1}} + \ldots + \frac{B_2}{(1+i)^{t_n}} = \frac{B_0 + B_1 + \ldots + B_n}{(1+i)^{\tau_B}} \tag{2.25}$$

Therefore, the eq. $PV(CF, i) = 0$ acquires the form:

$$\frac{A_0 + A_1 + \ldots + A_n}{(1+i)^{\tau_A}} = \frac{B_0 + B_1 + \ldots + B_n}{(1+i)^{\tau_B}} \tag{2.26}$$

From here

$$(1+i)^{\tau_A - \tau_B} = \frac{A_0 + A_1 + \ldots + A_n}{B_0 + B_1 + \ldots + B_n} \tag{2.27}$$

Consequently,

$$i = \left(\frac{A_0 + A_1 + \ldots + A_n}{B_0 + B_1 + \ldots + B_n}\right)^{\frac{1}{\tau_A - \tau_B}} - 1 \tag{2.28}$$

The formula (2.29) can be used to approximate the internal rate of return.

Example 2.2 Find the internal rate of return of the flow (2.20) using the formula (2.29).

Suppose,

$$A = \{(0,0), (1,2000), (2,3000)\},$$
$$B = \{(0,4000), (1,0), (2,0)\}.$$

Evidently, $\tau_B = 0$. We will find approximately the average term of the flow A as a weighted sum of the moments of payments:

$$\tau A \approx \frac{2000}{2000 + 3000} \cdot 1 + \frac{3000}{2000 + 3000} \cdot 2 = 1.6.$$

Consequently,

$$i_0 \approx \left(\frac{2000 + 3000}{4000}\right)^{\frac{1}{1.6}} - 1 \approx 14.97$$

Recall that, directly solving the equation, we obtained earlier $i_0 = 15.14$.

2.3.2 *The Internal Rate of Return of Typical Investment Flows*

In the previous paragraph, it was actually established that for a financial flow of the type (2.18), the internal rate of return i_0 is determined, and $i_0 > 0$ if the net amount of income exceeds the initial investment.

This result allows generalization to the so–called typical investment flows. A **typical investment flow** is understood as a flow of payments in which all negative payments precede positive ones, i.e. investments prevail over return only at the initial stages of the investment process. Thus, a typical investment flow has the form:

$$CF = \{(t_0, -C_0), \ldots, (t_m, -C_m), (t_{m+1}, C_{m+1}), \ldots, (t_n, C_n)\}, \tag{2.29}$$

where all C_k are positive, moreover C_0 and C_n are different from 0.

We prove that any typical investment flow has an unambiguously defined internal rate of return. Consider an arbitrary financial flow (2.30) satisfying the conditions of the theorem. The stream (2.30) can be represented as the difference of two streams:

$$CF = CF^{(+)} - CF^{(-)}, \tag{2.30}$$

where

$$\mathrm{CF}^{(+)} = \{(\mathrm{t_{m+1}}, \mathrm{C_{m+1}}), \ldots, (\mathrm{t_n}, \mathrm{C_n})\} \tag{2.31}$$

and

$$\mathrm{CF}^{(-)} = \{(\mathrm{t_0}, \mathrm{C_0}), \ldots, (\mathrm{t_n}, \mathrm{C_n})\}. \tag{2.32}$$

We shall bring the values of the flows (1.136) and (1.137) to a certain point in time τ between t_m and t_{m+1}. Let us suppose that

$$P^{(+)}(i) = PV_\tau\left(\mathrm{CF}^{(+)}, i\right) = \frac{C_{m+1}}{(1+i)^{t_{m+1}-\tau}} + \ldots + \frac{C_n}{(1+i)^{t_n-\tau}} \tag{2.33}$$

$$P^{(-)}(i) = PV_\tau\left(CF^{(-)}, i\right) = C_0(1+i)^{\tau - t_0} + \ldots + C_m(1+i)^{\tau - t_m} \tag{2.34}$$

When i runs from -1 to $+\infty$, the value of $P^{(+)}(i)$ decreases from $+\infty$ to 0, and the value of $P^{(-)}(i)$ increases from 0 to $+\infty$. Since

$$PV_\tau\ (CF, i) = P^{(+)}(i) - P^{(-)}(i), \tag{2.35}$$

$PV_\tau(CF, i)$—decreases from $+\infty$ to $-\infty$. Hence, the eq.

$$PV_\tau\ (CF, i) = 0 \tag{2.36}$$

has a unique solution $i_0 > -1$. Since

$$PV_\tau\ (CF, i) = (1+i)^\tau \cdot PV(CF, i), \tag{2.37}$$

the equation

$$PV\ (CF, i) = 0 \tag{2.38}$$

is equivalent to eq. (2.37). Therefore, the eq. $PV\,(CF, i) = 0$ for the flow (2.30) has a unique solution $i_0 > -1$, which is the internal rate of return of this flow.

The definition of the internal rate of return and the previous theorem can be extended to flows having a continuous component. So, if for the financial flow CF, the present value of $PV\ (CF,\ i)$ is defined and continuous with respect to i, and the equation

$$PV\ (CF, i) = 0 \tag{2.39}$$

has a unique solution i_0 for $i > -1$, then i_0 serves as the internal rate of return of the *CF* flow.

Consider the financial flow *CF* in the time interval from 0 to *T*, represented as a sum

$$CF = CF^{(d)} + CF^{(c)}, \tag{2.40}$$

where

$$CF^{(d)} = \{(t_1, C_1), \ldots, (t_n, C_n)\}, 0 \leq t_1 \leq \ldots \leq t_n \leq T \tag{2.41}$$

is a discrete stream, $CF^{(c)}$—continuous density flow $\mu(t)$, $0 \leq t \leq T$.

In this case, it is assumed that for a period of time from 0 to *T* there is such moment τ that all negative payments precede the moment τ, and positive ones follow it. More precisely, we assume that τ is different from the moments of payments of the discrete flow $CF^{(d)}$ and

$$\begin{aligned} &C_k \leq 0 \text{ when } t_k < \tau; C_k \geq 0 \text{ when } t_k > \tau; \\ &\mu(t) \leq 0 \text{ when } t < \tau; \mu(t) \geq 0 \text{ when } t > \tau. \end{aligned} \tag{2.42}$$

The equation $PV\ (CF, i) = 0$ has the form

$$\sum_{k=1}^{n} \frac{c_k}{(1+i)^{t_k}} + \int_0^T \frac{\mu}{(1+i)^t} dt = 0 \tag{2.43}$$

Discounting the cost of the flow to the moment τ and dividing the flows into positive and negative parts, we proceed from eq. (2.44) to the equation

$$\begin{aligned} &\sum_{t_k<\tau} |C_k|(1+\mathrm{i})^{\tau-t_k} + \int_0^T |\mu(t)|(1+\mathrm{i})^{\tau--t}\mathrm{d}t = \\ &= \sum_{t_1>\tau} \frac{C_k}{(1+i)^{t_k-\tau}} + \int_\tau^T \frac{\mu}{(1+\mathrm{i})^{t-\tau}}\mathrm{d}t \end{aligned} \tag{2.44}$$

When i runs through the interval from -1 to $+\infty$, the left side of equation (2.45) continuously increases from 0 to $+\infty$, and the right side continuously decreases from $+\infty$ to 0. Consequently, equation (2.45) has a unique solution: $i_0 > -1$, which is the internal rate of return of the financial flow *CF*.

Example 2.3 It is necessary to estimate the internal rate of return of the next project. One–time initial investments include expenses for the acquisition of the enterprise and the purchase of new equipment for it and amount to K. Over the next two years, there is a constant relative increase in daily income by q starting with daily income R_0. Production costs are constant for two years and amount to z per day. After T years, it is expected to sell the company for the amount of S.

It can be assumed that the financial flow associated with the described project contains discrete and continuous components. The discrete component has the form:

$$CF^{(d)} = \{(0, -K), (T, S)\}.$$

A continuous component is a financial flow with a density of

$$\mu(t) = 365 \cdot \left(R_0 \cdot (1+q)^t - z\right).$$

The internal rate of return is determined from the equation

$$-K + 365 \cdot \int_0^T \frac{R_0(1+q)^t - z}{(1+i)^t} dt + S = 0 \tag{2.45}$$

Suppose $\alpha = \ln(1+q)$ and $\delta = \ln(1+i)$. Then

$$\int_0^T \frac{R_0(1+q)^t - z}{(1+i)^t} dt = R_0 \int_0^T e^{(\alpha-\delta)t} dt - z \int_0^T e^{-\delta t} dt =$$
$$= \frac{R_0}{\alpha - \delta}\left(e^{(\alpha-\delta)T} - 1\right) - \frac{z}{\delta}\left(e^{-\delta T} - 1\right).$$

Accordingly, eq. (2.46) turns into the following equation with respect to δ:

$$-K + \frac{R_0}{\alpha - \delta}\left(e^{(\alpha-\delta)} - 1\right) - \frac{z}{\delta}\left(e^{-\delta T} - 1\right) + S = 0.$$

2.3.3 *Internal Rate of Return of Financial Flows with Alternating Positive and Negative Payments*

For an arbitrary financial flow in which negative payments alternate with positive ones, the internal rate of return may be uncertain. Consider, for example, such a financial flow:

$$CF = \{(0, -1000), (1, 3410), (2, -3856), (3, 1446)\}. \tag{2.46}$$

In this flow, the sum of negative payments is equal to the sum of positive ones, so that the present value is zero at $i_0 = 0$. In addition to the root i_0, the equation

$$-1000 + \frac{3410}{1+i} + \frac{3856}{(1+i)^2} - \frac{1446}{(1+i)^3} = 0$$

has two more roots: $i_1 = 0.1274$ and $i_2 = 0.2826$. Thus, two more rates claim to be the internal rate of return: 12.74% and 28.26%.

Another problem is related to the "instability" of the internal rate of return, even when it is clearly defined. So, relatively small changes in individual payments can lead to significant changes in the internal rate of return.

For example, consider along with the payment flow (1.151) flows

$$CF_1 = \{(0, -1000), (1, 3410), (2, -3856), (3, 1442)\} \tag{2.47}$$

$$CF_2 = \{(0, -1000), (1, 3410), (2, -3856), (3, 1452)\}, \tag{2.48}$$

The internal rate of return is determined unambiguously for flows (2.48) and (2.49). For the flow (2.48) it is negative and equal to -6.15%, for the flow (2.49) it is positive and equal to 35.62%.

Nevertheless, the theorem on the internal rate of return of a typical investment flow can be extended to some financial flows with alternating positive and negative payments.

We will call the **net amount** of the payment flow

$$CF = \{(t_0, C_0), (t_1, C_1), \ldots, (t_n, C_n)\}, t_0 \le t_1 \le \ldots \le t_n, \tag{2.49}$$

to the time moment τ the value $S(\tau) = \sum_{t_k \le t} C_k$.

We prove that if in the payment stream (2.50) all net amounts $S(t_k)$, $k = 1, \ldots, n$, are non–negative, then it has a non–negative present value for any positive conversion rate. If, in addition, the initial payment is positive, then the present value is positive for any positive conversion rate.

As $C_0 = S_0$ and $C_k = S_k - S_{k-1}$ when $k = 1, 2, \ldots, n$, then

$$\frac{C_0}{(1+i)^{t0}} \tag{2.50}$$

If $i > 0$, then all the terms of the last sum are non–negative, and the first term is positive, so that $PV(i) > 0$.

In proving the previous theorem, a discrete analogue of integration in parts was used. In fact, let CF be a continuous flow with continuous density $\mu(t)$, $t \in [0, T]$. Suppose all net amounts $S(t) = \int_0^t \mu(\tau)d\tau$ are non–negative. Denote by $P(t)$ the present value of the

CF flow over a period of time [0, *t*]. Then $dP(t) = e^{-\delta t} dS(t)$, where $\delta = \ln(1+i)$ is the intensity of growth rate. As $P(0) = 0$, then $P(t) = \int_0^t e^{-\delta\tau} dS(\tau)$ and $PV(i) = \int_0^T e^{-\delta\tau} dS(\tau)$. Integrating by parts, we get:

$$PV(i) = e^{-\delta\tau} S(\tau)|_0^T + \int_0^T \delta e^{-\delta\tau} S(\tau) d\tau. \tag{2.51}$$

It follows that $PV\ (i) \geq 0$. Definitely, $e^{-\delta\tau} S(\tau)|_0^T = e^{-\delta T} S(T) \geq 0$, a integrable function $\delta e^{-\delta\tau} S(\tau)$ is non–negative as $\delta > 0$ when $i > 0$ and $S(\tau) \geq 0$ by assumption. Assuming additionally that at some initial interval $[0, t_0]$ the density function is non–negative and $S(t_0) > 0$, it can be proved that $PV(i) > 0$.

Below we will indicate the class of financial flows for which a positive internal rate of return is determined.

We will say that the financial flow (2.50) is an investment—type payment flow if it has the following properties:

- the initial payment is negative;
- net–the sum of the total flow is positive;
- there is a point from which all payments are positive, and all previous net amounts are negative.

We show that for an investment–type flow a positive interest rate, relative to which the present value of the flow is zero, exists and is determined.

Let (2.50) be an investment—type flow. According to the definition, this means that $C_0 < 0$, $S\ (t_n) > 0$ and there exists an m, such that $C_k > 0$ for $k \geq m + 1$, and $S\ (t_k) < 0$ when $k \leq m$. For sufficiently large values of the conversion rate i the sign of the value $PV(i)$ coincides with the sign C_0, so that $PV(i)$ is negative. At the same time, $PV(0) = S(t_n) > 0$. Consequently, $PV(i_0) = 0$ for some $i_0 > 0$. Let us show that i_0 is uniquely determined, and at point i_0 the value of $PV(i)$ changes sign from plus to minus. To do this, it is enough to establish that $PV(i) < 0$ for all $i > i_0$.

Let us introduce into consideration a payment flow

$$CF' = \{(t_0, B_0), (t_1, B_1), \ldots, (t_n, B_n)\}, \tag{2.52}$$

in which (2.54)

$$B_k = \frac{C_k}{(1+i)^{tk}}, \quad k = 1, 2, \ldots, n \tag{2.53}$$

Let us denote by $S'(t_k)$ the net amount of payments of the CF' stream by the time t_k. Let us show that $S'(t_k) \leq 0$ for all k Definitely,

$$S'(t_k) = \sum_{l=1}^{k} \frac{C_l}{(1 + i_0)^{t_l}}$$

is a reduced cost of payments of the *CF* stream up to and including the t_k moment (relative to the i_0 rate). According to the definition of i_0, we have $S'(t_n) = 0$. When $k > m$, C_k payments are positive, and B_k payments are positive along with them. So, $S'(t_k) < 0$ when $k > m$. Finally, if $k \leq m$, then all the net amounts of the payment flow

$$\{(t_0, C_0), (t_1, C_1), \ldots., (t_k, C_k)\} \tag{2.54}$$

are negative by assumption. Then, according to the theorem proved above, the flow (2.55) has a negative present value relative to any positive bid. In particular, $S'(t_k) < 0$.

Since all the net amounts of the payment flow (2.53) (except the last one equal to zero) are negative, the flow (2.55) has a negative present value relative to any positive rate. Then for any rate $i > i_0$ supposing that $j = \frac{1+i}{1+i_0} - 1$, then

$$PV(i) = \sum_{k=1}^{n} \frac{C_k}{(1 + i)^{t_k}} = \sum_{k=1}^{n} \frac{B_k}{(1 + i)^{t_k}} < 0,$$

which is what we needed to prove.

2.4 Comparison of the Performance Indicators of an Investment Project

In order to identify the possibilities of practical use of investment performance evaluation methods, we will conduct a comparative analysis of them.

1. If the NPV and IRR criteria are applied to a single project in which, after the initial cash outlay, only cash receipts occur, then the results obtained using both methods are consistent with each other and lead to identical decisions.

   The NPV, IRR and PI criteria, most often used in investment analysis, are actually different versions of the same concept, as a result of which their results are related to each other. Therefore, for one project, the following mathematical relationships can be expected to be fulfilled:

   if NPV > 0, then PI > 1 and IRR > r;
   if NPV < 0, then PI < 1 and IRR < r;
   if NPV = 0, then PI = 1 and IRR = r,
   where r is the required rate of return.
2. For projects with different cash flow schedules, the values of the internal rate of return IRR can be as follows:*no IRR*:

(a) a project that does not spend cash always has a positive NPV, therefore the project has no IRR (where NPV $= 0$). In this case, you should reject IRR and use NPV. Since NPV > 0, this project should be accepted,
(b) a project that does not receive cash always has a negative NPV, and such a project also has no IRR. In this case, you should reject IRR and use NPV; since NPV < 0, this project should be rejected;

Opposite IRR values—a project that first receives cash and then spends it has an IRR value that is never consistent with NPV. A low IRR and a positive NPV will be observed simultaneously;

Several IRR values—a project that alternates between cash inflows and outflows will have as many internal rate of return values as there are changes in the direction of cash flows.

3. Project ranking is necessary:

(a) if the projects are alternative (to be able to choose one of them);
(b) if the amount of capital is limited and the company is unable to attract enough capital to implement all the "good" projects;
(c) when there is no agreement between NPV and IRR. When using two methods at the same time—NPV and IRR—different rankings often occur.

Let's consider the reasons for the discrepancy between the results of the IRR and NPV methods for several projects.

(a) Differences in the time of project implementation (see below for more details)—those projects that are implemented over a long period of time may have a low internal rate of return, but over time their net present value may be higher than that of short-term projects with a high rate of return.
(b) Differences in project size—large investments with low growth rates may generate higher net present value than small investments with high growth rates.
(c) Differences in the timing of cash inflows and outflows—for projects where positive cash flows occur over long periods of time, net present value is more sensitive to the cost of capital than for projects where cost recovery occurs quickly.

The choice between IRR and NPV is made as follows.

Using NPV as a criterion for choosing an investment project leads to maximizing the amount of cash, which is equivalent to maximizing value. If this is the goal of the company, then the net present value method should be used.

Using IRR as a criterion for choosing leads to maximizing the percentage of the company's growth. When the company's goal is growth, the most important characteristic of investment projects is the degree of return, the ability to earn cash for reinvestment.

The NPV indicator has a significant drawback, its dependence on the discount rate i, i.e. with different values of i, opposite results can be obtained. For a more accurate analysis of the effectiveness of investment projects, it is necessary to use a graph of the dependence of NPV on i. Dependence of the effectiveness of investment projects on the discount rate is shown in Fig. 2.1.

NPV allows you to choose the best project in any case.

2.5 Comparison of Investment Projects of Different Duration

The need to compare investment projects of different duration arises very often. There are several methods that allow comparing such projects:

1. the chain repetition method within the overall duration of the projects,
2. the method of infinite chain repetition of the compared projects,
3. the equivalent annuity method.

2.5.1 The Chain Repetition Method Within the Overall Duration of the Projects

1. The duration of one project is a multiple of the duration of the other
   In this case, the shorter project is repeated until its new duration equals the duration of the longer project. Then the NPV and IRR of the extended first and second projects are compared, on the basis of which the selection (or ranking) of the projects is made.

Example 2.1 Let us compare two projects of different durations (annual and biennial)

$$CF_1 = \{-50; 100\}$$

$$CF_2 = \{-30; 60; 60\}$$

at a discount rate of 10%.

In this case

$$NPV_1 = -50 + \frac{100}{1.1} = 40.91$$

$$NPV_2 = -30 + \frac{60}{1.1} + \frac{60}{1.1^2} = 74.13$$

The NPV of the second project is almost twice as large as the NPV of the first project, and it seems obvious that the second project is much more efficient and should be accepted unconditionally.

Let's repeat the first project once more (thus, the terms of both projects become equal)

$$CF_1' = \{-50; 100\} + \{0; -50; 100\} = \{-50; 50; 100\}$$

$$NPV_1' = -50 + \frac{50}{1.1} + \frac{100}{(1.1)^2} = 78.10$$

Now we calculate the IRR for all three projects (the two original ones and the extended first one)

$$NPV_1 = -50 + \frac{100}{(1 + IRR)} = 0; IRR = 100\%;$$

$$NPV_1' = -50 + \frac{50}{(1 + IRR)} + \frac{100}{(1 + IRR)^2} = 0; IRR = 100\%;$$

$$NPV_2 = -30 + \frac{60}{(1 + IRR)} + \frac{60}{(1 + IRR)^2} = 0; IRR = 173\%;$$

$$IRR_1 = 100\%; IRR_1' = 100\%; IRR_2 = 173\%.$$

We see that after the repeat operation, the NPV of the first project has become greater than the NPV of the second project and according to this criterion it becomes preferable. However, according to the IRR, the second project remains preferable. In conditions of conflicting indicators, the choice is made in accordance with the investor's preferences.

2. Project terms are not multiples

   In this case, the least common multiple (LCM) of the terms of both projects is found and each of the projects is repeated the required number of times until its term reaches the value of the least common multiple of the terms of both projects. For such projects with the required numbers of repetitions, but with the same terms, NPV and IRR are calculated and based on their comparison, a choice is made between the projects (or ranking of the projects).

Example 2.2 Let two projects be given—a 2-year and a 3-year

$$CF_1 = \{-50; 50; 70\}$$

$$CF_2 = \{-30; 40; 60; 60\}$$

Than

$$NPV_1 = -50 + \frac{50}{1.1} + \frac{70}{(1.1)^2} = 53.31$$

$$NPV_2 = -30 + \frac{40}{1.1} + \frac{60}{(1.1)^2} + \frac{60}{(1.1)^3} = 101.03.$$

The LCM of the terms of both projects is 6, therefore, repeating the first project three times, the second two times, we get the following 6-year projects

$$CF'_1 = \{-50; 50; 70\} + \{0; 0; -50; 50; 70\} + \{0; 0; 0; 0; -50; 50; 70\} = \{-50; 50; 20; 50; 20; 50; 70\}$$

$$CF'_2 = \{-30; 40; 60; 60\} + \{0; 0; 0; -30; 40; 60; 60\} = \{-30; 40; 60; 30; 40; 60; 60\}.$$

Then

$$NPV_1 = -50 + \frac{50}{1.1} + \frac{20}{(1.1)^2} + \frac{50}{(1.1)^3} + \frac{20}{(1.1)^4} + \frac{50}{(1.1)^5} + \frac{70}{(1.1)^6} = 133.76;$$

$$NPV_2 = -30 + \frac{40}{1.1} + \frac{60}{(1.1)^2} + \frac{30}{(1.1)^3} + \frac{40}{(1.1)^4} + \frac{60}{(1.1)^5} + \frac{60}{(1.1)^6} = 176.93$$

If before the repetition operation, the NPV of the second project was almost twice as large as the NPV of the first project, and it seemed obvious that the second project was significantly more effective and should be accepted unconditionally, then after the repetition operation, the NPV of the second project is already 32% greater than the NPV of the first project, so that the advantage of the second project, although remaining, becomes smaller. It is interesting to calculate the IRR for both projects, but this may not be so simple. Since the equations for finding the IRR become sixth-degree equations, they must be solved either by trial and error (iteration) or other approximate methods.

There is an alternative way to calculate the NPV of a repeating project. It does not require constructing a flow of the repeating project, as was done above. The calculations are performed for the original projects, and then the NPV of the repeating projects is calculated by summing the discounted values of their NPVs. The total NPV of the repeating project is found as follows

$$NPV(k, n, N) = NPV(k, n, N) \times \left[1 + \frac{1}{(1+r)^k} + \frac{1}{(1+r)^{2k}} + \frac{1}{(1+r)^{3k}} \cdots + \frac{1}{(1+r)^{N-k}}\right] \quad (2.55)$$

Here NPV (k) is the NPV of the original project with a duration of k years,

N = NOK (k_1, k_2, …, k_n), n is the number of repetitions of the original project (it is equal to the number of terms in square brackets).

Example 2.3 Let's continue Example 2.2, but we will calculate the NPV by summing the discounted NPV values of the repeating projects

$$NPV_1 = 53.31 \cdot \left(1 + \frac{1}{(1.1)^2} + \frac{1}{(1.1)^4}\right) = 133.76;$$

$$NPV_2 = 101.03 \cdot \left(1 + \frac{1}{(1.1)^3}\right) = 176.94.$$

Thus, we obtain the same results as when constructing flows of repeating projects, but such a scheme is simpler by as many times as the duration of each of the projects.

2.5.2 *Method of Infinite Chain Repetition of Projects*

The method of calculating NPV and IRR is significantly simplified when moving to an infinite chain repetition of projects. This is especially convenient for projects with significantly different terms.

Let us assume that each of the projects can be extended an unlimited number of times. In this case, we can move to the limit $n \to \infty$ in formula (2.57) and, using the formula for the sum of the terms of an infinitely decreasing geometric progression, obtain an expression for $NPV(k, \infty)$

$$NPV(k, \infty) = \lim_{n \to \infty} NPV(k, n, N) = NPV(k, N) \cdot \frac{(1 + r)^k}{(1 + r)^k - 1} \tag{2.56}$$

The project with the larger $NPV(k, \infty)$ value will be preferable.

Example 2.4 For the example we considered above, we have for both projects

$$NPV_1(2, \infty) = 53.31 \cdot \left(\frac{(1.1)^2}{(1.1)^2 - 1}\right) = 307.17;$$

$$NPV_2(3, \infty) = 101.03 \cdot \left(\frac{(1.1)^3}{(1.1)^3 - 1}\right) = 406.26.$$

Although with infinite extension of projects, their NPVs have grown significantly compared to the situation with finite repetitions of projects, the conclusion about the preference of the second project (with the larger NPV) remains valid.

2.5.3 Equivalent Annuity Method

This method correlates with the infinite chain repetition method of projects. The procedure for applying the method is as follows:

1. The NPV of each project is calculated.
2. For each project, an equivalent term annuity (EAA) is found, the present value of which is equal to the NPV of the project, i.e. the value of the annuity payment (P) is calculated using the formula

$$NPV = P \cdot \frac{1-(1+r)^{-n}}{r}. \tag{2.57}$$

3. Assuming that the found annuity can be replaced by a perpetual annuity with the same value of the annuity payment (P), the present value of the perpetual annuity is calculated using the formula

$$PV_{\infty} = \frac{P}{r}. \tag{2.58}$$

The project with the greater PV_{∞}value is preferable.

Example 2.5 For the example we considered above, we have for both projects

$$NPV_1 = P_1 \cdot \frac{1-(1+r)^{-n}}{r} = P_1 \cdot \frac{1-(1.1)^{-2}}{0.1} = 53.31;$$

$$NPV_2 = P_2 \cdot \frac{1-(1+r)^{-n}}{r} = P_2 \cdot \frac{1-(1.1)^{-3}}{0.1} = 101.03.$$

Hence,

$$P_1 = \frac{53.31}{1.74} = 30.64; P_2 = \frac{101.03}{2.49} = 40.57.$$

Next, we find the present value of perpetual annuities using the formula

$$PV_{\infty} = \frac{P}{r}$$

$$PV_{1,\infty} = 306.4; PV_{2,\infty} = 405.7.$$

The conclusion about the preference of the second project is confirmed by this method.

It should be noted that this conclusion could also be reached by comparing the amounts of annuity payments, ignoring the last procedure for calculating the present value of perpetual annuities. However, this can only be done if the discount rates for both projects are the same. The limitations of the above methods based on the repetition of the original projects are related to a number of points. Firstly, the initial conditions cannot always be extended into the future; they may change. Secondly, it is not always possible to accurately estimate the duration of the original project. Thirdly, it is not obvious that the project will be repeated n times, especially if it is quite long. Fourthly, such factors as inflation, scientific and technological progress, changes in the technologies of the original project, etc. are not taken into account in the calculations. Therefore, due to the limitations stated above, in cases where the initial parameters of the projects being compared are characterized by a sufficiently high uncertainty, it is apparently possible to ignore the difference in the duration of the projects and use standard efficiency criteria.

References

Brusov, P., & Filatova, T. (2025). *Business valuation and company value: New qualitative approaches considering business and financial risks* (850 pp). Springer Nature.

Brusov, P. N., Filatova, T. V., & Orekhova, N. P. (2023a). *Financial mathematics* (290 pp). Cambridge Scholars Publishing.

Brusov, P., Filatova, T., & Orekhova, N. (2023b). *The Brusov–Filatova–Orekhova theory of capital structure: Applications in corporate finance, investments, taxation and ratings* (769 pp). Springer Nature.

Brusov, P. N., Filatova, T. V., & Orekhova, N. P. (2024a). *Financial mathematics: Exercises and solutions* (Springer texts in business and economics) (286 pp). Springer Nature.

Brusov, P. N., Filatova, T. V., & Kulik, V. L. (2024b). Capital asset pricing model (CAPM) 2.0: Account of business and financial risk. *Finance: Theory and Practice, 28*(2), 128–142. https://doi.org/10.26794/2587-5671-2024-28-2-128-142

Brusov, P. N., Filatova, T. V., & Kulik, V. L. (2024c). Application of the company's "Golden age" effect in the economic practice. *Finance: Theory and Practice, 28*(3), 61–83. https://doi.org/10.26794/2587-5671-2024-28-3-61-83

Part II
Capital Structure Theories

One of the main parameters used in assessing the effectiveness of investment projects is the discount rate. This part deals with the state of the theory of capital structure and cost of capital, which are used for valuation the discount rates. Two main theories Brusov–Filatova–Orekhova (BFO) and Modigliani–Miller (MM) are discussed and analyzed, as well as their numerous modifications and generalizations. Also discussed is the latest stage in the development of the theory of capital structure, which began a couple of years ago and is associated with the adaptation of two main theories of capital structure (Brusov–Filatova–Orekhova and Modigliani–Miller) to the established practice of the functioning of companies. This generalization takes into account the real working conditions of companies.

Chapter 3
Capital Structure

3.1 Introduction

Under the capital structure, one understands the relationship between equity and debt capital of the company. Does capital structure affect the company's main settings, such as the cost of capital, profit, value of the company, and the others, and, if affects, how? Choice of an optimal capital structure, i.e., a capital structure, which minimizes the weighted average cost of capital, WACC, and maximizes the value of the company, *V*, is one of the most important tasks solved by financial manager and by the management of a company. The first serious study (and first quantitative study) of influence of capital structure of the company on its indicators of activities was the work by Modigliani and Miller (1958). Until this study, the approach existed (let us call it traditional), which was based on empirical data analysis.

3.2 Basic Theories of Capital Structure

3.2.1 A Historical Point of View

From a historical point of view, **five stages** in the development of the capital structure theory can be distinguished: **First** (before 1958)—the traditional approach, based on practical experience and existed before the appearance of the first quantitative theory by Modigliani and Miller (**the second stage**) (1958–1963). **The third stage** (1964–2008) is the numerous attempts of scientists to modify the Modigliani–Miller theory (Acharya et al. (2011), Arrow (1963), Asquith & Mullins (1986), Baumol et al. (1982), Harcourt Brace Jovanovich Black (1986), Brealey & Myers (1981), Brealey et al. (2013), Bhattacharya (1978), Brusov et al. (2020b), Derrig (1993), Diamond & He (2013), Donaldson (1961), Erel et al. (2015), Fairley (1979), Fama (1977), Fama (1996), Fisher & McGowan (1983), Filatova et al. (2022), Frank

P. Brusov, T. Filatova, *Efficiency of Investment Projects*, Contributions to Finance and Accounting, https://doi.org/10.1007/978-3-032-01028-5_3

& Goyal (2003), Fulghieri et al. (2013), Gordon & Shapiro (1956), Graham & Leary (2011), Hamada (1969), Hausman & Myers (2002), Healy et al. (2002), Hirshleifer (1965), Hirshleifer (1966), Jensen (1986), Jin & Myers (2006), Kaplan & Ruback (1995), Myer Annual Report (2015), Keynes (1936), Kolbe et al. (1993), Lambrecht & Myers (2007), Lambrecht & Myers (2008), Lambrecht & Myers (2012), Lambrecht & Myers (2015), Lemmon et al. (2008), Lemmon & Zender (2010), Lintner (1956), Lintner (1965), Majd & Myers (1987), Majluf (1978), Merton & Perold (1993), Miller & Modigliani (1961), Morck et al. (2000), Mullins Jr. (1976), Myers (1967), Myers (1968a), Myers (1968b), Myers (1972), Myers (1973a), Myers (1973b), Myers (1973), Myers (1977), Myers (1988), Myers (1989), Myers (1999), Myers (2000), Myers & Cohn (1987), Myers et al. (1976), Myers & Howe (1997), Myers et al. (1983), Myers et al. (1985), Myers & Majd (1990), Myers & Majluf (1983), Myers & Pogue (1973), Myers & Rajan (1998), Myers & Read Jr. (2001), Myers & Read Jr. (2013), Myers & Shyam-Sunder (1996), Myers & Turnbull (1977), Robichek & Myers (1965), Robichek & Myers (1966a), Robichek & Myers (1966b), Robichek & Myers (1966c), Rosenbaum & Pearl (2013), Schwartz (1959), Sharpe (1965), Shleifer & Summers (1988), Shyam-Sunder & Myers (1999), Skinner (2008), Solomon & Laya (1967), Van Horne (1966), Williams (1938). **The fourth stage** (2008–2019) is the appearance of Brusov–Filatova–Orekhova (BFO) theory, which removed the main restriction of the Modigliani–Miller (MM) theory associated with the infinite life–time of the company. And, finally, **the fifth stage** (2019 up to now), which began a few years ago and is associated with the adaptation of the two main theories of the capital structure (Brusov–Filatova–Orekhova and Modigliani–Miller) to the established financial practice of the company's functioning by taking into account the real conditions of their work (Brusov et al. 2021a, b, 2022a, b; Brusov and Filatova 2021, 2022a, b).

One of the most important assumptions of the Modigliani–Miller theory is that all financial flows and all companies are perpetuity. This limitation was lift out by Brusov–Filatova–Orekhova in 2008 (Filatova et al. 2008), who have created BFO (Brusov–Filatova–Orekhova) theory—modern theory of capital cost and capital structure for companies of arbitrary age (BFO–1 theory) and for companies of arbitrary life time (BFO–2 theory) (Brusov et al. 2018). Figure 3.1 shows the historical development of the theory of capital structure from the empirical traditional approach to the general theory of capital structure, BFO, through the perpetuity Modigliani–Miller theory.

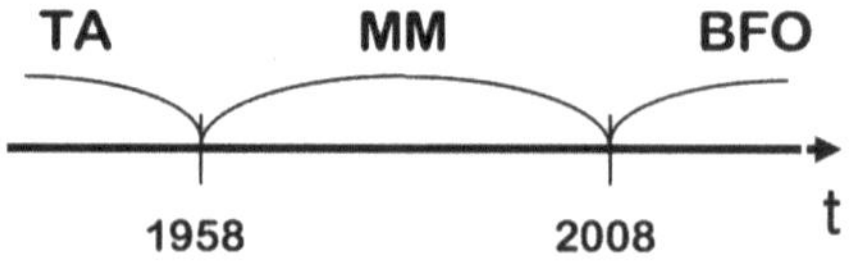

Fig. 3.1 The historical development of the theory of capital structure from the empirical traditional approach (TA) to the general theory of capital structure, BFO (Brusov–Filatova–Orekhova), through the perpetuity Modigliani–Miller theory

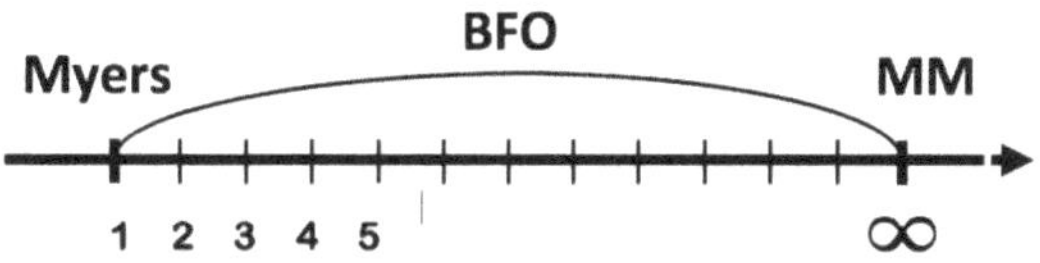

Fig. 3.2 Only two results for capital structure of the company were known before BFO theory: Modigliani–Miller (MM) for perpetuity companies and Myers for one–year company, while the created BFO theory filled the entire interval between n = 1 and $n = \infty$

One year companies were studied by Steve Myers in 2001, who showed that the weighted average cost of capital WACC is greater in this case than in the perpetual Modigliani–Miller case, and the value of company V is therefore less. Only two results for capital structure of the company were known up to 2008, when BFO theory has appeared: Modigliani–Miller for perpetuity companies and Myers for one–year company (see Fig. 3.2). The created BFO theory filled the entire interval between t = 1 and t = ∞. This expands capital structure theory for the companies of arbitrary ages and/or arbitrary life–time. Many new meaningful effects have been discovered in the BFO theory.

3.2.2 *The Empirical (Traditional) Approach*

In the traditional approach, based on practical experience and existing before the advent of the first quantitative theory of Modigliani and Miller, the weighted average cost of capital WACC and the associated capitalization of the company, $V =$ CF/WACC, depend on the capital structure, the level of debt load, L. The cost of debt is always lower than the cost of equity, because the former has less risk due to the fact that the claims of creditors are satisfied before the claims of shareholders in the event of bankruptcy. As a result, an increase in the share of cheaper borrowed capital in the total capital structure to the limit that does not cause a violation of financial stability and an increase in the risk of bankruptcy leads to a decrease in the weighted average cost of capital WACC.

The return required by investors (equal to the cost of equity) is growing; however, its growth did not offset the benefits of using cheaper borrowed capital. Therefore, the traditional approach welcomes the increase in leverage $L = D/S$and the associated increase in the value of the company $V =$ CF/WACC. The empirical traditional approach existed until the appearance of the first quantitative theory by Modigliani and Miller (1958).

Within the based on existing practical experience traditional approach the competition between the advantages of debt financing at a low leverage level and its disadvantages at a high leverage level forms the optimal capital structure, defined as the leverage level, at which WACC is minimal and company value, V, is maximum.

3.2.3 The Modigliani–Miller Theory

3.2.3.1 The Modigliani–Miller Theory with Taxes

There are two versions of the Modigliani–Miller theory: without taxes and with taxes. Without taxes the following expressions for *V*, *WACC* and k_e

$$V = V_0 = \frac{EBIT}{k_0}, \tag{3.1}$$

where V_0 stands for the unlevered company value, EBIT stands for earnings before interest and taxes and k_0 stands for the equity cost at zero leverage level L.

From (3.1) one gets for the weighted average cost of capital WACC:

$$WACC = k_0. \tag{3.2}$$

From the expression for WACC

$$\mathrm{WACC} = k_0 = k_e w_e + k_d w_d. \tag{3.3}$$

and accounting (3.1), one get for the equity cost, k_e,

$$k_e = \frac{k_0}{w_e} - k_d \frac{w_d}{w_e} = \frac{k_0(S+D)}{S} - k_d \frac{D}{S} = k_0 + (k_0 - k_d)\frac{D}{S} = k_0 + (k_0 - k_d)L \tag{3.4}$$

Here, D stands for debt capital value; S stands for equity capital value; k_d and w_d stand for the cost and share of the company's debt capital; k_e and w_e stand for the equity capital cost and share. It is seen from (3.4), that the equity increases linearly with the leverage level.

3.2.3.2 The Modigliani–Miller Theory with Taxes

Within the framework of the Modigliani–Miller theory with taxes [4], the following expression was postulated for the value of a company using borrowed funds, V,

$$V = V_0 + D \cdot t. \tag{3.5}$$

The expression for WACC immediately follows from (3.5)

$$WACC = k_0 \cdot (1 - w_d t) \tag{3.6}$$

The following formula for the cost of equity k_e can be obtained from (3.6) within the framework of the Modigliani–Miller theory with taxes

$$k_e = k_0 + L\cdot(k_0 - k_d)(1 - t) \tag{3.7}$$

Two formulas (3.7) (MM with taxes) and (3.4) (MM without taxes) differs by the multiplier (1 − t), called the tax corrector. It is less than unit, thus the k_e (L) curve slope decreases with accounting the taxes.

3.2.4 *Modifications of Modigliani–Miller Theory*

Since the creation of Modigliani–Miller theory, numerous attempts have been made to improve and develop this theory. We will discuss some of these modifications below. One of them is Hamada model.

3.2.4.1 Hamada Model

The Modigliani–Miller theory, with accounting taxes has been united with CAPM (Capital Asset Pricing Model) in 1961 by Hamada. For the cost of equity of leveraged company the below formula has been derived

$$k_e = k_F + (k_M - k_F)b_U + (k_M - k_F)b_U\frac{D}{S}(1 - T), \tag{3.8}$$

Here b_U is the β–coefficient of the unlevered company. First term represents risk–free profitability k_F, second term—business risk premium, $(k_M - k_F)b_U$,and third term—financial risk premium $(k_M - k_F)b_U\frac{D}{S}(1 - T)$.

In the case of a unlevered company (D = 0), the financial risk (the third term) is zero, and its shareholders receive only a business risk premium.

Equating CAPM formula to right side of (3.8), one gets:

$$k_F + (k_M - k_F)b_U = k_F + (k_M - k_F)b_U + (k_M - k_F)b_U\frac{D}{S}(1 - T) \tag{3.9}$$

or

$$b = b_U\left(1 + \frac{D}{S}(1 - T)\right). \tag{3.10}$$

Below are the formulas for the equity cost, k_e, debt cost, k_d and WACC in the CAPM model and (in parenthesis) in the Modigliani–Miller theory (Modigliani and Miller 1958, 1963, Miller and Modigliani 1966)].

The equity cost for unlevered company:

$$k_e = k_F + (\bar{k}_M - k_F)\beta_U, \quad (k_e = k_0). \tag{3.11}$$

The equity cost for levered company:

$$k_e = k_F + (\bar{k}_M - k_F)\beta_e, \quad (k_e = k_0 + (1 - T)(k_0 - k_d)L). \tag{3.12}$$

The debt cost:

$$k_d = k_F + (\bar{k}_M - k_F)\beta_d, (k_d = k_F; \beta_d = 0).$$

The weighted average cost of capital WACC

$$\mathrm{WACC} = k_e w_e + k_d w_d (1 - T), \quad (\mathrm{WACC} = k_0(1 - T w_d)). \tag{3.13}$$

3.2.4.2 The Cost of Capital Under Risky Debt

In the Modigliani—Miller theory there are two assets types: risky equity and risk—free debt). But assuming about the risk of bankruptcy of company and the ability to nonpayment of debt may change the situation with debt. It has been shown by Stiglitz (1969) and Rubinstein (1973) that the assumption concerning risky debt does not change the company value with respect to the Modigliani and Miller results under free of risk debt (Modigliani and Miller 1958, 1963; Miller and Modigliani 1966). However, the debt cost is changed from a constant value $k_d = k_F$ to variable one. As Hsia (1981), based on the models of pricing options, Modigliani–Miller and CAPM, has shown in the formula for the net discount profit, a term, reflecting tax shield should be discounted at the rate

$$k'_d = k_F + (k_0 - k_F)N(-d_1)\frac{1}{w_d}, \tag{3.14}$$

where

$$d_1 = \frac{-\ln w_d + k_F t}{\sigma\sqrt{t}} k_F + \frac{1}{2}\sigma\sqrt{t}, \tag{3.15}$$

$N(-d_1)$—cumulative normal distribution of probability of random value d_1 and t—a moment of payment a credit.

3.2.4.3 The Account of Corporate and Individual Taxes (Miller Model)

Modigliani and Miller (1963) have considered only corporate taxes, but they did not take into account the individual taxes of investors.

Miller (1977) has developed the model with account of the corporate and individual taxes and has studied the impact of leverage on the company value.

The model is described by the following formula for levered company value

$$V_{\mathrm{L}} = V_{\mathrm{U}} + \frac{I(1-T_{\mathrm{d}})}{k_{\mathrm{d}}}\left[1 - \frac{(1-T_{\mathrm{C}})(1-T_{\mathrm{S}})}{(1-T_{\mathrm{d}})}\right]. \tag{3.16}$$

Here first term means the unlevered company value

$$V_{\mathrm{U}} = \frac{\mathrm{EBIT}(1-T_{\mathrm{C}})(1-T_{\mathrm{S}})}{k_0}. \tag{3.17}$$

where, T_S stands for the tax on income of an individual investor from his ownership by corporation stock rate, T_C stands for the tax rate on corporate income, T_D stands for the tax rate on interest income from the provision of investor–individuals of credits to other investors and companies.

The Miller model estimates the value of levered company, accounting the corporate tax, as well as tax on individuals. Second term of sum in the Miller formula

$$\left[1 - \frac{(1-T_{\mathrm{C}})(1-T_{\mathrm{S}})}{(1-T_{\mathrm{d}})}\right]D, \tag{3.18}$$

represents the gains from use of debt capital. This term replaces the tax shield in the Modigliani–Miller model with corporate taxes:

$$V_{\mathrm{L}} = V_{\mathrm{U}} + TD. \tag{3.19}$$

3.2.4.4 Alternative Expression for WACC

Alternative formula for the WACC, different from Modigliani – Miller one has been derived in (Farber et al. 2006; Fernandez 2006; Berk and DeMarzo 2007; Harris and Pringle 1985) from the WACC definition and the balance identity (see as well Berk and DeMarzo 2007):

$$WACC = k_0(1 - w_d T) - k_d t w_d + k_{TS} t w_d, \tag{3.20}$$

where k_0, k_d and k_{TS} are the expected returns respectively on the unlevered company, the debt and the tax shield.

Some additional conditions are required eq. (3.20) practical applicability. If the WACC is constant over time, as it stated in (Farber et al. 2006) the levered company capitalization is found by discounting with the WACC of the unlevered company.

In textbooks (Berk and DeMarzo 2007; Harris and Pringle 1985) formulas for the special cases, where the WACC is constant, could be found.

In 1963 Modigliani and Miller assume that the debt value D is constant. Then, as the expected after–tax cash–flow of the unlevered firm is fixed, V_0 is constant as well. By assumption, $k_{TS} = k_D$ and the value of the tax shield is TS = tD. Thus, the capitalization of the company V is a constant and the alternative formula (3.20) becomes a formula for a constant WACC:

$$WACC = k_0(1 - w_d T) \tag{3.21}$$

Because the debt kd and the tax shield k_{TS} have debt nature it seems reasonable that the expected returns on they are equals as suggested by "classical" Modigliani–Miller (MM) theory, which has been modified by Brusov et al. for cases of practical meanings.

3.2.4.5 The Miles–Ezzell Model Versus the Modigliani–Miller Theory

Denis M. Becker (2021) discussed the differences between the Modigliani–Miller theory (1958, 1963) and the Miles–Ezzell model (1980, 1985), which deal with the stochasticity of free cash flows. The Modigliani–Miller theory considers a stationary process, while in the Miles–Ezzell model the process is stochastic. The author conducts a numerical experiment that allows you to determine the values and discount rates using a risk–neutral approach. He analyzes three formulas.

Modigliani–Miller theory (1958, 1963)

$$WACC = k_0 \cdot (1 - w_d t) \tag{3.22}$$

Miles–Ezzell model (1980, 1985),

$$WACC = k_0 - t \cdot w_d \cdot k_f \cdot \frac{1 + k_0}{1 + k_f} \tag{3.23}$$

Cooper and Nyborg (2006)

$$WACC = k_0 - k_f \cdot w_d t \tag{3.24}$$

where k_f stands for the risk–free rate, which equals the required return of the debt holders.

The author shows that in the Miles–Ezzell model, all cash flows and the values depend on the path, in contrast to the Modigliani–Miller theory. Also, in the Miles–Ezzell model, all discount rates are time independent, with the exception of the discount rate used to discount tax shields, which depends on the duration of the cash flows. Conversely, in the Modigliani–Miller theory, all discount rates change over time, except for the constant tax shield discount rate. This affects the applicability of the well–known formula for annuities and the development of models for estimating both finite and perpetual cash flows.

In this paper Becker (2021) raises the issue of paying the debt body together with payment of interest on the debt. Regarding this issue we would like to note that in both classical MM and BFO theories, the body of the debt is not paid. In the framework of the Modigliani–Miller theory, such an account is fundamentally impossible, while in the BFO theory it can be done and was done in the framework of the BFO–2 theory, where the amount of debt D decreases with time. This decrease in the value of debt D results in a decrease in the tax shield (see BFO–2 theory).

3.3 Trade–Off Theory

During decades one of the main theory of capital structure was the trade–off theory. There are two modifications of trade–off theory: the static and dynamic (Brennan et al., 1978, 1984; Leland, 1994; Kane et al., 1984).

3.3.1 Static Theory

The static trade–off theory accounts the income tax and cost of bankruptcy. Within this theory the optimal capital structure is formed by the balancing act between the benefits of debt financing at low leverage level (from the tax shield from interest deduction) and the disadvantage of debt financing at high leverage level (from the increased financial distress and expected bankruptcy costs). The tax shield benefit is equal to the product of corporate income tax rate and the market value of debt and the expected bankruptcy costs are equal to the product of probability of bankruptcy and the estimated bankruptcy costs.

The static version of the trade–off theory does not take into account the costs of the adapting of the capital structure to the optimal one, economic behavior of managers, owners, and other participants of economical process, as well as a number of other factors.

3.3.2 Dynamic Theory

The dynamic version of trade–off theory suggests that the costs of adjusting the capital structure are high, and therefore companies will only change their capital structure if the benefits outweigh the costs. Therefore, there is an optimal range that varies on the outside of each lever but remains the same on the inside. Companies try to adjust their leverage when it reaches the edge of the optimal range. Depending on the type of adaptation costs, companies reach the target ratio faster or slower. Proportional changes involve a small adjustment, while fixed changes imply significant costs.

In the dynamic version of the trade–off theory, company capital structure decision in current period depends on the waiting company income in the next period.

As it has been shown within BFO theory (Brusov et al. 2013a, b), under increased financial distress costs and bankruptcy risk, the optimal capital structure is absent. This means that the trade–off theory does NOT work in either the static or dynamic version.

3.3.3 *Proof of the Bankruptcy of the Trade–Off Theory*

Brusov et al. (2013a) tested whether assumptions about risky debt financing (and about rising lending rates in the run–up to bankruptcy) do not lead to an increase in the weighted average cost of capital, WACC. They used the following model

$$k_0 = 24\%; \quad \bar{k}_d = \left\{ \begin{array}{c} 0.07; \text{at } L \leq 2 \\ 0.07 + 0.1(L-2); \text{at } L > 2 \end{array} \right\}. \tag{3.25}$$

and showed that under the above conditions, WACC still decreases with leverage (Fig. 3.3).

This means there is no minimum for WACC with leverage change, and no maximum for company capitalization with leverage change. Thus, this means that there is no optimal capital structure in the famous theory of trade–offs. This fact has been explained by Brusov et al. (2013a, 2018). The explanation to this fact has been done within the Brusov–Filatova–Orekhova theory by studying the dependence of the equity cost k_e on leverage. The explanation for this fact is as follows. The dependence of the cost of equity k_e on the level of leverage undergoes significant

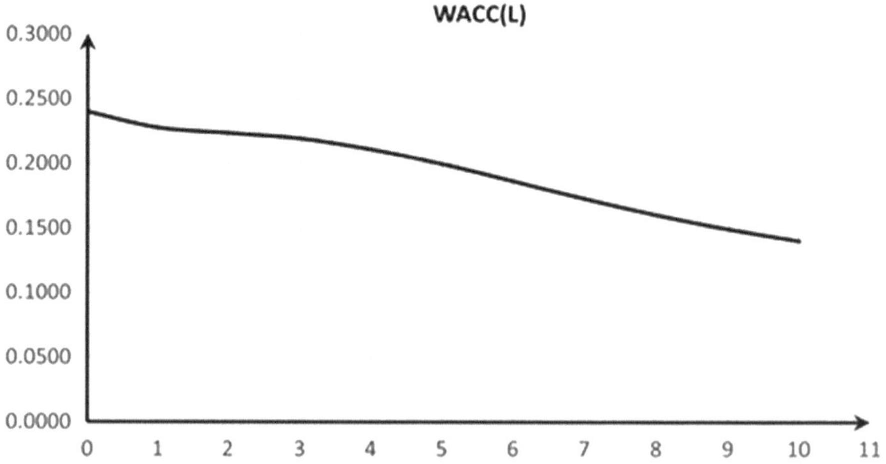

Fig. 3.3 Dependence of WACC on *L*

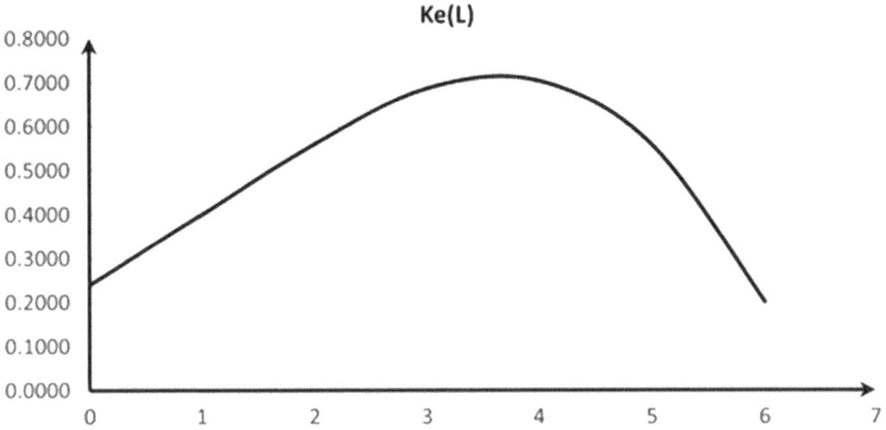

Fig. 3.4 Dependence of equity cost k_e on L

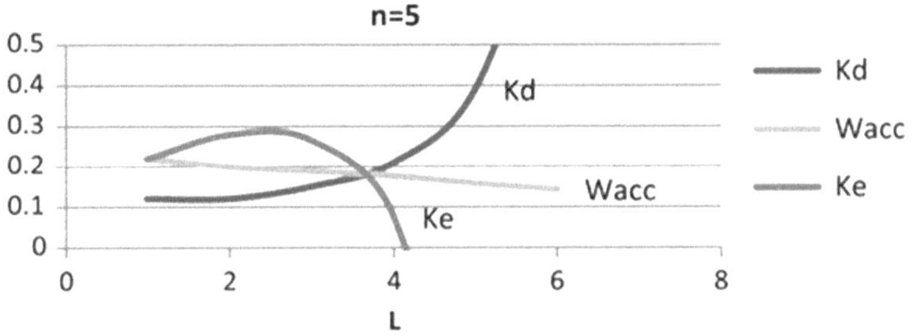

Fig. 3.5 Dependence of equity cost k_e, debt cost k_d, and WACC on leverage L

changes when the growth in the cost of debt capital k_d with the level of leverage is included. The linear growth of the cost of equity k_e at a low level of leverage is replaced by its fall, starting from a certain value L_0. The value of L_0 sometimes correlates exactly with the initial growth point k_d with leverage level, but sometimes it takes values much higher.

The rate of decline in the cost of equity k_e with leverage increases with an increase in the growth rate of the cost of debt k_d, as well as when moving from linear to quadratic growth and exponential growth. So, we come to the conclusion that the increase in the cost of debt capital k_d with leverage leads to the decrease of equity cost k_e with leverage, starting with some value L_0. **This is the cause of the absence of weighted average capital cost growth with leverage at all its values.** The conclusions made are independent of rate of growth of k_d with leverage (Figs. 3.4 and 3.5).

Thus, this means that there is no optimal capital structure in the famous theory of trade–offs. This paradoxical conclusion explains the absence of an optimal capital structure in the famous trade–off theory. This means that the competition between the benefits of leverage and the cost of financial hardship (or the cost of bankruptcy) is NOT balanced, and the hope that trade–off theory gives us an optimal capital structure is unfortunately not realized. Mechanism of formation of the company optimal capital structure, different from suggested by trade off theory has been developed by Brusov et al. (2014a).

3.4 Accounting of Transaction Cost

If the cost of changing the capital structure to its optimal value is high, the company may decide not to change the capital structure and maintain the current capital structure.

The company may decide that it is more cost effective not to change the capital structure, even if it is not optimal, for a certain period of time. Because of this, the actual and target capital structure may differ.

3.5 Accounting of Asymmetry of Information

In real financial markets, information is asymmetric (company managers have more reliable information than investors and creditors), and the rationality of economic entities is limited.

3.6 Signaling Theory

Information asymmetry can be reduced based on certain signals for lenders and investors related to the behavior of managers in the capital market. It should take into account the previous company development and the current and projected profitability of the activity (Myers 1984a, b).

3.7 Pecking Order Theory

The pecking order theory (Myers 1984a, b; Hovakimian et al., 2001; Fama and French 2005) describes a preferred sequence of funding types for raising capital. That is, companies first use financing from retained earnings (internal equity), the second source is debt, and the last source is the issuance of new shares of common stock (external equity). Empirical data on the leverage level of non–financial companies, combined with the decision–making process of top management and the board of directors, indicate a greater commitment to pecking order theory. Note that this theory contradicts both MM and BFO theory, which underline the quite importance of debt financing.

3.8 Behavioral Theories

Behavioral theories have been developed by Filbeck, Jenter (2005), Baker et al. (2002), Graham et al. (2011).

3.8.1 Manager Investment Autonomy

Company managers implement those decisions that, from their point of view, will be positively perceived by investors and, accordingly, will positively affect the market value of companies: when the market value of the company's shares and the degree of consensus between the expectations of managers and investors is high, the company conducts an additional share issue, and in the opposite situation uses debt instruments. Thus, the company capital structure is more influenced by investors, whose expectations are taken into account by company managers (Dittmar et al., 2007).

3.8.2 The Equity Market Timing Theory

The level of leverage is determined by market dynamics. Equity market timing theory means that a company should issue shares at a high price and repurchase them at a low price. The idea is to take advantage of the temporary fluctuations in the value of equity relative to the value of other forms of capital.

3.8.3 Information Cascades

To save costs and avoid mistakes, the company's capital structure can be formed not on the basis of calculations of the optimal capital structure (this is a non–trivial task and its correct solution could be found within the framework of the BFO theory) or depending on the available sources of financing of the company, but borrowed from other companies, having successful, proven managers (heads of companies), as well as using (following the majority) the most popular methods of managing the capital structure, or even simply by copying of the capital structure of successful companies in a similar industry (Bikhchandani et al., 1998).

3.9 Theories of Conflict of Interests

3.9.1 Theory of Agency Costs

The company management may make decisions that are contrary to the interests of shareholders or creditors, respectively; expenses are necessary to control its actions.

For solving the agency problem the correct choice of the compensation package (the agent's share in the property, bonus, stock options) is needed, which allows you to link the managers profit with the dynamics of equity capital and ensure the motivation of managers for preservation and growth of equity capital (Jensen et al., 1973).

3.9.2 Theory of Corporate Control and Costs Monitoring

In the presence of information asymmetry, creditors providing capital are interested in the possibility of self–control of the effectiveness of its use and return. Monitoring costs are usually passed on to company owners by being included in the loan rate. The level of monitoring costs depends on the scale of the business; therefore, with the increase in the scale of the business, the company weighted average capital cost increases, and the company market value decreases.

3.9.3 Theory of Stakeholders

Stakeholder theory is a theory that defines and models the groups that are the stakeholders of a company. The diversity and intersection of interests of stakeholders, their different assessment of acceptable risk give rise to conditions for a conflict of their interests, that is, they make adjustments to the process of the capital structure optimizing.

3.10 BFO Theory

The restriction associated with the infinite life of companies and the eternity of cash flows within the framework of the MM theory was removed in 2008 by Brusov, Filatova and Orekhova (Filatova et al. 2008), who created the modern theory of the cost of capital and capital structure—the BFO theory, which is valid both for companies of arbitrary age, as well as for companies with an arbitrary lifetime. A generalization of the assessment of the tax shield TS and the value of the company: without leverage V0 and with leverage V was required to modify the theory of MM (see the formulas below):

$$\mathrm{TS} = k_{\mathrm{d}} DT \sum_{t=1}^{n} (1 + k_{\mathrm{d}})^{-t} = DT[1 - (1 + k_{\mathrm{d}})^{-n}]. \tag{3.26}$$

$$V_0 = \mathrm{CF}[1 - (1 + k_0)^{-n}]/k_0; \quad V = \mathrm{CF}[1 - (1 + \mathrm{WACC})^{-n}]/\mathrm{WACC}. \tag{3.27}$$

$$\frac{1-(1+\text{WACC})^{-n}}{\text{WACC}} = \frac{1-(1+k_0)^{-n}}{k_0\left[1-\omega_d T\left(1-(1+k_d)^{-n}\right)\right]}. \tag{3.28}$$

Here, S stands for the equity capital value, $w_d = \frac{D}{D+S}$ stands for the debt capital share, $k_e, w_e = \frac{S}{D+S}$ stands for the equity cost and the equity capital share, and $L = D/S$ stands for the value of financial leverage, D stands for the debt capital value.

For a one–year company Steve Myers in 2001 has derived formula, which could be obtained easily from formula (3.28) by substituting n = 1

$$\text{WACC} = k_0 - \frac{(1+k_0)k_d}{1+k_d} w_d T$$

and by substituting $n = \infty$ we arrive to the Modigliani–Miller formula for WACC (Modigliani and Miller, 1963):

$$WACC = k_0 \cdot (1 - w_d t)$$

The methods and the conclusions of the Brusov–Filatova–Orekhova (BFO) theory are well–known in the scientific literature (for example, see references). In a number of works the BFO theory is used in practical calculations (Fig. 3.6).

3.10.1 Brusov–Filatova–Orekhova Theorem

3.10.1.1 Case of Absence of Corporate Taxes

Modigliani–Miller theory in case of absence of corporate taxes gives the following results for dependence of WACC and equity cost k_e on leverage:

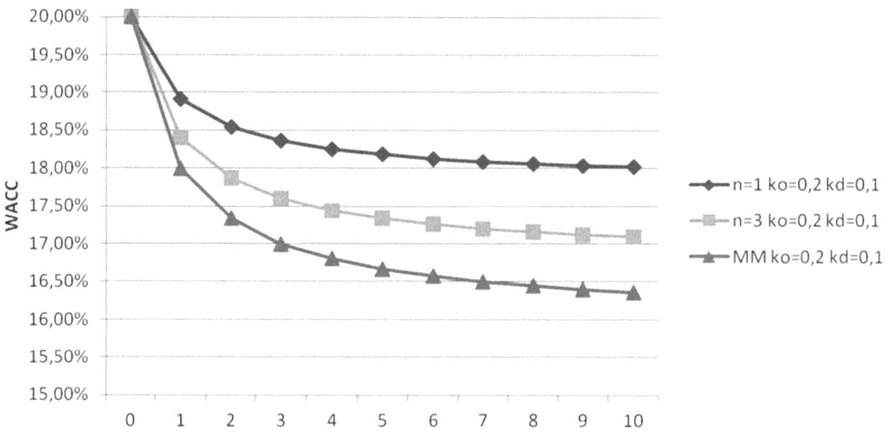

Fig. 3.6 Dependence of WACC on leverage level for $n = 1$, $n = 3$ and $n = \infty$

$$V_0 = V_L; \quad CF/k_0 = CF/WACC, \quad \text{and thus } WACC = k_0. \tag{3.29}$$

$$WACC = w_e \cdot k_e + w_d \cdot k_d \tag{3.30}$$

and thus

$$k_e = \frac{WACC - w_d \cdot k_d}{w_e} = \frac{k_0 - \frac{L}{1+L}k_d}{\frac{1}{1+L}} = k_0 + L(k_0 - k_d) \tag{3.31}$$

For the finite life–time (finite age) companies, Modigliani–Miller theorem about equality of value of financially independent and financially dependent companies ($V_0 = V_L$) has the following view (Brusov, 2012, 2013a, b, 2014a, b):

$$V_0 = V_L;$$
$$CF \cdot \frac{[1 - (1 + k_0)^{-n}]}{k_0} = CF \cdot \frac{[1 - (1 + WACC)^{-n}]}{WACC}. \tag{3.32}$$

Using this relation, we prove an important Brusov–Filatova–Orekhova theorem:

Under the absence of corporate taxes, the equity cost of the company, k_e, as well as its weighted average cost of capital, WACC, does not depend on the life–time (age) of the company and is equal, respectively, to

$$k_e = k_0 + L(k_0 - k_d); \quad WACC = k_0. \tag{3.33}$$

3.11 BFO Theory and Modigliani–Miller Theory Under Inflation

The influence of inflation on cost of raising capital and the company value has been studied within modern theory of capital cost and capital structure–Brusov–Filatova–Orekhova theory (BFO theory) (Brusov, 2012, 2013a, b, 2014a, b; Filatova et al. 2008; Brusova 2011) and within its perpetuity limit–Modigliani–Miller theory (Modigliani and Miller 1958, 1963; Miller and Modigliani 1966). For the first time it was shown (Brusov et al. 2014b) by the direct incorporation of inflation into both theories that inflation not only increases the cost of raising capital (the equity cost and the weighted average cost of capital), but as well it changes their dependence on leverage. In particular, it increases the growth rate of equity cost with leverage. The company value is decreased under inflation (Fig. 3.7).

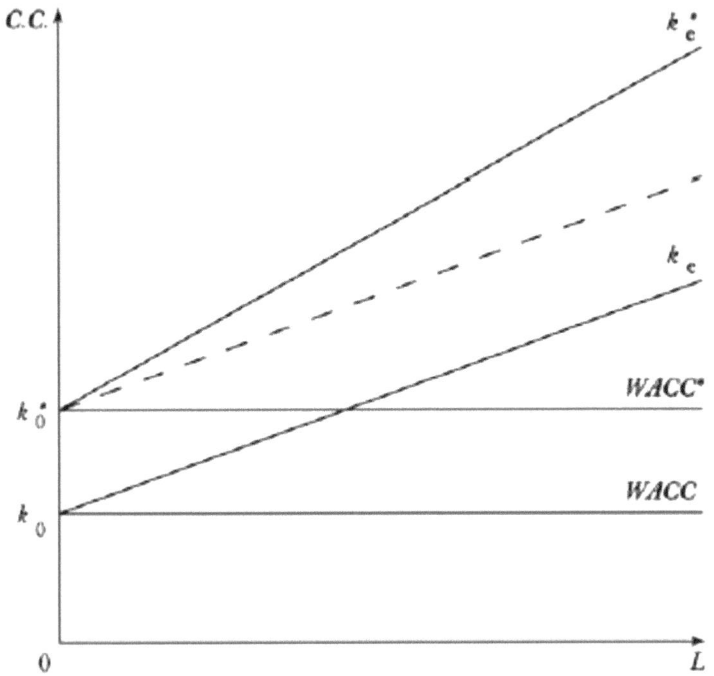

Fig. 3.7 Dependence of the equity cost and the weighted average cost of capital on leverage in the Modigliani–Miller theory without taxing under inflation. It is seen that growing rate of equity cost increases with leverage. Axis *y* means capital costs–CC

$$\frac{1-(1+\mathrm{WACC}^*)^{-n}}{\mathrm{WACC}^*} = \frac{1-[(1+k_0)(1+\alpha)]^{-n}}{(k_0(1+\alpha)+\alpha)\cdot[1-\omega_d T(1-((1+k_d)(1+\alpha))^{-n})]}. \tag{3.34}$$

Formula (3.34) is the generalized Brusov–Filatova–Orekhova formula under inflation.

Here, $\mathrm{WACC}^* = \mathrm{WACC}\cdot(1+\alpha)+\alpha$.

In the Modigliani–Miller theory (Fig. 3.8)

$$WACC^* = (k_0(1+\alpha)+\alpha)\cdot(1-w_d t)$$

$$\begin{aligned} k_e^* &= k_0^* + L(1-T)\left(k_0^* - k_d^*\right) \\ &= [k_0(1+\alpha)+\alpha] + L(1-T)(k_0-k_d)(1+\alpha). \end{aligned} \tag{3.35}$$

The modified equation for the weighted average cost of capital WACC, applicable to companies of arbitrary age under inflation, was derived in the framework of the modern theory of cost and capital structure—the theory of Brusov–Filatova–

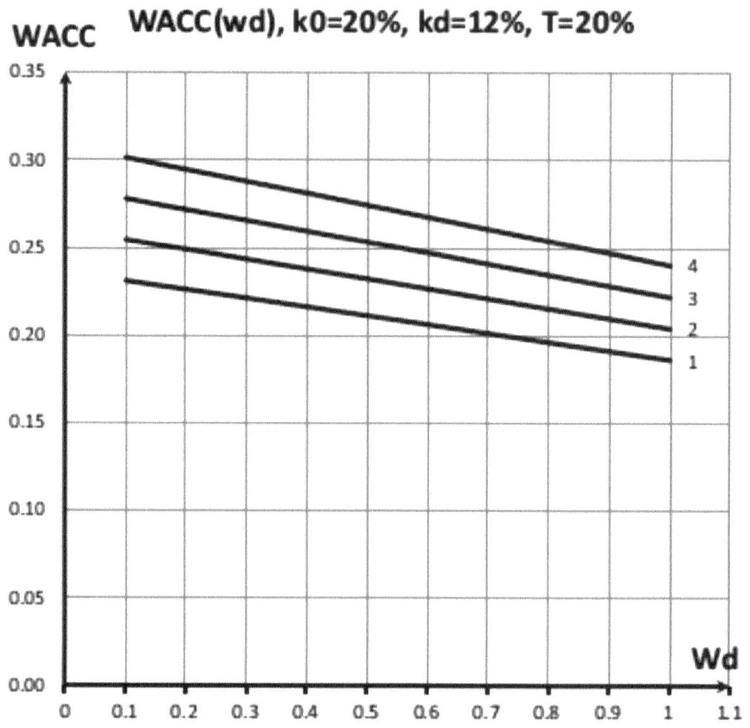

Fig. 3.8 Dependence of the weighted average cost of capital WACC on debt fraction w_d at different inflation rate α (1, $\alpha = 3\%$; 2, $\alpha = 5\%$; 3, $\alpha = 7\%$; and 4, $\alpha = 9\%$) for 5-year company

Orekhova (BFO theory). The modified BFO equation allows investigate the dependence of the weighted average cost of capital, WACC, and the cost of equity, k_e, on the level of financial leverage, L, on income tax rate, t, on the age of the company, n, on the cost of equity, k_0, and the cost of debt, k_d, for various inflation rates α.

Using the modified BFO equation, authors analyzed the dependence of the weighted average cost of capital WACC on the share of debt w_d at various income taxes t, as well as the inflation rate α.

It was shown that WACC decreases with an increase leverage level L (or the share of debt, w_d), and the faster, the higher the income tax rate t. The distance between the lines, corresponding to different values of the income tax rate at the same step (10%), increases with the growth of the inflation rate α. The area of change (with a change in the income tax rate t) of the weighted average cost of capital WACC increases with the inflation rate α, as well as with the age of the company n.

3.12 BFO Theory for the Companies Ceased to Exist at the Time Moment n (BFO–2 Theory)

The MM theory is only applicable to perpetual companies. As follows from the derivation of the BFO formula, the developed methodology is applied to companies of arbitrary age n (which have reached the age of n–years and continue to operate in the market). In other words, the BFO is applicable to the most interesting practical cases and allows you to analyze the financial condition of operating companies. However, the BFO theory also allows us to consider the financial condition of companies that have ceased to exist, i.e. those for which n means not age, but life–time, i.e. the time of existence. There can be many schemes for terminating a company: bankruptcy, merger, takeover, etc. Below we will consider one of these schemes, when the value of debt capital D becomes equal to zero at the time of the termination of the company n: in this case, the BFO theory requires minimal updates, as shown below (Brusov et al. 2018).

From the formula for the company's capitalization, it is easy to obtain an estimate of the "residual capitalization" of the company, discounted to the time moment *k:*

$$V_k = \sum_{t=k+1}^{n} \frac{\mathrm{CF}}{(1+\mathrm{WACC})^t} = \frac{\mathrm{CF}}{\mathrm{WACC}}\left[1-(1+\mathrm{WACC})^{-(n-k)}\right]. \tag{3.36}$$

Substituting

$$V_k = w_\mathrm{d} D, \tag{3.37}$$

one gets the expression for the tax shield for n years, subject to the termination of the company's activities at time moment n:

$$\begin{aligned} TS_n &= tk_\mathrm{d}w_\mathrm{d}\sum_{k=1}^{n}\frac{V_{k-1}}{(1+k_\mathrm{d})^k} = \frac{tk_\mathrm{d}w_\mathrm{d}CF}{\mathrm{WACC}}\sum_{k=1}^{n}\frac{1-(1+\mathrm{WACC})^{-(n-k+1)}}{(1+k_\mathrm{d})^k} = \\ &= \frac{tk_\mathrm{d}w_\mathrm{d}}{\mathrm{WACC}}\left[\frac{1-(1+k_\mathrm{d})^{-n}}{k_\mathrm{d}} - \frac{(1+k_\mathrm{d})^{-n}-(1+\mathrm{WACC})^{-n}}{\mathrm{WACC}-k_\mathrm{d}}\right] \end{aligned}. \tag{3.38}$$

Substituting this formula into equation

$$V_L = V_0 + (TS)_n \tag{3.39}$$

we arrive to the following equation (which is determines **the BFO–2 theory**)

$$\frac{1-(1+\text{WACC})^{-n}}{\text{WACC}}=\frac{1-(1+k_0)^{-n}}{k_0}+\frac{tk_\text{d}w_\text{d}}{\text{WACC}}\times\left[\frac{1-(1+k_\text{d})^{-n}}{k_\text{d}}-\frac{(1+k_\text{d})^{-n}-(1+\text{WACC})^{-n}}{\text{WACC}-k_\text{d}}\right], \tag{3.40}$$

whence it is possible to find WACC for companies with an arbitrary life–time n, under condition, that the company ceases to operate at time moment *n*.

3.13 The Modigliani–Miller Theory with Advance Payments of Tax on Profit

The first serious study (and first quantitative study) of influence of capital structure of the company on its indicators of activities was the work by Nobel Prize Winners Modigliani and Miller. Their theory has a lot of limitations. One of the most important and serious assumptions of the Modigliani – Miller theory is that all financial flows as well as all companies are perpetuity. This limitation was lift out by Brusov–Filatova–Orekhova in 2008 (Filatova et al. 2008), who have created BFO theory—modern theory of capital cost and capital structure for companies of arbitrary age. Despite the fact that the Modigliani–Miller theory is currently a particular case of the general theory of capital cost and capital structure—Brusov–Filatova–Orekhova (BFO) theory—it is still widely used at the West.

Brusov et al. (2020a) discuss another limitation of the Modigliani–Miller theory: the method of paying income tax. The Modigliani–Miller theory accounts for these payments as immediate annuities, while in practice these payments are often made in advance and therefore must be accounted for as annuity–due. The authors generalized the Modigliani–Miller theory to the case of advance payments on income tax, which is widely used in practice. This generalization leads to a serious change in all the main provisions of Modigliani and Miller. These changes are as follows: WACC begins to depend on the cost of debt k_d, WACC turns out to be lower than in the case of the classical Modigliani–Miller theory, and the value of the company becomes higher. The formula for WACC has the following form

$$WACC=k_0\left(1-w_d t(1+k_d)\right) \tag{3.41}$$

Although the dependence of the cost of equity k_e on the level of leverage L remains linear, the angle of inclination relative to the L axis turns out to be smaller: this may lead to a modification of the company's dividend policy. Correct accounting for the method of paying income tax shows that the shortcomings of the Modigliani–Miller theory are more serious than everyone expected: the underestimation of WACC is indeed greater, as is the overestimation of the company's capitalization. This means that the systematic risks arising from the use of the modified Modigliani–Miller theory (the MMM theory) (more correct than the "classical" theory) are in practice higher than it was supposed by the "classical" version of this theory.

3.14 The Modigliani–Miller Theory with Arbitrary Frequency of Payment of Tax on Profit

One of the conditions for the real functioning of companies is the payment of income tax with an arbitrary frequency (monthly, quarterly, every 6 months or annually). The return is not required more than once a year, but companies may be required to file estimated taxes based on profits earned. The authors of [1] generalized the Modigliani–Miller theory to the case of paying income tax with arbitrary periodicity. The formula for WACC has the following form Brusov and Filatova (2021)

$$WACC = k_0 \cdot \left(1 - \frac{k_d w_d t}{p} \frac{1}{(1 + k_d)^{1/p}} - 1 \right) \tag{3.42}$$

and for company value

$$V = \frac{WACC}{k_0 \cdot \left(1 - \frac{k_d w_d t}{p} \frac{1}{(1+k_d)^{1/p}} - 1 \right)} \tag{3.43}$$

They showed that the frequency of income tax payment affects all the main financial indicators of the company and leads to a number of important consequences. Theoretically, the authors derived all the main formulas of the modified Modigliani–Miller theory and used them to obtain all the main financial indicators of the company at arbitrary frequencies of income tax payment. They showed that: (1) all Modigliani–Miller theorems, assertions, and formulas have been changed; (2) all major financial indicators, such as the weighted average cost of capital (WACC), the value of the company, V and the cost of equity, k_e, depend on the frequency of income tax payments; (3) in the case of paying tax on profit more than once a year (with $p \neq 1$), as is the case in practice, WACC, company value, V and equity value, as well as equity cost k_e start depend on the cost of debt, k_d, and in the usual (classical) Modigliani–Miller theory, all these quantities do not depend on k_d; (4) the results obtained allow the company to choose the number of tax on profit payments per year (of course, within the framework of the current tax legislation): more frequent tax on profit payments are beneficial to both parties, the company and the tax regulator.

3.15 Generalization of the Modigliani–Miller Theory for the Case of Variable Profit

In (Brusov et al. 2021b), for the first time, the world–famous theory of Nobel laureates Modigliani and Miller was generalized to the case of variable income, which significantly expands the applicability of the theory in practice, in particular, in corporate finance, business valuation, investments, ratings etc. The following formulas for WACC and company value have been obtained

$$WACC - g = (k_0 - g) \cdot (1 - w_d t) \tag{3.44}$$

$$V = \frac{CF}{WACC - g} = \frac{CF}{(k_0 - g) \cdot (1 - w_d t)} \tag{3.45}$$

It was shown that all the theorems, assertions and formulas of Modigliani and Miller have been essentially changed. The authors obtained the following results: (1) WACC and k_0 are no longer the discount rates as it takes place in case of classical Modigliani–Miller theory with constant profit. The role of discount rate for leverage company pass from the weighted average cost of capital, WACC, to WACC–g (where g is growing rate), for a financially independent company from k_0 to k_0–g. The real discount rates WACC–g and k_0–g decrease with g, while WACC grows with g. This decrease leads to an increase of company value with g. (2) The slope of the equity cost k_e (L) grows with g. Via the fact, that the economically justified value of dividends is equal to equity cost this will modify the company's dividend policy. (3) It has been discovered the qualitatively new effect in corporate finance: the slope of the curve k_e (L) at rate $g < g^*$ turns out to be negative. This effect significantly alter the company's dividend policy principles.

3.16 The Generalization of the Brusov–Filatova–Orekhova Theory for the Case of Payments of Tax on Profit with Arbitrary Frequency

The Brusov–Filatova–Orekhova (BFO) theory and its limit of eternity, the Modigliani–Miller theory—both the main theories of the cost of capital and capital structure—consider income tax payments once a year. These payments in the real economy are made more frequently (every 6 months, quarter, month, etc.). In (Brusov et al. 2021b) the Modigliani–Miller theory was generalized by Brusov et al. for the case of paying income tax with arbitrary periodicity. In (Brusov et al. 2022b), for the first time, authors generalized the theory of Brusov–Filatova–Orekhova (BFO) to this case. Taking into account one of the features of the real functioning of companies—frequent payments of income tax—the authors brought the BFO theory closer to economic practice. Authors derived modified BFO formula

$$\frac{1 - (1 + WACC)^{-n}}{WACC} = \frac{1 - (1 + k_0)^{-n}}{k_0 \cdot \left(1 - \frac{k_d w_d t}{p} \frac{\left[1 - (1+k_d)^{-n}\right]}{(1+k_d)^{1/p} - 1} \right)} \tag{3.46}$$

and showed that: (1) All BFO formulas change; (2) all major financial parameters of a company, such as company capitalization, V, weighted average cost of capital, WACC, cost of equity, k_e, etc. depend on the frequency of tax on profit payments. The cost of raising capital WACC decreases with the frequency of tax on profit payments and the company value, V increases with the frequency of tax on profit payments. A qualitatively new anomalous effect takes place at some age of the company, n, and at some frequency of income tax payments p: the cost of equity k_e (L) decreases with leverage level L.

Since the economically justified amount of dividends is equal to the cost of equity, this fundamentally changes the company's dividend policy. More frequent income tax payments are beneficial to both parties—the company and the tax regulator: the company, because it increases the value of the company, and the tax regulator, because it benefits from earlier payments due to the time value of money.

3.17 Benefits of Advance Payments of Tax on Profit: Consideration within the Brusov–Filatova–Orekhova (BFO) Theory

Both theories of capital structure—the Brusov–Filatova–Orekhova (BFO) and its limit of eternity, the theory of Modigliani–Miller—consider the case of paying tax on profit at the end of the year. However, in practice, companies could make these payments in advance. In (Brusov et al. 2020a) the Modigliani–Miller theory was modified for the case of advance payments of tax on profit, and the results turned out to be completely different from the results of the usual Modigliani–Miller theory. In (Brusov et al. 2022a), the Brusov–Filatova–Orekhova (BFO) theory was modified for the first time for advance payments on income tax. The impact of such a transition to advance payments is much more significant than in the theory of MM, and even leads to a qualitatively new effect in the cost of equity dependence on leverage. Authors derived modified BFO formula for WACC

$$\frac{1-(1+WACC)^{-n}}{WACC}=\frac{1-(1+k_0)^{-n}}{k_0\cdot\left(1-w_d t[1-(1+k_d)^{-n}](1+k_d)^{1/p}\right)} \tag{3.47}$$

It is concluded that the tax shield and the way it is formed (payments at the end of the year or in advance) leads to very important consequences, changing all the company's financial indicators, such as the company's value and capital raising costs, and also significantly changing the company's dividend policy.

3.18 Influence of Method and Frequency of Profit Tax Payments on Company Financial Indicators

In practice, income tax payments may (1) be made in advance and (2) more frequently than once a year. In (Brusov et al. 2022a), the simultaneous influence of these two factors on the company's financial performance was studied. To this end, the Brusov–Filatova–Orekhova (BFO) theory was first generalized to the case of advance payments of income tax with arbitrary periodicity. Authors derived the following modified BFO formula for WACC

$$\frac{1-(1+WACC)^{-n}}{WACC}=\frac{1-(1+k_0)^{-n}}{k_0\cdot\left(1-\frac{k_d w_d t}{p}\frac{\left[1-(1+k_d)^{-n}\right]}{\left((1+k_d)^{1/p}-1\right)}(1+k_d)^{1/p}\right)} \quad (3.48)$$

It is concluded that all financial indicators of the company, such as the weighted average cost of capital (WACC) and the cost of equity (k_e), the value of the company (V) depend on the frequency of income tax payment. WACC increased with the frequency of payments, and the value of the company decreased. This meant that infrequent payments could be beneficial to the company in the case of advance income tax payments. The slope of the cost of equity, k_e (L), increased with the frequency of payments. Depending on the age of the company, k_e (L) either increased for some payment frequencies or decreased with L for all payment frequencies. The authors compared these results with their results for income tax payments at the end of the periods (see formula (3.47)) and found a huge difference: while in the latter case, WACC decreased with payment frequency, and the company capitalization increased with payment frequency, the value of WACC in this case remains bigger, and the company capitalization remains less than in the case of advance income tax payments of any frequency. The importance of advance income tax payments has been underlined by this fact. Authors developed the recommendations to regulator to increase the practice of advance payments of income tax due to the benefits of this for both parties: the companies and the state. A new effect—the decrease in equity cost with the level of leverage (L)—has been discovered.

3.19 The Brusov–Filatova–Orekhova (BFO) Theory with Variable Income

Two main theories of capital structure—Brusov–Filatova–Orekhova (BFO) and Modigliani–Miller consider the case of constant profit. But in practice, the company's income, of course, is variable. Both theories have recently been generalized to the case of variable income (Brusov et al. 2021b, 2022b), which significantly expands the applicability of both theories of capital structure in practice. The following generalized BFO formula for WACC has been discovered

$$\frac{1-\left(\frac{1+g}{1+WACC}\right)^n}{WACC-g}=\frac{1-\left(\frac{1+g}{1+k_0}\right)^n}{(k_0-g)\cdot(1-w_d t[1-(1+k_d)^{-n}])} \tag{3.49}$$

In the theory of Brusov–Filatova–Orekhova (BFO) (Brusov et al. 2022b), generalized to the case of variable income, income tax payments are made at the end of periods, while in practice these payments can also be made in advance. In (Brusov et al. 2022b) the authors considered two modifications of the Brusov–Filatova–Orekhova (BFO) theory with variable income: (1) with the payment of income tax at the end of periods and (2) with advance payments of income tax.

For these two cases, BFO formulas were derived for the weighted average cost of capital, WACC, for the value of the company, V, and within these formulas, a comprehensive analysis of the dependence of WACC, of the discount rate, WACC–g (here g is the growth rate), company capitalization, V, cost of equity, k_e, on debt financing at different values of the growth rate, g, at different values of the cost of debt capital, k_d, and at different values of the age of the company, n. The results for cases (1) and (2) are compared, which allows us to conclude that case (2) is always preferable for both the company and the regulator. This makes it possible to develop recommendations for both parties to expand the practice of advance income tax payments in the framework of real economic practice.

Brusov et al. emphasized an important observation. If in the classical versions of the Brusov–Filatova–Orekhova (BFO) theory and its perpetual limit—the theory of Nobel laureates Modigliani and Miller, where the case of constant profit was considered, and where the gap between these two theories is huge (many qualitative effects that take place in the first theory, missing in the second), when taking into account the variable profit, some effects of the BFO theory also take place in the Modigliani–Miller theory. This means that taking into account some effects that are present in economic practice (for example, variable income) brings both theories closer, and even the Modigliani–Miller theory, with all its many limitations, becomes more applicable in economic practice. However, it should be remembered that the Modigliani–Miller theory is only true for perpetual companies, while the BFO theory is valid for companies of any age and from this point of view they never coincide.

3.20 Qualitatively New Effects in the Theory of Capital Structure

Two Qualitatively new effects in the theory of capital structure.

Among the many new effects discovered in the framework of the BFO Theory, we will describe only two qualitatively new ones:

1. Golden and silver ages of the company
2. Anomalous dependence of the cost of equity on the leverage level

3.20.1 Golden and Silver Ages of the Company

To determine the minimum cost of raising capital depending on the age of the company, Brusov et al. In 2015, studied the dependence of the cost of raising capital on the age of the company, n, at different levels of leverage, L, at different values of capital costs. The authors used the Brusov–Filatov–Orekhova theory (Brusov & Filatova, 2011; Brusov et al., 2011b, c, 2012, 2013a, b, 2014a, b; Filatova et al., 2008; Brusova, 2011).

It was shown for the first time that the value of WACC in the Modigliani–Miller theory (Modigliani and Miller, 1958, 1963; Miller and Modigliani 1966) is not minimal, and the value of the company is not maximal, contrary to the opinion of all financiers: it turned out that at a certain age of the company, the value of WACC is lower than in the eternal Modigliani–Miller theory, and the value of company V is greater than the value of company V in the Modigliani–Miller theory.

Authors discovered two types of WACC dependences on the age of company n: monotonic decreasing with n and WACC decreasing with passage through minimum, followed by a limited growth. The latter type of dependence of WACC on the age of company n allows the companies take advantage of the benefits given at a certain stage of their development.

Moreover, since capital costs, k_e and k_d, affect the company's golden age by changing it (for example, changing the value of dividend payments reflecting the cost of equity, etc.), the company can postpone or extend its golden age. at which the cost of raising capital is minimal (and less than its perpetual limit), and the value of companies is maximum (above their perpetual value) (Fig. 3.9).

It has been shown that existence of the company "golden age" depends not on the capital costs values, but on the difference between equity k_0 and debt k_d costs. The effect of the company "golden age" takes place at small enough difference between k_0 and k_d costs, while at this difference high value this effect is absent: curve WACC (n) monotonic decreases with company age n. Curve WACC (L) for perpetuity limit ($n = \infty$) lies between curves WACC(L) for 1 year ($n = 1$) and 3 years ($n = 3$) companies (Fig. 3.10) for the companies with the "golden age", while curve WACC (L) for perpetuity limit is the lowest one for the companies without the "golden age" (Fig. 3.11).

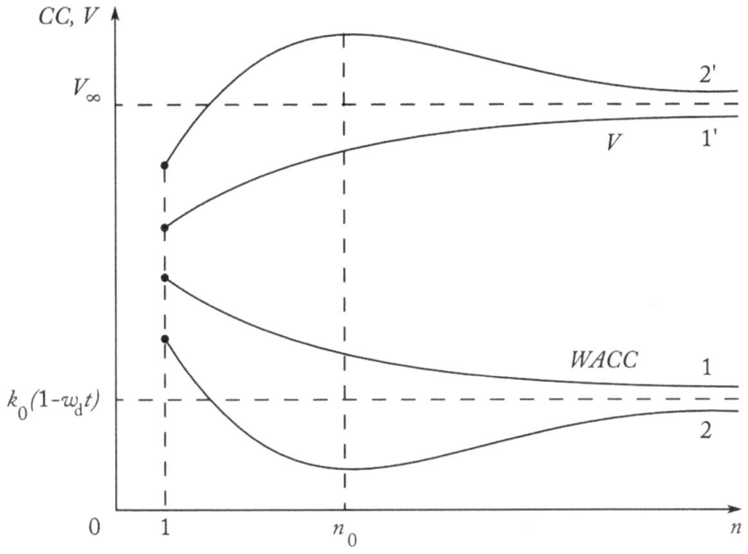

Fig. 3.9 Two types of dependences of WACC, and company value, V, on the company age n: 1–1′—monotonic decreasing of WACC, and company value, V, increasing with the company age n; 2–2′—WACC decreasing with n, passage through minimum(at n_0), followed by a limited growth and increase of V with the passage through a maximum (at n_0) and then a limited decreasing

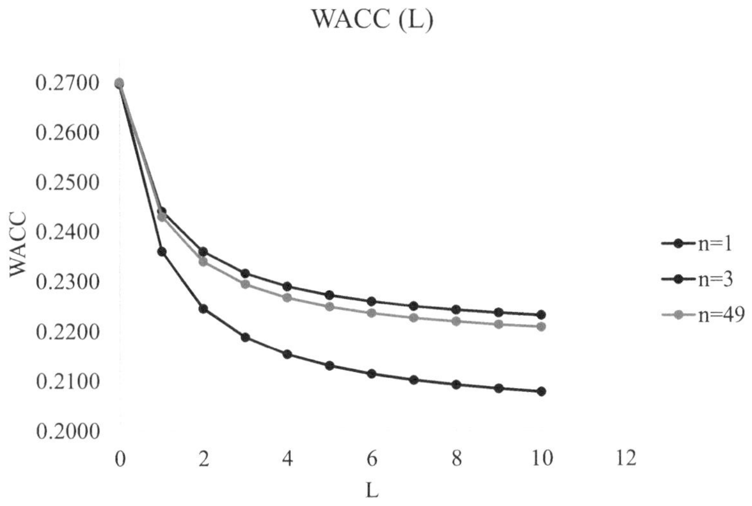

Fig. 3.10 The dependence of WACC on leverage level L at different the company age n ($n = 1{,}3{,}$ 49)

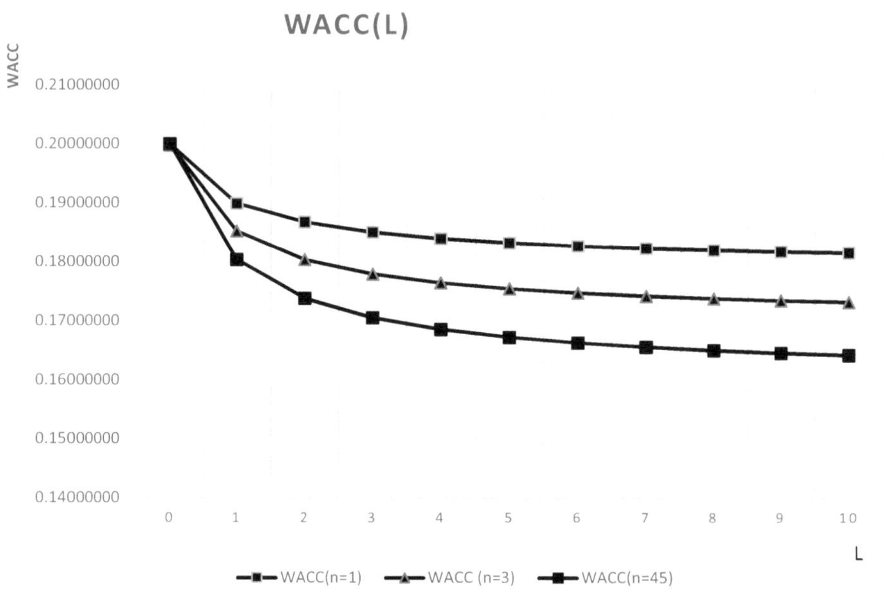

Fig. 3.11 The dependence of WACC on the leverage level, L, at different company age *n* (n = 1, 3, 45)

3.20.2 Silver Age of the Company (Fig. 3.12)

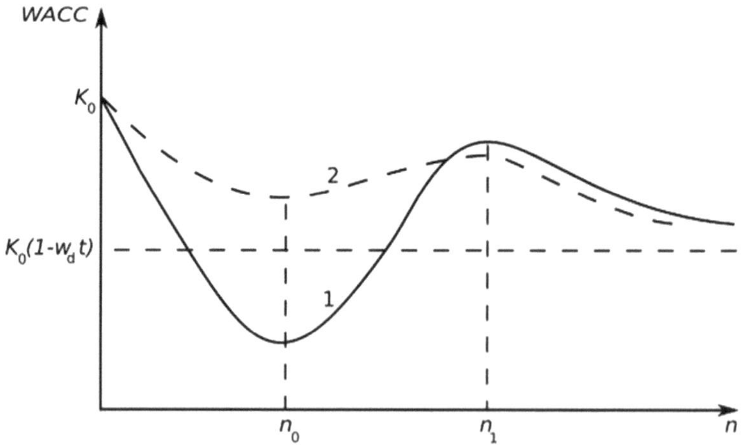

Fig. 3.12 Dependence of WACC on company age n: company "golden age" (curve 1) and company "silver age" (curve 2) under existence of "Kulik" effect, where n_0 is company "golden (silver) age" and n_1 is the company age of local maximum in the dependence of WACC (n)

3.20.3 *Anomalous Dependence of the Company's Equity Value on Leverage*

Brusov et al.(2013b) found a qualitatively new effect in the capital structure: a decrease in the cost of equity at a leverage level L. This effect is absent in the perpetual Modigliani–Miller limit, but appears in companies of finite age at an income tax rate exceeding a certain value T*.

For some values of the cost of equity and the cost of debt, this effect exists at the tax on income values existing in financial practice. This ensures the practical significance of the discovered effect. Accounting for this effect is important when changing tax legislation and can significantly change the company's dividend policy (Fig. 3.13).

3.21 A Stochastic Extension the Modigliani-Miller Theory

A Stochastic Extension of the Modigliani–Miller theory has been done in a few papers.

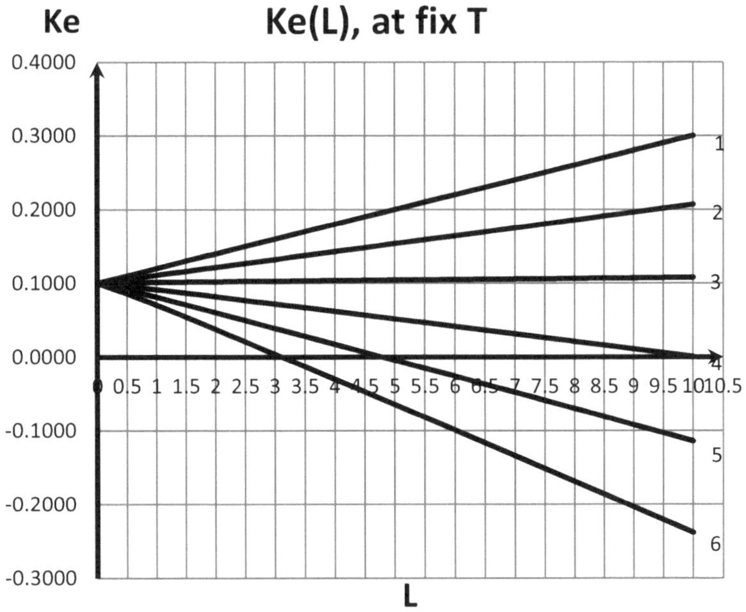

Fig. 3.13 Dependence of equity cost, k_e, on leverage level L at different tax on income rate T ($n = 5$, $k_0 = 10\ \%$, $k_d = 8\%$) (1—$T = 0$; 2—$T = 0.2$; 3—$T = 0.4$; 4—$T = 0.6$; 5—$T = 0.8$; 6—$T = 1$)

Demarzo (1988) showed that the Modigliani-Miller theory is correct in a multi-period stochastic economy with incomplete markets. In the model author developed, companies can trade securities, and stock prices and dividends become completely interdependent. Several corporate control mechanisms were considered, including maximizing the value of the company and shareholder voting. It is shown that in equilibrium companies have no incentive to trade securities, and the set of equilibrium distributions does not change as a result of such trading.

Sethi et al. (1991) study the company value valuation problem under a partial equilibrium, a time-dependent discount rate and a general stochastic environment in a discrete-time setting. The time path of share price and the company value are obtained upon the assumption of risk neutrality.

3.22 Conclusions

This Chapter analyzes all the main existing theories of capital structure, including Modigliani–Miller and BFO. Various modifications of these two and other theories, including the latest ones, related to the conditions of the real functioning of companies and developed in recent years are discussed in detail.

The generalization of Modigliani–Miller theory for the cases of practical interest lead to a much more significant change in Modigliani–Miller theory than all previous modifications, studied over the previous decades. Obtained results showed that under such a generalization there is some convergence of Modigliani–Miller theory and BFO theory: the former theory begins to have some new properties similar to those of BFO theory, which was not there in the classical Modigliani–Miller theory, between the properties of which and of the BFO theory was a huge gap. Most of the innovative effects existing in the BFO theory are absent in the classical Modigliani–Miller theory.

All of the above means that the considered effects, such as the variable income of the company, frequent paying income tax, advance payment of income tax, the combinations of this conditions and others, are more important than the effects studied earlier and having much less influence on the Modigliani–Miller theory. Taking into account these effects leads to some convergence between Modigliani–Miller theory and the BFO theory. At the same time, one must understand that Modigliani–Miller theory remains perpetual and will never describe a company of arbitrary age, which is extremely important for the practical application of the theory.

References

Acharya, V., Myers, S. C., & Rajan, R. (2011). The internal governance of firms. *The Journal of Finance, 66*(3), 689–720.

Arrow, K. J. (1963). The role of securities in the optimal allocation of risk–bearing. *The Review of Economic Studies, 31*(2), 91–96.
Asquith, P., & Mullins, D. W. (1986). Equity issues and offering dilution. *Journal of Financial Economics, 15*(1–2), 61–89.
Baker, J., Parasuraman, A., Grewal, D., & Voss, G. B. (2002). The influence of multiple store environment cues on perceived merchandise value and patronage intentions. *Journal of Marketing, 66*(2), 120–141. https://doi.org/10.1509/jmkg.66.2.120.18470
Baumol, W, Panzar, J. G., & Willig, R. D. (1982). *Contestable markets and the theory of industry structure.*
Becker, D. M. (2021). The difference between Modigliani–Miller and Miles–Ezzell and its consequences for the valuation of annuities. *Cogent Economics & Finance, 9*, 1. https://doi.org/10.1080/23322039.2020.1862446
Berk, J., & DeMarzo, P. (2007). *Corporate finance*. Pearson–Addison Wesley.
Bhattacharya, S. (1978). Project valuation with mean–reacting cash flows. *The Journal of Finance, 33*, 1317–1331.
Bikhchandani, S., Hirshleifer, D., & Welch, I. (1998). Learning from the behavior of others: Conformity, fads, and informational cascades. *The Journal of Economic Perspectives, 12*(3), 151–170.
Brealey, R. A., & Myers, S. C. (1981). *Principles of corporate finance* (1st ed.). McGraw–Hill.
Brealey, R. A., Myers, S. C., & Allen, F. (2013). *Principles of corporate finance* (11th ed.). McGraw–Hill.
Brennan, M., & Schwartz, E. (1978). Finite difference methods and jump processes arising in the pricing of contingent claims: A synthesis. *Journal of Financial and Quantitative Analysis, 13*, 461–474. https://doi.org/10.2307/2330152
Brusov, P., & Filatova, T. (2021). The Modigliani–Miller theory with arbitrary frequency of payment of tax on profit. *Mathematics, 9*(11), 1198.
Brusov, P., & Filatova, T. (2022a). Influence of method and frequency of profit tax payments on company financial indicators. *Mathematics, 10*(14), 2479. https://doi.org/10.3390/math10142479
Brusov, P., & Filatova, T. (2022b). Generalization of the Brusov–Filatova–Orekhova theory for the case of variable income. *Mathematics, 10*(19), 3661. https://doi.org/10.3390/math10193661
Brusov, P., Filatova, T., Eskindarov, M., & Orehova, N. (2012). Hidden global causes of the global financial crisis. *Journal of Reviews on Global Economics, 1*, 106–111.
Brusov, P., Filatova, P., & Orekhova, N. (2013a). Absence of an optimal capital structure in the famous tradeoff theory! *Journal of Reviews on Global Economics, 2*, 94–116.
Brusov, P., Filatova, T., & Orehova, N. (2013b). A qualitatively new effect in corporative finance: Abnormal dependence of cost of equity of company on leverage. *Journal of Reviews on Global Economics, 2*, 183–193.
Brusov, P., Filatova, P., & Orekhova, N. (2014a). Mechanism of formation of the company optimal capital structure, different from suggested by trade off theory. *Cogent Economics & Finance, 2*, 1–13. https://doi.org/10.1080/23322039.2013.946150
Brusov, P., Filatova, T., & Orehova, N. (2014b). Inflation in Brusov–Filatova–Orekhova theory and in its perpetuity limit–Modigliani–Miller theory. *Journal of Reviews on Global Economics, 3*, 175–185.
Brusov, P., Filatova, T., Orehova, N., & Eskindarov, M. (2018). *Modern corporate finance, investments, taxation and ratings* (2nd ed., pp. 1–571). Springer Nature Publishing.
Brusov, P. N., Filatova, T. V., Orekhova, N. P., Kulik, V. L., Chang, S. I., & Lin, Y. C. G. (2020a). Modification of the Modigliani–Miller theory for the case of advance payments of tax on profit. *Journal of Reviews on Global Economics, 9*, 257–267.
Brusov, P. N., Filatova, T. V., Orekhova, N. P., Kulik, V. L., Chang, S.-I., & Lin, Y. C. G. (2020b). Application of the Modigliani–Miller theory, modified for the case of advance payments of tax on profit, in rating methodologies. *Journal of Reviews on Global Economics, 9*, 282–292.

Brusov, P., Filatova, T., Chang, S.-I., & Lin, G. (2021a). Innovative investment models with frequent payments of tax on income and of interest on debt. *Mathematics, 9*(13), 1491.
Brusov, P., Filatova, T., Orekhova, N., Kulik, V., Chang, S.-I., & Lin, G. (2021b). Generalization of the Modigliani–Miller theory for the case of variable profit. *Mathematics, 9*(11), 1286.
Brusov, P., Filatova, T., Orekhova, N., Kulik, V., Chang, S.-I., & Lin, G. (2022a). The generalization of the Brusov–Filatova–Orekhova theory for the case of payments of tax on profit with arbitrary frequency. *Mathematics, 10*(8), 1343. https://doi.org/10.3390/math10081343
Brusov, P., Filatova, T., & Kulik, V. (2022b). Benefits of advance payments of tax on profit: Consideration within the Brusov–Filatova–Orekhova (BFO) theory. *Mathematics, 10*(12), 2013. https://doi.org/10.3390/math10122013
Brusova, A. (2011). A comparison of the three methods of estimation of weighted average cost of capital and equity cost of company. *Financial Analysis Problem Solving, 34*(76), 36–42.
DeMarzo, P. M. (1988). An extension of the Modigliani-Miller theorem to stochastic economies with incomplete markets and fully interdependent securities. *Journal of Economic Theory, 45*, 353–369.
Derrig, R. (1993). Theoretical considerations of the effect of federal income taxes on investment income in property–liability ratemaking. *The Journal of Risk and Insurance, 61*, 691–709.
Diamond, D. A., & He, Z. (2013). A theory of debt maturity: The long and short of debt overhang. *The Journal of Finance, 69*(2), 719–762.
Dittmar, H. (2007). The cost of consumers and the cage within: The impact of the material good life and body perfect ideals on individuals identity and Well-being. *Psychological Inquiry, 18*(1), 23–31.
Donaldson, G. (1961). *Debt capacity: A study of corporate debt policy and the determination of debt capacity*. Harvard University-Graduate School of Business Administration-Division of Research.
Erel, I., Myers, S. C., & Read, J. A., Jr. (2015). A theory of risk capital. *Journal of Financial Economics, 118*, 620–635. https://doi.org/10.1016/j.jfineco.2013.10.006
Fairley, W. B. (1979). Investment income and profit margins in property–liability insurance. *Bell Journal of Economics, 10*, 191–210.
Fama, E. F. (1977). Risk adjusted discount rates and capital budgeting under uncertainty. *Journal of Financial Economics, 5*(1), 3–24.
Fama, E. F. (1996). Discounting under uncertainty. *Journal of Business, 69*(4), 415–429.
Fama, E. F., & French, K. (2005). Financing decisions: Who issues stock? *Journal of Financial Economics, 76*(3), 549–582. Federal Power Commission et al. v. Hope Natural Gas Co., 320 U.S. 591 (1944).
Farber, A., Gillet, R., & Szafarz, A. (2006). A general formula for the WACC. *International Journal of Business, 11*(2), 211–218.
Fernandez, P. (2006). A general formula for the WACC: A comment. *International Journal of Business, 11*(2), 219.
Filatova, T., Orehova, N., & Brusova, A. (2008). Weighted average cost of capital in the theory of Modigliani–Miller, modified for a finite life–time company. *Bull FU, 48*, 68–77.
Filatova, T., Brusov, P., & Orekhova, N. (2022). Impact of advance payments of tax on profit on effectiveness of investments. *Mathematics, 10*(4), 666. https://doi.org/10.3390/math10040666
Fisher, F. M., & McGowan, J. J. (1983). On the misuse of accounting rates of return to infer monopoly profits. *The American Economic Review, 73*, 82–97.
Frank, M. Z., & Goyal, V. K. (2003). Testing the pecking order theory of capital structure. *Journal of Financial Economics, 67*, 217–248.
Fulghieri, P., Garcia, D., & Hackbarth, D. (2013). *Asymmetric information and the pecking (dis–) order*. Working Paper, Boston University, University of North Carolina.
Gordon, M. J., & Shapiro, E. (1956). Capital equipment analysis: The required rate of profit. *Management Science, 3*, 102–110.
Graham, J. R., & Leary, M. T. (2011). A review of empirical capital structure research and directions for the future. *Annual Review of Financial Economics, 3*, 309–345.

Hamada, R. (1969). Portfolio analysis, market equilibrium, and corporate finance. *The Journal of Finance, 24*, 13–31.
Harcourt Brace Jovanovich Black, F. (1986). Noise. *The Journal of Finance, 41*(3), 529–543.
Harris, R., & Pringle, J. (1985). Risk–adjusted discount rates—Extension form the average–risk case. *The Journal of Financial Research, 8*(3), 237–243.
Hausman, J., & Myers, S. C. (2002). Regulating U.S. railroads: The effects of sunk costs and asymmetric risk. *Journal of Regulatory Economics, 22*(3), 287–310.
Healy, P., Howe, C., & Myers, S. C. (2002). R&D accounting and the tradeoff between relevance and objectivity: A pharmaceutical industry simulation. *Journal of Accounting Research, 40*(3), 677–710.
Hirshleifer, J. (1965). Investment decision under uncertainty: Choice–theoretic approaches. *Quarterly Journal of Economics, 79*(4), 509–536.
Hirshleifer, J. (1966). Investment decision under certainty: Applications of the state–preference approach. *Quarterly Journal of Economics, 80*(2), 252–277.
Hovakimian, A., Opler, T., & Titman, S. (2001). The debt-equity choice. *Journal of Financial and Quantitative Analysis, 36*, 1–24. https://doi.org/10.2307/2676195
Hsia, C.-C. (1981). Optimal debt of a firm: An option pricing approach. *Journal of Financial Research, Southern Finance Association; Southwestern Finance Association, 4*(3), 221–231.
Jensen, M. C. (1986). Agency costs of free cash flow, corporate finance and takeovers. *The American Economic Review, 76*(2), 323–329.
Jenter, D. (2005). Market timing and managerial portfolio decisions. *Journal of Finance, American Finance Association, 60*(4), 1903–1949.
Jin, L., & Myers, S. C. (2006). R2 around the world: New theory and new tests. *Journal of Financial Economics, 79*(2), 257–292.
Kaplan, S., & Ruback, R. (1995). The valuation of cash flow forecasts: An empirical analysis. *The Journal of Finance, 50*(4), 1059–1093.
Kane, Alex & Marcus, Alan J & McDonald, Robert L, 1984. How big is the tax advantage to debt?, Journal of Finance, *American Finance Association,* 39(3).
Keynes, J. M. (1936). *The general theory of employment, interest and money*. Macmillan.
Kolbe, A. L., Tye, W. B., & Myers, S. C. (1993). *Regulatory risk: Economic principles and applications to natural gas pipelines and other industries*. Springer.
Lambrecht, B., & Myers, S. C. (2007). A theory of takeovers and disinvestment. *The Journal of Finance, 62*(2), 809–845.
Lambrecht, B., & Myers, S. C. (2008). Debt and managerial rents in a real–options model of the firm. *Journal of Financial Economics, 89*(2), 209–231.
Lambrecht, B., & Myers, S. C. (2012). A Lintner model of dividends and managerial rents. *The Journal of Finance, 67*(5), 1761–1810.
Lambrecht B, Myers SC. 2015. The dynamics of investment, payout and debt. Working Paper, Massachusetts Institute of Technology.
Leland, H. E. (1994). Corporate debt value, bond covenants, and optimal capital structure. *The Journal of Finance, 49*(4), 1213–1252.
Lemmon, M. L., & Zender, J. (2010). Debt capacity and tests of capital structure theories. *Journal of Financial and Quantitative Analysis, 45*, 1161–1187.
Lemmon, M. L., Roberts, M. R., & Zender, J. (2008). Back to the beginning: Persistence and the cross–section of corporate capital structure. *The Journal of Finance, 63*(4), 1575–1608.
Lintner, J. (1956). Distribution of incomes of corporations between dividends, retained earnings and taxes. *The American Economic Review, 46*, 97–113.
Lintner, J. (1965). Optimal dividends and corporate growth under uncertainty. *Quarterly Journal of Economics, 77*, 59–95.
Majd, S., & Myers, S. C. (1987). Tax asymmetries and corporate income tax reform. In M. Feldstein (Ed.), *The effects of taxation on capital accumulation* (pp. 343–373). The University of Chicago Press.

Majluf, N. S. (1978). *Study on mergers: A rationale for conglomerate mergers*. PhD Thesis, Massachusetts Institute of Technology.
Merton, R. C., & Perold, A. F. (1993). Theory of risk capital in financial firms. *Journal of Applied Corporate Finance, 6*, 16–32.
Miles, J., & Ezzell, R. (1980). The weighted average cost of capital, perfect capital markets and project life: A clarification. *Journal of Financial and Quantitative Analysis, 15*(3), 719–730.
Miller, M. H. (1977). Debt and taxes. *The Journal of Finance, 32*, 261–276.
Miller, M. H., & Modigliani, F. (1961). Dividend policy, growth and the valuation of shares. *Journal of Business, 34*, 411–433.
Miller, M. H., & Modigliani, F. (1966). Some estimates of the cost of capital to the electric utility industry, 1954–57. *The American Economic Review, 56*, 333–391.
Modigliani, F., & Miller, M. H. (1958). The cost of capital, corporation finance and the theory of investment. *The American Economic Review, 48*(3), 261–296.
Modigliani, F., & Miller, M. H. (1963). Corporate income taxes and the cost of capital: A correction. *The American Economic Review, 53*, 433–443.
Morck, R., Yeung, B. Y., & Yu, W. (2000). The information content on stock markets: Why do emerging markets have synchronous stock price movements? *Journal of Financial Economics, 58*, 215–260.
Mullins, D.W. Jr. (1976). Communications Satellite Corporation. Case Study 276195, Harvard Business School.
Myer Annual Report. (2015). *Review of Financial Economics. 7*:1–33. Downloaded from www.annualreviews.orgAccess provided by University of California—Los Angeles UCLA on 12/11/18. For personal use only.
Myers, S. C. (1967). *Effects of uncertainty on the valuation of securities and the financial decisions of the firm*. Ph.D. thesis, Stanford University.
Myers, S. C. (1968a). A time–state–preference model of security valuation. *Journal of Financial and Quantitative Analysis, 3*, 1–33.
Myers, S. C. (1968b). Procedures for capital budgeting under uncertainty. *Indian Management Review, 9*(3), 1–20.
Myers, S. C. (1972). Application of finance theory to public utility rates cases. *Bell Journal of Economics, 3*, 58–97.
Myers, S. C. (1973). Interactions of corporate financing and investment decisions: Implications for capital budgeting. *The Journal of Finance, 29*, 1–25.
Myers, S. C. (1973a). A simple model of firm behavior under regulation and uncertainty. *Bell Journal of Economics, 4*, 304–315.
Myers, S. C. (1973b). On public utility regulation under uncertainty. In R. H. Howard (Ed.), *Risk and regulated firms* (pp. 32–46). Div. Res. Grad. Sch. Bus., Mich. State Univ.
Myers, S. C. (1977). Determinants of corporate borrowing. *Journal of Financial Economics, 5*(2), 147–175.
Myers, S. C. (1984a). Finance theory and financial strategy. *Interfaces, 14*, 126–137.
Myers, S. C. (1984b). The capital structure puzzle. *The Journal of Finance, 39*(3), 575–592.
Myers SC. 1988. Fuzzy efficiency. Insta Invest. 8–9.
Myers, S. C. (1989). Still searching for optimal capital structure. In R. W. Kopke & E. S. Rosengren (Eds.), *Are the distinctions between debt and equity disappearing?* (pp. 80–95). Federal Reserve Bank.
Myers, S. C. (1999). Financial architecture. *European Financial Management, 5*(2), 133–141.
Myers, S. C. (2000). Outside equity. *The Journal of Finance, 55*(3), 1005–1037.
Myers, S. C., & Cohn, R. (1987). A discounted cash flow approach to property–liability insurance rate regulation. In J. D. Cummings & S. Harrington (Eds.), *Fair rate of return in property liability insurance* (pp. 55–78). Kluwer–Nijhoff.
Myers, S. C., & Howe, C. (1997). *A life cycle model of pharmaceutical R&D*. Work. Pap., Progr. Pharm. Ind., Mass. Inst. Technol.

Myers, S. C., & Majd, S. (1990). Abandonment value and project life. In F. Fabozzi (Ed.), *Advances in futures and options research* (Vol. 4, pp. 1–21). JAI.

Myers, S. C., & Majluf, N. S. (1983). Corporate financing and investment decisions when firms have information that investors do not have. *Journal of Financial Economics, 13*(2), 187–221.

Myers, S. C., & Pogue, G. A. (1973). A programming approach to corporate financial management. *The Journal of Finance, 29*(2), 579–599.

Myers, S. C., & Rajan, R. (1998). The paradox of liquidity. *Quarterly Journal of Economics, 113*(3), 733–771.

Myers, S. C., & Read, J. A., Jr. (2001). Capital allocation for insurance companies. *The Journal of Risk and Insurance, 68*(4), 545–580.

Myers, S. C., & Read, J. A. Jr. (2013). *Real options, taxes and leverage*. Work. Pap., Mass. Inst. Technol.

Myers, S. C., & Shyam-Sunder, L. (1996). Measuring pharmaceutical industry risk and the cost of capital. In R. B. Helms (Ed.), *Competitive strategies in the pharmaceutical industry* (pp. 208–237). American Enterprise Institute.

Myers, S. C., & Turnbull, S. M. (1977). Capital budgeting and the capital asset pricing model: Good news and bad news. *The Journal of Finance, 32*(2), 321–333.

Myers, S. C., Dill, D. A., & Bautista, A. J. (1976). Valuation of financial lease contracts. *The Journal of Finance, 31*(3), 799–819.

Myers, S. C., Kolbe, A. L., & Tye, W. B. (1983). Regulation and capital formation in the oil pipeline industry. *Transportation Journal, 23*(4), 25–49.

Myers, S. C., Kolbe, A. L., & Tye, W. B. (1985). Inflation and rate of return regulation. *Research in Transportation Economics, 2*, 83–119.

Robichek, A. A., & Myers, S. C. (1965). *Optimal financing decisions*. Prentice-Hall.

Robichek, A. A., & Myers, S. C. (1966a). Conceptual problems in the use of risk–adjusted discount rates. *The Journal of Finance, 21*(4), 727–730.

Robichek, A. A., & Myers, S. C. (1966b). Problems in the theory of optimal capital structure. *The Journal of Financial and Quantitative Analysis, 1*(2), 1–35.

Robichek, A. A., & Myers, S. C. (1966c). Valuation of the firm: Effects of uncertainty in a market context. *The Journal of Finance, 21*(2), 215–227.

Rosenbaum, J., & Pearl, J. (2013). *Investment banking* (2nd ed.). Wiley.

Rubinstein, M. E. (1973). A mean-variance synthesis of corporate financial theory. *The Journal of Finance, 28*, 167–181.

Schwartz, E. (1959). The theory of the capital structure of the firm. *The Journal of Finance, 14*, 18–39.

Sethi, S., Derzko, N., & Lehoczky, J. (1991). A stochastic extension of the Miller-Modigliani framework (1991). *Mathematical Finance, 1*(4), 57–76. SSRN: https://ssrn.com/abstract=1082940.

Sharpe, W. F. (1965). Capital asset prices: A theory of market equilibrium under conditions of risk. *The Journal of Finance, 19*(3), 425–442.

Shleifer, A., & Summers, L. H. (1988). Breach of trust in hostile takeovers. In A. J. Auerbach (Ed.), *Corporate takeovers: Causes and consequences* (pp. 33–56). The University of Chicago Press.

Shyam-Sunder, L., & Myers, S. C. (1999). Testing static tradeoff against pecking order models of capital structure. *Journal of Financial Economics, 51*(2), 219–244.

Skinner, D. J. (2008). The evolving relation between earnings, dividends and stock repurchases. *Journal of Financial Economics, 87*, 582–609.

Solomon, E., & Laya, J. (1967). Measurement of company profitability: Some systematic errors in the accounting rate of return. In A. A. Robichek (Ed.), *Financial research and management decisions* (pp. 152–183). Wiley.

Stiglitz, J. (1969). A re-examination of the Modigliani-miller theorem. *American Economic Review, 59*, 784–793.

Van Horne, J. (1966). Capital budgeting decisions involving combinations of risky assets. *Management Science, 19*, B84–B92.

Williams, J. B. (1938). *The theory of investment value*. Harvard University Press.

Chapter 4
Capital Structure: Modigliani–Miller Theory

4.1 Introduction

One of two main theories of capital cost and capital structure is the theory of Nobel Prize winners Modigliani and Miller (1958, 1963, 1966). In this Chapter we describe the main results of this theory. Under the capital structure, one understands the relationship between equity and debt capital of the company. Does capital structure affect the company's main settings, such as the cost of capital, profit, value of the company, and the others, and, if affects, how? Choice of an optimal capital structure, i.e., a capital structure, which minimizes the weighted average cost of capital, WACC, and maximizes the value of the company, *V*, is one of the most important tasks solved by financial manager and by the management of a company. The first serious study (and first quantitative study) of influence of capital structure of the company on its indicators of activities was the work by Modigliani and Miller (1958). Until this study, the approach existed (let us call it traditional), which was based on empirical data analysis.

One of the most important assumptions of the Modigliani–Miller theory is that all financial flows are perpetuity. This limitation was lift out by Brusov–Filatova–Orekhova in 2008 (Filatova et al., 2008), who have created BFO theory—modern theory of capital cost and capital structure for companies of arbitrary age (BFO–1 theory) and for companies of arbitrary life time (BFO–2 theory) (Brusov et al., 2015). In Fig. 4.1, the historical development of capital structure theory from the traditional (empirical) approach, through perpetuity Modigliani–Miller approach to general capital structure theory—Brusov–Filatova–Orekhova (BFO) theory is shown.

In 2001 Steve Myers has considered the case of one–year company and show that in this case the weighted average cost of capital, WACC, is higher than in Modigliani–Miller case, and the capitalization of the company, *V*, is less than in Modigliani–Miller case.

P. Brusov, T. Filatova, *Efficiency of Investment Projects*, Contributions to Finance and Accounting, https://doi.org/10.1007/978-3-032-01028-5_4

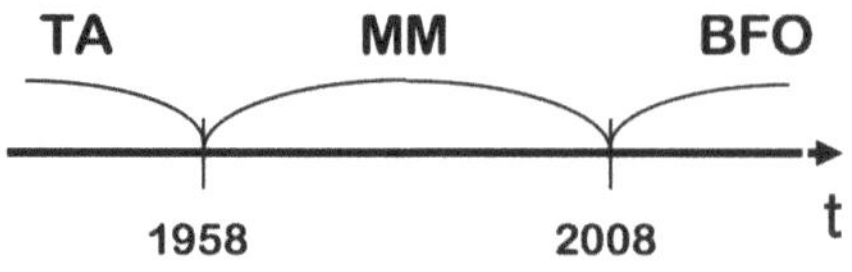

Fig. 4.1 Historical development of capital structure theory (here *TA* traditional (empirical) approach, *MM* Modigliani–Miller approach, *BFO* Brusov–Filatova–Orekhova theory)

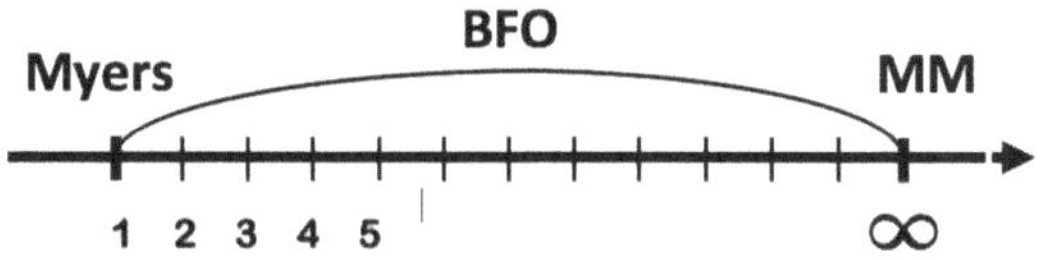

Fig. 4.2 MM theory describes perpetuity limit, Myers paper describes one–year company while BFO theory fills the whole numeric axis (from n = 1 up to perpetuity limit $n = \infty$)

So, before 2008 only two results for capital structure of the company were available: Modigliani–Miller for perpetuity company and Myers for one–year company (see Fig. 4.2). BFO theory has filled out whole interval between t = 1 and t = ∞. It gives the possibility to calculate the capitalization V, the weighted average cost of capital, *WACC*, equity cost k_e and other financial parameters for companies of arbitrary age and for companies of arbitrary life–time. BFO theory has lead to a lot of new meaningful effects in modern capital structure theory, discussed in this monograph.

4.2 The Traditional Approach

The traditional (empirical) approach told that weighted average cost of capital, WACC, and the associated company capitalization, V = CF/WACC, depend on the capital structure, the level of leverage, L. Debt cost always turns out to be lower than equity cost because first one has lower risk, via the fact, that in the event of bankruptcy creditor claims are met prior to shareholders claims.

As a result, an increase in the proportion of lower–cost debt capital in the overall capital structure up to the limit which does not cause violation of financial sustainability and growth of risk of bankruptcy leads to lower weighted average cost of capital, WACC.

The profitability required by investors (the equity cost) is growing; however, its growth has not led to compensation of benefits from use of lower–cost debt capital. Therefore, the traditional approach welcomes the increased leverage $L = D/S$ and the associated increase of company capitalization. The traditional (empirical) approach has existed up to appearance of the first quantitative theory by Modigliani and Miller (1958).

4.3 Modigliani–Miller Theory

4.3.1 Modigliani–Miller Theory Without Taxes

Modigliani and Miller (MM) in their first paper (Modigliani & Miller, 1958) have come to the conclusions which were fundamentally different from the conclusions of traditional approach. Under assumptions (see Sect. 4.3 for details) that there are no taxes, no transaction costs, no bankruptcy costs, perfect financial markets exist with symmetry information, equivalence in borrowing costs for both companies and investors, etc., they have showed that choosing of the ratio between the debt and equity capital does not affect company value as well as capital costs (Fig. 4.3).

Under above assumptions, Modigliani and Miller have analyzed the impact of financial leverage, supposing the absence of any taxes (on corporate profit as well as individual one). They have formulated and proven two following statements.

Without taxes, the total cost of any company is determined by the value of its EBIT–Earnings Before Interest and Taxes, discounted with fixed rate k_0, ***corresponding to group of business risk of this company:***

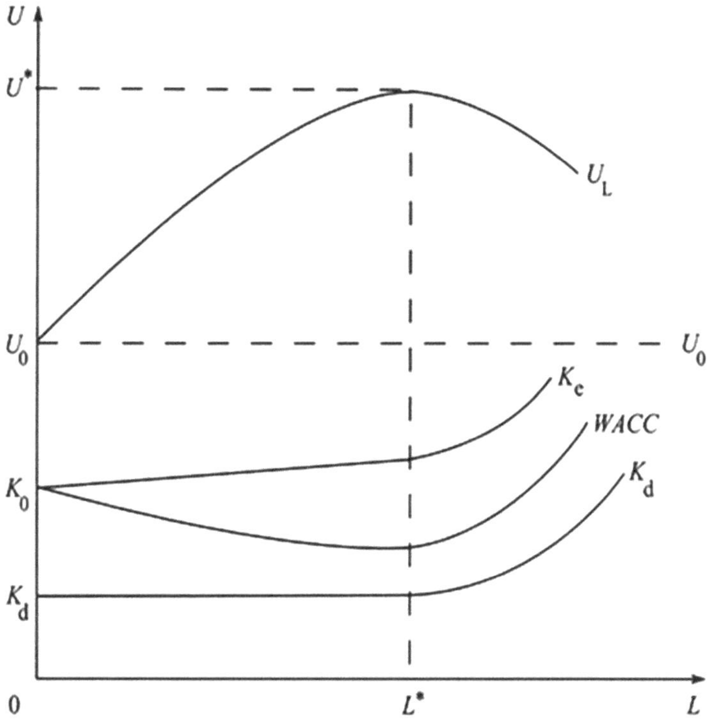

Fig. 4.3 Dependence of company capitalization, U_L, equity cost, k_e, debt cost, k_d, and weighted average cost of capital, WACC, in traditional (empirical) approach

$$V_L = V_U = \frac{\text{EBIT}}{k_0}. \tag{4.1}$$

Index L means financially dependent company (using debt financing), while index U means a financially independent company.

Authors supposed that both companies belong to the same group of business risk, and k_0 corresponds to required profitability of financially independent company, having the same business risk.

Because, as it follows from the formula (Eq. 4.1), value of the company does not depend on the value of debt, than according to Modigliani–Miller theorem (Modigliani & Miller, 1958), in the absence of taxes, value of the company is independent of the method of its funding. This means as well that weighted average cost of capital, WACC, of this company does not depend on its capital structure and is equal to the capital cost, which this company will have under the funding by equity capital only.

$$V_0 = V_L; \text{CF}/k_0 = \text{CF}/\text{WACC}, \quad \text{and thus WACC} = k_0.$$

Note that first Modigliani–Miller theorem is based on suggestion about independence of weighted average cost of capital and debt cost on leverage level.

From the first Modigliani–Miller theorem (Modigliani & Miller, 1958), it is easy to derive an expression for the equity capital cost

$$\text{WACC} = k_0 = k_e w_e + k_d w_d. \tag{4.2}$$

Finding from here k_e, one gets

$$k_e = \frac{k_0}{w_e} - k_d \frac{w_d}{w_e} = \frac{k_0(S+D)}{S} - k_d \frac{D}{S} = k_0 + (k_0 - k_d)\frac{D}{S} = k_0 + (k_0 - k_d)L \tag{4.3}$$

Here,

D	value of debt capital of the company
S	value of equity capital of the company
$k_d, w_d = \frac{D}{D+S}$	cost and fraction of debt capital of the company
$k_e, w_e = \frac{S}{D+S}$	cost and fraction of equity capital of the company
$L = D/S$	financial leverage

Thus, we come to second statement (theorem) of Modigliani–Miller theory about the equity cost of financially dependent (leverage) company (Modigliani & Miller, 1958).

Equity cost of leverage company k_e could be found as equity cost of financially independent company k_0 of the same group of risk, plus premium for risk, the value which is equal to production of difference $(k_0 - k_d)$ on leverage level L:

$$k_e = k_0 + (k_0 - k_d)L. \tag{4.4}$$

Formula (Eq. 4.4) shows that equity cost of the company increases linearly with leverage level (Fig. 4.3).

The combination of these two Modigliani–Miller statements implies that the increasing of level of debt in the capital structure of the company does not lead to increased value of firms, because the benefits gained from the use of more low–cost debt capital markets will be exactly offset by an increase in risk (we are speaking about the financial risk, the risk of bankruptcy) and, therefore, by an increase in cost of equity capital of firms: investors will increase the required level of profitability under increased risk, by which a higher level of debt in the capital structure is accompanied.

In this way, the Modigliani–Miller theorem argues that in the absence of the taxes, the capital structure of the company does not affect the value of the company and its weighted average cost of capital, WACC, and equity cost increases linearly with the increase of financial leverage.

Explanations, given by Modigliani and Miller under receiving of their conclusions, are the following (Modigliani & Miller, 1958). Value of the company depends on profitability and risk only and does not depend on the capital structure. Based on the principle of preservation of the value, they postulated that the value of the company, which is equal to the sum of the equity and debt funds, is not changed when the ratio between its parts is changed. An important role in justification of Modigliani–Miller statements an existence of an arbitral awards opportunities for the committed markets plays. Two identical companies, differing only by the leverage level, must have the same value. If this is not the case, the arbitration aligns business cost: investors of less cost company can invest capital in a company of more value. Selling of shares of the first company and buying of stock of the second company will continue until the values of both companies are not equalized.

Most of Modigliani and Miller assumptions (Modigliani & Miller, 1958), of course, are unrealistic. Some assumptions can be removed without changing the conclusions of the model. However, assuming no costs of bankruptcy and the absence of taxes (or the presence of corporate taxes only) are crucial–the change of these assumptions alters conclusions. The last two assumptions rule out the possibility of signaling theory and agency costs theory and, thus, also constitute a critical prerequisite (Fig. 4.4).

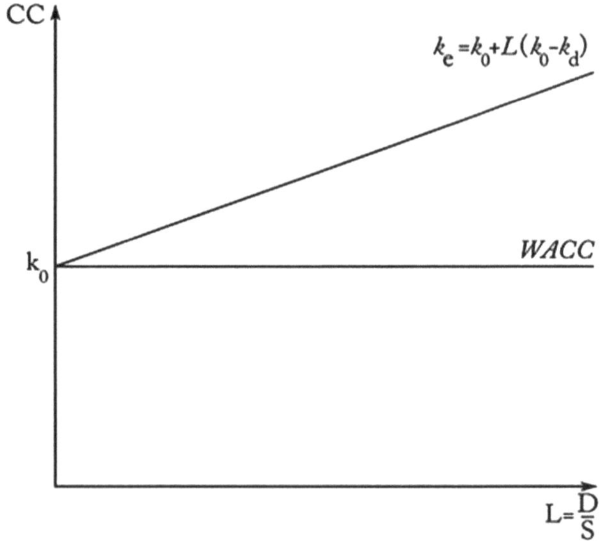

Fig. 4.4 Dependence of equity cost k_e and WACC on leverage level L within Modigliani–Miller theory without taxes

4.3.2 Modigliani–Miller Theory with Taxes

In the real situation, taxes on profit of companies always exist. Since the interest paid on debt are excluded from the tax base–it leads to the so–called effect of "tax shield": value of the company that used the borrowed capital (leverage company) is higher than the value of the company that financed entirely by the equity (non–leverage company). The value of the "tax shield" for 1 year is equal to $k_d\, D\, T$, where D–the value of debt, T–the income tax rate, and k_d–the interest on the debt (or debt capital cost) (Modigliani & Miller, 1963). The value of the "tax shield" for perpetuity company for all time of its existence is equal to (we used the formula for the sum of terms of an infinitely decreasing geometric progression).

$$(\mathrm{PV})_{\mathrm{TS}} = k_d DT \sum_{t=1}^{\infty} (1 + k_d)^{-t} = DT \tag{4.5}$$

and the cost of leverage company is equal to

$$V = V_0 + DT, \tag{4.6}$$

where V_0 is the value of financially independent company.

Thus, we obtain the following result obtained by Modigliani and Miller (1963): ***The value of financially dependent company is equal to the value of the company of the same risk group used no leverage, increased by the value of tax shield arising from financial leverage, and equal to the product of rate of corporate income tax T and the value of debt D.***

Let us now get the expression for the equity capital cost of the company under the existence of corporate taxes.

Accounting that $V_0 = \mathrm{CF}/k_0$ and that the ratio of debt capital $w_d = D/V$, one gets

$$V = \mathrm{CF}/k_0 + w_d VT. \tag{4.7}$$

Because the value of leverage company is $V = \mathrm{CF/WACC}$, for weighted average cost of capital, WACC, we get

$$\mathrm{WACC} = k_0(1 - w_d T). \tag{4.8}$$

From here the dependence of WACC on leverage $L = D/S$ becomes the following:

$$\mathrm{WACC} = k_0(1 - LT/(1 + L)). \tag{4.9}$$

On the other hand, on definition of the weighted average cost of capital with "tax shield" accounting, we have

$$\mathrm{WACC} = k_0 w_e + k_d w_d (1 - T). \tag{4.10}$$

Equating Eqs. (4.9) and (4.11), one gets

$$k_0(1 - w_d T) = k_0 w_e + k_d w_d (1 - T) \tag{4.11}$$

and from here, for equity cost, we get the following expression:

$$\begin{aligned} k_e &= k_0 \frac{(1 - w_d T)}{w_e} - k_d \frac{w_d}{w_e}(1 - T) = k_0 \frac{1}{w_e} - k_0 \frac{w_d}{w_e} T - k_d \frac{D}{S}(1 - T) \\ &= k_0 \frac{D + S}{S} - k_0 \frac{D}{S} T - k_d \frac{D}{S}(1 - T) = k_0 + L(1 - T)(k_0 - k_d). \end{aligned} \tag{4.12}$$

So, we get the following statement obtained by Modigliani and Miller (1963):

Equity cost of leverage company k_e paying tax on profit could be found as equity cost of financially independent company k_0 of the same group of risk, plus premium for risk, the value which is equal to production of difference ($k_0 - k_d$) on leverage level L and on tax shield *(1−T)*.

It should be noted that the formula (Eq. 4.12) is different from the formula (Eq. 4.4) without tax only by the multiplier $(1-T)$ in term, indicating a premium for risk. As the multiplier is less than unit, the corporate tax on profits leads to the fact that capital is growing with the increasing of financial leverage, slower than it would have been without them.

Analysis of formulas (Eqs. 4.4, 4.9, and 4.12) leads to following conclusions. When leverage grows:

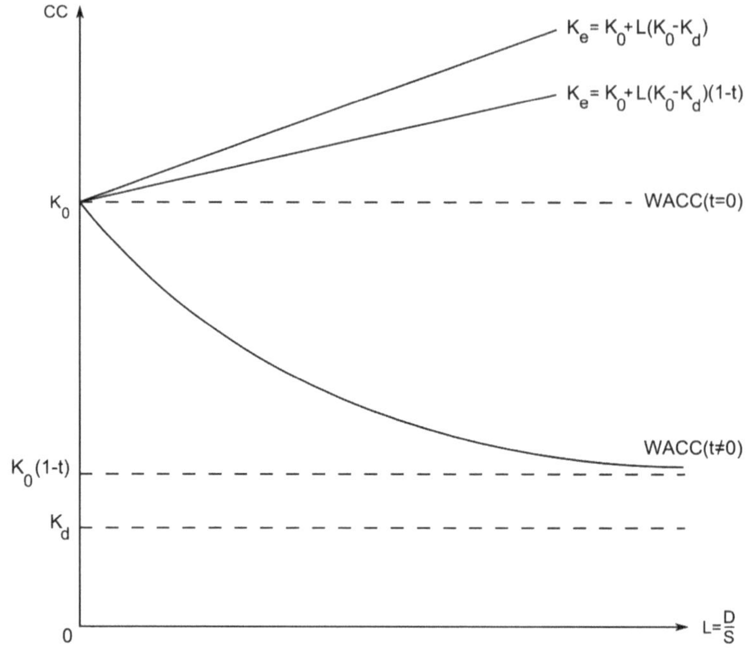

Fig. 4.5 Dependence of equity capital cost, debt cost, and WACC on leverage in Modigliani–Miller theory without taxes ($t = 0$) and with taxes ($t \neq 0$)

1. Value of company increases.
2. Weighted average cost of capital WACC decreases from k_0 (at $L = 0$) up to $k_0(1 - T)$ (at $L = \infty$) (when the company is funded solely by borrowed funds).
3. Equity cost increases linearly from k_0 (at $L = 0$) up to ∞ (at $L = \infty$).

Within their theory, Modigliani and Miller (1963) had come to the following conclusions. With the growth of financial leverage (Fig. 4.5):

1. The company value increases.
2. The weighted average cost of capital decreases from k_0 (for $L = 0$) up to $k_0(1 - T)$ (for $L = \infty$, when the company is financed entirely with borrowed funds).
3. The cost of equity capital increases linearly from k_0 (for $L = 0$) up to ∞ (for $L = \infty$).

4.3.3 Main Assumptions of Modigliani–Miller Theory

The most important assumptions of the Modigliani–Miller theory are as following:

1. Investors are behaving rationally and instantaneously, see profit opportunity, inadequate investment risk. Therefore, the possibility of a stable situation of the

arbitration, i.e., obtain the risk–free profit on the difference in prices for the same asset cannot be kept any length of time–reasonable investors quickly take advantage of it for their own purposes and equalize conditions in the market. This means that in a developed financial market capital, the same risk should be rewarded by the same rate of return.

2. Investment and financial market opportunities should be equally accessible to all categories of investors–whether institutional or individual investors, large or small, rapidly growing or stable, or experienced or relatively inexperienced.
3. Transaction costs associated with funding are very small. In practice, the magnitude of transaction costs is inversely proportional to the amount of finance involved, so this assumption is more consistent with reality than the large sums involved: i.e., in attracting small amounts, the transaction costs can be high, while, as in attracting large loans, as well as during placement of shares at a significant amount, the transaction costs can be ignored.
4. Investors get money and provide funds to borrowers at risk–free rate. In all probability, this assumption is due to the fact that the lender seeks to protect himself by using one or other guarantees, pledge of assets, the right to pay claims on third parties, and the treaty provisions restricting the freedom of the borrower to act to the detriment of the creditor. Lender's risk is really small, but its position can be considered risk free with respect to the position of the borrower and, accordingly, should be rewarded by a risk–free rate of return.
5. Companies have only two types of assets: risk–free debt capital and risky equity capital.
6. There is no possibility of bankruptcy, i.e., irrespective of what the level of financial leverage of the company–borrowers are reached–bankruptcy is not threatening them. Thus, bankruptcy costs are absent.
7. There are no corporate taxes and taxes on personal income of investors. If the personal income tax can indeed be neglected, because the assets of the company separated from the assets of shareholders, the corporate income taxes should be considered in the development of more realistic theories (which was done by Modigliani and Miller in their second paper devoted to the capital structure (Modigliani & Miller, 1963).
8. Companies are in the same class of risky companies.
9. All financial flows are perpetuity.
10. Companies have the same information.
11. Management of the company maximizes the capitalization of the company.

4.3.4 Modifications of Modigliani–Miller Theory

4.3.4.1 Taking into Account Market Risk: Hamada Model

Robert Hamada (1969) unites Capital Asset Pricing Model (CAPM) with Modigliani–Miller model taxation. As a result, he derived the following formula for calculation of the equity cost of financially dependent company, including both financial and business risk of company:

$$k_e = k_F + (k_M - k_F)b_U + (k_M - k_F)b_U \frac{D}{S}(1 - T), \quad (4.13)$$

where b_U is the β–coefficient of the company of the same group of business risk, that the company under consideration, but with zero financial leverage. The formula (Eq. 4.13) represents the desired profitability of equity capital k_e as a sum of three components: risk–free profitability k_F, compensating to shareholders a temporary value of their money, premium for business risk $(k_M - k_F)b_U$, and premium for financial risk $(k_M - k_F)b_U \frac{D}{S}(1 - T)$.

If the company does not have borrowing ($D = 0$), the financial risk factor will be equal to zero (the third term is drawn to zero), and its owners will only receive the premium for business risk.

To apply the Hamada equation, specialists in practice, in most cases, use book value of equity capital as its approach of market value. Nevertheless, the Hamada formula implies the use of market value of the assets.

It should be noted also that the formula (Eq. 4.13) can be used to derive other equation, using which you can analyze the impact of financial leverage on β–factor of company shares.

Equating CAPM formula to equity cost, we get:

$$k_F + (k_M - k_F)b_U = k_F + (k_M - k_F)b_U + (k_M - k_F)b_U \frac{D}{S}(1 - T) \quad (4.14)$$

or

$$b = b_U\left(1 + \frac{D}{S}(1 - T)\right). \quad (4.15)$$

In this way, the assumptions on which Modigliani–Miller theory and CAPM are based, β–factor of equity capital of financially dependent company is equal to β–factor of financially independent company, corrected on tax on profit rate and applied leverage level. Consequently, market risk of the company, measured by a factor b, depends on both the business risk of the company, a measure of which is b_U,, and on the financial risk b, which is calculated by the formula (Eq. 4.15).

In conclusion, here are the formulas for calculating the capital costs within the CAPM model [in parenthesis, there are formulas within the Modigliani–Miller theory (Modigliani and Miller 1958, 1963, 1966)].

The equity cost for company without debt capital:

$$k_e = k_F + (\overline{k}_M - k_F)\beta_U, \quad (k_e = k_0). \quad (4.16)$$

The equity cost for company with debt capital:

$$k_e = k_F + (\bar{k}_M - k_F)\beta_e, \quad (k_e = k_0 + (1 - T)(k_0 - k_d)L). \tag{4.17}$$

The debt cost:

$$k_d = k_F + (\bar{k}_M - k_F)\beta_d, (k_d = k_F; \beta_d = 0). \tag{4.18}$$

The weighted average cost of capital WACC

$$\mathrm{WACC} = k_e w_e + k_d w_d(1 - T), \quad (\mathrm{WACC} = k_0(1 - Tw_d)). \tag{4.19}$$

The Cost of Capital Under Risky Debt Another hypothesis of Modigliani and Miller was the suggestion about free of risk debt (in their theory, there are two types of assets: risky equity and free of risk debt). However, if we assume the risk of bankruptcy of company (and, accordingly, the ability to nonpayment of loans), the situation may change. Stiglitz (1969) and Rubinstein (1973) have shown that the conclusions concerning the total value of company do not change as compared to the findings derived by Modigliani and Miller under assumptions about free of risk debt (Modigliani and Miller 1958, 1963, 1966). However, the debt cost is changed. If previously, under assumption about the free of risk debt, it (debt cost) was regarded as a constant $k_d = k_F$, now it is not a constant. This claim is based on the work by Hsia (1981), where based on the models of pricing options, Modigliani–Miller and CAPM, it was shown that if one uses the formula for the net discount income, a term, reflecting tax protection on debt, should be discounted at the rate

$$k'_d = k_F + (k_0 - k_F)N(-d_1)\frac{1}{w_d}, \tag{4.20}$$

where

$$d_1 = \frac{-\ln w_d + k_F t}{\sigma\sqrt{t}} k_F + \frac{1}{2}\sigma\sqrt{t}, \tag{4.21}$$

here t–a moment of payment a credit, $N(-d_1)$–cumulative normal distribution of probability of random value d_1.

4.3.4.2 The Account of Corporate and Individual Taxes (Miller Model)

In the second article, Modigliani and Miller (1963) considered taxation of corporate profits, but did not take into account the presence in the economy of individual taxes of investors.

Merton Miller (1997) has introduced the model, demonstrating impact of leverage on the company value with account of the corporate and individual taxes (Miller, 1976).

To describe his model, we will enter the following legends: T_C–tax on corporate profits rate, T_S–the tax rate on income of an individual investor from his ownership by stock of corporation, T_D–tax rate on interest income from the provision of investor–individuals of credits to other investors and companies. Income from shares partly comes in the form of a dividend and, in part, as capital profits, so that T_S is a weighted average value of effective rates of tax on dividends and capital profits on shares, while the income from the provision of loans usually comes in the form of the interests. The last are usually taxed at a higher rate. In the light of the individual taxes, and with the same assumptions that have been made for Modigliani–Miller models previously, the financially independent company value can be determined as follows:

$$V_U = \frac{\mathrm{EBIT}(1 - T_C)(1 - T_S)}{k_0}. \tag{4.22}$$

A term $(1 - T_S)$ allows to take into account the individual taxes in the formula. In this way, numerator indicates which part of the operating company's profit remains in the possession of the investors, after the company will pay taxes on their profits, and its shareholders then will pay individual taxes on income from stock ownership. Since individual taxes reduce profits, remaining in the disposal of investors, the last, at other things being equal circumstances, also reduce and an overall assessment of the financially independent company value.

We will assess the financially dependent company under condition of a double taxation of income investors. To start, let us divide the annual cash flows of financially dependent company $\mathrm{CF_L}$ into flows sent to its shareholders $\mathrm{CF_e}$ and the flows belonging to debt owners $\mathrm{CF_d}$, with account of both corporation tax on profits and on the income of individuals:

$$\mathrm{CF_L} = \mathrm{CF_e} + \mathrm{CF_d} = (\mathrm{EBIT} - I)(1 - T_C)(1 - T_S) + I(1 - T_D), \tag{4.23}$$

where I is the annual interest payments on debt.

The formula (Eq. 4.23) can also be rewritten as follows:

$$\begin{aligned}\mathrm{CF_L} = \mathrm{CF_e} + \mathrm{CF_d} = \mathrm{EBIT}(1 - T_C)(1 - T_S) - I(1 - T_C)(1 - T_S)\\ + I(1 - T_D).\end{aligned} \tag{4.24}$$

The first term of the equation (Eq. 4.24) corresponds to cash flow after taxes for financially independent company, shown in equation (Eq. 4.22), which shows its present value. The second and the third terms of the equation, reflecting the financial dependence, corresponds to cash flows related to the debt financing, which, as previously, is considered as free of risk. Their present values are obtained by discounting by risk–free nominal rate on debt k_d.

By combining the present values of all three terms, we get the company value under using the debt financing and in the presence of all types of taxation:

$$V_U = \frac{\text{EBIT}(1 - T_C)(1 - T_S)}{k_0} - \frac{I(1 - T_C)(1 - T_S)}{k_d} + \frac{I(1 - T_d)}{k_d}. \tag{4.25}$$

First term in (Eq. 4.25) is identical to V_U in formula (Eq. 4.22). Accounting this and combining two last terms, we get the following formula:

$$V_L = V_U + \frac{I(1 - T_d)}{k_d}\left[1 - \frac{(1 - T_C)(1 - T_S)}{(1 - T_d)}\right]. \tag{4.26}$$

The amount of paid interests with taking into account the taxation, divided by the desired profitability of debt capital,

$$\frac{I(1 - T_d)}{k_d} \tag{4.27}$$

is equal to market value of the debt D. Substituting D in the previous formula, we get the final expression, which is known as a formula of a Miller model:

$$V_L = V_U + \left[1 - \frac{(1 - T_C)(1 - T_S)}{(1 - T_d)}\right]D. \tag{4.28}$$

The Miller model allows you to obtain an estimate of the value of financially dependent company, taking into account the corporate tax, as well as tax on individuals. The Miller formula (Eq. 4.28) has several important consequences:

1. Second term of sum,

$$\left[1 - \frac{(1 - T_C)(1 - T_S)}{(1 - T_d)}\right]D, \tag{4.29}$$

   represents the gains from use of debt capital. This term replaces the tax on profit of corporation rate in the Modigliani–Miller model with corporate taxes:

$$V_L = V_U + TD. \tag{4.30}$$

2. If we ignore taxes, a term (Eq. 4.29) will be equal to zero. Thus, in this case, the formula (Eq. 4.28) is transformed into the original version of the Modigliani–Miller model without taxes.
3. If we neglect taxes on individuals, the considering term becomes $1 - (1 - T_C) = T_C$, so, in this case, (Eq. 4.28) becomes a Modigliani–Miller model with corporate taxes (Eq. 4.30).
4. If the shareholder receives profit only in the form of dividend, and if effective tax rates on income from shares and bonds are equal ($T_S = T_D$), the terms $1 - T_S$ and $1 - T_D$ are shrinking, and the factor for D in (Eq. 4.29) again is equal to T_C.

5. If the shareholder receives dividends, and income from capital, the situation is changed.

   In this case, effective tax rates on income from shares and bonds are not equal. Let's take a look at common case, when individual taxes on income for the company shares are less than individual taxes of creditors. This encourages investors to purchase the shares of the company compared to purchasing the bonds of the company. In this case, $T_S < T_D$. Then factor in D in (Eq. 4.28) β has a look

$$\beta = 1 - \frac{(1 - T_C)(1 - T_S)}{(1 - T_d)} = 1 - (1 - T_C)\alpha = 1 - (1 - T_C)(1 + \gamma)$$
$$= T_C + \gamma(T_C - 1) < T_C. \tag{4.31}$$

   It is less than T_C, because $\gamma > 0$, $T_C < 1$, therefore, in this case, the effect of using of debt financing, although there is, but it is less, than in the absence of individual taxes. In other words, the effect of tax shields for the company in this case decreases, and it becomes less than the above individual taxes of creditors (individual taxes for the obligations of the company) in comparison with the individual income tax on shares.
6. Let's take a look at case $T_S > T_D$, when individual income taxes on shares are bigger than individual taxes creditors. The factor β takes view

$$\beta = 1 - \frac{(1 - T_C)(1 - T_S)}{(1 - T_d)} = 1 - (1 - T_C)\alpha = 1 - (1 - T_C)(1 - \gamma)$$
$$= T_C + \gamma(1 - T_C) > T_C. \tag{4.32}$$

   It is bigger than T_C, because $\gamma > 0$, $T_C < 1$; therefore, in this case, the effect of use of debt financing is increased compared with the case of the absence of individual taxes.
7. If $(1 - T_C)(1 - T_S) = 1 - T_D$, then this term is zero, and the effect of using debt financing will also be zero. This means that the benefits of the use of tax shields as a result of the application of debt financing will be fully offset by additional losses of investors, associated with a higher tax rate on interest on income of individuals. In this case, the capital structure will not affect the company value and its capital cost–in other words, you can apply Modigliani–Miller theory without tax (Modigliani & Miller, 1958).

In his paper, Miller (1976) claimed that companies on average will use issuance of shares and debt securities in such a way as to result in taxation of investors' income to be optimal. In such an equilibrium state will occur equality

$$(1 - T_C)(1 - T_S) = 1 - T_D, \tag{4.33}$$

and thus, as we have pointed out above, capital structure will not affect the market company value and its capital cost. Thus, by Miller, the conclusions on the irrelevance of the capital structure, made on the base of the original Modigliani–Miller model with zero taxes, remain in force.

Subsequently, researchers adapted and checked the Miller results. Their works, as a rule, have been devoted to the Miller's conclusion concerning the absence of the gains from the use of the debt capital by the company. In the United States, an effective tax rate on the income of shareholders is lower than the one on the income of creditors, but, nevertheless, the product $(1 - T_C)(1 - T_S)$ is less than $1 - T_D$.

Consequently, the companies may receive the benefit from use of debt financing. However, in the Miller's work, in fact, has been shown that the distinction of rates of individual taxes on income of shareholders and creditors to some extent compensates the advantages of use of debt financing, and, in this way, the tax benefits of debt are less than anticipated at a more earlier Modigliani–Miller model, where only corporate taxes have been taken into account.

4.3.4.3 The General WACC Formula

A more general formula for WACC, than in the famous Modigliani–Miller theory (MM) has been derived and discussed by a few authors in 2006–2007 (Farber et al. (2006); Fernandez (2006); Berk and DeMarzo (2007); Harris and Pringle (1985)). It takes the following form (Eqs. (18) in (Farber et al. (2006))

$$WACC = k_0(1 - w_d t) - k_d t w_d + k_{TS} t w_d, \qquad (4.34)$$

where k_0 is the required return on unleveraged company, k_d is the required return on its debt, k_{TS} is the expected return on the tax shield and t is the corporate taxes rate.

This formula is derived from the definition of the weighted average cost of capital and the balance sheet identity (for a similar presentation, see Berk and DeMarzo (2007)). At any point in time, it should therefore be verified, regardless of whether returns are annually or continuously compounded.

Practical applicability of Eq. (4.34) (while it is fairly general) requires additional conditions. Indeed, when the WACC is constant over time, the value of a leveraged company can be computed by discounting with the WACC of the unlevered free cash–flows. Therefore, it is interesting to consider the special cases when WACC is constant. The resulting formulas can also be found in textbooks (Brealey et al., 2005; Ross et al., 2005).

It was assumed by Modigliani and Miller in 1963, that the debt value D is constant. As the expected after-tax cash-flow of the unleveraged company is fixed,

V_0 is also constant. By this assumption, $k_{TS} = k_D$ and the value of the tax shield is TS = tD. Therefore, the capitalization of the leveraged company V is a constant and the general WACC Formula (4.34) simplifies to a constant WACC:

$$WACC = k_0(1 - w_d T) \tag{4.35}$$

However, our opinion is that "classical" Modigliani–Miller (MM) theory, which suggests that the expected returns on the debt k_d and the tax shield k_{TS} are equals (because both of them have debt nature), is much more reasonable and in our paper we modify the "classical" Modigliani–Miller (MM) theory, which is still widely used in practice.

4.3.4.4 Trade–Off Theory

The world-famous **trade-off theory** has been considered the cornerstone in the solution of the problem of optimal capital structure for a company for many decades and is still used today for decision analysis on capital structure. Below we give two examples.

Frank and Goyal (2009), "examines the relative importance of many factors in the capital structure decisions of publicly traded American firms from 1950 to 2003. The most reliable factors for explaining market leverage are: median industry leverage (+ effect on leverage), market-to-book assets ratio (−), tangibility (+), profits (−), log of assets (+), and expected inflation (+)". In addition, the authors have found that "dividend-paying firms tend to have lower leverage. When considering book leverage, somewhat similar effects are found. However, for book leverage, the impact of firm size, the market-to-book ratio, and the effect of inflation are not reliable". The empirical evidence seems to be reasonably consistent with some versions of the trade-off theory of capital structure.

Serrasqueiro and Caetano, in 2015, analyzed "to what extent decisions on the capital structure of small and medium–sized enterprises (SMEs) are closer to the assumptions of trade–off theory or to the assumptions of hierarchy theory. They used a sample of small and medium–sized enterprises located in the Portuguese hinterland, using dynamic LSDVC as the valuation method, and the empirical evidence suggests that the most profitable and oldest SMEs are less leveraged, confirming Pecking Order Theory 's forecasts. Larger SMEs are leveraging more borrowing, confirming the predictions of trade–off theory and hierarchy theory. In addition, SMEs are significantly adjusting their current debt levels towards the optimal debt ratio, which is consistent with the predictions of the compromise theory. It was concluded that theories of compromise and hierarchy are not mutually exclusive in explaining capital structure decisions of small and medium–sized enterprises".

However, the bankruptcy of trade-off theory has been proven by Brusov et al. in 2013b, 2013a. They have shown that risky debt financing (and growing credit rate near the bankruptcy) in contrast to waiting results does not lead to growing of weighted average cost of capital, WACC, which still decreases with leverage. This means the absence of minimum in the dependence of WACC on leverage level as well as the absence of maximum in the dependence of company capitalization on leverage. Thus, the well-known trade-off theory lacks an optimal capital structure. The explanation for this fact was made by Brusov et al. in 2013b, 2013a by analyzing the dependence of the cost of equity capital on the leverage level on the assumption that debt capital is risky.

Modigliani–Miller have considered tax shields from the interest on debt can increase the value of companies. In 1980, De Angelo and Masulis moved further in the theoretical examination of tax shields. They have noted that there are tax deductibles for companies other than debt to reduce their corporate tax burden and debt and non-debt tax shields should be accounted for. Depreciation, investment tax credits, or net-loss carry forwards could represents examples of such kind of non-debt tax shields. The first to test for these tax effects (suggested by DeAngelo and Masulis in 1980) has been carried out by Bradley, Jarrell and Kim in 1984. In contrast to the prediction in De Angelo and Masulis, 1980, by regressing company-specific debt-to-value ratios on non-debt tax shields, they have shown that debt is positively related to non-debt tax shields as measured by depreciation and investment tax credits. Titman and Wessels in 1988 found that "their results do not provide support for an effect on debt ratios arising from non-debt tax shields…". It was pointed out in 2003 by Graham, if a company invests heavily and uses debt financing to invest, a positive relation between such proxies for non-debt tax shield and debt may result. A mechanical positive relation of this type overwhelms and renders any substitution effects between debt and non-debt tax shields.

The original theory by Nobel Prize Winners Modigliani and Miller (1958, 1963, 1966) has been modified by many authors Brusov et al. (2018a), Brusov et al. (2018b), Brusov et al. (2018c), Brusov et al. (2018d), Sheridan & Wessels (1988), Harry & Masulis (1980) and above we shortly discussed some of them. In next paragraph we will generalize for the first time the Modigliani–Miller theory for the case of variable profit.

In conclusion, we present in Table 4.1 classification and summary of main theories of capital structures of company.

Table 4.1 Classification and summary of main theories of capital structures of company

Theory		Main thesis
Traditional theory		Empirical theory, existing before appearance of the first quantitative theory of capital structures (Modigliani–Miller theory) in 1958 (1958, 1963, 1966). Weighted average cost of capital depends on capital structures of company. There is an optimal dependence on capital structures of company
Modigliani–Miller theory (MM)	Without taxes	Capital cost and capitalization of the company are irrelevant on the capital structures of company
	With taxes	Weighted average cost of capital is decreased with leverage level, equity cost is increased linearly with leverage level, and capitalization of the company is increased with leverage level continuously
Brusov–Filatova–Orekhova theory (BFO–1)	For arbitrary age Without inflation	BFO theory (Brusov & Filatova, 2011; Brusov et al., 2011b, 2011a, 2011c, 2012a, 2012b, 2013b, 2013a, 2014b, 2014a; Filatova et al., 2008; Brusova, 2011) has replaced the famous theory of capital cost and capital structure by Nobel laureates, Modigliani and Miller (1958, 1963, 1966). The authors have moved from the assumption of Modigliani–Miller concerning the perpetuity (infinite time of life) of companies and further elaborated quantitative theory of valuation of core parameters of financial activities of companies with arbitrary age. Results of modern BFO theory turn out to be quite different from that of Modigliani–Miller theory. It shows that later, via its perpetuity, underestimates the assessment of weighted average cost of capital and the equity cost of the company and substantially overestimates the assessment of the capitalization of the company. Such an incorrect assessment of key performance indicators of financial activities of companies has led to an underestimation of risks involved, and impossibility, or serious difficulties in adequate managerial decision–making, which was one of the implicit reasons of global financial crisis of 2008 year. In the BFO theory, in investments at certain values of return–on–investment, there is an optimum investment structure. As well authors have developed a new mechanism of formation of the

(continued)

Table 4.1 (continued)

Theory		Main thesis
		company optimal capital structure, different from suggested by trade–off theory
	For arbitrary age With inflation	Inflation not only increases the equity cost and the weighted average cost of capital, but as well it changes their dependence on leverage. In particular, it increases growing rate of equity cost with leverage. Capitalization of the company is decreased under accounting of inflation
	For arbitrary age With increased financial distress costs and risk of bankruptcy	In BFO theory, with increased financial distress costs and risk of bankruptcy, the optimal capital structure is absent, which means that trade–off theory does NOT work
Brusov–Filatova–Orekhova theory (BFO–2)	For arbitrary lifetime	In perpetuity limit (Modigliani–Miller theory) time of life of company and company age turns out to be the same: both of them are infinite. When we move from the assumption of Modigliani–Miller concerning the perpetuity of companies these concepts (time of life of company and company age) become different and one should distinct them under generalization of Modigliani–Miller theory with respect to finite n Thus we have developed two kinds of finite n–theories: BFO–1 and BFO–4. BFO–1 theory is related to companies with arbitrary age and BFO–2 theory is related to companies with arbitrary life–time companies. By other words, BFO–1 is applicable for most interesting case of companies that reached the age of n–years and continue to exist on the market, and allows to analyze the financial condition of the operating companies. BFO–2 theory allows examine the financial status of the companies which ceased to exist, i.e., of those for which n means not age, but a life–time, i.e., the time of existence. A lot of schemes of termination of activities of the company can exist: bankruptcy, merger, acquisition, etc. One of those schemes, when the value of the debt capital D becomes zero at the time of termination of activity of company n, is considered in Chap. 3 and comparision of results of BFO–1 and BFO–2 has been done
Brusov–Filatova–Orekhova theory (BFO–3)	For rating needs	A new approach to rating methodology has been developed in Chaps. 21–23. Chapters 21 and 22 are devoted to rating of non–

(continued)

Table 4.1 (continued)

Theory		Main thesis
		financial issuers, while Chap. 23 is devoted to long–term project rating. The key factors of a new approach are: (1) The adequate use of discounting of financial flows virtually not used in existing rating methodologies, (2) The incorporation of rating parameters (financial "ratios") into the modern theory of capital structure (Brusov–Filatova–Orekhova (BFO–3 theory) (in Chaps. 21 into its perpetuity limit). This on the one hand allows use the powerful tools of this theory in the rating, and on the other hand it ensures the correct discount rates when discounting of financial flows. The interplay between rating ratios and leverage level which can be quite important in rating is discussed as well. All these create a new base for rating methodologies. New approach to ratings and rating methodologies allows to issue more correct ratings of issuers, makes the rating methodologies more understandable and transperant
Trade–off theory	Static	The static trade–off theory is developed with accounting of tax on profit and bankruptcy cost. It attempts to explain the optimal capital structure in terms of the balancing act between the benefits of debt (tax shield from interest deduction) and the disadvantage of debt (from the increased financial distress and expected bankruptcy costs). The tax shield benefit is the corporate income tax rate multiplied by the market value of debt and the expected bankruptcy costs are the probability of bankruptcy multiplied by the estimated bankruptcy costs
		Does not take into account the costs of the adaptation of financial capital structure to the optimal one, economic behavior of managers, owners, and other participants of economical process, as well as a number of other factors
		As it has been shown in BFO theory, the optimal capital structure is absent in trade–off theory
	Dynamic	The dynamic trade–off models assume that costs of constant capital adjustment are high and thus firms will change capital structure only if benefits exceed costs. Therefore,

(continued)

Table 4.1 (continued)

Theory		Main thesis
		there is an optimal range, outside of each leverage changes but remains unchanged inside. Companies try to adjust their leverage when it reaches the boundary of the optimal range. Subject to types of adjustment costs firms reach target ratio faster or slower. Proportional changes imply slight correction, whereas fixed changes imply considerable costs. In the dynamic model, correct decision on financial structure capital of the company in this period depends on the profit, which the company hopes to receive in the next period
		In BFO theory, with increased financial distress costs and risk of bankruptcy, the optimal capital structure is absent, which means that trade–off theory does NOT work: in static version as well as in dynamic one
Accounting of transaction cost		Accounting of the recapitalization transaction costs for the company, in which these costs are high, leads to the conclusion that a more cost–effective is not to modify financial capital structure, even if it is not optimal, during a certain period of time. The actual and target capital structure may vary because of the tool costs
Accounting of asymmetry of information		At the real financial markets, information is asymmetric (managers of the companies have owned more reliable information than investors and creditors), and rationality of economic subjects is limited
Signaling theory		Information asymmetry may be reduced on the basis of certain signals for creditors and investors, related to the behavior of managers on the capital market. It should take into account the previous development of the company and the current and projected cost–effectiveness of activities
Pecking order theory		The pecking order theory is the preferred, and empirically observed, sequence of financing type to raise capital. That is, firms first tap retained earnings (internal equity) finance, second source is debt, and the last source is issuing new common stock shares (external equity). The empirical evidence of nonfinancial firm debt ratios coupled with the decision–making process of top management and the board of directors point to

(continued)

Table 4.1 (continued)

Theory		Main thesis
		greater adherence to the pecking order theory
Theories of conflict of interests	Theory of agency costs	Management of the company may take decisions that are contrary to the interests of the shareholders or creditors, respectively; the costs are necessary to monitor its actions. An effective tool for resolving agent problem is the correct selection of compensation package (the share of participation of agent in property, bonus, stock options), allowing to link revenue of managers with the dynamics of equity capital and to provide motivation for managers to its (equity capital) conservation and growth
	Theory of corporate control and costs monitoring	If asymmetries of information exist, creditors, providing the capital, are interested in the possibility of the implementation of the self–monitoring of the effectiveness of its use and return. Costs for monitoring, as a rule, put on the company owners by their inclusion into credit rate. The level of monitoring costs depends on the scale of the business; therefore, with the increase of the business scale, the weighted average cost of capital of the company grows and company market value is reduced
	Theory of stakeholders	Stakeholder theory is a theory that identifies and models the groups that are stakeholders of a corporation or project. The diversity and the intersection of stakeholders' interests and different assessment by them of acceptable risk generate conditions for conflict of their interest, that is, making some corrections into the process of optimizing financial capital structure
Behavioral theories	Manager investment autonomy	Managers implement those decisions, which, from their point of view, will be positively perceived by investors and, respectively, positively affect the market value of companies: when the market value of shares of a company and the degree of consensus of expectations of managers and investors are high, the company has an additional issue of shares, and in the opposite situation, it uses debt instruments. In this way, the financial capital structure is more influenced by investors, the expectations of which are taken into account by managers

(continued)

Table 4.1 (continued)

Theory		Main thesis
	The equity market timing theory	Leverage level is determined by market dynamics. Equity market timing theory means that company should issue shares at high price and repurchase them at low price. The idea is to exploit temporary fluctuations in the equity cost relative to the cost of other forms of capital
	Information cascades	In order to save costs and to avoid errors, financial capital structure can be formed not on the basis of the calculations of optimal capital structure or depending on available in different periods of company life funding sources, but borrow from other companies that have successful, reputable managers (companies' leaders), as well as using (in the wake of the majority) the most popular methods of management of capital structure

References

Berk, J., & DeMarzo, P. (2007). *Corporate finance*. Pearson–Addison Wesley.
Bradley, M., Jarrell, G. A., & Kim, E. H. (1984). On the existence of an optimal capital structure: Theory and evidence. *Journal of Finance, 39*, 857–878.
Brealey, R., Myers, S., & Allen, F. (2005). *Principle of corporate finance* (7th ed.). McGraw Hill.
Brusova, A. (2011). A comparison of the three methods of estimation of weighted average cost of capital and equity cost of company. *Financial Analysis: Problems and Solutions, 34*(76), 36–42.
Brusov, P. N., & Filatova, T. V. (2011). From Modigliani–Miller to general theory of capital cost and capital structure of the company. *Finance Credit, 435*, 2–8.
Brusov, P., Filatova, T., Orehova, N., Brusov, P. P., & Brusova, N. (2011a). From Modigliani–Miller to general theory of capital cost and capital structure of the company. *Research Journal of Economics, Business and ICT, 2*, 16–21.
Brusov, P., Filatova, T., Orehova, N., & Brusova, A. (2011b). Weighted average cost of capital in the theory of Modigliani–Miller, modified for a finite life–time company. *Applied Financial Economics, 21*(11), 815–824.
Brusov, P., Filatova, T., Orehova, N., et al. (2011c). Influence of debt financing on the effectiveness of the investment project within the Modigliani–Miller theory. *Research Journal of Economics, Business and ICT, 2*, 11–15.
Brusov, P., Filatova, T., Eskindarov, M., & Orehova, N. (2012a). Influence of debt financing on the effectiveness of the finite duration investment project. *Applied Financial Economics, 22*(13), 1043–1052.
Brusov, P., Filatova, T., Eskindarov, M., & Orehova, N. (2012b). Hidden global causes of the global financial crisis. *Journal of Reviews on Global Economics, 1*, 106–111.
Brusov, P., Filatova, T., & Orehova, N. (2013a). A qualitatively new effect in corporative finance: Abnormal dependence of cost of equity of company on leverage. *Journal of Reviews on Global Economics, 2*, 183–193.
Brusov, P., Filatova, P., & Orekhova, N. (2013b). Absence of an optimal capital structure in the famous tradeoff theory! *Journal of Reviews on Global Economics, 2*, 94–116.
Brusov, P., Filatova, T., & Orehova, N. (2014a). Inflation in Brusov–Filatova–Orekhova theory and in its perpetuity limit–Modigliani–Miller theory. *Journal of Reviews on Global Economics, 3*, 175–185.
Brusov, P., Filatova, P., & Orekhova, N. (2014b). Mechanism of formation of the company optimal capital structure, different from suggested by trade off theory. *Cogent Economics & Finance, 2*, 1–13. https://doi.org/10.1080/23322039.2014.946150
Brusov, P., Filatova, T., Orehova, N., & Eskindarov, M. (2015). *Modern corporate finance, investment and taxation* (1st ed., pp. 1–368). Springer International Publishing.
Brusov, P., Filatova, T., Orehova, N., & Kulk, W. I. (2018a). New meaningful effects in modern capital structure theory. *Journal of Reviews on Global Economics, 7*, 104–122.
Brusov, P., Filatova, T., Orehova, N., & Kulk, V. (2018b). A "Golden age" of the companies: Conditions of its existence. *Journal of Reviews on Global Economics, 7*, 88–103.
Brusov, P., Filatova, T., Orehova, N., & Kulk, V. (2018c). Rating methodology: New look and new horizons. *Journal of Reviews on Global Economics, 7*, 63–87.
Brusov, P., Filatova, T., Orehova, N., & Kulk, V. (2018d). Rating: New approach. *Journal of Reviews on Global Economics, 7*, 37–62.
Farber, A., Gillet, R., & Szafarz, A. (2006). A general formula for the WACC. *International Journal of Business, 11*(2), 211–218.
Fernandez, P. (2006). A general formula for the WACC: A comment. *International Journal of Business, 11*(2), 219.
Filatova, T., Orehova, N., & Brusova, A. (2008). Weighted average cost of capital in the theory of Modigliani–Miller, modified for a finite life–time company. *Bull FU, 48*, 68–77.
Frank, M., & Goyal, V. (2009). Capital structure decisions: Which factors are reliably important? *Financial Management, 38*(1), 1–37. https://doi.org/10.1111/j.1755-053X.2009.01026.x

Graham, J. R. (2003). Taxes and corporate finance: A review. *Review of Financial Studies, 16*, 1075–1129.
Hamada, R. (1969). Portfolio analysis, market equilibrium, and corporate finance. *The Journal of Finance, 24*(1), 13–31.
Harris, R., & Pringle, J. (1985). Risk–adjusted discount rates—Extension form the average–risk case. *The Journal of Financial Research, 8*(3), 237–244.
Harry, D. A., & Masulis, R. W. (1980). Optimal capital structure under corporate and personal taxation. *Journal of Financial Economics, 8*, 3–29.
Hsia, C. (1981). Coherence of the modern theories of finance. *Financial Review, 16*(1), 27–42.
Miller, M. (1976). Debt and taxes. *Journal of Finance, 32*(2), 261–275.
Miller, M. (1997). *Merton Miller on derivatives*. Wiley.
Modigliani, F., & Miller, M. (1958). The cost of capital, corporate finance, and the theory of investment. *The American Economic Review, 48*, 261–297.
Modigliani, F., & Miller, M. (1963). Corporate income taxes and the cost of capital: A correction. *The American Economic Review, 53*, 147–175.
Modigliani, F., & Miller, M. (1966). Some estimates of the cost of capital to the electric utility industry 1954–1957. *The American Economic Review, 56*, 333–391.
Myers, S. (2001). Capital structure. *The Journal of Economic Perspectives, 15*(2), 81–102.
Ross, S., Westerfiel, R., & Jaffee, D. (2005). *Corporate finance* (7th ed.). McGraw Hill.
Rubinstein, M. (1973). A mean–variance synthesis of corporate financial theory. *The Journal of Finance, 28*, 167–181.
Sheridan, T., & Wessels, R. (1988). The determinants of capital structure choice. *Journal of Finance, 43*, 1–19.
Stiglitz, J. (1969). A re–examination of the Modigliani–Miller theorem. *The American Economic Review, 59*(5), 784–793.

Chapter 5
The Modigliani–Miller Theory with Arbitrary Frequency of Payment of Tax on Profit

5.1 Introduction

5.1.1 Literature Review

In their first paper Modigliani and Miller (Modigliani & Miller, 1958; Brusov et al., 2015, 2018a, 2020) consider the problem of capital cost and capital structure for perpetuity companies without any taxes. In the second article, Modigliani and Miller (Modigliani & Miller, 1963; Brusov et al., 2015, 2018a, 2020) considered taxation of corporate profits, but did not take into account the presence in the economy of individual taxes of investors. Merton Miller (Miller, 1977; Brusov et al., 2015, 2018a, 2020, Arzak and Glosten, 2005; Myers, 2001; Sethi et al., 1991) has introduced the model, demonstrating impact of leverage on the company value with account of the corporate and individual taxes. To describe his model, we will enter the following legends: T_C–tax on corporate profits rate, T_S–the tax rate on income of an individual investor from his ownership by stock of corporation, T_D–tax rate on interest income from the provision of investor–individuals of credits to other investors and companies. Income from shares partly comes in the form of a dividend and, in part, as capital profits, so that T_S is a weighted average value of effective rates of tax on dividends and capital profits on shares, while the income from the provision of loans usually comes in the form of the interests. The last are usually taxed at a higher rate. In the light of the individual taxes, and with the same assumptions that have been made for Modigliani–Miller models previously, the financially independent company value can be determined as follows:

$$V_U = \frac{\text{EBIT}(1 - T_C)(1 - T_S)}{k_0}. \tag{5.1}$$

P. Brusov, T. Filatova, *Efficiency of Investment Projects*, Contributions to Finance and Accounting, https://doi.org/10.1007/978-3-032-01028-5_5

A term $(1 - T_S)$ allows to take into account the individual taxes in the formula. In this way, numerator indicates which part of the operating company's profit remains in the possession of the investors, after the company will pay taxes on their profits, and its shareholders then will pay individual taxes on income from stock ownership. Since individual taxes reduce profits, remaining in the disposal of investors, the last, at other things being equal circumstances, also reduce and an overall assessment of the financially independent company value.

Robert Hamada (Hamada, 1969; Brusov et al., 2015, 2018a, 2020) unites Capital Asset Pricing Model (CAPM) with Modigliani–Miller model taxation. As a result, he derived the following formula for calculation of the equity cost of financially dependent company, including both financial and business risk of company:

$$k_e = k_F + (k_M - k_F)b_U + (k_M - k_F)b_U \frac{D}{S}(1 - T), \tag{5.2}$$

where b_U is the β–coefficient of the company of the same group of business risk, that the company under consideration, but with zero financial leverage. The formula (5.2) represents the desired profitability of equity capital k_e as a sum of three components: risk–free profitability k_F, compensating to shareholders a temporary value of their money, premium for business risk $(k_M - k_F)b_U$, and premium for financial risk $(k_M - k_F)b_U \frac{D}{S}(1 - T)$.

If the company does not have borrowing ($D = 0$), the financial risk factor will be equal to zero (the third term is drawn to zero), and its owners will only receive the premium for business risk.

More general formula for the WACC, then famous Modigliani–Miller (MM) one has been derived and discussed by a few authors (Farber et al., 2006; Fernandez, 2007; Harris & Pringle, 1985; Miles & Ezzel, 1980; DeMarzo 1988; Ross et al., 2005; Taggart, 1991). It takes the following form (Eq. (18) in Farber et al. (2006))

$$WACC = k_0(1 - w_d T) - k_d t w_d + k_{TS} t w_d \tag{5.3}$$

where k_0, k_d and k_{TS} are the expected returns respectively on the unlevered firm, the debt and the tax shield, V is the value of the levered firm, VTS is the value of the tax shield, D is the value of the debt and TC is the corporate tax rate.

This formula is derived from the balance sheet identity and the definition of the weighted average cost of capital (for a similar presentation, see Berk and De Marzo (2007)). It should therefore be verified at any point in time, whatever returns are annually or continuously compounded.

Although Eq. (5.3) is fully general, its practical applicability requires additional conditions. Indeed, when the WACC is constant over time, the value of a levered firm can be computed by discounting with the WACC the unlevered free cash–flows. Therefore, we pay a special attention to the special cases that make the WACC constant. The resulting particular formulas can also be found in textbooks (Brealey et al., 2014; Ross et al., 2005).

First, Modigliani and Miller (1963) assume that the level of debt D is constant. Then, as the expected after–tax cash–flow of the unlevered firm is fixed, V_0 is also constant. By assumption, $k_{TS} = k_D$ and the value of the tax shield is TS = tD. Therefore, the value of the firm V is a constant and the general WACC formula (15) simplifies to a constant WACC:

$$WACC = k_0(1 - w_d T) \tag{5.4}$$

But our opinion is that "classical" Modigliani–Miller (MM) theory, which coming from the suggestion that the expected returns on the debt k_d and the tax shield k_{TS} are equals (because both of them have debt nature), is much more reasonable and in our paper we modify namely "classical" Modigliani–Miller (MM) theory, which is still widely used in practice.

One of the most serious limitations of the Modigliani–Miller theory is the suggestion about perpetuity of the companies. In 2008, Brusov–Filatova–Orekhova (Brusov et al., 2015, 2018a, b, c, d, e, f, g, 2019, 2020; Brusov, 2018a, b; Filatova et al., 2008, 2018) have lifted up this limitation and have shown that the accounting of the finite life–time (or finite age) of the company leads to significant changes of all Modigliani–Miller results (Modigliani & Miller, 1958, 1963, 1966; Miller, 1977): capitalization of the company is changed, as well as the equity cost, k_e, and the weighted average cost of capital, WACC, in the presence of corporative taxes. Besides, a number of qualitatively new effects in corporate finance, obtained in Brusov–Filatova–Orekhova theory (Brusov et al., 2015, 2018a, b, c, d, e, f, g, 2019, 2020; Brusov, 2018a, b; Filatova et al., 2008, 2018), are absent in Modigliani–Miller theory (Modigliani & Miller, 1958, 1963; Miller, 1977).

The famous Brusov–Filatova–Orekhova formula for weighted average cost of capital, WACC, has the following form

$$\frac{1-(1+WACC)^{-n}}{WACC} = \frac{1-(1+k_0)^{-n}}{k_0[1-\omega_d T(1-(1+k_d)^{-n})]}. \tag{5.5}$$

Here, S—the value of own (equity) capital of the company, $w_d = \frac{D}{D+S}$—the share of debt capital; $k_e, w_e = \frac{S}{D+S}$—the cost and the share of the equity of the company, $L = D/S$—financial leverage.

A lot of meaningful effects has been discovered within BFO theory: these effects are absent within MM theory. BFO theory has destroyed some main existing principles of financial management: among them trade off theory, which was considered as keystone of formation of optimal capital structure of the company during many decades and **bankruptcy of which has been proven** by Brusov–Filatova–Orekhova (Brusov et al., 2015, 2018a, b, c, d, e, f, g, 2019, 2020; Brusov, 2018a, b; Filatova et al., 2008, 2018).

5.1.2 *Focus and Objective of the Chapter*

The tax shield plays a crucial role in both main capital structure theories: Brusov–Filatova–Orekhova (BFO theory) and its perpetuity limit Modigliani–Miller theory. How it is formed influences the results of both theories. Both of them describes the case of annual payments of tax of profit, but in practice these payments are made more frequent: semiannually, quarterly, monthly. While a return is not required more than once a year, businesses may be responsible for filing estimated taxes based on profits earned. This requirement is dependent on showing a profit. For example, sole proprietors must file estimated taxes on profits quarterly, on the 15th day of April, June, September and January. In Russia tax on profit payments could be made annually, quarterly, monthly. In current paper we show that the account of frequency of payment of tax on profit within the Modigliani–Miller theory turns out to be very important and leads to significant consequences.

To study this problem we modify the Modigliani–Miller theory for the case of arbitrary frequency of payment of tax on profit and show that:

- all Modigliani–Miller theorems, statements and all formulas change;
- all main financial indicators, such as the weighted average cost of capital, WACC, company value, V and equity cost, ke depend on the frequency of tax on profit payments. This allows to company manage WACC, V, ke etc. by choosing the number of payments of tax of profit p per year;
- increase the number of payments of tax of profit per year leads to decrease of the cost of attracting capital and increase of the company value;
- obtained results allow to company choose the number of payments of tax of profit per year, as many, as it is profitable to it (of course, within actual tax legislation);
- in case of income tax payments more than once per year (at $p \neq 1$), as it takes place in practice, the weighted average cost of capital, WACC, company value, V and equity cost, ke start depend on kd, while in ordinary (classical) Modigliani–Miller theory all these values DO NOT depend on kd.
- the tilt angle of the curve of equity cost, ke (L), decreases with the number of payments of tax of profit p, this modifies the dividend policy of the company, because the economically justified value of dividends is equal to equity cost.

5.2 Modification of the Modigliani–Miller Theory for the Case of Arbitrary Frequency of Payments of Tax on Profit

5.2.1 *Tax Shield*

Let us calculate first the tax shield in the Modigliani–Miller theory for the case of p payments of tax on profit per year

$$TS = \frac{k_d Dt}{p(1+k_d)^{1/p}} + \frac{k_d Dt}{p(1+k_d)^{2/p}} + \frac{k_d Dt}{p(1+k_d)^{3/p}} + \cdots \tag{5.6}$$

TS represents a geometric progression with denominator $q = \frac{1}{(1+k_d)^{1/p}}$.

Summing the progression, one gets

$$TS = \frac{k_d Dt}{p(1+k_d)^{1/p}\left(1-(1+k_d)^{-1/p}\right)} = \frac{k_d Dt}{p\left((1+k_d)^{1/p}-1\right)}. \tag{5.7}$$

Note, that in case of classical Modigliani–Miller theory

$$TS = Dt. \tag{5.8}$$

It is easy to get this result from (5.7), putting the frequency of payments of tax on profit p = 1.

5.2.2 The Weighted Average Cost of Capital, WACC

Let us now calculate the weighted average cost of capital, WACC, in the Modigliani–Miller theory for the case of p payments of tax on profit per year.

From the Modigliani–Miller theorem we have for the company value V

$$V = V_0 + TS. \tag{5.9}$$

Putting expression (5.7) for TS, we get

$$V = V_0 + \frac{k_d Dt}{p\left((1+k_d)^{1/p}-1\right)}. \tag{5.10}$$

Substituting $D = w_d V$, one gets

$$V\left(1 - \frac{k_d w_d t}{p\left((1+k_d)^{1/p}-1\right)}\right) = V_0. \tag{5.11}$$

Accounting, that

$$V = \frac{CF}{WACC}; V_0 = \frac{CF}{k_0}, \tag{5.12}$$

we have

$$\frac{CF}{WACC} \cdot \left(1 - \frac{k_d w_d t}{p\left((1 + k_d)^{1/p} - 1\right)}\right) = \frac{CF}{k_0}. \tag{5.13}$$

From here we arrive to the expression for the weighted average cost of capital, WACC, for the case of p payments of tax on profit per year.

$$WACC = k_0 \cdot \left(1 - \frac{k_d w_d t}{p\left((1 + k_d)^{1/p} - 1\right)}\right). \tag{5.14}$$

Note, that in case of classical Modigliani–Miller theory

$$WACC = k_0 \cdot (1 - w_d t). \tag{5.15}$$

It is easy to get this result from (5.14), putting the frequency of payments of tax on profit per year $p = 1$.

5.2.3 Company Value, V

Let us now calculate the company value, V, in the Modigliani–Miller theory for the case of p payments of tax on profit per year.

Putting WACC value into first formula of (5.12) one gets for company value, V

$$V = \frac{CF}{WACC} = \frac{CF}{k_0 \cdot \left(1 - \frac{k_d w_d t}{p\left((1+k_d)^{1/p} - 1\right)}\right)}. \tag{5.16}$$

In case of "classical" Modigliani–Miller theory one has

$$V = \frac{CF}{WACC} = \frac{CF}{k_0 \cdot (1 - w_d t)}. \tag{5.17}$$

It is easy to get this result from (5.16), putting the frequency of payments of tax on profit per year $p = 1$.

5.2.4 Equity Cost, k_e

Let us now calculate the equity cost, k_e, in the Modigliani–Miller theory for the case of p payments of tax on profit per year.

General expression for the weighted average cost of capital, WACC, upon definition has the following form

$$WACC = k_e w_e + k_d w_d(1-t). \tag{5.18}$$

We get from (5.18) the following formula for the equity cost, k_e,

$$k_e = \frac{WACC}{w_e} - \frac{k_d w_d(1-t)}{w_e}. \tag{5.19}$$

Accounting from (5.14) the expression for the weighted average cost of capital, WACC,

$$WACC = k_0 \cdot \left(1 - \frac{k_d w_d t}{p\left((1+k_d)^{1/p} - 1\right)}\right), \tag{5.20}$$

one gets

$$\begin{aligned} k_e &= \frac{k_0}{w_e} \cdot \left(1 - \frac{k_d w_d t}{p\left((1+k_d)^{1/p} - 1\right)}\right) - \frac{k_d w_d(1-t)}{w_e} = \\ &= k_0 \cdot \left(1 + L - \frac{k_d L t}{p\left((1+k_d)^{1/p} - 1\right)}\right) - k_d L(1-t) = \\ &= k_0 + L\left(k_0 - k_d(1-t) - \frac{k_0 k_d t}{p\left((1+k_d)^{1/p} - 1\right)}\right). \end{aligned} \tag{5.21}$$

Thus, we have the following formula for the equity cost, k_e, in case of arbitrary frequency of payment of tax on profit

$$k_e = k_0 + L\left(k_0 - k_d(1-t) - \frac{k_0 k_d t}{p\left((1+k_d)^{1/p} - 1\right)}\right). \tag{5.22}$$

From this formula it is easy to get "classical" Modigliani–Miller expression for the equity cost, k_e, in case of annual payment of tax on profit, putting $p = 1$

$$k_e = k_0 + L(k_0 - k_d)(1 - t) \quad (5.23)$$

Thus, from 5.21–5.24 we come to conclusion, that in case of income tax payments more than once per year (at $p \neq 1$), as it takes place in practice, the weighted average cost of capital, WACC, company value, V and equity cost, k_e start depend on k_d, while in ordinary (classical) Modigliani–Miller theory all these values DO NOT depend on k_d.

In subsequent paragraphs (5.3)–(5.5) we will study numerically with use of Microsoft Excel the dependence of the main financial indicators of the company (the weighted average cost of capital, WACC, the company value, V, the cost of equity, k_d, on leverage level L for cases of practical interest with respect to equity and debt costs and frequency of payment of tax on profit p.

5.3 Dependence of the Weighted Average Cost of Capital, WACC, on Leverage Level L

5.3.1 Dependence of the Weighted Average Cost of Capital, WACC, on Leverage Level L at Different p and Fixed k_d

Let us start from the study numerically with use of Microsoft Excel of the dependence of the weighted average cost of capital, WACC, on leverage level L at different p and fixed k_d, using obtained by us formula (5.14)

$$WACC = k_0 \cdot \left(1 - \frac{k_d w_d t}{p\left((1 + k_d)^{1/p} - 1\right)}\right).$$

From Tables 5.1, 5.2, 5.3, and 5.4 and Figs. 5.1 and 5.2 it is seen, that the weighted average cost of capital, WACC, decreases with leverage level L at all values of frequency of payments of tax on profit p. With increase of p WACC decreases. As we'll see below, this will lead to increase of company value, V, with p.

To demonstrate, that above conclusion takes place for all values of equity cost (at zero leverage level L) k_0, we consider one more example: $k_0 = 0.22$ and different $p = 2; 4; 12$ (Fig. 5.3; Table 5.5).

Table 5.1 Dependence of WACC on leverage level L at $k_0 = 0.2$, $k_d = 0.16$ and $p = 1$

L	T	k_0	k_d	p	WACC
0	0.2	0.2	0.16	1	0.200000
1	0.2	0.2	0.16	1	0.180000
2	0.2	0.2	0.16	1	0.173333
3	0.2	0.2	0.16	1	0.170000
4	0.2	0.2	0.16	1	0.168000
5	0.2	0.2	0.16	1	0.166667
6	0.2	0.2	0.16	1	0.165714
7	0.2	0.2	0.16	1	0.165000
8	0.2	0.2	0.16	1	0.164444
9	0.2	0.2	0.16	1	0.164000
10	0.2	0.2	0.16	1	0.163636

Table 5.2 Dependence of WACC on leverage level L at $k_0 = 0.2$, $k_d = 0.16$ and $p = 2$

L	T	k_0	k_d	p	WACC
0	0.2	0.2	0.16	2	0.200000
1	0.2	0.2	0.16	2	0.179230
2	0.2	0.2	0.16	2	0.172306
3	0.2	0.2	0.16	2	0.168845
4	0.2	0.2	0.16	2	0.166767
5	0.2	0.2	0.16	2	0.165383
6	0.2	0.2	0.16	2	0.164394
7	0.2	0.2	0.16	2	0.163652
8	0.2	0.2	0.16	2	0.163075
9	0.2	0.2	0.16	2	0.162613
10	0.2	0.2	0.16	2	0.162236

Table 5.3 Dependence of WACC on leverage level L at $k_0 = 0.2$, $k_d = 0.16$ and $p = 4$

L	T	k_0	k_d	p	WACC
0	0.2	0.2	0.16	4	0.200000
1	0.2	0.2	0.16	4	0.178837
2	0.2	0.2	0.16	4	0.171783
3	0.2	0.2	0.16	4	0.168256
4	0.2	0.2	0.16	4	0.166139
5	0.2	0.2	0.16	4	0.164728
6	0.2	0.2	0.16	4	0.163721
7	0.2	0.2	0.16	4	0.162965
8	0.2	0.2	0.16	4	0.162377
9	0.2	0.2	0.16	4	0.161907
10	0.2	0.2	0.16	4	0.161522

Table 5.4 Dependence of WACC on leverage level L at $k_0 = 0.2$, $k_d = 0.16$ and $p = 12$

L	T	k_0	k_d	p	WACC
0	0.2	0.2	0.16	12	0.200000
1	0.2	0.2	0.16	12	0.178573
2	0.2	0.2	0.16	12	0.171430
3	0.2	0.2	0.16	12	0.167859
4	0.2	0.2	0.16	12	0.165716
5	0.2	0.2	0.16	12	0.164288
6	0.2	0.2	0.16	12	0.163267
7	0.2	0.2	0.16	12	0.162502
8	0.2	0.2	0.16	12	0.161907
9	0.2	0.2	0.16	12	0.161431
10	0.2	0.2	0.16	12	0.161041

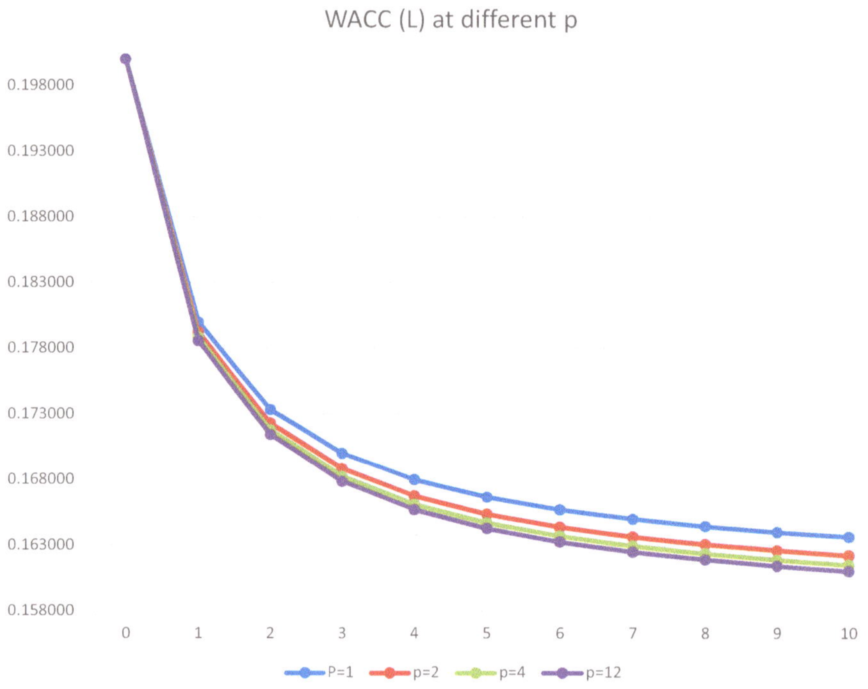

Fig. 5.1 Dependence of the weighted average cost of capital, WACC, on leverage level L at different p and fixed $k_d = 0.16$

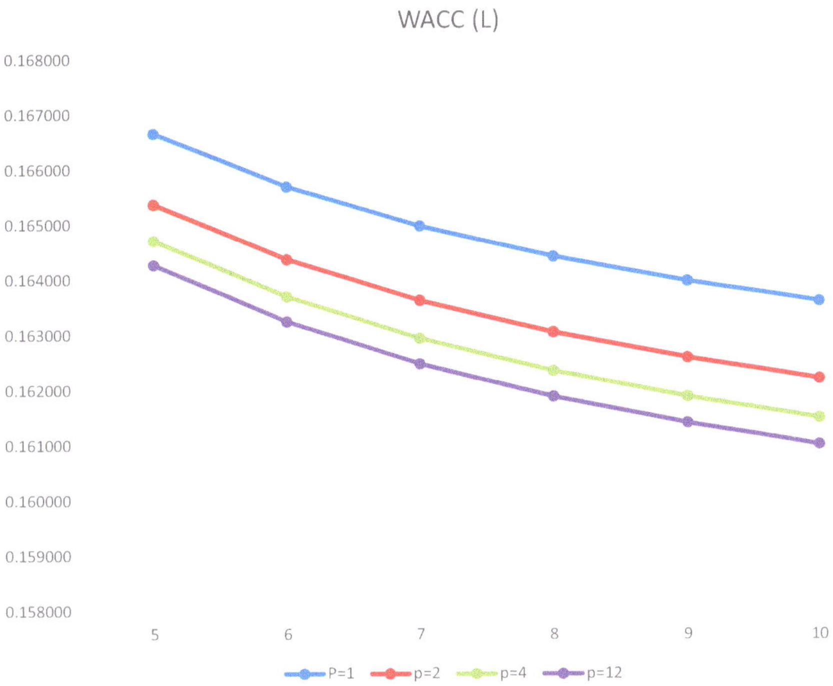

Fig. 5.2 Dependence of the weighted average cost of capital, WACC, on leverage level L at different p and fixed $k_d = 0.16$ (larger scale)

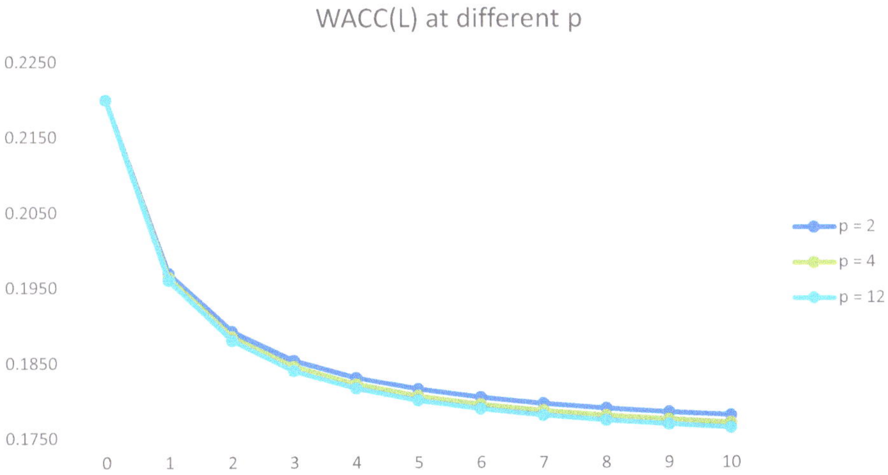

Fig. 5.3 Dependence of the weighted average cost of capital, WACC, on leverage level L at different p and fixed $k_0 = 0.22$

Table 5.5 Dependence of WACC on leverage level L at $k_0 = 0.22$ and different p = 2; 4; 12

L	WACC		
	p = 2	p = 4	p = 12
0	0.2200	0.2200	0.2200
1	0.1970	0.1964	0.1960
2	0.1893	0.1886	0.1881
3	0.1854	0.1846	0.1841
4	0.1831	0.1823	0.1817
5	0.1816	0.1807	0.1801
6	0.1805	0.1796	0.1789
7	0.1797	0.1787	0.1781
8	0.1790	0.1781	0.1774
9	0.1785	0.1775	0.1769
10	0.1781	0.1771	0.1765

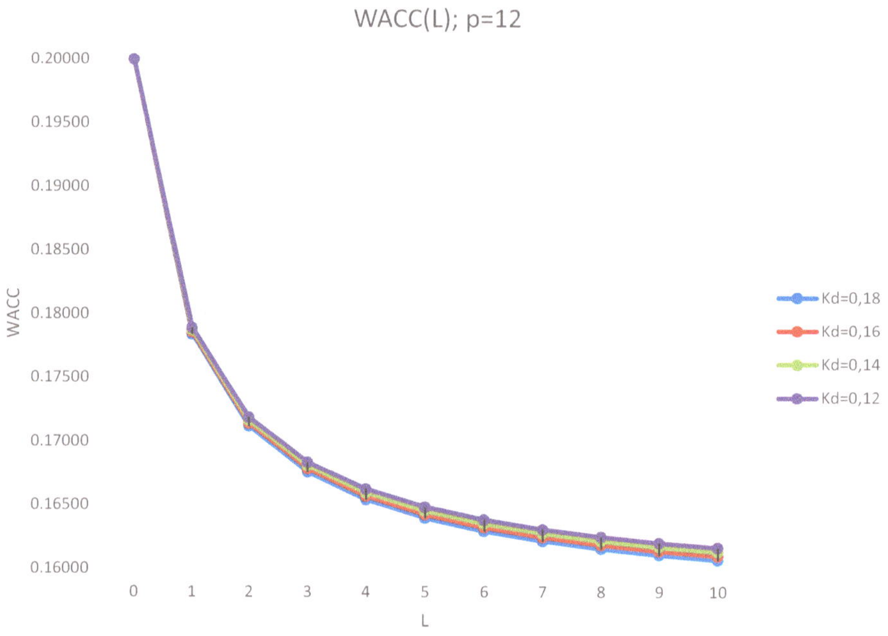

Fig. 5.4 Dependence of the weighted average cost of capital, WACC, on leverage level L at different k_d (12%; 14%; 16%; 18%) and fixed p = 12

5.3.2 *Dependence of the Weighted Average Cost of Capital, WACC, on Leverage Level L at Different k_d and Fixed p*

In this paragraph we investigate the dependence of the weighted average cost of capital, WACC, on leverage level L at different values of debt cost k_d and fixed p (Figs. 5.4 and 5.5).

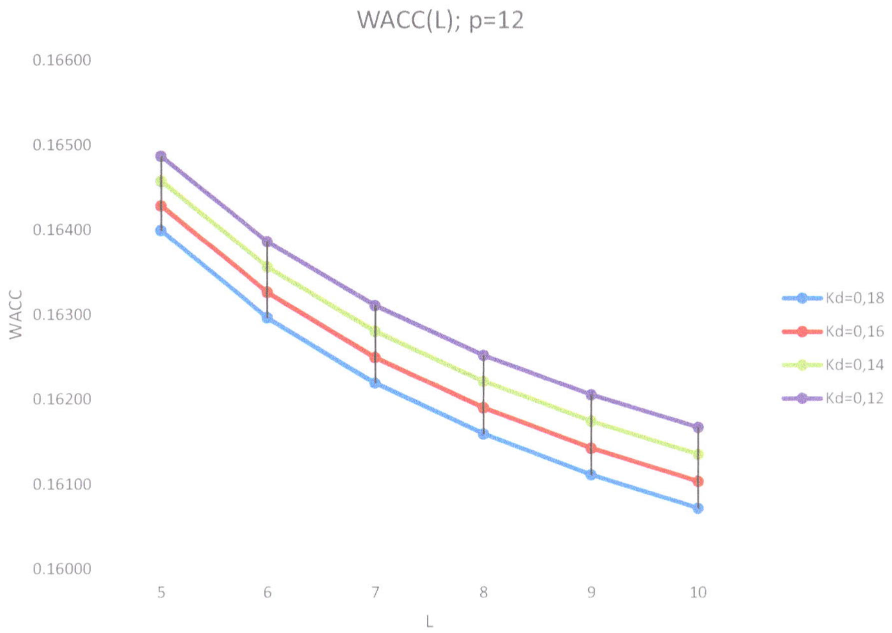

Fig. 5.5 Dependence of the weighted average cost of capital, WACC, on leverage level L at different k_d (12%; 14%; 16%; 18%) and fixed p = 12 (larger scale)

Table 5.6 Dependence of WACC on leverage level L at $k_0 = 0.2$, $k_d = 0.18$ and p = 1

L	T	k_0	k_d	p	WACC
0	0.2	0.2	0.18	12	0.20000
1	0.2	0.2	0.18	12	0.17840
2	0.2	0.2	0.18	12	0.17120
3	0.2	0.2	0.18	12	0.16760
4	0.2	0.2	0.18	12	0.16544
5	0.2	0.2	0.18	12	0.16400
6	0.2	0.2	0.18	12	0.16297
7	0.2	0.2	0.18	12	0.16220
8	0.2	0.2	0.18	12	0.16160
9	0.2	0.2	0.18	12	0.16112
10	0.2	0.2	0.18	12	0.16073

From Tables 5.6, 5.7, 5.8, 5.9, 5.10, 5.11, 5.12, and 5.13 and Figs. 5.6 and 5.7 it is seen, that in case of income tax payments more than once per year (at $p \neq 1$), as it takes place in practice, the weighted average cost of capital, WACC, (and as we'll see below, company value, V and equity cost, k_e start depend on debt cost, k_d, while in ordinary (classical) Modigliani–Miller theory all these values DO NOT depend on k_d). The weighted average cost of capital, WACC, decreases with leverage level L at all values of debt cost k_d and fixed frequency of payments of tax on profit p. With increase of k_d WACC decreases. This is connected to the fact that the increase of k_d

Table 5.7 Dependence of WACC on leverage level L at $k_0 = 0.2$, $k_d = 0.16$ and $p = 12$

L	T	k_0	k_d	p	WACC
0	0.2	0.2	0.16	12	0.20000
1	0.2	0.2	0.16	12	0.17857
2	0.2	0.2	0.16	12	0.17143
3	0.2	0.2	0.16	12	0.16786
4	0.2	0.2	0.16	12	0.16572
5	0.2	0.2	0.16	12	0.16429
6	0.2	0.2	0.16	12	0.16327
7	0.2	0.2	0.16	12	0.16250
8	0.2	0.2	0.16	12	0.16191
9	0.2	0.2	0.16	12	0.16143
10	0.2	0.2	0.16	12	0.16104

Table 5.8 Dependence of WACC on leverage level L at $k_0 = 0.2$, $k_d = 0.14$ and $p = 12$

L	T	k_0	k_d	p	WACC
0	0.2	0.2	0.14	12	0.20000
1	0.2	0.2	0.14	12	0.17875
2	0.2	0.2	0.14	12	0.17166
3	0.2	0.2	0.14	12	0.16812
4	0.2	0.2	0.14	12	0.16600
5	0.2	0.2	0.14	12	0.16458
6	0.2	0.2	0.14	12	0.16357
7	0.2	0.2	0.14	12	0.16281
8	0.2	0.2	0.14	12	0.16222
9	0.2	0.2	0.14	12	0.16174
10	0.2	0.2	0.14	12	0.16136

Table 5.9 Dependence of WACC on leverage level L at $k_0 = 0.2$, $k_d = 0.12$ and $p = 12$

L	T	k_0	k_d	p	WACC
0	0.2	0.2	0.12	12	0.20000
1	0.2	0.2	0.12	12	0.17892
2	0.2	0.2	0.12	12	0.17190
3	0.2	0.2	0.12	12	0.16838
4	0.2	0.2	0.12	12	0.16628
5	0.2	0.2	0.12	12	0.16487
6	0.2	0.2	0.12	12	0.16387
7	0.2	0.2	0.12	12	0.16311
8	0.2	0.2	0.12	12	0.16253
9	0.2	0.2	0.12	12	0.16206
10	0.2	0.2	0.12	12	0.16168

increases the value of tax shield in case of income tax payments more than once per year (at $p \neq 1$). Note, that in "classical" Modigliani–Miller theory with one income tax payment per year all financial indicators don't depend on debt cost, k_d.

Table 5.10 Dependence of WACC on leverage level L at $k_0 = 0.2$, $k_d = 0.18$ and $p = 4$

L	T	k_0	k_d	p	WACC
0	0.2	0.2	0.18	4	0.20000
1	0.2	0.2	0.18	4	0.17870
2	0.2	0.2	0.18	4	0.17160
3	0.2	0.2	0.18	4	0.16804
4	0.2	0.2	0.18	4	0.16591
5	0.2	0.2	0.18	4	0.16449
6	0.2	0.2	0.18	4	0.16348
7	0.2	0.2	0.18	4	0.16272
8	0.2	0.2	0.18	4	0.16213
9	0.2	0.2	0.18	4	0.16165
10	0.2	0.2	0.18	4	0.16127

Table 5.11 Dependence of WACC on leverage level L at $k_0 = 0.2$, $k_d = 0.16$ and $p = 4$

L	T	k_0	k_d	p	WACC
0	0.2	0.2	0.16	4	0.20000
1	0.2	0.2	0.16	4	0.17884
2	0.2	0.2	0.16	4	0.17178
3	0.2	0.2	0.16	4	0.16826
4	0.2	0.2	0.16	4	0.16614
5	0.2	0.2	0.16	4	0.16473
6	0.2	0.2	0.16	4	0.16372
7	0.2	0.2	0.16	4	0.16296
8	0.2	0.2	0.16	4	0.16238
9	0.2	0.2	0.16	4	0.16191
10	0.2	0.2	0.16	4	0.16152

Table 5.12 Dependence of WACC on leverage level L at $k_0 = 0.2$, $k_d = 0.14$ and $p = 4$

L	T	k_0	k_d	p	WACC
0	0.2	0.2	0.14	4	0.20000
1	0.2	0.2	0.14	4	0.17898
2	0.2	0.2	0.14	4	0.17197
3	0.2	0.2	0.14	4	0.16847
4	0.2	0.2	0.14	4	0.16637
5	0.2	0.2	0.14	4	0.16496
6	0.2	0.2	0.14	4	0.16396
7	0.2	0.2	0.14	4	0.16321
8	0.2	0.2	0.14	4	0.16263
9	0.2	0.2	0.14	4	0.16216
10	0.2	0.2	0.14	4	0.16178

Table 5.13 Dependence of WACC on leverage level L at $k_0 = 0.2$, $k_d = 0.12$ and $p = 4$

L	T	k_0	k_d	p	WACC
0	0.2	0.2	0.12	4	0.20000
1	0.2	0.2	0.12	4	0.17912
2	0.2	0.2	0.12	4	0.17216
3	0.2	0.2	0.12	4	0.16868
4	0.2	0.2	0.12	4	0.16659
5	0.2	0.2	0.12	4	0.16520
6	0.2	0.2	0.12	4	0.16421
7	0.2	0.2	0.12	4	0.16346
8	0.2	0.2	0.12	4	0.16288
9	0.2	0.2	0.12	4	0.16242
10	0.2	0.2	0.12	4	0.16204

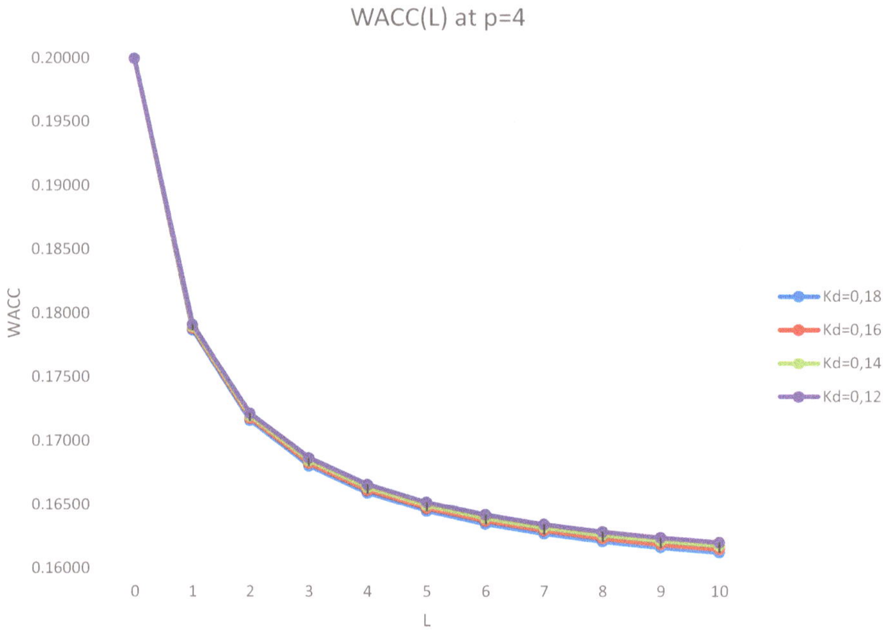

Fig. 5.6 Dependence of the weighted average cost of capital, WACC, on leverage level L at different k_d (12%; 14%; 16%; 18%) and fixed $p = 4$

5.4 Dependence of the Company Capitalization, V, on Leverage Level L

In this section we study the dependence of the company capitalization, V, on leverage level L for the following situations: (1) at different $p = 1;2;4;12$ and fixed $k_d = 0.16$; (2) for fixed p and at different values of debt cost $k_d = 0.18;0.16;0.14;012$.

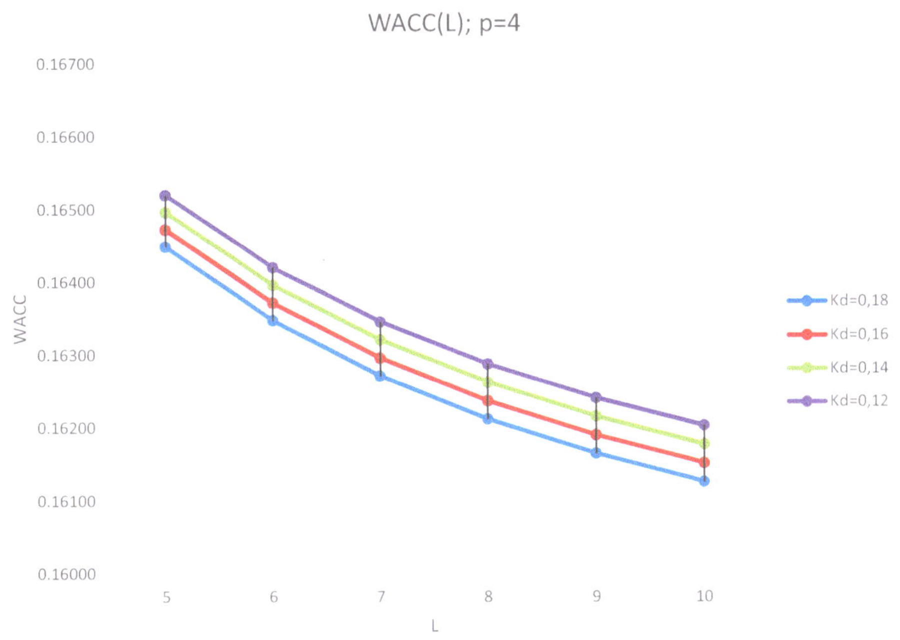

Fig. 5.7 Dependence of the weighted average cost of capital, WACC, on leverage level L at different k_d (12%; 14%; 16%; 18%) and fixed p = 4 (larger scale)

Table 5.14 Dependence of the company capitalization, V, on leverage level L at p = 1, $k_d = 0.16$

L	T	k_0	k_d	CF	p	V
0	0.2	0.2	0.16	1000	1	5000.00
1	0.2	0.2	0.16	1000	1	5555.56
2	0.2	0.2	0.16	1000	1	5769.23
3	0.2	0.2	0.16	1000	1	5882.35
4	0.2	0.2	0.16	1000	1	5952.38
5	0.2	0.2	0.16	1000	1	6000.00
6	0.2	0.2	0.16	1000	1	6034.48
7	0.2	0.2	0.16	1000	1	6060.61
8	0.2	0.2	0.16	1000	1	6081.08
9	0.2	0.2	0.16	1000	1	6097.56
10	0.2	0.2	0.16	1000	1	6111.11

5.4.1 Dependence of the Company Capitalization, V, on Leverage Level L at Different p and Fixed k_d

Obtained results on the dependence of the company capitalization, V, on leverage level L at different p = 1;2;4;12 and fixed $k_d = 0.16$ (Tables 5.14, 5.15, 5.16, and 5.17 and Figs. 5.8 and 5.9) show that the company capitalization, V, increases with p

Table 5.15 Dependence of the company capitalization, V, on leverage level L at p = 2, $k_d = 0.16$

L	T	k_0	k_d	CF	p	V
0	0.2	0.2	0.16	1000	2	5000.00
1	0.2	0.2	0.16	1000	2	5579.43
2	0.2	0.2	0.16	1000	2	5803.62
3	0.2	0.2	0.16	1000	2	5922.61
4	0.2	0.2	0.16	1000	2	5996.37
5	0.2	0.2	0.16	1000	2	6046.58
6	0.2	0.2	0.16	1000	2	6082.96
7	0.2	0.2	0.16	1000	2	6110.53
8	0.2	0.2	0.16	1000	2	6132.15
9	0.2	0.2	0.16	1000	2	6149.55
10	0.2	0.2	0.16	1000	2	6163.87

Table 5.16 Dependence of the company capitalization, V, on leverage level L at p = 4, $k_d = 0.16$

L	T	k_0	k_d	CF	p	V
0	0.2	0.2	0.16	1000	4	5000.00
1	0.2	0.2	0.16	1000	4	5591.68
2	0.2	0.2	0.16	1000	4	5821.31
3	0.2	0.2	0.16	1000	4	5943.34
4	0.2	0.2	0.16	1000	4	6019.04
5	0.2	0.2	0.16	1000	4	6070.60
6	0.2	0.2	0.16	1000	4	6107.96
7	0.2	0.2	0.16	1000	4	6136.29
8	0.2	0.2	0.16	1000	4	6158.51
9	0.2	0.2	0.16	1000	4	6176.39
10	0.2	0.2	0.16	1000	4	6191.11

Table 5.17 Dependence of the company capitalization, V, on leverage level L at p = 12, $k_d = 0.16$

L	T	k_0	k_d	CF	p	V
0	0.2	0.2	0.16	1000	12	5000.00
1	0.2	0.2	0.16	1000	12	5599.96
2	0.2	0.2	0.16	1000	12	5833.28
3	0.2	0.2	0.16	1000	12	5957.38
4	0.2	0.2	0.16	1000	12	6034.41
5	0.2	0.2	0.16	1000	12	6086.88
6	0.2	0.2	0.16	1000	12	6124.92
7	0.2	0.2	0.16	1000	12	6153.77
8	0.2	0.2	0.16	1000	12	6176.39
9	0.2	0.2	0.16	1000	12	6194.61
10	0.2	0.2	0.16	1000	12	6209.59

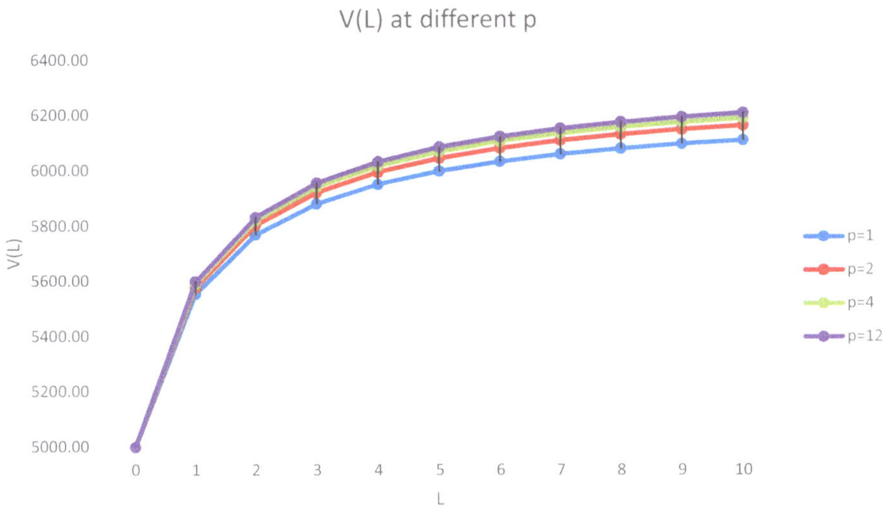

Fig. 5.8 Dependence of the company capitalization, V, on leverage level L at difference p (1; 2; 4; 12) and fixed $k_d = 0.16$

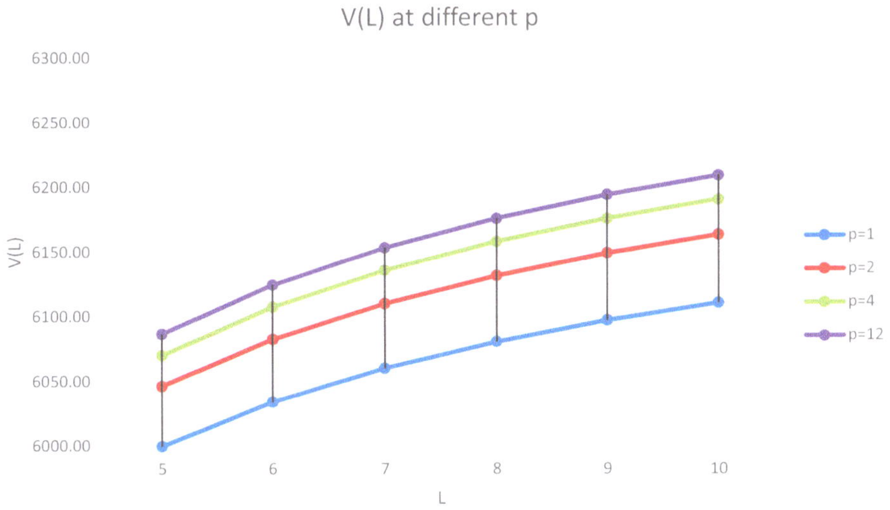

Fig. 5.9 Dependence of the company capitalization, V, on leverage level L at difference p (1; 2; 4; 12) and fixed $k_d = 0.16$ (larger scale)

and reach maximum at p = 12. This means, that for company is more profitable to pay tax on profit as much more frequently as tax legislation allows.

5.4.2 Dependence of the Company Capitalization, V, on Leverage Level L at Different k_d and Fixed p

Let us now study the dependence of the company capitalization, V, on leverage level L for fixed p and at different values of debt cost $k_d = 0.18; 0.16; 0.14; 0.12$.

Obtained results on the dependence of the company capitalization, V, on leverage level L at fixed p (but for p = 1; 2; 4; 12) and different $k_d = 12\%; 14\%; 16\%; 18\%$ (Tables 5.18, 5.19, 5.20, 5.21, 5.22, 5.23, 5.24, 5.25, 5.26, 5.27, 5.28, and 5.29 and Figs. 5.10, 5.11, 5.12, 5.13, 5.14, and 5.15) show that the company capitalization, V, increases with k_d at each value of fixed p and reach maximum at $k_d = 18\%$. This

Table 5.18 Dependence of the company capitalization, V, on leverage level L at p = 2 and $k_d = 0.18$

L	T	k_0	k_d	CF	p	V
0	0.2	0.2	0.18	1000	2	5000.00
1	0.2	0.2	0.18	1000	2	5582.31
2	0.2	0.2	0.18	1000	2	5807.78
3	0.2	0.2	0.18	1000	2	5927.48
4	0.2	0.2	0.18	1000	2	6001.70
5	0.2	0.2	0.18	1000	2	6052.22
6	0.2	0.2	0.18	1000	2	6088.83
7	0.2	0.2	0.18	1000	2	6116.58
8	0.2	0.2	0.18	1000	2	6138.34
9	0.2	0.2	0.18	1000	2	6155.85
10	0.2	0.2	0.18	1000	2	6170.26

Table 5.19 Dependence of the company capitalization, V, on leverage level L at p = 2 and $k_d = 0.16$

L	T	k_0	k_d	CF	p	V
0	0.2	0.2	0.16	1000	2	5000.00
1	0.2	0.2	0.16	1000	2	5579.43
2	0.2	0.2	0.16	1000	2	5803.62
3	0.2	0.2	0.16	1000	2	5922.61
4	0.2	0.2	0.16	1000	2	5996.37
5	0.2	0.2	0.16	1000	2	6046.58
6	0.2	0.2	0.16	1000	2	6082.96
7	0.2	0.2	0.16	1000	2	6110.53
8	0.2	0.2	0.16	1000	2	6132.15
9	0.2	0.2	0.16	1000	2	6149.55
10	0.2	0.2	0.16	1000	2	6163.87

Table 5.20 Dependence of the company capitalization, V, on leverage level L at p = 2 and $k_d = 0.14$

L	T	k_0	k_d	CF	p	V
0	0.2	0.2	0.14	1000	2	5000.00
1	0.2	0.2	0.14	1000	2	5576.53
2	0.2	0.2	0.14	1000	2	5799.44
3	0.2	0.2	0.14	1000	2	5917.71
4	0.2	0.2	0.14	1000	2	5991.01
5	0.2	0.2	0.14	1000	2	6040.90
6	0.2	0.2	0.14	1000	2	6077.05
7	0.2	0.2	0.14	1000	2	6104.44
8	0.2	0.2	0.14	1000	2	6125.92
9	0.2	0.2	0.14	1000	2	6143.21
10	0.2	0.2	0.14	1000	2	6157.43

Table 5.21 Dependence of the company capitalization, V, on leverage level L at p = 2 and $k_d = 0.12$

L	T	k_0	k_d	CF	p	V
0	0.2	0.2	0.12	1000	2	5000.00
1	0.2	0.2	0.12	1000	2	5573.61
2	0.2	0.2	0.12	1000	2	5795.22
3	0.2	0.2	0.12	1000	2	5912.77
4	0.2	0.2	0.12	1000	2	5985.62
5	0.2	0.2	0.12	1000	2	6035.19
6	0.2	0.2	0.12	1000	2	6071.10
7	0.2	0.2	0.12	1000	2	6098.31
8	0.2	0.2	0.12	1000	2	6119.65
9	0.2	0.2	0.12	1000	2	6136.83
10	0.2	0.2	0.12	1000	2	6150.96

Table 5.22 Dependence of the company capitalization, V, on leverage level L at $k_d = 0.18$ and fixed p = 4

L	T	k_0	k_d	CF	p	V
0	0.2	0.2	0.18	1000	4	5000.00
1	0.2	0.2	0.18	1000	4	5596.08
2	0.2	0.2	0.18	1000	4	5827.66
3	0.2	0.2	0.18	1000	4	5950.79
4	0.2	0.2	0.18	1000	4	6027.20
5	0.2	0.2	0.18	1000	4	6079.24
6	0.2	0.2	0.18	1000	4	6116.97
7	0.2	0.2	0.18	1000	4	6145.57
8	0.2	0.2	0.18	1000	4	6168.00
9	0.2	0.2	0.18	1000	4	6186.06
10	0.2	0.2	0.18	1000	4	6200.92

Table 5.23 Dependence of the company capitalization, V, on leverage level L at $k_d = 0.16$ and fixed $p = 4$

L	T	k_0	k_d	CF	p	V
0	0.2	0.2	0.16	1000	4	5000.00
1	0.2	0.2	0.16	1000	4	5591.68
2	0.2	0.2	0.16	1000	4	5821.31
3	0.2	0.2	0.16	1000	4	5943.34
4	0.2	0.2	0.16	1000	4	6019.04
5	0.2	0.2	0.16	1000	4	6070.60
6	0.2	0.2	0.16	1000	4	6107.96
7	0.2	0.2	0.16	1000	4	6136.29
8	0.2	0.2	0.16	1000	4	6158.51
9	0.2	0.2	0.16	1000	4	6176.39
10	0.2	0.2	0.16	1000	4	6191.11

Table 5.24 Dependence of the company capitalization, V, on leverage level L at $k_d = 0.14$ and fixed $p = 4$

L	T	k_0	k_d	CF	p	V
0	0.2	0.2	0.14	1000	4	5000.00
1	0.2	0.2	0.14	1000	4	5587.26
2	0.2	0.2	0.14	1000	4	5814.92
3	0.2	0.2	0.14	1000	4	5935.85
4	0.2	0.2	0.14	1000	4	6010.85
5	0.2	0.2	0.14	1000	4	6061.91
6	0.2	0.2	0.14	1000	4	6098.92
7	0.2	0.2	0.14	1000	4	6126.98
8	0.2	0.2	0.14	1000	4	6148.98
9	0.2	0.2	0.14	1000	4	6166.69
10	0.2	0.2	0.14	1000	4	6181.26

Table 5.25 Dependence of the company capitalization, V, on leverage level L at different $k_d = 0.12$ and fixed $p = 4$

L	T	k_0	k_d	CF	p	V
0	0.2	0.2	0.12	1000	4	5000.00
1	0.2	0.2	0.12	1000	4	5582.81
2	0.2	0.2	0.12	1000	4	5808.49
3	0.2	0.2	0.12	1000	4	5928.32
4	0.2	0.2	0.12	1000	4	6002.62
5	0.2	0.2	0.12	1000	4	6053.19
6	0.2	0.2	0.12	1000	4	6089.84
7	0.2	0.2	0.12	1000	4	6117.62
8	0.2	0.2	0.12	1000	4	6139.41
9	0.2	0.2	0.12	1000	4	6156.94
10	0.2	0.2	0.12	1000	4	6171.37

Table 5.26 Dependence of the company capitalization, V, on leverage level L at $k_d = 0.18$ and fixed $p = 12$

L	T	k_0	k_d	CF	p	V
0	0.2	0.2	0.18	1000	12	5000.00
1	0.2	0.2	0.18	1000	12	5605.40
2	0.2	0.2	0.18	1000	12	5841.15
3	0.2	0.2	0.18	1000	12	5966.63
4	0.2	0.2	0.18	1000	12	6044.53
5	0.2	0.2	0.18	1000	12	6097.61
6	0.2	0.2	0.18	1000	12	6136.09
7	0.2	0.2	0.18	1000	12	6165.28
8	0.2	0.2	0.18	1000	12	6188.17
9	0.2	0.2	0.18	1000	12	6206.60
10	0.2	0.2	0.18	1000	12	6221.77

Table 5.27 Dependence of the company capitalization, V, on leverage level L at $k_d = 0.16$ and fixed $p = 12$

L	T	k_0	k_d	CF	p	V
0	0.2	0.2	0.16	1000	12	5000.00
1	0.2	0.2	0.16	1000	12	5599.96
2	0.2	0.2	0.16	1000	12	5833.28
3	0.2	0.2	0.16	1000	12	5957.38
4	0.2	0.2	0.16	1000	12	6034.41
5	0.2	0.2	0.16	1000	12	6086.88
6	0.2	0.2	0.16	1000	12	6124.92
7	0.2	0.2	0.16	1000	12	6153.77
8	0.2	0.2	0.16	1000	12	6176.39
9	0.2	0.2	0.16	1000	12	6194.61
10	0.2	0.2	0.16	1000	12	6209.59

Table 5.28 Dependence of the company capitalization, V, on leverage level L at $k_d = 0.14$ and fixed $p = 12$

L	T	k_0	k_d	CF	p	V
0	0.2	0.2	0.14	1000	12	5000.00
1	0.2	0.2	0.14	1000	12	5594.50
2	0.2	0.2	0.14	1000	12	5825.38
3	0.2	0.2	0.14	1000	12	5948.11
4	0.2	0.2	0.14	1000	12	6024.27
5	0.2	0.2	0.14	1000	12	6076.13
6	0.2	0.2	0.14	1000	12	6113.73
7	0.2	0.2	0.14	1000	12	6142.23
8	0.2	0.2	0.14	1000	12	6164.58
9	0.2	0.2	0.14	1000	12	6182.59
10	0.2	0.2	0.14	1000	12	6197.39

Table 5.29 Dependence of the company capitalization, V, on leverage level L at $k_d = 0.12$ and fixed $p = 12$

L	T	k_0	k_d	CF	p	V
0	0.2	0.2	0.12	1000	12	5000.00
1	0.2	0.2	0.12	1000	12	5589.01
2	0.2	0.2	0.12	1000	12	5817.45
3	0.2	0.2	0.12	1000	12	5938.82
4	0.2	0.2	0.12	1000	12	6014.10
5	0.2	0.2	0.12	1000	12	6065.35
6	0.2	0.2	0.12	1000	12	6102.50
7	0.2	0.2	0.12	1000	12	6130.67
8	0.2	0.2	0.12	1000	12	6152.75
9	0.2	0.2	0.12	1000	12	6170.53
10	0.2	0.2	0.12	1000	12	6185.16

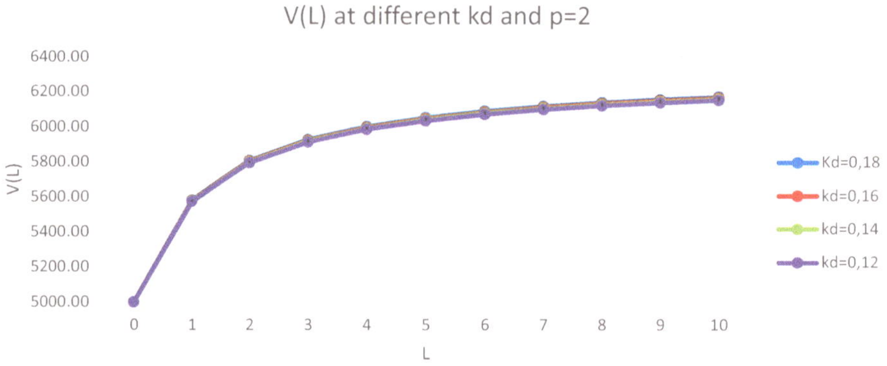

Fig. 5.10 Dependence of the company capitalization, V, on leverage level L at different k_d (12%; 14%; 16%; 18%) and fixed $p = 2$

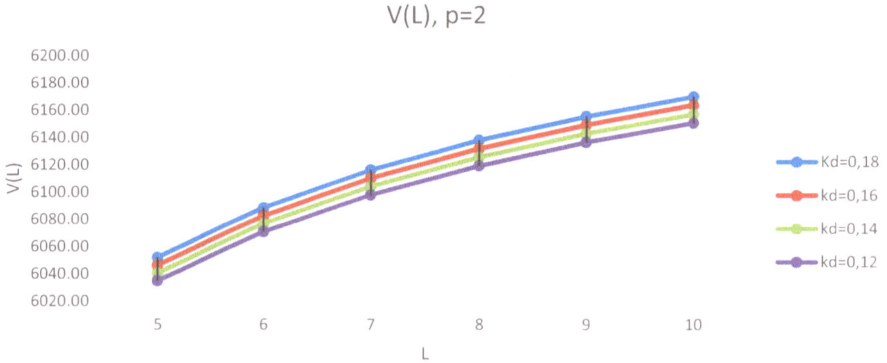

Fig. 5.11 Dependence of the company capitalization, V, on leverage level L at different k_d (12%; 14%; 16%; 18%) and fixed $p = 2$ (larger scale)

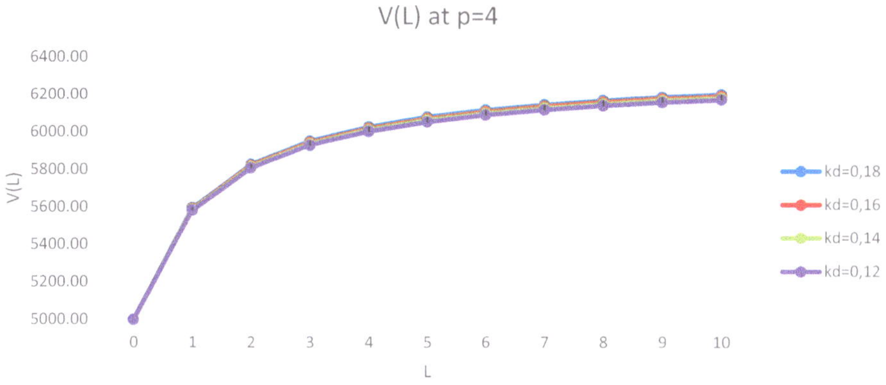

Fig. 5.12 Dependence of the company capitalization, V, on leverage level L at different k_d (12%; 14%; 16%; 18%) and fixed p = 4

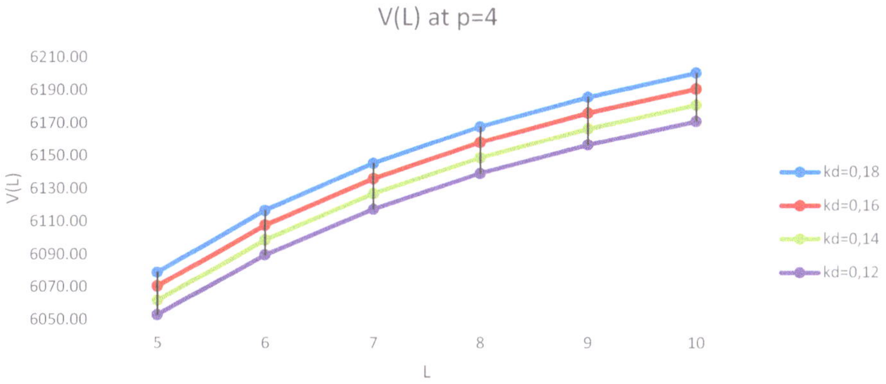

Fig. 5.13 Dependence of the company capitalization, V, on leverage level L at different k_d (12%; 14%; 16%; 18%) and fixed p = 4 (larger scale)

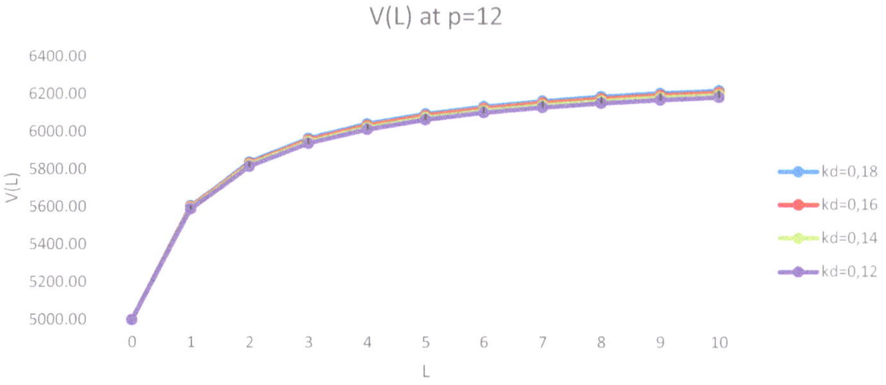

Fig. 5.14 Dependence of the company capitalization, V, on leverage level L at different k_d (12%; 14%; 16%; 18%) and fixed p = 12

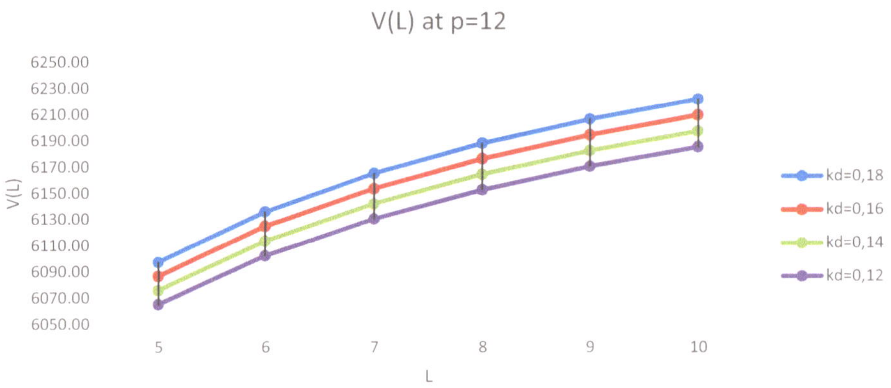

Fig. 5.15 Dependence of the company capitalization, V, on leverage level L at different k_d (12%; 14%; 16%; 18%) and fixed p = 12 (larger scale)

corresponds to decrease of the weighted average cost of capital, WACC, with k_d at fixed p, obtained by us in Sect. 5.3. This is connected to the fact that the increase of k_d increases the value of tax shield in case of income tax payments more than once per year (at $p \neq 1$). Note, that in "classical" Modigliani–Miller theory with one income tax payment per year all financial indicators, including company capitalization, V, don't depend on debt cost, k_d.

5.5 Dependence of the Equity Cost, k_e, on Leverage Level L

In this section we study the dependence of the equity cost, k_e, on leverage level L for the following situations: (1) at different p = 1; 2; 4; 12 and fixed $k_d = 0.16$; (2) for fixed p and at different values of debt cost $k_d = 0.18; 0.16; 0.14; 0.12$.

5.5.1 Dependence of the Equity Cost, k_e, on Leverage Level L at Different p and Fixed $k_d = 0.16$

Let us start from the study of the dependence of the equity cost, k_e, on leverage level L at different p and fixed $k_d = 0.16$.

Obtained results on the dependence of the equity cost, k_e, on leverage level L at fixed $k_d = 16\%$ and different p = 1; 2; 4; 12 (Tables 5.30, 5.31, 5.32, and 5.33 and Figs. 5.16 and 5.17) show that the tilt angle of the curve of equity cost, k_e (L), decreases with the number of payments of tax of profit p, this modifies the dividend policy of the company, because the economically justified value of dividends is equal to equity cost. Note, that in "classical" Modigliani–Miller theory with one

Table 5.30 Dependence of the equity cost, k_e, on leverage level L at $p = 1$ and fixed $k_d = 0.16$

L	T	k_0	k_d	p	k_e
0	0.2	0.2	0.16	1	0.2000
1	0.2	0.2	0.16	1	0.2320
2	0.2	0.2	0.16	1	0.2640
3	0.2	0.2	0.16	1	0.2960
4	0.2	0.2	0.16	1	0.3280
5	0.2	0.2	0.16	1	0.3600
6	0.2	0.2	0.16	1	0.3920
7	0.2	0.2	0.16	1	0.4240
8	0.2	0.2	0.16	1	0.4560
9	0.2	0.2	0.16	1	0.4880
10	0.2	0.2	0.16	1	0.5200

Table 5.31 Dependence of the equity cost, k_e, on leverage level L at $p = 2$ and fixed $k_d = 0.16$

L	T	k_0	k_d	p	k_e
0	0.2	0.2	0.16	2	0.2000
1	0.2	0.2	0.16	2	0.2305
2	0.2	0.2	0.16	2	0.2609
3	0.2	0.2	0.16	2	0.2914
4	0.2	0.2	0.16	2	0.3218
5	0.2	0.2	0.16	2	0.3523
6	0.2	0.2	0.16	2	0.3828
7	0.2	0.2	0.16	2	0.4132
8	0.2	0.2	0.16	2	0.4437
9	0.2	0.2	0.16	2	0.4741
10	0.2	0.2	0.16	2	0.5046

Table 5.32 Dependence of the equity cost, k_e, on leverage level L at $p = 4$ and fixed $k_d = 0.16$

L	T	k_0	k_d	p	k_e
0	0.2	0.2	0.16	4	0.2000
1	0.2	0.2	0.16	4	0.2297
2	0.2	0.2	0.16	4	0.2593
3	0.2	0.2	0.16	4	0.2890
4	0.2	0.2	0.16	4	0.3187
5	0.2	0.2	0.16	4	0.3484
6	0.2	0.2	0.16	4	0.3780
7	0.2	0.2	0.16	4	0.4077
8	0.2	0.2	0.16	4	0.4374
9	0.2	0.2	0.16	4	0.4671
10	0.2	0.2	0.16	4	0.4967

Table 5.33 Dependence of the equity cost, k_e, on leverage level L at p = 12 and fixed $k_d = 0.16$

L	T	k_0	k_d	p	k_e
0	0.2	0.2	0.16	12	0.2000
1	0.2	0.2	0.16	12	0.2291
2	0.2	0.2	0.16	12	0.2583
3	0.2	0.2	0.16	12	0.2874
4	0.2	0.2	0.16	12	0.3166
5	0.2	0.2	0.16	12	0.3457
6	0.2	0.2	0.16	12	0.3749
7	0.2	0.2	0.16	12	0.4040
8	0.2	0.2	0.16	12	0.4332
9	0.2	0.2	0.16	12	0.4623
10	0.2	0.2	0.16	12	0.4915

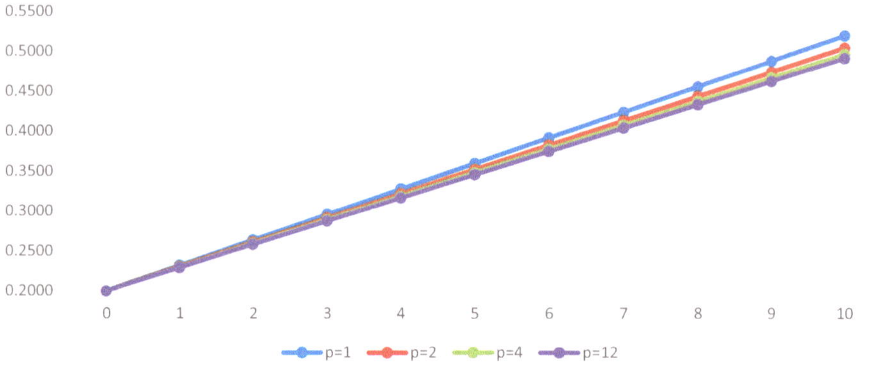

Fig. 5.16 Dependence of the equity cost, k_e, on leverage level L at different p (1; 2; 4; 12) and fixed $k_d = 0.16$

income tax payment per year all financial indicators, including the equity cost, k_e, don't depend on debt cost, k_d.

5.5.2 *Dependence of the Equity Cost, k_e, on Leverage Level L at Different k_d and Fixed p = 2*

In this section we study the dependence of the equity cost, k_e, on leverage level L at different k_d = 18%; 16%; 14%; 12% and fixed number of income tax payments per year p = 2.

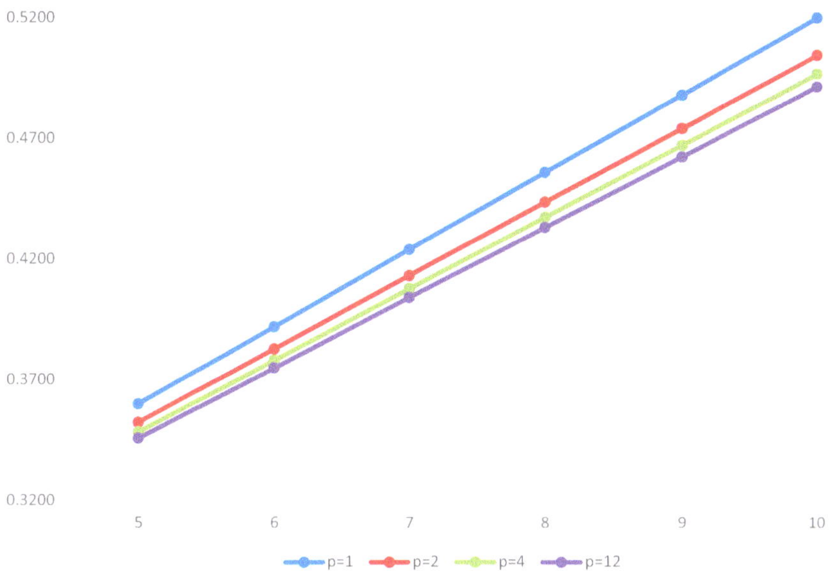

Fig. 5.17 Dependence of the equity cost, k_e, on leverage level L at different p (1; 2; 4; 12) and fixed kd = 0.16 (lager scale)

Table 5.34 Dependence of the equity cost, k_e, on leverage level L at p = 2 and fixed $k_d = 0.18$

L	T	k_0	k_d	p	k_e
0	0.2	0.2	0.18	2	0.2000
1	0.2	0.2	0.18	2	0.2143
2	0.2	0.2	0.18	2	0.2285
3	0.2	0.2	0.18	2	0.2428
4	0.2	0.2	0.18	2	0.2571
5	0.2	0.2	0.18	2	0.2714
6	0.2	0.2	0.18	2	0.2856
7	0.2	0.2	0.18	2	0.2999
8	0.2	0.2	0.18	2	0.3142
9	0.2	0.2	0.18	2	0.3285
10	0.2	0.2	0.18	2	0.3427

Conclusions. Obtained results on the dependence of the equity cost, k_e, on leverage level L at fixed but different p = 1; 2; 4; 12 and different values of debt cost k_d (Tables 5.34, 5.35, 5.36, 5.37, 5.38, 5.39, 5.40, 5.41, 5.42, 5.43, 5.44, and 5.45 and Figs. 5.18, 5.19, and 5.20) show that the tilt angle of the curve of equity cost, k_e (L), decreases with the increase of values of debt cost k_d: this modifies the dividend policy of the company, because the economically justified value of dividends is equal to equity cost. Note, that in "classical" Modigliani–Miller theory with

Table 5.35 Dependence of the equity cost, k_e, on leverage level L at p = 2 and fixed $k_d = 0.16$

L	T	k_0	k_d	p	k_e
0	0.2	0.2	0.16	2	0.2000
1	0.2	0.2	0.16	2	0.2305
2	0.2	0.2	0.16	2	0.2609
3	0.2	0.2	0.16	2	0.2914
4	0.2	0.2	0.16	2	0.3218
5	0.2	0.2	0.16	2	0.3523
6	0.2	0.2	0.16	2	0.3828
7	0.2	0.2	0.16	2	0.4132
8	0.2	0.2	0.16	2	0.4437
9	0.2	0.2	0.16	2	0.4741
10	0.2	0.2	0.16	2	0.5046

Table 5.36 Dependence of the equity cost, k_e, on leverage level L at p = 2 and fixed $k_d = 0.14$

L	T	k_0	k_d	p	k_e
0	0.2	0.2	0.14	2	0.2000
1	0.2	0.2	0.14	2	0.2466
2	0.2	0.2	0.14	2	0.2933
3	0.2	0.2	0.14	2	0.3399
4	0.2	0.2	0.14	2	0.3866
5	0.2	0.2	0.14	2	0.4332
6	0.2	0.2	0.14	2	0.4799
7	0.2	0.2	0.14	2	0.5265
8	0.2	0.2	0.14	2	0.5732
9	0.2	0.2	0.14	2	0.6198
10	0.2	0.2	0.14	2	0.6665

Table 5.37 Dependence of the equity cost, k_e, on leverage level L at p = 2 and fixed $k_d = 0.12$

L	T	k_0	k_d	p	k_e
0	0.2	0.2	0.12	2	0.2000
1	0.2	0.2	0.12	2	0.2628
2	0.2	0.2	0.12	2	0.3257
3	0.2	0.2	0.12	2	0.3885
4	0.2	0.2	0.12	2	0.4513
5	0.2	0.2	0.12	2	0.5142
6	0.2	0.2	0.12	2	0.5770
7	0.2	0.2	0.12	2	0.6398
8	0.2	0.2	0.12	2	0.7027
9	0.2	0.2	0.12	2	0.7655
10	0.2	0.2	0.12	2	0.8283

Table 5.38 Dependence of the equity cost, k_e, on leverage level L at $k_d = 0.18$ and fixed $p = 4$

L	T	k_0	k_d	p	k_e
0	0.2	0.2	0.18	4	0.2000
1	0.2	0.2	0.18	4	0.2134
2	0.2	0.2	0.18	4	0.2268
3	0.2	0.2	0.18	4	0.2402
4	0.2	0.2	0.18	4	0.2536
5	0.2	0.2	0.18	4	0.2670
6	0.2	0.2	0.18	4	0.2804
7	0.2	0.2	0.18	4	0.2938
8	0.2	0.2	0.18	4	0.3071
9	0.2	0.2	0.18	4	0.3205
10	0.2	0.2	0.18	4	0.3339

Table 5.39 Dependence of the equity cost, k_e, on leverage level L at $k_d = 0.16$ and fixed $p = 4$

L	T	k_0	k_d	p	k_e
0	0.2	0.2	0.16	4	0.2000
1	0.2	0.2	0.16	4	0.2297
2	0.2	0.2	0.16	4	0.2593
3	0.2	0.2	0.16	4	0.2890
4	0.2	0.2	0.16	4	0.3187
5	0.2	0.2	0.16	4	0.3484
6	0.2	0.2	0.16	4	0.3780
7	0.2	0.2	0.16	4	0.4077
8	0.2	0.2	0.16	4	0.4374
9	0.2	0.2	0.16	4	0.4671
10	0.2	0.2	0.16	4	0.4967

Table 5.40 Dependence of the equity cost, k_e, on leverage level L at $k_d = 0.14$ and fixed $p = 4$

L	T	k_0	k_d	p	k_e
0	0.2	0.2	0.14	4	0.2000
1	0.2	0.2	0.14	4	0.2460
2	0.2	0.2	0.14	4	0.2919
3	0.2	0.2	0.14	4	0.3379
4	0.2	0.2	0.14	4	0.3838
5	0.2	0.2	0.14	4	0.4298
6	0.2	0.2	0.14	4	0.4757
7	0.2	0.2	0.14	4	0.5217
8	0.2	0.2	0.14	4	0.5677
9	0.2	0.2	0.14	4	0.6136
10	0.2	0.2	0.14	4	0.6596

Table 5.41 Dependence of the equity cost, k_e, on leverage level L at $k_d = 0.12$ and fixed $p = 4$

L	T	k_0	k_d	p	k_e
0	0.2	0.2	0.12	4	0.2000
1	0.2	0.2	0.12	4	0.2622
2	0.2	0.2	0.12	4	0.3245
3	0.2	0.2	0.12	4	0.3867
4	0.2	0.2	0.12	4	0.4490
5	0.2	0.2	0.12	4	0.5112
6	0.2	0.2	0.12	4	0.5735
7	0.2	0.2	0.12	4	0.6357
8	0.2	0.2	0.12	4	0.6979
9	0.2	0.2	0.12	4	0.7602
10	0.2	0.2	0.12	4	0.8224

Table 5.42 Dependence of the equity cost, k_e, on leverage level L at $k_d = 0.18$ and fixed $p = 12$

L	T	k_0	k_d	p	k_e
0	0.2	0.2	0.18	12	0.2000
1	0.2	0.2	0.18	12	0.2128
2	0.2	0.2	0.18	12	0.2256
3	0.2	0.2	0.18	12	0.2384
4	0.2	0.2	0.18	12	0.2512
5	0.2	0.2	0.18	12	0.2640
6	0.2	0.2	0.18	12	0.2768
7	0.2	0.2	0.18	12	0.2896
8	0.2	0.2	0.18	12	0.3024
9	0.2	0.2	0.18	12	0.3152
10	0.2	0.2	0.18	12	0.3280

Table 5.43 Dependence of the equity cost, k_e, on leverage level L at $k_d = 0.16$ and fixed $p = 12$

L	T	k_0	k_d	p	k_e
0	0.2	0.2	0.16	12	0.2000
1	0.2	0.2	0.16	12	0.2291
2	0.2	0.2	0.16	12	0.2583
3	0.2	0.2	0.16	12	0.2874
4	0.2	0.2	0.16	12	0.3166
5	0.2	0.2	0.16	12	0.3457
6	0.2	0.2	0.16	12	0.3749
7	0.2	0.2	0.16	12	0.4040
8	0.2	0.2	0.16	12	0.4332
9	0.2	0.2	0.16	12	0.4623
10	0.2	0.2	0.16	12	0.4915

Table 5.44 Dependence of the equity cost, k_e, on leverage level L at $k_d = 0.12$ and fixed $p = 12$

L	T	k_0	k_d	p	k_e
0	0.2	0.2	0.14	12	0.2000
1	0.2	0.2	0.14	12	0.2455
2	0.2	0.2	0.14	12	0.2910
3	0.2	0.2	0.14	12	0.3365
4	0.2	0.2	0.14	12	0.3820
5	0.2	0.2	0.14	12	0.4275
6	0.2	0.2	0.14	12	0.4730
7	0.2	0.2	0.14	12	0.5185
8	0.2	0.2	0.14	12	0.5640
9	0.2	0.2	0.14	12	0.6094
10	0.2	0.2	0.14	12	0.6549

Table 5.45 Dependence of the equity cost, k_e, on leverage level L at $k_d = 0.12$ and fixed $p = 12$

L	T	k_0	k_d	p	k_e
0	0.2	0.2	0.12	12	0.2000
1	0.2	0.2	0.12	12	0.2618
2	0.2	0.2	0.12	12	0.3237
3	0.2	0.2	0.12	12	0.3855
4	0.2	0.2	0.12	12	0.4474
5	0.2	0.2	0.12	12	0.5092
6	0.2	0.2	0.12	12	0.5711
7	0.2	0.2	0.12	12	0.6329
8	0.2	0.2	0.12	12	0.6948
9	0.2	0.2	0.12	12	0.7566
10	0.2	0.2	0.12	12	0.8185

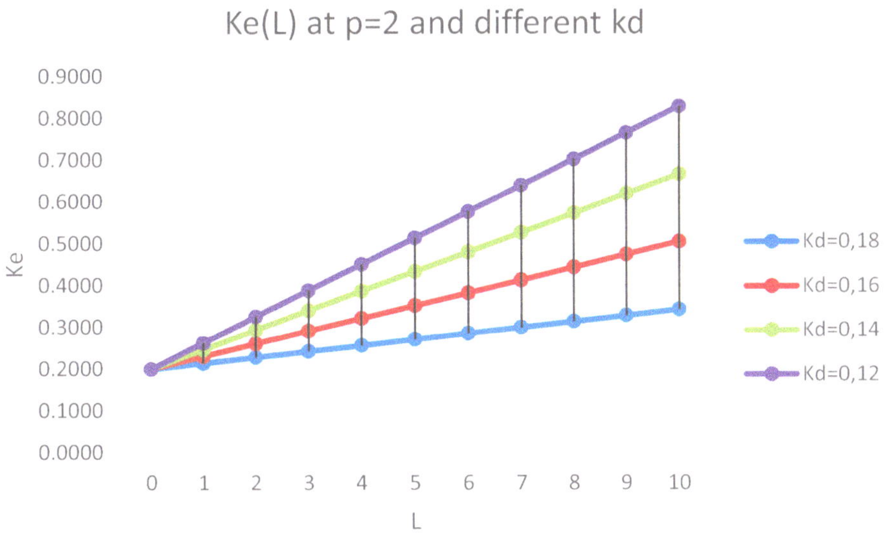

Fig. 5.18 Dependence of the equity cost, k_e, on leverage level L at different k_d (12%; 14%; 16%; 18%) and fixed $p = 2$

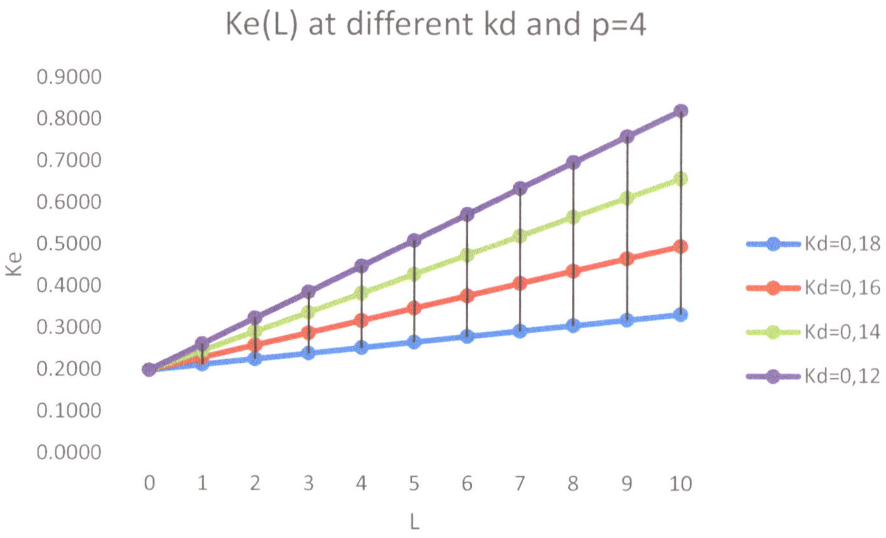

Fig. 5.19 Dependence of the equity cost, k_e, on leverage level L at different k_d (12%; 14%; 16%; 18%) and fixed $p = 4$

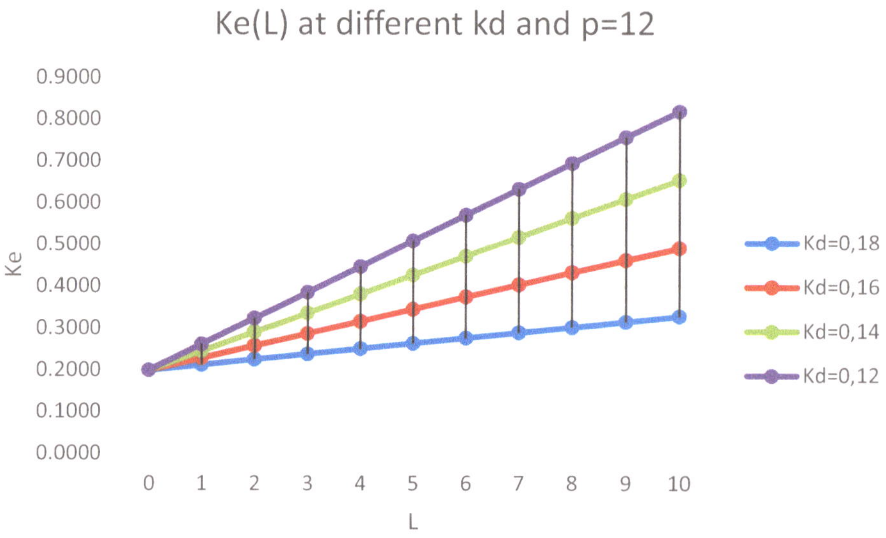

Fig. 5.20 Dependence of the equity cost, k_e, on leverage level L at different k_d (12%; 14%; 16%; 18%) and fixed $p = 12$

one income tax payment per year all financial indicators, including the equity cost, k_e, don't depend on debt cost, k_d.

5.6 Conclusions

The tax shield plays a crucial role in both main capital structure theories: Brusov–Filatova–Orekhova (BFO theory) and its perpetuity limit Modigliani–Miller theory. The way it is formed affects the results of the theories. Both of them describes the case of annual payments of tax of profit, but in practice these payments are made more frequent: semiannually, quarterly, monthly. While a return is not required more than once a year, businesses may be responsible for filing estimated taxes based on profits earned. This requirement is dependent on showing a profit. For example, sole proprietors must file estimated taxes on profits quarterly, on the 15th day of April, June, September and January. In Russia tax on profit payments could be made annually, quarterly, monthly. We show in current paper that the account of frequency of payment of tax on profit turns out to be very important and leads to significant consequences.

To study this problem we modify the Modigliani–Miller theory for the case of arbitrary frequency of payment of tax on profit and show that:

- all Modigliani–Miller statements and all formulas change,
- in case of income tax payments more than once per year (at $p \neq 1$), as it takes place in practice, the weighted average cost of capital, WACC, company value, V and equity cost, k_e start depend on k_d, while in ordinary (classical) Modigliani–Miller theory all these values DO NOT depend on k_d.
- obtained results allow to company choose the number of payments of tax of profit per year, as many, as it is profitable to it (of course, within actual tax legislation);
- the tilt angle of the curve of equity cost, k_e (L), decreases with p, this modifies the dividend policy of the company, because the economically justified value of dividends is equal to equity cost.
- one more influence of taxes within Modigliani–Miller theory is found: under accounting the number of payments of the tax of profit influent on all main indicators of the company: the weighted average cost of capital, WACC, company value, V, equity cost, k_e etc. This allows to company manage WACC, V, k_e etc. by choosing the number of payments of tax of profit p.

It turns out that increase the number of payments of tax of profit per year leads to decrease of the cost of attracting capital and increase of the company value. More frequent payments of income tax are beneficial for both parties: for the company and for the tax regulator: for the company, this leads to an increase in the value of the company, and for the tax regulator, earlier payments are beneficial due to the time value of money.

References

Arzac, E. R., & Glosten, L. R. (2005). A reconsideration of tax shield valuation. *European Financial Management, 11*(4), 453–461. International Journal of Business, 12(3), 2007 411.

Berk, J., & DeMarzo, P. (2007). *Corporate finance*. Pearson–Addison Wesley.

Brealey, R. A., Myers, S. C., & Allen, F. (2014). Principles of corporate finance (11th ed.). McGraw-Hill.

Brusov, P. (2018a). Editorial: Introduction on special issue on the banking system and financial markets of Russia and other countries: Problems and prospects. *Journal of Reviews on Global Economics, 7*, i–vi.

Brusov, P. (2018b). Editorial. *Journal of Reviews on Global Economics, 7*, i–v, SCOPUS.

Brusov, P., Filatova, T., Orehova, N., & Eskindarov, M. (2015). *Modern corporate finance, investments and taxation* (373 p). Springer International Publishing, monograph. Retrieved from https://www.springer.com/gp/book/9783319147314.

Brusov, P., Filatova, T., Orehova, N., & Eskindarov, M. (2018a). *Modern corporate finance, investments, taxation and ratings* (571 p). Springer Nature Publishing, monograph.

Brusov, P. N., Filatova, T. V., Orekhova, N. P., & Kulik, V. L. (2018b). Rating: New approach. *Journal of Reviews on Global Economics, 7*, 37–62. https://doi.org/10.6000/1929-7092.2018.07.05

Brusov, P. N., Filatova, T. V., Orekhova, N. P., & Kulik, V. L. (2018c). A "golden age" of the companies: Conditions of its existence. *Journal of Reviews on Global Economics, 7*, 88–103. https://doi.org/10.6000/1929-7092.2018.07.07

Brusov, P. N., Filatova, T. V., Orekhova, N. P., & Kulik, V. L. (2018d). Rating methodology: New look and new horizons. *Journal of Reviews on Global Economics, 7*, 63–87. https://doi.org/10.6000/1929-7092.2018.07.06

Brusov, P. N., Filatova, T. V., Orekhova, N. P., Kulik, V. L., & Weil, I. (2018e). New meaningful effects in modern capital structure theory. *Journal of Reviews on Global Economics, 7*, 104–122. https://doi.org/10.6000/1929-7092.2018.07.08

Brusov, P., Filatova, T., Orekhova, N., Kulik, V., Weil, I., & Brailov, A. (2018f). The impact of the central bank key rate and commercial banks credit rates on creating and maintaining of a favorable investment climate in the country. *Journal of Reviews on Global Economics, 7*, 360–376. SCOPUS. https://doi.org/10.6000/1929-7092.2018.07.31

Brusov, P. N., Filatova, T. V., & Orekhova, N. P. (2018g). *Modern corporate finance and investments*, monograph (517 p). Knorus Publishing House.

Brusov, P. N., Filatova, T. V., Orekhova, N. P., Kulik, V. L., & Weil, I. (2019). Ratings of the investment projects of arbitrary durations: New methodology. *Journal of Reviews on Global Economics, 8*, 437–448. https://doi.org/10.6000/1929-7092.2019.08.37

Brusov, P., Filatova, T., & Orehova, N. (2020). *Ratings: Critical analysis and new approaches of quantitative and qualitative methodology*. Springer Nature Publishing, monograph. Retrieved from https://www.springer.com/de/book/9783030562427.

DeMarzo, P. M. (1988). An extension of the Modigliani–Miller theorem to stochastic economies with incomplete markets and interdependent securities. *Journal of Economic Theory, 45*(2), 353–369.

Farber, A., Gillet, R., & Szafarz, A. (2006). A general formula for the WACC. *International Journal of Business, 11*(2), 211–218.

Fernandez, P. (2007). A general formula for the WACC: A comment. *International Journal of Business*, this issue.

Filatova, T., Orehova, N., & Brusova, A. (2008). Weighted average cost of capital in the theory of Modigliani–Miller, modified for a finite life–time company. *Bull FU, 48*, 68–77.

Filatova, T. V., Brusov, P. N., Orekhova, N. P., & Kulik, V. L. (2018a). Ratings of the long–term projects: New approach. *Journal of Reviews on Global Economics, 7*, 645–661. https://doi.org/10.6000/1929-7092.2018.07.59

Hamada, R. (1969). Portfolio analysis, market equilibrium, and corporate finance. *Journal of Finance, 24*, 13–31.

Harris, R., & Pringle, J. (1985). Risk–adjusted discount rates—extension form the average–risk case. *The Journal of Financial Research, 8*(3), 237–244.

Miles, J. A., & Ezzel, J. R. (1980). The weighted average cost of capital, perfect capital markets and project life: A clarification. *Journal of Financial and Quantitative Analysis, 15*(3), 719–730.

Miller, M. (1977). Debt and taxes. *The Journal of Finance, 32*(2), 261–275. https://doi.org/10.1111/j.1540-6261.1977.tb03267.x

Modigliani, F., & Miller, M. (1958). The cost of capital, corporate finance, and the theory of investment. *The American Economic Review, 48*, 261–297.

Modigliani, F., & Miller, M. (1963). Corporate income taxes and the cost of capital: A correction. *The American Economic Review, 53*, 147–175.

Modigliani, F., & Miller, M. (1966). Some estimates of the cost of capital to the electric utility industry 1954–1957. *The American Economic Review, 56*, 333–391.

Myers, S. (2001). Capital structure. *The Journal of Economic Perspectives, 15*(2), 81–102.

Ross, S., Westerfiel, R., & Jaffee, D. (2005). *Corporate finance* (7th ed.). McGraw Hill.

Sethi, S. P., Derzko, N. A., & Lehoczky, J. P. (1991). A stochastic extension of the Miller-Modigliani framework. *Mathematical Finance, 1*(4), 57–76.

Taggart, R. A. (1991). Consistent valuation and cost of capital expressions with corporate and personal taxes. *Financial Management, 20*, 8–20.

Chapter 6
Modification of the Modigliani–Miller Theory for the Case of Advance Tax on Profit Payments

In current Chapter we discuss one more limitation of Modigliani–Miller theory: a method of tax on profit payments. Modigliani–Miller theory accounts these payments as annuity–immediate while in practice these payments are often making in advance and thus should be accounting as annuity–due. We generalize the Modigliani–Miller theory for the case of advance tax on profit payments, which is widely used in practice, and show that this leads to some important consequences, which change seriously all the main statements by Modigliani and Miller (Brusov et al. 2020). The modified Modigliani–Miller theory (MMM theory) is the second of the three main theories which consists the base for conducting of modification of existing rating methodologies. In this Chapter we describe the main results of MMM theory.

6.1 Introduction

The first serious study (and first quantitative study) of influence of capital structure of the company on its indicators of activities was the work by Nobel Prize Winners Modigliani and Miller. Their theory has a lot of limitations. One of the most important and seriouse assumptions of the Modigliani–Miller theory is that all financial flows as well as all companies are perpetuity. This limitation was lift out by Brusov–Filatova–Orekhova in 2008 (Filatova et al. 2008), who have created BFO theory—modern theory of capital cost and capital structure for companies of arbitrary age.

Despite the fact that the Modigliani–Miller theory is currently a particular case of the general theory of capital cost and capital structure—Brusov–Filatova–Orekhova (BFO) theory—it is still widely used at the West.

P. Brusov, T. Filatova, *Efficiency of Investment Projects*, Contributions to Finance and Accounting, https://doi.org/10.1007/978-3-032-01028-5_6

In current Chapter we discuss one more limitation of Modigliani–Miller theory: a method of tax on profit payments. Modigliani–Miller theory accounts these payments as annuity–immediate while in practice these payments are making in advance and thus should be accounting as annuity–due.

We generalize the Modigliani–Miller theory for the case of advance tax on profit payments, which is widely used in practice, and show that this leads to some important consequences, which change seriously all the main statements by Modigliani and Miller. These consequences are as following: WACC starts depend on debt cost k_d, WACC turns out to be lower than in case of classical Modigliani–Miller theory and thus company capitalization becomes higher than in ordinary Modigliani–Miller theory. We show that equity dependence on leverage level L is still linear, but the tilt angle with respect to L–axis turns out to be smaller: this could lead to modification of the divident policy of the company.

Correct account of a method of tax on profit payments demonstrates that shortcomings of Modigliani–Miller theory are dipper, than everybody suggested: the underestimation of WACC really turns out to be bigger, as well as overestimation of the capitalization of the company. This means that systematic risks arising from the use of modified Modigliani–Miller theory (which is more correct than "classical" one) in practice is higher than it was suggested by the "classical" version of this theory.

6.2 Modified Modigliani–Miller Theory in Case of Advance Tax Payments

6.2.1 Tax Shield in Case of Advance Tax Payments

Modigliani–Miller theory accounts these payments as annuity–immediate while in practice these payments are making in advance and thus should be accounting as annuity–due.

To calculate tax shield *TS* in case of advance tax payments one should use annuity–due (Fig. 6.1)

$$TS = k_d Dt + \frac{k_d Dt}{(1+k_d)} + \frac{k_d Dt}{(1+k_d)^2} + \cdots = \frac{k_d Dt}{\left(1-(1+k_d)^{-1}\right)} = Dt(1+k_d) \quad (6.1)$$

This expression is different from the case of classical Modigliani–Miller theory (which used annuity–immediate) (Fig. 6.2)

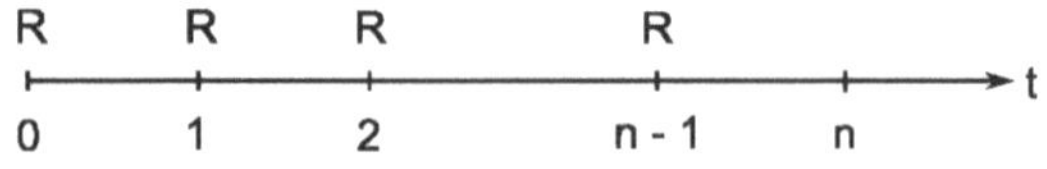

Fig. 6.1 Annuity–due

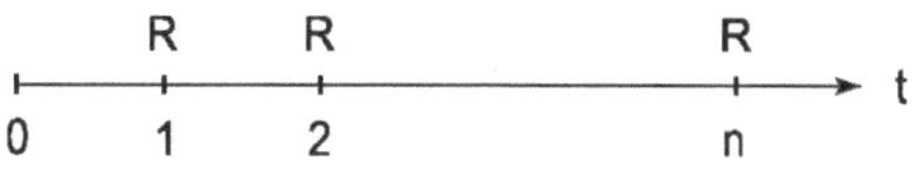

Fig. 6.2 Annuity–immediate

$$TS = \frac{k_d Dt}{(1+k_d)} + \frac{k_d Dt}{(1+k_d)^2} + \cdots = \frac{k_d Dt}{(1+k_d)\left(1-(1+k_d)^{-1}\right)} = Dt \tag{6.2}$$

Thus in the former case tax shield *TS* is bigger by multiplyer (1 + k_d). This is connected with the time value of money: money today is more expensive than money tomorrow due to the possibility of their alternative investment.

6.2.2 Capitalization of the Company

Modigliani–Miller theorem for capitalization of the company

$$V = V_0 + TS \tag{6.3}$$

takes the following form

$$V = V_0 + Dt(1+k_d). \tag{6.4}$$

Thus, we arrive to the following statement, which modifies the original Modigliani and Miller.

The value of financially dependent company making tax on profit payments in advance is equal to the value of the company of the same risk group used no leverage, increased by the value of tax shield arising from financial leverage, and equal to the product of rate of corporate income tax t, the value of debt D and multiplyer (1 + k_d).

Substituting $D = w_d \cdot V$

one has

$$V\left(1 - w_d t(1+k_d)\right) = V_0$$

$$\frac{CF}{WACC} \cdot \left(1 - w_d t(1+k_d)\right) = \frac{CF}{k_0}$$

And for WACC we have the following formula:

$$WACC = k_0\left(1 - w_d t(1+k_d)\right). \tag{6.5}$$

$$\text{At } L \to \infty \quad WACC = k_0(1 - t(1+k_d)). \tag{6.6}$$

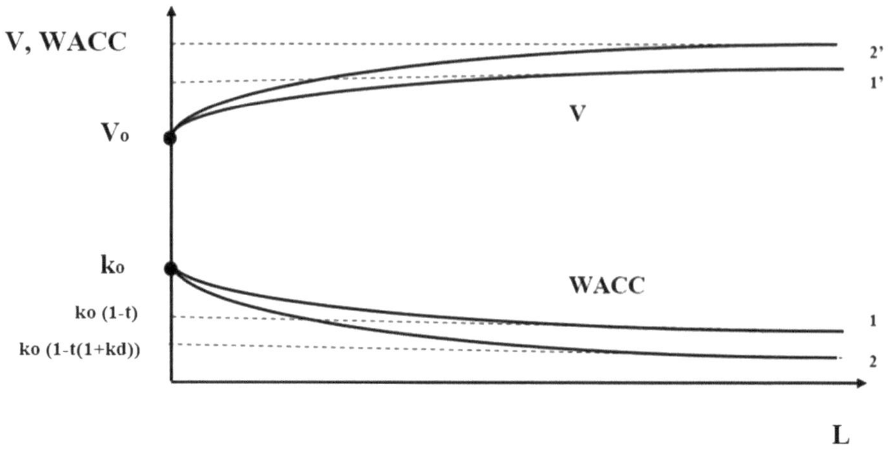

Fig. 6.3 Dependence of company capitalization, V and WACC on leverage level in "classical" Modigliani–Miller theory (curve 1′ and curve 1) and in "modified" Modigliani–Miller theory (curve 2′ and curve 2)

This expression is different from the similar one in classical Modigliani–Miller theory

$$\text{At } L \to \infty \quad WACC = k_0(1 - t).$$

From these expressions it is seen that WACC decreases with L, achieving lower value $WACC = k_0(1 - t(1 + k_d))$ at $L \to \infty$ in considering case comparing with the classical Modigliani–Miller theory $WACC = k_0(1 - t)$.

This means also, that company capitalization becomes higher than in ordinary Modigliani–Miller theory (Fig. 6.3).

6.2.3 Equity Cost

Let us find equity cost

$$WACC = k_e w_e + k_d w_d (1 - t). \tag{6.7}$$

Equating (6.5) and (6.7), we obtain

$$k_0(1 - w_d t(1 + k_d)) = k_e w_e + k_d w_d (1 - t), \tag{6.8}$$

whence we get the following expression for the equity cost:

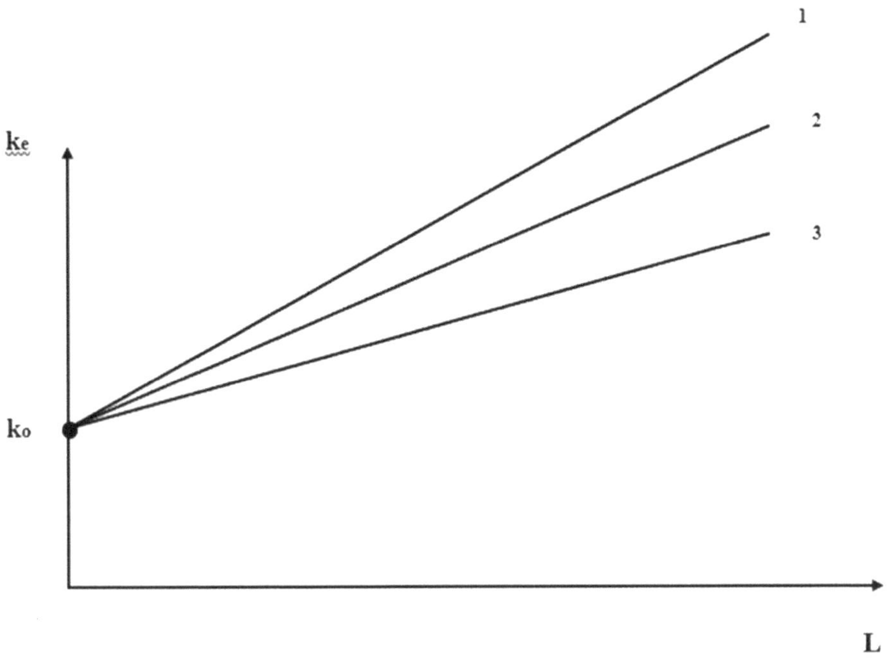

Fig. 6.4 Dependence of equity cost of the company, ke, on leverage level in "classical" Modigliani–Miller theory (curve 2), in "modified" Modigliani–Miller theory (curve 3) and for financially independent company (t = 0) (curve 1)

$$\begin{aligned} k_e &= k_0 \frac{(1 - w_d t(1 + k_d))}{w_e} - k_d \frac{w_d}{w_e}(1 - t) = \\ &= k_0 \frac{1}{w_e} - k_0 \frac{w_d}{w_e}(1 + k_d) \cdot t - k_d \frac{D}{S}(1 - t) = \\ &= k_0 \frac{D + S}{S} - k_0 \frac{D}{S}(1 + k_d) \cdot t - k_d \frac{D}{S}(1 - t) = \\ &= k_0 + L[(1 - t)(k_0 - k_d) - k_0 k_d t]. \end{aligned} \tag{6.9}$$

Finally, we have for the equity cost

$$k_e = k_0 + L[(1 - t)(k_0 - k_d) - k_0 k_d t]. \tag{6.10}$$

Thus, we arrive to the following statement, which modifies the original Modigliani and Miller.

Equity cost of leverage company k_e making tax on profit payments in advance could be found as equity cost of financially independent company k_0 of the same group of risk, plus premium for risk, the value which is equal to production of leverage level L on production of difference $(k_0 - k_d)$ and tax shield $(1 - t)$, decreasing by the value $k_0 k_d t$ (Fig. 6.4).

This means, that equity cost dependence on leverage level L is still linear, but the tilt angle with respect to L–axis turns out to be smaller $tg\alpha = (k_0 - k_d) \cdot (1 - t) - k_0 k_d t$.

This could lead to modification of the dividend policy of the company, because the equity cost represents itself economically sound value of dividends. Thus, company could decreases the value of dividends.

6.3 The Dependence of the Weighted Average Cost of Capital, WACC, on Leverage Level in "Classical" Modigliani–Miller Theory (MM Theory) and in Modified Modigliani–Miller Theory (MMM Theory)

Let us compare the dependence of the weighted average cost of capital, WACC, on leverage level in "classical" Modigliani–Miller theory (MM theory) and in modified Modigliani–Miller theory (MMM theory). Study of such dependence is very important, because, the weighted average cost of capital, WACC, plays the role of discount rate in operating financial flows discounting as well as of financial flows in rating methodologies. WACC's value determines as well the capitalization of the company $V = CF/WACC$. We use Microsoft Excel for calculations.

From the Tables 6.1, 6.2, 6.3, and 6.4 and from Fig. 6.5 it is seen that WACC in MMM theory turns out to be lower than in case of classical Modigliani–Miller theory and thus company capitalization becomes higher than in ordinary Modigliani–Miller theory. It is seen, that WACC decreases with debt cost k_d.

Correct account of a method of tax on profit payments demonstrates that shortcomings of Modigliani–Miller theory are dipper, than everybody suggested: the underestimation of WACC really turns out to be bigger, as well as overestimation of the capitalization of the company. This means that systematic risks arising from the

Table 6.1 Dependence of WACC on leverage level in "classical" Modigliani–Miller theory

k_0	L	t	WACC
0.2	0	0.2	0.2000
0.2	1	0.2	0.1800
0.2	2	0.2	0.1733
0.2	3	0.2	0.1700
0.2	4	0.2	0.1680
0.2	5	0.2	0.1667
0.2	6	0.2	0.1657
0.2	7	0.2	0.1650
0.2	8	0.2	0.1644
0.2	9	0.2	0.1640
0.2	10	0.2	0.1636

Table 6.2 Dependence of WACC on leverage level in modified Modigliani–Miller theory (MMM theory) at $k_d = 0.18$

k_0	L	t	k_d	WACC
0.2	0	0.2	0.18	0.2000
0.2	1	0.2	0.18	0.1764
0.2	2	0.2	0.18	0.1685
0.2	3	0.2	0.18	0.1646
0.2	4	0.2	0.18	0.1622
0.2	5	0.2	0.18	0.1607
0.2	6	0.2	0.18	0.1595
0.2	7	0.2	0.18	0.1587
0.2	8	0.2	0.18	0.1580
0.2	9	0.2	0.18	0.1575
0.2	10	0.2	0.18	0.1571

Table 6.3 Dependence of WACC on leverage level in modified Modigliani–Miller theory (MMM theory) at $k_d = 0.14$

k_0	L	t	k_d	WACC
0.2	0	0.2	0.14	0.2000
0.2	1	0.2	0.14	0.1772
0.2	2	0.2	0.14	0.1696
0.2	3	0.2	0.14	0.1658
0.2	4	0.2	0.14	0.1635
0.2	5	0.2	0.14	0.1620
0.2	6	0.2	0.14	0.1609
0.2	7	0.2	0.14	0.1601
0.2	8	0.2	0.14	0.1595
0.2	9	0.2	0.14	0.1590
0.2	10	0.2	0.14	0.1585

Table 6.4 Dependence of WACC on leverage level in modified Modigliani–Miller theory (MMM theory) at $k_d = 0.1$

k_0	L	t	k_d	WACC
0.2	0	0.2	0.1	0.2000
0.2	1	0.2	0.1	0.1780
0.2	2	0.2	0.1	0.1707
0.2	3	0.2	0.1	0.1670
0.2	4	0.2	0.1	0.1648
0.2	5	0.2	0.1	0.1633
0.2	6	0.2	0.1	0.1623
0.2	7	0.2	0.1	0.1615
0.2	8	0.2	0.1	0.1609
0.2	9	0.2	0.1	0.1604
0.2	10	0.2	0.1	0.1600

use of modified Modigliani–Miller theory (MMM theory) (which is more correct than "classical" one) in practice are higher than it was suggested by the "classical" version of this theory.

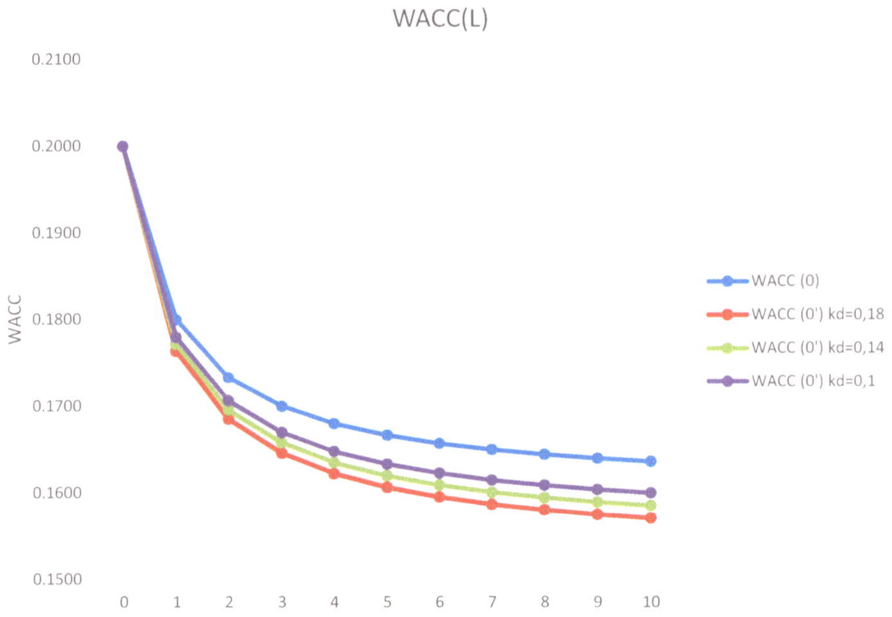

Fig. 6.5 Dependence of WACC on leverage level L: (1) in "classical" Modigliani–Miller theory (curve WACC(0)) and (2) in modified Modigliani–Miller theory (MMM theory) at different values of debt cost: $k_d = 0.18$; $k_d = 0.14$; $k_d = 0.1$ (curves WACC(0′))

6.4 Conclusions

Despite the fact that the Modigliani–Miller theory (Modigliani & Miller (1958), Modigliani & Miller (1963), Modigliani & Miller (1966), Myers (2001)) is currently a particular case of the general theory of capital cost and capital structure—Brusov–Filatova–Orekhova (BFO) theory—it is still widely used at the West. We generalize the Modigliani–Miller theory for the case of advance tax on profit payments, which is widely used in practice, and show that this leads to some important consequences, which change seriously all the main statements by Modigliani and Miller. These consequences are as following: WACC starts depend on debt cost k_d, WACC turns out to be lower than in case of classical Modigliani–Miller theory and thus company capitalization becomes higher than in ordinary Modigliani–Miller theory.

We study as well the equity cost dependence on leverage level L and show that it is still linear, but the tilt angle with respect to L–axis turns out to be smaller in MMM theory $tg\alpha = (k_0 - k_d) \cdot (1 - t) - k_0 k_d t$, than in classical Modigliani–Miller theory $tg\alpha = (k_0 - k_d) \cdot (1 - t)$. This could lead to modification of the dividend policy of the company, because the equity cost represents itself economically sound value of dividends. Thus, company could decreases the value of dividends, which they should pay to shareholders.

Correct account of a method of tax on profit payments demonstrates that shortcomings of Modigliani–Miller theory are dipper, than everybody suggested: the underestimation of WACC really turns out to be bigger, as well as overestimation of the capitalization of the company. This means that systematic risks arising from the use of modified Modigliani–Miller theory (which is more correct than "classical" one) in practice is higher than it was suggested by the "classical" version of this theory.

References

Brusov, P. (2018a). Editorial: Introduction on special issue on the banking system and financial markets of Russia and other countries: Problems and prospects. *Journal of Reviews on Global Economics, 7*, i–vi, SCOPUS.

Brusov, P. (2018b). Editorial. *Journal of Reviews on Global Economics, 7*, i–v, SCOPUS.

Brusov, P., Filatova, T., Orehova, N., & Eskindarov, M. (2015). *Modern corporate finance, investments and taxation* (373 p). Springer International Publishing. monograph, SCOPUS. Retrieved from https://www.springer.com/gp/book/9783319147314.

Brusov, P. N., Filatova, T. V., Orekhova, N. P., Kulik, V. L., & Weil, I. (2018a). New meaningful effects in modern capital structure theory. *Journal of Reviews on Global Economics, 7*, 104–122. SCOPUS. https://doi.org/10.6000/1929-7092.2016.07.08

Brusov, P., Filatova, T., Orehova, N., & Eskindarov, M. (2018b). *Modern corporate finance, investments, taxation and ratings* (571 p). Springer Nature Publishing. monograph.

Brusov, P. N., Filatova, T. V., & Orekhova, N. P. (2018c). *Modern corporate finance and investments*, monograph (517 p). Knorus Publishing House.

Brusov, P., Filatova, T., Orekhova, N., Kulik, V., Weil, I., & Brailov, A. (2018d). The impact of the central bank key rate and commercial banks credit rates on creating and maintaining of a favorable investment climate in the country. *Journal of Reviews on Global Economics, 7*, 360–376. SCOPUS. https://doi.org/10.6000/1929-7092.2016.07.31

Brusov, P. N., Filatova, T. V., Orekhova, N. P., & Kulik, V. L. (2018e). Rating: New approach. *Journal of Reviews on Global Economics, 7*, 37–62, SCOPUS. https://doi.org/10.6000/1929-7092.2016.07.05

Brusov, P. N., Filatova, T. V., Orekhova, N. P., & Kulik, V. L. (2018f). A "golden age" of the companies: Conditions of its existence. *Journal of Reviews on Global Economics, 7*, 88–103. SCOPUS. https://doi.org/10.6000/1929-7092.2016.07.07

Brusov, P. N., Filatova, T. V., Orekhova, N. P., & Kulik, V. L. (2018g). Rating methodology: New look and new horizons. *Journal of Reviews on Global Economics, 7*, 63–87. SCOPUS. https://doi.org/10.6000/1929-7092.2016.07.06

Brusov, P. N., Filatova, T. V., Orekhova, N. P., Kulik, V. L., & Weil, I. (2019). Ratings of the investment projects of arbitrary durations: New methodology. *Journal of Reviews on Global Economics, 8*, 437–446. SCOPUS. https://doi.org/10.6000/1929-7092.2019.06.37

Brusov, P. N., Filatova, T. V., Orekhova, N. P., Kulik, V. L., Chang, S.-I., & Lin, Y. C. G. (2020). Modification of the Modigliani–Miller theory for the case of advance payments of tax on profit. *Journal of Reviews on Global Economics, 9*, 257–266.

Filatova, T., Orehova, N., & Brusova, A. (2008). Weighted average cost of capital in the theory of Modigliani–Miller, modified for a finite life-time company. *Bull FU, 48*, 68–77.

Filatova, T. V., Brusov, P. N., Orekhova, N. P., & Kulik, V. L. (2018). Ratings of the long-term projects: New approach. *Journal of Reviews on Global Economics, 7*, 645–661, SCOPUS. https://doi.org/10.6000/1929-7092.2016.07.59

Modigliani, F., & Miller, M. (1958). The cost of capital, corporate finance, and the theory of investment. *The American Economic Review, 48*, 261–297.
Modigliani, F., & Miller, M. (1963). Corporate income taxes and the cost of capital: A correction. *The American Economic Review, 53*, 147–175.
Modigliani, F., & Miller, M. (1966). Some estimates of the cost of capital to the electric utility industry 1954–1957. *The American Economic Review, 56*, 333–391.
Myers, S. (2001). Capital structure. *The Journal of Economic Perspectives, 15*(2), 81–102.

Chapter 7
How Frequently Should Companies Pay Tax on Profit

7.1 Introduction

Brusov and Filatova (2021) have modified the Modigliani–Miller theory for the case of arbitrary frequency of payment of tax on profit and have shown that the account of frequency of payment of tax on profit within the Modigliani–Miller theory turns out to be very important and leads to significant consequences:

- all Modigliani–Miller theorems, statements and all formulas change;
- all main financial indicators, such as the weighted average cost of capital, WACC, company value, V and equity cost, k_e depend on the frequency of tax on profit payments. This allows to company manage WACC, V, k_e etc. by choosing the number of payments of tax of profit p per year;
- obtained results allow to company choose the number of payments of tax of profit per year, as many, as it is profitable to it (of course, within actual tax legislation);
- in case of income tax payments more than once per year (at $p \neq 1$), as it takes place in practice, the weighted average cost of capital, WACC, company value, V and equity cost, k_e start depend on k_d, while in ordinary (classical) Modigliani–Miller theory all these values **DO NOT depend on k_d**.
- the tilt angle of the curve of equity cost, k_e (L), decreases with the number of payments of tax of profit p, this modifies the dividend policy of the company, because the economically justified value of dividends is equal to equity cost.

As it was shown in that paper (Brusov & Filatova, 2021), more frequent payments of income tax are beneficial for both parties: for the company and for the tax regulator: for the company, this leads to decrease of the cost of attracting capital and thus to an increase in the value of the company, and for the tax regulator, earlier payments are beneficial due to the time value of money.

But companies could pay tax of profit in advance and as we show in current paper this fact changes drastically the main conclusion: more frequent payments of income tax are **NOT** beneficial for the company: this leads to increase of the cost of

P. Brusov, T. Filatova, *Efficiency of Investment Projects*, Contributions to Finance and Accounting, https://doi.org/10.1007/978-3-032-01028-5_7

attracting capital and thus to decrease in the value of the company, but for the tax regulator it remains beneficial: earlier payments are beneficial due to the time value of money.

The structure of the Chapter is as following:

1. We give a literature review for Modigliani–Miller theory and for its modifications.
2. We modify the Modigliani–Miller theory for the case of frequent payment of tax on profit (in case of advanced payments) and get modified Modigliani–Miller theorems as well as new formulas for the weighted average cost of capital, WACC, for the equity cost, ke, and for the capitalization of the company, V.
3. Within new Modified Modigliani–Miller theory (MMM theory) we study the dependence of the main financial indicators of the company (WACC, k_e, V) on leverage level L, on debt cost k_d and on tax on profit t at different frequency of payments of tax on profit (at the end of period and in advance).
4. Based on obtained results we come to some very important conclusions, which allow company choose the method of payments of tax of profit (at the end of period or in advance) and number of payments of tax of profit per year, as many, as it is profitable to it (of course, within actual tax legislation).

7.2 Literature Review

In their first paper Modigliani and Miller (1958) considered the problem of capital cost and capital structure for perpetuity companies without any taxes. In the second article, Modigliani and Miller (1963) considered taxation of corporate profits, but did not take into account the presence in the economy of individual taxes of investors. Merton Miller (1977) has introduced the model, demonstrating impact of leverage on the company value with account of the corporate and individual taxes. To describe his model, we will enter the following legends: T_C–tax on corporate profits rate, T_S–the tax rate on income of an individual investor from his ownership by stock of corporation, T_D–tax rate on interest income from the provision of investor–individuals of credits to other investors and companies. Income from shares partly comes in the form of a dividend and, in part, as capital profits, so that T_S is a weighted average value of effective rates of tax on dividends and capital profits on shares, while the income from the provision of loans usually comes in the form of the interests. The last are usually taxed at a higher rate. In the light of the individual taxes, and with the same assumptions that have been made for Modigliani–Miller models previously, the financially independent company value can be determined as follows:

$$V_U = \frac{\text{EBIT}(1 - T_C)(1 - T_S)}{k_0}. \tag{7.1}$$

A term $(1 - T_S)$ allows to take into account the individual taxes in the formula. In this way, numerator indicates which part of the operating company's profit remains in the possession of the investors, after the company will pay taxes on their profits, and its shareholders then will pay individual taxes on income from stock ownership. Since individual taxes reduce profits, remaining in the disposal of investors, the last, at other things being equal circumstances, also reduce and an overall assessment of the financially independent company value.

Robert Hamada (1969) unites Capital Asset Pricing Model (CAPM) with Modigliani–Miller model taxation. As a result, he derived the following formula for calculation of the equity cost of financially dependent company, including both financial and business risk of company:

$$k_e = k_F + (k_M - k_F)b_U + (k_M - k_F)b_U \frac{D}{S}(1 - T), \quad (7.2)$$

where b_U is the β–coefficient of the company of the same group of business risk, that the company under consideration, but with zero financial leverage. The formula (7.2) represents the desired profitability of equity capital k_e as a sum of three components: risk–free profitability k_F, compensating to shareholders a temporary value of their money, premium for business risk $(k_M - k_F)b_U$, and premium for financial risk $(k_M - k_F)b_U \frac{D}{S}(1 - T)$.

If the company does not have borrowing $(D = 0)$, the financial risk factor will be equal to zero (the third term is drawn to zero), and its owners will only receive the premium for business risk.

More general formula for the WACC, then famous Modigliani–Miller (MM) one has been derived and discussed by a few authors (Farber et al., 2007; Fernandez, 2007; Harris & Pringle, 1985; Miles & Ezzel, 1980; De Marzo, 1988). It takes the following form (Eq. (18) in FGS, 2006)

$$WACC = k_0(1 - w_d T) - k_d t w_d + k_{TS} t w_d \quad (7.3)$$

where k_0, k_d and k_{TS} are the expected returns respectively on the unlevered firm, the debt and the tax shield, V is the value of the levered firm, VTS is the value of the tax shield, D is the value of the debt and TC is the corporate tax rate.

This formula is derived from the balance sheet identity and the definition of the weighted average cost of capital (for a similar presentation, see Berk & De Marzo, 2007). It should therefore be verified at any point in time, whatever returns are annually or continuously compounded.

Although Eq. (7.3) is fully general, its practical applicability requires additional conditions. Indeed, when the WACC is constant over time, the value of a levered firm can be computed by discounting with the WACC the unlevered free cash–flows. Therefore, we pay a special attention to the special cases that make the WACC constant. The resulting particular formulas can also be found in textbooks (Brealey et al., 2006; Ross et al., 2005).

First, Modigliani and Miller (1963) assume that the level of debt D is constant. Then, as the expected after–tax cash–flow of the unlevered firm is fixed, V_0 is also constant. By assumption, $k_{TS} = k_D$ and the value of the tax shield is TS = tD. Therefore, the value of the firm V is a constant and the general WACC formula (7.3) simplifies to a constant WACC:

$$WACC = k_0(1 - w_d T) \tag{7.4}$$

But our opinion is that "classical" Modigliani–Miller (MM) theory, which coming from the suggestion that the expected returns on the debt k_d and the tax shield k_{TS} are equals (because both of them have debt nature), is much more reasonable and in our paper we modify namely "classical" Modigliani–Miller (MM) theory, which is still widely used in practice.

Recently the Modigliani–Miller theory has been modified by Brusov et al. (2019, 2021) into two ways: it was applied for rating methodologies needs and it was generalized for the case of advance payments of tax on profit, which are widely used in practice (MMM theory). As well Brusov et al. (2019) used the Modified Modigliani–Miller theory (MMM theory) and apply it for rating methodologies needs. The financial "ratios" (main rating parameters) were introduced into MMM theory. The dependence of the weighted average cost of capital (WACC), which plays the role of discount rate in financial flows discounting in rating methodologies, on coverage and leverage ratios is analyzed. Obtained results will help improve the existing rating methodologies.

One of the most serious limitations of the Modigliani–Miller theory is the suggestion about perpetuity of the companies. In 2008, Brusov–Filatova–Orekhova (Filatova et al., 2008) have lifted up this limitation and have shown that the accounting of the finite age (or finite life–time) of the company leads to significant changes of all Modigliani–Miller results (Modigliani & Miller, 1958, 1963, 1966): capitalization of the company is changed, as well as the equity cost, k_e, and the weighted average cost of capital, WACC, in the presence of corporative taxes. Besides, a number of qualitatively new effects in corporate finance, obtained within Brusov–Filatova–Orekhova theory (BFO theory) (Brusov et al., 2018a, b, c, d, e, f, g, 2019, 2020a, b, c), are absent in Modigliani–Miller theory (Modigliani & Miller, 1958, 1963).

The famous Brusov–Filatova–Orekhova formula for weighted average cost of capital, WACC, has the following form

$$\frac{1-(1+WACC)^{-n}}{WACC} = \frac{1-(1+k_0)^{-n}}{k_0\left[1-\omega_d T\left(1-(1+k_d)^{-n}\right)\right]}. \tag{7.5}$$

Here, S—the value of own (equity) capital of the company, $w_d = \frac{D}{D+S}$—the share of debt capital, $k_e, w_e = \frac{S}{D+S}$—the cost and the share of the equity of the company, $L = D/S$—financial leverage.

A lot of meaningful effects has been discovered within BFO theory: these effects are absent within Modigliani–Miller theory. BFO theory has destroyed some main existing principles of financial management: among them trade off theory, which was considered as keystone of formation of optimal capital structure of the company during many decades and **bankruptcy of which has been proven** by Brusov–Filatova–Orekhova.

7.3 Focus and Objective of the Chapter

The tax shield plays a crucial role in both main capital structure theories: Brusov–Filatova–Orekhova (BFO theory) and its perpetuity limit Modigliani–Miller theory. How it is formed influences the results of both theories. Both of them describes the case of annual payments of tax of profit, but in practice these payments are made more frequent: semiannually, quarterly, monthly. While a return is not required more than once a year, businesses may be responsible for filing estimated taxes based on profits earned. This requirement is dependent on showing a profit. For example, sole proprietors must file estimated taxes on profits quarterly, on the 15th day of April, June, September and January. In Russia tax on profit payments could be made annually, quarterly, monthly. Brusov and Filatova (2021) have shown that the account of frequency of payment of tax on profit within the Modigliani–Miller theory turns out to be very important and leads to significant consequences.

To study this problem they have modified the Modigliani–Miller theory for the case of arbitrary frequency of payment of tax on profit and show that:

- all Modigliani–Miller theorems, statements and all formulas change;
- all main financial indicators, such as the weighted average cost of capital, WACC, company value, V and equity cost, k_e depend on the frequency of tax on profit payments. This allows to company manage WACC, V, k_e etc. by choosing the number of payments of tax of profit p per year;
- increase the number of payments of tax of profit per year leads to decrease of the cost of attracting capital and increase of the company value;
- obtained results allow to company choose the number of payments of tax of profit per year, as many, as it is profitable to it (of course, within actual tax legislation);
- in case of income tax payments more than once per year (at $p \neq 1$), as it takes place in practice, the weighted average cost of capital, WACC, company value, V and equity cost, k_e start depend on k_d, while in ordinary (classical) Modigliani–Miller theory all these values DO NOT depend on k_d.
- the tilt angle of the curve of equity cost, k_e (L), decreases with the number of payments of tax of profit p, this modifies the dividend policy of the company, because the economically justified value of dividends is equal to equity cost.

As it was shown in that paper (Brusov & Filatova, 2021), more frequent payments of income tax are beneficial for both parties: for the company and for the tax regulator: for the company, this leads to decrease of the cost of attracting capital and thus to an increase in the value of the company, and for the tax regulator, earlier payments are beneficial due to the time value of money.

But companies could pay tax of profit in advance and as we show in current paper this fact changes drastically the main conclusion: more frequent payments of income tax are **NOT** beneficial for the company: this leads to increase of the cost of attracting capital and thus to decrease in the value of the company, but for the tax regulator it remains beneficial: earlier payments are beneficial due to the time value of money.

7.4 Modification of the Modigliani–Miller Theory for the Case of Arbitrary Frequency of Advanced Payments of Tax on Profit

7.4.1 Tax Shield

Let us calculate first the tax shield *TS* in the Modigliani–Miller theory in the case of arbitrary frequency (p per year) of advanced payments of tax on profit

$$TS = \frac{k_d Dt}{p} + \frac{k_d Dt}{p(1+k_d)^{1/p}} + \frac{k_d Dt}{p(1+k_d)^{2/p}} + \cdots \tag{7.6}$$

(TS) represents a geometric progression with denominator $q = \frac{1}{(1+k_d)^{\frac{1}{p}}}$. Here D is debt value, k_d is debt cost, t is tax on profit rate, p is the frequency of payments of tax on profit.

Summing the progression, one gets

$$TS = \frac{k_d Dt}{p\left(1-(1+k_d)^{-1/p}\right)} = \frac{k_d Dt \cdot (1+k_d)^{1/p}}{p\left((1+k_d)^{1/p}-1\right)}. \tag{7.7}$$

Note, that in case of classical Modigliani–Miller theory, modified by us for the case of one advanced payment of tax on profit per year

$$TS = Dt(1+k_d). \tag{7.8}$$

It is easy to get this result from (7.7), putting p = 1.

7.4.2 The Weighted Average Cost of Capital, WACC

Let us calculate now the weighted average cost of capital, WACC in the Modigliani–Miller theory in the case of arbitrary frequency (p per year) of advanced payments of tax on profit.

From the Modigliani–Miller theorem we have for the company value V

$$V = V_0 + TS. \tag{7.9}$$

Here V_0 is the financially independent company value.

Putting expression for tax shield TS, we get

$$V = V_0 + \frac{k_d D t \cdot (1 + k_d)^{1/p}}{p\left((1 + k_d)^{1/p} - 1\right)}. \tag{7.10}$$

Substituting $D = w_d V$, one gets

$$V\left(1 - \frac{k_d w_d t \cdot (1 + k_d)^{1/p}}{p\left((1 + k_d)^{1/p} - 1\right)}\right) = V_0. \tag{7.11}$$

Accounting, that

$$V = \frac{CF}{WACC}; V_0 = \frac{CF}{k_0}, \tag{7.12}$$

we have

$$\frac{CF}{WACC} \cdot \left(1 - \frac{k_d w_d t \cdot (1 + k_d)^{1/p}}{p\left((1 + k_d)^{1/p} - 1\right)}\right) = \frac{CF}{k_0}. \tag{7.13}$$

From here we arrive to the expression for the weighted average cost of capital, WACC,

$$WACC = k_0 \cdot \left(1 - \frac{k_d w_d t \cdot (1 + k_d)^{1/p}}{p\left((1 + k_d)^{1/p} - 1\right)}\right). \tag{7.14}$$

Note, that in case of classical Modigliani–Miller theory, modified by us for the case of one advanced payment of tax on profit per year

$$WACC = k_0 \cdot (1 - w_d t \cdot (1 + k_d)). \tag{7.15}$$

It is easy to get this result from (7.14), putting p = 1.

7.4.3 Company Value, V

Let us calculate now the company value, V, in the Modigliani–Miller theory in the case of arbitrary frequency (p per year) of advanced payments of tax on profit.

Putting WACC value into first formula of (7.12) one gets for company value, V

$$V = \frac{CF}{WACC} = \frac{CF}{k_0 \cdot \left(1 - \frac{k_d w_d t \cdot (1+k_d)^{1/p}}{p\left((1+k_d)^{1/p} - 1\right)}\right)}. \tag{7.16}$$

Here CF is the profit of the company per year, w_d is the debt share.

In case of classical Modigliani–Miller theory, modified by us for the case of one advanced payment of tax on profit per year (Brusov et al., 2020a, b, c), one has

$$V = \frac{CF}{WACC} = \frac{CF}{k_0 \cdot (1 - w_d t (1 + k_d))}. \tag{7.17}$$

It is easy to get this result from (7.16), putting p = 1.

7.4.4 Equity Cost

Let us calculate now the equity cost, k_d, in the Modigliani–Miller theory in the case of arbitrary frequency (p per year) of advanced payments of tax on profit.

For the weighted average cost of capital, WACC, we have from formula (7.14)

$$WACC = k_0 \cdot \left(1 - \frac{k_d w_d t \cdot (1 + k_d)^{1/p}}{p\left((1 + k_d)^{1/p} - 1\right)}\right). \tag{7.18}$$

From other side, upon definition one has

$$WACC = k_e w_e + k_d w_d (1 - t). \tag{7.19}$$

From (7.19) we get

$$k_e = \frac{WACC}{w_e} - \frac{k_d w_d (1-t)}{w_e} \tag{7.20}$$

Putting (7.18) into (7.20), we arrive to expression for equity cost k_e

$$\begin{aligned} k_e &= \frac{k_0}{w_e} \cdot \left(1 - \frac{k_d w_d t \cdot (1+k_d)^{1/p}}{p\left((1+k_d)^{1/p} - 1\right)}\right) - \frac{k_d w_d (1-t)}{w_e} = \\ &= k_0 \cdot \left(1 + L - \frac{k_d L t \cdot (1+k_d)^{1/p}}{p\left((1+k_d)^{1/p} - 1\right)}\right) - k_d L (1-t) = \\ &= k_0 + L\left(k_0 - k_d(1-t) - \frac{k_0 k_d t \cdot (1+k_d)^{1/p}}{p\left((1+k_d)^{1/p} - 1\right)}\right) \end{aligned} \tag{7.21}$$

Finally, one has

$$k_e = k_0 + L\left(k_0 - k_d(1-t) - \frac{k_0 k_d t \cdot (1+k_d)^{1/p}}{p\left((1+k_d)^{1/p} - 1\right)}\right) \tag{7.22}$$

In case of classical Modigliani–Miller theory, modified by us for the case of one advanced payment of tax on profit per year (Brusov et al., 2020a, b, c), one has

$$k_e = k_0 + L(k_0 - k_d \cdot (1-t) - k_0 t (1+k_d)). \tag{7.23}$$

Putting in formula (7.22) p = 1, it is easy to get formula (7.23).

7.5 Dependence of the Weighted Average Cost of Capital, WACC, on Leverage Level L

7.5.1 Dependence of the Weighted Average Cost of Capital, WACC, on Leverage Level L at Different p and Fixed k_d

Let us start from the study numerically with use of Microsoft Excel of the dependence of the weighted average cost of capital, WACC, on leverage level L at different frequency of payments p and fixed debt cost k_d, using obtained by us formulas

$$WACC = k_0 \cdot \left(1 - \frac{k_d w_d t}{p\left((1+k_d)^{1/p} - 1\right)}\right) \text{ and } WACC = k_0 \cdot \left(1 - \frac{k_d w_d t \cdot (1+k_d)^{1/p}}{p\left((1+k_d)^{1/p} - 1\right)}\right)$$

From Tables 7.1, 7.2, 7.3, and 7.4 and Figs. 7.1 and 7.2 it is seen, that the weighted average cost of capital, WACC, decreases with leverage level L at all values of frequency of payments of tax on profit p. With increase of the frequency of payments of tax on profit per year p WACC (1) decreases in the case of payments at the ends of the periods and WACC (2) increases in the case of advanced payments of tax on profit. As we'll see below, this will lead to increase of company value, V, with increase of p in the case (1) and to decrease of company value, V, with increase of p in the case (2).

Note, that the weighted average cost of capital, WACC, in the case of advanced payments of tax on profit for all values of frequency of payments turns out to be low than for any values of frequency of payments in the case of payments at the ends of the periods.

Table 7.1 Dependence of WACC on leverage level L at $k_0 = 0.2$, $k_d = 0.16$ and $p = 1$

k_0	k_d	w_d	L	t	p	i	WACC
0.2	0.16	0	0	0.2	1	1.16	0.2
0.2	0.16	0.5	1	0.2	1	1.16	0.18
0.2	0.16	0.666667	2	0.2	1	1.16	0.173333
0.2	0.16	0.75	3	0.2	1	1.16	0.17
0.2	0.16	0.8	4	0.2	1	1.16	0.168
0.2	0.16	0.833333	5	0.2	1	1.16	0.166667
0.2	0.16	0.857143	6	0.2	1	1.16	0.165714
0.2	0.16	0.875	7	0.2	1	1.16	0.165
0.2	0.16	0.888889	8	0.2	1	1.16	0.164444
0.2	0.16	0.9	9	0.2	1	1.16	0.164
0.2	0.16	0.909091	10	0.2	1	1.16	0.163636

Table 7.2 Dependence of WACC on leverage level L at $k_0 = 0.2$, $k_d = 0.16$ and $p = 12$

k_0	k_d	w_d	L	t	p	i	WACC
0.2	0.16	0	0	0.2	12	1.012445	0.2
0.2	0.16	0.5	1	0.2	12	1.012445	0.178573
0.2	0.16	0.666667	2	0.2	12	1.012445	0.17143
0.2	0.16	0.75	3	0.2	12	1.012445	0.167859
0.2	0.16	0.8	4	0.2	12	1.012445	0.165716
0.2	0.16	0.833333	5	0.2	12	1.012445	0.164288
0.2	0.16	0.857143	6	0.2	12	1.012445	0.163267
0.2	0.16	0.875	7	0.2	12	1.012445	0.162502
0.2	0.16	0.888889	8	0.2	12	1.012445	0.161907
0.2	0.16	0.9	9	0.2	12	1.012445	0.161431
0.2	0.16	0.909091	10	0.2	12	1.012445	0.161041

Table 7.3 Dependence of WACC on leverage level L at $k_0 = 0.2$, $k_d = 0.16$ and $p = 1$ (advanced payment of tax of profit)

k_0	k_d	w_d	L	t	p	i	WACC
0.2	0.16	0	0	0.2	1	1.16	0.2
0.2	0.16	0.5	1	0.2	1	1.16	0.1768
0.2	0.16	0.666667	2	0.2	1	1.16	0.169067
0.2	0.16	0.75	3	0.2	1	1.16	0.1652
0.2	0.16	0.8	4	0.2	1	1.16	0.16288
0.2	0.16	0.833333	5	0.2	1	1.16	0.161333
0.2	0.16	0.857143	6	0.2	1	1.16	0.160229
0.2	0.16	0.875	7	0.2	1	1.16	0.1594
0.2	0.16	0.888889	8	0.2	1	1.16	0.158756
0.2	0.16	0.9	9	0.2	1	1.16	0.15824
0.2	0.16	0.909091	10	0.2	1	1.16	0.157818

Table 7.4 Dependence of WACC on leverage level L at $k_0 = 0.2$, $k_d = 0.16$ and $p = 12$ (advanced payment of tax of profit)

k_0	k_d	w_d	L	t	p	i	WACC
0.2	0.16	0	0	0.2	12	1.012445	0.2
0.2	0.16	0.5	1	0.2	12	1.012445	0.178306
0.2	0.16	0.666667	2	0.2	12	1.012445	0.171075
0.2	0.16	0.75	3	0.2	12	1.012445	0.167459
0.2	0.16	0.8	4	0.2	12	1.012445	0.16529
0.2	0.16	0.833333	5	0.2	12	1.012445	0.163843
0.2	0.16	0.857143	6	0.2	12	1.012445	0.16281
0.2	0.16	0.875	7	0.2	12	1.012445	0.162035
0.2	0.16	0.888889	8	0.2	12	1.012445	0.161433
0.2	0.16	0.9	9	0.2	12	1.012445	0.160951
0.2	0.16	0.909091	10	0.2	12	1.012445	0.160556

7.5.2 Dependence of the Weighted Average Cost of Capital, WACC, and of Company Value, V, on Debt Cost Value k_d at Fixed Value of Leverage Level L and Fixed p

In this paragraph we investigate the dependence of the weighted average cost of capital, WACC, and of company value, V on debt cost value kd at fixed value of leverage level L and fixed p.

From Tables 7.5, 7.6, 7.7, and 7.8 and Fig. 7.3 it is seen, that the weighted average cost of capital, WACC, decreases with increase of debt cost value k_d, except the classical MM case $p = 1$ and payment at the end of the period. It supports obtained by us result that account of each of two effects (either advanced payments of tax on profit, or more frequent than once per year payments) leads to dependence of all main financial indicators, such as the weighted average cost of capital, WACC,

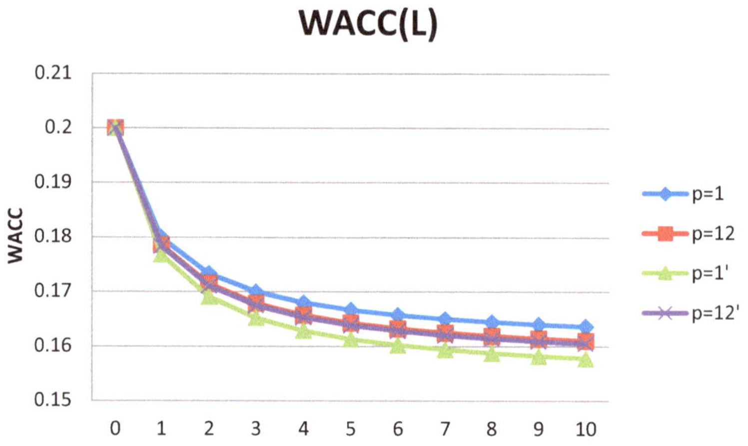

Fig. 7.1 Dependence of the Weighted Average Cost of Capital, WACC, on Leverage Level L at Different p and Fixed $k_d = 0.16$

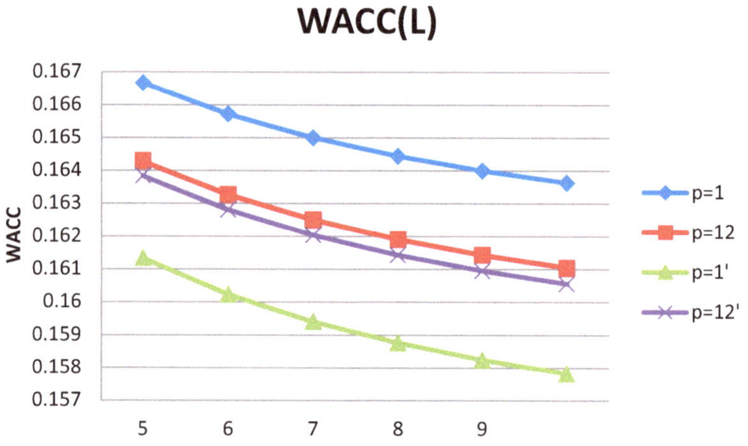

Fig. 7.2 Dependence of the Weighted Average Cost of Capital, WACC, on Leverage Level L at Different p and Fixed $k_d = 0.16$ (Larger Scale)

company value, V, equity cost, k_e, etc. on debt cost value k_d, while in the classical MM case these financial indicators DO NOT depend on debt cost value k_d.

Similar to previous paragraph it is seen, that with increase of the frequency of payments of tax on profit per year p WACC (1) decreases in the case of payments at the ends of the periods and WACC (2) increases in the case of advanced payments of tax on profit. As we'll see below, this will lead to increase of company value, V, with increase of p in the case (1) and to decrease of company value, V, with increase of p in the case (2).

Table 7.5 Dependence of WACC on debt cost value k_d at $k_0 = 0.2$, at $L = 1$ and $p = 1$

k_0	k_d	w_d	L	t	p	i	WACC	CF	V	k_e
0.2	0.08	0.5	1	0.2	1	1.08	0.18	1000	5555.556	0.296
0.2	0.09	0.5	1	0.2	1	1.09	0.18	1000	5555.556	0.288
0.2	0.1	0.5	1	0.2	1	1.1	0.18	1000	5555.556	0.28
0.2	0.11	0.5	1	0.2	1	1.11	0.18	1000	5555.556	0.272
0.2	0.12	0.5	1	0.2	1	1.12	0.18	1000	5555.556	0.264
0.2	0.13	0.5	1	0.2	1	1.13	0.18	1000	5555.556	0.256
0.2	0.14	0.5	1	0.2	1	1.14	0.18	1000	5555.556	0.248
0.2	0.15	0.5	1	0.2	1	1.15	0.18	1000	5555.556	0.24
0.2	0.16	0.5	1	0.2	1	1.16	0.18	1000	5555.556	0.232
0.2	0.17	0.5	1	0.2	1	1.17	0.18	1000	5555.556	0.224
0.2	0.18	0.5	1	0.2	1	1.18	0.18	1000	5555.556	0.216

Table 7.6 Dependence of WACC on debt cost value k_d at $k_0 = 0.2$, at $L = 1$ and $p = 12$

k_0	k_d	w_d	L	t	p	i	WACC	CF	V	k_e
0.2	0.08	0.5	1	0.2	12	1.006434	0.179277	1000	5577.965	0.294554
0.2	0.09	0.5	1	0.2	12	1.007207	0.179188	1000	5580.736	0.286376
0.2	0.1	0.5	1	0.2	12	1.007974	0.179099	1000	5583.501	0.278198
0.2	0.11	0.5	1	0.2	12	1.008735	0.179011	1000	5586.259	0.270021
0.2	0.12	0.5	1	0.2	12	1.009489	0.178923	1000	5589.012	0.261845
0.2	0.13	0.5	1	0.2	12	1.010237	0.178835	1000	5591.758	0.253669
0.2	0.14	0.5	1	0.2	12	1.010979	0.178747	1000	5594.499	0.245494
0.2	0.15	0.5	1	0.2	12	1.011715	0.17866	1000	5597.234	0.237319
0.2	0.16	0.5	1	0.2	12	1.012445	0.178573	1000	5599.963	0.229145
0.2	0.17	0.5	1	0.2	12	1.01317	0.178486	1000	5602.686	0.220972
0.2	0.18	0.5	1	0.2	12	1.013888	0.178399	1000	5605.404	0.212799

Table 7.7 Dependence of WACC on debt cost value k_d at $k_0 = 0.2$, at $L = 1$ and $p = 1$ (for advanced payments of tax of profit)

k_0	k_d	w_d	L	t	p	i	WACC	CF	V	k_e
0.2	0.08	0.5	1	0.2	1	1.08	0.1784	1000	5605.381	0.2928
0.2	0.09	0.5	1	0.2	1	1.09	0.1782	1000	5611.672	0.2844
0.2	0.1	0.5	1	0.2	1	1.1	0.178	1000	5617.978	0.276
0.2	0.11	0.5	1	0.2	1	1.11	0.1778	1000	5624.297	0.2676
0.2	0.12	0.5	1	0.2	1	1.12	0.1776	1000	5630.631	0.2592
0.2	0.13	0.5	1	0.2	1	1.13	0.1774	1000	5636.979	0.2508
0.2	0.14	0.5	1	0.2	1	1.14	0.1772	1000	5643.341	0.2424
0.2	0.15	0.5	1	0.2	1	1.15	0.177	1000	5649.718	0.234
0.2	0.16	0.5	1	0.2	1	1.16	0.1768	1000	5656.109	0.2256
0.2	0.17	0.5	1	0.2	1	1.17	0.1766	1000	5662.514	0.2172
0.2	0.18	0.5	1	0.2	1	1.18	0.1764	1000	5668.934	0.2088

Table 7.8 Dependence of WACC on debt cost value k_d at $k_0 = 0.2$, at $L = 1$ and $p = 12$ (for advanced payments of tax of profit)

k_0	k_d	w_d	L	t	p	i	WACC	CF	V	k_e
0.2	0.08	0.5	1	0.2	12	1.006434	0.179144	1000	5582.116	0.294287
0.2	0.09	0.5	1	0.2	12	1.007207	0.179038	1000	5585.412	0.286076
0.2	0.1	0.5	1	0.2	12	1.007974	0.178932	1000	5588.702	0.277865
0.2	0.11	0.5	1	0.2	12	1.008735	0.178827	1000	5591.986	0.269655
0.2	0.12	0.5	1	0.2	12	1.009489	0.178723	1000	5595.266	0.261445
0.2	0.13	0.5	1	0.2	12	1.010237	0.178618	1000	5598.541	0.253236
0.2	0.14	0.5	1	0.2	12	1.010979	0.178514	1000	5601.811	0.245027
0.2	0.15	0.5	1	0.2	12	1.011715	0.17841	1000	5605.077	0.236819
0.2	0.16	0.5	1	0.2	12	1.012445	0.178306	1000	5608.338	0.228612
0.2	0.17	0.5	1	0.2	12	1.01317	0.178202	1000	5611.594	0.220405
0.2	0.18	0.5	1	0.2	12	1.013888	0.178099	1000	5614.846	0.212199

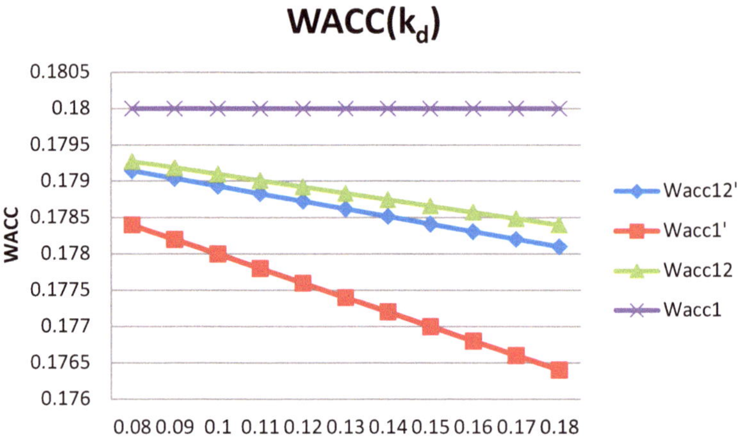

Fig. 7.3 Dependence of WACC on Debt Cost Value k_d at $k_0 = 0.2$, at $L = 1$ and $p = 1$ and $p = 12$ (' Means Advanced Payments of Tax of Profit)

Note, that the weighted average cost of capital, WACC, in the case of advanced payments of tax on profit for all values of frequency of payments turns out to be low than for any values of frequency of payments in the case of payments at the ends of the periods.

From Tables 7.5, 7.6, 7.7, and 7.8 and Fig. 7.4 it is seen, that the company value, V, increases with increase of debt cost value kd, except the classical Modigliani–Miller theory case $p = 1$ and payment at the end of the period, when V remains constant. It supports obtained by us result that account of each of two effects (either advanced payments of tax on profit, or more frequent than once per year) leads to dependence of all main financial indicators, such as the weighted average cost of capital, WACC, company value, V, equity cost, k_e, etc. on debt cost

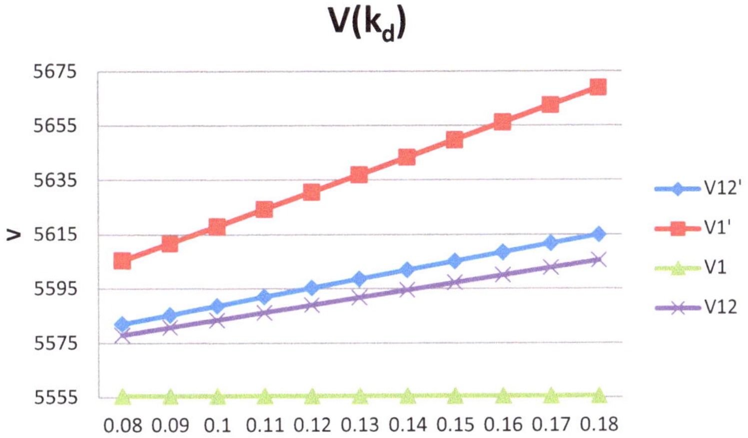

Fig. 7.4 Dependence of Company Value V on Debt Cost Value k_d at $k_0 = 0.2$, at $L = 1$ and $p = 1$ and $p = 12$ (' Means Advanced Payments of Tax of Profit)

value k_d, while in the classical Modigliani–Miller theory case these financial indicators DO NOT depend on debt cost value kd.

Similar to previous paragraph it is seen, that with increase of the frequency of payments of tax on profit per year p the company value, V, (1) increases in the case of payments at the ends of the periods and WACC (2) decreases in the case of advanced payments of tax on profit.

Note, that the company value, V, in the case of advanced payments of tax on profit for all values of frequency of payments turns out to be higher than for any values of frequency of payments in the case of payments at the ends of the periods.

7.6 Dependence of the Company Capitalization, V, on Leverage Level L

7.6.1 Dependence of the Company Capitalization, V, on Leverage Level L at Different p and Fixed k_d

Let us now calculate the dependence of the company capitalization, V, on leverage level L at different p and fixed k_d.

From Tables 7.9, 7.10, 7.11, and 7.12 and Figs. 7.5 and 7.6 it is seen, that the company value, V, increases with the increase of leverage level L for all frequencies of payment of tax on profit per year p.

While in case of payments at the end of the periods the company value, V, increases with the increase of the frequency of payment of tax on profit per year p, in case of advanced payments of tax on profit the company value, V, decreases with the increase of the frequency of payment of tax on profit per year p. These facts

Table 7.9 Dependence of the company capitalization on leverage level L at $k_d = 0.16$ and at $p = 1$

k_0	k_d	w_d	L	t	p	i	WACC	CF	V
0.2	0.16	0	0	0.2	1	1.16	0.2	1000	5000
0.2	0.16	0.5	1	0.2	1	1.16	0.18	1000	5555.556
0.2	0.16	0.666667	2	0.2	1	1.16	0.173333	1000	5769.231
0.2	0.16	0.75	3	0.2	1	1.16	0.17	1000	5882.353
0.2	0.16	0.8	4	0.2	1	1.16	0.168	1000	5952.381
0.2	0.16	0.833333	5	0.2	1	1.16	0.166667	1000	6000
0.2	0.16	0.857143	6	0.2	1	1.16	0.165714	1000	6034.483
0.2	0.16	0.875	7	0.2	1	1.16	0.165	1000	6060.606
0.2	0.16	0.888889	8	0.2	1	1.16	0.164444	1000	6081.081
0.2	0.16	0.9	9	0.2	1	1.16	0.164	1000	6097.561
0.2	0.16	0.909091	10	0.2	1	1.16	0.163636	1000	6111.111

Table 7.10 Dependence of the company capitalization on leverage level L at $k_d = 0.16$ and at $p = 12$

k_0	k_d	w_d	L	t	p	i	WACC	CF	V
0.2	0.16	0	0	0.2	12	1.012445	0.2	1000	5000
0.2	0.16	0.5	1	0.2	12	1.012445	0.178573	1000	5599.963
0.2	0.16	0.666667	2	0.2	12	1.012445	0.17143	1000	5833.279
0.2	0.16	0.75	3	0.2	12	1.012445	0.167859	1000	5957.383
0.2	0.16	0.8	4	0.2	12	1.012445	0.165716	1000	6034.413
0.2	0.16	0.833333	5	0.2	12	1.012445	0.164288	1000	6086.883
0.2	0.16	0.857143	6	0.2	12	1.012445	0.163267	1000	6124.923
0.2	0.16	0.875	7	0.2	12	1.012445	0.162502	1000	6153.767
0.2	0.16	0.888889	8	0.2	12	1.012445	0.161907	1000	6176.39
0.2	0.16	0.9	9	0.2	12	1.012445	0.161431	1000	6194.608
0.2	0.16	0.909091	10	0.2	12	1.012445	0.161041	1000	6209.594

Table 7.11 Dependence of the company capitalization on leverage level L at $k_d = 0.16$ and at $p = 1$ (for advanced payments of tax of profit)

k_0	k_d	w_d	L	t	p	i	WACC	CF	V
0.2	0.16	0	0	0.2	1	1.16	0.2	1000	5000
0.2	0.16	0.5	1	0.2	1	1.16	0.1768	1000	5656.109
0.2	0.16	0.666667	2	0.2	1	1.16	0.169067	1000	5914.826
0.2	0.16	0.75	3	0.2	1	1.16	0.1652	1000	6053.269
0.2	0.16	0.8	4	0.2	1	1.16	0.16288	1000	6139.489
0.2	0.16	0.833333	5	0.2	1	1.16	0.161333	1000	6198.347
0.2	0.16	0.857143	6	0.2	1	1.16	0.160229	1000	6241.084
0.2	0.16	0.875	7	0.2	1	1.16	0.1594	1000	6273.526
0.2	0.16	0.888889	8	0.2	1	1.16	0.158756	1000	6298.992
0.2	0.16	0.9	9	0.2	1	1.16	0.15824	1000	6319.515
0.2	0.16	0.909091	10	0.2	1	1.16	0.157818	1000	6336.406

Table 7.12 Dependence of the company capitalization on leverage level L at $k_d = 0.16$ and at $p = 12$ (for advanced payments of tax of profit)

k_0	k_d	w_d	L	t	p	i	WACC	CF	V
0.2	0.16	0	0	0.2	12	1.012445	0.2	1000	5000
0.2	0.16	0.5	1	0.2	12	1.012445	0.178306	1000	5608.338
0.2	0.16	0.666667	2	0.2	12	1.012445	0.171075	1000	5845.403
0.2	0.16	0.75	3	0.2	12	1.012445	0.167459	1000	5971.613
0.2	0.16	0.8	4	0.2	12	1.012445	0.16529	1000	6049.99
0.2	0.16	0.833333	5	0.2	12	1.012445	0.163843	1000	6103.394
0.2	0.16	0.857143	6	0.2	12	1.012445	0.16281	1000	6142.121
0.2	0.16	0.875	7	0.2	12	1.012445	0.162035	1000	6171.49
0.2	0.16	0.888889	8	0.2	12	1.012445	0.161433	1000	6194.528
0.2	0.16	0.9	9	0.2	12	1.012445	0.160951	1000	6213.082
0.2	0.16	0.909091	10	0.2	12	1.012445	0.160556	1000	6228.345

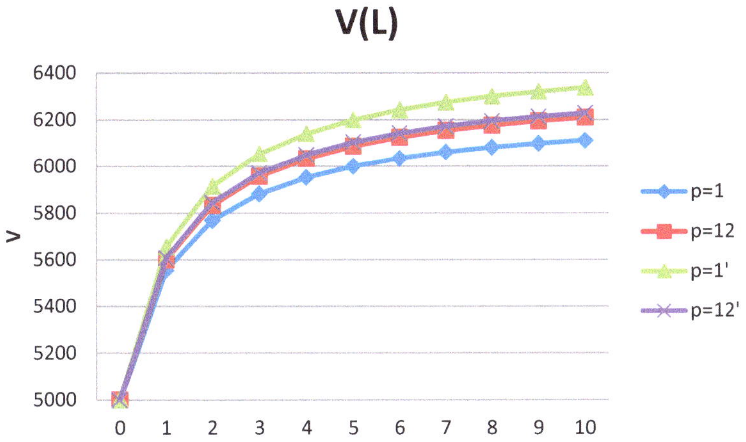

Fig. 7.5 Dependence of the Company Capitalization, V, on Leverage Level L at $k_d = 0.16$ and $p = 1$ and 12 (' Means Advanced Payment of Tax of Profit)

consistent with the dependence of the weighted average cost of capital, WACC, on the leverage level L.

All the company values, V, in case of advanced payments of tax on profit turns out to be bigger than in case of payments at the end of the periods. Thus if the companies tend to increase their capitalization, they should pay tax on profit (1) in advance; (2) as seldom as tax legislation allows.

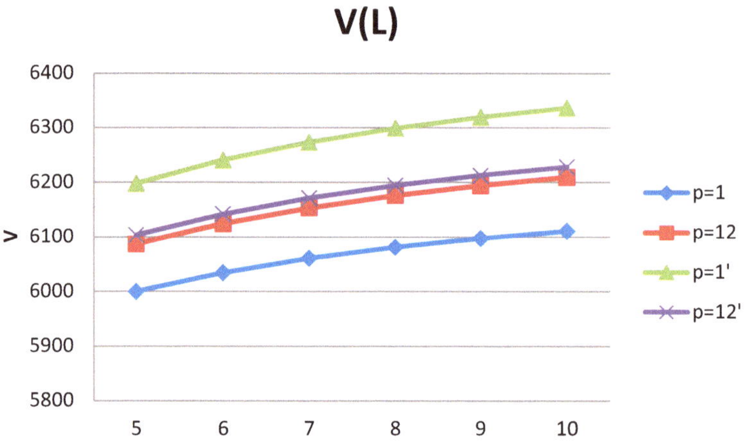

Fig. 7.6 Dependence of the Company Capitalization, V, on Leverage Level L at $k_d = 0.16$ and $p = 1$ and 12 (' Means Advanced Payment of Tax of Profit) (Larger Scale)

Table 7.13 Dependence of the equity cost, k_e, on leverage level L at $k_d = 0.16$ and at $p = 1$

k_0	k_d	w_d	L	t	p	i	WACC	CF	V	k_e
0.2	0.16	0	0	0.2	1	1.16	0.2	1000	5000	0.2
0.2	0.16	0.5	1	0.2	1	1.16	0.18	1000	5555.556	0.232
0.2	0.16	0.666667	2	0.2	1	1.16	0.173333	1000	5769.231	0.264
0.2	0.16	0.75	3	0.2	1	1.16	0.17	1000	5882.353	0.296
0.2	0.16	0.8	4	0.2	1	1.16	0.168	1000	5952.381	0.328
0.2	0.16	0.833333	5	0.2	1	1.16	0.166667	1000	6000	0.36
0.2	0.16	0.857143	6	0.2	1	1.16	0.165714	1000	6034.483	0.392
0.2	0.16	0.875	7	0.2	1	1.16	0.165	1000	6060.606	0.424
0.2	0.16	0.888889	8	0.2	1	1.16	0.164444	1000	6081.081	0.456
0.2	0.16	0.9	9	0.2	1	1.16	0.164	1000	6097.561	0.488
0.2	0.16	0.909091	10	0.2	1	1.16	0.163636	1000	6111.111	0.52

7.7 Dependence of the Equity Cost, k_e, on Leverage Level L

Let us now calculate the dependence of the equity cost, ke, on leverage level L at different p and fixed k_d.

From Tables 7.13, 7.14, 7.15, and 7.16 and Figs. 7.7, 7.8, and 7.9 it is seen, that the equity cost, k_e, linearly increases with the increase of leverage level L for all frequencies of payment of tax on profit per year p.

While in the case of tax on profit payments at the end of the periods the tilt of the curve of the equity cost, k_e, decreases with the increase of the frequency of payment of tax on profit per year p, in case of the advanced payments of tax on profit, the tilt of the curve of the equity cost, k_e, increases with the increase of the frequency of

Table 7.14 Dependence of the equity cost, k_e, on leverage level L at $k_d = 0.16$ and at $p = 12$

k0	kd	wd	L	t	p	i	WACC	CF	V	ke
0.2	0.16	0	0	0.2	12	1.012445	0.2	1000	5000	0.2
0.2	0.16	0.5	1	0.2	12	1.012445	0.178573	1000	5599.963	0.229145
0.2	0.16	0.666667	2	0.2	12	1.012445	0.17143	1000	5833.279	0.25829
0.2	0.16	0.75	3	0.2	12	1.012445	0.167859	1000	5957.383	0.287436
0.2	0.16	0.8	4	0.2	12	1.012445	0.165716	1000	6034.413	0.316581
0.2	0.16	0.833333	5	0.2	12	1.012445	0.164288	1000	6086.883	0.345726
0.2	0.16	0.857143	6	0.2	12	1.012445	0.163267	1000	6124.923	0.374871
0.2	0.16	0.875	7	0.2	12	1.012445	0.162502	1000	6153.767	0.404017
0.2	0.16	0.888889	8	0.2	12	1.012445	0.161907	1000	6176.39	0.433162
0.2	0.16	0.9	9	0.2	12	1.012445	0.161431	1000	6194.608	0.462307
0.2	0.16	0.909091	10	0.2	12	1.012445	0.161041	1000	6209.594	0.491452

Table 7.15 Dependence of the equity cost, k_e, on leverage level L at $k_d = 0.16$ and at $p = 1$ (for advanced payments of tax of profit)

k_0	k_d	w_d	L	t	p	i	WACC	CF	V	k_e
0.2	0.16	0	0	0.2	1	1.16	0.2	1000	5000	0.2
0.2	0.16	0.5	1	0.2	1	1.16	0.1768	1000	5656.109	0.2256
0.2	0.16	0.666667	2	0.2	1	1.16	0.169067	1000	5914.826	0.2512
0.2	0.16	0.75	3	0.2	1	1.16	0.1652	1000	6053.269	0.2768
0.2	0.16	0.8	4	0.2	1	1.16	0.16288	1000	6139.489	0.3024
0.2	0.16	0.833333	5	0.2	1	1.16	0.161333	1000	6198.347	0.328
0.2	0.16	0.857143	6	0.2	1	1.16	0.160229	1000	6241.084	0.3536
0.2	0.16	0.875	7	0.2	1	1.16	0.1594	1000	6273.526	0.3792
0.2	0.16	0.888889	8	0.2	1	1.16	0.158756	1000	6298.992	0.4048
0.2	0.16	0.9	9	0.2	1	1.16	0.15824	1000	6319.515	0.4304
0.2	0.16	0.909091	10	0.2	1	1.16	0.157818	1000	6336.406	0.456

Table 7.16 Dependence of the equity cost, k_e, on leverage level L at $k_d = 0.16$ and at $p = 12$ (for advanced payments of tax of profit)

k_0	k_d	w_d	L	t	p	i	WACC	CF	V	k_e
0.2	0.16	0	0	0.2	12	1.012445	0.2	1000	5000	0.2
0.2	0.16	0.5	1	0.2	12	1.012445	0.178306	1000	5608.338	0.228612
0.2	0.16	0.666667	2	0.2	12	1.012445	0.171075	1000	5845.403	0.257224
0.2	0.16	0.75	3	0.2	12	1.012445	0.167459	1000	5971.613	0.285836
0.2	0.16	0.8	4	0.2	12	1.012445	0.16529	1000	6049.99	0.314448
0.2	0.16	0.833333	5	0.2	12	1.012445	0.163843	1000	6103.394	0.34306
0.2	0.16	0.857143	6	0.2	12	1.012445	0.16281	1000	6142.121	0.371671
0.2	0.16	0.875	7	0.2	12	1.012445	0.162035	1000	6171.49	0.400283
0.2	0.16	0.888889	8	0.2	12	1.012445	0.161433	1000	6194.528	0.428895
0.2	0.16	0.9	9	0.2	12	1.012445	0.160951	1000	6213.082	0.457507
0.2	0.16	0.909091	10	0.2	12	1.012445	0.160556	1000	6228.345	0.486119

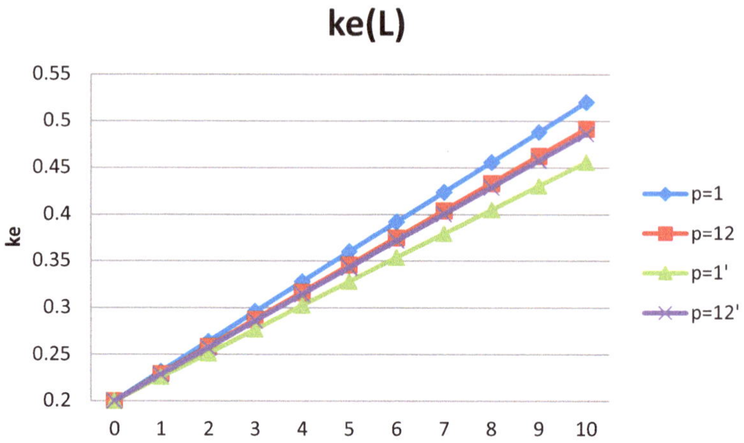

Fig. 7.7 Dependence of the Dependence of the Equity Cost, k_e, on Leverage Level L at Different p and Fixed $k_d = 0.16$ (' Means Advanced Payment of Tax of Profit)

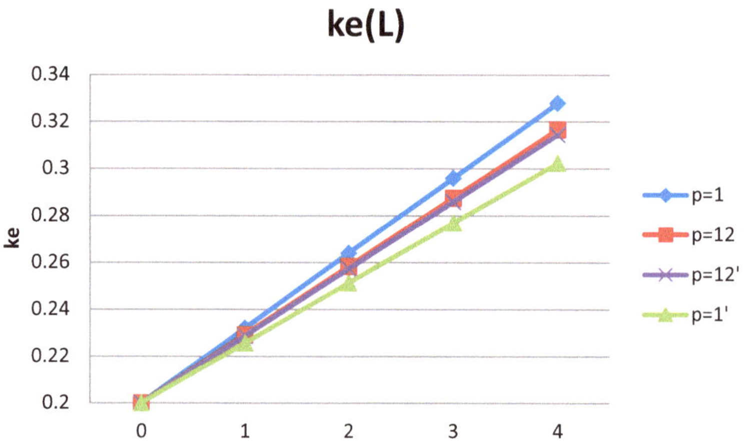

Fig. 7.8 Dependence of the Dependence of the Equity Cost, k_e, on Leverage Level L at Different p and Fixed $k_d = 0.16$ (' Means Advanced Payment of Tax of Profit) (Larger Scale)

payment of tax on profit per year p. These facts should be accounted when developing the company's dividend policy, because the value of the equity cost, k_e, is the economically justified amount of dividends.

The company's dividend policy should be as following:

1. *If the company would like **to minimize the dividend payments**:*

   – in case of payments of tax on profit at the ends of periods they should be as frequently as tax legislation allows.

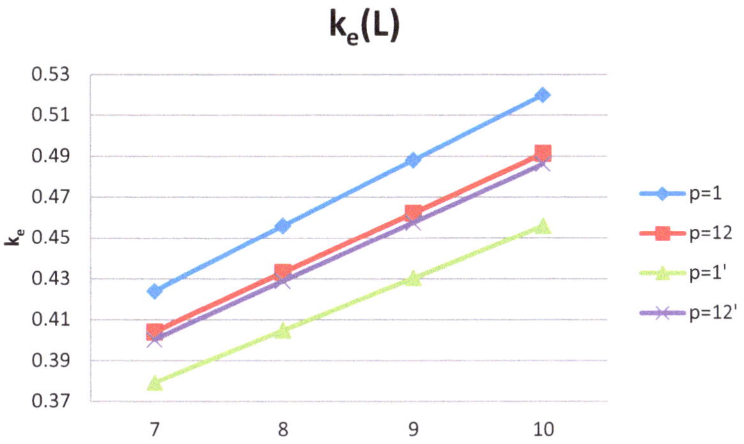

Fig. 7.9 Dependence of the Dependence of the Equity Cost, k_e, on Leverage Level L at Different p and Fixed $k_d = 0.16$ (' Means Advanced Payment of Tax of Profit) (Larger Scale)

– in case of advanced payments of tax on profit they should be as seldom as tax legislation allows.

2. *If the company would like* ***to maximize the dividend payments****:*

– in case of payments of tax on profit at the ends of periods they should be as seldom as tax legislation allows.
– in case of advanced payments of tax on profit they should be as frequently as tax legislation allows.

7.8 Dependence of the Equity Cost, k_e, and the Weighted Average Cost of Capital, WACC, on Tax on Profit Value t

In this paragraph we study the dependence of the equity cost, k_e, and the weighted average cost of capital, WACC, on tax on profit value.

Let us first analyze the dependence of the equity cost, k_e, on tax on profit value t at different p and fixed debt cost k_d.

From Tables 7.17, 7.18, 7.19, and 7.20 and Figs. 7.10 and 7.11 it is seen, that the equity cost, k_e, decreases with the increase of on tax on profit value t for all frequencies of payment of tax on profit per year p.

While in case of payments at the end of the periods the equity cost, k_e, decreases with the increase of the frequency of payments of tax on profit per year p (negative tilt of the curve $k_e(t)$ increases on modulus), in case of advanced payments of tax on

Table 7.17 Dependence of the equity cost, k_e, and the weighted average cost of capital, WACC, on tax on profit value at p = 1

k_0	k_d	w_d	L	t	p	i	WACC	CF	V	k_e
0.2	0.16	0.5	1	0	1	1.16	0.2	1000	5000	0.24
0.2	0.16	0.5	1	0.1	1	1.16	0.19	1000	5263.158	0.236
0.2	0.16	0.5	1	0.2	1	1.16	0.18	1000	5555.556	0.232
0.2	0.16	0.5	1	0.3	1	1.16	0.17	1000	5882.353	0.228
0.2	0.16	0.5	1	0.4	1	1.16	0.16	1000	6250	0.224
0.2	0.16	0.5	1	0.5	1	1.16	0.15	1000	6666.667	0.22
0.2	0.16	0.5	1	0.6	1	1.16	0.14	1000	7142.857	0.216
0.2	0.16	0.5	1	0.7	1	1.16	0.13	1000	7692.308	0.212
0.2	0.16	0.5	1	0.8	1	1.16	0.12	1000	8333.333	0.208
0.2	0.16	0.5	1	0.9	1	1.16	0.11	1000	9090.909	0.204
0.2	0.16	0.5	1	1	1	1.16	0.1	1000	10,000	0.2

Table 7.18 Dependence of the equity cost, k_e, and the weighted average cost of capital, WACC, on tax on profit value at p = 12

k_0	k_d	w_d	L	t	p	i	WACC	CF	V	k_e
0.2	0.16	0.5	1	0	12	1.012445	0.2	1000	5000	0.24
0.2	0.16	0.5	1	0.1	12	1.012445	0.189286	1000	5283.002	0.234573
0.2	0.16	0.5	1	0.2	12	1.012445	0.178573	1000	5599.963	0.229145
0.2	0.16	0.5	1	0.3	12	1.012445	0.167859	1000	5957.383	0.223718
0.2	0.16	0.5	1	0.4	12	1.012445	0.157145	1000	6363.54	0.21829
0.2	0.16	0.5	1	0.5	12	1.012445	0.146432	1000	6829.129	0.212863
0.2	0.16	0.5	1	0.6	12	1.012445	0.135718	1000	7368.227	0.207436
0.2	0.16	0.5	1	0.7	12	1.012445	0.125004	1000	7999.733	0.202008
0.2	0.16	0.5	1	0.8	12	1.012445	0.11429	1000	8749.634	0.196581
0.2	0.16	0.5	1	0.9	12	1.012445	0.103577	1000	9654.672	0.191154
0.2	0.16	0.5	1	1	12	1.012445	0.092863	1000	10,768.54	0.185726

Table 7.19 Dependence of the equity cost, k_e, and the weighted average cost of capital, WACC, on tax on profit value at p = 1 (for advanced payments of tax of profit)

k_0	k_d	w_d	L	t	p	i	WACC	CF	V	k_e
0.2	0.16	0.5	1	0	1	1.16	0.2	1000	5000	0.24
0.2	0.16	0.5	1	0.1	1	1.16	0.1884	1000	5307.856	0.2328
0.2	0.16	0.5	1	0.2	1	1.16	0.1768	1000	5656.109	0.2256
0.2	0.16	0.5	1	0.3	1	1.16	0.1652	1000	6053.269	0.2184
0.2	0.16	0.5	1	0.4	1	1.16	0.1536	1000	657.417	0.2112
0.2	0.16	0.5	1	0.5	1	1.16	0.142	1000	7042.254	0.204
0.2	0.16	0.5	1	0.6	1	1.16	0.1304	1000	7668.712	0.1968
0.2	0.16	0.5	1	0.7	1	1.16	0.1188	1000	8417.508	0.1896
0.2	0.16	0.5	1	0.8	1	1.16	0.1072	1000	9328.358	0.1824
0.2	0.16	0.5	1	0.9	1	1.16	0.0956	1000	10,460.25	0.1752
0.2	0.16	0.5	1	1	1	1.16	0.084	1000	11,904.76	0.168

Table 7.20 Dependence of the equity cost, k_e, and the weighted average cost of capital, WACC, on tax on profit value at p = 12 (for advanced payments of tax of profit)

k_0	k_d	w_d	L	t	p	i	WACC	CF	V	k_e
0.2	0.16	0.5	1	0	12	1.012445	0.2	1000	5000	0.24
0.2	0.16	0.5	1	0.1	12	1.012445	0.189153	1000	5286.726	0.234306
0.2	0.16	0.5	1	0.2	12	1.012445	0.178306	1000	5608.338	0.228612
0.2	0.16	0.5	1	0.3	12	1.012445	0.167459	1000	5971.613	0.222918
0.2	0.16	0.5	1	0.4	12	1.012445	0.156612	1000	6385.21	0.217224
0.2	0.16	0.5	1	0.5	12	1.012445	0.145765	1000	6860.363	0.21153
0.2	0.16	0.5	1	0.6	12	1.012445	0.134918	1000	7411.917	0.205836
0.2	0.16	0.5	1	0.7	12	1.012445	0.124071	1000	8059.911	0.200142
0.2	0.16	0.5	1	0.8	12	1.012445	0.113224	1000	8832.063	0.194448
0.2	0.16	0.5	1	0.9	12	1.012445	0.102377	1000	9767.838	0.188754
0.2	0.16	0.5	1	1	12	1.012445	0.09153	1000	10,925.41	0.18306

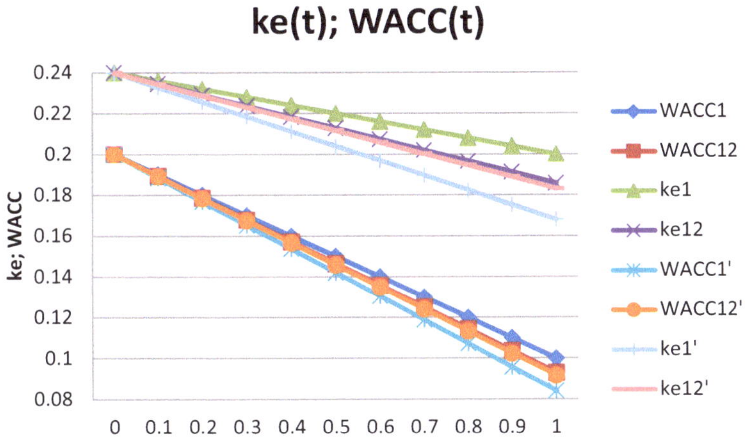

Fig. 7.10 Dependence of the Equity Cost, k_e, and the Weighted Average Cost of Capital, WACC, on Tax on Profit Value at Different p and Fixed $k_d = 0.16$ (' Means Advanced Payment of Tax of Profit)

profit t the equity cost, k_e, increases with the increase of the frequency of payment of tax on profit per year p (negative tilt of the curve $k_e(t)$ increases on modulus).

All the equity cost, k_e, values in case of advanced payments of tax on profit turns out to be lower than in case of payments at the end of the periods.

Let us now analyze the dependence of the weighted average cost of capital, WACC, on tax on profit value at different p and fixed debt cost k_d.

From Tables 7.17, 7.18, 7.19, and 7.20 and Figs. 7.10 and 7.11 it is seen, that the weighted average cost of capital, WACC, decreases with the increase of on tax on profit value t for all frequencies of payment of tax on profit per year p.

While in case of payments at the end of the periods the weighted average cost of capital, WACC, decreases with the increase of the frequency of payments of tax on

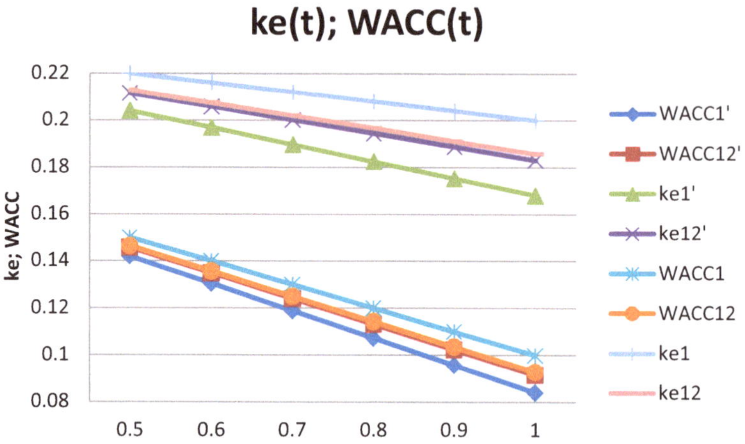

Fig. 7.11 Dependence of the Dependence of the Equity Cost, k_e, and the Weighted Average Cost of Capital, WACC, on Tax on Profit Value at Different p and Fixed $k_d = 0.16$ (' Means Advanced Payment of Tax of Profit) (Larger Scale)

profit per year p (negative tilt of the curve WACC(t) increases on modulus), in case of advanced payments of tax on profit t the weighted average cost of capital, WACC, increases with the increase of the frequency of payment of tax on profit per year p (negative tilt of the curve WACC(t) increases on modulus).

All the weighted average cost of capital, WACC, values in case of advanced payments of tax on profit turns out to be lower than in case of payments at the end of the periods.

7.9 Discussions

Let us discuss the main obtained results.

1. The weighted average cost of capital, WACC, decreases with leverage level L at all values of frequency of payments of tax on profit p. With increase of the frequency of payments of tax on profit per year p the weighted average cost of capital, WACC, (1) decreases in the case of payments at the ends of the periods and WACC (2) increases in the case of advanced payments of tax on profit. Note, that the weighted average cost of capital, WACC, in the case of advanced payments of tax on profit for all values of frequency of payments turns out to be low than for any values of frequency of payments in the case of payments at the ends of the periods.
2. The weighted average cost of capital, WACC, decreases with increase of debt cost value k_d, except the classical Modigliani–Miller case p = 1 and payment at the end of the period. It supports obtained by us result that account of each of two

effects (either advanced payments of tax on profit, or more frequent than once per year) leads to dependence of all main financial indicators, such as the weighted average cost of capital, WACC, company value, V, equity cost, k_e, etc. on debt cost value k_d, while in the classical Modigliani–Miller case these financial indicators DO NOT depend on debt cost value k_d.

Similar to previous paragraph it is seen, that with increase of the frequency of payments of tax on profit per year p WACC (1) decreases in the case of payments at the ends of the periods and WACC (2) increases in the case of advanced payments of tax on profit. As we'll see below, this will lead to increase of company value, V, with increase of p in the case (1) and to decrease of company value, V, with increase of p in the case (2).

Note, that the weighted average cost of capital, WACC, in the case of advanced payments of tax on profit for all values of frequency of payments turns out to be low than for any values of frequency of payments in the case of payments at the ends of the periods.

3. The company value, V, increases with the increase of leverage level L for all frequencies of payment of tax on profit per year p.

   While in case of payments at the end of the periods the company value, V, increases with the increase of the frequency of payment of tax on profit per year p, in case of advanced payments of tax on profit the company value, V, decreases with the increase of the frequency of payment of tax on profit per year p. These facts consistent with the dependence of the weighted average cost of capital, WACC, on the leverage level L.

   All the company values, V, in case of advanced payments of tax on profit turns out to be bigger than in case of payments at the end of the periods. Thus if the companies tend to increase their capitalization, they should pay tax on profit (1) in advance; (2) as seldom as tax legislation allows.

4. The equity cost, k_e, linearly increases with the increase of leverage level L for all frequencies of payment of tax on profit per year p.

   While in the case of tax on profit payments at the end of the periods the tilt of the curve of the equity cost, k_e, decreases with the increase of the frequency of payment of tax on profit per year p, in case of the advanced payments of tax on profit, the tilt of the curve of the equity cost, k_e, increases with the increase of the frequency of payment of tax on profit per year p. These facts should be accounted when developing the company's dividend policy, because the value of the equity cost, k_e, is the economically justified amount of dividends.

   The company's dividend policy should be as following:

   (a) *If the company would like* ***to minimize the dividend payments****:*

      – in case of payments of tax on profit at the ends of periods they should be as frequently as tax legislation allows.
      – in case of advanced payments of tax on profit they should be as seldom as tax legislation allows.

   (b) *If the company would like* ***to maximize the dividend payments****:*

– in case of payments of tax on profit at the ends of periods they should be as seldom as tax legislation allows.
– in case of advanced payments of tax on profit they should be as frequently as tax legislation allows.

5. The equity cost, k_e, decreases with the increase of on tax on profit value t for all frequencies of payment of tax on profit per year p.

   While in case of payments at the end of the periods the equity cost, k_e, decreases with the increase of the frequency of payments of tax on profit per year p (negative tilt of the curve $k_e(t)$ increases on modulus), in case of advanced payments of tax on profit t the equity cost, ke, increases with the increase of the frequency of payment of tax on profit per year p (negative tilt of the curve $k_e(t)$ increases on modulus).

   All the equity cost, k_e, values in case of advanced payments of tax on profit turns out to be lower than in case of payments at the end of the periods.

   The weighted average cost of capital, WACC, decreases with the increase of on tax on profit value t for all frequencies of payment of tax on profit per year p.

   While in case of payments at the end of the periods the weighted average cost of capital, WACC, decreases with the increase of the frequency of payments of tax on profit per year p (negative tilt of the curve WACC(t) increases on modulus), in case of advanced payments of tax on profit t the weighted average cost of capital, WACC, increases with the increase of the frequency of payment of tax on profit per year p (negative tilt of the curve WACC(t) increases on modulus).

   All the weighted average cost of capital, WACC, values in case of advanced payments of tax on profit turns out to be lower than in case of payments at the end of the periods.

7.10 Conclusions

Brusov and Filatova (2021) have modified the Modigliani–Miller theory for the case of arbitrary frequency of payment of tax on profit t and show that the account of frequency of payment of tax on profit within the Modigliani–Miller theory turns out to be very important and leads to significant consequences: (1) all Modigliani–Miller theorems, statements and all formulas change; (2) all main financial indicators, such as the weighted average cost of capital, WACC, company value, V and equity cost, k_e depend on the frequency of tax on profit payments p; (3) this allows to company choose the number of payments of tax of profit per year, as many, as it is profitable to it (of course, within actual tax legislation).

As it was shown in that paper (Brusov & Filatova, 2021), more frequent payments of income tax are beneficial for both parties: for the company and for the tax regulator: for the company, this leads to decrease of the cost of attracting capital and thus to an increase in the value of the company, and for the tax regulator, earlier payments are beneficial due to the time value of money.

But companies could (and sometimes should) pay tax of profit in advance. In current paper we use two modifications of the Modigliani–Miller theory simultaneously: (1) frequent payments of income tax; (2) advance payments of income tax and show that this fact changes drastically the main conclusions of the paper (Brusov & Filatova, 2021): more frequent payments of income tax are **NOT** beneficial for the company: this leads to increase of the cost of attracting capital (the weighted average cost of capital, WACC, and the equity cost, ke,) and thus to decrease in the value of the company, but for the tax regulator it remains beneficial: earlier payments are beneficial due to the time value of money.

Thus, the main conclusions of current Chapter are as follows:

1. for payments of tax of profit at the end of periods more frequent payments of income tax are beneficial for both parties: for the company and for the tax regulator: (1) for the company, this leads to decrease of the cost of attracting capital, WACC, and thus to an increase in the value of the company, V, and (2) for the tax regulator, more frequent payments are beneficial due to the time value of money.
2. for payments of tax of profit in advance more frequent payments of income tax are NOT beneficial for the company: this leads to increase of the cost of attracting capital, WACC, and thus to decrease in the value of the company, V, but for the tax regulator it remains beneficial: earlier payments are beneficial due to the time value of money.

References

Berk, J., & DeMarzo, P. (2007). Corporate finance. pearson international edition, Addison-Wesley, 988.

Brealey, R., Myers S., Allen F. (2006). Corporate finance, McGrow Hill, New York.

Brusov, P., & Filatova, T. (2021). The Modigliani–Miller theory with arbitrary frequency of payment of tax on profit. *Mathematics, 9*(11), 1198. https://doi.org/10.3390/math9111198

Brusov, P., Filatova, T., Orehova, N., & Eskindarov, M. (2018a). *Modern corporate finance, investments, taxation and ratings* (571 p), Springer Nature Publishing. Monograph.

Brusov, P. N., Filatova, T. V., Orekhova, N. P., & Kulik, V. L. (2018c). A "golden age" of the companies: Conditions of its existence. *Journal of Reviews on Global Economics, 7*, 88–103. https://doi.org/10.6000/1929-7092.2018.07.07

Brusov, P. N., Filatova, T. V., Orekhova, N. P., & Kulik, V. L. (2018d). Rating methodology: New look and new horizons. *Journal of Reviews on Global Economics, 7*, 63–87. https://doi.org/10.6000/1929-7092.2018.07.06

Brusov, P. N., Filatova, T. V., Orekhova, N. P., Kulik, V. L., & Weil, I. (2018e). New meaningful effects in modern capital structure theory. *Journal of Reviews on Global Economics, 7*, 104–122. https://doi.org/10.6000/1929-7092.2018.07.08

Brusov, P., Filatova, T., Orekhova, N., Kulik, V., Weil, I., & Brailov, A. (2018f). The impact of the central bank key rate and commercial banks credit rates on creating and maintaining of a favorable investment climate in the country. *Journal of Reviews on Global Economics, 7*, 360–376. https://doi.org/10.6000/1929-7092.2018.07.31

Brusov, P. N., Filatova, T. V., & Orekhova, N. P. (2018g). *Modern corporate finance and investments* (517 p). Monograph. Knorus Publishing House.

Brusov, P. N., Filatova, T. V., Orekhova, N. P., Kulik, V. L., & Weil, I. (2019). Ratings of the investment projects of arbitrary durations: New methodology. *Journal of Reviews on Global Economics, 8*, 437–448. https://doi.org/10.6000/1929-7092.2019.08.37

Brusov, P., Filatova, T., & Orehova, N. (2020a). *Ratings: Critical analysis and new approaches of quantitative and qualitative methodology*. Springer Nature Publishing. Monograph. Retrieved from https://www.springer.com/de/book/9783030562427.

Brusov, P. N., Filatova, T. V., Orekhova, N. P., Kulik, V. L., Chang, S.-I., & Lin, Y. C. G. (2020b). Application of the Modigliani–Miller theory, modified for the case of advance payments of tax on profit, in rating methodologies. *Journal of Reviews on Global Economics, 9*, 282–292. https://doi.org/10.6000/1929-7092.2020.09.28

Brusov, P. N., Filatova, T. V., Orekhova, N. P., Kulik, V. L., Chang, S.-I., & Lin, Y. C. G. (2020c). Modification of the Modigliani–Miller theory for the case of advance payments of tax on profit. *Journal of Reviews on Global Economics, 9*, 257–267. https://doi.org/10.6000/1929-7092.2020.09.25

De Marzo, P. M. (1988). An extension of the Modigliani–Miller theorem to stochastic economies with incomplete markets and interdependent securities. *Journal of Economic Theory, 45*(2), 353–369.

Farber, A., Gillet, R., & Szafarz, A. (2007). A general formula for the WACC. *International Journal of Business, 11*(2), 211–218.

Fernandez, P. (2007). A general formula for the WACC: A comment. *International Journal of Business*, this issue.

Filatova, T., Orehova, N., & Brusova, A. (2008). Weighted average cost of capital in the theory of Modigliani–Miller, modified for a finite life–time company. *Bull FU, 48*, 68–77.

Hamada, R. (1969). Portfolio analysis, market equilibrium, and corporate finance. *Journal of Finance, 24*, 13–31.

Harris, R., & Pringle, J. (1985). Risk–adjusted discount rates—Extension form the average–risk case. *The Journal of Financial Research, 8*(3), 237–244.

Miles, J. A., & Ezzel, J. R. (1980). The weighted average cost of capital, perfect capital markets and project life: A clarification. *Journal of Financial and Quantitative Analysis, 15*(3), 719–730.

Miller, M. (1977). Debt and taxes. *The Journal of Finance, 32*(2), 261–275. https://doi.org/10.1111/j.1540-6261.1977.tb03267.x

Modigliani, F., & Miller, M. (1958). The cost of capital, corporate finance, and the theory of investment. *The American Economic Review, 48*, 261–297.

Modigliani, F., & Miller, M. (1963). Corporate income taxes and the cost of capital: A correction. *The American Economic Review, 53*, 147–175.

Modigliani, F., & Miller, M. (1966). Some estimates of the cost of capital to the electric utility industry 1954–1957. *The American Economic Review, 56*, 333–391.

Ross, S., Westerfiel, R., & Jaffee, D. (2005). *Corporate finance* (7th ed.). McGraw Hill.

Chapter 8
Generalization of the Modigliani–Miller Theory for the Case of Variable Profit

For the first time we have generalized the world–famous theory by Nobel Prize winners Modigliani and Miller for the case of variable profit, which significantly extends the application of the theory in practice, specifically in business valuation, ratings, corporate finance, etc. We demonstrate that all the theorems, statements and formulae of Modigliani and Miller are changed significantly. We combine theoretical and numerical (by MS Excel) considerations. The following results are obtained:

1. Discount rate for leverage company changes from the weighted average cost of capital, WACC, to WACC–g (where g is growing rate), for a financially independent company from k_0 to k_0–g. This means that WACC and k_0 are no longer the discount rates as it takes place in case of classical Modigliani–Miller theory with constant profit. WACC grows with g, while real discount rates WACC–g and k_0–g decrease with g. This leads to an increase of company capitalization with g.
2. The tilt angle of the equity cost k_e (L) grows with g. This should change the dividend policy of the company, because the economically justified value of dividends is equal to equity cost.
3. A qualitatively new effect in corporate finance has been discovered: at rate $g < g^*$ the slope of the curve k_e (L) turns out to be negative, which could significantly alter the principles of the company's dividend policy.

8.1 Introduction

The original theory by Nobel Prize Winners Modigliani and Miller (1958, 1963, 1966) has been modified by many authors and we shortly discuss some of these. A few important modifications have been done by the authors of this monograph (Brusov et al., 2015, 2018; Filatova et al., 2008), who created the general theory of capital cost and capital structure, the Brusov–Filatova–Orekhova (BFO) theory,

P. Brusov, T. Filatova, *Efficiency of Investment Projects*, Contributions to Finance and Accounting, https://doi.org/10.1007/978-3-032-01028-5_8

which generalized the Modigliani–Miller theory for the case of companies of arbitrary age (and arbitrary lifetime), as well as for the case of advance payments of tax on profit (Brusov et al. 2020a), for rating needs (Brusov et al. 2020b), as well as for variable debt cost (Brusov et al. 2021b). Note that a stochastic extension of the Miller–Modigliani theory has been created by some authors (Sethi et al. 1991; DeMarzo 1988).

In the current chapter for the first time we have generalized the world–famous theory by Nobel Prize winners Modigliani and Miller for the case of variable profit (Brusov et al. 2021b), which significantly extends the application of the theory in practice, specifically in business valuation, ratings, investments and in other areas of the economy and of finance. We consider the case of growing profit as well as decreasing profit and show that all theorems, all statements by Modigliani and Miller (and all their main formulas) are changed significantly.

Within the new Generalized Modigliani–Miller theory (GMM theory), we study the dependence of the weighted average cost of capital, WACC, the equity cost, k_e, the discount rate, *i*, and the capitalization of the company, V, on leverage level L. Some important results have been obtained, which allows for the development of a new approach to financial policy and financial strategy for the company. Some of these are as follows:

- The discount rate for a leveraged company changes from the weighted average cost of capital, WACC, to WACC–g (where g is growth rate), for an unleveraged company from k_0 to k_0–g. This means that WACC and k_0 are no longer the discount rates they are in the case of classical Modigliani–Miller theory with constant profit.
- All curves WACC(L) for different values of g start from one point, k_0. They decrease with L at $g < k_0$ and increase at $g > k_0$. It turns out that WACC grows with g, while real discount rates WACC–g and k_0–g decrease with g. This leads to an increase of company capitalization along with g: V = CF/WACC–g. Knowing the correct value of the discount rate allows for management of companies' financial flows.
- The equity cost, k_e, which grows linearly with the leverage level, increases with g: the tilt angle k_e (L) grows along with the growth rate g.

A qualitatively new effect in corporate finance has been discovered: at rate $g < g^*$ the slope of the curve k_e (L) turns out to be negative. Two these effects, which are absent in classical Modigliani–Miller theory, could significantly alter the principles behind the company's dividend policy, because the economically justified value of dividends is equal to equity cost.

The final effect is similar to the qualitatively new effect in corporate finance that was discovered by Brusov–Filatova–Orekhova within the BFO theory: the abnormal dependence of equity cost on leverage level at tax on profit rate T, which exceeds rate value T*. This discovery also significantly alters the principles behind the company's dividend policy.

The current paper, contrary to papers on (stochastically) growing cash flows, can be directly applied for calculation of all the main indicators for companies. The results obtained will have applications in corporate finance, business valuation, ratings, etc.

The structure of the Chapter is as follows:

1. We give an introduction to the traditional approach of the Modigliani–Miller theory and to its modifications.

   - We generalize the Modigliani–Miller theory to the case of variable profit and obtain generalized Modigliani–Miller theorems, as well as new formulae for the weighted average cost of capital, WACC, equity cost, k_e, discount rate, i, and capitalization of the company, V.
   - Within the new Generalized Modigliani–Miller theory (GMM theory), we numerically study (with MS Excel) the dependence of the main financial indicators of the company (WACC, k_e, i, V) on leverage level L.
   - We discuss the results obtained and based on these, we arrive at some important conclusions.

Below we discuss the problem of capital structure.

Capital structure is the relationship between the debt and the equity capital of the company. Does capital structure impact the main financial indicators of the company, such as the cost of capital, profit, value of the company, etc., and, if so, how? The choice of an optimal capital structure, i.e., a capital structure that maximizes the company's capitalization, *V*, and minimizes the weighted average cost of capital, WACC, is one of the most important problems to be solved by the financial manager and senior management of a company. The first quantitative study of the impact of company capital structure on its financial indicators was performed by Modigliani and Miller (1958). Before 1958, the traditional approach, based on empirical data analysis, was used.

8.1.1 The Traditional Approach

The traditional approach supposes that the weighted average cost of capital, WACC, and the associated company capitalization value, $V = \text{CF/WACC}$, depend on the capital structure (the level of leverage (L)). The reason for this is that the debt cost always turns out to be lower than the equity cost because the first one has lower risk, due to the fact that, in the event of bankruptcy, creditor claims are met prior to shareholders claims.

Thus, if the firm increases the share of lower–cost debt capital in the overall capital structure, this will lead to a lower weighted average cost of capital and WACC up to the limit, which does not cause violation of financial sustainability or result in increase of bankruptcy risk.

Investor profitability is required and the cost of equity grows with the leverage level; however, its growth has not lead to compensation of benefits from the increasing use of lower–cost debt capital. Therefore, at a low leverage level

WACC decreases with the increase of leverage $L = D/S$ and company capitalization increases. At a high leverage level when the risk of bankruptcy becomes higher, WACC may increase with the increase of leverage L and company capitalization decreases. Thus, the tradeoff between advantages of debt financing at a low leverage level and its shortcomings at a high leverage level forms an optimal capital structure, which maximizes the company capitalization, V, and minimizes the weighted average cost of capital, WACC.

The traditional approach has existed up to 1958, when the first quantitative theory by Modigliani and Miller has appeared (Modigliani and Miller 1958).

8.1.2 Modigliani–Miller Theory Without Taxes

In their first paper, Modigliani and Miller (MM) (1958) under a lot of assumptions (there are no taxes, no bankruptcy costs, no transaction costs, perfect financial markets exist with symmetry information, equivalence in borrowing costs for both companies and investors, etc.), came to the conclusions that choosing of the ratio between the debt and equity capital does not affect company value as well as capital

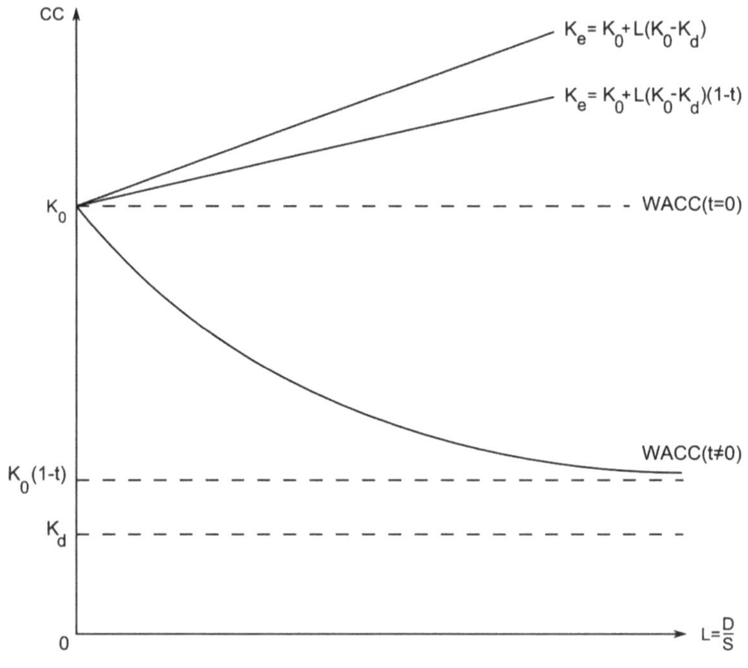

Fig. 8.1 Dependence of equity capital cost, debt cost and WACC on leverage in Modigliani–Miller theory without taxes ($t = 0$) and with taxes ($t \neq 0$)

costs (Fig. 8.1). These conclusions were fundamentally different from the conclusions of the traditional approach.

Modigliani and Miller, under the above assumptions, have analyzed the impact of financial leverage, assuming the absence of any taxes (on corporate profit as well as individual one). They have formulated and proven two following theorems.

Without taxes, the total cost of any company is determined by the value of its EBIT (Earnings Before Interest and Taxes), discounted with fixed rate k_0, corresponding to group of business risk of this company:

$$V = V_0 = \frac{EBIT}{k_0} \tag{8.1}$$

This leads to the following expressions for WACC:

$$WACC = k_0 \tag{8.2}$$

Note, that k_0 here and below is the weighted average cost of capital, WACC for an unleveraged company. For a leveraged company, k_0 is the equity cost (and weighted average cost of capital, WACC) at zero leverage level ($L = 0$).

From the first Modigliani–Miller theorem [1], it is easy to derive an expression for the equity capital cost:

$$\text{WACC} = k_0 = k_e w_e + k_d w_d. \tag{8.3}$$

Finding from here k_e, one gets

$$k_e = \frac{k_0}{w_e} - k_d \frac{w_d}{w_e} = \frac{k_0(S+D)}{S} - k_d \frac{D}{S} = k_0 + (k_0 - k_d)\frac{D}{S} = k_0 + (k_0 - k_d)L \tag{8.4}$$

Here,

D	value of debt capital of the company;
S	value of equity capital of the company;
$k_d, w_d = \frac{D}{D+S}$	cost and fraction of debt capital of the company;
$k_e, w_e = \frac{S}{D+S}$	cost and fraction of equity capital of the company;
$L = D/S$	financial leverage;
WACC	weighted average cost of capital.

Thus, we come to second theorem of Modigliani–Miller theory regarding the equity cost of a leveraged company (Modigliani and Miller 1958).

The equity cost of a leveraged company (k_e) could be found as k_0 (the cost of equity of an unleveraged company with the same asset risk), plus premium for risk, which is equal to the product of difference ($k_0 - k_d$) and leverage level L:

$$k_e = k_0 + L(k_0 - k_d) \tag{8.5}$$

Formula (8.5) shows that equity cost of the company increases linearly with is leverage level.

8.2 Some Modifications of Modigliani–Miller Theory

8.2.1 Modigliani–Miller Theory with Taxes

In 1963 Modigliani and Miller (1963) accounted for the effect of corporate taxes and obtained the following result for the capitalization value of a leveraged company, V:

$$V = V_0 + DT, \tag{8.6}$$

where V_0 is the value of unleveraged company, D is the debt value and T is the corporate taxes on profit rate.

The value of a leveraged company is equal to the value of the unleveraged company with the same asset risk, plus the value of tax shield arising from financial leverage, which is equal to the product of corporate income tax rate (T) and the debt value (D).

The following section presents the formal derivation result for WACC and the equity capital cost (k_e) of the company with consideration of corporate taxes.

8.2.1.1 Weighted Average Cost of Capital, WACC

$$\begin{aligned} V &= V_0 + DT \\ V(1 - w_d T) &= V_0 \end{aligned} \tag{8.7}$$

$$\frac{CF}{WACC}(1 - w_d T) = \frac{CF}{k_0} \tag{8.8}$$

$$WACC = k_0 \cdot (1 - w_d t) \tag{8.9}$$

The above is a very important formula for weighted average cost of capital (WACC) and is one of the main results of the Modigliani–Miller theory with taxes.

8.2.1.2 Equity Cost

Let us derive formula for equity cost.

On definition of the weighted average cost of capital with “tax shield”, we have:

$$\mathrm{WACC} = k_0 w_e + k_d w_d (1 - T). \tag{8.10}$$

Equating Eqs. (8.9) and (8.10), one gets:

$$k_0 (1 - w_d T) = k_0 w_e + k_d w_d (1 - T) \tag{8.11}$$

and from here, for equity cost, we develop the following expression:

$$\begin{aligned} k_e &= k_0 \frac{(1 - w_d T)}{w_e} - k_d \frac{w_d}{w_e}(1 - T) = k_0 \frac{1}{w_e} - k_0 \frac{w_d}{w_e} T - k_d \frac{D}{S}(1 - T) \\ &= k_0 \frac{D + S}{S} - k_0 \frac{D}{S} T - k_d \frac{D}{S}(1 - T) = k_0 + L(1 - T)(k_0 - k_d). \end{aligned} \tag{8.12}$$

Therefore, we see the following theorem obtained by Modigliani and Miller in 1963:

Equity cost of leveraged company (k_e) could be found as equity cost of unleveraged company (k_0) with the same business risk, plus premium for risk, the value of which is equal to the triple product of difference between the cost of capital for an unleveraged firm and cost of debt ($k_0 - k_d$), leverage level ($L = \frac{D}{S}$)) and tax corrector (*1* − *T*).

It should be noted that Formula (8.12) is different from Formula (8.5) without tax only by the multiplier (1 − T), suggesting the tax benefit of debt will lower the cost of equity as well.

Analysis of Formulas (8.5), (8.9) and (8.12) leads to the following conclusions. When leverage grows:

1. The value of company increases.

- The weighted average cost of capital WACC decreases from k_0 (at $L = 0$) up to $k_0(1 - T)$ (at $L = \infty$) (when the company is funded solely by borrowed funds).
- Equity cost increases linearly from k_0 (at $L = 0$) up to ∞ (at $L = \infty$).

Within their theory, Modigliani and Miller (1963) came to the following conclusions regarding the growth of financial leverage (Fig. 8.1).

8.2.2 *Taking into Account Market Risk: Hamada Model*

In 1969 Hamada united CAPM (Capital Asset Pricing Model) and Modigliani–Miller theory with taxes. For the equity cost of a leveraged company, the following formula has been derived, which includes financial and business risk of company:

$$k_e = k_F + (k_M - k_F)b_U + (k_M - k_F)b_U \frac{D}{S}(1 - T), \quad (8.13)$$

here b_U is the CAPM βeta of the unleveraged company with the same asset risk as the leveraged company under consideration. Formula (8.13) represents the cost of equity (k_e) for a leveraged firm as a sum of three components: risk–free rate (k_F), risk premium for business/asset risk $(k_M - k_F)b_U$ and risk premium for financial risk $(k_M - k_F)b_U \frac{D}{S}(1 - T)$.

If the company does not have borrowing ($D = 0$), the financial risk factor will be equal to zero (the third term is drawn to zero), and equity holder will only require the premium for business/asset risk.

8.2.3 The Account of Corporate and Individual Taxes (Miller Model)

In their second article, Modigliani and Miller (1963) only considered corporate taxes benefit of debt but did not take into account the effect of individual investors' income taxes.

In 1977 Merton Miller (1977) have developed such a model, showing the influence of financial leverage on the capitalization of the company when the effects of both corporate and individual taxes are accounted for, i.e., there is double taxation on corporate earnings. We will use the following definitions: T_C–corporate taxes rate and T_S–a weighted average value of effective taxes rates on dividends and capital gains on shares. With the same other assumptions that have been made for Modigliani–Miller models previously, the unleveraged company value can be determined as follows:

$$V_U = \frac{\text{EBIT}(1 - T_C)(1 - T_S)}{k_0}.$$

A term $(1 - T_S)$ accounts for the individual taxes. The numerator in (1) indicates which part of the operating company's profit remains in the possession of the investors, after the company earnings are taxed twice: by corporate taxes and individual taxes. Since individual taxes reduce investors' residual income, it also reduces the overall assessment of the unleveraged company value.

8.2.4 Brusov–Filatova–Orekhova (BFO) Theory

One of the most serious limitations of the Modigliani–Miller theory is the suggestion about the perpetuity of the companies. In 2008, Brusov–Filatova–Orekhova (Filatova et al., 2008) have considered this limitation and showed that accounting for the finite lifetime length (or arbitrary age n) of the company leads to significant changes of all Modigliani–Miller results (Modigliani & Miller 1958, 1963, 1966): capitalization of the company V is changed, as well as the equity cost, k_e, and the weighted average cost of capital, WACC, in the presence of corporative taxes. Moreover, a number of new findings for corporate finance, in the Brusov–Filatova–Orekhova theory (Brusov et al., 2015, 2018; Filatova et al. 2008), are absent in Modigliani–Miller theory.

The formula for weighted average cost of capital (WACC) for the company with arbitrary age n, as derived by Brusov–Filatova–Orekhova, has the following form (Brusov et al., 2015, 2018; Filatova et al. 2008):

$$\frac{1-(1+WACC)^{-n}}{WACC}=\frac{1-(1+k_0)^{-n}}{k_0[1-w_d T(1-(1+k_d)^{-n})]} \tag{8.14}$$

Here, w_d $(=\frac{D}{D+S})$ is the debt share; k_0 is the cost of capital for an unleveraged firm with the same asset risk; k_d the cost of debt;T denotes the corporate taxes rate; and n stands for the firm's lifetime length (age).

A perpetuity (Modigliani–Miller) limit could be easily obtained from (8.14), substituting $n \to \infty$.

A lot of meaningful effects have been discovered within the BFO theory: these effects are absent within MM theory. BFO theory has challenged some main existing principles of financial management: among them the trade–off theory, which has been prevailing for many decades and established the foundation to claim the existence of an optimal capital structure for firms. However, BFO theory has proven the bankruptcy of trade–off theory (Brusov et al., 2015, 2018).

8.2.5 The General WACC Formula

A more general formula for WACC, than in the famous Modigliani–Miller theory (MM) has been derived and discussed by a few authors in 2006–2007 (Farber et al. 2006; Fernandez 2006; Berk and DeMarzo 2007; Harris and Pringle 1985). It takes the following form (Eq. (18) in (Farber et al. 2006)

$$WACC=k_0(1-w_d t)-k_d t w_d+k_{TS} t w_d \tag{8.15}$$

where k_0 is the required return on unleveraged company, k_d is the required return on its debt, k_{TS} is the expected return on the tax shield and t is the corporate taxes rate.

This formula is derived from the definition of the weighted average cost of capital and the balance sheet identity (for a similar presentation, see (Berk and DeMarzo 2007). At any point in time, it should therefore be verified, regardless of whether returns are annually or continuously compounded.

Practical applicability of Eq. (8.15) (while it is fairly general) requires additional conditions. Indeed, when the WACC is constant over time, the value of a leveraged company can be computed by discounting with the WACC of the unlevered free cash–flows. Therefore, it is interesting to consider the special cases when WACC is constant. The resulting formulas can also be found in textbooks (Brealey et al. 2005; Ross et al. 2005).

It was assumed by Modigliani and Miller in 1963 that the debt value D is constant. As the expected after–tax cash–flow of the unleveraged company is fixed, V_0 is also constant. By this assumption, $k_{TS} = k_D$ and the value of the tax shield is TS = tD. Therefore, the capitalization of the leveraged company V is a constant and the general WACC Formula (8.15) simplifies to a constant WACC:

$$WACC = k_0(1 - w_d T)$$

However, our opinion is that "classical" Modigliani–Miller (MM) theory, which suggests that the expected returns on the debt k_d and the tax shield k_{TS} are equals (because both of them have debt nature), is much more reasonable and in our paper we modify the "classical" Modigliani–Miller (MM) theory, which is still widely used in practice.

8.2.6 Trade–Off Theory

The world–famous **trade–off theory** has been considered the cornerstone in the solution of the problem of optimal capital structure for a company for many decades and is still used today for decision analysis on capital structure. Below we give two examples.

Frank, M., & Goyal, V., in their 2009 paper, "examines the relative importance of many factors in the capital structure decisions of publicly traded American firms from 1950 to 2003. The most reliable factors for explaining market leverage are: median industry leverage (+ effect on leverage), market–to–book assets ratio (−), tangibility (+), profits (−), log of assets (+), and expected inflation (+)". In addition, the authors have found that "dividend–paying firms tend to have lower leverage. When considering book leverage, somewhat similar effects are found. However, for book leverage, the impact of firm size, the market–to–book ratio, and the effect of inflation are not reliable". The empirical evidence seems to be reasonably consistent with some versions of the trade–off theory of capital structure.

Serrasqueiro, Z., & Caetano, A., in 2015, analyzed "to what extent decisions on the capital structure of small and medium–sized enterprises (SMEs) are closer to the assumptions of trade–off theory or to the assumptions of hierarchy theory. They used a sample of small and medium–sized enterprises located in the Portuguese hinterland, using dynamic LSDVC as the valuation method, and the empirical evidence suggests that the most profitable and oldest SMEs are less leveraged, confirming Pecking Order Theory 's forecasts. Larger SMEs are leveraging more borrowing, confirming the predictions of trade–off theory and hierarchy theory. In addition, SMEs are significantly adjusting their current debt levels towards the optimal debt ratio, which is consistent with the predictions of the compromise theory. It was concluded that theories of compromise and hierarchy are not mutually exclusive in explaining capital structure decisions of small and medium–sized enterprises".

However, the bankruptcy of trade–off theory has been proven by Brusov et al. in 2013 (Brusov et al. 2018). They have shown that risky debt financing (and growing credit rate near the bankruptcy) in contrast to waiting results does not lead to growing of weighted average cost of capital, WACC, which still decreases with leverage. This means the absence of minimum in the dependence of WACC on leverage level as well as the absence of maximum in the dependence of company capitalization on leverage. Thus, the well–known trade–off theory lacks an optimal capital structure. The explanation for this fact was made by Brusov et al. in 2013 (Brusov et al. 2018) by analyzing the dependence of the cost of equity capital on the leverage level on the assumption that debt capital is risky.

Modigliani–Miller have considered tax shields from the interest on debt can increase the value of companies. In 1980, De Angelo and Masulis moved further in the theoretical examination of tax shields. They have noted that there are tax deductibles for companies other than debt to reduce their corporate tax burden and debt and non–debt tax shields should be accounted for. Depreciation, investment tax credits, or net–loss carry forwards could represents examples of such kind of non–debt tax shields. The first to test for these tax effects (suggested by DeAngelo and Masulis in 1980) has been carried out by Bradley, Jarrell and Kim in 1984. In contrast to the prediction in De Angelo and Masulis (1980), by regressing company–specific debt–to–value ratios on non–debt tax shields, they have shown that debt is positively related to non–debt tax shields as measured by depreciation and investment tax credits. Titman and Wessels in 1988 found that "their results do not provide support for an effect on debt ratios arising from non–debt tax shields…". It was pointed out in 2003 by Graham, if a company invests heavily and uses debt financing to invest, a positive relation between such proxies for non–debt tax shield and debt may result. A mechanical positive relation of this type overwhelms and renders any substitution effects between debt and non–debt tax shields.

The original theory by Nobel Prize Winners Modigliani and Miller (Modigliani & Miller 1958, 1963, 1966) has been modified by many authors and above we shortly discussed some of them. In next paragraph we will generalize for the first time the Modigliani–Miller theory for the case of variable profit.

8.3 Generalization of the Modigliani–Miller Theory for the Case of Variable Profit

8.3.1 Modigliani–Miller Theory Without Taxes

8.3.1.1 Company Value, V

In this part we generalize for the first time the world–famous theory by Nobel Prize winners Modigliani and Miller for the case of variable profit. Let us calculate capitalization for leverage company, assuming profit per period grows with growth rate g.

$$V = \frac{CF}{1 + WACC} + \frac{CF(1+g)}{(1+WACC)^2} + \frac{CF(1+g)^2}{(1+WACC)^3} + \ldots + \frac{CF(1+g)^{n-1}}{(1+WACC)^n} \quad (8.16)$$

Here, CF is the annual profit of company, $WACC$ is the weighted average cost of capital.

This is geometric progression with denominator

$$\hat{g} = \frac{1+g}{1+WACC} \quad (8.17)$$

Summing (8.16), one gets

$$V = \frac{CF}{1+WACC} \cdot \frac{1}{1 - \frac{1+g}{1+WACC}} = \frac{CF}{WACC - g} \quad (8.18)$$

Let us calculate the capitalization for a financially independent company

$$V_0 = \frac{CF}{1+k_0} + \frac{CF(1+g)}{(1+k_0)^2} + \frac{CF(1+g)^2}{(1+k_0)^3} + \ldots + \frac{CF(1+g)^{n-1}}{(1+k_0)^n} \quad (8.19)$$

This is geometric progression with denominator

$$\hat{g} = \frac{1+g}{1+k_0} \quad (8.20)$$

Summing (8.19), one gets

$$V_0 = \frac{CF}{1+k_0} \cdot \frac{1}{1 - \frac{1+g}{1+k_0}} = \frac{CF}{k_0 - g} \quad (8.21)$$

The original first theorem by Modigliani–Miller changes by the following one:

Without taxes, the total cost of any company with growing profit with rate g is determined by the value of its EBIT—Earnings Before Interest and Taxes—discounted with fixed rate $k_0 - g$, corresponding to group of business risk of this company:

$$V = V_0 = \frac{EBIT}{k_0 - g} \tag{8.22}$$

Note, that this formula for company capitalization is similar to one of the Gordon Growth Model (GGM), which assumes a company is perpetuity and pays dividends per share that increase at a constant rate. To estimate the value of a stock, the model takes the infinite series of dividends per share and discounts them back into the present using the required rate of return.

$$P = \frac{DPS_1}{k_e - g} \tag{8.23}$$

where p—value of stock; DPS_1—expected Dividends one year from now (next period); k_e—required rate of return for equity investors (equity cost); g—growth rate in dividends forever.

Note that, despite some formal similarities, these are not the same model. In our model, CF is the company's annual income (not dividends), and we study all the financial indicators of the company (WACC, V, k_e, i, etc.), all of them (except V) are absent in Gordon's growth model.

8.3.1.2 The Weighted Average Cost of Capital, WACC

Below we calculate the WACC value.

$$\begin{aligned} &V = V_0 \\ &\frac{CF}{WACC - g} = \frac{CF}{k_0 - g} \\ &WACC = k_0 \end{aligned} \tag{8.24}$$

It is seen, that the WACC value turns out to be the same as in case of classical Modigliani–Miller theory with constant profit.

8.3.1.3 Equity Cost, k_e

It is easy to show that the expression for equity cost turns out to be the same as in case of classical Modigliani–Miller theory with constant profit

$$k_e = k_0 + (k_0 - k_d)L. \tag{8.25}$$

Therefore, as we show in the case of the Modigliani–Miller theory without taxes accounting of variable profit changes the first of three Modigliani–Miller theorems only. As we show below, in the case of the Modigliani–Miller theory with taxes (practically interesting case) all three Modigliani–Miller theorems (concerning value of the company, V, its WACC and its equity cost, k_e) change significantly.

8.3.2 *Modigliani–Miller Theory with Taxes*

8.3.2.1 The Weighted Average Cost of Capital, WACC

The following result is obtained by Modigliani and Miller (1963):

The value of leveraged company is equal to the value of the company of the same risk group used no leverage, increased by the value of tax shield arising from financial leverage, and equal to the product of rate of corporate income tax T and the value of debt D.

$$V = V_0 + DT \tag{8.26}$$

From here it is easy to get expression for WACC (see above).

$$V(1 - w_d T) = V_0 \tag{8.27}$$

$$\frac{CF}{WACC}(1 - w_d T) = \frac{CF}{k_0} \tag{8.28}$$

$$WACC = k_0 \cdot (1 - w_d t) \tag{8.29}$$

Under variable profit, as we have seen above in the case of Modigliani–Miller theory without taxes, one has the same statement for capitalization, but discount rates are different:

$$\begin{aligned} WACC &\rightarrow WACC - g \\ k_0 &\rightarrow k_0 - g \end{aligned} \tag{8.30}$$

$$\frac{CF}{WACC - g}(1 - w_d T) = \frac{CF}{k_0 - g} \tag{8.31}$$

$$WACC - g = (k_0 - g) \cdot (1 - w_d T) \tag{8.32}$$

$$WACC = (k_0 - g) \cdot (1 - w_d T) + g \tag{8.33}$$

This expression for WACC is different from the original result for WACC by Modigliani and Miller (8.29):

$$WACC = k_0 \cdot (1 - w_d T)$$

In order to understand the character of dependence of WACC (L) let us calculate the derivative of WACC with respect to L using Eq. (8.33):

$$WACC = (k_0 - g) \cdot (1 - w_d T) + g = k_0 - (k_0 - g) \cdot T \cdot \frac{L}{1 + L} \tag{8.34}$$

$$WACC'_L = -\frac{(k_0 - g) \cdot T}{(1 + L)^2} \tag{8.35}$$

At $g < k_0$ the derivative $WACC'_L$ is negative, thus $WACC\ (L)$ decreases with L, while at the $g > k_0$ derivative $WACC'_L$ is positive and $WACC\ (L)$ increases with L. We will see these effects in part (8.4) under numerical calculations.

8.3.2.2 The Company Value, V

For the company value from the Formulas (8.30) and (8.32) we get the following formula

$$V = \frac{CF}{WACC - g} = \frac{CF}{(k_0 - g) \cdot (1 - w_d t)} \tag{8.36}$$

We should note that in case of growing rate g equals to equity cost (at L = 0) k_0, the company value V becomes infinite. This is the limitation of the perpetuity Modigliani–Miller theory. As we will see in our future publications such kind of restriction is absent for companies of finite ages when we consider the generalization of BFO theory for the case of variable profit.

8.3.2.3 Equity Cost, k_e

Let us calculate the equity cost k_e. From (8.33) we have for the weighted average cost of capital, WACC,

$$WACC = (k_0 - g) \cdot (1 - w_d T) + g \tag{8.37}$$

As well according to the definition of the weighted average cost of capital, considering the "tax shield" we have

$$WACC = k_e w_e + k_d w_d (1 - T) \tag{8.38}$$

Equating (8.37) and (8.38), we obtain

$$(k_0 - g)\cdot(1 - w_d T) + g = k_e w_e + k_d w_d (1 - T) \tag{8.39}$$

whence we get the following expression for the cost of equity:

$$\begin{aligned} k_e &= WACC(1 + L) - L k_d (1 - T) = \\ & k_0 + L\cdot[(k_0 - k_d)\cdot(1 - T) + gT] \end{aligned} \tag{8.40}$$

This expression is different from Formula (8.12) in case of the classical Modigliani–Miller theory with tax. Thus, we have the following modified statement for equity cost within Generalized Modigliani–Miller theory.

Equity cost of leverage company k_e could be found as equity cost of unleveraged company k_0 of the same group of risk, plus premium for risk, the value which is equal to production of leverage level L on sum of production of difference $(k_0 - k_d)$on tax corrector and gT:

$$k_e = k_0 + L[(k_0 - k_d)(1 - T) + gT] \tag{8.41}$$

The Qualitatively New Effect in Corporate Finance

From the Formula (8.41) it is seen that with decrease of g tilt angle of straight $k_e(L)$ decreases and at some value of g = g*

$$k_e = k_0 \tag{8.42}$$

Let us find g* value. Equating $k_e$tok_0 in Formula (8.41), one has

$$L[(k_0 - k_d)(1 - T) + gT] = 0$$

and from here we find g*

$$g^* = -\frac{(k_0 - k_d)(1 - T)}{T} \tag{8.43}$$

It is clear that at with growth rate g < g* the slope of the straight $k_e(L)$ turns out to be negative.

Thus, the qualitatively new effect in corporate finance has been discovered: at growth rate g < g* the slope of the straight $k_e(L)$ turns out to be negative. This effect, which is absent in the classical Modigliani–Miller theory, could significantly alter the principles of the company's dividend policy. Rating methodology takes into account the company's dividend policy, thus the results obtained could change the credit rating of an issuer.

We have shown in Sect. 8.3, that in case of the Modigliani–Miller theory with taxes, generalized for the case of variable profit all three Modigliani–Miller statements (concerning value of the company, V, the weighted average cost of capital, WACC, and its equity cost, k_e) change significantly. The consequences of such changes will be investigated numerically in next part.

8.4 Results and Discussions

In this section we study numerically (within Microsoft Excel) the dependence of the weighted average cost of capital, WACC, discount rate, i, company value, V, and equity cost, k_e, on leverage level L in the Generalized Modigliani–Miller theory (GMM theory) at two values of equity cost k_0 (0.2 and 0.3) and different values of growth rate, g. The obtained results are discussed.

8.4.1 Dependence of WACC on Leverage Level L in Generalized Modigliani–Miller Theory (GMM Theory) at $k_0 = 0.2$ and Different Values of g (0.4; 0.3; 0.2; 0.0; −0.2; −0.3; −0.4)

We study first the dependence of the weighted average cost of capital, WACC, on leverage level L in Generalized Modigliani–Miller theory (GMM theory) at $k_0 = 0.2$ and different values of g (0.4; 0.3; 0.2; 0.0; −0.2; −0.3; −0.4).

The results of Table 8.1 are shown in Fig. 8.2.

Table 8.1 Dependence of WACC on leverage level L in generalized Modigliani–Miller theory (GMM theory) at $k_0 = 0.2$ and $g = 0.4; 0.3; 0.2; 0.0; -0.2; -0.3; -0.4$

			WACC						
L	T	k_0	$g = 0.4$	$g = 0.3$	$g = 0.2$	$g = 0$	$g = -0.2$	$g = -0.3$	$g = -0.4$
0	0.2	0.2	0.20000	0.20000	0.20000	0.20000	0.20000	0.20000	0.20000
1	0.2	0.2	0.22000	0.21000	0.20000	0.18000	0.16000	0.15000	0.14000
2	0.2	0.2	0.22667	0.21333	0.20000	0.17333	0.14667	0.13333	0.12000
3	0.2	0.2	0.23000	0.21500	0.20000	0.17000	0.14000	0.12500	0.11000
4	0.2	0.2	0.23200	0.21600	0.20000	0.16800	0.13600	0.12000	0.10400
5	0.2	0.2	0.23333	0.21667	0.20000	0.16667	0.13333	0.11667	0.10000
6	0.2	0.2	0.23429	0.21714	0.20000	0.16571	0.13143	0.11429	0.09714
7	0.2	0.2	0.23500	0.21750	0.20000	0.16500	0.13000	0.11250	0.09500
8	0.2	0.2	0.23556	0.21778	0.20000	0.16444	0.12889	0.11111	0.09333
9	0.2	0.2	0.23600	0.21800	0.20000	0.16400	0.12800	0.11000	0.09200
10	0.2	0.2	0.23636	0.21818	0.20000	0.16364	0.12727	0.10909	0.09091

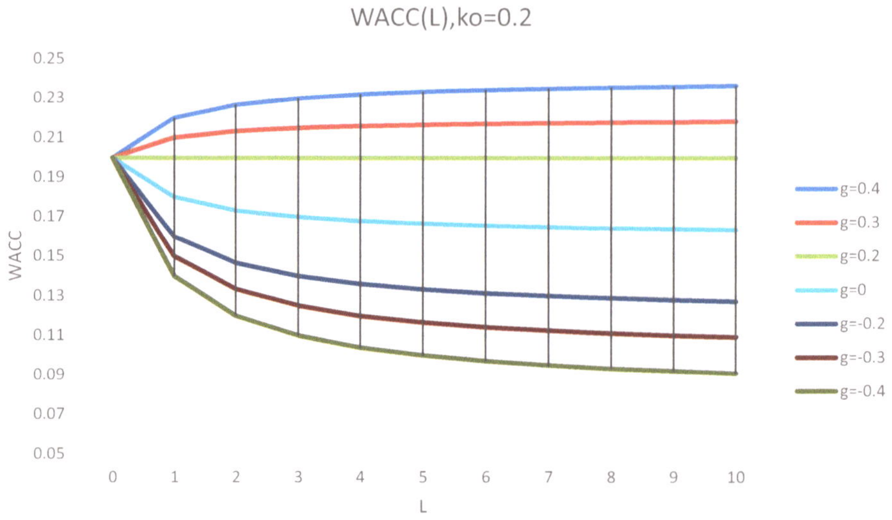

Fig. 8.2 Dependence of WACC on leverage level L in Generalized Modigliani–Miller theory (GMM theory) at $k_0 = 0.2$ and $g = 0; \pm 0.2; \pm 0.3; \pm 0.4$

From Fig. 8.2 it is seen that all curves WACC (L) for different g start from one point k_0, in this case from point (0; 0.2). They decrease with leverage level L at $g < 0.2$ (at $g = 0; \pm 0.2; -0.3; -0.4$) and increase at $g > 0.2$ (at $g = 0.3; 0.4$). The curves WACC(L) increase with growth rate, g. Note, that cut–off value of g, which separate increasing curves WACC(L) from decreasing ones is equal to $k_0 = 0.2$, and WACC is constant at $g = k_0$ and equal to k_0. Below we check this observation at different value of k_0 ($k_0 = 0.3$).

8.4.2 Dependence of the Weighted Average Cost of Capital WACC on Leverage Level L in Generalized Modigliani–Miller Theory (GMM Theory) at $k_0 = 0.3$ and Different Values of g

Let us study the dependence of WACC on leverage level L in Generalized Modigliani–Miller theory (GMM theory) at $k_0 = 0.3$ and different values of g.

The results of Table 8.2 are shown in Fig. 8.3.

From Fig. 8.3 it is seen that all curves WACC (L) for different g start from one point, k_0, in this case from (0; 0.3). They decrease with L at $g < 0.3$ (at $g = 0; \pm 0.2; \pm 0.3; -0.4$) and increase at $g > 0.3$ (at $g = 0.4$). Note, that the cut–off value of g, which separate increasing curves WACC(L) from decreasing ones as in above case is equal to k_0 ($k_0 = 0.3$), and WACC is constant at $g = k_0$ and equal to k_0. Therefore, our observation that this conclusion is valid at different value of k_0 is right.

Table 8.2 Dependence of WACC on leverage level L in generalized Modigliani–Miller theory (GMM theory) at $k_0 = 0.3$ and $g = 0; \pm 0.2; -0.3; -0.4$

L	T	k_0	WACC						
			$g = 0.4$	$g = 0.3$	$g = 0.2$	$g = 0$	$g = -0.2$	$g = -0.3$	$g = -0.4$
0	0.2	0.3	0.300	0.300	0.300	0.300	0.300	0.300	0.300
1	0.2	0.3	0.310	0.300	0.290	0.270	0.250	0.240	0.230
2	0.2	0.3	0.313	0.300	0.287	0.260	0.233	0.220	0.207
3	0.2	0.3	0.315	0.300	0.285	0.255	0.225	0.210	0.195
4	0.2	0.3	0.316	0.300	0.284	0.252	0.220	0.204	0.188
5	0.2	0.3	0.317	0.300	0.283	0.250	0.217	0.200	0.183
6	0.2	0.3	0.317	0.300	0.283	0.249	0.214	0.197	0.180
7	0.2	0.3	0.318	0.300	0.283	0.248	0.213	0.195	0.178
8	0.2	0.3	0.318	0.300	0.282	0.247	0.211	0.193	0.176
9	0.2	0.3	0.318	0.300	0.282	0.246	0.210	0.192	0.174
10	0.2	0.3	0.318	0.300	0.282	0.245	0.209	0.191	0.173

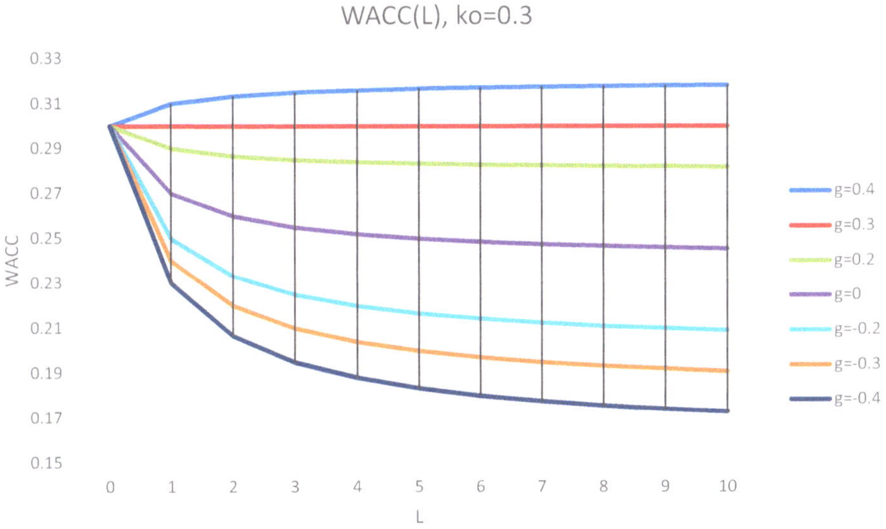

Fig. 8.3 Dependence of WACC on leverage level L in Generalized Modigliani–Miller theory (GMM theory) at $k_0 = 0.3$ and $g = 0; \pm 0.2; \pm 0.3; \pm 0.4$

From Figs. 8.2 and 8.3 it is seen that all curves WACC (L) for different g start from one point, k_0. They decrease with L at $g < k_0$ and increase at $g > k_0$. It turns out that WACC grows with g, while, as we will see below, real discount rates WACC–g and k_0–g decrease with g. This leads to increase of company capitalization with g: V = CF/WACC–g.

8.4.3 Dependence of Discount Rate i on Leverage Level L in Generalized Modigliani–Miller Theory (GMM Theory) at $k_0 = 0.2$ and Different Values of g

Discount rate for the leverage company change from the weighted average cost of capital, WACC, to WACC–g (where g is growth rate), for a financially independent company from k_0 to k_0–g. This means that WACC and k_0 are no longer the discount rates as it takes place in case of the classical Modigliani–Miller theory with constant profit. Below we study the dependence of discount rate i on leverage level L in Generalized Modigliani–Miller theory (GMM theory) at two values of k_0 (0.2; 0.3) and different values of g (−0.4; −0.3; −0.2; 0.0; 0.1; 0.15; 0.2; 0.3; 0.4). Let us start from $k_0 = 0.2$.

The results of Table 8.3 are shown in Fig. 8.4.

From Fig. 8.4 it is seen that discount rate i decreases with leverage level L at growth values $g < k_0$ (at g = 0; 0.1; 0.15; ±0.2; −0.3; −0.4). Discount rate i in contrast to WACC decreases with g: this provides the increase of company value V with g.

At $g > k_0$ discount rate i increases with L, but being negative it is not shown in Fig. 8.4.

8.4.4 Dependence of Discount Rate i on Leverage Level L in Generalized Modigliani–Miller Theory (GMM Theory) at $k_0 = 0.3$ and Different Values of g

Let us study the dependence of discount rate i on leverage level L in Generalized Modigliani–Miller theory (GMM theory) at $k_0 = 0.3$ and different values of g.

The results of Table 8.4 are shown in Fig. 8.5.

From Fig. 8.5 it is seen that discount rate i decreases with leverage level L at growth values at $g < k_0$ (at g = 0; 0.1; 0.15; ±0.2; −0.3; −0.4). Discount rate i in opposite to WACC decreases with g: this provides the increase of company value V with g.

At $g > k_0$ discount rate i increases with L, but being negative it is not shown in Fig. 8.5.

From Figs. 8.4 and 8.5 it is seen that for $k_0 = 0.2$ and 0.3 discount rate i decreases with g: this provides the increase of company value V with g. Discount rate i decreases with L at $g < k_0$ (at g = 0; 0.1; 0.15; ±0.2; −0.3; −0.4). At $g > k_0$ discount rate i increases with L, but being negative it is not shown in Figs. 8.4 and 8.5.

Table 8.3 Dependence of discount rate i on leverage level L in generalized Modigliani–Miller theory (GMM theory) at $k_0 = 0.2$ and $g = -0.4; -0.3; -0.2; 0.0; 0.1; 0.15; 0.2; 0.3; 0.4$

			i								
L	t	k_0	g = 0.4	g = 0.3	g = 0.2	g = 0.15	g = 0.1	g = 0	g = −0.2	g = −0.3	g = −0.4
0	0.2	0.2	−0.200	−0.100	0.000	0.050	0.100	0.200	0.400	0.500	0.600
1	0.2	0.2	−0.180	−0.090	0.000	0.045	0.090	0.180	0.360	0.450	0.540
2	0.2	0.2	−0.173	−0.087	0.000	0.043	0.087	0.173	0.347	0.433	0.520
3	0.2	0.2	−0.170	−0.085	0.000	0.043	0.085	0.170	0.340	0.425	0.510
4	0.2	0.2	−0.168	−0.084	0.000	0.042	0.084	0.168	0.336	0.420	0.504
5	0.2	0.2	−0.167	−0.083	0.000	0.042	0.083	0.167	0.333	0.417	0.500
6	0.2	0.2	−0.166	−0.083	0.000	0.041	0.083	0.166	0.331	0.414	0.497
7	0.2	0.2	−0.165	−0.083	0.000	0.041	0.083	0.165	0.330	0.413	0.495
8	0.2	0.2	−0.164	−0.082	0.000	0.041	0.082	0.164	0.329	0.411	0.493
9	0.2	0.2	−0.164	−0.082	0.000	0.041	0.082	0.164	0.328	0.410	0.492
10	0.2	0.2	−0.164	−0.082	0.000	0.041	0.082	0.164	0.327	0.409	0.491

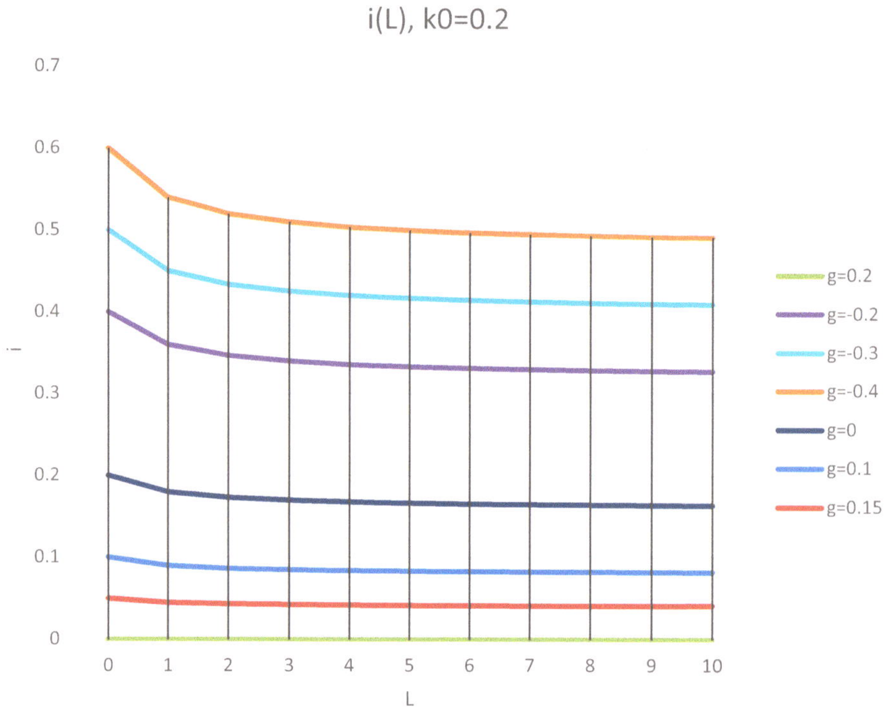

Fig. 8.4 Dependence of discount rate i on leverage level L in Generalized Modigliani–Miller theory (GMM theory) at $k_0 = 0.2$ and $g = 0; 0.1; 0.15; \pm 0.2; -0.3; -0.4$

8.4.5 *Dependence of Company Value V on Leverage Level L in Generalized Modigliani–Miller Theory (GMM Theory) at $k_0 = 0.2$ and Different Values of g*

Let us study the dependence of company value V on leverage level L in Generalized Modigliani–Miller theory (GMM theory) at $k_0 = 0.2$ and different values of g.

The results of Table 8.5 are shown in Fig. 8.6.

From Fig. 8.6 it is seen, that at $k_0 = 0.2$ and $g = 0; 0.1; 0.15; -0.2; -0.3; -0.4$ the company value V at fixed growth rate g increases with leverage level L in Generalized Modigliani–Miller theory (GMM theory). The company value V also increases with growth rate g.

Table 8.4 Dependence of discount rate i on leverage level L in generalized Modigliani–Miller theory (GMM theory) at $k_0 = 0.3$ and $g = 0; 0.1; 0.15; \pm 0.2; \pm 0.3; \pm 0.4$

			i								
L	T	k_0	g = 0.4	g = 0.3	g = 0.2	g = 0.15	g = 0.1	g = 0	g = −0.2	g = −0.3	g = −0.4
0	0.2	0.3	−0.100	0.000	0.100	0.150	0.200	0.300	0.500	0.600	0.700
1	0.2	0.3	−0.090	0.000	0.090	0.135	0.180	0.270	0.450	0.540	0.630
2	0.2	0.3	−0.087	0.000	0.087	0.130	0.173	0.260	0.433	0.520	0.607
3	0.2	0.3	−0.085	0.000	0.085	0.128	0.170	0.255	0.425	0.510	0.595
4	0.2	0.3	−0.084	0.000	0.084	0.126	0.168	0.252	0.420	0.504	0.588
5	0.2	0.3	−0.083	0.000	0.083	0.125	0.167	0.250	0.417	0.500	0.583
6	0.2	0.3	−0.083	0.000	0.083	0.124	0.166	0.249	0.414	0.497	0.580
7	0.2	0.3	−0.083	0.000	0.083	0.124	0.165	0.248	0.413	0.495	0.578
8	0.2	0.3	−0.082	0.000	0.082	0.123	0.164	0.247	0.411	0.493	0.576
9	0.2	0.3	−0.082	0.000	0.082	0.123	0.164	0.246	0.410	0.492	0.574
10	0.2	0.3	−0.082	0.000	0.082	0.123	0.164	0.245	0.409	0.491	0.573

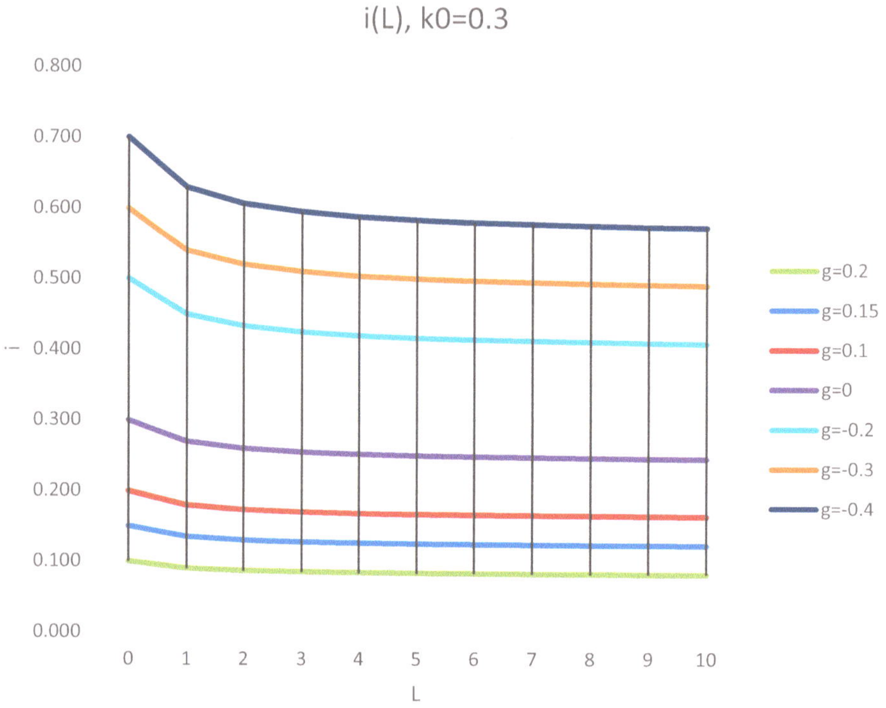

Fig. 8.5 Dependence of discount rate i on leverage level L in Generalized Modigliani–Miller theory (GMM theory) at $k_0 = 0.3$ and $g = 0; 0.1; 0.15; \pm 0.2; -0.3; -0.4$

8.4.6 *Dependence of Company Value V on Leverage Level L in Generalized Modigliani–Miller Theory (GMM Theory) at $k_0 = 0.3$ and Different Values of g*

Let us study the dependence of company value V on leverage level L in Generalized Modigliani–Miller theory (GMM theory) at $k_0 = 0.3$ and different values of g.

The results of Table 8.6 are shown in Fig. 8.7.

From Fig. 8.7 it is seen, that at $k_0 = 0.3$ and $g = 0; 0.1; 0.15; -0.2; -0.3; -0.4$ the company value V at fixed growth rate g increases with leverage level L in Generalized Modigliani–Miller theory (GMM theory). The company value V as well increases with growth rate g.

From Figs. 8.6 and 8.7 it is seen, that at two values of k_0 (0.2; 0.3) and $g = 0; 0.1; 0.15; -0.2; -0.3; -0.4$ the company value V at fixed growth rate g increases with leverage level L in Generalized Modigliani–Miller theory (GMM theory). The company value V as well increases with growth rate g.

We should note that in case of growing rate g equals to equity cost (at $L = 0$) k_0, the company value V becomes infinite. This is the limitation of the perpetuity Modigliani–Miller theory. As one can see in our future publications where we will

Table 8.5 Dependence of company value V on leverage level L in generalized Modigliani–Miller theory (GMM theory) at $k_0 = 0.2$ and different values of $g = 0.0; 0.1; 0.15; -0.2; -0.3; -0.4$

					V					
L	t	k_0	k_d	CF	$g = 0.1$	$g = 0.15$	$g = 0$	$g = -0.2$	$g = -0.3$	$g = -0.4$
0	0.2	0.2	0.16	100	1000	2000	500	250	200	166.6667
1	0.2	0.2	0.16	100	118.111	2222.222	555.5556	277.7778	222.2222	185.1852
2	0.2	0.2	0.16	100	1153.846	2307.692	576.9231	288.4615	230.7692	192.3077
3	0.2	0.2	0.16	100	1176.471	2352.941	588.2353	294.1176	235.2941	196.0784
4	0.2	0.2	0.16	100	1190.476	2380.952	595.2381	297.619	238.0952	198.4127
5	0.2	0.2	0.16	100	1200	2400	600	300	240	200
6	0.2	0.2	0.16	100	1206.897	2413.793	603.4483	301.7241	241.3793	201.1494
7	0.2	0.2	0.16	100	1212.121	2424.242	606.0606	303.0303	242.4242	202.0202
8	0.2	0.2	0.16	100	1216.216	2432.432	608.1081	304.0541	243.2432	202.7027
9	0.2	0.2	0.16	100	1219.512	2439.024	609.7561	304.878	243.9024	203.252
10	0.2	0.2	0.16	100	1222.222	2444.444	68.1111	305.5556	244.4444	203.7037

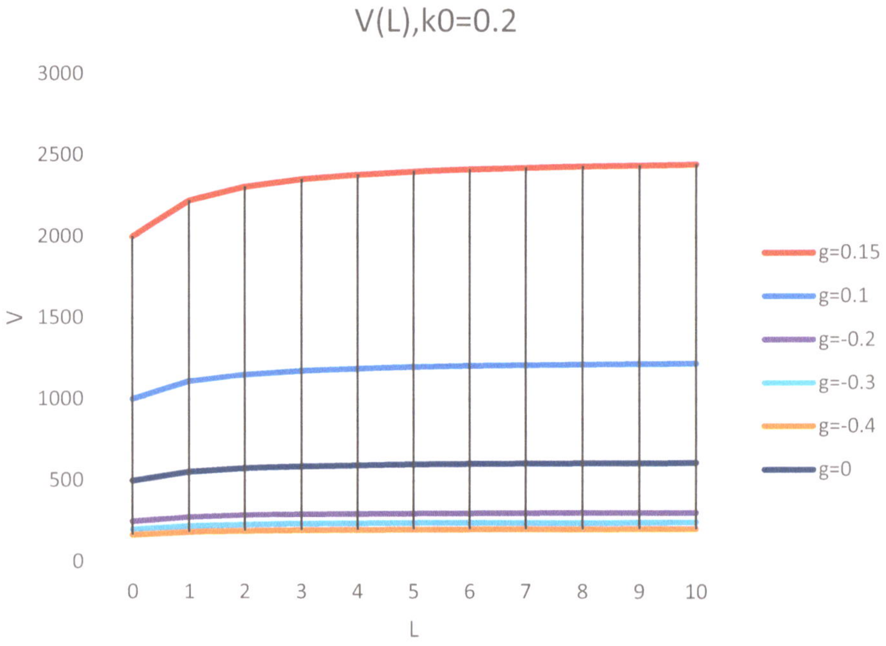

Fig. 8.6 Dependence of company value V on leverage level L in Generalized Modigliani–Miller theory (GMM theory) at $k_0 = 0.2$ and $g = 0; 0.1; 0.15; -0.2; -0.3; -0.4$

consider the generalization of BFO theory for the case of variable profit such kind of restriction is absent for companies of finite ages.

8.4.7 Dependence of Equity Cost k_e on Leverage Level L in Generalized Modigliani–Miller Theory (GMM Theory) at $k_0 = 0.2$ and Different Values of g (0; ±0.2; ±0.3; ±0.4)

The economically justified value of dividends is equal to equity cost. Thus, knowing the dependence of equity cost of company on leverage level L, on growth rate g is very important, because it could impact the dividend policy of the company and help to company management to develop reasonable dividend policy.

Below we study the dependence of equity cost k_e on leverage level L in Generalized Modigliani–Miller theory (GMM theory) at $k_0 = 0.2$ and different values of g (0; ±0.2; ±0.3; ±0.4).

The results of Table 8.7 are shown in Fig. 8.8.

We have investigated the dependence of equity cost k_e on leverage level L in the Generalized Modigliani–Miller theory (GMM theory) at $k_0 = 0.2$ and $g = 0; \pm 0.2;$

Table 8.6 Dependence of company value V on leverage level L in generalized Modigliani–Miller theory (GMM theory) at $k_0 = 0.3$ and $g = 0.25$

L	t	k_0	k_d	CF	V						
					$g = 0.1$	$g = 0.15$	$g = 0.25$	$g = 0$	$g = -0.2$	$g = -0.3$	$g = -0.4$
0	0.2	0.3	0.16	100	500	666.6667	2000	333.3333	200	166.6667	142.8571
1	0.2	0.3	0.16	100	555.5556	740.7407	2222.222	370.3704	222.2222	185.1852	158.7302
2	0.2	0.3	0.16	100	576.9231	769.2308	2307.692	384.6154	230.7692	192.3077	164.8352
3	0.2	0.3	0.16	100	588.2353	784.3137	2352.941	392.1569	235.2941	196.0784	168.0672
4	0.2	0.3	0.16	100	595.2381	793.6508	2380.952	396.8254	238.0952	198.4127	170.068
5	0.2	0.3	0.16	100	600	800	2400	400	240	200	171.4286
6	0.2	0.3	0.16	100	603.4483	804.5977	2413.793	402.2989	241.3793	201.1494	172.4138
7	0.2	0.3	0.16	100	606.0606	808.0808	2424.242	404.0404	242.4242	202.0202	173.1602
8	0.2	0.3	0.16	100	608.1081	810.8108	2432.432	405.4054	243.2432	202.7027	173.7452
9	0.2	0.3	0.16	100	609.7561	813.0081	2439.024	406.5041	243.9024	203.252	174.216
10	0.2	0.3	0.16	100	68.1111	814.8148	2444.444	407.4074	244.4444	203.7037	174.6032

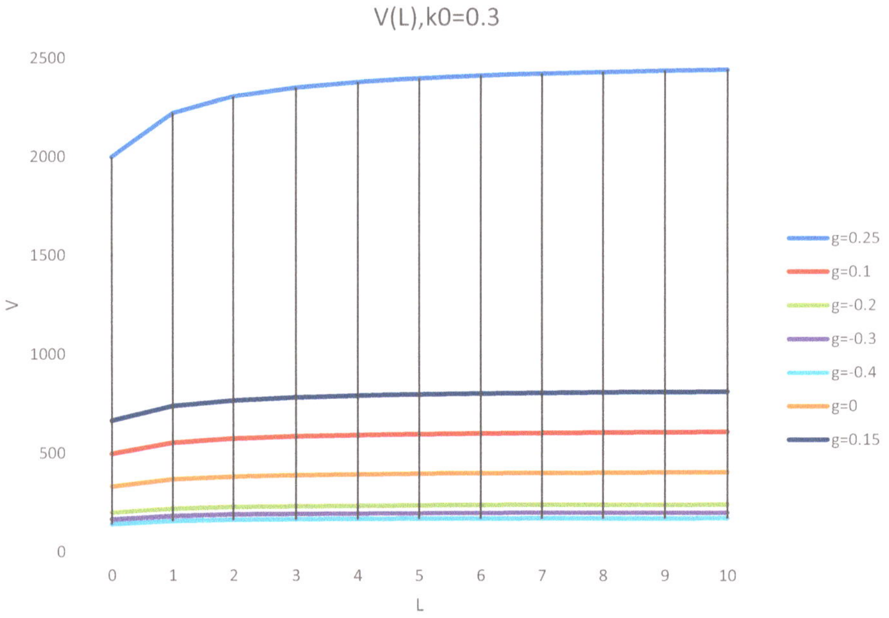

Fig. 8.7 Dependence of company value V on leverage level L in Generalized Modigliani–Miller theory (GMM theory) at $k_0 = 0.3$ and $g = 0; 0.1; 0.15; 0.25; -0.2; -0.3; -0.4$

Table 8.7 Dependence of equity cost k_e on leverage level L in generalized Modigliani–Miller theory (GMM theory) at $k_0 = 0.2$ and different values of g (0; ±0.2; ±0.3; ±0.4)

L	T	k_0	k_d	k_e						
				$g = 0.4$	$g = 0.3$	$g = 0.2$	$g = 0$	$g = -0.2$	$g = -0.3$	$g = -0.4$
0	0.2	0.2	0.16	0.2	0.2	0.2	0.2	0.2	0.2	0.2
1	0.2	0.2	0.16	0.312	0.292	0.272	0.232	0.192	0.172	0.152
2	0.2	0.2	0.16	0.424	0.384	0.344	0.264	0.184	0.144	0.104
3	0.2	0.2	0.16	0.536	0.476	0.416	0.296	0.176	0.116	0.056
4	0.2	0.2	0.16	0.648	0.568	0.488	0.328	0.168	0.088	0.008
5	0.2	0.2	0.16	0.76	0.66	0.560	0.360	0.16	0.060	−0.040
6	0.2	0.2	0.16	0.872	0.752	0.632	0.392	0.152	0.032	−0.088
7	0.2	0.2	0.16	0.984	0.844	0.704	0.424	0.144	0.004	−0.136
8	0.2	0.2	0.16	1.096	0.936	0.776	0.456	0.136	−0.024	−0.184
9	0.2	0.2	0.16	1.208	1.028	0.848	0.488	0.128	−0.052	−0.232
10	0.2	0.2	0.16	1.320	1.120	0.920	0.520	0.120	−0.080	−0.280

±0.3; ±0.4. From Fig. 8.8 it is seen that the equity cost, ke, which linearly grows with leverage level L increases with g: the tilt angle k_e (L) grows with g. It is interesting, that at $k_0 = 0.2$; $k_d = 0.16$ and at $g^* = -0.16$ in accordance with formula

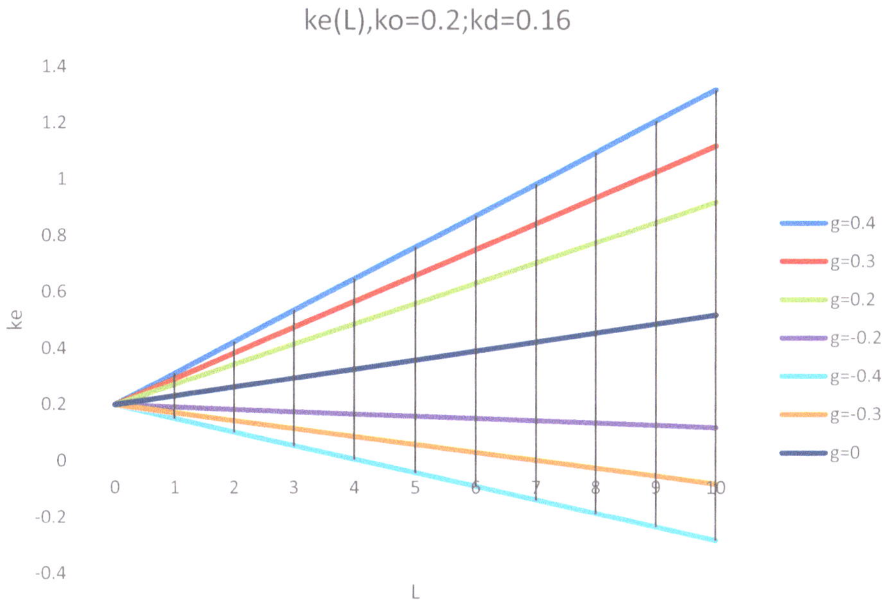

Fig. 8.8 Dependence of equity cost k_e on leverage level L in Generalized Modigliani–Miller theory (GMM theory) at $k_0 = 0.2$; $k_d = 0.16$ and $g = 0; \pm 0.2; \pm 0.3; \pm 0.4$

$g^* = -\frac{(k_0 - k_d)\cdot(1-T)}{T}$ the equity cost k_e turns out to be equal to k_0 and does not change with leverage level L.

This should change the dividend policy of the company, because the economically justified value of dividends is equal to equity cost.

8.4.8 *Dependence of Equity Cost k_e on Leverage Level L in Generalized Modigliani–Miller Theory (GMM Theory) at $k_0 = 0.3$ and Different Values of g*

Presented below is the dependence of equity cost k_e on leverage level L in the Generalized Modigliani–Miller theory (GMM theory) at $k_0 = 0.3$ and different values of g (0; ±0.2; ±0.3; ±0.4).

The results of Table 8.8 are shown in Fig. 8.9.

We have investigated the dependence of equity cost k_e on leverage level L in the Generalized Modigliani–Miller theory (GMM theory) at $k_0 = 0.3$ and $g = 0; \pm 0.2; \pm 0.3; \pm 0.4$. From Fig. 8.9 it is seen that the equity cost, k_e, which linearly grows with leverage level increases with g: the tilt angle k_e (L) grows with g. It is interesting, that at $k_0 = 0.3$; $k_d = 0.22$ and at $g^* = -0.32$ in accordance with Formula (8.43) the equity cost k_e turns out to be equal to k_0 and does not change with leverage level L.

Table 8.8 Dependence of equity cost k_e on leverage level L in Generalized Modigliani–Miller theory (GMM theory) at $k_0 = 0.3$ and different g (0; ±0.2; ±0.3; ±0.4)

L	T	k_0	k_d	k_e						
				g = 0.4	g = 0.3	g = 0.2	g = 0	g = −0.2	g = −0.3	g = −0.4
0	0.2	0.3	0.16	0.3	0.3	0.3	0.3	0.3	0.3	0.3
1	0.2	0.3	0.16	0.444	0.424	0.404	0.364	0.324	0.304	0.284
2	0.2	0.3	0.16	0.588	0.548	0.508	0.428	0.348	0.308	0.268
3	0.2	0.3	0.16	0.732	0.672	0.612	0.492	0.372	0.312	0.252
4	0.2	0.3	0.16	0.876	0.796	0.716	0.556	0.396	0.316	0.236
5	0.2	0.3	0.16	1.020	0.920	0.820	0.620	0.420	0.32	0.220
6	0.2	0.3	0.16	1.164	1.044	0.924	0.684	0.444	0.324	0.204
7	0.2	0.3	0.16	1.308	1.168	1.028	0.748	0.468	0.328	0.188
8	0.2	0.3	0.16	1.452	1.292	1.132	0.812	0.492	0.332	0.172
9	0.2	0.3	0.16	1.596	1.416	1.236	0.876	0.516	0.336	0.156
10	0.2	0.3	0.16	1.740	1.540	1.340	0.940	0.540	0.340	0.140

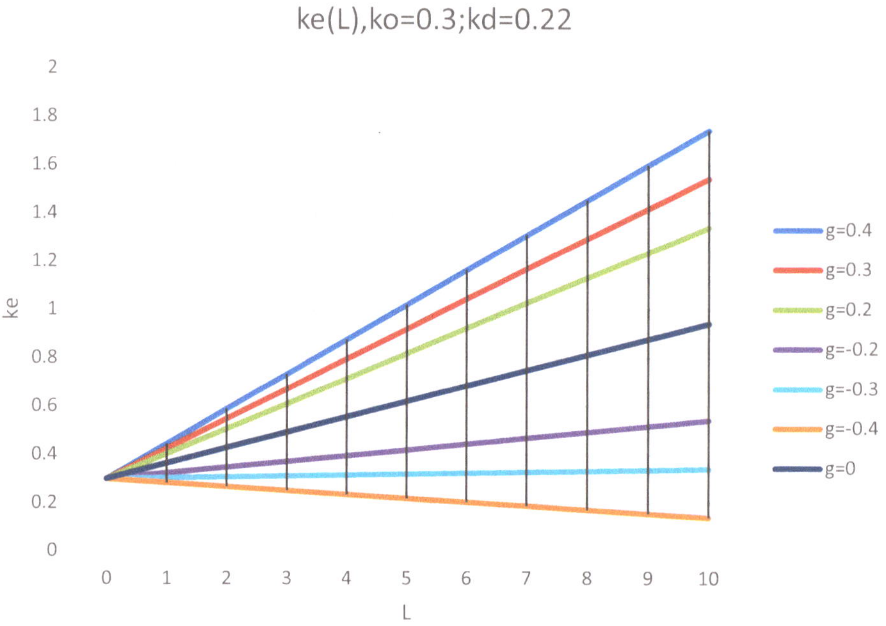

Fig. 8.9 Dependence of equity cost k_e on leverage level L in Generalized Modigliani–Miller theory (GMM theory) at $k_0 = 0.3$; $k_d = 0.22$ and g = 0; ±0.2; ±0.3; ±0.4

Here, we have investigated the dependence of equity cost k_e on leverage level L in Generalized Modigliani–Miller theory (GMM theory) at two values of k_0 (0.2 and 0.3) and different values of g. The equity cost, k_e, which linearly grows with leverage level increases with g: the tilt angle k_e (L) grows with g. This should change the dividend policy of the company, because the economically justified value of dividends is equal to equity cost.

It is interesting, that at $k_0 = 0.2$; $k_d = 0.16$ and at $g^* = -0.16$ and at $k_0 = 0.3$; $k_d = 0.22$ and at $g^* = -0.32$ in accordance with Formula (8.43) the equity cost k_e turns out to be equal to k_0 and does not change with leverage level L.

The qualitatively new effect in corporate finance has been discovered: at rate $g < g^*$ the slope of the curve k_e (L) turns out to be negative. This effect, which is absent in classical Modigliani–Miller theory, could significantly alter the principles of the company's dividend policy.

The last effect is similar to the qualitatively new effect in corporate finance, which has been discovered by Brusov–Filatova–Orekhova within the BFO theory (Brusov et al., 2021b): abnormal dependence of equity cost k_e on leverage level at tax on profit rate T, which exceed some rate value T*: this discovery also significantly alters the principles of the company's dividend policy.

8.5 Conclusions

The main purpose of the current study is to generalize the Modigliani–Miller theory by taking into account one of the most important conditions of companies real functioning: variable profit. We use analytical and numerical methods: we derive all main formulas of the Generalized Modigliani–Miller theory theoretically and then use them to obtain all the main financial indicators of company and their dependences on different parameters by MS Excel. The generalized Modigliani–Miller theory significantly extends its application in practice, especially in business valuation, in ratings and in other areas of economy and finance. We consider the case of growing profit as well as decreasing of profit and show, that all the statements by Modigliani and Miller change significantly.

Within the new Generalized Modigliani–Miller theory (GMM theory), we study the dependence of the weighted average cost of capital, WACC, the cost of equity, k_e, the discount rate *i* and the capitalization of the company, V, on leverage level L at different values of growth rate g and obtained the following results:

1. Discount rate for leverage company changes from the weighted average cost of capital, WACC, to WACC–g (where g is growing rate), for a financially independent company from k_0 to k_0–g. This means that WACC and k_0 are no longer the discount rates as it takes place in case of classical Modigliani–Miller theory with constant profit.
2. All curves WACC (L) for different g start from one point, k_0. They decrease with L at $g < k_0$ and increase at $g > k_0$. It turns out that WACC grows with g, while real discount rates WACC–g and k_0–g decrease with g. This leads to increase of company capitalization with g: V = CF/WACC–g.
3. We have investigated the dependence of equity cost k_e on leverage level L in the Generalized Modigliani–Miller theory (GMM theory) at two values of k_0 (0.2 and

0.3) and different values of g. The equity cost, k_e, which linearly grows with leverage level L increases with g: the tilt angle k_e (L) grows with g. This should change the dividend policy of the company, because the economically justified value of dividends is equal to equity cost.

It is interesting, that at $k_0 = 0.2$; $k_d = 0.16$ and at $g^* = -0.16$ and at $k_0 = 0.3$; $k_d = 0.22$ and at $g^* = -0.32$ in accordance with Formula (8.43) the equity cost k_e turns out to be equal to k_0 and does not change with leverage level L.

A new qualitative finding related to corporate finance has been discovered in this study: at rate $g < g^*$ the slope of the curve k_e (L) turns out to be negative. This effect, which is absent in the classical Modigliani–Miller theory, could significantly alter the principles of the company's dividend policy. Rating methodology should take into account the company's dividend policy (Brusov and Filatova 2021), thus such results obtained could change the credit rating of an issuer.

The last effect found in this study is similar to the qualitatively new effect in corporate finance, discussed above, which has been discovered by Brusov–Filatova–Orekhova within BFO theory: abnormal dependence of equity cost on leverage level when corporate taxes rate T exceeds some rate value T*; this discovery also significantly alters the principles of the company's dividend policy.

Thus, within the new Generalized Modigliani–Miller theory, a lot of significant implications are obtained, allowing us to develop new approach to financial policy and financial strategy of the company.

Note that the limitations of the Generalized Modigliani–Miller theory are similar to the limitations of the classical Modigliani–Miller theory and are connected with the numerous assumptions of classical Modigliani–Miller theory (Brusov et al. 2020a, b, 2021a). Therefore, the direction of future research is clear: we will consider the generalization of BFO theory, which describes the companies of arbitrary age, for the case of variable profit. The limitations of the Generalized Modigliani–Miller theory will be absent in the generalized BFO theory, which is valid for companies of an arbitrary age.

The Generalized Modigliani–Miller theory could and should be used in corporate finance and corporate management, in investments, business valuation, taxation, ratings, etc., for more correct assessments and more qualified management decisions.

Authors planning further development of the Modigliani–Miller theory should account for practical conditions of business and its application in all mentioned above areas. In particular, more frequent payments of tax of profit will be considered, combination of this case with advance payments of tax of profit, modification of BFO theory (which is valid for the companies of arbitrary ages) for the case of variable profit.

References

Berk, J., & DeMarzo, P. (2007). *Corporate finance*. Pearson–Addison Wesley.

Bradley, M., Jarrell, G. A., & Kim, E. H. (1984). On the existence of an optimal capital structure: Theory and evidence. *Journal of Finance, 39*, 857–878.

Brealey, R., Myers, S., & Allen, F. (2005). *Principle of corporate finance* (7th ed.). McGraw Hill, USA.

Brusov et al. (2015). Modern theory of capital cost and capital structure by Brusov–Filatova–Orekhova (BFO Theory) for companies, which ceased to exist at the time moment n. *Journal of Reviews on Global Economics*, *4*, 87–95.

Brusov, P., & Filatova, T. (2021). The Modigliani–Miller theory with arbitrary frequency of payment of tax on profit. *Mathematics, 9*(11), 1198. (WoS Q1).

Brusov, P., Filatova, T., Orehova, N., & Eskindarov, M. (2018). *Modern corporate finance, investments, taxation and ratings* (2nd ed., pp. 1–571). Springer Nature Publishing.

Brusov, P. N., Filatova, T. V., Orekhova, N. P., Kulik, V. L., Chang, S.-I., & Lin, Y. C. G. (2020a). Modification of the Modigliani–Miller theory for the case of advance payments of tax on profit. *Journal of Reviews on Global Economics, 9*, 257–267.

Brusov, P. N., Filatova, T. V., Orekhova, N. P., Kulik, V. L., Chang, S.-I., & Lin, Y. C. G. (2020b). Application of the Modigliani–Miller theory, modified for the case of advance payments of tax on profit, in rating methodologies. *Journal of Reviews on Global Economics, 9*, 282–292.

Brusov, P., Filatova, T., Chang, S.-I., & Lin, G. (2021a). Innovative investment models with frequent payments of tax on income and of interest on debt. *Mathematics, 9*(13), 1491. (WoS Q1).

Brusov, P., Filatova, T., Orekhova, N., Kulik, V., Chang, S.-I., & Lin, G. (2021b). Generalization of the Modigliani–Miller theory for the case of variable profit. *Mathematics, 9*(11), 1286. (WoS Q1).

DeAngelo, H. & Masulis, R. W., (1980). Optimal capital structure under corporate and personal taxation. *Journal of Financial Economics*, 3(1) 3–27.

DeMarzo, P. M. (1988). An extension of the Modigliani–Miller theorem to stochastic economies with incomplete markets and interdependent securities. *Journal of Economic Theory, 45*(2), 353–369.

Farber, A., Gillet, R., & Szafarz, A. (2006). A general formula for the WACC. *International Journal of Business, 11*(2), 211–218.

Fernandez, P. (2006). A general formula for the WACC: A comment. *International Journal of Business, 11*(2), 219.

Filatova, T., Orehova, & N., Brusova, A. (2008). Weighted average cost of capital in the theory of Modigliani–Miller, modified for a finite life–time company. Bull FU, 48, 68–77.

Frank, M., & Goyal, V. (2009). Capital structure decisions: Which factors are reliably important? *Financial Management, 38*(1), 1–37. https://doi.org/10.1111/j.1755-053X.2009.01026.x

Graham, J. R. (2003). Taxes and corporate finance: A review. *Review of Financial Studies, 16*, 1075–1129.

Hamada, R. (1969). Portfolio analysis, market equilibrium, and corporate finance. *Journal of Finance, 24*(1), 13–31.

Harris, R., & Pringle, J. (1985). Risk–adjusted discount rates—Extension form the average–risk case. *The Journal of Financial Research, 8*(3), 237–244.

Miller, M. (1977). Debt and taxes. *The Journal of Finance, 32*(2), 261–275. https://doi.org/10.1111/j.1540-6261.1977.tb03267.x

Modigliani, F., & Miller, M. (1958). The cost of capital, corporate finance, and the theory of investment. *The American Economic Review, 48*, 261–297.

Modigliani, F., & Miller, M. (1963). Corporate income taxes and the cost of capital: A correction. *The American Economic Review, 53*, 147–175.

Modigliani, F., & Miller, M. (1966). Some estimates of the cost of capital to the electric utility industry 1954–1957. *The American Economic Review, 56*, 333–391.

Ross, S., Westerfiel, R., & Jaffee, D. (2005). *Corporate finance* (7th ed.). McGraw Hill, USA.

Sethi, S. P., Derzko, N. A., & Lehoczky, J. P. (1991). A stochastic extension of the Miller–Modigliani framework. *Mathematical Finance, 1*(4), 57–76.

Chapter 9
Modern Theory of Capital Cost and Capital Structure: Brusov–Filatova–Orekhova Theory (BFO Theory)

As we discussed in Chap. 4, one of the serious limitations of the Modigliani–Miller theory is the suggestion about perpetuity of the companies. In 2008, Brusov–Filatova–Orekhova (Filatova et al. 2008) have lifted up this limitation and have shown that the accounting of the finite life–time (or finite age) of the company leads to significant changes of all Modigliani–Miller results (Modigliani and Miller 1958, 1963, 1966): capitalization of the company is changed, as well as the equity cost, k_e, and the weighted average cost of capital, WACC, in the presence of corporative taxes. Besides, a number of qualitatively new effects in corporate finance, obtained in Brusov–Filatova–Orekhova theory, are absent in Modigliani–Miller theory.

Only in the absence of corporate taxes, we give a rigorous proof of the Brusov–Filatova–Orekhova theorem that equity cost, k_e, as well as its weighted average cost, WACC, does not depend on the life–time (or age) of the company, so the Modigliani–Miller theory could be generalized for arbitrary life–time (arbitrary age) companies.

Until recently (before 2008, when the first paper by Brusov–Filatova–Orekhova (Filatova et al. 2008) has appeared), the basic theory (and the first quantitative one) of the cost of capital and capital structure of companies was the theory by Nobel Prize winners Modigliani and Miller (Modigliani and Miller 1958, 1963, 1966). One of the serious limitations of the Modigliani–Miller theory is the suggestion about perpetuity of the companies. We lift up this limitation and show that the accounting of the finite life–time (finite age) of the company leads to change of the equity cost, k_e, as well as of the weighted average cost of capital, WACC, in the presence of corporate taxes. The effect of leverage on the cost of equity capital of the company, k_e, with an arbitrary life–time, and its weighted average cost of WACC is investigated. We give a rigorous proof of the Brusov–Filatova–Orekhova theorem that in the absence of corporate taxes, cost of company equity, k_e, as well as its weighted average cost, WACC, does not depend on the age (life–time) of the company.

P. Brusov, T. Filatova, *Efficiency of Investment Projects*, Contributions to Finance and Accounting, https://doi.org/10.1007/978-3-032-01028-5_9

9.1 Companies of Arbitrary Age and Companies with Arbitrary Life–Time: Brusov–Filatova–Orekhova Equation

Let us consider companies of arbitrary age and companies with arbitrary life–time.

In perpetuity limit (Modigliani–Miller theory) ***time of life of company*** and ***company age*** turn out to be the same: both of them are infinite. When we move from the assumption of Modigliani–Miller concerning the perpetuity of companies these concepts (time of life of company and company age) become different and one should distinct them under generalization of Modigliani–Miller theory with respect to finite n.

Thus we have developed two kinds of finite n—theories: BFO–1 and BFO–2. BFO–1 theory is related to companies with arbitrary age and BFO–2 theory is related to companies with arbitrary life–time companies. By other words, BFO–1 is applicable for the most interesting case of companies that reached the age of n–years and continue to exist on the market, and allows to analyze the financial state of the operating companies. BFO–2 theory allows examine the financial status of the companies which ceased to exist, i.e., of those companies for which n means not age, but a life–time, i.e., the time of existence. A lot of schemes of termination of activities of the company can exist: bankruptcy, merger, acquisition, etc. One of those schemes, when the value of the debt capital D becomes zero at the time of termination of activity of company n, is considered in this chapter below (Sect. 9.5) and comparision of results of BFO–1 and BFO–2 has been done as well.

To start with the case of finite n let us first of all, find the value of tax shield, TS, of the company for n years

$$\mathrm{TS} = k_\mathrm{d} DT \sum_{t=1}^{n} (1 + k_\mathrm{d})^{-t} = DT[1 - (1 + k_\mathrm{d})^{-n}]. \tag{9.1}$$

(We used the formula for the sum of n terms of a geometric progression).

Here, D is the value of debt capital; k_d the cost of debt capital; and T the tax on profit rate.

Next, we use the Modigliani–Miller theorem (Modigliani and Miller 1958, 1963, 1966):

The value of financially dependent company is equal to the value of the company of the same risk group used no leverage, increased by the value of tax shield arising from financial leverage, and equal to the product of rate of corporate income tax T and the value of debt D.

$$V = V_0 + DT. \tag{9.2}$$

This theorem was formulated by Modigliani and Miller for perpetuity companies, but we modify it for a company of arbitrary age.

$$\begin{aligned} V &= V_0 + \mathrm{TS} = V_0 + k_d DT \sum_{t=1}^{\infty} (1 + k_d)^{-t} \\ &= V_0 + w_d VT[1 - (1 + k_d)^{-n}], \end{aligned} \tag{9.3}$$

$$V(1 - w_d VT[1 - (1 + k_d)^{-n}]) = V_0. \tag{9.4}$$

There is a common use of the following two formulas for the cost of the financially independent and financially dependent companies (Modigliani and Miller 1958, 1963, 1966):

$$V_0 = \mathrm{CF}/k_0 \text{ and } V = \mathrm{CF}/\mathrm{WACC}. \tag{9.5}$$

However, these almost always used formulas were derived for perpetuity company, and in case of a company of finite age (or with a finite life–time), they must be modified in the same manner as the value of tax shields (Brusov & Filatova, 2011; Brusov et al., 2011a, b, c, 2012a, b, 2013a, b, 2014a, b; Filatova et al. 2008):

$$V_0 = \mathrm{CF}[1 - (1 + k_0)^{-n}]/k_0; \quad V = \mathrm{CF}[1 - (1 + \mathrm{WACC})^{-n}]/\mathrm{WACC}. \tag{9.6}$$

From formula (Eq. 9.4), we get Brusov–Filatova–Orekhova equation for WACC (Brusov et al., 2011a, b, c, 2012a, b, 2013a, b, 2014a, b; Filatova et al. 2008):

$$\frac{1 - (1 + \mathrm{WACC})^{-n}}{\mathrm{WACC}} = \frac{1 - (1 + k_0)^{-n}}{k_0[1 - \omega_d T(1 - (1 + k_d)^{-n})]}. \tag{9.7}$$

Here, *S*—the value of equity capital of the company, $w_d = \frac{D}{D+S}$—the share of debt capital, $k_e, w_e = \frac{S}{D+S}$—the cost and the share of the equity capital of the company, and $L = D/S$—financial leverage.

At $n = 1$, we get Myers (Myers 2001) formula for 1-year company

$$\mathrm{WACC} = k_0 - \frac{(1 + k_0)k_d}{1 + k_d} w_d T \tag{9.8}$$

For $n = 2$, one has

$$\frac{1 - (1 + \mathrm{WACC})^{-2}}{\mathrm{WACC}} = \frac{1 - (1 + k_0)^{-2}}{k_0\left[1 - \omega_d T\left(1 - (1 + k_d)^{-2}\right)\right]}. \tag{9.9}$$

This equation can be solved for WACC analytically:

$$\mathrm{WACC} = \frac{1 - 2\alpha \pm \sqrt{4\alpha + 1}}{2\alpha}, \tag{9.10}$$

where

$$\alpha = \frac{2 + k_0}{(1 + k_0)^2 \left[1 - \omega_d T \frac{2k_d + k_d^2}{(1 + k_d)^2}\right]}. \tag{9.11}$$

For $n = 3$ and $n = 4$, equation for the WACC becomes more complicated, but it still can be solved analytically, while for $n > 4$, it can be solved only numerically.

We would like to make **an important methodological notice**: taking into account the finite life–time of the company, all formulas, without exception, should be received with use formulas (Eq. 9.6) instead of their perpetuity limits (Eq. 9.5).

Below, we will describe the algorithm for the numerical solution of the equation (Eq. 9.7).

9.1.1 Algorithm for Finding WACC in Case of Companies of Arbitrary Age

Let us return back to *n*–year project (*n*–year company). We have the following equation for WACC in *n*–year case:

$$\frac{1 - (1 + \mathrm{WACC})^{-n}}{\mathrm{WACC}} - A(n) = 0, \tag{9.12}$$

Where,

$$A(n) = \frac{1 - (1 + k_0)^{-n}}{k_0[1 - \omega_d T(1 - (1 + k_d)^{-n})]}. \tag{9.13}$$

The algorithm of the solving of the Eq. (9.12) should be as follows:

1. Putting the values of parameters k_0, ω_d, T and given n, we calculate $A(n)$;
2. We determine two WACC values, for which the left part of the eq. (9.12) has opposite signs. It is obvious that as these two values we can use WACC_1 and WACC_∞, because $\mathrm{WACC}_1 > \mathrm{WACC}_n > \mathrm{WACC}_\infty$ for finite $n \geq 2$.
3. Using, for example, the bisection method, we can solve the Eq. (9.12) numerically.

In MS Excel, it is possible to solve the Eq. (9.7) much easily by using the option "matching of parameter": we will use it through the monograph.

9.2 Comparison of Modigliani–Miller Results (Perpetuity Company) with Myers Results (1-Year Company) and Brusov–Filatova–Orekhova Ones (Company of Arbitrary Age)

Myers (2001) has compared his result for 1-year company (project) (Eq. 9.8) with Modigliani and Miller's result for perpetuity limits (Eq. 9.8). He has used the following values of parameters:

$$k_0 = 8\% \div 24\%; k_d = 7\%; T = 50\%; w_d = 0\% \div 60\%$$

and estimated the difference in the WACC values following from the formulas (Eqs. 9.8). We did make the similar calculations for 2-, 3-,5-, and 10-year project for the same set of parameters, and we have gotten the following results, shown in Tables [Table 9.1 (second line (bulk)), Table 9.2 (second line (bulk)), and Table 9.3)] and corresponding figures (Figs. 9.1, 9.2, and 9.3).

Note that data for equity cost $k_0 = 8\%$ turn out to be a little bit uncertain: this could be related to the fact that this value of equity cost is quite close to value of interest rate of the debt $k_d = 7\%$. For all other values of equity cost, the results are reproducible and very informative and are discussed below.

For a graphic illustration of the results, we use data for $n = 1, 2, \infty$, which adequately reflect the results we have obtained.

9.2.1 Discussion of Results

1. From Table 9.1 and Fig. 9.1, it is obvious that WACC is maximum for 1-year company (project) and decreases with the life–time (age) of the company (project) and reaches the minimum in the Modigliani–Miller perpetuity case. Dependence of all WACC values on debt share w_d turns out to be linear at any equity cost k_0 for all considered durations of the project (life–time values of the companies). It is natural for 1-year project because it is described by Myers linear formula (9.8) as well as, in the Modigliani–Miller perpetuity case, described by the formula (2.8), which is linear too, but it is surprise for 2-year project, where formula for WACC (9.7) is obviously nonlinear.
   The negative slope in WACC increases with the equity cost k_0.
2. As it follows from the Table 9.2 and Fig. 9.3, the dependence of the average ratios $\bar{r} = \langle \Delta_1 / \Delta_2 \rangle$ on debt share w_d is quite weak and can be considered as almost constant. The value of this constant increases practically linear with the equity cost k_0 from 1.22 at $k_0 = 10\%$ up to 9.69 at $k_0 = 24\%$ (see Fig. 9.4).
3. The relative difference between 1-year and 2-year projects increases when the equity cost k_0 decreases. At the same time, the relative difference between 2-year project and perpetuity MM project increases with the equity cost k_0.

Table 9.1 WACC dependence on debt share w_d for different values of equity cost k_0 for companies with different life–time n

k_0	n	$w_d = 10\%$	20%	30%	40%	50%	60%
$k_0 = 8\%$	$n = 1$	7.6	7.3	6.9	6.6	6.2	5.9
	$n = 2$	**7.52**	**7.08**	**6.6**	**6.17**	**5.67**	**5.21**
	$n = \infty$	7.6	7.2	6.8	6.4	6.0	5.6
$k_0 = 10\%$	$n = 1$	9.7	9.3	8.9	8.6	8.2	7.8
	$n = 2$	**9.51**	**9.05**	**8.59**	**8.13**	**7.64**	**7.16**
	$n = \infty$	9.5	9.0	8.5	8.0	7.5	7.0
$k_0 = 12\%$	$n = 1$	11.6	11.3	10.9	10.5	10.2	9.8
	$n = 2$	**11.51**	**11.02**	**10.54**	**10.07**	**9.6**	**9.09**
	$n = 3$	**11.46**	**10.93**	**10.39**	**9.85**	**9.31**	**8.77**
	$n = 5$	**11.42**	**10.83**	**10.25**	**9.66**	**9.06**	**8.46**
	$n = 10$	**11.396**	**10.786**	**10.1695**	**9.5455**	**8.914**	**8.2745**
	$n = \infty$	11.4	10.8	10.2	9.6	9.0	8.4
$k_0 = 16\%$	$n = 1$	15.62	15.2	14.9	14.5	14.1	13.7
	$n = 2$	**15.52**	**14.99**	**14.5**	**13.98**	**13.47**	**12.96**
	$n = 3$	**15.44**	**14.88**	**14.31**	**13.75**	**13.18**	**12.61**
	$n = 5$	**15.38**	**14.76**	**14.14**	**13.51**	**12.88**	**12.24**
	$n = 10$	**15.34**	**14.67**	**13.99**	**13.31**	**12.62**	**11.92**
	$n = \infty$	15.2	14.4	13.6	12.8	12.0	11.2
$k_0 = 20\%$	$n = 1$	19.6	19.2	18.8	18.4	18.1	17.7
	$n = 2$	**19.45**	**18.97**	**18.45**	**17.93**	**17.37**	**16.86**
	$n = 3$	**19.41**	**18.82**	**18.23**	**17.64**	**17.05**	**16.45**
	$n = 5$	**19.35**	**18.69**	**18.03**	**17.36**	**16.70**	**16.03**
	$n = 10$	**19.27**	**18.54**	**17.80**	**17.05**	**16.30**	**15.54**
	$n = \infty$	19.0	18.0	17.0	16.0	15.0	14.0
$k_0 = 24\%$	$n = 1$	23.6	23.2	22.8	22.4	22.0	21.6
	$n = 2$	**23.46**	**22.94**	**22.37**	**21.80**	**21.30**	**20.75**
	$n = 3$	**23.39**	**22.77**	**22.15**	**21.54**	**20.91**	**20.29**
	$n = 5$	**23.31**	**22.61**	**21.91**	**21.21**	**20.51**	**19.80**
	$n = 10$	**23.21**	**22.40**	**21.60**	**20.78**	**19.96**	**19.13**
	$n = \infty$	22.8	21.6	20.4	19.2	18.0	16.8

We can also show below at Table 9.4 and Fig. 9.4 the dependence of the WACC on leverage level L for $n = 1$, $n = 3$, and $n = \infty$.

From Table 9.4 and Fig. 9.5 it is obviously that WACC has a maximum for one year company and decreases with the age (life–time) of the company, reaching the minimum in the Modigliani–Miller perpetuity case. (Note, however, that this not always be so via the effect of "golden age" of the company (Brusov et al. 2018c).

Table 9.2 Dependence of the differences $\Delta_1 = \text{WACC}_1 - \text{WACC}_\infty$ (first line), $\Delta_2 = \text{WACC}_1 - \text{WACC}_2$ [second line (bulk)], and their ratio $r = \Delta_1/\Delta_2$ (third line) on debt share w_d for different values of equity cost k_0

	$w_d = 10\%$	20%	30%	40%	50%	60%
$k_0 = 10\%$	0.20	0.30	0.4	0.60	0.7	0.8
	0.19	**0.25**	**0.31**	**0.47**	**0.56**	**0.64**
	1.05	1.2	1.29	1.28	1.25	1.25
$k_0 = 12\%$	0.2	0.5	0.7	0.9	1.2	1.4
	0.09	**0.28**	**0.36**	**0.43**	**0.6**	**0.71**
	2.22	1.76	1.94	2.09	2	1.97
$k_0 = 16\%$	0.4	0.8	1.3	1.7	2.1	2.5
	0.08	**0.21**	**0.4**	**0.52**	**0.63**	**0.74**
	5.0	3.81	3.25	3.27	3.33	3.38
$k_0 = 20\%$	0.6	1.2	1.8	2.4	3.1	3.7
	0.15	**0.23**	**0.35**	**0.47**	**0.73**	**0.84**
	4.0	5.22	5.14	5.11	4.25	4.4
$k_0 = 24\%$	0.8	1.6	2.4	3.2	4.0	4.8
	0.14	**0.26**	**0.43**	**0.6**	**0.7**	**0.85**
	5.7	6.15	5.58	5.33	5.71	5.65

Table 9.3 Average (by debt share w_d) values of ratios $\bar{r} = \langle \Delta_1/\Delta_2 \rangle$ for $k_0 = 10\%$; 12%; 16%; 20%; and 24%

k_0	10%	12%	16%	20%	24%
$\bar{r} = <\Delta_1/\Delta_2>$	1.22	2.00	3.67	4.69	3.69

9.3 Brusov–Filatova–Orekhova Theorem

9.3.1 Case of Absence of Corporate Taxes

Modigliani–Miller theory in case of absence of corporate taxes gives the following results for dependence of WACC and equity cost k_e on leverage:

1.

$$V_0 = V_L; \quad \text{CF}/k_0 = \text{CF}/\text{WACC}, \quad \text{and thus WACC} = k_0. \tag{9.14}$$

2. $\text{WACC} = w_e \cdot k_e + w_d \cdot k_d$; and thus

$$k_e = \frac{\text{WACC} - w_d \cdot k_d}{w_e} = \frac{k_0 - \dfrac{L}{1+L} k_d}{\dfrac{1}{1+L}} = k_0 + L(k_0 - k_d). \tag{9.15}$$

For the finite life–time (finite age) companies, Modigliani–Miller theorem about equality of value of financially independent and financially dependent companies

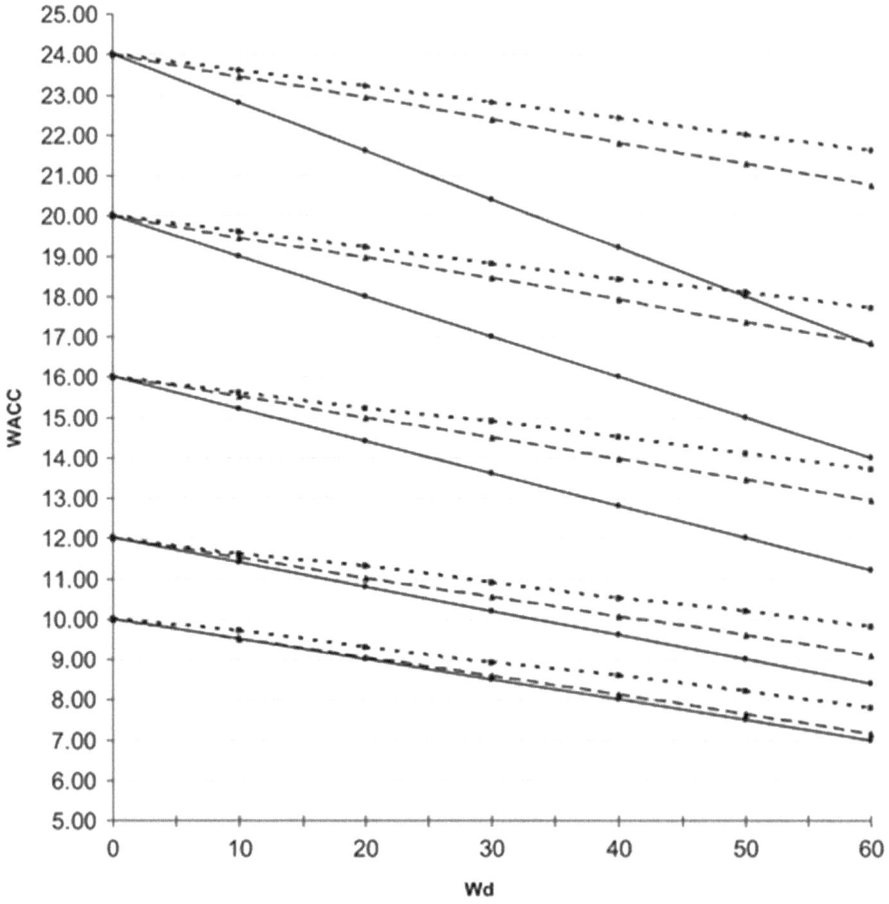

Fig. 9.1 The dependence of the WACC on debt share w_d for companies with different life–time n for different cost of equity, k_0 (from Table 9.1)

($V_0 = V_L$) has the following view (Brusov & Filatova, 2011; Brusov et al., 2011a, b, c, 2012a, b, 2013a, b, 2014a, b):

$$V_0 = V_L;$$
$$\mathrm{CF}\cdot\frac{\left[1-(1+k_0)^{-n}\right]}{k_0} = \mathrm{CF}\cdot\frac{\left[1-(1+\mathrm{WACC})^{-n}\right]}{\mathrm{WACC}}. \tag{9.16}$$

Using this relation, we prove an important Brusov–Filatova–Orekhova theorem:

Under the absence of corporate taxes, the equity cost of the company, k_e, as well as its weighted average cost of capital, WACC, does not depend on the life–time (age) of the company and is equal, respectively, to

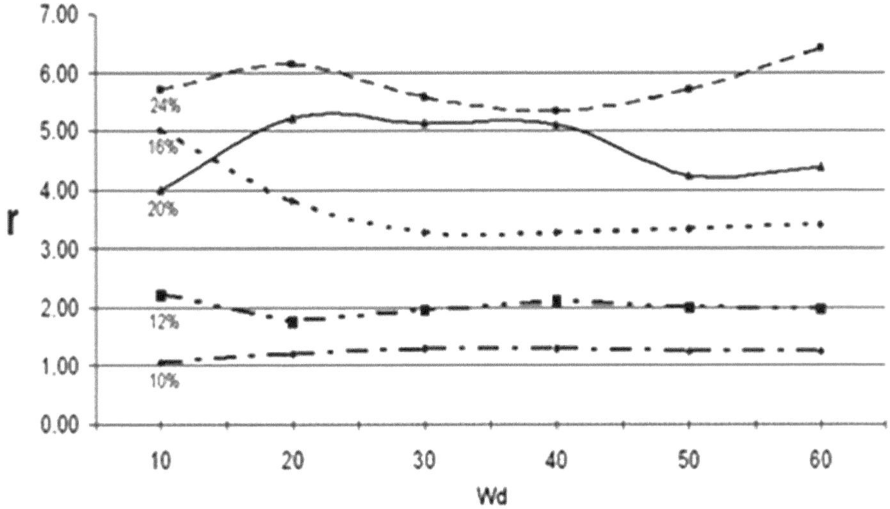

Fig. 9.2 Dependence of the ratio $r = \Delta_1/\Delta_2$ of differences $\Delta_1 = \mathrm{WACC}_1 - \mathrm{WACC}_\infty$ and $\Delta_2 = \mathrm{WACC}_1 - \mathrm{WACC}_2$ on debt share w_d for different values of equity cost k_0 (from Table 9.2)

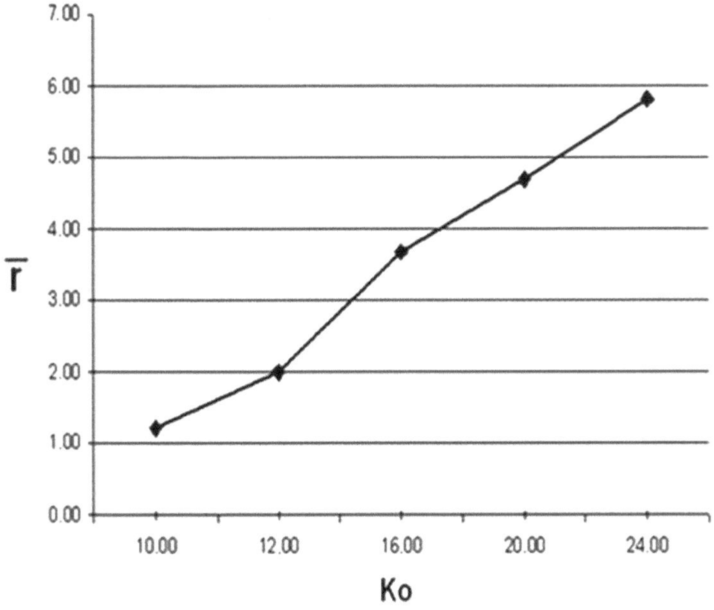

Fig. 9.3 Dependence of the average values of ratio $\bar{r} = \langle \Delta_1/\Delta_2 \rangle$ on the equity cost, k_0

$$k_\mathrm{e} = k_0 + L(k_0 - k_\mathrm{d}); \quad \mathrm{WACC} = k_0. \tag{9.17}$$

Let us consider first the 1- and 2-year companies

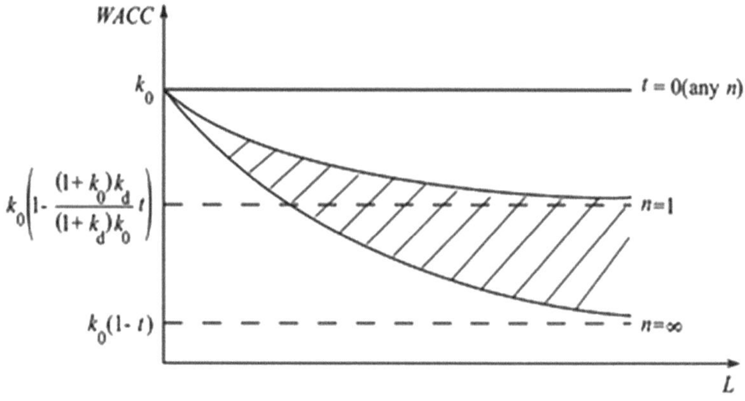

Fig. 9.4 The dependence of the WACC on leverage in the absence of corporate taxes [the horizontal line ($t = 0$)], as well as in the presence of corporate taxes [for 1-year ($n = 1$) and perpetuity companies ($n = \infty$)]. Curves for the WACC of companies with an intermediate life–time (age) ($1 < n < \infty$) lie within the shaded region

Table 9.4 The dependence of the WACC on leverage level L for $n = 1$, $n = 3$, and $n = \infty$

L	k_0	k_d	t	w_d	WACC n = 1	WACC n = 3	WACC (MM)
0	0.2	0.1	0.2	0.00	20.00%	20.00%	20.00%
1	0.2	0.1	0.2	0.50	18.91%	18.41%	18.00%
2	0.2	0.1	0.2	0.67	18.55%	17.87%	17.33%
3	0.2	0.1	0.2	0.75	18.36%	17.61%	17.00%
4	0.2	0.1	0.2	0.80	18.25%	17.44%	16.80%
5	0.2	0.1	0.2	0.83	18.18%	17.34%	16.67%
6	0.2	0.1	0.2	0.86	18.13%	17.26%	16.57%
7	0.2	0.1	0.2	0.88	18.09%	17.20%	16.50%
8	0.2	0.1	0.2	0.89	18.06%	17.16%	16.44%
9	0.2	0.1	0.2	0.90	18.04%	17.12%	16.40%
10	0.2	0.1	0.2	0.91	18.02%	17.09%	16.36%

(a) For 1-year company, one has from (9.15)

$$\frac{1-(1+k_0)^{-1}}{k_0} = \frac{1-(1+\text{WACC})^{-1}}{\text{WACC}}, \tag{9.18}$$

and thus

$$\frac{1}{1+k_0} = \frac{1}{1+\text{WACC}}. \tag{9.19}$$

Hence

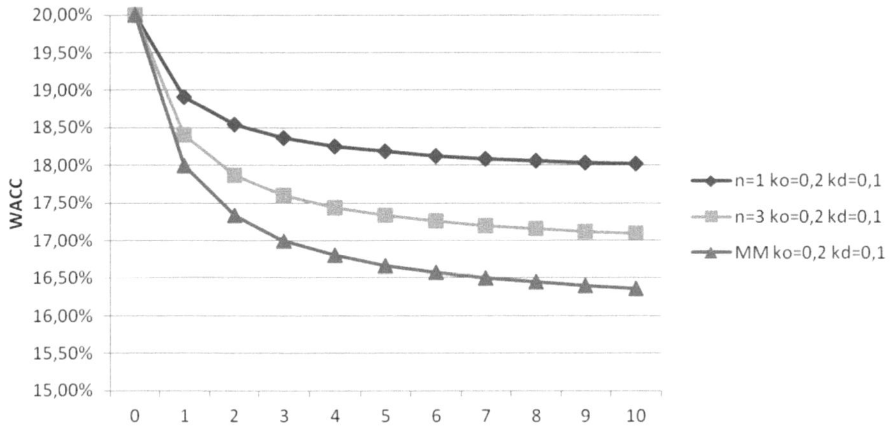

Fig. 9.5 Dependence of WACC on leverage level for $n = 1$, $n = 3$ and $n = \infty$

$$\text{WACC} = k_0. \tag{9.20}$$

(b) Formula for equity cost $k_e = k_0 + L(k_0 - k_d)$ now obtained by substituting $\text{WACC} = k_0$ into (9.14).

For 2-year company, one has from (9.15)

$$\frac{\left[1-(1+k_0)^{-2}\right]}{k_0} = \frac{\left[1-(1+\text{WACC})^{-2}\right]}{\text{WACC}},$$

and thus

$$\frac{2+k_0}{(1+k_0)^2} = \frac{2+\text{WACC}}{(1+\text{WACC})^2}. \tag{9.21}$$

Denoting $\alpha = \frac{2+k_0}{(1+k_0)^2}$, we get the following quadratic equation for WACC:

$$\alpha \cdot \text{WACC}^2 + (2\alpha - 1) \cdot \text{WACC} + (\alpha - 2) = 0. \tag{9.22}$$

It has two solutions

$$\text{WACC}_{1,2} = \frac{1 - 2\alpha \pm \sqrt{4\alpha + 1}}{2\alpha}. \tag{9.23}$$

Substituting $\alpha = \frac{2+k_0}{(1+k_0)^2}$, we get

$$\mathrm{WACC}_{1,2} = \frac{(k_0^2 - 3) \pm (k_0 + 3)(1 + k_0)}{2(2 + k_0)}. \tag{9.24}$$

$$\mathrm{WACC}_1 = k_0; \quad \mathrm{WACC}_2 = -\frac{2k_0 + 3}{k_0 + 2} < 0. \tag{9.25}$$

The second root is negative, but the weighted average cost of capital can only be positive, so only one value remains

$$\mathrm{WACC}_1 = k_0.$$

(c) For company with arbitrary life–time, n, Brusov–Filatova–Orekhova formula (9.15) gives

$$\frac{1 - (1 + k_0)^{-n}}{k_0} = \frac{1 - (1 + \mathrm{WACC})^{-n}}{\mathrm{WACC}}. \tag{9.26}$$

For a fixed k_0, (Eq. 9.25) is an equation of $(n + 1)$–degree relative to WACC. It has $n + 1$ roots (in general complex). One of the roots, as a direct substitution shows, is always WACC = k_0.. Investigation of the remaining roots is difficult and not a part of our problem.

Formula for equity cost $k_e = k_0 + L(k_0 - k_d)$ is now obtained by substituting WACC = k_0 into (Eq. 9.14).

Thus, we have proved the Brusov–Filatova–Orekhova theorem.

9.3.2 Case of the Presence of Corporate Taxes

Modigliani–Miller theory in case of presence of corporate taxes gives the following results for dependence of WACC and equity cost k_e on leverage:

1. WACC

$$V_L = V_0 + Dt; D = w_d V_L; \tag{9.27}$$

$$\mathrm{CF}/\mathrm{WACC} = \mathrm{CF}/k_0 + Dt = \mathrm{CF}/k_0 + w_d t\mathrm{CF}/\mathrm{WACC}, \tag{9.28}$$

$$\frac{1 - w_d t}{\mathrm{WACC}} = \frac{1}{k_0}; \tag{9.29}$$

$$\mathrm{WACC} = k_0(1 - w_d t) = k_0\left(1 - \frac{L}{1 + L}t\right). \tag{9.30}$$

Thus, WACC decreases with leverage from k_0 [in the absence of debt financing ($L = 0$)] up to $k_0(1 - t)$ (at $L = \infty$).

2. The Equity Cost k_e

$$\mathrm{WACC} = k_0(1 - w_d t) = w_e \cdot k_e + w_d \cdot k_d(1 - t);$$

and thus

$$k_e = \frac{\mathrm{WACC} - w_d \cdot k_d \cdot (1 - t)}{w_e} = \frac{k_0(1 - w_d t) - \frac{L}{1 + L} k_d(1 - t)}{\frac{1}{1 + L}} = k_0 + L(k_0 - k_d)(1 - t). \tag{9.31}$$

9.4 From Modigliani–Miller to General Theory of Capital Cost and Capital Structure

Let us consider, how the weighted average cost of capital, WACC, and the cost of equity capital, k_e, will be changed when taking into account the finite age of the company.

(a) 1-year company

From (9.7), one has

$$\frac{1 - (1 + \mathrm{WACC})^{-n}}{\mathrm{WACC}} = \frac{1 - (1 + k_0)^{-n}}{k_0[1 - w_d t(1 - (1 + k_d)^{-n})]}. \tag{9.32}$$

For 1-year company, we get

$$\frac{1 - (1 + \mathrm{WACC})^{-1}}{\mathrm{WACC}} = \frac{1 - (1 + k_0)^{-1}}{k_0\left[1 - w_d t\left(1 - (1 + k_d)^{-1}\right)\right]} \tag{9.33}$$

From (Eq. 9.33), we obtain the well–known Myers formula (Eq. 9.8), which is the particular case of Brusov–Filatova–Orekhova formula (Eq. 9.7).

$$\mathrm{WACC} = k_0 - \frac{1 + k_0}{1 + k_d} k_d w_d t.$$

Thus

$$\mathrm{WACC}=k_0\left(1-\frac{(1+k_0)\cdot k_\mathrm{d}}{(1+k_\mathrm{d})\cdot k_0}\cdot\frac{L}{1+L}t\right). \tag{9.34}$$

Thus, WACC decreases with leverage from k_0 [in the absence of debt financing ($L=0$)] up to $k_0\left(1-\frac{(1+k_0)\cdot k_\mathrm{d}}{(1+k_\mathrm{d})\cdot k_0}t\right)$ (at $L=\infty$)..

Equating the right part of Eq. (9.34) to general expression for WACC

$$WACC=w_\mathrm{e}\cdot k_\mathrm{e}+w_\mathrm{d}\cdot k_\mathrm{d}(1-t), \tag{9.35}$$

one gets

$$k_0-\frac{1+k_0}{1+k_\mathrm{d}}k_\mathrm{d}w_\mathrm{d}t=w_\mathrm{e}\cdot k_\mathrm{e}+w_\mathrm{d}\cdot k_\mathrm{d}(1-t). \tag{9.36}$$

Thus

$$\begin{aligned}k_\mathrm{e}&=\frac{1}{w_\mathrm{e}}\left[k_0-\frac{1+k_0}{1+k_\mathrm{d}}k_\mathrm{d}w_\mathrm{d}t-k_\mathrm{d}w_\mathrm{d}(1-t)\right]\\&=(1+L)k_0-L\frac{k_\mathrm{d}}{1+k_\mathrm{d}}\left[(1+k_0)t+(1+k_\mathrm{d})(1-t)\right].\\&=k_0+L(k_0-k_\mathrm{d})\left(1-\frac{k_\mathrm{d}}{1+k_\mathrm{d}}t\right)\end{aligned}$$

$$k_\mathrm{e}=k_0+L(k_0-k_\mathrm{d})\left(1-\frac{k_\mathrm{d}}{1+k_\mathrm{d}}t\right). \tag{9.37}$$

So we see that in case of 1-year company, the perpetuity limit
$k_\mathrm{e}=k_0+L(k_0-k_\mathrm{d})(1-t)$ is replaced by (Eq. 9.37).

Difference is due to different values of the tax shield for a one–year company and perpetuity one (Figs. 9.5 and 9.6).

Let us investigate the question of the tax shield value for companies with different life–time (age) in more detail.

9.4.1 Tax Shield

General expression for the tax shield for n-year company has the form (Brusov–Filatova–Orekhova)

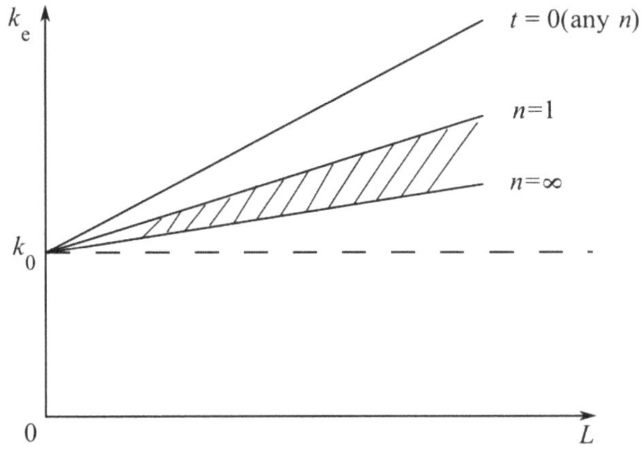

Fig. 9.6 Dependence of the equity cost, k_e, on leverage in the absence of corporate taxes [the upper line ($t = 0$)], as well as in the presence of corporate taxes [for 1-year ($n = 1$) and perpetuity companies ($n = \infty$)]. Dependences of the cost of equity capital of companies, k_e, of an intermediate age ($1 < n < \infty$) lie within the shaded region

$$\mathrm{TS} = \sum_{i=1}^{n} \frac{k_d Dt}{(1+k_d)^i} = \frac{k_d Dt[1-(1+k_d)^{-n}]}{(1+k_d)\left(1-(1+k_d)^{-1}\right)} = Dt[1-(1+k_d)^{-n}]. \quad (9.38)$$

1. In perpetuity limit ($n \to \infty$), tax shield is equal to $\mathrm{TS}_\infty = Dt$, which leads to the so–called effect of the tax shield associated with the appearance of a factor (tax corrector) $(1 - t)$ in the equity cost $k_e = k_0 + L(k_0 - k_d)(1 - t)$.
2. For the 1-year company, tax shield value is equal to

$$\mathrm{TS}_1 = Dt\left(1-(1+k_d)^{-1}\right) = Dtk_d/(1+k_d). \quad (9.39)$$

This leads to appearance of a factor $\left(1 - \frac{k_d}{1+k_d}t\right)$ in the equity cost (Eq. 9.36) $k_e = k_0 + L(k_0 - k_d)\left(1 - \frac{k_d}{1+k_d}t\right)$.

3. Tax shield for a 2-year company is equal to

$$\mathrm{TS}_2 = Dt\left(1-(1+k_d)^{-2}\right) = Dtk_d(2+k_d)/(1+k_d)^2 \quad (9.40)$$

and if the analogy with 1-year company will keep, then factor $(1 - t)$ in the Modigliani–Miller theory would be replaced by the factor

$$\left(1 - \frac{k_d(2+k_d)}{(1+k_d)^2}t\right). \quad (9.41)$$

However, due to a nonlinear relation between WACC and k_0 and k_d in Brusov–Filatova–Orekhova formula (Eqs. 9.9, 9.10, and 9.11) for 2-year company (and companies of bigger age), such a simple analogy is no longer observed, and the calculations become more complex.

9.5 BFO Theory in the Case, When the Company Ceased to Exist at the Time Moment *n* (BFO–2 Theory)

From the output of the BFO formula it follows that developed ideology is applied to companies which have reached the age of n-years and continue to exist on the market, while the theory of MM is only applicable to an infinitely old (perpetuity) companies. By other words, BFO is applicable for most interesting case of companies that reached the age of n-years and continue to exist on the market, and allows to analyze the financial condition of the operating companies.

However, the BFO theory allows also to examine the financial status of the companies which ceased to exist, i.e., of those for which n means not age, but a life–time, i.e., the time of existence. A lot of schemes of termination of activities of the company can exist: bankruptcy, merger, acquisition, etc. Below we consider one of those schemes, when the value of the debt capital D becomes zero at the time of termination of activity of company n: in this case the BFO theory requires minimal upgrades, showed below.

From the formula for the capitalization of the company (9.1), it is easy to get an estimation for the "residual capitalization" of the company, discounted to the time moment k:

$$V_k = \sum_{t=k+1}^{n} \frac{\text{CF}}{(1+\text{WACC})^t} = \frac{\text{CF}}{\text{WACC}}\left[1-(1+\text{WACC})^{-(n-k)}\right]. \tag{9.42}$$

Using the formula

$$V_k = w_d D, \tag{9.43}$$

we obtain an expression for the tax shield for n years subject to the termination of the activities of the company at the moment n:

$$\begin{aligned} TS_n &= tk_d w_d \sum_{k=1}^{n} \frac{V_{k-1}}{(1+k_d)^k} = \frac{tk_d w_d CF}{\text{WACC}} \sum_{k=1}^{n} \frac{1-(1+\text{WACC})^{-(n-k+1)}}{(1+k_d)^k} = \\ &= \frac{tk_d w_d}{\text{WACC}}\left[\frac{1-(1+k_d)^{-n}}{k_d} - \frac{(1+k_d)^{-n}-(1+\text{WACC})^{-n}}{\text{WACC}-k_d}\right] \end{aligned} . \tag{9.44}$$

Substituting this expression into Eq. (9.3)

$$V_L = V_0 + (TS)_n$$

one gets the equation (let us call it BFO–2)

$$\frac{1-(1+\mathrm{WACC})^{-n}}{\mathrm{WACC}}=\frac{1-(1+k_0)^{-n}}{k_0}+\frac{tk_\mathrm{d}w_\mathrm{d}}{\mathrm{WACC}}\times\left[\frac{1-(1+k_\mathrm{d})^{-n}}{k_\mathrm{d}}-\frac{(1+k_\mathrm{d})^{-n}-(1+\mathrm{WACC})^{-n}}{\mathrm{WACC}-k_\mathrm{d}}\right], \tag{9.45}$$

from which one can find the WACC for companies with arbitrary life–time n, provided that the company ceases to function at the time moment n.

Below in the monograph, we investigate the companies that have reached the age of n–years and continue to exist on the market, i.e., we will use formula BFO (9.7), but in this paragraph we present some results obtained from the formula BFO–2 (9.45).

9.5.1 Application of Formula BFO–2

Formula BFO–2 (9.45) in MS Excel takes a following form:

$$\begin{aligned}&\left(\left(1-(1+C4)^{(-H4)}\right)/C4\right)-\left(\left(\left(1-(1+D4)^{(-H4)}\right)/D4\right)\right.\\&+((G4^{*}E4^{*}F4)/C4)*\left(\left(1-(1+E4)^{(-H4)}\right)/E4\right)\\&-\left.\left(\left((1+E4)^{(-H4)}-(1+C4)^{(-H4)}\right)/(C4-E4)\right)\right)\\&=0.\end{aligned} \tag{9.46}$$

Using it we get the following results for dependence of WACC on leverage level L, life–time n, and on tax on profit rate t (Figs. 9.7, 9.8, and 9.9).

9.5.2 Comparison of Results Obtained from Formulas BFO and BFO–2

Let us compare results obtained from formulas BFO and BFO–2 (Figs. 9.10, 9.11, 9.12, and 9.13).

Comparison of results obtained from formulas BFO and BFO–2 shows that WACC values (at the same values of other parameters) turn out to be higher for the companies which ceased to exist at the time moment n, than for companies which have reached the age of n-years and continue to exist on the market. By other words, the companies which ceased to exist at the time moment n can attract a capital at higher rate, than for companies which have reached the age of n-years and continue to exist on the market.

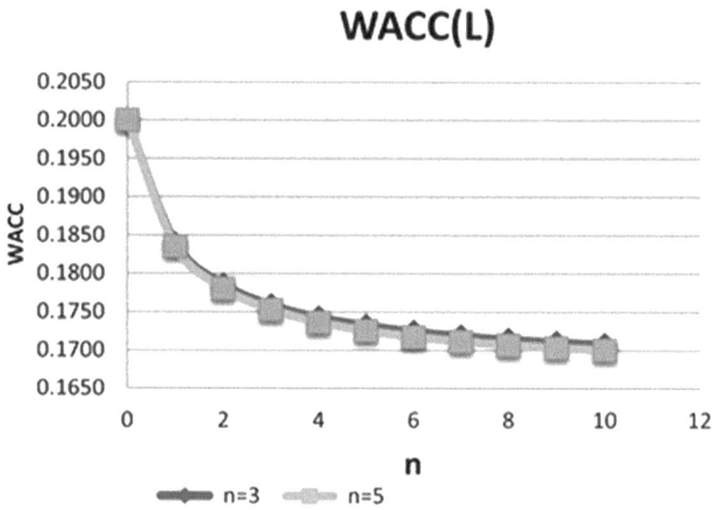

Fig. 9.7 The dependence of the WACC on leverage level L for $n = 3$ and $n = 5$; $k_0 = 0.2$; $k_d = 0.15$

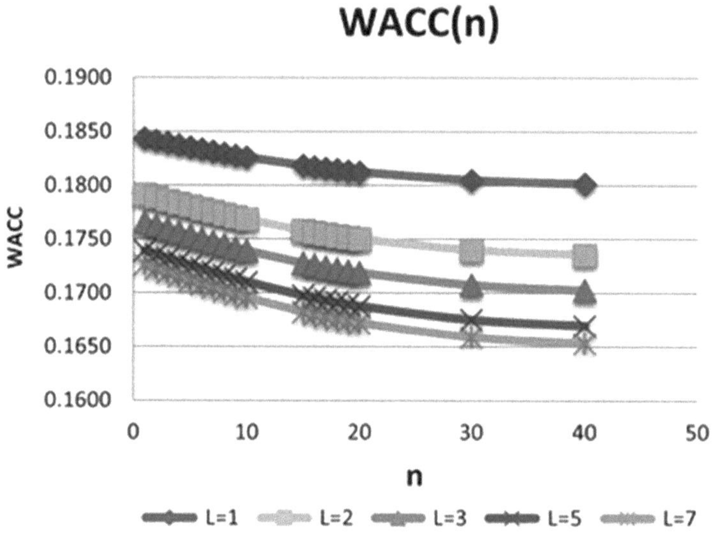

Fig. 9.8 The dependence of the WACC on life–time n at different leverage level L

We will develop the detailed investigation of the companies which ceased to exist at the time moment n (described by formula BFO–2) somewhere also and in this monograph we will limit ourselves by consideration of the companies which have reached the age of n-years and continue to exist on the market (described by formula BFO).

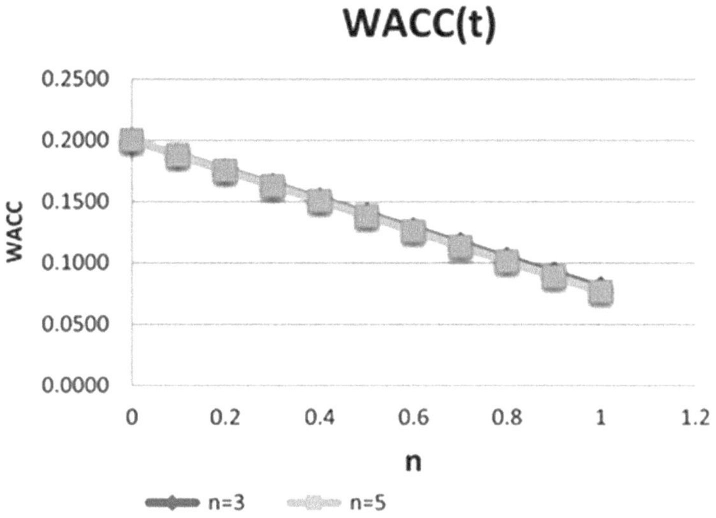

Fig. 9.9 The dependence of the WACC on tax on profit rate t for $n = 3$ and $n = 5$; $k_0 = 0.2$; $k_d = 0.15$

Fig. 9.10 Comparison of the dependence of the WACC on leverage level L for $n = 3$ and $n = 5$ from formulas BFO and BFO–2

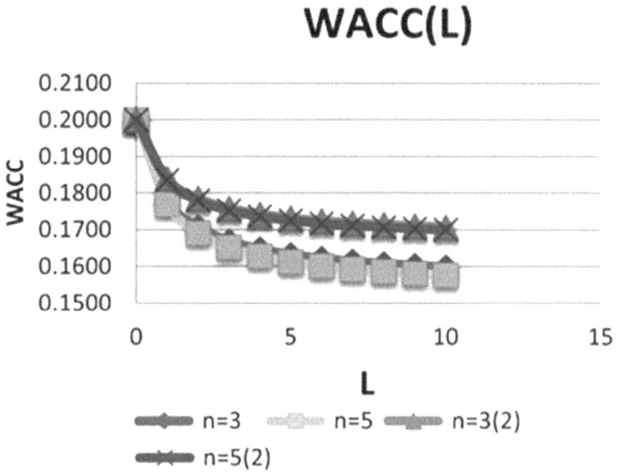

9.6 Conclusions

In this Chapter, an important step towards a general theory of capital cost and capital structure of the company has been done. For this, perpetuity theory of Nobel Prize winners Modigliani and Miller, which was until recently (until 2008) the main and the basic theory of capital cost and capital structure of companies, has been extended to the case of companies of arbitrary age and of companies with an arbitrary age (life–time), as well as for projects of arbitrary duration.

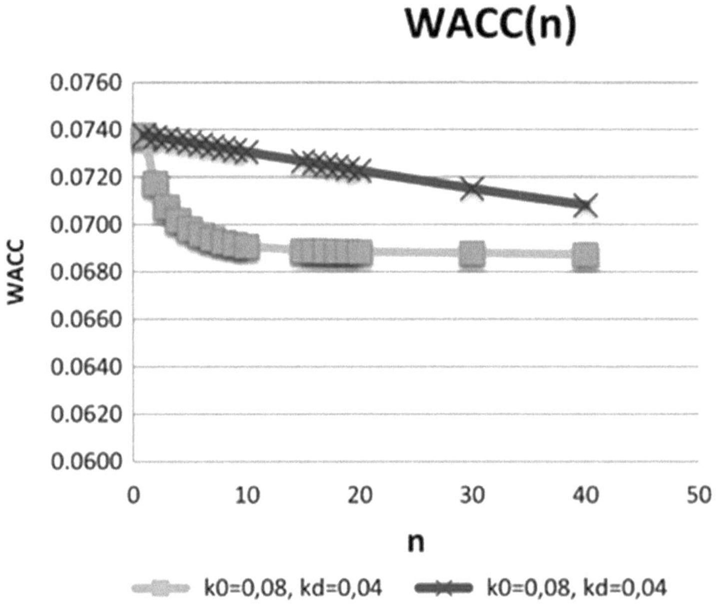

Fig. 9.11 Comparison of the dependence of the WACC on life–time *n* from formulas BFO (*lower curve*) and BFO–2 (*upper curve*); $k_0 = 0.08$; $k_d = 0.04$

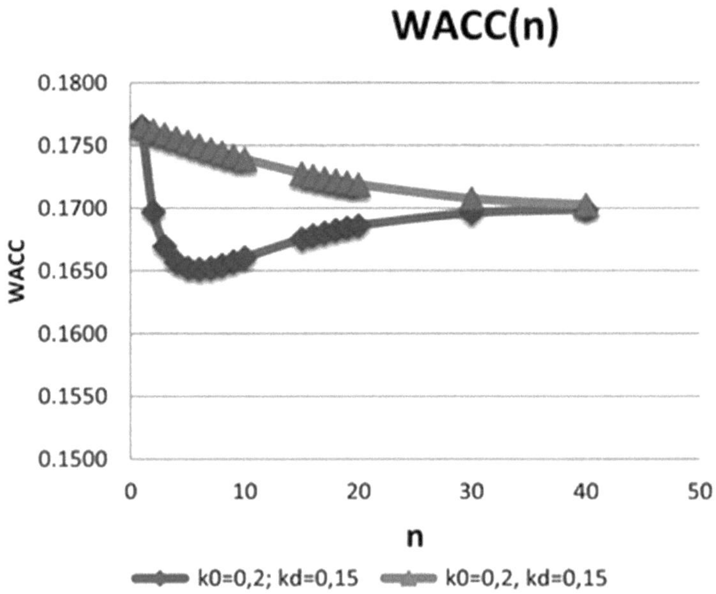

Fig. 9.12 Comparison of the dependence of the WACC on life–time *n* from formulas BFO (*lower curve*) and BFO–2 (*upper curve*); $k_0 = 0.2$; $k_d = 0.15$; $L = 3$

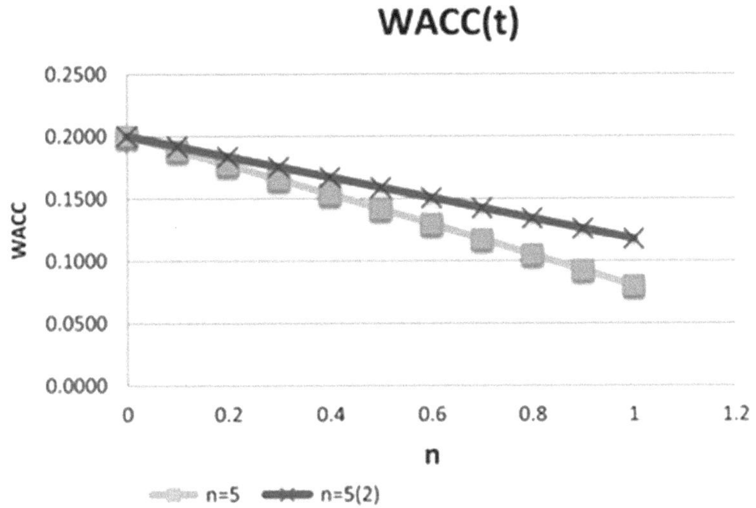

Fig. 9.13 Comparison of the dependence of the WACC on tax on profit rate t from formulas BFO (*lower curve*) and BFO–2 (*upper curve*); $k_0 = 0.2$; $k_d = 0.15$; $n = 5$

We show that taking into account the finite age of the company or the finite life–time of the company in the presence of corporate taxes leads to a change in the equity cost of the company, k_e, as well as in its weighted average cost, WACC, and company capitalization, V.

Thus, we have removed one of the most serious limitations of the theory of Modigliani–Miller, connected with the assumption of perpetuity of the companies. The effect of leverage on the cost of equity capital, k_e, of the company of an arbitrary age or with arbitrary life–time and its weighted average cost, WACC, is investigated. We give a rigorous proof of an important Brusov–Filatova–Orekhova theorem that in the absence of corporate tax, equity cost of companies, k_e, as well as its weighted average cost, WACC, does not depend on the life–time (age) of the company.

We summarize the difference in results obtained within modern Brusov–Filatova–Orekhova theory and within classical Modigliani–Miller one in Table 9.5.

The first four formulas from the right–hand column are sometimes used in practice, but there are several significant nuances. First–these formulas do not take account of the residual value of the company, and only take into account the operating flows and this must be borne in mind. Second–these formulas contain the weighted average cost of capital of the company, WACC. If it is estimated within the traditional approach or the theory of Modigliani–Miller, it gives a lower WACC value, than the real value, and, therefore, overestimates the capitalization of both financially dependent and financially independent companies. Therefore, in order to assess a company's capitalization by the first two formulas, one needs to use Brusov–Filatova–Orekhova formulas for weighted average cost of capital, WACC, and equity cost, k_e.

Table 9.5 Comparison of results, obtained within Modigliani–Miller theory and within general Brusov–Filatova–Orekhova theory

Financial parameter	Modigliani–Miller (MM) results	Brusov–Filatova–Orekhova (BFO) results
Capitalization of financially independent company	$V_0 = \mathrm{CF}/k_0$	$V_0 = \frac{\mathrm{CF}}{k_0}\left[1-(1+k_0)^{-n}\right]$
Capitalization of leverage (financially dependent) company	$V = \mathrm{CF}/\mathrm{WACC}$	$V = \frac{\mathrm{CF}}{\mathrm{WACC}}\left[1-(1+\mathrm{WACC})^{-n}\right]$
Tax shield	$(TS)_\infty = DT$	BFO–1: $TS_n = DT[1-(1+k_d)^{-n}]$
		BFO–2: $\mathrm{TS}_n = \frac{T \cdot k_d \cdot w_d}{\mathrm{WACC}}\left[\frac{1-(1+k_d)^{-n}}{k_d} - -\frac{(1+k_d)^{-n}-(1+\mathrm{WACC})^{-n}}{\mathrm{WACC}-k_d}\right]$
Modigliani–Miller theorem with taxes	$V = V_0 + DT$	$V = V_0 + DT[1-(1+k_d)^{-n}]$
Weighted average cost of capital, WACC	$\mathrm{WACC} = k_0(1-w_d t)$	BFO–1: $\frac{1-(1+\mathrm{WACC})^{-n}}{\mathrm{WACC}} = = \frac{1-(1+k_0)^{-n}}{k_0\left[1-\omega_d T\left(1-(1+k_d)^{-n}\right)\right]}$
		BFO–2: $\frac{1-(1+\mathrm{WACC})^{-n}}{\mathrm{WACC}} = \frac{1-(1+k_0)^{-n}}{k_0} + \frac{T k_d w_d}{\mathrm{WACC}} \cdot \cdot\left[\frac{1-(1+k_d)^{-n}}{k_d} - \frac{(1+k_d)^{-n}-(1+\mathrm{WACC})^{-n}}{\mathrm{WACC}-k_d}\right]$
Equity cost, k_e	$k_e = k_0 + +L(k_0-k_d)(1-t)$	$k_e = (1+L)\cdot\mathrm{WACC} - -k_d L(1-T)$

To calculate the equity cost in BFO approximation (last line in Table 9.4), one first needs to use Brusov–Filatova–Orekhova formula for weighted average cost of capital, WACC (Brusov & Filatova, 2011; Brusov et al., 2011a, b, c, 2012a, b, 2013a, b, 2014a, b; Filatova et al. 2008; Brusov (2018a), Brusov (2018b), Brusov et al. (2015), Brusov et al. (2018a), Brusov et al. (2018b), Brusov et al. (2018d), Brusov et al. (2018e), Brusov et al. (2018f), Brusov et al. (2018g), Brusov et al. (2019), Brusov et al. (2020), Filatova et al. (2018)).

References

Brusov, P. (2018a). Editorial: Introduction on special issue on the banking system and financial markets of Russia and other countries: Problems and prospects. *Journal of Reviews on Global Economics, 7*, i–vi, SCOPUS.

Brusov, P. (2018b). Editorial. *Journal of Reviews on Global Economics, 7*, i–v, SCOPUS.

Brusov, P., Filatova, T., Orehova, N., & Eskindarov, M. (2015). *Modern corporate finance, investments and taxation* (373 p). Springer International Publishing. Monograph, SCOPUS. Retrieved from https://www.springer.com/gp/book/9783319147314.

Brusov, P., Filatova, T., Orehova, N., & Eskindarov, M. (2018a). *Modern corporate finance, investments, taxation and ratings* (571 p), Springer Nature Publishing. Monograph.

Brusov, P. N., Filatova, T. V., Orekhova, N. P., & Kulik, V. L. (2018b). Rating: New approach. *Journal of Reviews on Global Economics, 7*, 37–62. SCOPUS. https://doi.org/10.6000/1929-7092.2018.07.05

Brusov, P. N., Filatova, T. V., Orekhova, N. P., & Kulik, V. L. (2018c). A "golden age" of the companies: Conditions of its existence. *Journal of Reviews on Global Economics, 7*, 88–109. SCOPUS. https://doi.org/10.6000/1929-7092.2018.07.07

Brusov, P. N., Filatova, T. V., Orekhova, N. P., & Kulik, V. L. (2018d). Rating methodology: New look and new horizons. *Journal of Reviews on Global Economics, 7*, 63–87. SCOPUS. https://doi.org/10.6000/1929-7092.2018.07.06

Brusov, P. N., Filatova, T. V., Orekhova, N. P., Kulik, V. L., & Weil, I. (2018e). New meaningful effects in modern capital structure theory. *Journal of Reviews on Global Economics, 7*, 104–122. SCOPUS. https://doi.org/10.6000/1929-7092.2018.07.08

Brusov, P., Filatova, T., Orekhova, N., Kulik, V., Weil, I., & Brailov, A. (2018f). The impact of the central bank key rate and commercial banks credit rates on creating and maintaining of a favorable investment climate in the country. *Journal of Reviews on Global Economics, 7*, 360–376. SCOPUS. https://doi.org/10.6000/1929-7092.2018.07.31

Brusov, P. N., Filatova, T. V., & Orekhova, N. P. (2018g). *Modern corporate finance and investments* (517 p), monograph. Knorus Publishing House.

Brusov, P. N., Filatova, T. V., Orekhova, N. P., Kulik, V. L., & Weil, I. (2019). Ratings of the investment projects of arbitrary durations: New methodology. *Journal of Reviews on Global Economics, 8*, 437–448. SCOPUS. https://doi.org/10.6000/1929-7092.2019.08.37

Brusov, P. N., Filatova, T. V., Orekhova, N. P., Kulik, V. L., Chang, S.-I., & Lin, Y. C. G. (2020). Modification of the Modigliani–Miller theory for the case of advance payments of tax on profit. *Journal of Reviews on Global Economics, 9*, 257–268.

Filatova, T., Orehova, N., & Brusova, A. (2008). Weighted average cost of capital in the theory of Modigliani–Miller, modified for a finite life–time company. *Bull FU, 48*, 68–77.

Filatova, T. V., Brusov, P. N., Orekhova, N. P., & Kulik, V. L. (2018). Ratings of the long–term projects: New approach. *Journal of Reviews on Global Economics, 7*, 645–661, SCOPUS. https://doi.org/10.6000/1929-7092.2018.07.59

Modigliani, F., & Miller, M. (1958). The cost of capital, corporate finance, and the theory of investment. *The American Economic Review, 48*, 261–297.
Modigliani, F., & Miller, M. (1963). Corporate income taxes and the cost of capital: A correction. *The American Economic Review, 53*, 147–175.
Modigliani, F., & Miller, M. (1966). Some estimates of the cost of capital to the electric utility industry 1954–1957. *The American Economic Review, 56*, 333–391.
Myers, S. (2001). Capital structure. *The Journal of Economic Perspectives, 15*(2), 81–102.

Chapter 10
The Generalization of the Brusov–Filatova–Orekhova Theory for the Case of Payments of Tax on Profit with Arbitrary Frequency

10.1 Introduction

The Chapter has the following structure:

In Introduction we describe some approaches to the capital structure theory: traditional approach, Modigliani–Miller theory and some of its modifications.

We generalize the Brusov–Filatova–Orekhova (BFO) theory for the case of arbitrary frequency of payments of income tax and get generalized Brusov–Filatova–Orekhova (BFO) formulas for the weighted average cost of capital, WACC, for the equity cost, k_e, and for the capitalization of the company, V.

Within the generalized Brusov–Filatova–Orekhova (BFO) theory the dependence of the main financial parameters of the company (the weighted average cost of capital, WACC, equity cost, k_e, company capitalization, V) on leverage level L at different frequency of payments of income tax p is studied.

Obtained results allow come to some very important conclusions and give recommendations how frequently company should pay income tax in order to decrease the cost of attracting capital and to increase its capitalization.

10.1.1 Capital Structure of the Company

This Chapter is devoted to one of the most important problem of financial management—problem of capital cost and capital structure of the company. The management by capital structure of the company (the relationship between equity and debt capital of the company) allows to company's management solve the main task—increase the company value. This relates as well with the problem of the optimal capital structure, i.e., with the capital structure, which minimizes the weighted average cost of capital, WACC, and maximizes the value of the company, V, which is one of the most important problem in financial management.

P. Brusov, T. Filatova, *Efficiency of Investment Projects*, Contributions to Finance and Accounting, https://doi.org/10.1007/978-3-032-01028-5_10

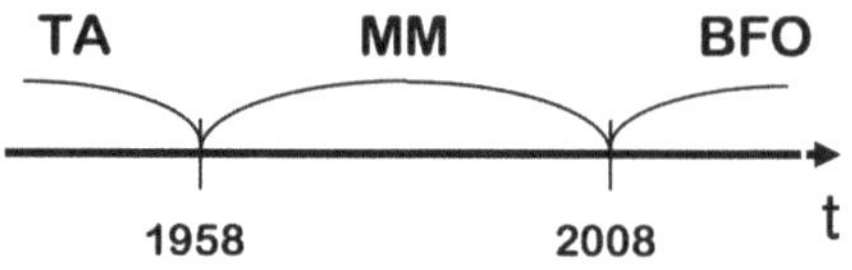

Fig. 10.1 Historical aspects of capital cost and capital structure theory: here empirical traditional approach—TA, Modigliani—Miller theory—MM, Brusov–Filatova–Orekhova theory—BFO

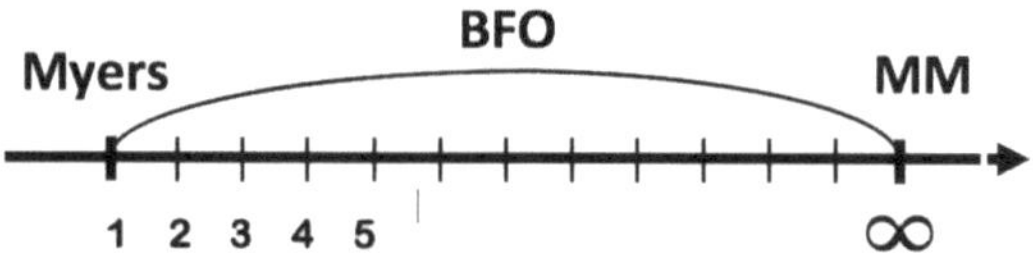

Fig. 10.2 Myers result describes 1-year company, Modigliani–Miller theory (MM) corresponds to perpetuity limit, while Brusov–Filatova–Orekhova (BFO) theory fills the whole numeric axis (from n = 1 up to perpetuity limit $n = \infty$)

The first quantitative study of influence of capital structure of the company on its (company) financial indicators was the work by Nobel Prize winners Modigliani and Miller (1958). Before their work, the traditional approach existed, based on empirical data analysis. In 2008 the modern capital cost and capital structure theory—Brusov–Filatova–Orekhova (BFO) theory has been developed (Filatova et al., 2008), which made Modigliani–Miller theory its particular case. Within BFO theory many qualitatively new effects has been discovered, which are absent within Modigliani–Miller theory. Some main existing principles of financial management have been destroyed by BFO theory, among them the world famous trade off theory (Brennan & Schwartz, 1978, 1984; Leland, 1994), which was considered as keystone of formation of optimal capital structure of the company during many decades. Brusov–Filatova–Orekhova have proven the bankruptcy of trade off theory and have found the cause of this (Chap. 4 in monograph (Brusov et al., 2018)). Below we briefly describe the historical development of the theory of the capital cost and the capital structure.

Figure 10.1 shows the historical development of capital cost and capital structure theory (from the empirical traditional approach, through perpetuity Modigliani–Miller theory to general capital cost and capital structure theory—Brusov–Filatova–Orekhova (BFO) theory (Brusov et al., 2018).

Myers in 2001 has considered the 1-year company case and show that the weighted average cost of capital, WACC, in this case is higher than in Modigliani–Miller case, and the capitalization of the company, *V*, is less than in Modigliani–Miller case.

So, before 2008 only two results for capital structure of the company were available: Modigliani–Miller for perpetuity company (Modigliani & Miller, 1958, 1963, 1966) and Myers (2001) for 1-year company (see Fig. 10.2). Wherein Brusov–Filatova–Orekhova theory has filled out whole interval between t = 1 and t = ∞. Brusov–Filatova–Orekhova theory allows calculate the company value, *V*, the weighted average cost of capital, *WACC*, the equity cost k_e and other financial parameters for arbitrary age companies and for arbitrary life–time companies. BFO theory has lead to a lot of new meaningful effects in modern capital structure theory, absent in Modigliani–Miller theory (Brusov et al., 2018).

The main purpose of this Chapter is to generalize and further develop the Brusov–Filatova–Orekhova (BFO) theory, taking into account the case of paying income tax with an arbitrary frequency (monthly, quarterly, semi–annual or annual payments), which takes place in real economic practice. Note that both major theories on the cost of capital and capital structure of the Brusov–Filatova–Orekhova (BFO) theory and its perpetuity limit, the Modigliani–Miller theory, take into account annual income tax payments. The latter was recently generalized by us to the case of an arbitrary periodicity in the payment of income tax (Filatova et al., 2022), which showed a significant change in the results of the classical Modigliani–Miller theory. Here we make this generalization for the Brusov–Filatova–Orekhova (BFO) theory.

Below we discuss the traditional approach, the Modigliani–Miller theory and its different modifications, the Brusov–Filatova–Orekhova (BFO) theory.

10.1.2 The Modigliani–Miller Theory

Before the Modigliani–Miller work, the traditional approach existed, based on empirical data analysis. In the traditional approach weighted average cost of capital, WACC, and the company value, V, depend on the capital structure, on the leverage level, L. The cost of debt is lower than cost of equity via the fact that first one has lower risk, because creditor claims are met prior to shareholders claims in the event of bankruptcy. Thus the increase of the share of lower–cost debt capital up to the values which does not cause violation of financial sustainability and growth of risk of bankruptcy leads to lower weighted average cost of capital, WACC. The required by investors profitability, which is equal to the equity cost, is growing; but its growth does not compensate the benefits from use of lower–cost debt capital. Therefore, in the traditional approach the increase of leverage level (at low leverage level) leads to decrease of WACC and to the associated increase of the value of company V. At high leverage level, financial difficulties and the risk of bankruptcy appear. The debt capital cost increases as well as the equity cost (which always increases with leverage level) and these lead to increase of WACC and decrease of company value V. Thus, competition between the advantages of debt financing at a low level of leverage and its disadvantages at a high level of leverage forms the optimal capital structure, when WACC reaches minimum and capital value V reaches maximum. This is the main point of the traditional approach and as well as trade off theory. This empirical approach has existed up to creation of the first quantitative theory by Nobel Prize Winners Modigliani and Miller (1958).

Modigliani and Miller (MM) in their first paper (Modigliani & Miller, 1958) have come to the conclusions which were fundamentally different from the conclusions of traditional approach. Under assumptions that there are no taxes, no transaction costs, no bankruptcy costs, perfect financial markets exist with symmetry information, equivalence in borrowing costs for both companies and investors, etc., they have showed that choosing of the ratio between the debt and equity capital does not affect company value V, as well as weighted average cost of capital, WACC.

Modigliani and Miller analyzing the influence of financial leverage on V and V_0, weighted average cost of capital, WACC and k_e, get the following results: (here, EBIT–earnings before interest and taxes, k_0 is discount rate, V and V_0 are the values of levered and unlevered companies, k_e is the equity cost)

$$V = V_0 = \frac{EBIT}{k_0} \tag{10.1}$$

This leads to the following expressions for weighted average cost of capital, WACC:

$$WACC = k_0. \tag{10.2}$$

Note, that k_0 here and below is the equity cost for financially independent company. For financially dependent company k_0 is the equity cost at zero leverage level ($L = 0$).

From (10.1), and formula for WACC definition

$$WACC = k_0 = k_e w_e + k_d w_d \tag{10.3}$$

it is easy to derive an expression for the equity capital cost.

Finding k_e, one gets

$$k_e = \frac{k_0}{w_e} - k_d \frac{w_d}{w_e} = \frac{k_0(S+D)}{S} - k_d \frac{D}{S} = k_0 + (k_0 - k_d)\frac{D}{S} = k_0 + (k_0 - k_d)L \tag{10.4}$$

Here,

D	value of debt capital of the company
S	value of equity capital of the company
$k_d, w_d = \frac{D}{D+S}$	cost and fraction of debt capital of the company
$k_e, w_e = \frac{S}{D+S}$	cost and fraction of equity capital of the company
$L = D/S$	financial leverage
WACC	weighted average cost of capital

Thus, we have the following formula for equity cost

$$k_e = k_0 + L(k_0 - k_d) \tag{10.5}$$

Formula (10.5) shows that equity cost of the company increases linearly with leverage level.

In 1963 Modigliani and Miller have accounted the tax on profit and get the following formula for the cost of financially dependent company, V,

$$V = V_0 + DT, \tag{10.6}$$

where V_0 is the value of financially independent company, D is debt value and T is the tax on profit.

Below we get the expression for WACC and the equity capital cost k_e of the company under the existence of corporate taxes.

10.1.2.1 Weighted Average Cost of Capital, WACC

Substituting into (10.6) $D = w_d V$, one gets

$$V(1 - w_d T) = V_0. \tag{10.7}$$

Putting $V = \frac{CF}{WACC}$; $V_0 = \frac{CF}{k_0}$ into formula (10.7), we get

$$\frac{CF}{WACC}(1 - w_d T) = \frac{CF}{k_0}. \tag{10.8}$$

Here CF is the income of the company for one period.

And from (10.8) we arrive to the expression for weighted average cost of capital, WACC

$$WACC = k_0 \cdot (1 - w_d t) \tag{10.9}$$

This formula for WACC is one of the main result of Modigliani–Miller theory with taxes.

10.1.2.2 The Equity Cost

Let us derive formula for equity cost.

On definition of the weighted average cost of capital WACC in the presence of corporate taxes, we have

$$\mathrm{WACC} = k_0 w_\mathrm{e} + k_\mathrm{d} w_\mathrm{d}(1 - T). \tag{10.10}$$

Equating Eqs. (10.9) and (10.10), one gets

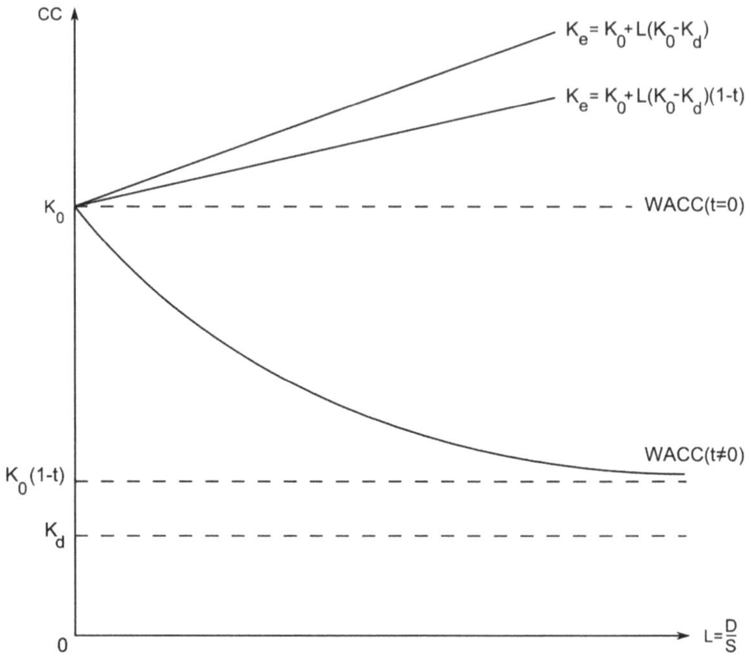

Fig. 10.3 Impact of leverage level on equity capital cost k_e, and WACC within the Modigliani–Miller theory without taxes ($t = 0$) and with taxes ($t \neq 0$)

$$k_0(1 - w_d T) = k_0 w_e + k_d w_d (1 - T) \tag{10.11}$$

and from here, for equity cost, k_e we get the following expression (Modigliani & Miller, 1963):

$$\begin{aligned} k_e &= k_0 \frac{(1 - w_d T)}{w_e} - k_d \frac{w_d}{w_e}(1 - T) = k_0 \frac{1}{w_e} - k_0 \frac{w_d}{w_e} T - k_d \frac{D}{S}(1 - T) \\ &= k_0 \frac{D + S}{S} - k_0 \frac{D}{S} T - k_d \frac{D}{S}(1 - T) \; = k_0 + L(1 - T)(k_0 - k_d) \end{aligned} . \tag{10.12}$$

The formula (10.12) is different from the formula (10.5) without tax only by the multiplier $(1-T)$ in term, indicating a premium for risk. As the multiplier is less than unit, the corporate income tax leads to decrease of the slope of the curve k_e (L) with respect to L–axis.

From formulas (10.5), (10.9) and (10.12) we get the following conclusions (see Fig. 10.3).

When leverage level grows:

1. Value of company, V, increases;
2. Weighted average cost of capital, WACC decreases from k_0 up to $k_0(1 - T)$ (at $L = \infty$);
3. Equity cost increases linearly from k_0 up to infinity.

10.2 Some Modifications of Modigliani–Miller Theory

10.2.1 Hamada Model: Accounting Market Risk

The Modigliani–Miller theory were united with Capital Asset Pricing Model (CAPM) in 1969 (Hamada, 1969) by R. Hamada, who has derived the following formula for the leveraged company equity cost, accounting both financial and business risk of company:

$$k_e = k_F + (k_M - k_F)b_U + (k_M - k_F)b_U \frac{D}{S}(1 - T), \qquad (10.13)$$

where b_U is the β–coefficient of the same group of business risk company, that the considering company, but with L = 0. The formula (10.13) represents the desired equity capital k_e profitability as a sum of three components: risk–free profitability k_F, compensating to shareholders a temporary value of their money, premium for business risk $(k_M - k_F)b_U$, and premium for financial risk $(k_M - k_F)b_U \frac{D}{S}(1 - T)$. If the company does not have debt financing, the financial risk factor will be equal to zero (the third term is drawn to zero), and only the premium for business risk will be received by shareholders.

10.2.2 The Account of Corporate and Individual Taxes (Miller Model)

In 1963 Modigliani and Miller have accounted the taxation of corporate income, but the individual taxes of investors were not take into account.

The account of the corporate and individual taxes has been done by Merton Miller in 1977, who study the influence of debt financing on the company capitalization. In Miller model, the following definitions have been used: TC–the tax on corporate income rate, TS–the tax rate on income of an individual investor from his ownership by corporation stock, TD–tax rate on interest income from the provision of investor–individuals of credits to other investors and companies. Because the income from shares partly comes in the form of a dividend and, in part, as capital profits, so TS is a weighted average value of the rates of tax on capital profits on shares and on dividends, while the profit from the provision of loans usually comes in the form of the interests. The former are usually taxed at a higher rate. For the capitalization of the financially independent company one has:

$$V_{\mathrm{U}} = \frac{\mathrm{EBIT}(1 - T_{\mathrm{C}})(1 - T_{\mathrm{S}})}{k_0}. \tag{10.14}$$

A term $(1 - T_{\mathrm{S}})$ allows to take into account the individual taxes in the formula. In this way, numerator indicates which part of the operating company's profit remains in the possession of the investors, after the company will pay taxes on their profits, and its shareholders then will pay individual taxes on income from stock ownership. Since individual taxes reduce profits, remaining in the disposal of investors, the last, at other things being equal circumstances, also reduce and an overall assessment of the financially independent company value.

10.2.3 More General Case for WACC Formula

In (Farber et al., 2006; Fernandez, 2006; Berk & DeMarzo, 2007; Harris & Pringle, 1985) more general than Modigliani–Miller (MM) formula for the WACC has been derived: it is described by the following formula (Farber et al., 2006)

$$WACC = k_0(1 - w_d T) - k_d t w_d + k_{TS} t w_d. \tag{10.15}$$

Here k_0, k_d and k_{TS} are the expected returns respectively on the financially independent company, the debt and the tax shield, V is the capitalization of the financially dependent company, VTS is the tax shield value, D is the debt value and TC is the rate of corporate tax.

Formula (10.15) is derived from the WACC definition and the balance identity (see as well Berk and De Marzo (2007)). Although Eq. (10.15) is general enough, some additional conditions are required its practical applicability. If the WACC is constant over time, as it stated in (Farber et al., 2006) the levered company value can be found by discounting with the WACC for the unlevered free cash–flows. The resulting formulas, which describe the special cases, where the WACC is constant, can be found in textbooks (Berk & DeMarzo, 2007; Harris & Pringle, 1985).

First, Modigliani and Miller in 1963 assume that the level of debt D is constant. Then, as the expected after–tax cash–flow of the unlevered firm is fixed, V_0 is also constant. By assumption, $k_{TS} = k_D$ and the value of the tax shield is TS = tD. Therefore, the company value V is a constant and the general WACC formula (10.15) simplifies to a constant WACC:

$$WACC = k_0(1 - w_d T)$$

We think that "classical" Modigliani–Miller (MM) theory, which suggests that the expected returns on the debt k_d and the tax shield k_{TS} are equals (because both of them have debt nature), is much more reasonable and namely the "classical" Modigliani–Miller (MM) theory, which is still widely used in practice, has been modified by us in (Filatova et al., 2022).

10.2.4 Fiscal Pressure, Financial Liquidity, Financial Solvency and Financial Leverage

The problem of fiscal pressure is more current than ever in most countries. The article by Batrancea (2021a) was the first empirical study of the impact of fiscal pressure on the financial equilibrium of energy companies listed on the New York Stock Exchange. Empirical results obtained by econometric models showed that fiscal pressure had a stronger impact on the short–term and long–term equilibrium of electricity and oil companies than on the equilibrium of gas companies. The research could be useful for the managers of energy companies interested in estimating the evolution of company equilibrium state when considering different possible financial crises.

How financial liquidity and financial solvency influenced the performance of healthcare companies that are publicly traded on the New York Stock Exchange has been studied by Batrancea (2021b), who used econometric models with two–stage least squares (2SLS) panel and panel generalized method of moments (GMM). Empirical evidence showed that the financial indicators current liquidity ratio, quick liquidity ratio, and financial leverage significantly influenced company performance measured by return on assets, gross margin ratio, operating margin ratio, earnings before interest, tax, depreciation, and amortization. Strategies intended to improve business performance based on liquidity and solvency insights were also addressed.

10.2.5 Brusov–Filatova–Orekhova (BFO) Theory

The suggestion about perpetuity of the companies and of all financial flows is the most serious limitations of the Modigliani–Miller theory. This limitation have been lifted up in 2008 by Brusov, Filatova and Orekhova (Brennan & Schwartz, 1984), who have created the modern theory of capital cost and capital structure, generalizing the Modigliani–Miller theory for the companies of arbitrary age (arbitrary life–time). They have shown that in this case all Modigliani–Miller results (Modigliani & Miller, 1958; Filatova et al., 2008; Brennan & Schwartz, 1978) significant changes: in the presence of corporative taxes capitalization of the company V is changed, as well as the equity cost, k_e, and the weighted average cost of capital, WACC. A number of qualitatively new effects in corporate finance, obtained in Brusov–

Filatova–Orekhova theory (Brennan & Schwartz, 1984), are absent in Modigliani–Miller theory (Modigliani & Miller, 1958, 1963, 1966).

The Brusov–Filatova–Orekhova formula for weighted average cost of capital, WACC for the company of arbitrary age n, takes the following form (Brennan & Schwartz, 1984)

$$\frac{1-(1+WACC)^{-n}}{WACC}=\frac{1-(1+k_0)^{-n}}{k_0[1-w_dT(1-(1+k_d)^{-n})]}. \tag{10.16}$$

Here, S—the equity capital of the company value, $w_d=\frac{D}{D+S}$—the share of debt capital; $k_e, w_e=\frac{S}{D+S}$—the cost and the share of the equity capital of the company, $L=D/S$—leverage level.

Substituting in (10.14) $n\to\infty$, we easy arrive to the perpetuity (Modigliani–Miller) limit formula for weighted average cost of capital, WACC.

$$WACC=k_0(1-w_dT). \tag{10.17}$$

Results of BFO theory are well known in the world literature (for example, see references (Dimitropoulos, 2014; Luiz et al., 2015; Massimiliano, 2011; Pavel, 2018; Franc-Dąbrowska et al., 2021; Angotti et al., 2018; Cristian et al., 2018)). Some Chapters (Cristian et al., 2018) use the BFO theory in practical calculations.

The similarities and differences between BFO and MM results are shown in Table 10.1.

10.2.6 Trade–Off Theory

As we mentioned above, the main theory of optimal capital structure of the company during many decades was the world famous trade–off theory (Brennan & Schwartz, 1978, 1984; Leland, 1994), which is still used now for decision on capital structure making. But, in 2013 Brusov et al. (Brusov et al., 2018) have proven the trade–off theory's bankruptcy. In opposite to waiting result, it was shown that suggestion of risky debt financing which causes the growing of the credit rate near the bankruptcy, does not lead to the growing of weighted average cost of capital, WACC, which still decreases with leverage level. Thus, the minimum in the dependence of WACC on leverage level is absent as well. This means, that the world famous trade–off theory lacks an optimal capital structure. Brusov et al. in 2013 (Brusov et al., 2018) have given the explanation of this fact by analyzing the equity capital cost dependence on the leverage level on the risky debt capital assumption.

Table 10.1 The similarities and differences between BFO and MM results

	BFO	MM
Time parameter, n	Arbitrary age of companies, Arbitrary life–time of companies	Perpetuity companies, perpetuity financial flows; $n = \infty$
Company age and life–time	Are different	Are the same and equal infinity
Dependence of all main financial indicators: WACC, V, k_e on time parameter, n	Allow study this dependences	Time parameter, n is absent
All main financial indicators: WACC, V, k_e depend on debt cost k_d	Yes	None
WACC	Correct estimation; Decreases with leverage level L	Underestimation; Decreases with leverage level L
V	Correct estimation; Increases with leverage level L	Overestimation; Increases with leverage level L
Equity cost k_e	Increases linearly with leverage level L	Increases linearly with leverage level L
Equity cost k_e: the slope of the curve k_e (L)	Less than in MM theory	Bigger than in BFO theory
k_e: the slope of the curve k_e (L)	**Could be negative at T > T* (qualitatively new effect)**	Always is positive **New effect is absent**
Golden and silver age effects (Brusov et al., 2018)	Exist	Absent

10.3 Modification of the Brusov–Filatova–Orekhova (BFO) Theory for Companies with Frequent Payments of Tax on Income

Below we use the following definitions:

D	value of debt capital of the company
S	value of equity capital of the company
$k_d, w_d = \frac{D}{D+S}$	cost and fraction of debt capital of the company
$k_e, w_e = \frac{S}{D+S}$	cost and fraction of equity capital of the company
$L = D/S$	financial leverage
WACC	weighted average cost of capital

k_0 is the equity cost for financially independent company; p is the number of payments of tax on profit per year; T is tax on profit.

10.3.1 Calculation of the Tax Shield

We start from the calculation of the tax shield within Brusov–Filatova–Orekhova theory for the case of p payments of tax on profit per year (payments are made at the end of periods). The tax shield, *TS*, for period of n-years is equal to sum of discounted values of benefits from the use of tax incentives

$$(TS)_n = \frac{k_d Dt}{p(1+k_d)^{1/p}} + \frac{k_d Dt}{p(1+k_d)^{2/p}} + \cdots + \frac{k_d Dt}{p(1+k_d)^{np/p}} \tag{10.18}$$

We have a geometric progression with denominator $q = \frac{1}{(1+k_d)^{\frac{1}{p}}}$.

After summing the progression, one obtains:

$$(TS)_n = \frac{k_d Dt(1-(1+k_d)^{-n})}{p(1+k_d)^{\frac{1}{p}}\left(1-(1+k_d)^{-1/p}\right)} = \frac{k_d Dt(1-(1+k_d)^{-n})}{p\left((1+k_d)^{1/p}-1\right)} \tag{10.19}$$

In classical Brusov–Filatova–Orekhova theory (at p = 1):

$$(TS)_n = Dt(1-(1+k_d)^{-n}) \tag{10.20}$$

It is easy to obtain this result from (10.19), putting the frequency of payments of tax on profit $p = 1$.

10.3.2 Derivation of the Modified BFO Formula for Weighted Average Cost of Capital (WACC)

We derive now the modified formula BFO for weighted average cost of capital (WACC) for the case of p payments of tax on profit per year (payments at the end of periods).

One has for the financially dependent company value V:

$$V = V_0 + (TS)_n \tag{10.21}$$

Here V_0 is the value of a financially independent company. Putting expression (10.19) for TS, one has:

$$V = V_0 + \frac{k_d Dt(1-(1+k_d)^{-n})}{p\left((1+k_d)^{1/p-1}\right)} \tag{10.22}$$

After substituting $D = w_d V$, we get:

$$V\left(1 - \frac{k_d w_d t(1-(1+k_d)^{-n})}{p\left((1+k_d)^{1/p}-1\right)}\right) = V_0 \tag{10.23}$$

Accounting, that the values of financially dependent company, V, and financially independent company, V_0, are respectively equal to

$$V = \frac{CF(1-(1+WACC)^{-n})}{WACC};\, V_0 = \frac{CF(1-(1+k_0)^{-n})}{k_0} \tag{10.24}$$

one gets:

$$\frac{CF(1-(1+WACC)^{-n})}{WACC}\cdot\left(1 - \frac{k_d w_d t(1-(1+k_d)^{-n})}{p\left((1+k_d)^{1/p}-1\right)}\right)$$
$$= \frac{CF}{k_0}\cdot(1-(1+k_0)^{-n}) \tag{10.25}$$

From here we derive now the modified formula BFO for weighted average cost of capital (WACC) for company of age n years for the case of p payments of tax on profit per year (payments at the end of periods)

$$\frac{1-(1+WACC)^{-n}}{WACC} = \frac{CF(1-(1+k_0)^{-n})}{k_0\left(1 - \frac{k_d w_d t(1-(1+k_d)^{-n})}{p\left((1+k_d)^{1/p}-1\right)}\right)} \tag{10.26}$$

At p = 1 we get **classical** BFO formula

$$\frac{1-(1+WACC)^{-n}}{WACC} = \frac{1-(1+k_0)^{-n}}{k_0\cdot(1-w_d t[1-(1+k_d)^{-n}])} \tag{10.27}$$

10.3.3 Formulas for Capital Value, V, and Equity Cost, k_e

Below in Sect. 10.3 we will investigate the dependence of the weighted average cost of capital, WACC, capital value, V, equity cost, k_e, on leverage level L at different frequency of payment of tax on profit p for 3-year and 6-year companies, using Microsoft Excel. For WACC we will use formula (10.10) and for capital value, V, and equity cost, k_e, we will use formulas (10.28) and (10.29) respectively (see below).

Company of age n capitalization could be calculated by the following formula

$$V = \frac{CF \cdot (1 - (1 + WACC)^{-n})}{WACC}. \tag{10.28}$$

k_e should be found from the equation

$$WACC = k_e w_e + k_d w_d (1 - t)$$

$$k_e = \frac{WACC}{w_e} - \frac{k_d(1 - t)w_d}{w_e} = WACC(1 + L) - Lk_d(1 - t) \tag{10.29}$$

where one should substitute WACC from the formula (10.26).

10.4 Results

In this section we study the dependence of the weighted average cost of capital, WACC, capital value, V, equity cost, k_e, on leverage level L at different frequency of payment of tax on profit p for 3-year and 6-year companies, using Microsoft Excel. As we mentioned above, for WACC we use formula (10.10) and for capital value, V, and equity cost, k_e, we use formulas (10.11) and (10.13) respectively.

We use the following parameters: $k_0 = 0.2$; $k_d = 0.18$; $t = 0.2$; $n = 3; 6$; $CF = 100$.

10.4.1 Dependence of the Weighted Average Cost of Capital, WACC, Capital Value, V, Equity Cost, k_e, on Leverage Level L at Different Frequency of Payment of Tax on Profit p for 3-Year Company

10.4.1.1 Dependence of the Weighted Average Cost of Capital, WACC, on Leverage Level L at Different Frequency of Payment of Tax on Profit p for 3-Year Company

Dependence of the weighted average cost of capital, WACC, on leverage level L at different frequency of payments of tax on profit *p* **for 3-year company** is shown **in** Table 10.2 and at Fig. 10.4.

The weighted average cost of capital, WACC, decreases with leverage level L at any frequency of payments of tax on profit p. The difference between the WACC

Table 10.2 Dependence of the weighted average cost of capital, WACC, on leverage level L at different frequency of payments of tax on profit p for 3-year company

		WACC				
L	wd	p = 1	p = 2	p = 4	p = 6	p = 12
1	0.5	0.1749	0.1738	0.1732	0.1730	0.1728
2	0.666667	0.1664	0.1650	0.1642	0.1640	0.1637
3	0.75	0.1622	0.1605	0.1597	0.1594	0.1591
4	0.8	0.1596	0.1579	0.1570	0.1567	0.1564
5	0.833333	0.1579	0.1561	0.1552	0.1549	0.1545
6	0.857143	0.1567	0.1548	0.1539	0.1536	0.1532
7	0.875	0.1558	0.1539	0.1529	0.1526	0.1522
8	0.888889	0.1551	0.1532	0.1522	0.1518	0.1515
9	0.9	0.1545	0.1526	0.1515	0.1512	0.1509
10	0.909091	0.1541	0.1521	0.1511	0.1507	0.1504

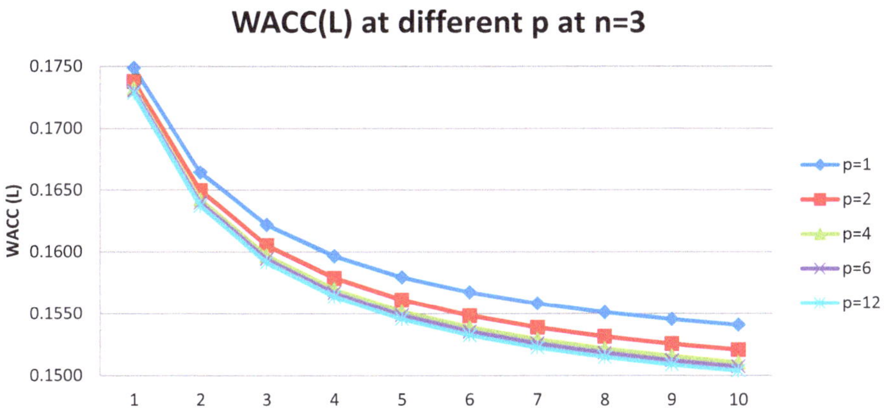

Fig. 10.4 Dependence of the weighted average cost of capital, WACC, on leverage level L at different frequency of payments of tax on profit p for 3-year company

(L) curves is maximum when moving from annual (p = 1) to semi–annual (p = 2) income tax payments and decreases when moving from semi–annual (p = 2) to quarterly (p = 4) payments and from quarterly (p = 4) to monthly (p = 12) payments.

10.4.1.2 Dependence of the Company Value, on Leverage Level L at Different Frequency of Payment of Tax on Profit p for 3-Year Company

Dependence of the company value, V, on leverage level L at different frequency of payments of tax on profit *p* for 3-year company is shown in Table 10.3 and at Fig. 10.5.

Table 10.3 Dependence of the company value, V, on leverage level L at different frequency of payments of tax on profit p for 3-year company

		V				
L	w_d	p = 1	p = 2	p = 4	p = 6	p = 12
1	0.5	219.2281	219.6139	219.8116	219.8781	219.9451
2	0.666667	222.2455	222.7746	223.0459	223.1373	223.2292
3	0.75	223.7856	224.3893	224.6989	224.8033	224.9082
4	0.8	224.7199	225.3694	225.7025	225.8149	225.9278
5	0.833333	225.3471	226.0275	226.3766	226.4944	226.6127
6	0.857143	225.7973	226.5000	226.8606	226.9822	227.1045
7	0.875	226.1361	226.8557	227.2250	227.3495	227.4747
8	0.888889	226.4003	227.1331	227.5091	227.6360	227.7635
9	0.9	226.6122	227.3555	227.7370	227.8657	227.9951
10	0.909091	226.7858	227.5378	227.9238	228.0540	228.1849

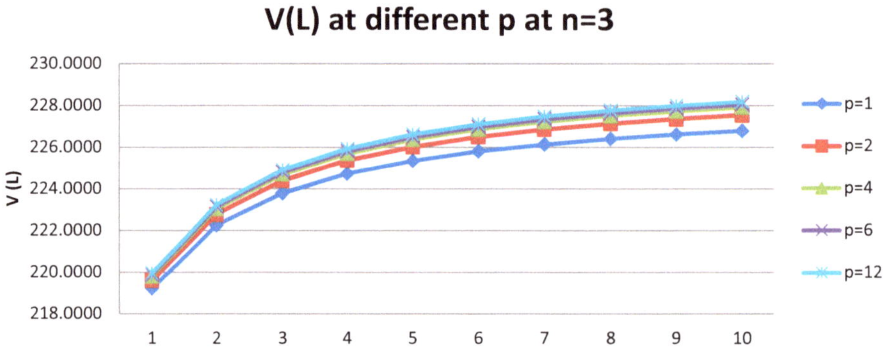

Fig. 10.5 Dependence of the company value, V, on leverage level L at different frequency of payments of tax on profit p for 3-year company

The value of company V increases with the frequency p. The largest increase occurs when moving from annual (p = 1) to semi–annual (p = 2) income tax payments and decreases when moving from semi–annual (p = 2) to quarterly (p = 4) payments and from quarterly (p = 4) to monthly (p = 12) payments.

10.4.1.3 Dependence of the Equity Cost, k_e, on Leverage Level L at Different Frequency of Payment of Tax on Profit p for 3-Year Company

Dependence of the equity cost, k_e, on leverage level L at different frequency of payments of tax on profit *p* for 3-year company is shown in Table 10.4 and at Fig. 10.6.

The cost of equity k_e increases linearly with the level of leverage L. The slope of the curve k_e (L) depends on the frequency of paying income tax: it decreases with

Table 10.4 Dependence of the equity cost, k_e, on leverage level L at different frequency of payments of tax on profit p for 3-year company

		k_e				
L	w_d	p = 1	p = 2	p = 4	p = 6	p = 12
1	0.5	0.2057	0.2036	0.2024	0.2021	0.2017
2	0.666667	0.2113	0.2069	0.2046	0.2039	0.2031
3	0.75	0.2168	0.2102	0.2068	0.2057	0.2045
4	0.8	0.2222	0.2134	0.2089	0.2074	0.2059
5	0.833333	0.2277	0.2167	0.2111	0.2092	0.2073
6	0.857143	0.2331	0.2199	0.2132	0.2109	0.2086
7	0.875	0.2386	0.2232	0.2153	0.2126	0.2100
8	0.888889	0.2440	0.2264	0.2174	0.2144	0.2113
9	0.9	0.2495	0.2296	0.2195	0.2161	0.2127
10	0.909091	0.2549	0.2329	0.2216	0.2178	0.2140

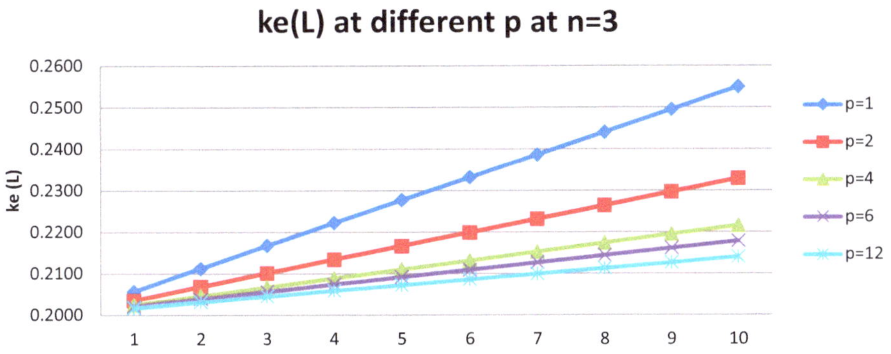

Fig. 10.6 Dependence of the equity cost, k_e, on leverage level L at different frequency of payments of tax on profit p for 3-year company

increasing p, most rapidly when moving from annual (p = 1) to semi–annual (p = 2) payments of income tax and slower in the transition from semi–annual (p = 2) to quarterly (p = 4) payments and from quarterly (p = 4) to monthly (p = 12) payments.

10.4.2 Dependence of the Weighted Average Cost of Capital, WACC, Capital Value, V, Equity Cost, k_e, on Leverage Level L at Different Frequency of Payment of Tax on Profit p for 6-Year Company

10.4.2.1 Dependence of the Weighted Average Cost of Capital, WACC, on Leverage Level L at Different Frequency of Payment of Tax on Profit p for 6-Year Company

Dependence of the weighted average cost of capital, WACC, on leverage level L at different frequency of payments of tax on profit p for 6-year company is shown in Table 10.5 and at Fig. 10.7.

Table 10.5 Dependence of the weighted average cost of capital, WACC, on leverage level L at different frequency of payments of tax on profit p for 6-year company

		WACC				
L	w_d	p = 1	p = 2	p = 4	p = 6	p = 12
1	0.5	0.1743	0.1732	0.1726	0.1725	0.1723
2	0.666667	0.1657	0.1641	0.1634	0.1631	0.1629
3	0.75	0.1613	0.1596	0.1587	0.1584	0.1581
4	0.8	0.1586	0.1568	0.1559	0.1556	0.1553
5	0.833333	0.1569	0.1550	0.1540	0.1537	0.1534
6	0.857143	0.1556	0.1537	0.1527	0.1523	0.1520
7	0.875	0.1547	0.1527	0.1517	0.1513	0.1510
8	0.888889	0.1539	0.1519	0.1509	0.1505	0.1502
9	0.9	0.1534	0.1513	0.1502	0.1499	0.1495
10	0.909091	0.1529	0.1508	0.1497	0.1494	0.1490

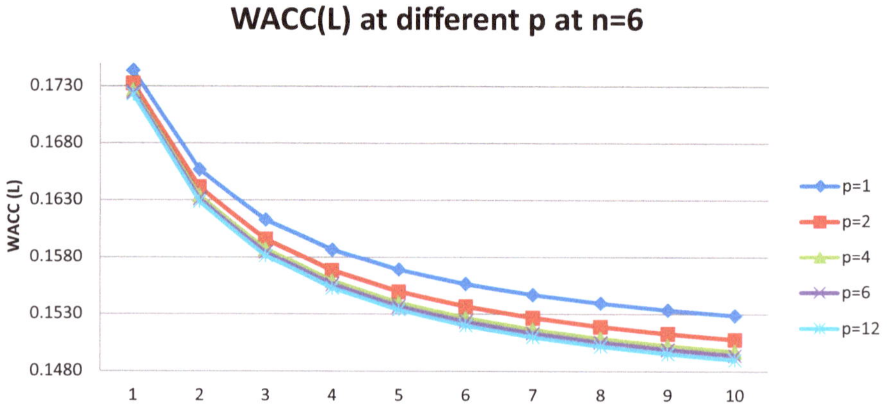

Fig. 10.7 Dependence of the weighted average cost of capital, WACC, on leverage level L at different frequency of payments of tax on profit p for 6-year company

We use the following parameters: $k_0 = 0.2$; $k_d = 0.18$; $t = 0.2$; $n = 6$;

The weighted average cost of capital, WACC, decreases with leverage level L at any frequency of payments of tax on profit p. The difference between the WACC (L) curves is maximum when moving from annual ($p = 1$) to semi–annual ($p = 2$) income tax payments and decreases when moving from semi–annual ($p = 2$) to quarterly ($p = 4$) payments and from quarterly ($p = 4$) to monthly ($p = 12$) payments.

Table 10.6 Dependence of the company value, V, on leverage level L at different frequency of payments of tax on profit p for 6-year company

		V				
L	w_d	$p = 1$	$p = 2$	$p = 4$	$p = 6$	$p = 12$
1	0.5	354.8940	355.9256	356.4549	356.6333	356.8127
2	0.666667	363.0242	364.4649	365.2052	365.4550	365.7063
3	0.75	367.2305	368.8900	369.7435	370.0315	370.3214
4	0.8	369.8015	371.5971	372.5210	372.8329	373.1468
5	0.833333	371.5355	373.4240	374.3960	374.7242	375.0545
6	0.857143	372.7841	374.7399	375.7468	376.0869	376.4291
7	0.875	373.7261	375.7330	376.7664	377.1154	377.4667
8	0.888889	374.4620	376.5090	377.5632	377.9193	378.2776
9	0.9	375.0529	377.1322	378.2031	378.5648	378.9289
10	0.909091	375.5377	377.6435	378.7283	379.0947	379.4635

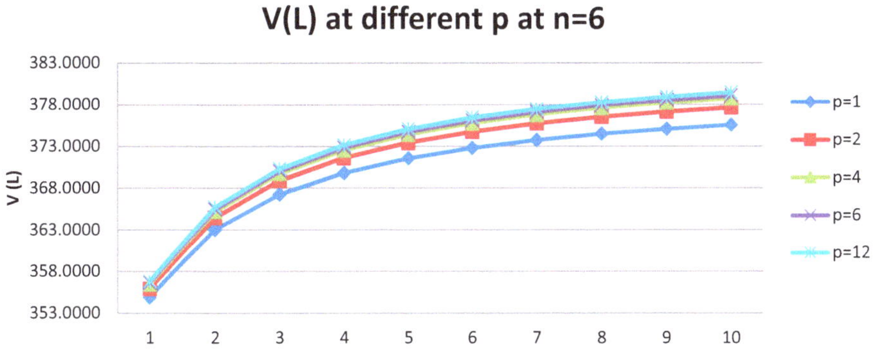

Fig. 10.8 Dependence of the company value, V, on leverage level L at different frequency of payments of tax on profit p for 6-year company

10.4.2.2 Dependence of the Company Value, V, on Leverage Level L at Different Frequency of Payment of Tax on Profit p for 6-Year Company

Dependence of the company value, V, on leverage level L at different frequency of payments of tax on profit *p* for 6-year company is shown in Table 10.6 and at Fig. 10.8.

The value of company V increases with the frequency p. The largest increase occurs when moving from annual ($p = 1$) to semi–annual ($p = 2$) income tax payments and decreases when moving from semi–annual ($p = 2$) to quarterly ($p = 4$) payments and from quarterly ($p = 4$) to monthly ($p = 12$) payments.

10.4.2.3 Dependence of the Equity Cost, k_e, on Leverage Level L at Different Frequency of Payment of Tax on Profit p for 6-Year Company

Dependence of the equity cost, k_e, on leverage level L at different frequency of payments of tax on profit *p* for 6-year company is shown in Table 10.7 and at Fig. 10.9.

The cost of equity k_e increases linearly with the level of leverage L. The slope of the curve k_e (L) depends on the frequency of paying income tax: it decreases with increasing p, most rapidly when moving from annual ($p = 1$) to semi–annual ($p = 2$) payments of income tax and slower in the transition from semi–annual ($p = 2$) to quarterly ($p = 4$) payments and from quarterly ($p = 4$) to monthly ($p = 12$) payments.

One can see, that for a 6-year-old company with monthly income tax payments **a qualitatively new anomalous effect takes place**: k_e (L) decreases with an increase in the level of leverage L. This radically changes the company's dividend policy,

Table 10.7 Dependence of the equity cost, k_e, on leverage level L at different frequency of payments of tax on profit p for 6-year company

		k_e				
L	w_d	$p = 1$	$p = 2$	$p = 4$	$p = 6$	$p = 12$
1	0.5	0.2047	0.2024	0.2013	0.2009	0.2005
2	0.666667	0.2090	0.2044	0.2021	0.2013	0.2006
3	0.75	0.2131	0.2063	0.2028	0.2016	0.2005
4	0.8	0.2172	0.2081	0.2035	0.2019	0.2003
5	0.833333	0.2213	0.2099	0.2041	0.2021	0.2001
6	0.857143	0.2254	0.2117	0.2047	0.2023	0.1999
7	0.875	0.2294	0.2134	0.2052	0.2025	0.1997
8	0.888889	0.2335	0.2152	0.2058	0.2027	0.1995
9	0.9	0.2376	0.2169	0.2064	0.2029	0.1993
10	0.909091	0.2416	0.2187	0.2070	0.2030	0.1991

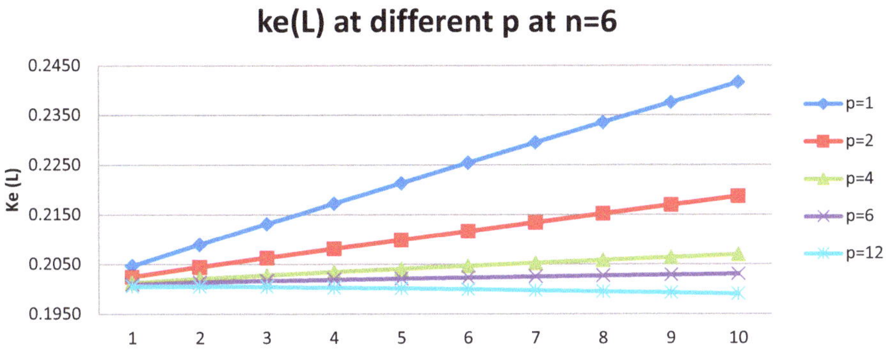

Fig. 10.9 Dependence of the equity cost, k_e, on leverage level L at different frequency of payments of tax on profit p for 6-year company

since the economically justified amount of the dividends is equal to the cost of equity.

10.5 The Discussion and Conclusions

The main purpose of the Chapter is bring the Brusov–Filatova–Orekhova (BFO) theory closer to economic practice, taking into account one of the features of the real functioning of companies—the frequent payments of tax on profit. We generalize for the first time the Brusov–Filatova–Orekhova (BFO) theory for the case of tax on profit payments with an arbitrary frequency. We derive modified BFO formulas and show that: (1) all BFO formulas change; (2) all main financial parameters of the company, such as company value, V, the weighted average cost of capital, WACC, and equity cost, k_e, depend on the frequency of tax on profit payments. It turns out that the increase of the number of payments of tax of profit per year leads to decrease of the cost of attracting capital and increase of the company value. More frequent payments of income tax are beneficial for both parties—for the company and for the tax regulator: for the company, this leads to an increase in the value of the company, and for the tax regulator, earlier payments are beneficial due to the time value of money.

We numerically studied the dependence of the weighted average cost of capital WACC, value of company, V, cost of equity k_e on the level of leverage L at different frequency of income tax payments p for 3- and 6-year companies and obtained the following results. The calculation of financial indicators are made for some typical parameters of enterprises (equity and debt costs etc.).

The weighted average cost of capital, WACC, decreases with leverage level L at any frequency of payments of tax on profit p. The difference between the WACC (L) curves is maximum when moving from annual ($p = 1$) to semi–annual ($p = 2$) income tax payments and decreases when moving from semi–annual ($p = 2$) to quarterly (p = 4) payments and from quarterly (p = 4) to monthly (p = 12) payments.

The value of company V increases with the frequency p. The largest increase occurs when moving from annual (p = 1) to semi–annual (p = 2) income tax payments and decreases when moving from semi–annual (p = 2) to quarterly (p = 4) payments and from quarterly (p = 4) to monthly (p = 12) payments.

The cost of equity k_e increases linearly with the level of leverage L. The slope of the curve k_e (L) depends on the frequency of paying income tax: it decreases with increasing p, most rapidly when moving from annual (p = 1) to semi–annual (p = 2) payments of income tax and slower in the transition from semi–annual (p = 2) to quarterly (p = 4) payments and from quarterly (p = 4) to monthly (p = 12) payments.

At a certain age of the company, **a qualitatively new anomalous effect** takes place: k_e (L) decreases with an increase in the level of leverage L. We obtain this result for a 6-year-old company with monthly income tax payments, but it is clear that this effect should take place for the companies of different age and at different frequency of income tax payment. This radically changes the company's dividend policy, since the economically justified amount of the dividends is equal to the cost of equity.

Note, that this result remind discovery, made by us earlier (Chap. 8 in monograph (Brusov et al., 2018)): the abnormal dependence of equity cost, k_e, on leverage level L: above some value of tax on income, T^*, equity cost k_e decreases with the leverage level L.

Thus, it is interesting to note, that **increase of frequency of income tax payments p and increase of tax on profit *T* lead to the similar abnormal effect: decrease of equity cost k_e with leverage level L**. If in the first case there are threshold values for the frequency of paying income tax, p*, and the age of the company, n*, then in the second case there is a threshold value for income tax T^*.

In Table 10.8 we compare the impact of the frequency of income tax payments in BFO theory and in MM theory.

10.5.1 Recommendations

From the analysis of the results of current Chapter the following recommendations could be done:

For company: company should pay tax on profit as frequently as it is possible, because in this case company value increases.

For regulator: should encourage more frequent payments of tax on profit, because earlier payments are beneficial for budget due to the time value of money.

Table 10.8 Comparison of the impact of the frequency of income tax payments in BFO theory and in MM theory

	BFO	MM
Theorems, statements and formulas	Change	Change
All main financial indicators: WACC, V, k_e	Depend on the frequency of tax on profit payments p	Depend on the frequency of tax on profit payments p
All main financial indicators: WACC, V, k_e	Value of the effect of p on all indicators is higher, than in MM	Value of the effect of p on all indicators is smaller, than in BFO
All main financial indicators: WACC, V, k_e depend on debt cost k_d	At all p	At $p \neq 1$ only
WACC	Decreases with p	Decreases with p
V	Increases with p	Increases with p
k_e: the slope of the curve k_e (L)	Decreases with increasing p	Decreases with increasing p
k_e: the slope of the curve k_e (L)	**Could be negative (qualitatively new effect)**	Always is positive

References

Angotti, M., de Lacerda Moreira, R., do Nascimento, J. H. B., & de Almeida Bispo, O. N. (2018). Analysis of an equity investment strategy based on accounting and financial reports in Latin American markets. *Reficont, 5*(2), 22–40.

Batrancea, L. (2021a). An econometric approach regarding the impact of fiscal pressure on equilibrium: Evidence from electricity, gas and oil companies listed on the new york stock exchange. *Mathematics, 9*(6), 630. https://doi.org/10.3390/math906063

Batrancea, L. (2021b). The influence of liquidity and solvency on performance within the healthcare industry: Evidence from publicly listed companies. *Mathematics, 9*, 2231. https://doi.org/10.3390/math9182231

Berk, J., & DeMarzo, P. (2007). *Corporate finance*. Pearson–Addison Wesley.

Brennan, M., & Schwartz, E. (1978). Corporate income taxes, valuation, and the problem of optimal capital structure. *Journal of Business, 51*, 103–114.

Brennan, M., & Schwartz, E. (1984). Optimal financial policy and firm valuation. *The Journal of Finance, 39*, 593–607.

Brusov, P., Filatova, T., Orehova, N., & Eskindarov, M. (2018). *Modern corporate finance, investments, taxation and ratings*. Springer Nature Publishing.

Cristian, V.-N., Juan, P. S.-R., Alfaro Miguel, D., & Nicolás, R. (2018). Cost of capital estimation for highway concessionaires in Chile. *Journal of Advanced Transportation, 2018*, 153536. https://doi.org/10.1155/2018/2153536

Dimitropoulos, P. (2014). Capital structure and corporate governance of soccer clubs: European evidence. *Management Research Review, 37*(7), 658–678. https://doi.org/10.1108/MRR-09-2012-0207

Farber, A., Gillet, R., & Szafarz, A. (2006). A general formula for the WACC. *International Journal of Business, 11*(2), 211–218.

Fernandez, P. (2006). A general formula for the WACC: A comment. *International Journal of Business, 11*(2), 219.

Filatova, T., Orehova, N., & Brusova, A. (2008). Weighted average cost of capital in the theory of Modigliani–Miller, modified for a finite life–time company. *Bull FU, 48*, 68–77.

Filatova, T., Brusov, P., & Orekhova, N. (2022). Impact of advance payments of tax on profit on effectiveness of investments. *Mathematics, 10*(4), 666. https://doi.org/10.3390/math10040666

Franc-Dąbrowska, J., et al. (2021). Energy sector risk and cost of capital assessment—Companies and investors perspective. *Energies, 14*(6), 1613. https://doi.org/10.3390/en14061613

Hamada, R. (1969). Portfolio analysis, market equilibrium, and corporate finance. *The Journal of Finance, 24*(1), 13–31.

Harris, R., & Pringle, J. (1985). Risk–adjusted discount rates—Extension form the average–risk case. *The Journal of Financial Research, 8*(3), 237–244.

Leland, H. (1994). Corporate debt value, bond covenants, and optimal capital structure. *The Journal of Finance, 49*(4), 1213–1252.

Luiz, K., Cruz, M., et al. (2015). The relevance of capital structure on firm performance: a multivariate analysis of publicly traded Brazilian companies. *REPeC, Brasília, 9*(4), 384–401. https://doi.org/10.17524/repec.v9i4.1313

Massimiliano, B. (2011). On the risk–neutral value of debt tax shields. *Applied Financial Economics, 22*(3), 251–258. https://doi.org/10.1080/09603107.2010.613754

Miller, M. (1977). Debt and taxes. *The Journal of Finance, 32*(2), 261–275. https://doi.org/10.1111/j.1540-6261.1977.tb03267.x

Modigliani, F., & Miller, M. (1958). The cost of capital, corporate finance, and the theory of investment. *American Economic Review, 48*, 261–297.

Modigliani, F., & Miller, M. (1963). Corporate income taxes and the cost of capital: A correction. *American Economic Review, 53*, 147–175.

Modigliani, F., & Miller, M. (1966). Some estimates of the cost of capital to the electric utility industry 1954–1957. *American Economic Review, 56*, 333–391.

Myers, S. (2001). Capital structure. *The Journal of Economic Perspectives, 15*(2), 81–102.

Pavel, Z. (2018). The impact of cash flows and weighted average cost of capital to enterprise value in the oil and gas sector. *Journal of Reviews on Global Economics, 7*, 138–145.

Chapter 11
Benefits of Advance Payments of Tax on Profit: Consideration Within Brusov–Filatova–Orekhova (BFO) Theory

11.1 Introduction

The importance of accounting for upfront payments rests on two points: (1) companies can pay tax payments upfront with subsequent adjustments if they earn a stable and predictable income; (2) the development of a methodology for advance payments allows the regulator to expand this practice, which ensures an increase in the stability of budget revenues. As results of this chapter show the advance payments of tax on income are quite important for both parties: for companies and for regulator.

We give below the short review of evolution of the theory of capital cost and capital structure.

The Modigliani-Miller (MM) theory, created by Nobel Prize winners in 1958 (Modigliani and Miller 1958), was the first quantitative theory of capital structure. It was based on many restrictions, the main of which were the absence of taxes and the perpetual nature of all financial flows and companies. The first limitation was removed by the MMs themselves (Modigliani and Miller 1963, 1966) and the following formulas were obtained:

$$\text{For tax shield, TS} : \text{TS} = \text{DT}; \tag{11.1}$$

For company value, V (financially independent company) V_0, leverage company, V : $V_0 = \text{CF}/k_0$ and $V = \text{CF}/\text{WACC}$;

$$\tag{11.2}$$

$$\text{For weighted average cost of capital, WACC} : WACC = k_0 \cdot (1 - w_d t); \tag{11.3}$$

$$\text{For equity cost, k}_\text{e} : k_e = k_0 + L(k_e - k_d) \cdot (1 - t). \tag{11.4}$$

Hamada (1969) has derived the following formula for the leveraged company equity cost, accounting both financial and business risk of company, united the Modigliani–Miller theory with Capital Asset Pricing Model (CAPM):

P. Brusov, T. Filatova, *Efficiency of Investment Projects*, Contributions to Finance and Accounting, https://doi.org/10.1007/978-3-032-01028-5_11

$$k_e = k_F + (k_M - k_F)b_U + (k_M - k_F)b_U \frac{D}{S}(1-T), \tag{11.5}$$

where b_U is the β-coefficient of the company of the same group of business risk, that the considering company, but with L = 0.

Merton Miller (1977) took into account the corporate and individual taxes He got for the capitalization of the financially independent company the following formula:

$$V = V_0 + \left[1 - \frac{(1-T_C)(1-T_S)}{(1-T_D)}\right] \cdot D \tag{11.6}$$

Here T_C—the tax on corporate income rate, T_S—the tax rate on income of an individual investor from his ownership by corporation stock, T_D—tax rate on interest income from the provision of investor-individuals of credits to other investors and companies.

In 2008, Brusov–Filatova–Orekhova (Brusov et al. 2018, 2022) have lifted up the limitation concerning the perpetuity of flows and companies and created the modern theory of capital cost and capital structure—BFO theory—which describes the companies of arbitrary age (and arbitrary life-time). The generalization of MM theory for the companies of arbitrary age (and arbitrary life-time) required the modification of the valuation of the tax shield TS, as well as of the valuation of the company capitalization: financially independent, V_0 as well as financially dependent, V.

The following formulas were obtained: for the tax shield, TS, company value, V, and for weighted average cost of capital WACC:

$$\mathrm{TS} = k_d DT \sum_{t=1}^{n} (1+k_d)^{-t} = DT[1-(1+k_d)^{-n}]. \tag{11.7}$$

$$V_0 = \mathrm{CF}[1-(1+k_0)^{-n}]/k_0; \quad V = \mathrm{CF}[1-(1+\mathrm{WACC})^{-n}]/\mathrm{WACC}. \tag{11.8}$$

$$\frac{1-(1+\mathrm{WACC})^{-n}}{\mathrm{WACC}} = \frac{1-(1+k_0)^{-n}}{k_0[1-\omega_d T(1-(1+k_d)^{-n})]}. \tag{11.9}$$

Here, S—the value of equity capital of the company, $w_d = \frac{D}{D+S}$—the share of debt capital, $k_e, w_e = \frac{S}{D+S}$—the cost and the share of the equity capital of the company, and $L = D/S$—financial leverage, D—the value of debt capital.

At $n = 1$, from BFO formula for WACC we get Myers (2001) formula for 1-year company

$$\mathrm{WACC} = k_0 - \frac{(1 + k_0)k_\mathrm{d}}{1 + k_\mathrm{d}} w_\mathrm{d} T; \tag{11.10}$$

and at $n = \infty$ we get Modigliani–Miller formula for WACC (Modigliani and Miller 1963):

$$WACC = k_0 \cdot (1 - w_d t) \tag{11.11}$$

Note, that before BFO discovery in 2008 there were two results only: Modigliani–Miller for perpetuity companies (Modigliani and Miller 1958, 1963, 1966) and Myers one for a year companies (Myers 2001). The accounting of the company's finite age, as it was shown by BFO authors leads to significant changes of all Modigliani–Miller results (Modigliani and Miller 1958, 1963, 1966). Besides, a number of qualitatively new effects in corporate finance, obtained in Brusov–Filatova–Orekhova theory (Brusov et al. 2018, 2022) are absent in Modigliani–Miller theory.

Recently we have modified the Modigliani–Miller theory for the case of advanced payments of income tax (Brusov et al. 2022, 2020) and have shown that obtained results are quite different from ones in "classical" Modigliani–Miller theory.

11.2 Modification of the Brusov–Filatova–Orekhova (BFO) Theory for Companies with Frequent Payments of Tax on Income

11.2.1 Calculation of the Tax Shield

Let us calculate the tax shield within Brusov–Filatova–Orekhova theory for the case of advance payments of tax on profit. The tax shield, *TS*, for period of n-years is equal to sum of discounted values of benefits from the use of tax incentives

$$(TS)_n = k_d Dt + \frac{k_d Dt}{(1 + k_d)} + \frac{k_d Dt}{(1 + k_d)^2} + \ldots + \frac{k_d Dt}{(1 + k_d)^{n-1}} \tag{11.12}$$

This expression represents a geometric progression with denominator

$$q = \frac{1}{(1 + k_d)} \tag{11.13}$$

After summing the progression, one obtains:

$$(TS)_n = \frac{k_d Dt(1-(1+k_d)^{-n})}{\left(1-(1+k_d)^{-1}\right)} = Dt(1-(1+k_d)^{-n})(1+k_d) \tag{11.14}$$

In classical Brusov–Filatova–Orekhova theory under payments at the end of periods one has

$$(TS)_n = Dt(1-(1+k_d)^{-n}) \tag{11.15}$$

11.2.1.1 Company Value

For the financially dependent company value V we have:

$$V = V_0 + (TS)_n \tag{11.16}$$

$$V = V_0 + Dt(1-(1+k_d)^{-n})\cdot(1+k_d). \tag{11.17}$$

This formula differs from the equation for company value V in case of payments of income tax at the end of the year

$$V = V_0 + Dt(1-(1+k_d)^{-n}). \tag{11.18}$$

11.2.1.2 The Weighted Average Cost of Capital, WACC

Substituting

$$D = w_d V \tag{11.19}$$

into (11.17), we get

$$V\left(1 - w_d t(1-(1+k_d)^{-n})\cdot(1+k_d)\right) = V_0 \tag{11.20}$$

Using the values of financially dependent company, V, and financially independent company, V_0,

$$V = \frac{CF(1-(1+WACC)^{-n})}{WACC}; V_0 = \frac{CF}{k_0}\cdot(1-(1+k_0)^{-n}), \tag{11.21}$$

we come to the following intermediate equation

$$\frac{CF(1-(1+WACC)^{-n})}{WACC}\cdot(1-w_d t\cdot(1-(1+k_d)^{-n}))\cdot(1+k_d)$$
$$=\frac{CF}{k_0}\cdot(1-(1+k_0)^{-n}) \tag{11.22}$$

and then to the final Brusov–Filatova–Orekhova equation for WACC for the case of advanced payments of income tax

$$\frac{(1-(1+WACC)^{-n})}{WACC}=\frac{(1-(1+k_0)^{-n})}{k_0(1-w_d t\cdot(1-(1+k_d)^{-n}))\cdot(1+k_d)}. \tag{11.23}$$

This formula differs from the classical Brusov–Filatova–Orekhova equation by the factor $(1 + k_d)$ in the left denominator

$$\frac{(1-(1+WACC)^{-n})}{WACC}=\frac{(1-(1+k_0)^{-n})}{k_0(1-w_d t\cdot(1-(1+k_d)^{-n}))} \tag{11.24}$$

11.2.2 Calculation of the Equity Cost

To calculate the equity cost ke one should use the equation of WACC definition

$$WACC=k_e w_e+k_d w_d(1-t) \tag{11.25}$$

from where one gets the expression for ke

$$k_e=\frac{WACC}{w_e}-\frac{k_d(1-t)w_d}{w_e}=WACC(1+L)-Lk_d(1-t). \tag{11.26}$$

One should substitute WACC from the formula (11.12), calculating the equity cost ke for the case of advanced payments of income tax and from the formula (11.13), calculating the equity cost ke for the case of payments of income tax at the end of the year.

11.3 Results

Here the dependence of the weighted average cost of capital, WACC, capital value, V, equity cost, ke, on leverage level L for 3-year and 6-year companies, using Microsoft Excel is being studied. We consider two types of payments of income tax: (1) at the end of the year and (2) in advance. As we mentioned above, for WACC we use formulas (11.12) and (11.13), for capital value, V, we use formulas (11.6) and (11.7) and for equity cost, ke, we use formula (11.15).

We use the following parameters: $k_0 = 0.2$; $k_d = 0.18$; $t = 0.2$; $n = 3$; 6; $CF = 100$.

11.3.1 Dependence of the Weighted Average Cost of Capital, WACC, Capital Value, V, Equity Cost, ke, on Leverage Level L for 3-Year Company

11.3.2 Dependence of the Weighted Average Cost of Capital, WACC, Capital Value, V, Equity Cost, ke, on Leverage Level L for 6-Year Company

11.4 Comparison of Results for 3-Year and 6-Year Companies

11.5 Discussion

We derive BFO formulas for WACC, V, ke for the case of advanced payments of income tax. Making the calculations using these formulas within Microsoft Excel we get the following results (see Tables 11.1 and 11.2 and Figs. 11.1, 11.2, 11.3, 11.4, 11.5, 11.6, 11.7, 11.8, and 11.9).

1. WACC(L) decreases with L in both cases: advanced payments of income tax and payments at the ends of years. This means that debt financing is important and should be used by company—it leads to decrease of attracting capital cost with L.
2. WACC turns out to be lower in the case of advanced payments of income tax, this tells about importance of the use of advanced payments of income tax for companies.
3. WACC decreases with company age: this is one of the important results of classical BFO theory.

Table 11.1 Dependence of the weighted average cost of capital, WACC, equity cost, ke and company value, V, on leverage level L in the cases of payments of tax on profit at the end of the year (1) and in the beginning of the year (2) for 3-year company

n = 3	WACC		ke		V	
L	1	2	1	2	1	2
1	0.1749	0.1703	0.2057	0.1966	219.2281	220.8472
2	0.1664	0.1603	0.2113	0.1930	222.2455	224.4700
3	0.1622	0.1553	0.2168	0.1892	223.7856	226.3263
4	0.1596	0.1523	0.2222	0.1855	224.7199	227.4549
5	0.1579	0.1503	0.2277	0.1817	225.3471	228.2136
6	0.1567	0.1489	0.2331	0.1780	225.7973	228.7586
7	0.1558	0.1478	0.2386	0.1742	226.1361	229.1691
8	0.1551	0.1469	0.2440	0.1704	226.4003	229.4893
9	0.1545	0.1463	0.2495	0.1666	226.6122	229.7462
10	0.1541	0.1457	0.2549	0.1628	226.7858	229.9568

Table 11.2 Dependence of the weighted average cost of capital, WACC, equity cost, ke and company value, V, on leverage level L in the cases of payments of tax on profit at the end of the year (1) and in the beginning of the year (2) for 6-year company

n = 6	WACC		ke		V	
L	1	2	1	2	1	2
1	0.1743	0.1697	0.2047	0.1953	354.8940	359.2385
2	0.1657	0.1593	0.2090	0.1900	363.0242	369.1124
3	0.1613	0.1542	0.2131	0.1846	367.2305	374.2557
4	0.1586	0.1510	0.2172	0.1791	369.8015	377.4110
5	0.1569	0.1489	0.2213	0.1736	371.5355	379.5443
6	0.1556	0.1474	0.2254	0.1680	372.7841	381.0830
7	0.1547	0.1463	0.2294	0.1624	373.7261	382.2451
8	0.1539	0.1454	0.2335	0.1569	374.4620	383.1540
9	0.1534	0.1447	0.2376	0.1513	375.0529	383.8841
10	0.1529	0.1442	0.2416	0.1457	375.5377	384.4836

4. Company value, V, increases with L in both cases, this follows from the decrease of attracting capital cost with L.
5. Company value, V, turns out to be bigger in the case of advanced payments of income tax: this tells about importance of the use of advanced payments of income tax for companies.
6. Company value, V, increases with company age.
7. Equity cost ke decreases with company age in both cases: this is one of the important results of classical BFO theory.
8. Equity cost ke increases with leverage level L in the case of payments of income tax at the end of the year.
9. Equity cost ke decreases with leverage level L in the case of advanced payments of income tax. This means the appearance of a qualitatively new effect that can

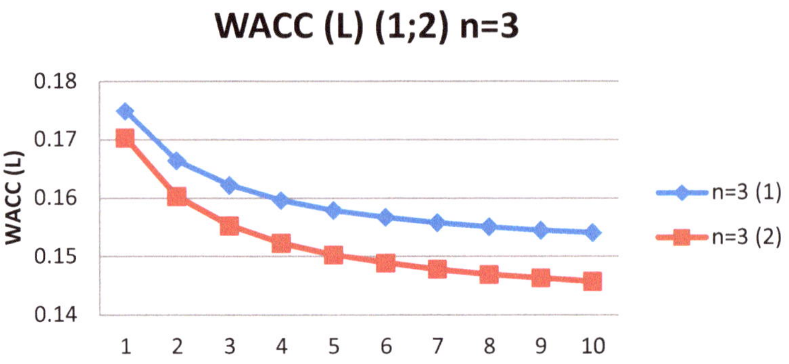

Fig. 11.1 Dependence of the weighted average cost of capital, WACC, on leverage level L in the cases of payments of tax on profit at the end of the year (1) and in the beginning of the year (2) for 3-year company

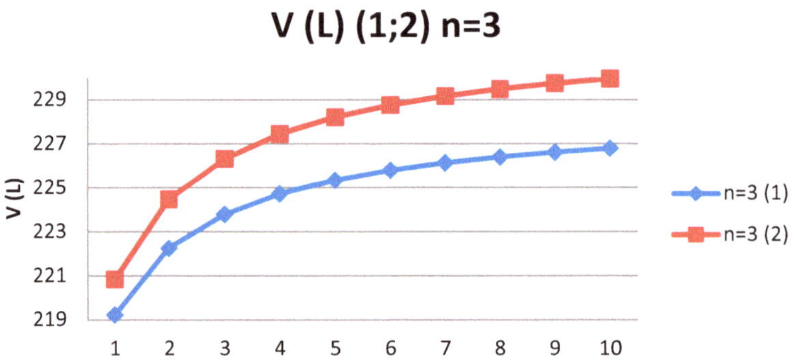

Fig. 11.2 Dependence of the company value, V, on leverage level L in the cases of payments of tax on profit at the end of the year (1) and in the beginning of the year (2) for 3-year company

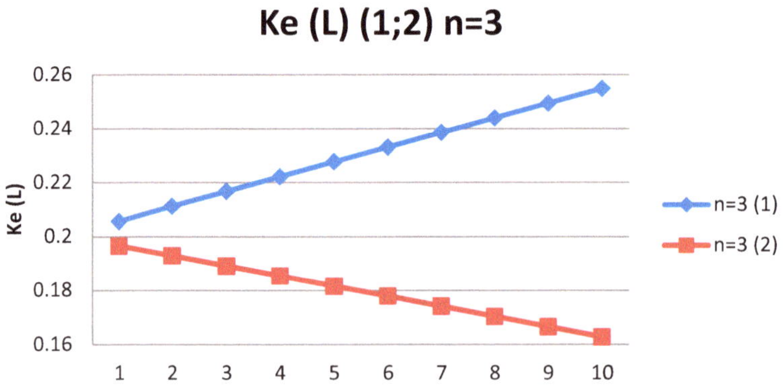

Fig. 11.3 Dependence of the equity cost, ke on leverage level L in the cases of payments of tax on profit at the end of the year (1) and in the beginning of the year (2) for 3-year company

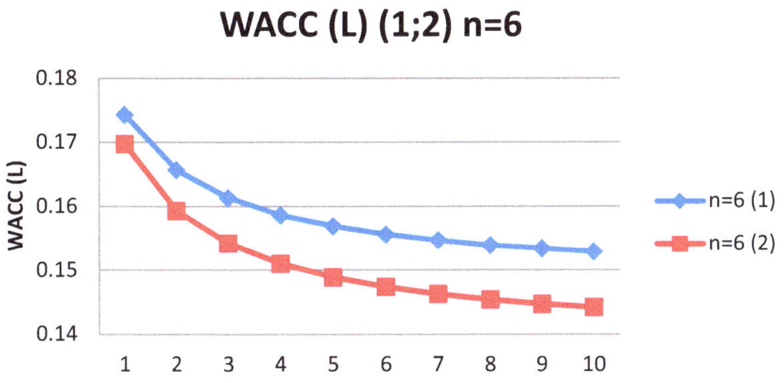

Fig. 11.4 Dependence of the weighted average cost of capital, WACC, on leverage level L in the cases of payments of tax on profit at the end of the year (1) and in the beginning of the year (2) for 6-year company

Fig. 11.5 Dependence of the company value, V, on leverage level L in the cases of payments of tax on profit at the end of the year (1) and in the beginning of the year (2) for 6-year company

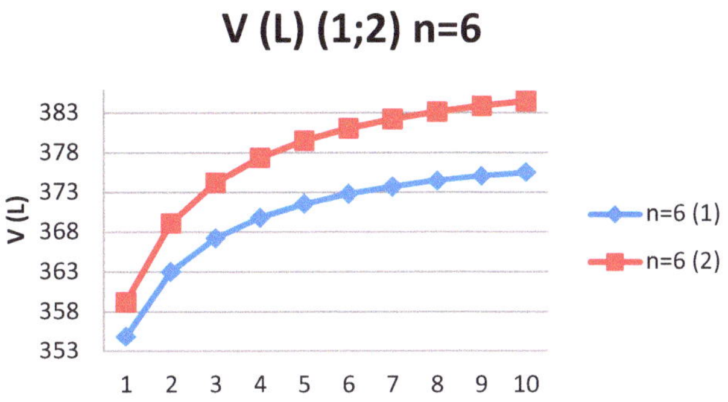

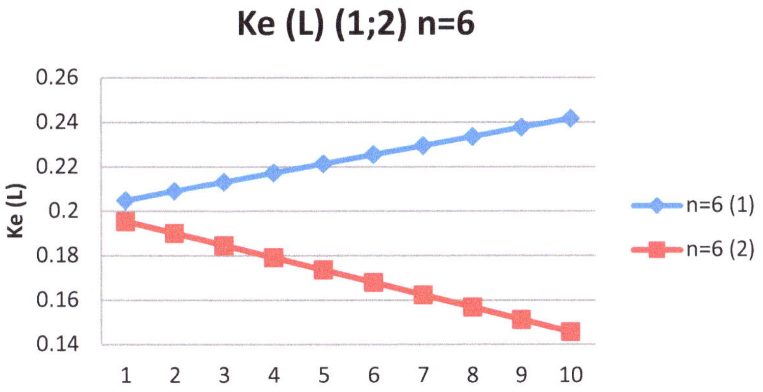

Fig. 11.6 Dependence of the equity cost, ke on leverage level L in the cases of payments of tax on profit at the end of the year (1) and in the beginning of the year (2) for 3-year company

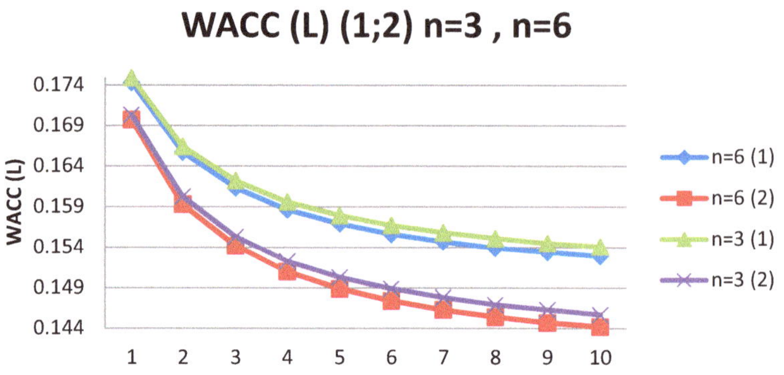

Fig. 11.7 Dependence of the weighted average cost of capital, WACC, on leverage level L in the cases of payments of tax on profit at the end of the year (1) and in the beginning of the year (2) for 3-year and 6-year companies

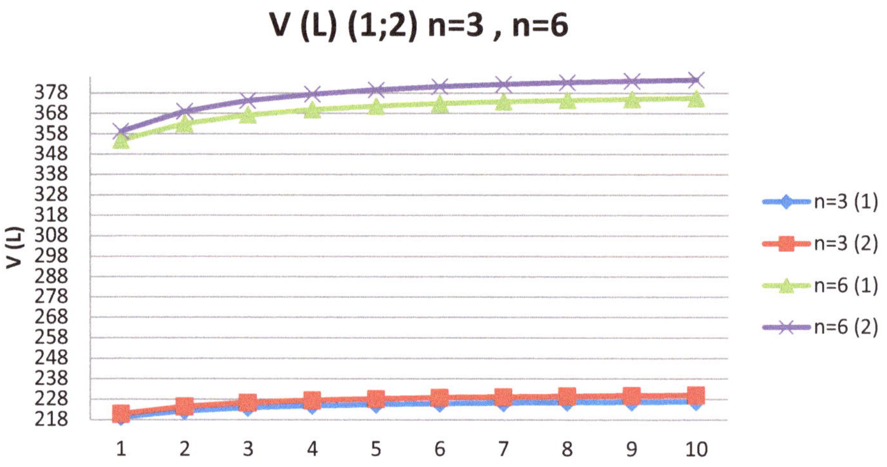

Fig. 11.8 Dependence of the company value, V, on leverage level L in the cases of payments of tax on profit at the end of the year (1) and in the beginning of the year (2) for 3-year and 6-year companies

greatly change the company's dividend policy, because the economically justified amount of dividends is equal to the equity cost.

11.6 Summary and Conclusions

Advance income tax payments are beneficial to both parties: to companies, because they lead to decrease of cost of attracting capital and increase of company values; to regulator, because earlier replenishment of the budget ensures an increase in the

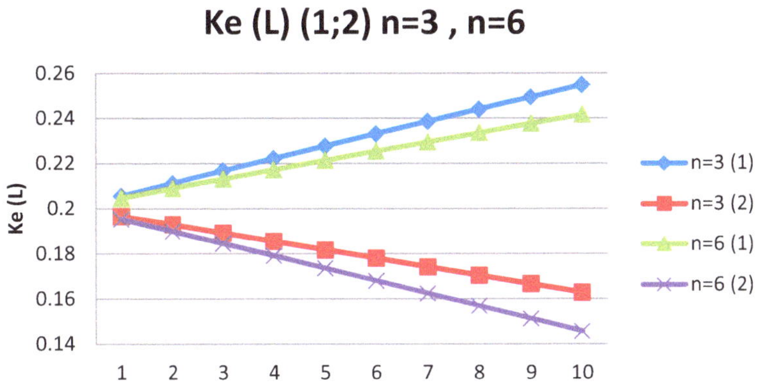

Fig. 11.9 Dependence of the equity cost, ke, on leverage level L in the cases of payments of tax on profit at the end of the year (1) and in the beginning of the year (2) for 3-year and 6-year companies

stability of budget revenues. Thus the regulator should extend the practice of tax payments in advance by the companies.

An important conclusion drawn in this chapter is that the tax shield is very important and the way it is formed (payments at the end of the year or in advance) leads to very important consequences, changing all the financial indicators of the company, such as the cost of raising capital and company value and radically changing the company's dividend policy. If with payments at the end of the year, the amount of dividends should increase with the increase in the use of debt financing, with advance payments of income tax the amount of dividends should decrease with the increase in the use of debt financing: this is a pioneering result that radically changes the company's dividend policy.

References

Brusov, P., Filatova, T., Orehova, N., & Eskindarov, M. (2018). *Modern corporate finance, investments, taxation and ratings* (2nd ed., pp. 1–571). Springer Nature Publishing.

Brusov, P. N., Filatova, T. V., Orekhova, N. P., Kulik, V. L., Chang, S.-I., & Lin, Y. C. G. (2020). Modification of the Modigliani–Miller theory for the case of advance payments of tax on profit. *Journal of Reviews on Global Economics, 2020*(9), 257–267.

Brusov, P., Filatova, T., & Orehova, N. (2022). *Generalized Modigliani–Miller theory: Applications in corporate finance, investments, taxation and ratings* (pp. 1–362). Springer Nature Publishing. Retrieved from https://rd.springer.com/book/11.1007/978-3-030-93893-2

Hamada, R. (1969). Portfolio analysis, market equilibrium, and corporate finance. *The Journal of Finance, 24*, 13–31.

Miller, M. (1977). Debt and taxes. *The Journal of Finance, 32*(2), 261–275. https://doi.org/10.1111/j.1540-6261.1977.tb03267.x

Modigliani, F., & Miller, M. (1958). The cost of capital, corporate finance, and the theory of investment. *The American Economic Review, 48*, 261–297.

Modigliani, F., & Miller, M. (1963). Corporate income taxes and the cost of capital: A correction. *The American Economic Review, 53*, 147–175.

Modigliani, F., & Miller, M. (1966). Some estimates of the cost of capital to the electric utility industry 1954–1957. *The American Economic Review, 56*, 333–391.

Myers, S. (2001). Capital structure. *The Journal of Economic Perspectives, 15*(2), 81–102.

Chapter 12
Influence of Method and Frequency of Profit Tax Payments on Company Financial Indicators

12.1 Introduction

In practice, profit tax payments are (1) made more frequently than annually and (2) can be made in advance. Recently, the Brusov–Filatova–Orekhova (BFO) theory (Brusov et al. 2022) was generalized for the case of profit tax payments with an arbitrary frequency. Using modified BFO formulae, it was shown that all financial indicators of a company depend on the frequency of profit tax payments. The weighted average cost of capital (WACC) decreases with the payments and the company value increases with the payments. A new effect was thus discovered: the decrease in equity cost with an increase in the level of leverage (L). Due to the fact that the economically justified amount of dividends is equal to the equity cost, this can significantly change the dividend policy of a company.

More frequent profit tax payments are beneficial for both parties: the company and the tax regulator. For the company, this practice increases the company value and it helps the tax regulator due to the time value of money.

In current chapter, for the first time, the Brusov–Filatova–Orekhova (BFO) theory was generalized for the case of advance profit tax payments with an arbitrary frequency. We compared the obtained results to those that we described recently for profit tax payments at the end of the financial period (Brusov et al. 2022) and found them to be totally different.

We showed that in spite the fact that the WACC decreases with the payments and the company value increases with the payments in this case (Brusov et al. 2022), the WACC values turned out to be bigger and the company value turned out to be smaller than in case of advance profit tax payments of any frequency. This underlined the importance of advance profit tax payments.

The novelty and originality of this study lie in the fact that the simultaneous influence of two effects was investigated for the first time: (1) advance profit tax payments and (2) frequent profit tax payments.

P. Brusov, T. Filatova, *Efficiency of Investment Projects*, Contributions to Finance and Accounting, https://doi.org/10.1007/978-3-032-01028-5_12

Before this study, we generalized the Brusov–Filatova–Orekhova (BFO) theory to (1) the case of frequent profit tax payments at the end of the financial period and (2) the case of advance annual profit tax payments.

In this article, we also compared the obtained results to the results of consideration (1) and found a large difference between the results. Thus, in this chapter, an in-depth study of frequent profit tax payments was carried out for both possible cases of payments: advance and at the end of the financial period.

The robustness of the outcomes was provided by the use of the modern Brusov–Filatova–Orekhova (BFO) theory of capital cost and capital structure, which has been tested in real economic practice and its correct generalization for the case of frequent advance profit tax payments has been proven.

The structure of the chapter is as follows.

In the introduction, we present a literature review of the development of the capital structure and capital cost theory (Sect. 12.1.1). The Brusov–Filatova–Orekhova (BFO) theory is generalized for frequent advance profit tax payments for the first time in Sect. 12.2, The modified formulae are derived for the tax shield in Sect. 12.2.1, company value in Sect. 12.2.2 and the equity cost and WACC in Sect. 12.2.3.

In Sect. 12.3, using the generalized BFO formulae that were obtained in Sect. 12.2 and Microsoft Excel, we numerically calculate the dependence of a company's financial parameters, such as WACC, company value and equity cost, on the level of leverage for an arbitrary frequency of profit tax payment for 3- and 6-year-old companies (Sects. 12.3.1 and 12.3.2, respectively).

We considered frequent advance profit tax payments. In Sect. 12.3, we also compare the obtained results to those that were found previously for frequent profit tax payments tax at the end of the financial period (Brusov et al. 2022). This section also compares the results for companies of different ages (3- and 6-year-old companies). In Sect. 12.4, the discussion and conclusions are presented.

12.1.1 A Literature Review on the Development of the Capital Cost and Capital Structure Theory

In this section, we present an overview of the development of the capital cost and capital structure theory. Following the traditional empirical approaches, the first quantitative theory of capital cost and capital structure was proposed the Nobel Prize winners Modigliani and Miller (the MM theory) (1958). Among the many limitations of this theory, two main shortcomings can be distinguished: the non-accounting of taxes and the infinite lifetime of companies. The first constraint was removed by Modigliani and Miller themselves (Modigliani and Miller 1963, 1966), who obtained the following formulae.

The tax shield (*TS*) had the following form:

$$TS = DT \tag{12.1}$$

Company value (V) and financially independent companies (V_0) were expressed as:

$$V_0 = \mathrm{CF}/k_0 \quad \text{and} \quad V = \mathrm{CF}/\mathrm{WACC}. \tag{12.2}$$

For *WACC*, Modigliani and Miller obtained the following formula:

$$WACC = k_0 \cdot (1 - w_d t) \tag{12.3}$$

For equity cost (k_e), they had:

$$k_e = k_0 + L(k_e - k_d) \cdot (1 - t) \tag{12.4}$$

In 1969, the capital asset pricing model (CAPM) was combined with the Modigliani–Miller theory by Hamada (1969), which accounted for both the business and financial risk of companies and used the below formula for the equity cost (k_e) of financially dependent companies:

$$k_\mathrm{e} = k_\mathrm{F} + (k_\mathrm{M} - k_\mathrm{F})b_\mathrm{U} + (k_\mathrm{M} - k_\mathrm{F})b_\mathrm{U}\frac{D}{S}(1 - T), \tag{12.5}$$

where b_U is the β-coefficient of a company with $L = 0$.

Corporate and individual taxes were then taken into account by Miller (1977). The following formula was obtained for unleveraged company value:

$$V = V_0 + \left[1 - \frac{(1 - T_C)(1 - T_S)}{(1 - T_D)}\right] \cdot D \tag{12.6}$$

where T_S is the profit tax rate for an individual investor for their ownership of corporation stock, T_C is the tax rate for corporate profit and T_D is the profit tax interest rate for the provision of credit to other investors or companies.

In Farber et al. (2006), Fernandez (2006), Berk and De Marzo (2007), and Harris and Pringle (1985), a more general expression for *WACC* than the Modigliani–Miller (MM) version was obtained (Farber et al. 2006):

$$WACC = k_0(1 - w_d t) - k_d w_d t + k_{TS} t w_d \tag{12.7}$$

where k_0 is the equity cost of the unleveraged company, k_d is the debt cost, k_{TS} is the expected return on the tax shield, V is the leveraged company value, V_{TS} is the value of the tax shield and D is the debt value.

Derived in Berk and De Marzo (2007), formula (12.7) is more general than the Modigliani–Miller (MM) expression, so some additional conditions are necessary for its use in practice. When the WACC value is constant over time, the leveraged company value can be found by discounting the *WACC* from the unleveraged company cash flows. For this specific case, the formulae can be found in textbooks (Berk and De Marzo 2007; Harris and Pringle 1985).

In 1963, Modigliani and Miller supposed that the level of debt was constant. The expected after-tax cash flow of a financially independent company was also fixed, thus V_0 was constant. By that assumption, $kTS = kd$ and the tax shield value was $TS = tD$. Thus, for the financially dependent company value, we used the ordinary Modigliani–Miller (MM) formula for *WACC* instead of formula (12.7):

$$WACC = k_0(1 - w_d t) \tag{12.8}$$

From our point of view, the "classical" Modigliani–Miller (MM) theory supposed that the debt cost and expected return on the tax shield were equal (due to the fact that both of them are debt-like in nature), which is much more reasonable and so we modified the "classical" Modigliani–Miller (MM) in Brusov et al. (2020).

A study on the influence of tax pressure on the financial balance of energy companies was carried out in Batrancea (2021a). The results showed that tax pressure has a stronger influence on the financial equilibrium (both in the short term and long term) of oil and electricity companies than gas companies. These results are useful for the managers of energy companies who underestimate the evolution of the financial equilibrium of the company because they take into account various possible financial crises.

A study on the influence of financial liquidity and financial solvency on the performance of healthcare companies was conducted in Batrancea (2021b) using econometric models. From the empirical evidence, it follows that certain financial parameters such as current liquidity ratio, quick liquidity ratio and financial leverage significantly influence company performance, as measured by the return on assets, gross margin ratio, operating margin ratio, taxes, earnings before interest, amortization and depreciation.

The limitation that was related to the infinite lifetime of companies was removed in 2008 by Brusov, Filatova and Orekhova (Brusov et al. 2018), who developed the modern theory of capital cost and capital structure (the BFO theory), which is applicable to companies of all ages.

Modifications to the calculations for the tax shield and company value (unleveraged, V_0; leveraged, V) were required for the generalization of the MM theory (see the formulae below):

$$\mathrm{TS} = k_\mathrm{d} DT \sum_{t=1}^{n} (1 + k_\mathrm{d})^{-t} = DT[1 - (1 + k_\mathrm{d})^{-n}], \tag{12.9}$$

$$V_0 = \mathrm{CF}[1 - (1 + k_0)^{-n}]/k_0; \quad V = \mathrm{CF}[1 - (1 + \mathrm{WACC})^{-n}]/\mathrm{WACC}, \tag{12.10}$$

$$\frac{1 - (1 + \mathrm{WACC})^{-n}}{\mathrm{WACC}} = \frac{1 - (1 + k_0)^{-n}}{k_0[1 - \omega_\mathrm{d} T(1 - (1 + k_\mathrm{d})^{-n})]}, \tag{12.11}$$

where S is the equity capital value, $w_d = \frac{D}{D+S}$ is the debt capital, $k_e, w_e = \frac{S}{D+S}$ is the equity capital, $L = D/S$ is the value of financial leverage and D is the debt value.

The Myers formula for a 1-year company (Myers 2001) can be easily obtained from formula (12.11) by substituting $n = 1$:

$$\mathrm{WACC} = k_0 - \frac{(1 + k_0)k_d}{1 + k_d} w_d T \tag{12.12}$$

Then, by substituting $n = \infty$, we arrive at the Modigliani–Miller formula for *WACC* (Modigliani and Miller 1958):

$$WACC = k_0 \cdot (1 - w_d t) \tag{12.13}$$

Until 2008, when the BFO theory was created, there were only two options: the Modigliani–Miller theory for perpetual companies (Modigliani and Miller 1958, 1963, 1966) and the Myers theory for 1-year companies (Myers 2001). As the authors of the BFO showed, taking the finite age of a company into account leads to significant changes in all Modigliani–Miller results (Modigliani and Miller 1958, 1963, 1966). In addition, a number of the new effects that are observed within corporate finance using the Brusov–Filatov–Orekhova theory (Brusov et al. 2018) are absent in the Modigliani–Miller theory.

In Brusov et al. (2020), the Modigliani-Miller theory was modified for the case of advance profit tax payments and significant differences were found between the modified theory and the "classical" Modigliani-Miller theory.

In the literature, the methodology and results of the Brusov–Filatova–Orekhova (BFO) theory are well known (for example, see Dimitropoulos 2014; Luiz and Cruz 2015; Barbi 2011; Zhukov 2018; Franc-Dąbrowska et al. 2021; Angotti et al. 2018; Vergara-Novoa et al. 2018; Mundi et al. 2022; Sadiq et al. 2022; Becker 2022; El-Chaarani et al. 2022; Huang et al. 2020). A number of chapters (see, for example Vergara-Novoa et al. 2018; Mundi et al. 2022; Sadiq et al. 2022) have used the BFO theory in practical calculations.

The authors of Franc-Dąbrowska et al. (2021) valuated the capital costs of energy companies by including an investor and market risk approach. The study also concerned a *WACC* intra-industry analysis the companies, which was related to the sector characteristics of revenue, total assets, company age and market capitalization.

In Mundi et al. (2022), the influence of the overconfidence of finance managers on the capital structure decisions of family-run businesses in India was studied. It was shown that the capital structure decisions of managers could be explained by measurable managerial characteristics. In Sadiq et al. (2022), the correlation between company risk and capital structure was explored using datasets from sugar and cement companies in Pakistan. It was shown that the role of risk assessment and capital structure is vital for the sustainable growth of companies and the increase in the wealth of shareholders.

The authors of Becker (2022) considered the adjusted present value method, the free cash flow (FCF) method, the flow-to-equity method and the relationships between these methods. The authors departed from Modigliani and Miller method and used a stationary FCF method and the Miles and Ezzell method to consider FCF as a first-order autoregressive possession. The authors derived DCF valuation formulae for annuities. The authors found (1) the correct discount rate for the tax shield when the free cash flow is a first-order autoregressive annuity and (2) the tax shield valuation of a first-order autoregressive free cash flow annuity.

The author of El-Chaarani et al. (2022) studied the impacts of internal and external corporate governance mechanisms on the financial performance of banks in the MENA region during the COVID-19 pandemic. The results showed that the corporate governance measures of the presence of independent members on the board of directors, strong legal protection, a high concentration of ownership and the lack of political pressure on board members had positive effects on the financial indicators of banks.

In Huang et al. (2020), the influence of leverage level on the expected equity returns of listed companies was studied. The existence of an optimal capital structure to maximize the expected equity returns was discussed. The authors thought that this problem should be discussed from the perspective of maximizing shareholder wealth and not from the perspective of maximizing company value due to the existence of financial distress costs and agency costs. Using Gebhardt–Lee–Swaminathan (GLS) model to measure the implied capital costs and to measure the expected equity returns, the authors empirically found that the leverage level had a positive correlation with the equity returns of a listed company. The relationship between the leverage level and the expected equity returns appeared to an inverse "U":

- When the leverage level fell below the optimal level, increasing the leverage could increase the expected equity returns;
- When the leverage level was higher than the optimal level, deleveraging was beneficial to increase the expected equity returns.

Within a standard capital structure model, the authors of Islam and Khandaker (2015) investigated the leverage decisions of over 1500 companies that were listed in the ASX for 2000–2012 and divided the sample into mining and non-mining industries. Applying a dummy variable approach, the authors showed that there were fundamental differences between two these types of companies when making leverage decisions. It turned out that mining companies were more sensitive to profitability and asset tangibility, whereas neither profitability nor asset tangibility were significant for non-mining companies. The obtained results suggested that industry type matters for companies when making leverage decisions.

In Barbi (2011), the authors discussed the disagreement in the financial literature about the meaning of the "value of tax shields". Although it is accepted that the tax deductibility of interest increases the value of a firm, the correct valuation of this added value is controversial. Using a risk-neutral approach, the authors derived a general formula for the valuation of tax shields, which shows that this value equals the summation of the discounted future tax savings. As soon as the authors specified a leverage level policy and cash flow dynamics, they obtained some well-known formulae for the calculation of tax shields.

12.2 The Modified Brusov–Filatova–Orekhova (BFO) Theory for the Case of Frequent Advance Profit Tax Payments

In this section, we generalize the Brusov–Filatova–Orekhova (BFO) theory for the case of frequent advance profit tax payments. The expressions for the tax shield, company value, *WACC* and equity cost are presented in Sects. 12.2.1, 12.2.2 and 12.2.3, respectively.

In this section, we use the following definitions.

D	Debt Capital Value
S	Equity Capital Value
$k_{\mathrm{d}}, w_{\mathrm{d}} = \frac{D}{D+S}$	Debt Capital Cost and Share
$k_{\mathrm{e}}, w_{\mathrm{e}} = \frac{S}{D+S}$	Equity Capital Cost and Share
$L = D/S$	Leverage Level
$WACC$	Weighted Average Cost of Capital

k_0, the equity cost for a financially independent company; p, the number of payments per year; T, profit tax.

12.2.1 The Tax Shield Calculation

We started with the calculation of the tax shield for the case of frequent advance profit tax payments using the Brusov–Filatova–Orekhova theory (where p is the number of payments per year when payments are made at the beginning of the financial period). For a period of n years, the tax shield (*TS*) was equal to the sum of the discounted values of benefits from the use of tax incentives:

$$(TS)_n = \frac{k_d Dt}{p} + \frac{k_d Dt}{p(1+k_d)^{1/p}} + \frac{k_d Dt}{p(1+k_d)^{2/p}} + \dots + \frac{k_d Dt}{p(1+k_d)^{np/p}} \tag{12.14}$$

We used a geometric progression with the denominator $q = \frac{1}{(1+k_d)^{\frac{1}{p}}}$.

After summation, we obtained:

$$(TS)_n = \frac{k_d Dt(1-(1+k_d)^{-n})}{p\left(1-(1+k_d)^{-1/p}\right)} = \frac{k_d Dt(1-(1+k_d)^{-n})\cdot(1+k_d)^{1/p}}{p\left((1+k_d)^{1/p}-1\right)} \tag{12.15}$$

Using the classical Brusov–Filatova–Orekhova theory at $p = 1$, we obtained:

$$(TS)_n = Dt(1-(1+k_d)^{-n})\cdot(1+k_d) \tag{12.16}$$

It was easy to obtain this result from formula (12.15) by inputting the value of $p = 1$ for the number of profit tax payments.

12.2.2 Derivation of the Modified BFO Formula for the Weighted Average Cost of Capital (WACC)

We now present the modified BFO formula for the weighted average cost of capital (*WACC*) for the case of frequent advance profit tax payments (where p is the number of payments per year when the payments are made at the beginning of the financial period).

For the value of a leveraged company (V), we obtained:

$$V = V_0 + (TS)_n \tag{12.17}$$

where V_0 is the value of an unleveraged company. By substituting formula (12.19) for TS, we obtained:

$$V = V_0 + \frac{k_d Dt(1-(1+k_d)^{-n})\cdot(1+k_d)^{1/p}}{p\left((1+k_d)^{1/p}-1\right)} \tag{12.18}$$

After substituting $D = w_d V$, we obtained:

$$V\left(1-\frac{k_d w_d t(1-(1+k_d)^{-n})\cdot(1+k_d)^{1/p}}{p\left((1+k_d)^{1/p}-1\right)}\right) = V_0 \tag{12.19}$$

Assuming that the values of a leveraged company (V) and an unleveraged company (V_0) were equal to:

$$V = \frac{CF(1-(1+WACC)^{-n})}{WACC};\ V_0 = \frac{CF(1-(1+k_0)^{-n})}{k_0} \tag{12.20}$$

respectively, we obtained:

$$\frac{CF(1-(1+WACC)^{-n})}{WACC}\cdot\left(1-\frac{k_d w_d t(1-(1+k_d)^{-n})\cdot(1+k_d)^{1/p}}{p\left((1+k_d)^{1/p}-1\right)}\right)=\frac{CF}{k_0}\cdot(1-(1+k_0)^{-n}) \tag{12.21}$$

From there, we could derive the modified BFO formula for the WACC of a company of age n years for the case of p profit tax payments per year (payments at the beginning of the financial period):

$$\frac{1-(1+WACC)^{-n}}{WACC}=\frac{CF(1-(1+k_0)^{-n})}{k_0\left(1-\frac{k_d w_d t(1-(1+k_d)^{-n})\cdot(1+k_d)^{1/p}}{p\left((1+k_d)^{1/p}-1\right)}\right)} \tag{12.22}$$

For $p=1$, we obtained the modified BFO formula for advance annual profit tax payments:

$$\frac{1-(1+WACC)^{-n}}{WACC}=\frac{1-(1+k_0)^{-n}}{k_0\cdot(1-w_d t[1-(1+k_d)^{-n}]\cdot(1+k_d))} \tag{12.23}$$

12.2.3 Formulae for the Capital Value and Equity Cost

In Sect. 12.3, we present our investigation into the impact of the frequency of profit tax payments on the dependence of the WACC, capital value and equity cost on the leverage level of companies of different ages (3 or 6 years) using Microsoft Excel. For the WACC, we used formula (12.10) and for capital value and equity cost, we used formulae (12.24) and (12.25), respectively (see below).

The value of a company of age n could be calculated using the following formula:

$$V=\frac{CF\cdot(1-(1+WACC)^{-n})}{WACC} \tag{12.24}$$

Then, the equity cost could be found using the following equation:

$$WACC=k_e w_e+k_d w_d(1-t)$$

$$k_e=\frac{WACC}{w_e}-\frac{k_d(1-t)w_d}{w_e}=WACC(1+L)-Lk_d(1-t) \tag{12.25}$$

where the *WACC* value is found using formula (12.22).

12.3 Results and Discussions

We studied the impact of the frequency of advance profit tax payments on the dependence of the WACC, capital value and equity cost on the leverage level of companies of different ages (3 or 6 years) using Microsoft Excel. We considered the advance profit tax payments using the formulae that were obtained above: formula (12.22) was used for WACC, formula (12.24) was used for capital value and formula (12.25) was used for equity cost. By using the BFO theory, we could study companies of any age. Three- and six-year-old companies were considered as examples to estimate the impact of age on the financial parameters of a company. We used a large database, which can be made available to interested readers upon request. We used data that showed the influence of advance profit tax payments with an arbitrary frequency on the financial parameters of the selected companies. Then, we compared the results of the calculations of two types of profit tax payments: (1) payments at the end of the financial period (Brusov et al. 2022) and (2) at the beginning of the financial period (i.e., in advance).

The following parameters were used: $k_0 = 0.2$; $k_d = 0.18$; $t = 0.2$; $n = 3$ or 6; $CF = 100$. The chosen parameters were typical for companies and the results that were obtained using other parameters were similar and did not show any qualitative differences.

12.3.1 *The Impact of the Frequency of Profit Tax Payments on the Dependence of the Weighted Average Cost of Capital, Capital Value and Equity Cost on the Leverage Level of a 3-Year-Old Company*

In Tables 12.1, 12.2, and 12.3, we present the results of our study on the impact of the frequency of profit tax payments on the dependence of the *WACC*, capital value and equity cost on the leverage level of a 3-year-old company with payment frequencies of $p = 1, 2, 4, 6$ and 12.

As can be seen from Fig. 12.1, the WACC decreased with L for all frequencies (annual, semi-annual, quarterly, monthly, etc.) of advance profit tax payments. So, the debt financing led to a decrease in the corresponding capital cost with L and thus, its use by the company was important. The *WACC* values increased with the frequency of payments, which meant that more infrequent payments were beneficial for the company. The influence of the payment frequency was stronger for annual, semi-annual and quarterly payments and decreased for monthly payments.

As can be seen from Fig. 12.2, the company value (V) increased with L for all frequencies (annual, semi-annual, quarterly, monthly, etc.) of advance profit tax payments. So, the debt financing led to an increase in the company value with L and thus, its use by the company was important. This was consistent with the decrease in *WACC* with the leverage level, as shown in the previous section. The company value

Table 12.1 The dependence of the *WACC* on the leverage level (L) of a 3-year-old company for advance profit tax payments with frequencies of $p = 1, 2, 4, 6$ and 12

L	Wd	*WACC*, $n = 3$ (2)				
		$p = 1$	$p = 2$	$p = 4$	$p = 6$	$p = 12$
1	0.5	0.1703	0.1715	0.1721	0.1723	0.1725
2	0.666667	0.1603	0.1619	0.1627	0.1629	0.1632
3	0.75	0.1553	0.1571	0.1580	0.1583	0.1586
4	0.8	0.1523	0.1542	0.1552	0.1555	0.1558
5	0.833333	0.1503	0.1523	0.1533	0.1536	0.1539
6	0.857143	0.1489	0.1509	0.1519	0.1522	0.1526
7	0.875	0.1478	0.1499	0.1509	0.1512	0.1516
8	0.888889	0.1469	0.1491	0.1501	0.1505	0.1508
9	0.9	0.1463	0.1484	0.1495	0.1498	0.1502
10	0.909091	0.1457	0.1479	0.1490	0.1493	0.1497

Table 12.2 The dependence of the company value (V) on the leverage level (L) of a 3-year-old company for advance profit tax payments with frequencies of $p = 1, 2, 4, 6$ and 12

L	Wd	*V*				
		$p = 1$	$p = 2$	$p = 4$	$p = 6$	$p = 12$
1	0.5	220.8472	220.4234	220.2163	220.1479	220.0800
2	0.666667	224.4700	223.8866	223.6018	223.5079	223.4144
3	0.75	226.3263	225.6593	225.3339	225.2266	225.1199
4	0.8	227.4549	226.7365	226.3861	226.2705	226.1557
5	0.833333	228.2136	227.4603	227.0930	226.9719	226.8515
6	0.857143	228.7586	227.9802	227.6007	227.4756	227.3512
7	0.875	229.1691	228.3717	227.9829	227.8548	227.7274
8	0.888889	229.4893	228.6771	228.2811	228.1506	228.0208
9	0.9	229.7462	228.9220	228.5202	228.3878	228.2561
10	0.909091	229.9568	229.1228	228.7162	228.5823	228.4490

Table 12.3 The dependence of the equity cost (*ke*) on the leverage level (L) of a 3-year-old company for advance profit tax payments with frequencies of $p = 1, 2, 4, 6$ and 12

L	Wd	*ke*				
		$p = 1$	$p = 2$	$p = 4$	$p = 6$	$p = 12$
1	0.5	0.1966	0.1990	0.2002	0.2005	0.2009
2	0.666667	0.1930	0.1977	0.2001	0.2008	0.2016
3	0.75	0.1892	0.1964	0.1999	0.2011	0.2022
4	0.8	0.1855	0.1951	0.1998	0.2013	0.2028
5	0.833333	0.1817	0.1937	0.1996	0.2015	0.2034
6	0.857143	0.1780	0.1923	0.1994	0.2017	0.2040
7	0.875	0.1742	0.1910	0.1992	0.2019	0.2046
8	0.888889	0.1704	0.1896	0.1990	0.2021	0.2052
9	0.9	0.1666	0.1882	0.1988	0.2023	0.2058
10	0.909091	0.1628	0.1868	0.1986	0.2025	0.2063

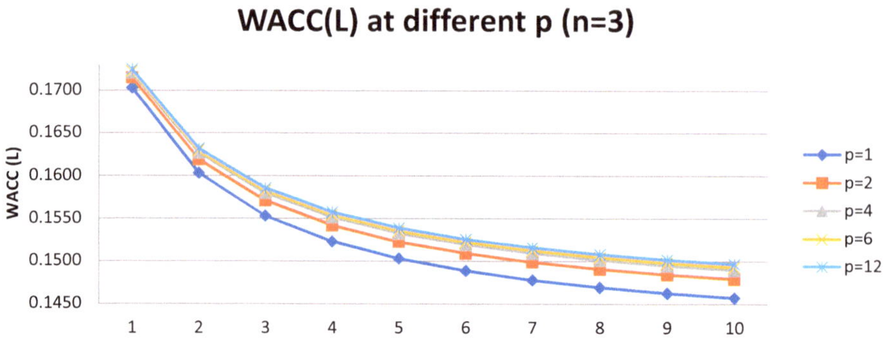

Fig. 12.1 The dependence of the WACC on the leverage level (L) of a 3-year-old company for advance profit tax payments with frequencies of $p = 1, 2, 4, 6$ and 12

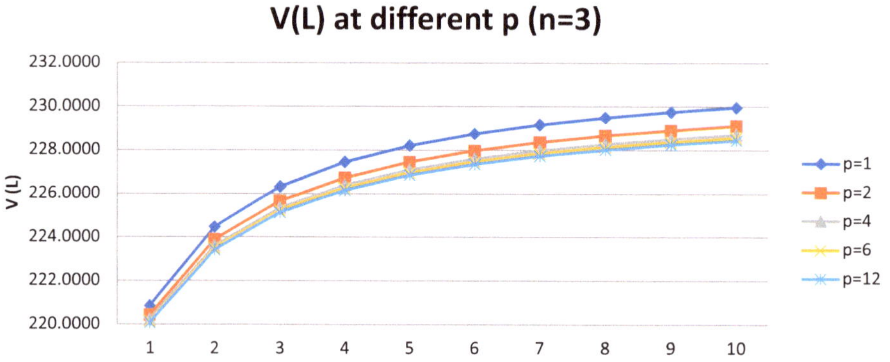

Fig. 12.2 The dependence of the company value (V) on the leverage level (L) of a 3-year-old company for advance profit tax payments with frequencies of $p = 1, 2, 4, 6$ and 12

decreased with the frequency of payments, which meant that more infrequent payments were beneficial for the company. The influence of the payment frequency was stronger for annual, semi-annual and quarterly payments and decreased for monthly payments. This was consistent with the *WACC* behavior that was shown in the previous section.

As can be seen from Fig. 12.3, the equity cost (*ke*) depended on the leverage level in a linear manner. It was not expected for *ke*(L) to be linear because the BFO equation for the WACC was an $n + 1$ power equation. However, as can be seen from the figures and tables for the equity cost results, this occurred with a high accuracy. We called the relationship quasi-linear (or practically linear). Our entire extensive database showed that this was the case. The equity cost decreased for annual ($p = 1$), semi-annual ($p = 2$) and quarterly ($p = 4$) payments and increased for monthly ($p = 12$) and bi-monthly payments ($p = 6$). The tilt angle of *ke*(L) increased with p and the transition from *ke* increasing with L to *ke* decreasing with L occurred at $p = 4$.

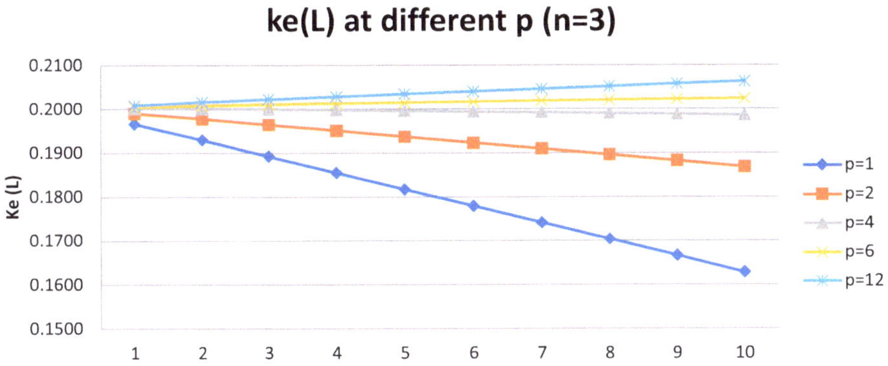

Fig. 12.3 The dependence of the equity cost (*ke*) on the leverage level (L) of a 3-year-old company for advance profit tax payments with frequencies of $p = 1, 2, 4, 6$ and 12

The economically justified amount of dividends was equal to the equity cost; therefore, the decrease in the equity cost with the leverage level signaled the discovery of a new effect that could significantly change the company's dividend policy.

Below, we compare the dependence of the WACC, company value and equity cost on the leverage level of a 3-year-old company for profit tax payments that were (1) made at the end of the financial period (data from Brusov et al. 2022) and (2) made at the beginning of the financial period (i.e., advance payments) with frequencies of $p = 2, 4, 6$ and 12.

The comparison of the dependence of the WACC on the leverage level (L) of a 3-year-old company for profit tax payments that were (1) made at the end of the financial period and (2) made at the beginning of the financial period (i.e., advance payments) with frequencies of $p = 2, 4, 6$ and 12 is shown in Fig. 12.4.

The $WACC(L)$ decreased with L in both cases: advance profit tax payments and payments at the end of the financial year. This meant that debt financing was important and should be used by the company as it led to a decrease in the corresponding capital cost with L.

In the case of profit tax payments that were made at the end of the financial year, the $WACC(L)$ decreased with p, while in the case of advance profit tax payments, the $WACC(L)$ increased with p. However, in spite of this, the $WACC(L)$ always turned out to be lower in the case of advance profit tax payments. From Fig. 12.4, it can be seen, for example, that $WACC(L = 10) = 0.1504$ at $p = 12$ for profit tax payments that were made at the end of the financial year while $WACC(L = 10) = 0.1497$ at $p = 12$ for advance profit tax payments. The $WACC(L)$ curves for two these cases never crossed, i.e., the $WACC$ values for advance profit tax payments always turned out to be lower than in the case of profit tax payments that were made at the end of the financial year. This demonstrated the importance of the use of advance profit tax payments for companies because the lower values of $WACC$ corresponded to higher company values.

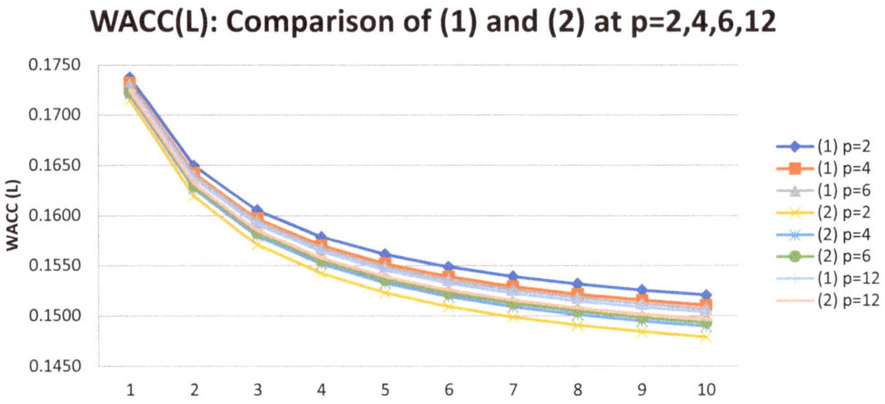

Fig. 12.4 The comparison of the dependence of the WACC on the leverage level (L) of a 3-year-old company for profit tax payments that were (1) made at the end of the financial period and (2) made at the beginning of the financial period (i.e., advance payments) with frequencies of $p = 2$, 4, 6 and 12

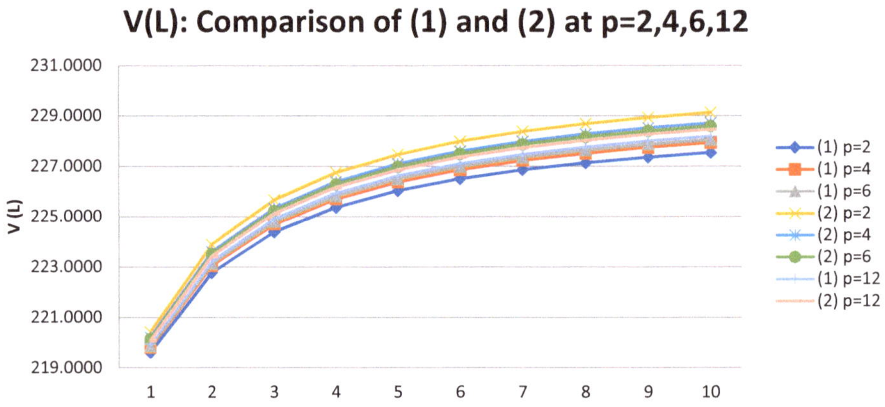

Fig. 12.5 The comparison of the dependence of the company value (V) on the leverage level (L) of a 3-year-old company for profit tax payments that were (1) made at the end of the financial period and (2) made at the beginning of the financial period (i.e., advance payments) with frequencies of $p = 2$, 4, 6 and 12 is shown in Fig. 12.4

The comparison of the dependence of the company value (V) on the leverage level (L) of a 3-year-old company for profit tax payments that were (1) made at the end of the financial period and (2) made at the beginning of the financial period (i.e., advance payments) with frequencies of $p = 2$, 4, 6 and 12 is shown in Fig. 12.5.

Figure 12.5 shows that the company value increased with L for all payment frequencies (annual, semi-annual, quarterly, monthly, etc.) in both cases: advance profit tax payments and payments the end of the financial period. So, the debt financing led to an increase in the company value with L and thus, it was important and should be used by the company. This was consistent with the decrease in the

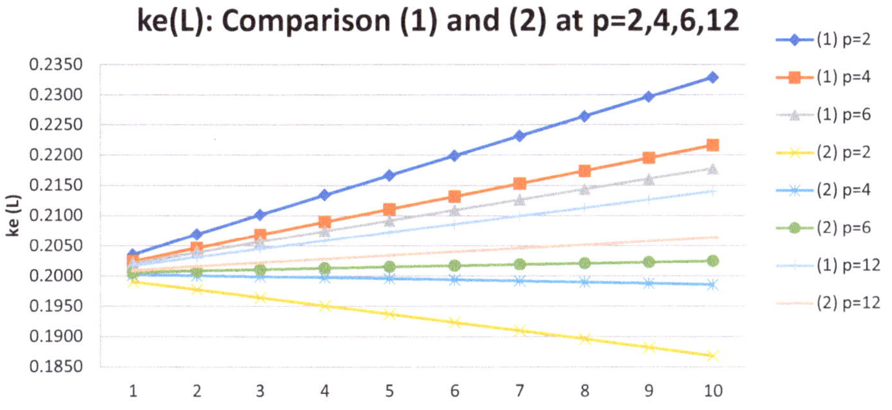

Fig. 12.6 The comparison of the dependence of the equity cost (*ke*) on the leverage level (*L*) of a 3-year-old company for profit tax payments that were (1) made at the end of the financial period and (2) made at the beginning of the financial period (i.e., advance payments) with frequencies of p = 2, 4, 6 and 12

WACC with *L*, as shown in the previous section. The company value decreased with the frequency of payments in the case of profit tax payments being made at the beginning of the financial period (i.e., advance payments), which meant that more infrequent payments were beneficial for the company. The company value increased with the frequency of payments in the case of profit tax payments being made at the end of the financial period. However, in spite of this, *V*(*L*) always turned out to be bigger in the case of advance profit tax payments. From Fig. 12.5, it can be seen, for example, that $V(L = 10) = 228.1849$ at $p = 12$ for profit tax payments that were made at the end of the financial year, while V(L = 10) = 228.4490 at $p = 12$ for advance profit tax payments. Note that the *V*(*L*) curves that corresponded to these two cases never crossed. The influence of the payment frequency was stronger for annual, semi-annual and quarterly payments and decreased for monthly payments. This was consistent with the WACC behavior that was shown in the previous section.

The comparison of the dependence of the equity cost (*ke*) on the leverage level (*L*) of a 3-year-old company for profit tax payments that were (1) made at the end of the financial period and (2) made at the beginning of the financial period (i.e., advance payments) with frequencies of $p = 2$, 4, 6 and 12 is shown in Fig. 12.6.

From Fig. 12.6, it can be seen that the equity cost (*ke*) depended on the leverage level (*L*) in a linear manner in both cases ((1) and (2)) for all values of p. In the case of profit tax payments being made at the end of the financial period, the tilt angle of *ke*(*L*) decreased with p but *ke*(L) continued to increase with L for all values of p. In the case of advance profit tax payments, the tilt angle of *ke*(*L*) increased with p and the transition from *ke* increasing with L to *ke* decreasing with L occurred at $p = 4$. Additionally, *ke*(*L*) decreased with L for annual ($p = 1$), semi-annual ($p = 2$) and quarterly ($p = 4$) payments and increased for monthly ($p = 12$) and bi-monthly payments ($p = 6$). The economically justified amount of dividends was equal to the equity cost; therefore, the decrease in the equity cost with the leverage level signaled the discovery of a new effect that could significantly change the dividend policy of a company. This effect only occurred in the case of advance profit tax payments. Note that with the increase in p, the $ke(L)$ curves for the two cases ((1) and (2)) approached each other but never crossed.

12.3.2 The Impact of the Frequency of Profit Tax Payments on the Dependence of the Weighted Average Cost of Capital, Capital Value and Equity Cost on the Leverage Level of a 6-Year-Old Company

In Tables 12.4, 12.5 and 12.6, we present the results of our study on the impact of the frequency of profit tax payments on the dependence of the WACC, capital value and equity cost on the leverage level of a 6-year-old company with payment frequencies of $p = 1, 2, 4, 6$ and 12.

As can be seen from Fig. 12.7, the WACC decreased with L in the case of advance profit tax payments for all payment frequencies (annual, semi-annual, quarterly, monthly, etc.). So, debt financing led to a decrease in the corresponding capital cost with L and thus, should be used by the company. The WACC values increased with the frequency of payments, which meant that more infrequent payments were beneficial for the company. The influence of the payment frequency was stronger for annual, semi-annual and quarterly payments and decreased for monthly payments.

As can be seen from Fig. 12.8, the company value (V) increased with L in the case of advance profit tax payments for all payment frequencies (annual, semi-annual, quarterly, monthly, etc.). This was consistent with the decrease in the WACC with the leverage level, as shown in the previous section. The company value decreased

Table 12.4 The dependence of the WACC on the leverage level (L) of a 6-year-old company for advance profit tax payments with frequencies of $p = 1, 2, 4, 6$ and 12

		WACC, $n = 6$ (2)				
L	Wd	$p = 1$	$p = 2$	$p = 4$	$p = 6$	$p = 12$
1	0.5	0.1697	0.1709	0.1715	0.1717	0.1719
2	0.666667	0.1593	0.1610	0.1618	0.1621	0.1623
3	0.75	0.1542	0.1560	0.1569	0.1572	0.1575
4	0.8	0.1510	0.1530	0.1540	0.1543	0.1546
5	0.833333	0.1489	0.1510	0.1520	0.1524	0.1527
6	0.857143	0.1474	0.1496	0.1506	0.1510	0.1513
7	0.875	0.1463	0.1485	0.1496	0.1499	0.1503
8	0.888889	0.1454	0.1477	0.1487	0.1491	0.1495
9	0.9	0.1447	0.1470	0.1481	0.1484	0.1488
10	0.909091	0.1442	0.1464	0.1475	0.1479	0.1483

Table 12.5 The dependence of the company value (V) on the leverage level (L) of a 6-year-old company for advance profit tax payments with frequencies of $p = 1, 2, 4, 6$ and 12

		V				
L	Wd	$p = 1$	$p = 2$	$p = 4$	$p = 6$	$p = 12$
1	0.5	359.2385	358.0973	357.5406	357.3571	357.1746
2	0.666667	369.1124	367.5078	366.7265	366.4692	366.2133
3	0.75	374.2557	372.4010	371.4988	371.2017	370.9064
4	0.8	377.4110	375.4000	374.4222	374.1003	373.7805
5	0.833333	379.5443	377.4263	376.3969	376.0581	375.7214
6	0.857143	381.0830	378.8871	377.8201	377.4690	377.1201
7	0.875	382.2451	379.9901	378.8947	378.5342	378.1761
8	0.888889	383.1540	380.8525	379.7347	379.3669	379.0014
9	0.9	383.8841	381.5452	380.4093	380.0356	379.6643
10	0.909091	384.4836	382.1139	380.9631	380.5845	380.2084

Table 12.6 The dependence of the equity cost (ke) on the leverage level (L) of a 6-year-old company for advance profit tax payments with frequencies of $p = 1, 2, 4, 6$ and 12

		ke				
L	Wd	$p = 1$	$p = 2$	$p = 4$	$p = 6$	$p = 12$
1	0.5	0.1953	0.1978	0.1990	0.1994	0.1997
2	0.666667	0.1900	0.1950	0.1974	0.1982	0.1990
3	0.75	0.1846	0.1920	0.1957	0.1969	0.1981
4	0.8	0.1791	0.1891	0.1939	0.1955	0.1971
5	0.833333	0.1736	0.1860	0.1921	0.1941	0.1961
6	0.857143	0.1680	0.1830	0.1903	0.1927	0.1951
7	0.875	0.1624	0.1799	0.1885	0.1913	0.1941
8	0.888889	0.1569	0.1769	0.1867	0.1899	0.1931
9	0.9	0.1513	0.1738	0.1848	0.1885	0.1921
10	0.909091	0.1457	0.1707	0.1830	0.1870	0.1911

with the frequency of payments, which meant that more infrequent payments were beneficial for the company. The influence of the payment frequency was stronger for annual, semi-annual and quarterly payments and decreased for monthly payments. This was consistent with the WACC behavior that was shown in the previous section.

As can be seen from Fig. 12.9, the equity cost (ke) depended on the leverage level (L) in a linear manner. It decreased for annual ($p = 1$), semi-annual ($p = 2$) and quarterly ($p = 4$) payments and increased for monthly ($p = 12$) and bi-monthly payments ($p = 6$). The economically justified amount of dividends was equal to the equity cost; therefore, the decrease in the equity cost with the leverage level signaled the discovery of a new effect that could significantly change the dividend policy of the company. Note that for the 3-year-old company, ke increased with L for monthly ($p = 12$) and bi-monthly payments ($p = 6$) and the transition from decreasing to increasing behavior occurred at $p = 4$.

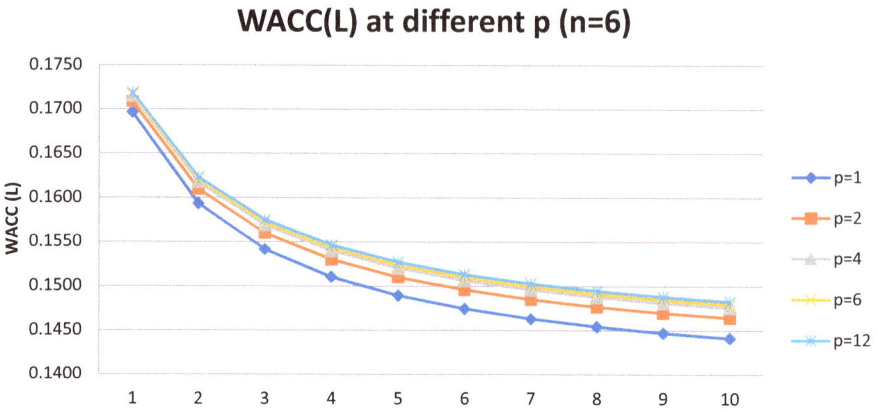

Fig. 12.7 The dependence of the *WACC* on the leverage level (L) of a 6-year-old company for advance profit tax payments with frequencies of $p = 1, 2, 4, 6$ and 12

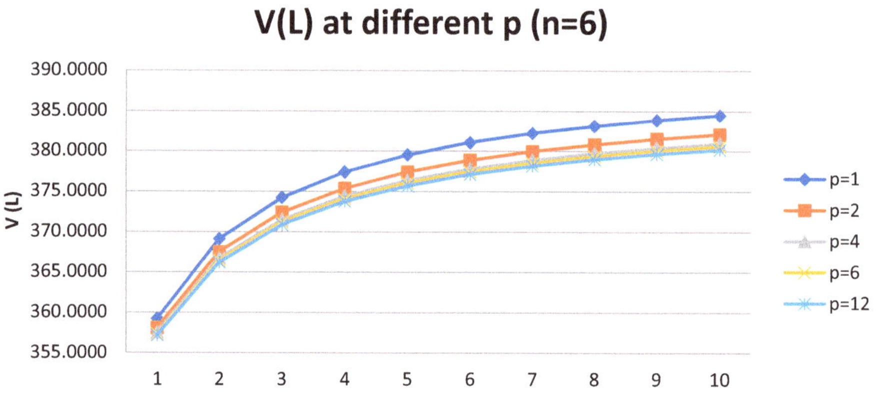

Fig. 12.8 The dependence of the company value (V) on the leverage level (L) of a 6-year-old company for advance profit tax payments with frequencies of $p = 1, 2, 4, 6$ and 12

The comparison of the dependence of the *WACC* on the leverage level (L) of a 6-year-old company for profit tax payments that were (1) made at the end of the financial period and (2) made at the beginning of the financial period (i.e., advance payments) with frequencies of $p = 2, 4, 6$ and 12 is shown in Fig. 12.10.

The WACC(L) decreased with L in both cases: profit tax payments that were made at the beginning of the financial period and payments that were made at the end of the financial period. This meant that the debt financing led to a decrease in the corresponding capital cost with L and should be used by the company.

The *WACC*(L) decreased with p in the case of profit tax payments being made at the end of the financial period, while in the case of profit tax payments being made at the beginning of the financial period (i.e., advance payments), the WACC (L) increased with p. However, in spite of this, the *WACC*(L) always turned out to

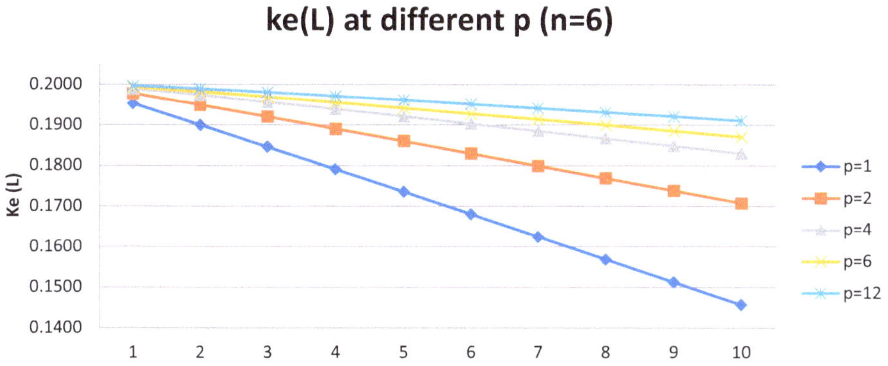

Fig. 12.9 The dependence of the equity cost (*ke*) on the leverage level (L) of a 6-year-old company for advance profit tax payments with frequencies of $p = 1, 2, 4, 6$ and 12

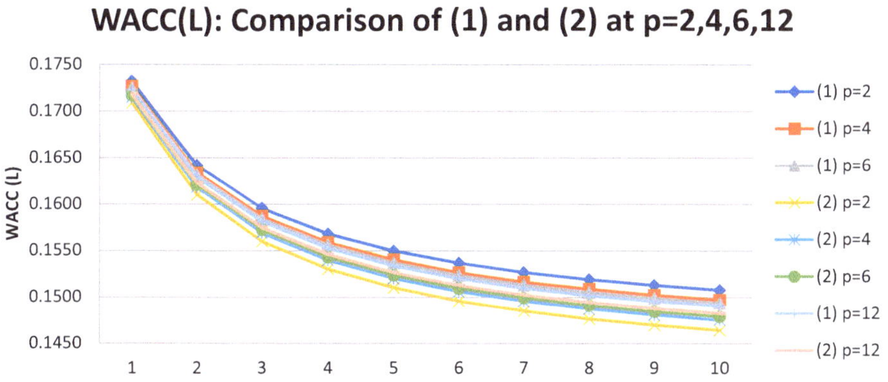

Fig. 12.10 The comparison of the dependence of the *WACC* on the leverage level (L) of a 6-year-old company for profit tax payments that were (1) made at the end of the financial period and (2) made at the beginning of the financial period (i.e., advance payments) with frequencies of $p = 2$, 4, 6 and 12

be lower in the latter case. From Fig. 12.10, it can be seen, for example, that $WACC(L = 10) = 0.1490$ at $p = 12$ for profit tax payments that were made at the end of the financial year while $WACC(L = 10) = 0.1483$ at $p = 12$ for advance profit tax payments. The $WACC(L)$ curves for two these cases never crossed, i.e., the *WACC* values for advance profit tax payments always turned out to be lower than in the case of profit tax payments that were made at the end of the financial year. Thus, the use of advance profit tax payments was quite important for the company because the lower values of *WACC* corresponded to higher company values. This conclusion is valid for companies of any age.

The comparison of the dependence of the company cost (V) on the leverage level (L) of a 6-year-old company for profit tax payments that were (1) made at the end of

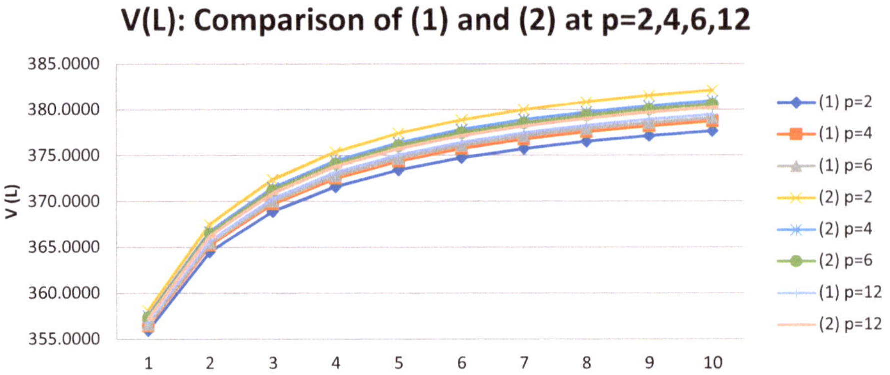

Fig. 12.11 The comparison of the dependence of the company value (V) on the leverage level (L) of a 6-year-old company for profit tax payments that were (1) made at the end of the financial period and (2) made at the beginning of the financial period (i.e., advance payments) with frequencies of $p = 2, 4, 6$ and 12

the financial period and (2) made at the beginning of the financial period (i.e., advance payments) with frequencies of $p = 2, 4, 6$ and 12 is shown in Fig. 12.11.

As can be seen from Fig. 12.11, the company value (V) increased with L for all payment frequencies (annual, semi-annual, quarterly, monthly, etc.) in both cases: profit tax payments that were made at the beginning of the financial period (i.e., advance payments) and profit tax payments that were made at the end of the financial period. So, the debt financing led to an increase in the company value with the leverage level and thus, should be used by the company. This was consistent with the decrease in the *WACC* with the leverage level, as shown in the previous section. The company value decreased with the frequency of payments in the case of advance profit tax payments, which meant that more infrequent payments were beneficial for the company, and increased with the frequency of payments in the case of profit tax payments that were made at the end of the financial period. However, in spite of this, the V(L) always turned out to be bigger in the case of advance profit tax payments. From Fig. 12.11, it can be seen, for example, that V (L = 10) = 379.4635 at $p = 12$ for profit tax payments that were made at the end of the financial year while V(L = 10) = 380.2084 at $p = 12$ for advance profit tax payments. Note that the V(L) curves that corresponded to these two cases never crossed. The influence of the payment frequency was stronger for annual, semi-annual and quarterly payments and decreased for monthly payments. This was consistent with the WACC behavior that was shown in the previous section.

The comparison of the dependence of the equity cost (ke) on the leverage level (L) of a 6-year-old company for profit tax payments that were (1) made at the end of the financial period and (2) made at the beginning of the financial period (i.e., advance payments) with frequencies of $p = 2, 4, 6$ and 12 is shown in Fig. 12.12.

From Fig. 12.12, it can be seen that the equity cost (ke) depended on the leverage level (L) in a linear manner in both cases ((1) and (2)) for all values of p (similar to

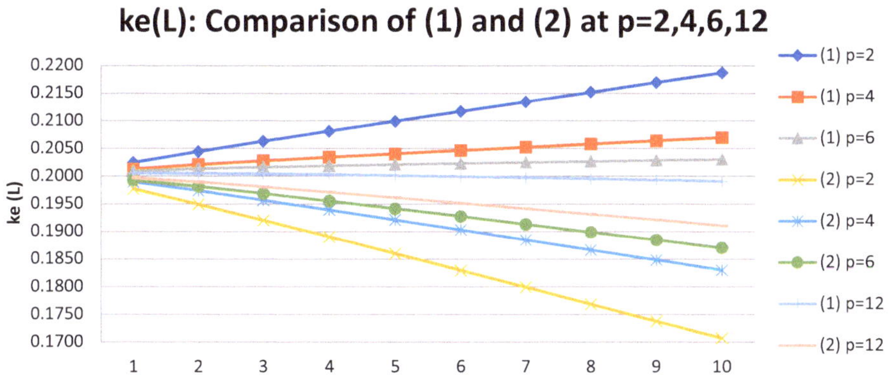

Fig. 12.12 The comparison of the dependence of the equity cost (*ke*) on the leverage level (*L*) of a 6-year-old company for profit tax payments that were (1) made at the end of the financial period and (2) made at the beginning of the financial period (i.e., advance payments) with frequencies of p = 2, 4, 6 and 12

the case of the 3-year-old company). In the case of profit tax payments that were made at the end of the financial period, the tilt angle of $ke(L)$ decreased with p, but $ke(L)$ continued to increase with L at $p = 2$, 4 and 6 and we only observed the transition to a slight decrease with leverage level at $p = 12$. Note that for the 3-year-old company, this transition was absent and $ke(L)$ increased with L for all values of p.

In the case of profit tax payments that were made at the beginning of the financial period (i.e., advance payments), the tilt angle of $ke(L)$ increased with p, but $ke(L)$ decreased with L for all payment frequencies. The economically justified amount of dividends was equal to the equity cost; thus, the discovery of the decrease in equity cost with the leverage level represented a new effect that could greatly change the company's dividend policy. This effect only occurred in the case of advance profit tax payments. Note that with the increase in p, the $ke(L)$ curves for the two cases ((1) and (2)) approached each other but never crossed.

12.4 Conclusions

In this chapter, we studied the influence of frequent advance profit tax payments on the main financial indicators of a company. For this, we generalized the Brusov–Filatova–Orekhova (BFO) theory for the case of advance profit tax payments with an arbitrary frequency for the first time. We derived modified BFO formulae and showed that (1) all BFO formulae could change and (2) all of the financial parameters of a company, such as *WACC*, company value and equity cost, depended on the frequency of profit tax payments. The weighted average cost of capital (*WACC*) increased with the payments and the company value decreased with the payments.

This meant that more infrequent payments were beneficial for the companies. The tilt angle of the equity cost (*ke*(*L*)) also increased with the payments. Depending on the company age, *ke*(*L*) either decreased with L for all payment frequencies or a transition from the decrease to an increase in ke with L occurred for some frequencies.

We compared the obtained results to those that we recently obtained for frequent profit tax payments that were made at the end of the financial period (Brusov et al. 2022) and found them to be totally different. We showed that in spite the fact that the *WACC* decreased with the payments and the company value increased with the payments in this case (Brusov et al. 2022), the *WACC* values turned out to be bigger and the company values turned out to be smaller than in the case of frequent advance profit tax payments for all payment frequencies. This underlined the importance of advance profit tax payments. Regulators should extend the practice of advance income tax payments to profit tax payments.

The economically justified amount of dividends was equal to the equity cost; thus, the discovered decrease in the equity cost with the leverage level represented a new effect that could greatly change a company's dividend policy. In the case of profit tax payments that were made at the end of the financial period, the tilt angle of *ke*(*L*) decreased with the payments, but *ke*(*L*) continued to increase with L at $p = 2$, 4 and 6 and we only observed the transition to a slight decrease with leverage level at $p = 12$. Note that for the 3-year-old company, this transition was absent and ke (L) increased with L for all values of p.

In the case of advance profit tax payments, the tilt angle of *ke*(*L*) increased with the payments, but *ke*(*L*) decreased with L for all payment frequencies for the 6-year-old company. For the 3-year-old company, *ke*(*L*) decreased at $p = 1$, 2 and 4 and increased at $p = 6$ and 12. Thus, for the 3-year-old company, the transition from the decreasing *ke*(*L*) behavior to the *ke*(*L*) increasing behavior occurred at $p = 6$, while for the 6-year-old company, this transition was absent.

The novelty of the chapter is as follows:

1. We generalized the BFO theory for the case of advance profit tax payments with an arbitrary frequency for the first time.
2. By comparing the obtained results to those for frequent profit tax payments at the end of the financial period, we showed that in spite the fact that the WACC decreased with the payments and the company value increased with the payments, the WACC value in this case turned out to be bigger and the company value turned out to be smaller than in case of frequent advance profit tax payments for all payment frequencies, which underlined the importance of advance profit tax payments.
3. A new effect of the dependence of equity cost on leverage level was discovered and this pioneering result could radically change the company dividend policies. In the case of advance profit tax payments, the tilt angle of *ke*(*L*) increased with the payments, but *ke*(*L*) decreased with L for all payment frequencies for the 6-year-old company and at $p = 1$, 2 and 4 for the 3-year-old company, although it increased at $p = 6$ and 12. Thus, for the 3-year-old company, the transition from

the decreasing $ke(L)$ behavior to the increasing $ke(L)$ behavior occurred at $p = 6$, while this transition was absent for the 6-year-old company.

4. Regulator recommendations were developed to expand the practice of advance profit tax payments, which would be beneficial to both parties, the companies and the state, (1) this practice would lead to a decrease in the cost of raising capital and an increase in the company value for companies and (2) ensure an increase in the stability of budget revenues for the regulators.

There was one limitation of the current consideration: the BFO theory was used with the assumption of a constant income. We will generalize the BFO theory for the case of variable incomes in the next chapter.

References

Angotti, M., de Lacerda Moreira, R., Hipólito Bernardes do Nascimento, J., & Neto de Almeida Bispo, O. (2018). Analysis of an equity investment strategy based on accounting and financial reports in Latin American markets. *Reficont, 5*, 22–40.

Barbi, M. (2011). On the risk-neutral value of debt tax shields. *Applied Financial Economics, 22*, 251–258.

Batrancea, L. (2021a). An econometric approach regarding the impact of fiscal pressure on equilibrium: Evidence from electricity, gas and oil companies listed on the New York stock exchange. *Mathematics, 9*, 630.

Batrancea, L. (2021b). The influence of liquidity and solvency on performance within the healthcare industry: Evidence from publicly listed companies. *Mathematics, 9*, 2231.

Becker, D. M. (2022). Getting the valuation formulas right when it comes to annuities. *Managerial Finance, 48*, 470.

Berk, J., & De Marzo, P. (2007). *Corporate finance*. Pearson–Addison Wesley.

Brusov, P., Filatova, T., Orekhova, N., & Eskindarov, M. (2018). *Modern corporate finance, investments, taxation and ratings* (2nd ed., pp. 1–571). Springer Nature Publishing.

Brusov, P. N., Filatova, T. V., Orekhova, N. P., Kulik, V. L., Chang, S. I., & Lin, Y. C. G. (2020). Modification of the Modigliani–Miller theory for the case of advance payments of tax on profit. *Journal of Reviews on Global Economics, 9*, 257–267.

Brusov, P. N., Filatova, T. V., Orekhova, N. P., Kulik, V. L., Chang, S. I., & Lin, Y. C. G. (2022). The generalization of the Brusov–Filatova–Orekhova theory for the case of payments of tax on profit with arbitrary frequency. *Mathematics, 10*, 1343. https://doi.org/10.3390/math10081343

Dimitropoulos, P. (2014). Capital structure and corporate governance of soccer clubs: European evidence. *Management Research Review, 37*, 658–678.

El-Chaarani, H., Abraham, R., & Skaf, Y. (2022). The impact of corporate governance on the financial performance of the banking sector in the MENA (Middle Eastern and North African) region: An immunity test of banks for COVID-19. *Journal of Risk and Financial Management, 15*, 82.

Farber, A., Gillet, R., & Szafarz, A. A. (2006). General formula for the WACC. *International Journal of Business, 11*, 211–218.

Fernandez, P. A. (2006). General formula for the WACC: A comment. *International Journal of Business, 11*, 219.

Franc-Dąbrowska, J., Mądra-Sawicka, M., & Milewska, A. (2021). Energy sector risk and cost of capital assessment—Companies and investors perspective. *Energies, 14*, 1613.

Hamada, R. (1969). Portfolio analysis, market equilibrium, and corporate finance. *The Journal of Finance, 24*, 13–31.

Harris, R., & Pringle, J. (1985). Risk-adjusted discount rates–extension from the average-risk case. *Journal of Financial Research, 8*, 237–244.
Huang, S., Sun, H., Zhao, H., & Zhang, Y. (2020). Influence of leverage on the return on equity. *Systems Engineering—Theory & Practice, 40*, 355.
Islam, S. Z., & Khandaker, S. (2015). Firm leverage decisions: Does industry matter? *North American Journal of Economics and Finance, 31*, 94.
Luiz, K., & Cruz, M. (2015). The relevance of capital structure on firm performance: A multivariate analysis of publicly traded Brazilian companies. *REPeC Brasília, 9*, 384–401.
Miller, M. (1977). Debt and taxes. *The Journal of Finance, 32*, 261–275. https://doi.org/10.1111/j.1540-6261.1977.tb03267.x
Modigliani, F., & Miller, M. (1958). The cost of capital, corporate finance, and the theory of investment. *The American Economic Review, 48*, 261–297.
Modigliani, F., & Miller, M. (1963). Corporate income taxes and the cost of capital: A correction. *The American Economic Review, 53*, 147–175.
Modigliani, F., & Miller, M. (1966). Some estimates of the cost of capital to the electric utility industry 1954–1957. *The American Economic Review, 56*, 333–391.
Mundi, H. S., Kaur, P., & Murty, R. L. N. (2022). A qualitative inquiry into the capital structure decisions of overconfident finance managers of family-owned businesses in India. *Qualitative Research in Financial Markets, 14*, 357–379.
Myers, S. (2001). Capital structure. *The Journal of Economic Perspectives, 15*, 81–102.
Sadiq, M., Alajlani, S., Hussain, M. S., Bashir, F., & Chupradit, S. (2022). Impact of credit, liquidity, and systematic risk on financial structure: Comparative investigation from sustainable production. *Environmental Science and Pollution Research International, 29*, 20963–20975.
Vergara-Novoa, C., Sepúlveda-Rojas, J. P., Miguel, D. A., & Nicolás, R. (2018). Cost of capital estimation for highway concessionaires in Chile. *Journal of Advanced Transportation, 2018*, 2153536.
Zhukov, P. (2018). The impact of cash flows and weighted average cost of capital to enterprise value in the oil and gas sector. *Journal of Reviews on Global Economics, 7*, 138–145.

Chapter 13
Generalization of the Brusov–Filatova–Orekhova Theory for the Case of Variable Income

13.1 Introduction

Among the theories of capital structure, the two main ones are the Brusov–Filatova–Orekhova (BFO) theory and the Modigliani–Miller (MM) theory, which is the eternal limit of the BFO theory. Both of them consider the case of constant income, although in practice, a company's income is, of course, variable. The generalization of these two theories of capital structure for the case of variable income is very important, since it allows them to expand their applicability in practice. Recently, we have generalized the case of variable income the Modigliani–Miller theory (Brusov et al. 2021a), and here we have generalized for the first time the case of variable income of the Brusov–Filatova–Orekhova theory. This generalization significantly expands the applicability of the modern capital structure theory, which is valid for companies of any age, and in practice, for investments, corporate finance, business valuation, banking, ratings, etc.

13.1.1 Literature Review

The Brusov–Filatova–Orekhova (BFO) theory and its perpetual limit—the theory of Nobel laureates Modigliani and Miller—study capital structure problems. Capital structure is the ratio between the company's own and borrowed capital, and asks whether capital structure affects key company metrics, such as weighted average cost of capital, WACC, cost of equity, ke, earnings, company value, and others, and if so, how? The determination of the optimal capital structure, that is, the determination of capital structure in which WACC is minimal and the company value is maximized, is one of the main problems, solved by the company's management. The Modigliani and Miller chapter (Modigliani and Miller 1958) was the first

P. Brusov, T. Filatova, *Efficiency of Investment Projects*, Contributions to Finance and Accounting, https://doi.org/10.1007/978-3-032-01028-5_13

quantitative study of the influence of capital structure on a company's financial parameters. Prior to this, there was a traditional approach based on empirical data analysis.

13.1.2 Before the Modigliani and Miller Work

In the traditional approach, the WACC and the value of the company, V, depend on the level of leverage, L (and, therefore, on the capital structure). Debt is always cheaper than equity, since the former has less risk, and because in the event of bankruptcy, the claims of creditors are satisfied earlier than the claims of shareholders.

Thus, an increase in the share of cheaper borrowed capital of the total capital structure to the limit that does not violate financial stability and does not increase the risk of bankruptcy leads to a decrease in the WACC and an increase in the value of the company, V.

Further increase of debt financing could lead to financial stability violation and an increase in the bankruptcy risk. WACC increases and the company value, V, decreases. The competition of advantages of debt financing and its shortcomings at low and at high leverage levels forms the optimal capital structure where WACC is minimal and the company value, V, is maximum. These traditional approach results have been used in the trade-off theory.

13.1.3 Modigliani–Miller Theory

13.1.3.1 Modigliani–Miller Theory Without Taxes

Modigliani and Miller (MM) (1958), under a lot of assumptions, including the absence of corporate and individual taxes, the perpetuity of all companies and all cash flows, etc., obtained results that are completely different from the results of the traditional approach: capital structure does not affect capital cost and company value.

Under the above restrictions, Modigliani and Miller have shown that without taxes, the company value, *V*, is equal

$$V = V_0 = \frac{EBIT}{k_0} \tag{13.1}$$

Here, EBIT is Earnings Before Interest and Taxes, k_0 is discount rate, and V_0 stands for the unlevered company value.

From (13.1) it is easy to obtain WACC:

$$WACC = k_0 \tag{13.2}$$

k_0 is the cost of equity for a company without borrowed funds, and for a company with borrowed capital, k_0 is the cost of equity with a zero level of borrowed funds ($L = 0$).

From (13.1) and the expression for WACC

$$\mathrm{WACC} = k_0 = k_e w_e + k_d w_d. \tag{13.3}$$

One can obtain the cost of equity, ke

$$k_e = \frac{k_0}{w_e} - k_d \frac{w_d}{w_e} = \frac{k_0(S + D)}{S} - k_d \frac{D}{S} = k_0 + (k_0 - k_d)\frac{D}{S} = k_0 + (k_0 - k_d)L \tag{13.4}$$

Here, WACC is weighted average cost of capital; L is leverage level; D is debt capital value; S is equity capital value; k_d and w_d are the cost and share of the company's debt capital; and k_e and w_e are the equity capital cost and share. From (13.4), it follows that the equity cost increases linearly with the leverage level.

13.1.3.2 Modigliani–Miller Theory with Taxes

Taking into account income tax, in 1963, Modigliani and Miller (1963, 1966) obtained the following result for the value of the levered company, V,

$$V = V_0 + D \cdot t \tag{13.5}$$

Here, V_0 stands for the unlevered company value, t is the tax on income, and D is debt value.

From (13.5), it is easy to derive the expression for the WACC

$$WACC = k_0 \cdot (1 - w_d t) \tag{13.6}$$

From (13.6), one can obtain the formula for equity cost, ke, within the Modigliani–Miller theory with taxes

$$k_e = k_0 + L \cdot (k_0 - k_d)(1 - t) \tag{13.7}$$

Formula (13.7) differs from formula (13.4) (MM without taxes) by the factor (1 − t), which is called the tax corrector. This is less than one, so the tilt of the ke(L) curve decreases with taxes.

13.1.4 Unification of Capital Asset Pricing Model (CAPM) with Modigliani–Miller Model

Unification of Capital Asset Pricing Model (CAPM) with the Modigliani–Miller model was done in 1969 (Hamada 1969). Hamada derived the below formula for the cost of equity of a levered company, which included the financial and business risks of a company:

$$k_e = k_F + (k_M - k_F)b_U + (k_M - k_F)b_U \frac{D}{S}(1 - T), \tag{13.8}$$

where b_U is the β-coefficient of the company of the same group of business risk, that the company under consideration, but with L = 0. The formula (13.8) consists of three terms: risk-free income ability k_F, compensating shareholders a time value of their money, business risk premium $(k_M - k_F)b_U$, and financial risk premium $(k_M - k_F)b_U \frac{D}{S}(1 - T)$.

For a financially independent company, the financial risk is equal to zero (the third term disappears), and its owners will only receive the business risk premium.

13.1.4.1 Miller Model

In Miller (1977), Miller has accounted for corporate and individual taxes to obtain the following formula for unlevered company value, V_U,

$$V_U = \frac{\text{EBIT}(1 - T_C)(1 - T_S)}{k_0}. \tag{13.9}$$

Here, T_C—tax rate on corporate income, T_S—the tax rate on incomes of an individual investor from his ownership through corporation stocks.

13.1.5 Brusov–Filatova–Orekhova (BFO) Theory

The perpetuity of all company cash flow and of a company's lifetime was one of the main restrictions of the Modigliani–Miller (MM) theory, which has been lifted up in 2008 by Brusov–Filatova–Orekhova (Brusov et al. 2018, 2021b). They generalized the MM theory for the case of the company of any age, n, and derived the following Brusov–Filatova–Orekhova formula for the WACC

$$\frac{1-(1+WACC)^{-n}}{WACC}=\frac{1-(1+k_0)^{-n}}{k_0[1-w_dT(1-(1+k_d)^{-n})]} \tag{13.10}$$

To obtain the Modigliani–Miller formula for the WACC from (13.10), one should substitute $n \rightarrow \infty$.

It was shown (Brusov et al. 2018, 2021b) that a number of innovative effects, discovered in the BFO theory, are absent in the MM theory (Modigliani and Miller 1958, 1963, 1966).

Some main existing principles of financial management spanning many decades have been destroyed by the BFO theory; among them is the keystone of optimal capital structure formation—trade-off theory, and the bankruptcy of this theory has been proven within the BFO theory (Brusov et al. 2018, 2021b).

13.1.6 Alternate WACC Formula

An alternate formula for the WACC has been suggested (Farber et al. 2006; Fernandez 2006; Berk and DeMarzo 2007; Harris and Pringle 1985). It has the form below (Eq. 13.18 in Farber et al. 2006)

$$WACC=k_0(1-w_dt)-k_dtw_d+k_{TS}tw_d \tag{13.11}$$

Here, k_0, k_d, and k_{TS} are the returns on the financially independent company, the debt, and the tax shield, respectively, t is the corporate tax rate, and w_d is the debt share.

Although Eq. (13.11) is quite general, additional conditions are needed for practical applicability. When the WACC remains constant over time, the value of a leveraged company can be found by discounting the unleveraged free cash flows using the WACC. In this case, specific formulas can be found in textbook (Berk and DeMarzo 2007).

In the Modigliani–Miller theory (Modigliani and Miller 1963), the debt value D is constant. V_0 is also constant, as the expected after-tax cash-flow of the financially independent company is fixed. By assumption, $k_{TS} = k_d$ and the tax shield value is TS = tD. Therefore, the company value V is a constant and the alternate WACC formula (13.11) simplifies the MM formula:

$$WACC=k_0(1-w_dt)$$

The "classical" MM theory, suggesting that the returns on the debt k_d and the tax shield k_{TS} are equals (both these values have debt nature), is much more reasonable, so this is why in Brusov et al. (2021a), we modify the "classical" MM theory, namely.

13.1.7 Trade-Off Theory

In the study of the problem of optimal capital structure of the company during many decades, the cornerstone was the world-famous trade-off theory. It is still widely used now for decisions on capital structure. In Frank and Goyal (2009), the relative importance of different factors in capital structure decisions of publicly traded American companies has been studied. The most important factors to explain leverage level are: median industry leverage (+ effect on leverage), log of assets (+), market-to-book assets ratio (−), inflation (+), tangibility (+), and incomes (−). It was noted that companies that pay dividends tend to have lower leverage levels. The related effects have been found under considering book leverage. Authors found empirical data consistent with some trade-off theory versions.

In Sheridan and Wessels (1988), authors compare the applicability of the trade-off theory and pecking order theory for small and medium-sized companies' decisions about capital structure. It was found that the most lucrative and oldest companies have smaller leverage levels, which confirms the forecasts of the pecking order theory. Larger companies have a higher level of leverage, which is consistent with the predictions of the trade-off theory and pecking order theory. It is concluded that the trade-off theory and pecking order theory for small and medium-sized companies are not mutually exclusive when explaining capital structure decisions.

However, in 2013, Brusov et al. (2018, 2021b) proved the inconsistency of the trade-off theory in the framework of the BFO theory they created. It is shown that the assumption of risky debt financing does not lead to an increase in the WACC, which still decreases with increasing leverage. Thus, there is no minimum depending on the level of WACC leverage and no maximum depending on the value of the company from the level of leverage. Therefore, in the world-famous theory of trade-offs, there is no optimal capital structure. Brusov et al., in 2013 (Brusov et al. 2018, 2021b), having analyzed the equity cost dependence on the level of leverage under the assumption that debt capital is risky, gave an explanation for this fact.

The Modigliani–Miller theory proved that tax shields provided substantial gains to the company. In Harry and Masulis (1980), the theoretical study of tax shields was continued. It was noted that companies may have tax deductibles other than debt. Such non-debt tax shields are investment tax credits, depreciation, and net-loss carry forwards. In Bradley et al. (1984), the tax effects suggested in Harry and Masulis (1980) have been tested. In contrast to the prediction in Harry and Masulis (1980), it was shown that debt is positively related to non-debt tax shields, as measured by investment tax credits and depreciation. The results of Sheridan and Wessels (1988) do not provide support for an effect on debt ratios arising from nondebt tax shields. In Graham (2003), it was pointed out that a positive relationship between such proxies for non-debt tax shield and debt may result if a company invests heavily and borrows to invest. Any substitution effects between debt and non-debt tax shields could be suppressed by a mechanical positive relation of this type.

The Brusov–Filatova–Orekhova (BFO) theory methodology and results are well known in the literature (for example, see Dimitropoulos 2014; Luiz and Cruz 2015;

Barbi 2011; Franc-Dąbrowska et al. 2021; Angotti et al. 2018; Vergara-Novoa et al. 2018; Mundi et al. 2022; Becker 2022; El-Chaarani et al. 2022; Sadiq et al. 2022; Singhal et al. 2022). Chapters (Angotti et al. 2018; Vergara-Novoa et al. 2018; Mundi et al. 2022; Singhal et al. 2022) use the BFO theory in practice. The impact on capital structure decisions of the overconfidence of finance managers of family-run businesses in India has been studied in Vergara-Novoa et al. (2018). The study concludes that manager decisions about capital structure could be explained by measurable managerial characteristics. In Mundi et al. (2022), the correlation between capital structure and company risk was studied using datasets from Pakistani companies. It was shown that the role of capital structure and risk valuation is vital for the increase in the wealth of shareholders and the sustainable growth of companies. In Becker (2022), the adjusted present value method, the free cash flow (FCF) method, the flow-to-equity method and the relationships between these methods have been considered. The authors used a stationary FCF method and the Miles and Ezzell method instead the Modigliani–Miller method to derive DCF valuation formulas for annuities. In El-Chaarani et al. (2022), the influence of internal and external corporate governance mechanisms on the financial performance of banks in the MENA region is studied. It was shown that the corporate governance had positive effects on the financial indicators of banks. The energy companies capital costs by including an investor and market risk approach have been evaluated in Sadiq et al. (2022). The WACC intra-industry analysis of the companies has been done. The connection of capitalization and income ability in the BRICS banking sector has been examined in Singhal et al. (2022) under the signaling theory, the bankruptcy cost theory, the agency theory, the pecking order theory, the Modigliani and Miller theory, and the general theory of the cost of capital and capital structure—the Brusov–Filatova–Orekhova (BFO) theory. Over the past 2 years, the theory of capital structure has received a new impetus. A large-scale modification of both main theories of the capital structure, BFO and MM, has been carried out and continues in order to better take into account the conditions for the real functioning of companies, such as variable income, advance income tax payments, frequent income tax payments, their combinations, etc. (Brusov et al. 2021a; Brusov and Filatova 2022, 2013). A study of different aspects of emerging markets was carried out in Durana et al. (2021, 2022), Virglerova et al. (2021), Xu and Liu (2020), Valaskova et al. (2020, 2021a, b), and Białek-Jaworska and Klapkiv (2021). In Xu and Liu (2020), the impact of intellectual capital on firm performance within a modified and extended VAIC model has been studied.

In the near future, the authors plan to publish a large review, which will examine in detail the problems of capital structure.

13.1.8 Materials and Methods

We combine analytical and numerical methods. First, we derive formulas for the company's leveraged value, V; the leverage less value of the company, V_0; the tax shield TS, and finally the $WACC$ in the case of variable income.

Then, using Microsoft Excel, we study the dependences of the following values: weighted average cost of capital, WACC, discount rate, WACC-g, company value, V, and the cost of equity, ke, on the level of leverage, L, at different values of the growth rate, g. We have created a large database for sets of cost of equity k_0 and cost of debt kd, which is available upon request.

13.2 Modification of the BFO Theory for the Case of Companies with Variable Incomes

13.2.1 The Levered Company Value, V

Below, for the first time, we generalize the modern theory of the capital cost and capital structure—the Brusov–Filatova–Orekhova theory—to the case of variable income. We start by deriving the capitalization formula for a financially dependent company, assuming that income for the period grows with the growth rate g.

When accounting for the cost of any asset being equal to the sum of discounted values of incomes generated by this asset, one could write the capitalization for a financially dependent company's V of age, n, as the following expression

$$V = \frac{CF}{1+WACC} + \frac{CF(1+g)}{(1+WACC)^2} + \frac{CF(1+g)^2}{(1+WACC)^3} + \dots + \frac{CF(1+g)^{n-1}}{(1+WACC)^n} \quad (13.12)$$

Here, $WACC$ is the weighted average cost of capital, CF is an annual income of company, and (13.12) is geometric progression with denominator

$$g = \frac{(1+g)}{(1+WACC)} \quad (13.13)$$

Summarizing (13.12), we get the expression for the capitalization, V, of the levered company of age n

$$V = \frac{CF}{1 + WACC} \cdot \frac{1 - \left(\frac{1+g}{1+WACC}\right)^n}{1 - \frac{1+g}{1+WACC}} = \frac{CF}{WACC - g} \cdot \left(1 - \left(\frac{1 + g}{1 + WACC}\right)^n\right) \tag{13.14}$$

In the perpetuity limit ($n \to \infty$), we get the following formula for the levered company value, V,

$$V = \frac{CF}{WACC - g} \tag{13.15}$$

This formula shows that the discount rate is WACC–g, and not WACC.

13.2.2 The Unlevered Company Value, V_0

Let us now derive the capitalization formula for a financially independent company, assuming that income for the period grows with the growth rate g.

$$V_0 = \frac{CF}{1 + k_0} + \frac{CF(1 + g)}{(1 + k_0)^2} + \frac{CF(1 + g)^2}{(1 + k_0)^3} + \ldots + \frac{CF(1 + g)^{n-1}}{(1 + k_0)^n} \tag{13.16}$$

Equation (13.16) is geometric progression with denominator

$$g = \frac{(1 + g)}{(1 + k_0)} \tag{13.17}$$

Summarizing (13.16), we get the expression for the value V_0 of unlevered company of age n

$$V_0 = \frac{CF}{1 + k_0} \cdot \frac{1 - \left(\frac{1+g}{1+k_0}\right)^n}{1 - \frac{1+g}{1+k_0}} = \frac{CF}{k_0 - g} \cdot \left(1 - \left(\frac{1 + g}{1 + k_0}\right)^n\right) \tag{13.18}$$

In the perpetuity limit ($n \to \infty$), we get the following formula for the unlevered company value, V_0,

$$V_0 = \frac{CF}{k_0 - g} \tag{13.19}$$

This formula shows that the discount rate is $k_0 - g$, and not k_0.

13.2.3 The Tax Shield Value

The tax shield for n-years is equal to

$$(TS)_n = \frac{tk_d D}{1+k_d} + \frac{tk_d D}{(1+k_d)^2} + \ldots + \frac{tk_d D}{(1+k_d)^n} \tag{13.20}$$

Equation (13.20) is geometric progression with denominator

$$g = \frac{1}{(1+k_d)} \tag{13.21}$$

Summarizing (13.20), for the tax shield, we have

$$(TS)_n = \frac{tk_d D}{1+k_d} \cdot \frac{1-(1+k_d)^{-n}}{1-\frac{1}{1+k_d}} = Dt(1-(1+k_d)^{-n}) \tag{13.22}$$

$$(TS)_n = Dt(1-(1+k_d)^{-n}) \tag{13.23}$$

Using an analog of the first theorem by Modigliani–Miller for finite time, one gets

$$V = V_0 + (TS)_n \tag{13.24}$$

Substituting

$$D = w_d V \tag{13.25}$$

we arrive to the following expression

$$V(1 - w_d t(1-(1+k_d)^{-n})) = V_0 \tag{13.26}$$

Substituting into this equation the values of an unlevered company, V_0 (13.18) and of levered company, V, (13.14) one gets the following expression

$$\frac{CF\cdot\left(1-\left(\frac{1+g}{1+WACC}\right)^n\right)\cdot(1-w_d t[1-(1+k_d)^{-n}])}{WACC-g} = \frac{CF\cdot\left(1-\left(\frac{1+g}{1+k_0}\right)^n\right)}{(k_0-g)} \tag{13.27}$$

Dividing both parts by $(1 - w_d t(1 - (1 + k_d)^{-n}))$, we get the BFO equation for the WACC of the company with variable income

$$\frac{1-\left(\frac{1+g}{1+WACC}\right)^n}{WACC-g}=\frac{1-\left(\frac{1+g}{1+k_0}\right)^n}{(k_0-g)\cdot(1-w_d t[1-(1+k_d)^{-n}])} \tag{13.28}$$

This is the main theoretical result of the current chapter.

In the perpetuity limit ($n \to \infty$), we get the following equation for WACC in the case of variable income (Brusov et al. 2021a)

$$WACC-g=(k_0-g)\cdot(1-w_d t) \tag{13.29}$$

$$WACC=(k_0-g)\cdot(1-w_d t)+g \tag{13.30}$$

13.3 Results and Discussions

In order to understand the impact of a variable growth rate g on the main financial indicators of a company within the framework of the generalized theory of Brusov–Filatova–Orekhova (GBFO), we study the dependences of the following values: weighted average cost of capital, WACC, discount rate, WACC − g, company value, V, and the cost of equity, ke, on the level of leverage L at different values of the growth rate, g. We have created a large database for sets of cost of equity k_0 and cost of debt k_d, which is available upon request. To illustrate the results obtained and the conclusions drawn below, we present below the results for the following financial parameters of the company:

$k_0=0.18; k_d=0.16; t=0.2; CF=100; n=2$ and $n=4; g=0.2; 0.1; 0.0; -0.1; -0.2$

Here, k_0 is the equity cost at zero leverage level; k_d is the debt cost; t is the tax on income; *CF* is income per period; n is the company age; L is the leverage level; and g is the growth rate.

Note that if the results for different parameters could be and are numerically different, then the qualitative effect of the variable growth rate g on the main financial indicators is similar.

13.3.1 Calculations for 2-Year Company

13.3.1.1 Calculations of Weighted Average Cost of Capital, WACC

As can be seen from Table 13.1 and Fig. 13.1, for different values of g all curves WACC(L) start from one point (0; $k_0 = 0.18$). These curves WACC(L) demonstrate

Table 13.1 The WACC depending on the level of leverage L at different growth rates g in Generalized Brusov–Filatova–Orekhova theory (GBFO theory)

	WACC, n = 2				
L	g = 0.2	g = −0.1	g = 0	g = 0.1	g = 0.2
0	0.1799864	0.1799858	0.179995	0.1799849	0.1799844
1	0.1583502	0.1583545	0.1591545	0.1594954	0.1597882
2	0.15084	0.1513935	0.1521805	0.1523026	0.1526805
3	0.1472744	0.1478992	0.148688	0.1489255	0.1493521
4	0.1451307	0.1457982	0.1465907	0.1468947	0.1473505
5	0.1436999	0.1443958	0.1451917	0.1455391	0.1460143
6	0.142677	0.1433932	0.1441921	0.1445699	0.145059
7	0.1419093	0.1426408	0.1434421	0.1438425	0.144342
8	0.141312	0.1420553	0.1428587	0.1432764	0.1437841
9	0.140834	0.1415867	0.1423919	0.1428234	0.1433376
10	0.1404427	0.1412033	0.1420099	0.1424527	0.1429721

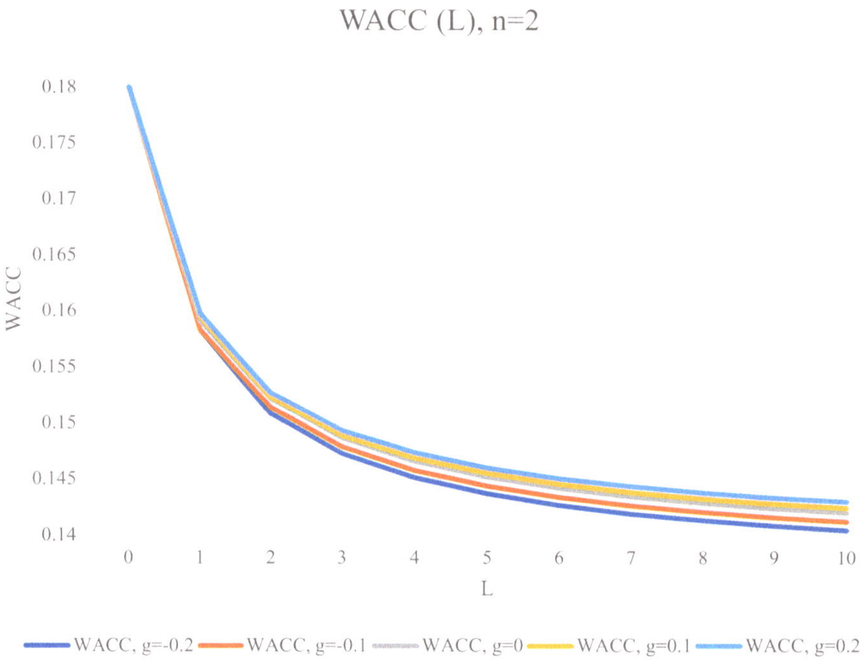

Fig. 13.1 The WACC depending on the level of leverage L at different growth rates g in Generalized Brusov–Filatova–Orekhova theory (GBFO theory)

the decrease of WACC with leverage level L at all g values. The curves WACC (L) increase with growth rate, g.

The results for a 2-year company differ from the results for a perpetual limit—the theory of Modigliani and Miller (Brusov et al. 2021a). In the latter case, the WACC

(L) curves decrease with the level of leverage L at $g < k_0$ and increase at $g > k_0$. k_0 is the threshold value g separating the increasing WACC(L) curves from the decreasing ones, and for $g = k_0$ WACC = const = k_0. In the first case (BFO theory), the WACC(L) curves decrease with increasing leverage L for all values of the growth rate g. The WACC(L) curves increase with the rate g both in the Brusov–Filatova–Orekhova theory and in the Modigliani and Miller theory. This is the first indication that WACC is no longer a discount rate, since it is intuitive that the discount rate must decrease in order to increase the value of company V. As we will see below, WACC–g and WACC–k_0 play the role of the discount rate.

The discount rate moves for a financially dependent company from the weighted average cost of capital, WACC, to WACC–g (where g is the growth rate), for a financially independent company from k_0 to k_0–g. As we intuitively understood above, this means that WACC and k_0 are no longer discount rates, as is the case for the classical Brusov–Filatova–Orekhova constant income theory. Below, we study the dependence of the discount rate WACC–g on the level of leverage L in the Generalized theory of Brusov–Filatova–Orekhova (the GBFO theory) at $k_0 = 0.18$ and different values of g (−0.2; −0.1; 0.0; 0.1; 0.2) for a 2- and 4-year company.

13.3.1.2 Calculations of the Discount Rate, WACC–g

As can be seen from Table 13.2 and Fig. 13.2, the curves (WACC–g)(L) demonstrate the decrease of WACC–g with leverage level L at all g values. The curves (WACC–g)(L) decrease with growth rate, g.

This behavior of the (WACC–g)(L) curves can be explained as follows: all WACC(L) curves originate from the same point (L = 0; WACC = 0.18). The (WACC–g)(L) curves will be ordered as follows for L = 0: the larger g, the lower the starting point and hence the entire graph lies, since the curves do not intersect. As

Table 13.2 The discount rate WACC–g depending on the level of leverage L at different growth rates g in Generalized Brusov–Filatova–Orekhova theory (GBFO theory)

	(WACC–g), n = 2				
L	g = −0.2	g = −0.1	g = 0	g = 0.1	g = 0.2
0	0.3799864	0.2799858	0.179995	0.0799849	−0.020016
1	0.3583502	0.2583545	0.1591545	0.0594954	−0.040212
2	0.35084	0.2513935	0.1521805	0.0523026	−0.047319
3	0.3472744	0.2478992	0.148688	0.0489255	−0.050648
4	0.3451307	0.2457982	0.1465907	0.0468947	−0.052649
5	0.3436999	0.2443958	0.1451917	0.0455391	−0.053986
6	0.342677	0.2433932	0.1441921	0.0445699	−0.054941
7	0.3419093	0.2426408	0.1434421	0.0438425	−0.055658
8	0.341312	0.2420553	0.1428587	0.0432764	−0.056216
9	0.340834	0.2415867	0.1423919	0.0428234	−0.056662
10	0.3404427	0.2412033	0.1420099	0.0424527	−0.057028

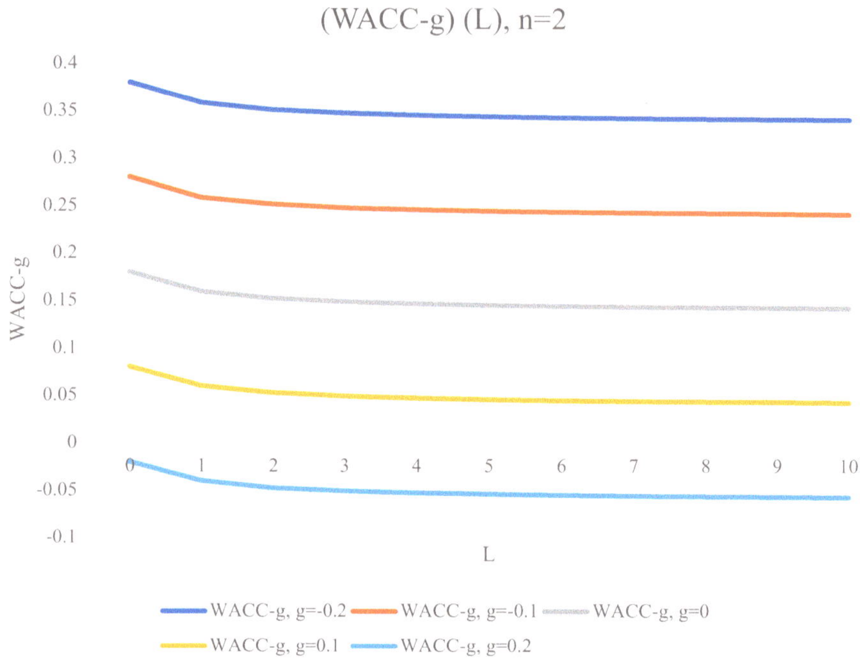

Fig. 13.2 The discount rate WACC–g depending on the level of leverage L at different growth rates g in Generalized Brusov–Filatova–Orekhova theory (GBFO theory

we will see below, the decrease of (WACC–g)(L) with growth rate, g, will lead to an increase of the company value, V, with g.

13.3.1.3 Calculations of the Company Value, V

It is seen from Table 13.3 and Fig. 13.3 that the company value V at fixed growth rate g increases with leverage level L in the Generalized Brusov–Filatova–Orekhova theory (GBFO theory). The company value V as well increases with growth rate g. This is a consequence of a decrease in the discount rate (WACC–g)(L) with an increase in the growth rate g.

Below, we have studied the dependence of cost of equity, k_e, on leverage level L and on growth rate, g, in the Generalized Brusov–Filatova–Orekhova theory (GBFO theory) at $k_0 = 0.18$; $k_d = 0.16$; and $g = 0;\pm0.1;\pm0.2$.

Table 13.3 The company value, V, depending on the level of leverage L at different growth rates g in Generalized Brusov–Filatova–Orekhova theory (GBFO theory)

	V, n = 2				
L	g = −0.2	g = −0.1	g = 0	g = 0.1	g = 0.2
0	142.20282	149.38493	156.56518	163.74916	170.93128
1	145.95218	153.40417	160.69451	168.06349	175.43476
2	147.29627	154.73958	162.12039	169.62647	177.07002
3	147.94237	155.41792	162.84303	170.36931	177.84511
4	148.3333	155.82837	163.27978	170.81878	178.31411
5	148.5953	156.10344	163.57228	171.12001	178.62842
6	148.78312	156.30063	163.78186	171.33596	178.85374
7	148.92435	156.44891	163.93941	171.49835	179.02318
8	149.03442	156.56448	164.06217	171.6249	179.15523
9	149.12261	156.65707	164.16051	171.7263	179.26104
10	149.19486	156.73293	164.24106	171.80938	179.34772

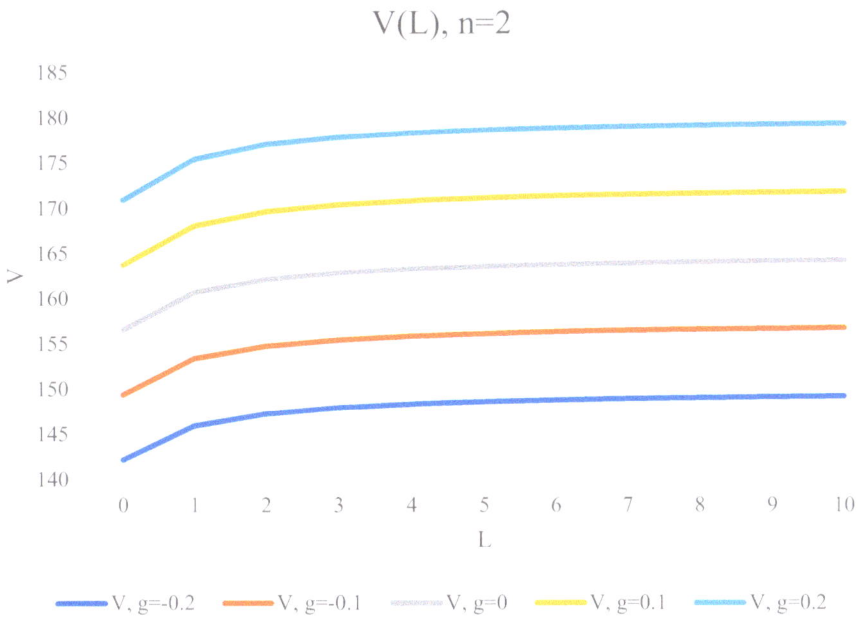

Fig. 13.3 The company value, V, depending on the level of leverage L at different growth rates g in Generalized Brusov–Filatova–Orekhova theory (GBFO theory)

13.3.1.4 Calculations of the Equity Cost, ke

As is seen from Table 13.4 and Fig. 13.4, the equity cost, ke, practically linearly grows with leverage level L at all growth rate g values. The tilt angle ke(L) grows with g.

Table 13.4 The equity cost, ke, depending on the level of leverage L at different growth rates g in Generalized Brusov–Filatova–Orekhova theory (GBFO theory)

L	ke, n = 2				
	g = −0.2	g = −0.1	g = 0	g = 0.1	g = 0.2
0	0.1799864	0.1799858	0.179995	0.1799849	0.1799844
1	0.1887005	0.1887091	0.1903091	0.1909909	0.1915764
2	0.19652	0.1981805	0.2005415	0.2009079	0.2020416
3	0.2050974	0.2075966	0.2107521	0.2117018	0.2134083
4	0.2136535	0.2169911	0.2209535	0.2224736	0.2247527
5	0.2221991	0.2263748	0.2311504	0.2332344	0.236086
6	0.2307387	0.2357525	0.2413446	0.243989	0.2474131
7	0.2392746	0.2451264	0.2515371	0.2547397	0.2587362
8	0.247808	0.2544978	0.2617285	0.2654878	0.2700568
9	0.2563397	0.2638674	0.271919	0.2762342	0.2813755
10	0.2648701	0.2732358	0.282109	0.2869792	0.292693

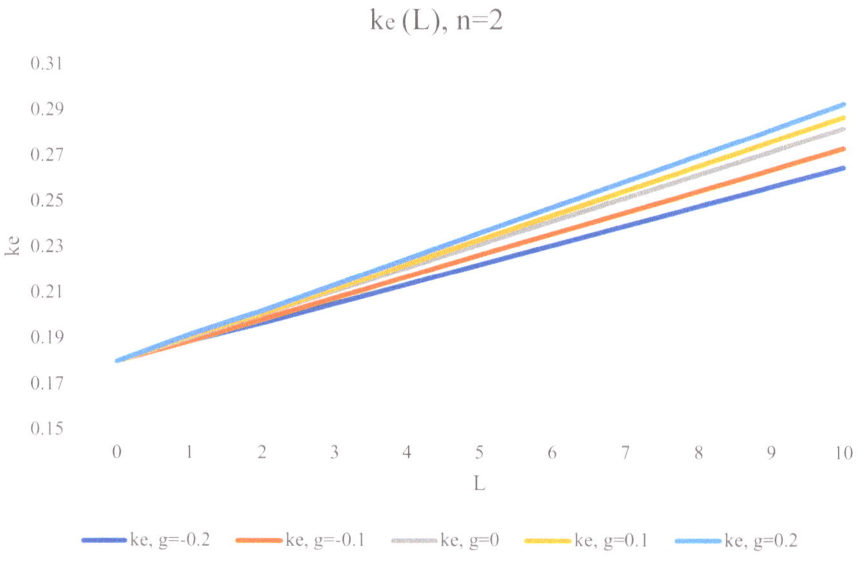

Fig. 13.4 The equity cost, ke, depending on the level of leverage L at different growth rates g in Generalized Brusov–Filatova–Orekhova theory (GBFO theory)

13.3.2 Calculations for 4-Year Company

Below, we study the dependence of WACC, WACC–g, V, and ke on leverage level L in the Generalized Brusov–Filatova–Orekhova theory (GBFO theory) at $k_0 = 0.18$; $k_d = 0.16$; $t = 0.2$; $g = 0.2$; 0.1; 0.0; −0.1; −0.2 for a 4-year company and compare obtained results with ones for a 2-year company (see above).

13.3.2.1 Calculations of Weighted Average Cost of Capital, WACC

From Table 13.5 and Fig. 13.5, it follows that different values of g all curves WACC (L) start from one point (0; $k_0 = 0.18$). These curves WACC(L) demonstrate the

Table 13.5 The WACC depending on the level of leverage L at different growth rates g in Generalized Brusov–Filatova–Orekhova theory (GBFO theory)

	WACC				
L	g = −0.2	g = −0.1	g = 0	g = 0.1	g = 0.2
0	0.1799864	0.1799858	0.179995	0.1799849	0.1799844
1	0.1538597	0.1554667	0.1567983	0.1579005	0.1588493
2	0.1450677	0.1472039	0.1489781	0.1504635	0.1517081
3	0.1406528	0.1430555	0.1450509	0.1467206	0.1481196
4	0.1379983	0.1405609	0.1426891	0.1444691	0.145877
5	0.1362263	0.1388955	0.1411122	0.1429656	0.1445542
6	0.1349595	0.1377048	0.1399846	0.1418905	0.1434877
7	0.1340088	0.1368112	0.1391383	0.1410836	0.1427137
8	0.1332689	0.1361157	0.1384797	0.1404555	0.1421113
9	0.1326768	0.1355592	0.1379526	0.1399529	0.1416291
10	0.1321922	0.1351036	0.1375212	0.1395414	0.1412344

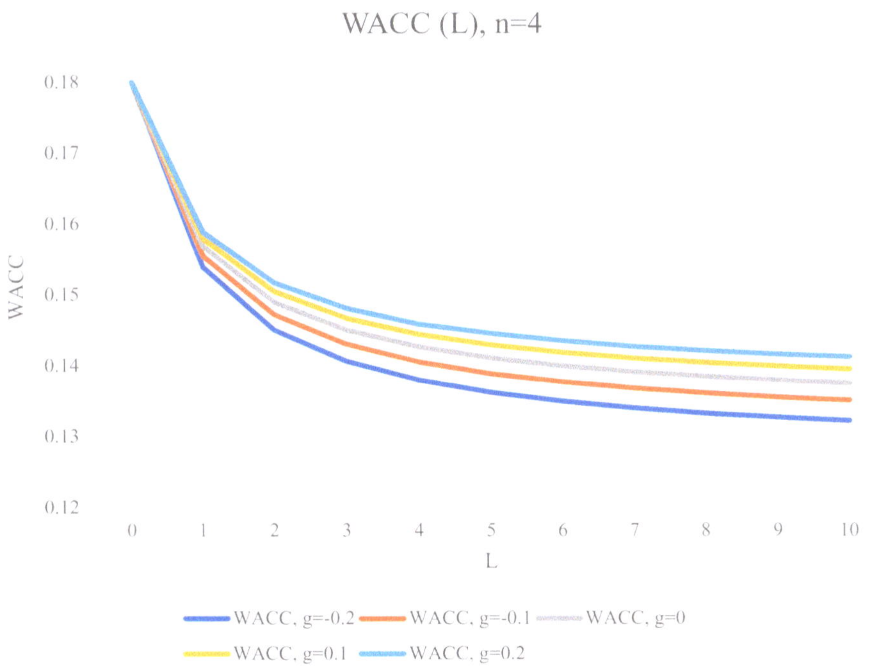

Fig. 13.5 The WACC depending on the level of leverage L at different growth rates g in Generalized Brusov–Filatova–Orekhova theory (GBFO theory

decrease of WACC with leverage level L at all g values. The curves WACC (L) increase with growth rate, g. In this part, the results for a 4-year company are similar to ones for a 2-year company. There is quantitative difference between them; the splitting of the curves WACC(L) Δ for g = 0.2 and g = −0.2 increases with g (at L = 10 Δ = 0.0018 for n = 2, while for n = 4 Δ = 0.009). In the perpetual case, ($n = \infty$) Δ = 0.073) (Brusov et al. 2021a).

The results for a 4-year company, as it was seen for a 2-year company, differ from the results for the perpetual limit—the theory of Modigliani and Miller (Brusov et al. 2021a); the WACC(L) curves decrease with increasing leverage L for all values of the growth rate g. As is known from Brusov et al. (2021a), in the case of Modigliani and Miller, the WACC(L) curves decrease with the level of leverage L for g < k_0 and increase for g > k_0. Therefore, k_0 is the threshold value g separating the increasing WACC(L) curves from the decreasing ones, and for g = k_0 WACC = const = k_0. As we mentioned above, the increase in WACC(L) with the growth rate of g is the first indication that WACC is no longer a discount rate, since it is intuitive that the discount rate must decrease with growth rate g in order to increase the value of company V. As we will see below, WACC–g and WACC–k_0 play the role of the discount rates and satisfy the above condition.

13.3.2.2 Calculations of the Discount Rate, WACC–g

Table 13.6 and Fig. 13.6 show all the curves (WACC–g)(L) with leverage level L at all g values. The (WACC–g) values at fixed leverage level L decrease with growth rate g. This means that WACC–g is a suitable candidate for the discount rate.

As in the case of the 2-year company, the behavior of (WACC–g)(L) with g growth can be explained as follows: all WACC(L) curves originate from the same point (L = 0; WACC = 0.18). The (WACC–g)(L) curves will be ordered as follows

Table 13.6 The discount rate, WACC–g, depending on the level of leverage L at different growth rates g in Generalized Brusov–Filatova–Orekhova theory (GBFO theory)

	(WACC–g), n = 4				
L	g = −0.2	g = −0.1	g = 0	g = 0.1	g = 0.2
0	0.3799864	0.2799858	0.179995	0.0799849	−0.020016
1	0.3538597	0.2554667	0.1567983	0.0579005	−0.041151
2	0.3450677	0.2472039	0.1489781	0.0504635	−0.048292
3	0.3406528	0.2430555	0.1450509	0.0467206	−0.05188
4	0.3379983	0.2405609	0.1426891	0.0444691	−0.054123
5	0.3362263	0.2388955	0.1411122	0.0429656	−0.055446
6	0.3349595	0.2377048	0.1399846	0.0418905	−0.056512
7	0.3340088	0.2368112	0.1391383	0.0410836	−0.057286
8	0.3332689	0.2361157	0.1384797	0.0404555	−0.057889
9	0.3326768	0.2355592	0.1379526	0.0399529	−0.058371
10	0.3321922	0.2351036	0.1375212	0.0395414	−0.058766

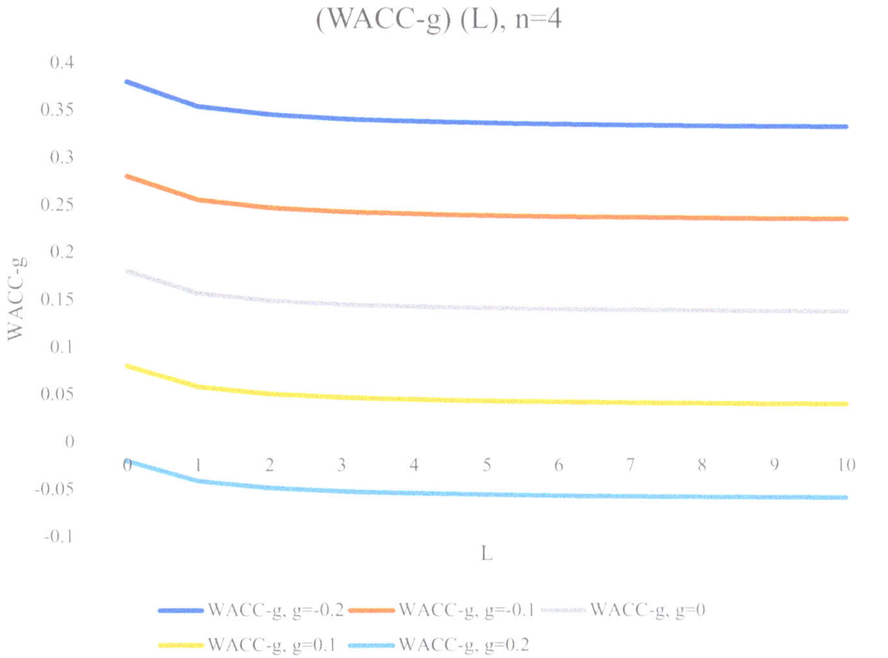

Fig. 13.6 The discount rate, WACC–g, depending on the level of leverage L at different growth rates g in Generalized Brusov–Filatova–Orekhova theory (GBFO theory)

for L = 0: the larger g, the lower the starting point and hence the entire graph lies, since the curves do not intersect. As we will see below, the decrease of (WACC–g)(L) with growth rate, g, will lead to an increase of the company value, V, with g.

13.3.2.3 Calculations of the Company Value, V

As can be seen from Table 13.7 and Fig. 13.7, the company value V at fixed growth rate g increases with leverage level L in Generalized Brusov–Filatova–Orekhova theory (GBFO theory). The company value V increases at fixed leverage level L with growth rate g at fixed as well. This is the consequence of a decrease in the discount rate (WACC–g)(L) with an increase in the growth rate g. Comparing with the results for the 2-year-old company, we see that the value of the company V increases with the age of the company: we have a range from 149 to 157 with L = 1 for g from −0.2 to 0.2 for the 2-year-old company and a range from 217 to 364 with L = 1 for g from −0.2 to 0.2 for a 4 year old company. This is the obvious conclusion, because it is well known that the value of any asset (company, stock, bond, etc.) is equal to the sum of the discounted returns generated by this asset. Since this value is proportional to the lifetime of this asset, the capitalization of the company will grow with its age.

Table 13.7 The company value, V, depending on the level of leverage L at different growth rates g in Generalized Brusov–Filatova–Orekhova theory (GBFO theory)

	V, n = 4				
L	g = −0.2	g = −0.1	g = 0	g = 0.1	g = 0.2
0	207.56615	236.28862	269.00881	306.05136	347.71061
1	217.29735	247.3597	281.61599	320.39442	363.9946
2	220.75339	251.2914	286.08901	325.47139	369.77167
3	222.5253	253.30558	288.37987	328.07604	372.72979
4	223.60263	254.53001	289.7723	329.65911	374.59742
5	224.32687	255.35305	290.70818	330.72309	375.70606
6	224.84716	255.94428	291.38042	331.48732	376.60369
7	225.23902	256.38954	291.88667	332.06283	377.25722
8	225.54478	256.73695	292.28165	332.51185	377.7671
9	225.79	257.01557	292.59842	332.87195	378.176
10	225.99105	257.244	292.85811	333.16716	378.51122

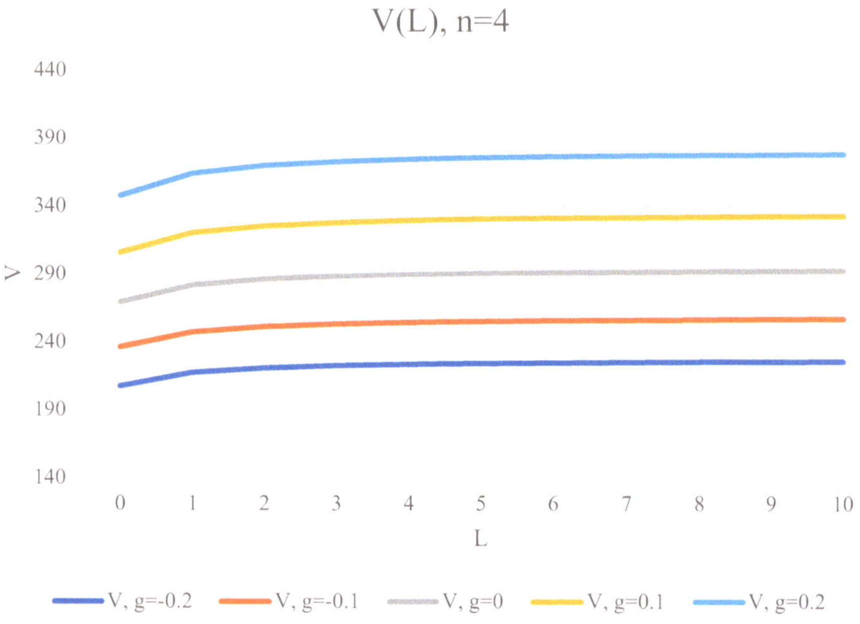

Fig. 13.7 The company value, V, depending on the level of leverage L at different growth rates g in Generalized Brusov–Filatova–Orekhova theory (GBFO theory)

The dependence of the cost of equity ke on the level of leverage L in the Generalized theory of Brusov–Filatova–Orekhova (GBFO theory) was studied below for a 4-year company with growth rates g = 0; ±0.1; and ± 0.2.

13.3.2.4 Calculations of the Cost of Equity ke

From Table 13.8 and Fig. 13.8, it can be seen that the cost of equity ke increases practically linearly with leverage level L at all growth rates g (except for $g = -0.2$ where we see a decrease in ke with leverage level L). The slope angle ke(L) increases with g. This should change the dividend policy of an enterprise with variable income

Table 13.8 The equity cost, ke, depending on the level of leverage L at different growth rates g in Generalized Brusov–Filatova–Orekhova theory (GBFO theory)

	ke, n = 4				
L	$g = -0.2$	$g = -0.1$	$g = 0$	$g = 0.1$	$g = 0.2$
0	0.1799864	0.1799858	0.179995	0.1799849	0.1799844
1	0.1797193	0.1829333	0.1855967	0.187801	0.1896985
2	0.1792032	0.1856116	0.1909343	0.1953906	0.1991244
3	0.1786111	0.1882219	0.1962038	0.2028824	0.2084783
4	0.1779915	0.1908046	0.2014454	0.2103453	0.2173852
5	0.1773581	0.1933733	0.2066729	0.2177938	0.2273252
6	0.1767166	0.1959339	0.2118923	0.2252338	0.2364136
7	0.1760702	0.1984895	0.2171066	0.2326686	0.2457093
8	0.1754204	0.2010417	0.2223176	0.2400999	0.2550015
9	0.1747682	0.2035915	0.2275261	0.2475287	0.2642911
10	0.1741143	0.2061396	0.2327328	0.2549558	0.2735789

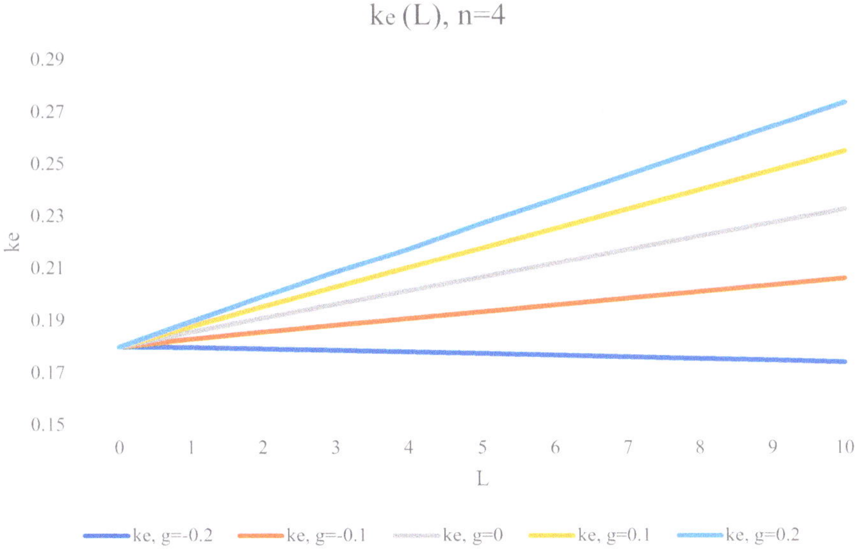

Fig. 13.8 The equity cost, ke, depending on the level of leverage L at different growth rates g in Generalized Brusov–Filatova–Orekhova theory (GBFO theory)

since, economically, the reasonable amount of dividends is equal to the cost of equity.

However, the biggest change in the company's dividend policy is related to the discovery of a qualitatively new effect in corporate finance: at a rate $g < g^*$, the slope of the ke(L) curve turns out to be negative (one can observe this effect here for $g = -0.2$ where a decrease in ke with leverage level L takes place). This effect, which is absent in the classical Modigliani–Miller theory and the classical Brusov–Filatova–Orekhova theory with constant income, exists in the Modigliani–Miller theory with variable income and in the Brusov–Filatova–Orekhova theory with variable income at a certain age of the company, n, which exceeds some cutoff age value, n^*.

The latter effect is similar to a qualitatively new effect in corporate finance, discovered by Brusov–Filatova–Orekhova within the framework of the BFO theory (Brusov et al. 2018, 2021b): anomalous dependences of the cost of equity ke on the leverage level L when income tax T exceeds a certain value T^*. This discovery also significantly changes the principles of the company's dividend policy.

13.3.3 Comparison with the Theory of Modigliani and Miller with Variable Income

The main difference between the Modigliani–Miller (MM) theory and the Brusov–Filatova–Orekhova (BFO) theory is that the latter describes companies of an arbitrary age, whereas the former represents the eternal limit of the BFO theory and is valid only for perpetual companies (with an infinite lifetime). If there is no time factor in the MM theory, then within the framework of the modern BFO theory, it is possible to analyze the financial condition of a company of any age; in this chapter, this is done for two company ages—2 and 4 years.

Since the MM theory is the eternal BFO limit, it is important to note that some of the results, such as the change in the discount rate from WACC to WACC–g, etc., are general, and this emphasizes the correctness of both theories and the connection between them as a connection between a general theory and its limiting case.

Along with the fundamental differences between the results of the two theories indicated above, there are also qualitative and quantitative differences. One of them is the next. The BFO results for a 4-year company, as it was seen for a 2-year company as well, differ from the results for the perpetual limit—the theory of Modigliani and Miller (Brusov et al. 2021a), where the WACC(L) curves decrease with increasing leverage L for all values of the growth rate g. As is known from Brusov et al. (2021a), in the case of Modigliani and Miller, the WACC(L) curves decrease with the level of leverage L for $g < k_0$ and increase for $g > k_0$. Therefore, k_0 is the threshold value g separating the increasing WACC(L) curves from the decreasing ones, and for $g = k_0$ WACC = const = k_0 (see Fig. 13.9 below).

We would like to emphasize an important observation. If in the classical versions of the Brusov–Filatova–Orekhova (BFO) theory and its perpetual limit, the theory of

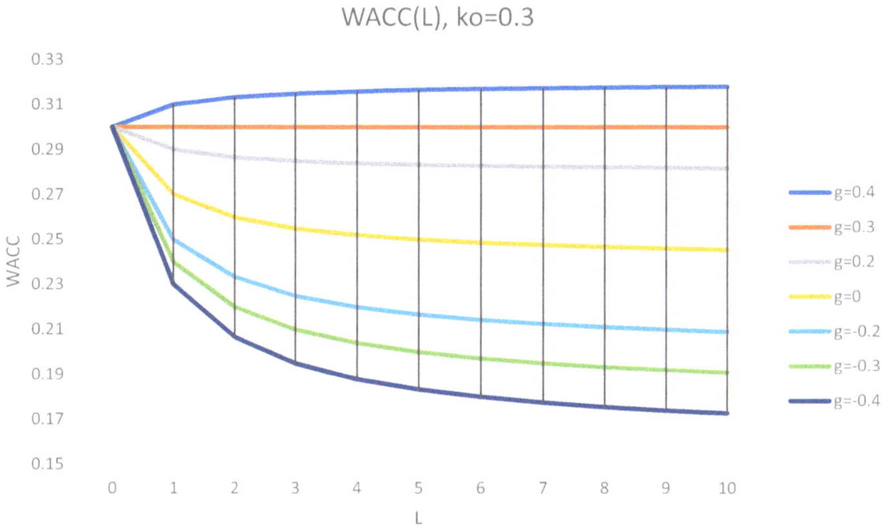

Fig. 13.9 WACC depending on leverage level L in Generalized Modigliani–Miller theory (GMM theory) at $k_0 = 0.3$ and $g = 0;\pm0.2;\pm0.3;\pm0.4$. (From Brusov et al. 2021a)

Nobel laureates Modigliani and Miller, where the case of constant income was considered, and where the gap between these two theories is huge (many qualitative effects that take place in the first theory, missing in the second), when taking into account the variable income, some effects of the BFO theory also take place in the Modigliani–Miller theory. This means that taking into account some effects that are present in economic practice (for example, variable income) brings both theories closer, and even the Modigliani–Miller theory, with all its many limitations, becomes more applicable in economic practice. However, it should be remembered that the Modigliani–Miller theory is only true for perpetual companies, whereas the BFO theory is valid for companies of any age, and from this point of view, they never coincide.

13.4 Conclusions

The Brusov–Filatova–Orekhova (BFO) theory of capital cost and capital structure as well as the theory by Nobel Prize winners Modigliani and Miller (perpetuity limit of BFO theory) consider the case of constant income, whereas in practice, the income of a company is, of course, variable. Recently, we have generalized the latter for the case of variable income, and here we have generalized for the first time the Brusov–Filatova–Orekhova theory for the case of variable income.

This generalization significantly expands the applicability of this modern capital structure theory, which is valid for companies of any age, and in practice for corporate finance, business valuation, investments, banking, ratings, etc.

We have derived the generalized BFO formula for WACC and this consists of a main theoretical result of a current chapter.

From this formula and as well from using this formula in MS Excel, we show that the role of the discount rate shifts from WACC to WACC–g (where g is the growth rate) for financially dependent companies and from k_0 to k_0–g (for financially independent companies). Whereas the WACC increases with g, the actual discount rates WACC–g and k_0–g decrease with g and, accordingly, the company value, V, increases with g. For the cost of equity ke, the slope of the curve ke(L) increases with g. Since the cost of equity determines the economically justified amount of dividends, this should change the company's dividend policy. It turns out that at the rate $g < g^*$, the slope of the curve ke(L) becomes negative, which can significantly change the company's dividend policy principles. This means a qualitatively new effect on discoveries in corporate finance.

The novelty of the work is the generalization of the theory of Brusov–Filatova–Orekhova to the case of variable income, the derivation of generalized BFO formulas for the weighted average cost of capital, WACC, cost of equity, ke, company value, V, and the use of these formulas to study the influence of the growth rate g on the dependence of the main financial performance of the company on debt financing.

The importance of the current consideration is due to a couple of points:

- gives an idea of the behavior of the main financial indicators of the company with a change in the growth rate g for both cases—an increase in g and a decrease in g;
- creates a developed methodology that allows the researcher to analyze the main financial indicators of the company (capital costs, company value, etc.) for the actual conditions of the company's functioning.

Limitations of the study: a generalization of the BFO theory was carried out for the case of paying income tax at the end of reporting periods. However, in practice, these payments may also be made in advance. This determines the further direction of the study: consideration of the case of variable income when income tax is paid in advance.

References

Angotti, M., de Lacerda Moreira, R., Hipólito Bernardes do Nascimento, J., & Neto de Almeida Bispo, O. (2018). Analysis of an equity investment strategy based on accounting and financial reports in Latin American markets. *Reficont, 5*, 22–40.

Barbi, M. (2011). On the risk–neutral value of debt tax shields. *Applied Financial Economics, 22*, 251–258.

Becker, D. M. (2022). Getting the valuation formulas right when it comes to annuities. *Managerial Finance, 48*, 470.

Berk, J., & DeMarzo, P. (2007). *Corporate finance*. Pearson–Addison Wesley.

Białek-Jaworska, A., & Klapkiv, L. (2021). Does withholding tax on interest limit international profit-shifting by FDI? *Equilibrium: Quarterly Journal of Economics and Economic Policy, 16*, 11–44. https://doi.org/10.24136/eq.2021.001

Bradley, M., Jarrell, G. A., & Kim, E. H. (1984). On the existence of an optimal capital structure: Theory and evidence. *The Journal of Finance, 39*, 857–878.

Brusov, P., & Filatova, T. (2013). Veniamin Kulik benefits of advance payments of tax on profit: Consideration within the Brusov–Filatova–Orekhova (BFO) theory. *Mathematics, 2022*, 10. https://doi.org/10.3390/math10122013

Brusov, P., & Filatova, T. (2022). Influence of method and frequency of profit tax payments on company financial indicators. *Mathematics, 10*, 2479. https://doi.org/10.3390/math10142479

Brusov, P., Filatova, T., Orehova, N., & Eskindarov, M. (2018). *Modern corporate finance, investments, taxation and ratings* (2nd ed., pp. 1–571). Springer Nature Publishing.

Brusov, P. N., Filatova, T. V., Orekhova, N. P., Kulik, V. L., Chang, S. I., & Lin, Y. C. G. (2021a). Generalization of the Modigliani–Miller theory for the case of variable income. *Mathematics, 9*, 1286. https://doi.org/10.3390/math9111286

Brusov, P., Filatova, T., & Orehova, N. (2021b). *Ratings: Critical analysis and new approaches of quantitative and qualitative methodology* (pp. 1–369). Springer Nature Publishing. Retrieved from https://www.springer.com/de/book/9783030562427

Dimitropoulos, P. (2014). Capital structure and corporate governance of soccer clubs: European evidence. *Management Research Review, 37*, 658–678.

Durana, P., Ginevicius, R., Urbanski, M., Podhorska, I., & Tumpach, M. (2021). Parallels and differences in earnings management of the Visegrad Four and the Baltics. *Journal of Competitiveness, 13*, 39–55. https://doi.org/10.7441/joc.2021.03.03

Durana, P., Valaskova, K., Blazek, R., & Palo, J. (2022). Metamorphoses of earnings in the transport sector of the V4 region. *Mathematics, 10*, 1204. https://doi.org/10.3390/math10081204

El-Chaarani, H., Abraham, R., & Skaf, Y. (2022). The impact of corporate governance on the financial performance of the banking sector in the MENA (Middle Eastern and North African) region: An immunity test of banks for COVID–19. *Journal of Risk and Financial Management, 15*, 82.

Farber, A., Gillet, R., & Szafarz, A. (2006). A general formula for the WACC. *International Journal of Business, 11*, 211–218.

Fernandez, P. (2006). A general formula for the WACC: A comment. *International Journal of Business, 11*, 219.

Franc-Dąbrowska, J., Mądra-Sawicka, M., & Milewska, A. (2021). Energy sector risk and cost of capital assessment—Companies and investors perspective. *Energies, 14*, 1613.

Frank, M., & Goyal, V. (2009). Capital structure decisions: Which factors are reliably important? *Financial Management, 38*, 1–37. https://doi.org/10.1111/j.1755-053X.2009.01026.x

Graham, J. R. (2003). Taxes and corporate finance: A review. *Review of Financial Studies, 16*, 1075–1129.

Hamada, R. (1969). Portfolio analysis, market equilibrium, and corporate finance. *The Journal of Finance, 24*, 13–31.

Harris, R., & Pringle, J. (1985). Risk–adjusted discount rates—Extension form the average–risk case. *Journal of Financial Research, 8*, 237–244.

Harry, D. A., & Masulis, R. W. (1980). Optimal capital structure under corporate and personal taxation. *Journal of Financial Economics, 8*, 3–29.

Luiz, K., & Cruz, M. (2015). The relevance of capital structure on firm performance: A multivariate analysis of publicly traded Brazilian companies. *REPeC Brasília, 9*, 384–401.

Miller, M. (1977). Debt and taxes. *The Journal of Finance, 32*, 261–275. https://doi.org/10.1111/j.1540-6261.1977.tb03267.x

Modigliani, F., & Miller, M. (1958). The cost of capital, corporate finance, and the theory of investment. *The American Economic Review, 48*, 261–297.

Modigliani, F., & Miller, M. (1963). Corporate income taxes and the cost of capital: A correction. *The American Economic Review, 53*, 147–175.

Modigliani, F., & Miller, M. (1966). Some estimates of the cost of capital to the electric utility industry 1954–1957. *The American Economic Review, 56*, 333–391.

Mundi, H. S., Kaur, P., & Murty, R. L. N. (2022). A qualitative inquiry into the capital structure decisions of overconfident finance managers of family-owned businesses in India. *Qualitative Research in Financial Markets, 14*, 357–379.

Sadiq, M., Alajlani, S., Hussain, M. S., Bashir, F., & Chupradit, S. (2022). Impact of credit, liquidity, and systematic risk on financial structure: Comparative investigation from sustainable production. *Environmental Science and Pollution Research International, 29*, 20963–20975.

Sheridan, T., & Wessels, R. (1988). The determinants of capital structure choice. *The Journal of Finance, 43*, 1–19.

Singhal, N., Goyal, S., Sharma, D., Kumari, S., & Nagar, S. (2022). Capitalization and profit ability: Applicability of capital theories in BRICS banking sector. *Future Business Journal, 8*, 30. https://doi.org/10.1186/s43093-022-00140-w

Valaskova, K., Gavurova, B., Durana, P., & Kovacova, M. (2020). Alter ego only four times? The case study of business profits in the Visegrad Group. *Economics and Management, 23*, 101–119. https://doi.org/10.15240/tul/001/2020-3-007

Valaskova, K., Androniceanu, A.-M., Zvarikova, K., & Olah, J. (2021a). Bonds between earnings management and corporate financial stability in the context of the competitive ability of enterprises. *Journal of Competitiveness, 13*, 167–184. https://doi.org/10.7441/joc.2021.04.10

Valaskova, K., Kliestik, T., & Gajdosikova, D. (2021b). Distinctive determinants of financial indebtedness: Evidence from Slovak and Czech enterprises. *Equilibrium: Quarterly Journal of Economics and Economic Policy, 16*, 639–659. https://doi.org/10.24136/eq.2021.023

Vergara-Novoa, C., Sepúlveda-Rojas, J. P., Miguel, D. A., & Nicolás, R. (2018). Cost of capital estimation for highway concessionaires in Chile. *Journal of Advanced Transportation, 2018*, 2153536.

Virglerova, Z., Ivanova, E., Dvorsky, J., Belas, J., & Krulický, T. (2021). Selected factors of internationalisation and their impact on the SME perception of the market risk. *Oeconomia Copernicana, 12*, 1011–1032. https://doi.org/10.24136/oc.2021.033

Xu, J., & Liu, F. (2020). The impact of intellectual capital on firm performance: A modified and extended VAIC model. *Journal of Competitiveness, 12*, 161–176. https://doi.org/10.7441/joc.2010.01.10

Chapter 14
Inflation in Brusov–Filatova–Orekhova Theory and in Its Perpetuity Limit: Modigliani–Miller Theory

In this chapter the influence of inflation on capital cost and capitalization of the company within modern theory of capital cost and capital structure—Brusov–Filatova–Orekhova theory (BFO theory) (Brusov and Filatova 2011; Brusov et al. 2011a, b, c, 2012a, b, 2013a, b, 2014a, b; Filatova et al. 2008) and within its perpetuity limit—Modigliani–Miller theory (Modigliani and Miller 1958, 1963, 1966) is investigated. By direct incorporation of inflation into both theories, it is shown for the first time, that inflation not only increases the equity cost and the weighted average cost of capital, but as well it changes their dependence on leverage. In particular, it increases growing rate of equity cost with leverage. Capitalization of the company is decreased under accounting of inflation.

14.1 Introduction

Created more than half a century ago by Nobel Prize winners Modigliani and Miller theory of capital cost and capital structure (Modigliani and Miller 1958, 1963, 1966) did not account a lot of factors of a real economy, such as taxing, bankruptcy, unperfected capital markets, inflation and many others. But while taxing has been included into consideration by authors themselves and some other limitations have been taken off by their followers, direct incorporation of inflation to Modigliani–Miller theory is absent still now.

The influence of inflation on valuation of capital cost of company and its capitalization is investigated within Modigliani–Miller theory (MM) (Modigliani and Miller 1958, 1963, 1966), which is now outdate, but still widely used at the West, as well as within modern theory of capital cost and capital structure—Brusov–Filatova–Orekhova theory (BFO theory) (Brusov and Filatova 2011; Brusov et al. 2011a, b, c, 2012a, b, 2013a, b, 2014a, b; Filatova et al. 2008), which should replaced Modigliani–Miller theory (Modigliani and Miller 1958, 1963, 1966). It is shown, that inflation not only increases the equity cost and the weighted average cost

P. Brusov, T. Filatova, *Efficiency of Investment Projects*, Contributions to Finance and Accounting, https://doi.org/10.1007/978-3-032-01028-5_14

of capital, but as well it changes their dependence on leverage. In particular, it increases growing rate of equity cost with leverage. Capitalization of the company is decreased under accounting of inflation.

We start from the study of inflation within Modigliani–Miller theory without taxing (Modigliani and Miller 1958), than with taxing (Modigliani and Miller 1963) and finally within modern theory of capital cost and capital structure—Brusov–Filatova–Orekhova theory (BFO theory) (Brusov and Filatova 2011; Brusov et al. 2011a, b, c, 2012a, b, 2013a, b, 2014a, b; Filatova et al. 2008).

14.2 Accounting of Inflation in Modigliani–Miller Theory Without Taxes

Note, that any modification of Modigliani–Miller theory, as well as of any other one, requires going behind the frame of modified theory. Thus, in current case we should go behind the frame of perpetuity of the company (remind to reader, that Modigliani–Miller theory describes only perpetuity companies—companies with infinite lifetime), come to the companies with finite lifetime, make necessary calculations and then use the perpetuity limit.

As known, in profit approach capitalization of the company is equal to discounted sum of profits of the company. Suppose that profit is constant for all periods and equal to *CF*, one gets for capitalization of the financially independent company V_0, existing n years at market

$$V_0 = \frac{CF}{1+k_0} + \frac{CF}{(1+k_0)^2} + \ldots + \frac{CF}{(1+k_0)^n}. \tag{14.1}$$

Here k_0 capital cost of the financially independent company.

Under inflation with rate α the capitalization of the financially independent company V_0^* becomes equal to

$$V_0^* = \frac{CF}{(1+k_0)(1+\alpha)} + \frac{CF}{[(1+k_0)(1+\alpha)]^2} + \ldots + \frac{CF}{[(1+k_0)(1+\alpha)]^n}. \tag{14.2}$$

Using the formula for sum of the terms of indefinitely diminishing geometrical progression with the first term

$$a_1 = \frac{CF}{(1+k_0)(1+\alpha)} \tag{14.3}$$

and denominator

$$q = \frac{1}{(1+k_0)(1+\alpha)} \tag{14.4}$$

one gets for capitalization of the financially independent company V_0^* the following expression

$$V_0^* = \frac{a_1}{1-q} = \frac{CF}{(1+k_0)(1+\alpha)\left[1-((1+k_0)(1+\alpha))^{-1}\right]} =$$
$$= \frac{CF}{(1+k_0)(1+\alpha)-1} = \frac{CF}{k_0(1+\alpha)+\alpha}.$$
$$V_0^* = \frac{CF}{k_0(1+\alpha)+\alpha}. \tag{14.5}$$

It is seen, that under accounting of inflation the capitalization of the company decreases.

At discount rate $k_0 = 10\%$ and inflation rate $\alpha = 3\%$ the decrease is equal to 5.7%, and at discount rate $k_0 = 15\%$ and inflation rate $\alpha = 7\%$ the decrease is equal to 35%. One can see that influence of inflation on the company capitalization could be significant enough and negative.

For leverage company, using debt capital one has without inflation

$$V_L = \frac{CF}{1+WACC} + \frac{CF}{(1+WACC)^2} + \ldots + \frac{CF}{(1+WACC)^n}, \tag{14.6}$$

and in perpetuity limit

$$V_L = \frac{CF}{WACC}. \tag{14.7}$$

Under accounting of inflation the capitalization of the company is equal to

$$V_L^* = \frac{CF}{(1+WACC)(1+\alpha)} + \frac{CF}{[(1+WACC)(1+\alpha)]^2} + \ldots + \frac{CF}{[(1+WACC)(1+\alpha)]^n}. \tag{14.8}$$

Summing the infinite set, we get for leverage company capitalization under accounting of inflation in Modigliani–Miller limit

$$V_L^* = \frac{a_1}{1-q} = \frac{CF}{(1+WACC)(1+\alpha)\left[1-((1+WACC)(1+\alpha))^{-1}\right]} =$$
$$= \frac{CF}{(1+WACC)(1+\alpha)-1} = \frac{CF}{WACC(1+\alpha)+\alpha}$$
$$V_L^* = \frac{CF}{WACC(1+\alpha)+\alpha}. \tag{14.9}$$

It is seen, that similar to the case of the financially independent company inflation decreases the company capitalization and the decrease could be significant. From the formulas (14.7) and (14.9) it follows that effective values of capital costs (equity and WACC) are equal to

$$k_0^* = k_0(1+\alpha)+\alpha \tag{14.10}$$

$$WACC^* = WACC\cdot(1+\alpha)+\alpha \tag{14.11}$$

Note, that both capital costs increases under inflation.

We can compare obtained results with Fisher formula for inflation.

$$i^* = \frac{i-\alpha}{1+\alpha}. \tag{14.12}$$

Solving this equation with respect to nominal rate i, one gets equation, similar Eqs. (14.10) and (14.11)

$$i = i^*\cdot(1+\alpha)+\alpha. \tag{14.13}$$

Thus, effective capital costs in our case have meaning of nominal ones, accounting inflation.

From the Modigliani–Miller theorem, that the weighted average cost of capital WACC does not depends on leverage level (without taxing), formulating under accounting of inflation, it is easy to get expression for the equity cost:

$$WACC^* = k_0^* = k_e^* w_e + k_d^* w_d. \tag{14.14}$$

Finding from here k_e^*, one gets:

$$\begin{aligned} k_e^* &= \frac{k_0^*}{w_e} - k_d^*\frac{w_d}{w_e} = \frac{k_0^*(S+D)}{S} - k_d^*\frac{D}{S} = k_0^* + \left(k_0^* - k_d^*\right)\frac{D}{S} \\ &= k_0^* + \left(k_0^* - k_d^*\right)L \end{aligned} \tag{14.15}$$

Putting instead of k_0^*, k_d^* their expressions, one gets finally

$$\begin{aligned} k_e^* = k_0^* + \left(k_0^* - k_d^*\right)L = k_0(1+\alpha)+\alpha+ \\ +L(k_0-k_d)(1+\alpha) = (1+\alpha)[k_0+\alpha+L(k_0-k_d)] \end{aligned}$$

$$k_e^* = k_0(1+\alpha)+\alpha+L(k_0-k_d)(1+\alpha). \tag{14.16}$$

It is seen, that inflation not only increases the equity cost, but as well it changes its dependence on leverage. In particular, it increases growing rate of equity cost with

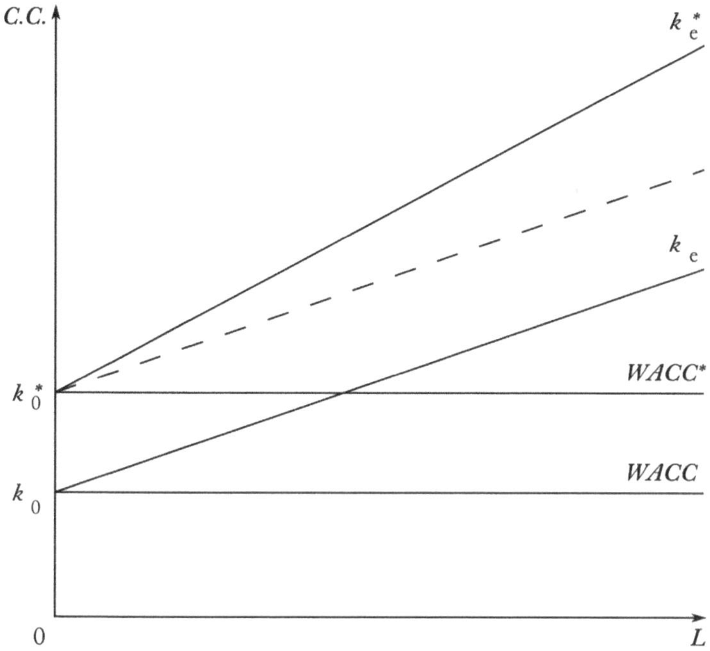

Fig. 14.1 Dependence of the equity cost and the weighted average cost of capital on leverage in the Modigliani–Miller theory without taxing under accounting of inflation. It is seen, that growing rate of equity cost increases with leverage. Axis y means capital costs—C.C.

leverage by multiplier (1+ α). The growing rate of equity cost with leverage, which is equal to $(k_0 - k_d)$ without inflation becomes equal to $(k_0 - k_d)(1 + \alpha)$ under accounting of inflation (Fig. 14.1).

Thus, we come to conclusion, that it is necessary to modified the second statement of the Modigliani–Miller theory (Modigliani and Miller 1958) concerning the equity cost of leverage company.

Second Original MM Statement

Equity cost of leverage company k_e could be found as equity cost of financially independent company k_0 of the same group of risk, plus premium for risk, which value is equal to production of difference $(k_0 - k_d)$ on leverage level L.

Second Modified MM-BFO Statement

Under existing of inflation with rate α equity cost of leverage company k_e could be found as equity cost of financially independent company k_0 of the same group of risk, multiplied by (1+ α), plus inflation rate α and plus premium for risk, which value is equal to production of difference $(k_0 - k_d)$ on leverage level L and on multiplier $(1 + \alpha)$.

14.3 Accounting of Inflation in Modigliani–Miller Theory with Corporate Taxes

Let us calculate first the tax shield for perpetuity company under accounting of inflation

$$(PV)_{TS} = k_d^* DT \sum_{t=1}^{\infty} \left(1 + k_d^*\right)^{-t} = DT \tag{14.17}$$

It is interesting to note, that inspite of dependence of each term of set on effective credit rate.

k_d^* tax shield turns out to be independent of it and equal to "inflationless" value DT and Modigliani–Miller theorem under accounting of inflation takes the form (Modigliani and Miller 1963)

$$V_L^* = V_0^* + DT. \tag{14.18}$$

Substituting $D = w_d V_L^*$, one gets

$$V_L^* = CF/k_0^* + w_d V_L^* T \tag{14.19}$$

or

$$V_L^*(1 - w_d T) = CF/k_0^*. \tag{14.20}$$

Because leverage company capitalization is equal to $V_L^* = CF/WACC^*$ for the weighted average cost of capital one has

$$WACC^* = k_0^*(1 - w_d T). \tag{14.21}$$

From Eq. (14.21) we get the dependence of $WACC^*$ on leverage level $L = D/S$:

$$WACC^* = k_0^*(1 - LT/(1 + L))$$

$$WACC^* = [k_0(1 + \alpha) + \alpha]\cdot(1 - w_d T) \tag{14.22}$$

On definition of the weighted average cost of capital with accounting of the tax shield one has

$$WACC^* = k_0^* w_e + k_d^* w_d (1 - T). \tag{14.23}$$

Equating right hand parts of expressions (Eq. 14.21) and (Eq. 14.23), we get

$$k_0^*(1 - w_d T) = k_0^* w_e + k_d^* w_d (1 - T), \tag{14.24}$$

from where one obtains the following expression for equity cost:

$$\begin{aligned} k_e^* &= k_0^* \frac{(1 - w_d T)}{w_e} - k_d^* \frac{w_d}{w_e}(1 - T) = k_e^* \frac{1}{w_e} - k_0^* \frac{w_d}{w_e} T - k_d^* \frac{D}{S}(1 - T) = \\ &= k_0^* \frac{D + S}{S} - k_0^* \frac{D}{S} T - k_d^* \frac{D}{S}(1 - T) = k_0^* + L(1 - T)\left(k_0^* - k_d^*\right). \\ k_e^* &= k_0^* + L(1 - T)\left(k_0^* - k_d^*\right) = [k_0(1 + \alpha) + \alpha] + L(1 - T)(k_0 - k_d)(1 + \alpha) \end{aligned} \tag{14.25}$$

It is seen, that similar to the case without taxes inflation not only increases the equity cost, but as well it changes its dependence on leverage (Fig. 14.2). In particular, it increases growing rate of equity cost with leverage by multiplier (1+ α). The growing rate of equity cost with leverage, which is equal to $(k_0 - k_d)(1 - T)$ without inflation becomes equal to $(k_0 - k_d)(1 + \alpha)(1 - T)$ under accounting of inflation.

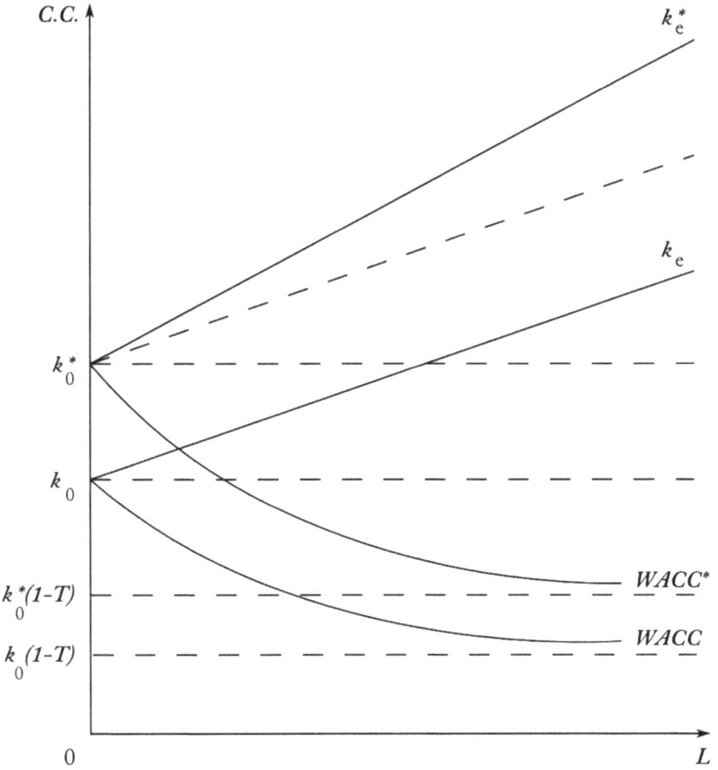

Fig. 14.2 Dependence of the equity cost and the weighted average cost of capital on leverage in the Modigliani–Miller theory with taxing under accounting of inflation. It is seen, that growing rate of equity cost increases with leverage. Axis y means capital costs—C.C.

We can now reformulate the fourth statement of the Modigliani–Miller theory (Modigliani and Miller 1963) concerning the equity cost of leverage company for case of accounting of inflation.

Fourth Original MM Statement
Equity cost of leverage company k_e paying tax on profit could be found as equity cost of financially independent company k_0 of the same group of risk, plus premium for risk, which value is equal to production of difference $(k_0 - k_d)$ on leverage level L and on tax shield $(1\text{-}T)$ and on multiplier $(1 + \alpha)$.

Fourth Modified MM-BFO Statement
Equity cost of leverage company k_e paying tax on profit under existing of inflation with rate α could be found as equity cost of financially independent company k_0 of the same group of risk, multiplied by $(1 + \alpha)$, plus inflation rate α and plus premium for risk, which value is equal to production of difference $(k_0 - k_d)$ on leverage level L, on tax shield $(1\text{-}T)$ and on multiplier $(1 + \alpha)$.

14.4 Accounting of Inflation in Brusov–Filatova–Orekhova Theory with Corporate Taxes

14.4.1 Generalized Brusov–Filatova–Orekhova Theorem

Brusov–Filatova–Orekhova, generalized the Modigliani–Miller theory for the case of the companies with arbitrary lifetime (Brusov and Filatova 2011; Brusov et al. 2011a, b, c, 2012a, b, 2013a, b, 2014a, b; Filatova et al. 2008), have proved the following important theorem in case of absence of corporate taxing:

Without corporate taxing the equity cost k_{e0}, ***as well as the weighted average cost of capital WACC do not depend on company lifetime and are equal to***

$$k_e = k_0 + L(k_0 - k_d) \quad \text{and} \quad WACC = k_0. \tag{14.26}$$

consequently.

Thus, the theorem has been proved, that without corporate taxes (say, in offshore zones) the Modigliani–Miller results for capital costs, in spite of the fact, that they have been obtained in perpetuity limit, remain in force for companies with arbitrary lifetime, describing by Brusov–Filatova–Orekhova theory (BFO theory). To prove this theorem Brusov, Filatova and Orekhova, of course, had to go behind Modigliani–Miller approximation.

Under accounting of inflation we can generalize this theorem (Brusov and Filatova 2011; Brusov et al. 2011a, b, c, 2012a, b, 2013a, b, 2014a, b; Brusova 2011; Filatova et al. 2008):

Generalized Brusov–Filatova–Orekhova Theorem

Under accounting of inflation without corporate taxing the equity cost k_0^*, as well as the weighted average cost of capital $WACC^*$ do not depend on company lifetime and are equal to

$$k_e^* = k_0^* + L(k_0^* - k_d^*) = k_0(1+\alpha) + \alpha + L(k_0 - k_d)(1+\alpha) \quad \text{and}$$
$$WACC^* = k_0^* = k_0(1+\alpha) + \alpha. \tag{14.27}$$

consequently.

Following to Brusov–Filatova–Orekhova (Brusov and Filatova 2011; Brusov et al. 2011a, b, c, 2012a, b, 2013a, b, 2014a, b; Filatova et al. 2008), let us consider the situation for arbitrary lifetime companies with account of corporate taxing. They have derived the famous formula for weighted average cost of capital of companies with arbitrary lifetime

$$\frac{1-(1+WACC)^{-n}}{WACC} = \frac{1-(1+k_0)^{-n}}{k_0[1-\omega_d T(1-(1+k_d)^{-n})]} \tag{14.28}$$

The application of BFO formula (14.28) is very wide: authors have applied it in corporate finance, in investments, in taxing, in business valuation, in banking and some other areas (Brusov et al. 2011a, b, 2013a***). Using this formula (14.28), one can study the dependence of the weighted average cost of capital, WACC, as well as the equity cost k_e on leverage level L, on tax on profit rate t, on lifetime of the company n and on relation between equity and debt cost. The qualitatively new effect in corporate finance has been discovered: decrease of the equity cost k_e on leverage level L, which is quite important for corporate finance in general and, in particular, for creating the adequate dividend policy.***

Below we generalize formula (14.28) under existing of inflation.

14.5 Generalized Brusov–Filatova–Orekhova Formula Under Existing of Inflation

Under existing of inflation it is necessary to replace all capital costs: the equity, the debt and the weighted average cost of capital k_0, k_d, $WACC$ by effective ones $k_0^*, k_d^*, WACC^*$, where

$$k_0^* = k_0(1+\alpha) + \alpha$$
$$k_d^* = k_d(1+\alpha) + \alpha$$
$$WACC^* = WACC \cdot (1+\alpha) + \alpha$$

Rewriting the equations for tax shield $(TS)_n$, capitalization of financially independent company V_0^* as well as for financially dependent company V_L^* for the case of existing of inflation, one gets

$$(PV)_{TS} = (TS)_n = k_d^* DT \sum_{t=1}^{n} \left(1 + k_d^*\right)^{-t} = DT\left[1 - \left(1 + k_d^*\right)^{-n}\right] \quad (14.29)$$

$$V_0^* = CF \sum_{t=1}^{n} \left(1 + k_0^*\right)^{-t} = CF\left[1 - \left(1 + k_0^*\right)^{-n}\right]/k_0^* \quad (14.30)$$

$$V_L^* = CF \sum_{t=1}^{n} (1 + WACC^*)^{-t} = CF[1 - (1 + WACC^*)^{-n}]/WACC^* \quad (14.31)$$

$$V_L^* = V_0^* + (TS)_n. \quad (14.32)$$

After substitution $D = w_d V_L^*$ we have

$$V_L^* = CF/k_0^* + w_d V_L^* T \quad (14.33)$$

From here after some transformations we get generalized Brusov–Filatova–Orekhova formula under existing of inflation

$$\frac{1 - (1 + WACC^*)^{-n}}{WACC^*} = \frac{1 - \left(1 + k_0^*\right)^{-n}}{k_0^*\left[1 - \omega_d T\left(1 - \left(1 + k_d^*\right)^{-n}\right)\right]}, \quad (14.34)$$

or after substitutions

$$k_0^* = k_0(1 + \alpha) + \alpha; \quad k_d^* = k_d(1 + \alpha) + \alpha$$

one gets finally

$$\frac{1-(1+WACC^*)^{-n}}{WACC^*} = \frac{1-[(1+k_0)(1+\alpha)]^{-n}}{(k_0(1+\alpha)+\alpha)\cdot[1-\omega_d T(1-((1+k_d)(1+\alpha))^{-n})]} \quad (14.35)$$

Formula (14.35) is the generalized Brusov–Filatova–Orekhova formula under existing of inflation.

Let us show some figures, illustrating obtained results.

At Figs. 14.3 and 14.4 the dependence of the weighted average cost of capital WACC on debt ratio w_d at different inflation rate α (1 – α = 3%; 2 – α = 5%; 3 – α = 7%; 4 – α = 9%) for 5-year company as well as for 2-year company. It is seen that with increase of inflation rate lines, showing the dependence WACC (w_d) shift practically homogeneously to higher values.

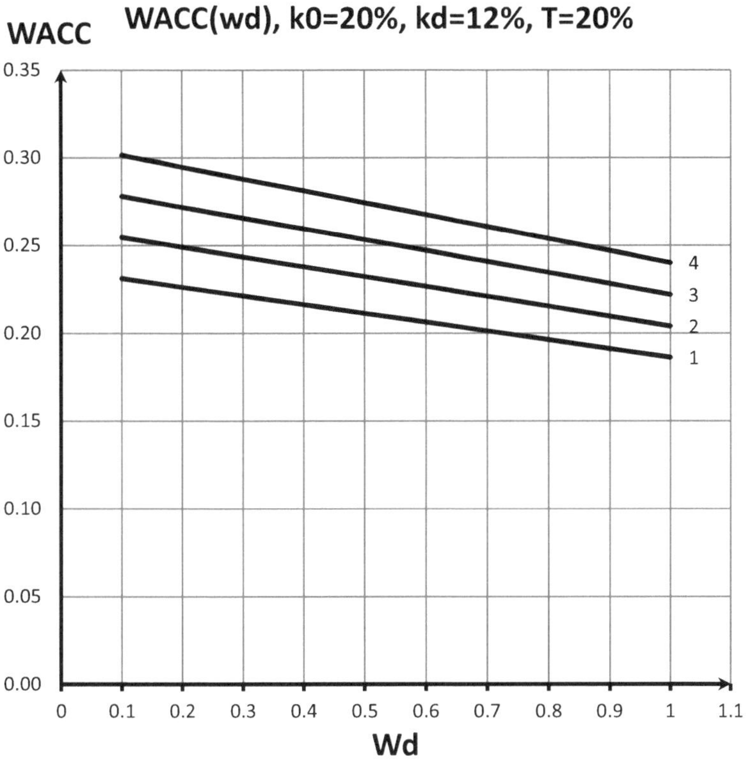

Fig. 14.3 Dependence of the weighted average cost of capital WACC on debt ratio w_d at different inflation rate α (1 − α = 3%; 2 − α = 5%; 3 − α = 7%; 4 − α = 9%) for 5-year company

It is seen, that difference in results for 2-year company and 5-year company is very small. More obviously it could be observed from Tables 14.1 and 14.2.

Below we show the dependences of the weighted average cost of capital WACC on debt ratio w_d at different tax on profit rate from $T = 10\%$ up to $T = 100\%$ at different inflation rate $\alpha = 3\%, 5\%, 7\%, 9\%$ for 5-year company (Figs. 14.5, 14.6, 14.7, and 14.8) as well as for 2-year company (Figs. 14.9, 14.10, 14.11, and 14.12). Tax on profit rate increases from $T = 0.1$ (upper line) up to $T = 1$ (lowest line) with the step 0.1.

The analysis of Figs. 14.5, 14.6, 14.7, 14.8, 14.9, 14.10, 14.11, and 14.12 shows, that the weighted average cost of capital WACC decreases with debt ratio w_d faster with increase of tax on profit rate. The space between lines, corresponding to different tax on profit rates, increases with inflation rate. The variation range of WACC increases with inflation rate as well as with lifetime of the company.

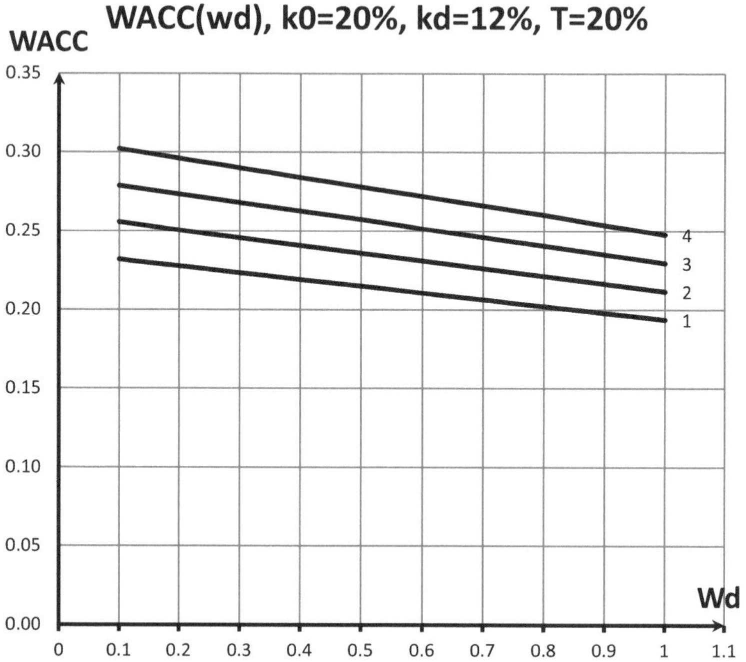

Fig. 14.4 Dependence of the weighted average cost of capital WACC on debt ratio w_d at different inflation rate α (1 – α = 3%; 2 – α = 5%; 3 – α = 7%; 4 – α = 9%) for 2-year company

14.6 Irregular Inflation

Above we considered inflation rate as constant. Really, as a rule, the inflation rate is a variable. It is possible to generalize all above consideration for the case of non-homogeneous inflation, introducing effective inflation for a few periods.

The effective inflation rate for a few periods $t = t_1 + t_2 + \ldots + t_n$ is equal to

$$\alpha = (1 + \alpha_1)(1 + \alpha_2)\cdot\ldots\cdot(1 + \alpha_n) - 1, \tag{14.36}$$

where $\alpha_1, \alpha_2, \ldots, \alpha_n$ are inflation rates for periods $t_1, t_2, \ldots, t_n$.The proof of the formula (14.36) will be done below in (14.6).

In the case of non-homogeneous inflation it could be accounted in both theories: Modigliani–Miller and Brusov–Filatova–Orekhova theory (BFO theory) either through effective inflation rate, or directly upon discounting of financial flow.

Table 14.1 Dependence of the weighted average cost of capital WACC on debt ratio w_d at different inflation rate $\alpha = 3\%; 5\%; 7\%; 9\%$ for 2-year company

α\wd	0.1	0.2	0.3	0.4	0.5	0.6	0.7	0.8	0.9	1
0.03	0.2318	0.2276	0.2233	0.2191	0.2149	0.2106	0.2064	0.2021	0.1979	0.1937
0.05	0.2557	0.2503	0.2455	0.2406	0.2358	0.2309	0.2261	0.2212	0.2164	0.2115
0.07	0.2786	0.2733	0.2679	0.2626	0.2573	0.2514	0.2459	0.2404	0.2350	0.2295
0.09	0.3020	0.2960	0.2900	0.2839	0.2779	0.2720	0.2661	0.2602	0.2537	0.2476

Table 14.2 Dependence of the weighted average cost of capital WACC on debt ratio w_d at different inflation rate $\alpha = 3\%; 5\%; 7\%; 9\%$ for 5-year company

α\wd	0.1	0.2	0.3	0.4	0.5	0.6	0.7	0.8	0.9	1
0.03	0.2311	0.2262	0.2213	0.2163	0.2113	0.2064	0.2013	0.1963	0.1912	0.1863
0.05	0.2546	0.2491	0.2434	0.2379	0.2323	0.2267	0.2210	0.2154	0.2097	0.2040
0.07	0.2781	0.2718	0.2657	0.2595	0.2534	0.2472	0.2408	0.2346	0.2283	0.2219
0.09	0.3015	0.2947	0.2879	0.2812	0.2744	0.2676	0.2608	0.2539	0.2471	0.2400

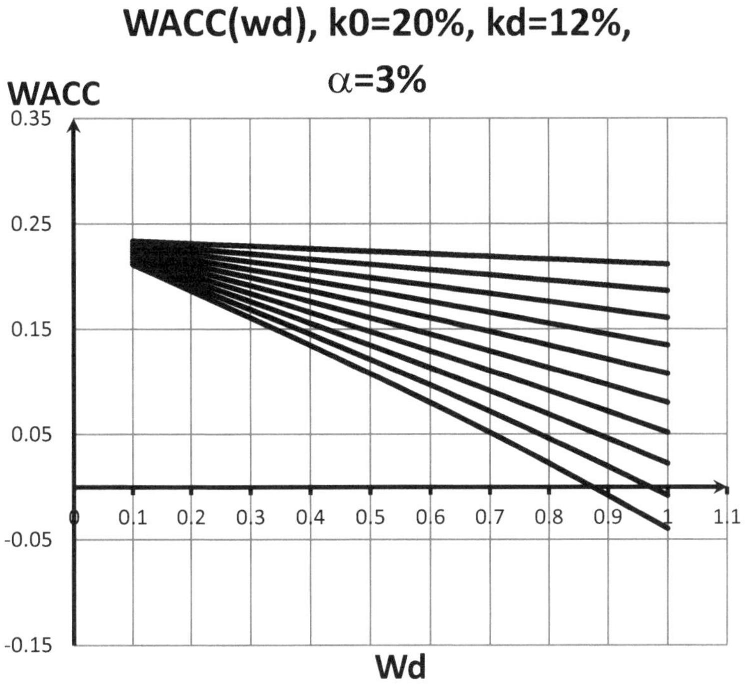

Fig. 14.5 Dependence of the weighted average cost of capital WACC on debt ratio w_d at different tax on profit rate at inflation rate $\alpha = 3\%$ for 5-year company. Tax on profit rate increases from $T = 0.1$ (upper line) up to $T = 1$ (lowest line) with step 0.1

14.7 Conclusions

In this chapter the influence of inflation on capital cost and capitalization of the company within modern theory of capital cost and capital structure—Brusov–Filatova–Orekhova theory (BFO theory) (Brusov and Filatova 2011; Brusov et al. 2011a, b, c, 2012a, b, 2013a, b, 2014a, b; Filatova et al. 2008) and in its perpetuity limit—Modigliani–Miller theory (Modigliani and Miller 1958, 1963, 1966), which is now outdate, but still widely used at the West, is investigated. All basic results of Modigliani–Miller theory were modified. It is shown, that inflation not only increases the equity cost and the weighted average cost of capital, but as well it changes their dependence on leverage. In particular, it increases growing rate of equity cost with leverage. Capitalization of the company is decreased under accounting of inflation.

Within modern theory of capital cost and capital structure—Brusov–Filatova–Orekhova theory (BFO theory) the modified equation for the weighted average cost of capital, WACC, applicable to companies with arbitrary lifetime under accounting of inflation has been derived. Modified BFO equation allow to investigate the

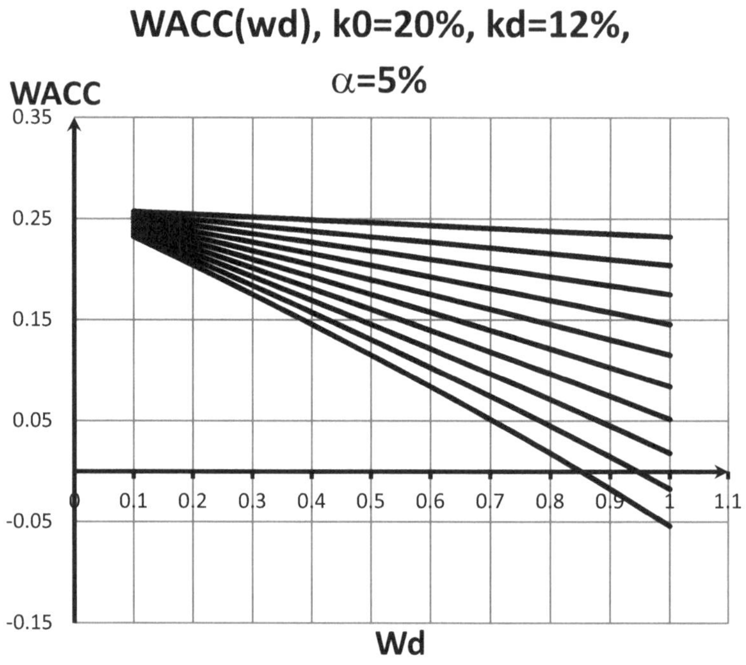

Fig. 14.6 Dependence of the weighted average cost of capital WACC on debt ratio w_d at different tax on profit rate at inflation rate $\alpha = 5\%$ for 5-year company. Tax on profit rate increases from $T = 0.1$ (upper line) up to $T = 1$ (lowest line) with step 0.1

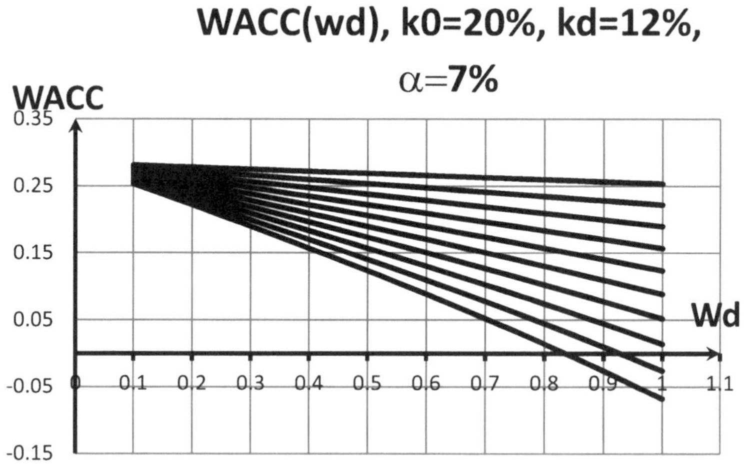

Fig. 14.7 Dependence of the weighted average cost of capital WACC on debt ratio w_d at different tax on profit rate at inflation rate $\alpha = 7\%$ for 5-year company. Tax on profit rate increases from $T = 0.1$ (upper line) up to $T = 1$ (lowest line) with step 0.1

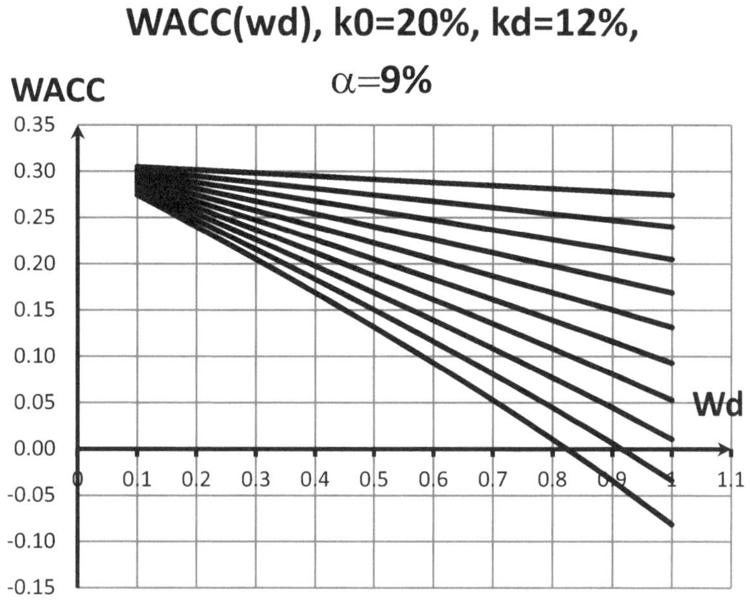

Fig. 14.8 Dependence of the weighted average cost of capital WACC on debt ratio w_d at different tax on profit rate at inflation rate $\alpha = 9\%$ for 5-year company. Tax on profit rate increases from $T = 0.1$ (upper line) up to $T = 1$ (lowest line) with step 0.1

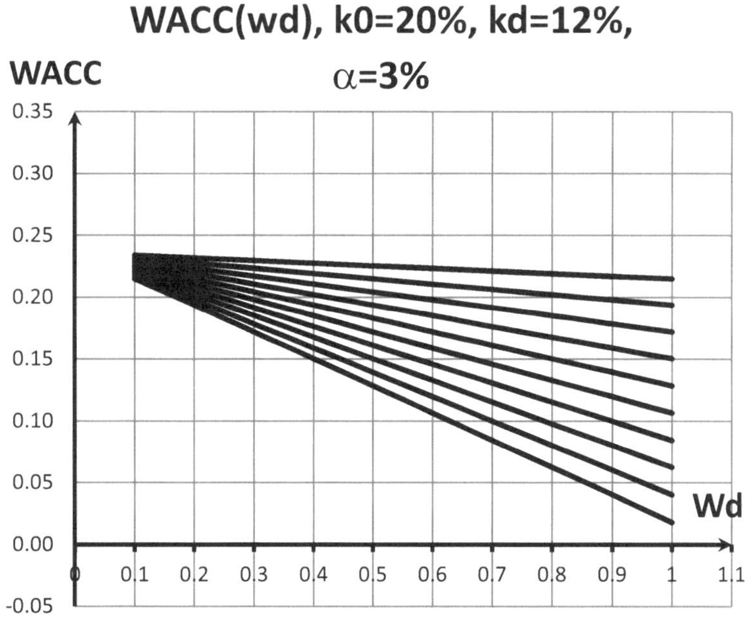

Fig. 14.9 Dependence of the weighted average cost of capital WACC on debt ratio w_d at different tax on profit rate at inflation rate $\alpha = 3\%$ for 2-year company. Tax on profit rate increases from $T = 0.1$ (upper line) up to $T = 1$ (lowest line) with step 0.1

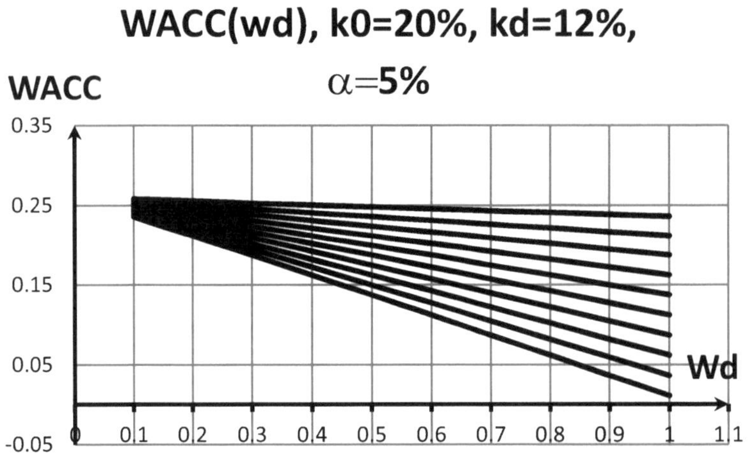

Fig. 14.10 Dependence of the weighted average cost of capital WACC on debt ratio w_d at different tax on profit rate at inflation rate $\alpha = 5\%$ for 2-year company. Tax on profit rate increases from $T = 0.1$ (upper line) up to $T = 1$ (lowest line) with step 0.1

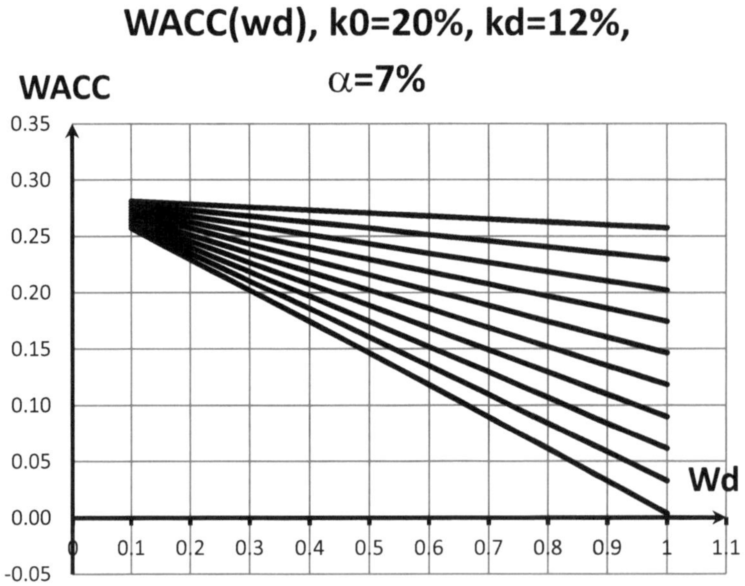

Fig. 14.11 Dependence of the weighted average cost of capital WACC on debt ratio w_d at different tax on profit rate at inflation rate $\alpha = 7\%$ for 2-year company. Tax on profit rate increases from $T = 0.1$ (upper line) up to $T = 1$ (lowest line) with step 0.1

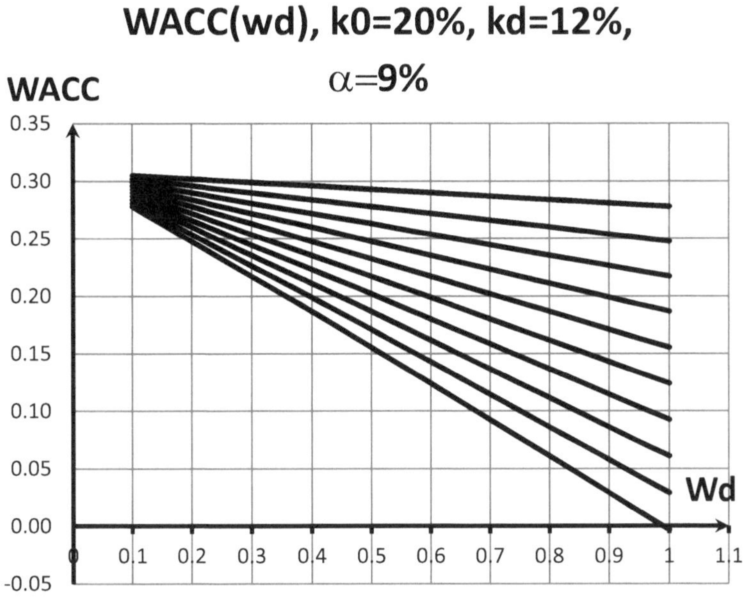

Fig. 14.12 Dependence of the weighted average cost of capital WACC on debt ratio w_d at different tax on profit rate at inflation rate $\alpha = 9\%$ for 2-year company. Tax on profit rate increases from $T = 0.1$ (upper line) up to $T = 1$ (lowest line) with step 0.1

dependence of the weighted average cost of capital, WACC, and equity cost, k_e, on leverage level L, on tax on profit rate t, on lifetime of the company n, on equity cost of financially independent company, k_0, and debt cost, k_d, as well as on inflation rate α.

Using modified BFO equation the analysis of the dependence of the weighted average cost of capital, WACC, on debt ratio, w_d, at different tax on profit rate t, as well as inflation rate α has been done.

It has been shown, that WACC decreases with debt ratio, w_d, faster at bigger tax on profit rate t. The space between lines, corresponding to different values of tax on profit rate at the same step (10%), increases with inflation rate α. The variation region (with change of tax on profit rate t) of the weighted average cost of capital, WACC, increases with inflation rate α, as well as with lifetime of the company n.

14.8 Inflation Rate for a Few Periods

Suppose that the inflation rate for the consistent time periods $t_1, t_2, \ldots, t_n$ are equal to $\alpha_1, \alpha_2, \ldots, \alpha_n$ consequently. Let us find the inflation rate α for total time period $t = t_1 + t_2 + \ldots + t_n$.

Common sense dictates that inflation rate is an additive value, so that α, at least approximately, is equal to the sum of the inflation rates $\alpha_1, \alpha_2, \ldots, \alpha_n$

$$\alpha \approx \alpha_1 + \alpha_2 + \ldots + \alpha_n. \tag{14.37}$$

Below we will get an exact expression for inflation rate for the total period of time t, and will see how it is different from an intuitive result (14.37).

At the end of the first commitment period the gained sum will be equal to the amount $S_1 = S_0(1 + i)$, and with the account of inflation $S_{1\alpha} = S_0(1 + i)^{t_1}/(1 + \alpha_1)$. At the end of the second commitment period the gained sum will be equal to the amount $S_2 = S_0(1 + i)^{t_1+t_2}$, and with the account of inflation $S_{2\alpha} = S_0(1 + i)^{t_1+t_2}/(1 + \alpha_1)(1 + \alpha_2)$. At the end of the n-th commitment period the gained sum will be equal to the amount $S_n = S_0(1 + i)^{t_1+t_2+\ldots+t_n}$, and with the account of inflation

$$S_{n\alpha} = S_0(1 + i)^{t_1+t_2+\ldots+t_n}/(1 + \alpha_1)(1 + \alpha_2)\cdot\ldots\cdot(1 + \alpha_n). \tag{14.38}$$

On the other hand, at inflation rate α for the total period at $t = t_1 + t_2 + \ldots + t_n$ at the end of this period t gained sum will be equal to

$$S_{n\alpha} = S_0(1 + i)^t/(1 + \alpha). \tag{14.39}$$

Equating the right-hand part of (14.38) and (14.39), we get

$$(1 + \alpha_1)(1 + \alpha_2)\cdot\ldots\cdot(1 + \alpha_n) = 1 + \alpha \tag{14.40}$$

From where

$$\alpha = (1 + \alpha_1)(1 + \alpha_2)\cdot\ldots\cdot(1 + \alpha_n) - 1 \tag{14.41}$$

It is easy to get a strict proof of this formula by the method of mathematical induction. Note that inflation rate for the n-periods does not depend on both of the length of constituting periods and on the period t.

For equal inflation rates $\alpha_1 = \alpha_2 = \ldots = \alpha_n$ (it is interesting to note, that herewith the time intervals $t_1, t_2, \ldots, t_n$ can be arbitrary and do not equal each other) one has

$$\alpha = (1 + \alpha_1)^n - 1 \tag{14.42}$$

References

Brusov, P. N., & Filatova, T. V. (2011). From Modigliani–Miller to general theory of capital cost and capital structure of the company. *Finance and Credit, 435*, 2–8.

Brusov, P., Filatova, T., Orehova, N., & Brusova, A. (2011a). Weighted average cost of capital in the theory of Modigliani–Miller, modified for a finite life-time company. *Applied Financial Economics, 21*(11), 815–824.

Brusov, P., Filatova, T., Orehova, N., et al. (2011b). From Modigliani–Miller to general theory of capital cost and capital structure of the company. *Research Journal of Economics, Business and ICT, 2*, 16–21.
Brusov, P., Filatova, T., Orehova, N., et al. (2011c). Influence of debt financing on the effectiveness of the investment project within the Modigliani–Miller theory. *Research Journal of Economics, Business and ICT (UK), 2*, 11–15.
Brusov, P., Filatova, T., Eskindarov, M., & Orehova, N. (2012a). Influence of debt financing on the effectiveness of the finite duration investment project. *Applied Financial Economics, 22*(13), 1043–1052.
Brusov, P., Filatova, T., Eskindarov, M., & Orehova, N. (2012b). Hidden global causes of the global financial crisis. *Journal of Reviews on Global Economics, 1*, 106–111.
Brusov, P., Filatova, P., & Orekhova, N. (2013a). Absence of an optimal capital structure in the famous tradeoff theory! *Journal of Reviews on Global Economics, 2*, 94–116.
Brusov, P., Filatova, T., & Orehova, N. (2013b). A qualitatively new effect in corporative finance: Abnormal dependence of cost of equity of company on leverage. *Journal of Reviews on Global Economics, 2*, 183–193.
Brusov, P., Filatova, P., & Orekhova, N. (2014a). Mechanism of formation of the company optimal capital structure, different from suggested by trade off theory. *Cogent Economics and Finance, 2*, 1–13. https://doi.org/10.1080/233220314.2014.946150
Brusov, P., Filatova, T., & Orehova, N. (2014b). Inflation in Brusov–Filatova–Orekhova theory and in its perpetuity limit—Modigliani–Miller theory. *Journal of Reviews on Global Economics, 3*, 175–185.
Brusova, A. (2011). A comparison of the three methods of estimation of weighted average cost of capital and equity cost of company. *Financial Analysis: Problems and Solutions, 34*(76), 36–42.
Filatova, T., Orehova, N., & Brusova, A. (2008). Weighted average cost of capital in the theory of Modigliani–Miller, modified for a finite life–time company. *Bulletin of the FU, 48*, 68–77.
Modigliani, F., & Miller, M. (1958). The cost of capital, corporate finance, and the theory of investment. *American Economic Review, 48*, 261–297.
Modigliani, F., & Miller, M. (1963). Corporate income taxes and the cost of capital: A correction. *American Economic Review, 53*, 147–175.
Modigliani, F., & Miller, M. (1966). Some estimates of the cost of capital to the electric utility industry 1954–1957. *American Economic Review, 56*, 333–391.

Part III
Modern Investment Models

In this part (this and the next chapters) we describe modern investment models created by the authors and well tested in the real economy. These models are based on the first principles and are used in assessing the effectiveness of investment projects and to study various investment problems, such as the impact of debt financing, leverage, taxation, project duration, financing method and some other parameters on investment efficiency and other problems. But we use them in Chap. 18 to modify the methodology of project ratings. In this chapter we look at investment models with debt repayment at the end of the project and their application, and in Chap. 16 we look at investment models with equal debt repayment and their application.

Chapter 15
Investment Models with Debt Repayment at the End of the Project and Their Application

In this and next chapter, we describe the modern investment models, created by the authors and well tested in the real economy. These models used by us for investigation of different problems of investments, such as influence of debt financing, leverage level, taxing, project duration, method of financing, and some other parameters on efficiency of investments and other problems. But we use them in Chap. 18 for modification of methodology of project ratings. In this chapter we consider the investment models with debt repayment at the end of the project and their application while in Chap.18 we consider the investment models with uniform debt repayment and their application.

15.1 Investment Models

The effectiveness of the investment project is considered from two perspectives: the owners of equity and debt and the equity holders only. For each of these cases, NPV is calculated in two ways: with the division of credit and investment flows (and thus discounting of the payments using two different rates) and without such a division (in this case, both flows are discounted using the same rate, as which can be, obviously, chosen WACC). For each of the four situations, two cases are considered: (1) a constant value of equity S and (2) a constant value of the total invested capital $I = S + D$ (D is value of debt funds).

As it was stated above, the effectiveness of the investment project is considered from two perspectives: the owners of equity and debt and the equity holders only. In the first case, the interest and duty paid by owners of equity (negative flows) returned to the project because they are exactly equal to the flow (positive), obtained by owners of debt capital. The only effect of leverage in this case is—the effect of the tax shield, generated from the tax relief: interest on the loan is entirely included into the cost and thus reduces the tax base. After-tax flow of capital for each period in this case is equal to

P. Brusov, T. Filatova, *Efficiency of Investment Projects*, Contributions to Finance and Accounting, https://doi.org/10.1007/978-3-032-01028-5_15

$$\mathrm{NOI}(1-t)+k_{\mathrm{d}}Dt \tag{15.1}$$

and the value of investments at the initial time moment $T=0$ is equal to $-I=-S-D$.

Here NOI stands for net operating income (before taxes).

In the second case, investments at the initial time moment $T=0$ are equal to $-S$ and the flow of capital for the period (in addition to the tax shields $k_{\mathrm{d}}Dt$ it includes a payment of interest on a loan $-k_{\mathrm{d}}D$):

$$(\mathrm{NOI}-k_{\mathrm{d}}D)(1-t). \tag{15.2}$$

Here, for simplicity, we suppose that interest on the loan will be paid in equal shares $k_{\mathrm{d}}\,D$ during all periods. Note that principal repayment is made at the end of the last period.

Some variety of repayment of long-term loans will be considered below (see in Chap. 14).

We will consider two different ways of discounting:

1. Operating and financial flows are not separated and both are discounted, using the general rate (as which, obviously, the weighted average cost of capital (WACC) can be selected). In this case for perpetuity projects, the Modigliani–Miller formula (Modigliani and Miller 1958, 1963, 1966) for WACC will be used and for projects of finite (arbitrary) duration Brusov–Filatova–Orekhova formula (Brusov et al. 2015, 2018a, b, c, d, e, f, g, 2019, 2020; Filatova et al. 2008; Filatova et al. 2018; Brusov 2018).
2. Operating and financial flows are separated and are discounted at different rates: the operating flow at the rate which is equal to the equity cost k_{e}, depending on leverage, and credit flow—at the rate which is equal to the debt cost k_{d}, which until fairly large values of leverage remain constant and start to grow only at high values of leverage L, when there is a danger of bankruptcy.

Note that loan capital is the least risky, because interest on credit is paid after taxes in the first place. Therefore, the cost of credit will always be less than the equity cost, whether of ordinary or of preference shares $k_{\mathrm{e}}>k_{\mathrm{d}}$; $k_{\mathrm{p}}>k_{\mathrm{d}}$. Here k_{e}; k_{p} is the equity cost of ordinary or of preference shares consequently.

15.2 The Effectiveness of the Investment Project from the Perspective of the Equity Holders Only

15.2.1 With the Division of Credit and Investment Flows

15.2.1.1 Projects of Finite (Arbitrary) Duration

In this case, the expression for NPV has a view

$$
\begin{aligned}
\mathrm{NPV} &= -S + \sum_{i=1}^{n} \frac{\mathrm{NOI}(1-t)}{(1+k_e)^i} + \sum_{i=1}^{n} \frac{-k_d D(1-t)}{(1+k_d)^i} - \frac{D}{(1+k_d)^n} \\
&= -S + \frac{\mathrm{NOI}(1-t)}{k_e}\left(1 - \frac{1}{(1+k_e)^n}\right) - D(1-t)\left(1 - \frac{1}{(1+k_d)^n}\right) - \frac{D}{(1+k_d)^n}.
\end{aligned} \tag{15.3}
$$

The last term in the first line-discounted (present) value of credit, extinguished a one-off payment at the end of the last period n.

Below we will look at two cases:

1. A constant value of the invested capital $I = S + D$ (D—debt value)
2. A constant value of equity capital S

We will start with the first case.

At a Constant Value of the Invested Capital (I = const)

In the case of a constant value of the invested capital (I = const), taking into account $D = IL/(1 + L)$, $S = I/(1 + L)$, one gets

$$
\begin{aligned}
\mathrm{NPV} = &-\frac{I}{1+L}\left[1 + L\left((1-t)\left(1 - \frac{1}{(1+k_d)^n}\right) + \frac{1}{(1+k_d)^n}\right)\right] \\
&+ \frac{\mathrm{NOI}(1-t)}{k_e}\left(1 - \frac{1}{(1+k_e)^n}\right).
\end{aligned} \tag{15.4}
$$

For 1-Year Project

Putting at Eq. (15.4) $n = 1$, one gets for NPV

$$
\mathrm{NPV} = -\frac{I}{1+L}\left[1 + L\left(\frac{1 + k_d(1-t)}{(1+k_d)}\right)\right] + \frac{\mathrm{NOI}(1-t)}{1+k_e}. \tag{15.5}
$$

At a Constant Value of Equity Capital (S = const)

Accounting that in the case S = const NOI is proportional to the invested capital, $\mathrm{NOI} = \beta I = \beta S(1 + L)$, we get

$$
\begin{aligned}
\mathrm{NPV} = &-S\left[1 + L\left((1-t)\left(1 - \frac{1}{(1+k_d)^n}\right) + \frac{1}{(1+k_d)^n}\right)\right] \\
&+ \frac{\beta S(1+L)(1-t)}{k_e}\left(1 - \frac{1}{(1+k_e)^n}\right).
\end{aligned} \tag{15.6}
$$

For 1-Year Project Putting at Eq. (15.6) $n = 1$, one gets for NPV

$$\mathrm{NPV} = -S\left(1 + L\left(\frac{1 + k_\mathrm{d}(1-t)}{1 + k_\mathrm{d}}\right)\right) + \frac{\beta S(1+L)(1-t)}{1+k_\mathrm{e}}. \tag{15.7}$$

15.3 Without Flows Separation

In this case operating and financial flows are not separated and are discounted, using the general rate (as which, obviously, WACC can be selected).

The credit reimbursable at the end of the project (at the end of the period (n)) can be discounted either at the same rate WACC or at the debt cost rate k_d. Now we choose a uniform rate and the first option.

$$\begin{aligned}\mathrm{NPV} &= -S + \sum_{i=1}^{n} \frac{\mathrm{NOI}(1-t) - k_\mathrm{d}D(1-t)}{(1+\mathrm{WACC})^i} - \frac{D}{(1+\mathrm{WACC})^n} \\ &= -S + \frac{\mathrm{NOI}(1-t) - k_\mathrm{d}D(1-t)}{\mathrm{WACC}}\left(1 - \frac{1}{(1+\mathrm{WACC})^n}\right) - \frac{D}{(1+\mathrm{WACC})^n}.\end{aligned} \tag{15.8}$$

At a Constant Value of the Invested Capital (I = const)

In case of a constant value of the invested capital (I = const), taking into account $D = IL/(1 + L)$, $S = I/(1 + L)$, one gets

$$\begin{aligned}\mathrm{NPV} = &-\frac{I}{1+L}\left[1 + L\frac{k_\mathrm{d}(1-t)}{\mathrm{WACC}}\left(1 - \frac{1}{(1+\mathrm{WACC})^n}\right) + \frac{L}{(1+\mathrm{WACC})^n}\right] \\ &+ \frac{\mathrm{NOI}(1-t)}{\mathrm{WACC}}\left(1 - \frac{1}{(1+\mathrm{WACC})^n}\right).\end{aligned} \tag{15.9}$$

For 1-Year Project

Putting into Eq. (15.9) $n = 1$, one gets for NPV

$$\mathrm{NPV} = -\frac{I}{1+L}\left[1 + L\frac{1 + k_\mathrm{d}(1-t)}{1 + \mathrm{WACC}}\right] + \frac{\mathrm{NOI}(1-t)}{1+\mathrm{WACC}}. \tag{15.10}$$

At a Constant Value of Equity Capital (S = const)

Accounting that in the case S = const NOI is proportional to the invested capital, I, $\mathrm{NOI} = \beta I = \beta S(1 + L)$, and substituting $D = LS$, we get

$$\mathrm{NPV} = -S + \frac{\mathrm{NOI}(1-t) - k_{\mathrm{d}}D(1-t)}{\mathrm{WACC}}\left(1 - \frac{1}{(1+\mathrm{WACC})^n}\right) - \frac{D}{(1+\mathrm{WACC})^n}, \tag{15.11}$$

$$\mathrm{NPV} = -S\left[1 + \frac{Lk_{\mathrm{d}}(1-t)}{\mathrm{WACC}}\left(1 - \frac{1}{(1+\mathrm{WACC})^n}\right) + \frac{L}{(1+\mathrm{WACC})^n}\right] + \frac{\beta S(1+L)(1-t)}{\mathrm{WACC}}\left(1 - \frac{1}{(1+\mathrm{WACC})^n}\right). \tag{15.12}$$

For 1-Year Project
Putting into Eq. (15.12) $n = 1$, one gets for NPV

$$\mathrm{NPV} = -S\left[1 + \frac{Lk_{\mathrm{d}}(1-t)}{\mathrm{WACC}}\left(1 - \frac{1}{(1+\mathrm{WACC})^n}\right) + \frac{L}{(1+\mathrm{WACC})^n}\right] + \frac{\beta S(1+L)(1-t)}{\mathrm{WACC}}\left(1 - \frac{1}{(1+\mathrm{WACC})^n}\right).$$

$$\mathrm{NPV} = -S + \frac{\mathrm{NOI}(1-t) - k_{\mathrm{d}}D(1-t) - D}{1+\mathrm{WACC}}. \tag{15.13}$$

Substituting $D = LS$, $\mathrm{NOI} = \beta I = \beta S(1 + L)$, we get

$$\mathrm{NPV} = -S\left[1 + \frac{L(k_{\mathrm{d}}(1-t) - 1)}{1+\mathrm{WACC}}\right] + \frac{\beta S(1+L)(1-t)}{1+\mathrm{WACC}}. \tag{15.14}$$

15.4 Modigliani–Miller Limit (Perpetuity Projects)

15.4.1 With Flows Separation

In perpetuity limit (Modigliani–Miller limit) (turning to the limit $n \to \infty$ in the relevant equations), we have

$$\mathrm{NPV} = -S + \frac{\mathrm{NOI}(1-t)}{k_{\mathrm{e}}} - D(1-t). \tag{15.15}$$

At a Constant Value of the Invested Capital (I = const)
At a constant value of the invested capital (I = const), accounting $D = IL/(1 + L)$, $S = I/(1 + L)$, we get

$$\mathrm{NPV} = -\frac{I}{1+L}(1+L(1-t)) + \frac{\mathrm{NOI}(1-t)}{k_e}. \tag{15.16}$$

$$\mathrm{NPV} = -\frac{I}{1+L}(1+L(1-t)) + \frac{\mathrm{NOI}(1-t)}{k_0 + (k_0 - k_d)L(1-t)}. \tag{15.17}$$

In order to obtain Eqs. (15.17) from (15.16), we used the Modigliani–Miller formula (Modigliani and Miller 1963) for equity cost k_e for perpetuity projects:

$$k_e = k_0 + (k_0 - k_d)L(1-t). \tag{15.18}$$

At a Constant Value of Equity Capital (S = const)
Accounting $D = LS$, we get in perpetuity limit ($n \to \infty$) (Modigliani–Miller limit)

$$\mathrm{NPV} = -S(1+L(1-t)) + \frac{\beta S(1+L)(1-t)}{k_0 + (k_0 - k_d)Lt}. \tag{15.19}$$

15.4.2 Without Flows Separation

In perpetuity limit ($n \to \infty$) (Modigliani–Miller limit) (turning to the limit $n \to \infty$ in the relevant equations), we have

$$\mathrm{NPV} = -S + \frac{\mathrm{NOI}(1-t) - k_d D(1-t)}{\mathrm{WACC}}. \tag{15.20}$$

At a Constant Value of the Invested Capital (I = const)
At a constant value of the invested capital (I = const), accounting $D = IL/(1 + L)$, $S = I/(1 + L)$, we get

$$\begin{aligned}\mathrm{NPV} &= -I\cdot\frac{1}{1+L} + \frac{\mathrm{NOI}(1-t) - I\frac{L}{1+L}k_d(1-t)}{\mathrm{WACC}} \\ &= -I\cdot\frac{1}{1+L}\left[1 + \frac{Lk_d(1-t)}{k_0(1-Lt/(1+L))}\right] + \frac{\mathrm{NOI}(1-t)}{k_0(1-Lt/(1+L))}.\end{aligned} \tag{15.21}$$

At a Constant Value of Equity Capital (S = const)

$$\mathrm{NPV} = -S + \frac{\mathrm{NOI}(1-t) - k_d D(1-t)}{\mathrm{WACC}} \tag{15.22}$$

Substituting $D = LS$, we get

$$\begin{aligned}\mathrm{NPV} &= -S\left[1+\frac{Lk_{\mathrm{d}}(1-t)}{\mathrm{WACC}}\right]+\frac{\mathrm{NOI}(1-t)}{\mathrm{WACC}} \\ &= -S\left[1+\frac{Lk_{\mathrm{d}}(1-t)}{k_0(1-Lt/(1+L))}\right]+\frac{\beta S(1+L)(1-t)}{k_0(1-Lt/(1+L))}.\end{aligned} \tag{15.23}$$

15.5 The Effectiveness of the Investment Project from the Perspective of the Owners of Equity and Debt

15.5.1 With Flows Separation

15.5.1.1 Projects of Arbitrary (Finite) Duration

In this case, operating and financial flows are separated and are discounted, using different rates: the operating flow at the rate equal to the equity cost k_e, depending on leverage, and credit flow—at the rate equal to the debt cost k_d, which until fairly large values of leverage remain constant and start to grow only at high values of leverage L, when there is a danger of bankruptcy.

$$\begin{aligned}\mathrm{NPV} &= -I+\sum_{i=1}^{n}\frac{\mathrm{NOI}(1-t)}{(1+k_e)^i}+\sum_{i=1}^{n}\frac{k_d Dt}{(1+k_d)^i} \\ &= -I+\frac{\mathrm{NOI}(1-t)}{k_e}\left(1-\frac{1}{(1+k_e)^n}\right)+Dt\left(1-\frac{1}{(1+k_d)^n}\right).\end{aligned} \tag{15.24}$$

Below we will consider two cases:

1. At a constant value of the invested capital $I = S + D$ (D is the debt value)
2. At a constant value of equity capital S

We will start with the first case.

At a Constant Value of the Invested Capital (I = const)

At a constant value of the invested capital (I = const), accounting $D = IL/(1 + L)$, $S = I/(1 + L)$, we get

$$\begin{aligned}\mathrm{NPV} &= -I+\frac{\mathrm{NOI}(1-t)}{k_e}\left(1-\frac{1}{(1+k_e)^n}\right)+\frac{ILt}{1+L}\left(1-\frac{1}{(1+k_d)^n}\right) \\ &= -I\left[1-\frac{Lt}{1+L}\left(1-\frac{1}{(1+k_d)^n}\right)\right]+\frac{\mathrm{NOI}(1-t)}{k_e}\left(1-\frac{1}{(1+k_e)^n}\right).\end{aligned} \tag{15.25}$$

At a Constant Value of Equity Capital (S = const)
Accounting $D = LS$, $I = S(1 + L)$, we get

$$\mathrm{NPV} = -S - LS + \frac{\mathrm{NOI}(1-t)}{k_\mathrm{e}}\left(1 - \frac{1}{(1+k_\mathrm{e})^n}\right) + Dt\left(1 - \frac{1}{(1+k_\mathrm{d})^n}\right). \quad (15.26)$$

Accounting that in the case S = const NOI is not a constant, but is proportional to the invested capital, $\mathrm{NOI} = \beta I = \beta S(1 + L)$, we get

$$\mathrm{NPV} = -S\left[1 + L - tL\left(1 - \frac{1}{(1+k_\mathrm{d})^n}\right)\right] + \frac{\beta S(1+L)(1-t)}{k_\mathrm{e}} \times \left(1 - \frac{1}{(1+k_\mathrm{e})^n}\right). \quad (15.27)$$

For 1-Year Project

$$\mathrm{NPV} = -S\left[1 + L - tL\frac{k_\mathrm{d}}{1+k_\mathrm{d}}\right] + \frac{\beta S(1+L)(1-t)}{1+k_\mathrm{e}}. \quad (15.28)$$

15.5.2 Without Flows Separation

In this case operating and financial flows are not separated and both are discounted, using the general rate (as which, obviously, WACC can be selected):

$$\begin{aligned}\mathrm{NPV} &= -I + \sum_{i=1}^{n} \frac{\mathrm{NOI}(1-t) + k_\mathrm{d}Dt}{(1+\mathrm{WACC})^i} \\ &= -I + \frac{\mathrm{NOI}(1-t) + k_\mathrm{d}Dt}{\mathrm{WACC}}\left(1 - \frac{1}{(1+\mathrm{WACC})^n}\right).\end{aligned} \quad (15.29)$$

At a Constant Value of the Invested Capital (I = const)
At a constant value of the invested capital (I = const), we have

$$\mathrm{NPV} = -I + \frac{\mathrm{NOI}(1-t) + k_\mathrm{d}Dt}{\mathrm{WACC}}\left(1 - \frac{1}{(1+\mathrm{WACC})^n}\right).$$

Accounting $D = IL/(1 + L)$, $S = I/(1 + L)$, we get

$$\mathrm{NPV} = -I\left[1 - \frac{k_\mathrm{d} t \frac{L}{1+L}}{k_0\left(1-\gamma\frac{L}{1+L}t\right)}\left(1 - \frac{1}{\left(1+k_0\left(1-\gamma\frac{L}{1+L}t\right)\right)^n}\right)\right] + \frac{\mathrm{NOI}(1-t)}{k_0\left(1-\gamma\frac{L}{1+L}t\right)}\left(1 - \frac{1}{\left(1+k_0\left(1-\gamma\frac{L}{1+L}t\right)\right)^n}\right). \tag{15.30}$$

For 1-Year Project
Putting at the Eq. (15.30) $n = 1$, one gets for NPV

$$\mathrm{NPV} = -I\left[1 - \frac{k_\mathrm{d} t \frac{L}{1+L}}{1+\mathrm{WACC}}\right] + \frac{\mathrm{NOI}(1-t)}{1+\mathrm{WACC}}. \tag{15.31}$$

At a Constant Value of Equity Capital ($S = \mathrm{const}$)

$$\begin{aligned}\mathrm{NPV} &= -I + \frac{\mathrm{NOI}(1-t) + k_\mathrm{d} D t}{\mathrm{WACC}}\left(1 - \frac{1}{(1+\mathrm{WACC})^n}\right)\\ &= -S\left[1 + L - \frac{k_\mathrm{d} L t}{\mathrm{WACC}}\left(1 - \frac{1}{(1+\mathrm{WACC})^n}\right)\right]\\ &\quad + \frac{\beta S(1+L)(1-t)}{\mathrm{WACC}}\left(1 - \frac{1}{(1+\mathrm{WACC})^n}\right).\end{aligned} \tag{15.32}$$

For 1-Year Project

$$\begin{aligned}\mathrm{NPV} &= -I + \frac{\mathrm{NOI}(1-t) + k_\mathrm{d} D t}{1+\mathrm{WACC}}\\ &= -S\left[1 + L - \frac{k_\mathrm{d} L t}{1+\mathrm{WACC}}\right] + \frac{\mathrm{NOI}(1-t)}{1+\mathrm{WACC}}.\end{aligned} \tag{15.33}$$

15.6 Modigliani–Miller Limit

15.6.1 With Flows Separation

In perpetuity limit ($n \to \infty$) (Modigliani–Miller limit), we have

$$\mathrm{NPV} = -I + \frac{\mathrm{NOI}(1-t)}{k_\mathrm{e}} + Dt. \tag{15.34}$$

At a constant value of the invested capital ($I = \text{const}$), accounting $D = IL/(1 + L)$, we have

$$\text{NPV} = -I\left(1 - t\frac{L}{1+L}\right) + \frac{\text{NOI}(1-t)}{k_e}. \tag{15.35}$$

For equity cost k_e and WACC in Modigliani–Miller theory, we have consequently

$$k_e = k_0 + (k_0 - k_d)L(1-t), \tag{15.36}$$

$$\text{WACC} = k_0(1 - w_d t) = k_0(1 - Lt/(1+L)). \tag{15.37}$$

Putting Eqs. (15.36) into (15.37), we get

$$\text{NPV} = -I\left(1 - t\frac{L}{1+L}\right) + \frac{\text{NOI}(1-t)}{k_0 + (k_0 - k_d)L(1-t)}. \tag{15.38}$$

At a Constant Value of Equity Capital (S = const)
Accounting $D = LS$, $I = S(1 + L)$, in perpetuity limit ($n \to \infty$) (Modigliani–Miller limit), we have

$$\text{NPV} = -S(1 + L(1-t)) + \frac{\text{NOI}(1-t)}{k_0 + (k_0 - k_d)Lt}. \tag{15.39}$$

Note that in the case S = const NOI is not a constant, but is proportional to the invested capital, $\text{NOI} = \beta I = \beta S(1 + L)$.

In this case, Eq. (15.38) is replaced by

$$\text{NPV} = -S(1 + L(1-t)) + \frac{\beta S(1+L)(1-t)}{k_0 + (k_0 - k_d)Lt}, \tag{15.40}$$

15.6.2 Without Flows Separation

In perpetuity limit ($n \to \infty$) (Modigliani–Miller limit), we have

$$\text{NPV} = -I + \frac{\text{NOI}(1-t) + k_d D t}{\text{WACC}}. \tag{15.41}$$

At a constant value of the invested capital (I = const), we have

$$\mathrm{NPV} = -I + \frac{\mathrm{NOI}(1-t) + k_{\mathrm{d}}Dt}{\mathrm{WACC}}$$
$$= -I\left(1 - \frac{k_{\mathrm{d}}t\dfrac{L}{1+L}}{k_0\left(1 - \dfrac{L}{1+L}t\right)}\right) + \frac{\mathrm{NOI}(1-t)}{k_0\left(1 - \dfrac{L}{1+L}t\right)}. \tag{15.42}$$

At a Constant Value of Equity Capital (S = const)
In perpetuity limit ($n \to \infty$) (Modigliani–Miller limit), we have

$$\mathrm{NPV} = -S\left[1 + L - \frac{k_{\mathrm{d}}Lt}{\mathrm{WACC}}\right] + \frac{\mathrm{NOI}(1-t)}{\mathrm{WACC}}. \tag{15.43}$$

$$\mathrm{NPV} = -S\left[1 + L - \frac{k_{\mathrm{d}}Lt}{k_0\left(1 - \frac{L}{1+L}t\right)}\right] + \frac{\beta S(1+L)(1-t)}{k_0\left(1 - \frac{L}{1+L}t\right)}. \tag{15.44}$$

References

Brusov, P. (2018). Editorial: Introduction on special issue on the Banking System and financial markets of Russia and other countries: Problems and prospects. *Journal of Reviews on Global Economics, 7*, i–vi.

Brusov, P., Filatova, T., Orehova, N., & Eskindarov, M. (2015). *Modern corporate finance, investments and taxation*. Springer International Publishing, 373 p. Retrieved from https://www.springer.com/gp/book/9783319147314

Brusov, P., Filatova, T., Orehova, N., & Eskindarov, M. (2018a). *Modern corporate finance, investments, taxation and ratings*. Springer Nature Publishing, 571 p.

Brusov, P. N., Filatova, T. V., & Orekhova, N. P. (2018b). *Modern corporate finance and investments, monograph*. Knorus Publishing House, 517 p.

Brusov, P. N., Filatova, T. V., Orekhova, N. P., & Kulik, V. L. (2018c). Rating: New approach. *Journal of Reviews on Global Economics, 7*, 37–62. https://doi.org/10.6000/1929-7092.2018.07.05

Brusov, P. N., Filatova, T. V., Orekhova, N. P., & Kulik, V. L. (2018d). A "golden age" of the companies: Conditions of its existence. *Journal of Reviews on Global Economics, 7*, 88–103. https://doi.org/10.6000/1929-7092.2018.07.07

Brusov, P. N., Filatova, T. V., Orekhova, N. P., & Kulik, V. L. (2018e). Rating methodology: New look and new horizons. *Journal of Reviews on Global Economics, 7*, 63–87. https://doi.org/10.6000/1929-7092.2018.07.06

Brusov, P., Filatova, T., Orekhova, N., Kulik, V., Weil, I., & Brailov, A. (2018f). The impact of the central bank key rate and commercial banks credit rates on creating and maintaining of a favorable investment climate in the country. *Journal of Reviews on Global Economics, 7*, 360–376. https://doi.org/10.6000/1929-7092.2018.07.31

Brusov, P. N., Filatova, T. V., Orekhova, N. P., Kulik, V. L., & Weil, I. (2018g). New meaningful effects in modern capital structure theory. *Journal of Reviews on Global Economics, 7*, 104–122. https://doi.org/10.6000/1929-7092.2018.07.08

Brusov, P. N., Filatova, T. V., Orekhova, N. P., Kulik, V. L., & Weil, I. (2019). Ratings of the investment projects of arbitrary durations: New methodology. *Journal of Reviews on Global Economics, 8*, 437–448. https://doi.org/10.6000/1929-7092.20115.08.37

Brusov, P. N., Filatova, T. V., Orekhova, N. P., Kulik, V. L., Chang, S.-I., & Lin, Y. C. G. (2020). Modification of the Modigliani–Miller theory for the case of advance payments of tax on profit. *Journal of Reviews on Global Economics, 9*, 257–268.

Filatova, T., Orehova, N., & Brusova, A. (2008). Weighted average cost of capital in the theory of Modigliani–Miller, modified for a finite life–time company. *Bulletin of the FU, 48*, 68–77.

Filatova, T. V., Brusov, P. N., Orekhova, N. P., & Kulik, V. L. (2018). Ratings of the long-term projects: New approach. *Journal of Reviews on Global Economics, 7*, 645–661. https://doi.org/10.6000/1929-7092.2018.07.59

Modigliani, F., & Miller, M. (1958). The cost of capital, corporate finance, and the theory of investment. *The American Economic Review, 48*, 261–297.

Modigliani, F., & Miller, M. (1963). Corporate income taxes and the cost of capital: A correction. *The American Economic Review, 53*, 147–175.

Modigliani, F., & Miller, M. (1966). Some estimates of the cost of capital to the electric utility industry 1954–1957. *The American Economic Review, 56*, 333–391.

Chapter 16
Investment Models with Uniform Debt Repayment and Their Application

In previous chapter, we have established investment models with debt repayment at the end of the project, well proven in the analysis of real investment projects. In practice, however, a scheme of uniform debt repayment during the duration of the project is more extended. In this chapter, we describe new investment models with uniform debt repayment during the duration of the investment project, quite adequately describing real investment projects. Within these models it is possible, in particular, to analyze the dependence of effectiveness of investment projects on debt financing and taxation. We will work on the modern theory of capital cost and capital structure developed by Brusov–Filatova–Orekhova (Brusov et al. 2015, 2018a, b, c, d, e, f, g, 2019, 2020; Filatova et al. 2008; Brusov 2018) as well as on perpetuity limit (Modigliani and Miller 1958, 1963, 1966).

In this chapter we consider the application of the investment models with uniform debt repayment.

16.1 Investment Models with Uniform Debt Repayment

As in the case of debt repayment at the end of the project, the effectiveness of the investment project is considered from two perspectives: the owners of equity and debt and the equity holders only. In the first case, the interest and duty paid by owners of equity (negative flows) returned to the project because they are exactly equal to the flow (positive), obtained by owners of debt capital. The only effect of leverage in this case is the effect of tax shield, generated from the tax relief: interest on the loan is entirely included into the net cost and, thus, reduces the tax base. For each of these cases, NPV is calculated in two ways: with the division of credit and investment flows (and thus discounting the payments, using two different rates) and without such a division (in this case, both flows are discounted at the same rate as which, obviously, WACC can be chosen). For each of the four situations, two cases

P. Brusov, T. Filatova, *Efficiency of Investment Projects*, Contributions to Finance and Accounting, https://doi.org/10.1007/978-3-032-01028-5_16

Table 16.1 The sequence of debt and interest values and credit values

Period number	1	2	3	…	n
Debt	D	$D\cdot\frac{n-1}{n}$	$D\cdot\frac{n-2}{n}$	…	$D\cdot\frac{1}{n}$
Interest	k_dD	$k_dD\cdot\frac{n-1}{n}$	$k_dD\cdot\frac{n-2}{n}$	…	$k_dD\cdot\frac{1}{n}$

are considered: (1) a constant value of equity S and (2) a constant value of the total invested capital $I = S + D$ (D is value of debt funds).

The main debt repayment occurs evenly (by equal parts) at the end of each period, and the remaining debt at the end of the each period is an arithmetic progression with the difference $-D/n$

$$\left\{D, D-\frac{D}{n}, D-\frac{2D}{n}, \ldots, \frac{D}{n}\right\} = \left\{D, D\left(\frac{n-1}{n}\right), D\left(\frac{n-2}{n}\right), \ldots, \frac{D}{n}\right\} \tag{16.1}$$

Interest constitutes a sequence:

$$\left\{k_dD, k_dD\left(\frac{n-1}{n}\right), k_dD\left(\frac{n-2}{n}\right), \ldots, k_d\frac{D}{n}\right\}. \tag{16.2}$$

In the case of consideration from the point of view of equity owners and debt owners, the after-tax flow of capital for each period is equal to

$$\mathrm{NOI}(1-t) + k_dD_it, \tag{16.3}$$

where

$$D_i = D\frac{n-(i-1)}{n}, \tag{16.4}$$

and investments at time moment $T = 0$ are equal to $-I = -S - D$. Here NOI stands for net operating income (before tax).

In the second case (from the point of view of equity owners only), investments at the initial moment $T = 0$ are equal to $-S$, and the flow of capital for the ith period (apart from tax shields k_dDt it includes payment of interest on the loan $-k_dD_i$) is equal to

$$(\mathrm{NOI} - k_dD_i)(1-t) - \frac{D_i}{n}. \tag{16.5}$$

We suppose that the interest on the loan and the loans itself are paid in tranches k_dDi and $D\cdot\frac{n-i}{n}$ consequently during the all ith periods. We cite in Table 16.1 the sequence of debt and interest values and credit values.

As in the case of debt repayment at the end of the project, we will consider two different ways of discounting:

1. Operating and financial flows are not separated and both are discounted, using the general rate (as which, obviously, the weighted average cost of capital (WACC) can be selected). For perpetuity projects, the Modigliani–Miller formula (Modigliani and Miller 1963) for WACC will be used and for projects of finite duration Brusov–Filatova–Orekhova formula for WACC (Brusov and Filatova 2011; Brusov et al. 2015, 2018a, b, c, d, e, f, g, 2019, 2020; Filatova et al. 2008).
2. Operating and financial flows are separated and are discounted at different rates: the operating flow at the rate equal to the equity cost k_e, depending on leverage, and credit flow–at the rate equal to the debt cost k_d, which until fairly large values of leverage remain constant and start to grow only at high values of leverage L, when there is a danger of bankruptcy.

Note once again that loan capital is the least risky, because interest on credit is paid after tax in the first place. Therefore, the cost of credit will always be less than the equity cost, whether of ordinary or of preference shares $k_e > k_d$; $k_p > k_d$. Here $k_e; k_p$ is equity cost of ordinary or of preference shares consequently.

One can show that the present value of interest can be calculated by using the following formula, which we have been able to derive:

$$\frac{1}{a}+\frac{2}{a^2}+\frac{3}{a^3}+\cdots+\frac{n}{a^n}=\frac{a(1-a^{-n})}{(a-1)^2}-\frac{n}{(a-1)a^n}. \tag{16.6}$$

Here $a = 1 + i$.

We will use this formula in the further calculations.

16.2 The Effectiveness of the Investment Project from the Perspective of the Equity Holders Only

16.2.1 With the Division of Credit and Investment Flows

To obtain an expression for NPV, the discounted flow values for one period, given by formulas (16.3) and (16.5), must be summed, using our obtained formula (16.6), in which $a = 1 + i$, where i is the discount rate. Its accurate assessment is one of the most important advantages of BFO theory (Brusov–Filatova–Orekhova) (Brusov and Filatova 2011; Brusov et al. 2015, 2018a, b, c, d, e, f, g, 2019, 2020; Filatova et al. 2018; Filatova et al. 2008) over its perpetuity limit–Modigliani–Miller theory (Modigliani and Miller 1958, 1963, 1966).

In this case, the expression for NPV has a view

$$
\begin{aligned}
\mathrm{NPV} = & -S + \sum_{i=1}^{n} \frac{\mathrm{NOI}(1-t)}{(1+k_\mathrm{e})^i} + \sum_{i=1}^{n} \frac{-k_\mathrm{d} D \frac{n+1-i}{n}(1-t) - \frac{D}{n}}{(1+k_\mathrm{d})^i} \\
= & -S + \frac{\mathrm{NOI}(1-t)(1-(1+k_\mathrm{e})^{-n})}{k_\mathrm{e}} \\
& - \left(\frac{D}{n} + k_\mathrm{d} D \frac{n+1}{n}(1-t)\right) \frac{1-(1+k_\mathrm{d})^{-n}}{k_\mathrm{d}} \\
& + k_\mathrm{d} \frac{D}{n}(1-t) \left\{ \frac{(1+k_\mathrm{d})[1-(1+k_\mathrm{d})^{-n}]}{k_\mathrm{d}^2} - \frac{n}{k_\mathrm{d}(1+k_\mathrm{d})^n} \right\}
\end{aligned}
\tag{16.7}
$$

In perpetuity limit (let us call it Modigliani–Miller limit), one has

$$
\mathrm{NPV} = -S + \frac{\mathrm{NOI}(1-t)}{k_\mathrm{e}} - D(1-t). \tag{16.8}
$$

16.2.2 Without Flows Separation

In this case, operating and financial flows are not separated and are discounted, using the general rate (as which, obviously, WACC can be selected).

The main debt repayment, which occurs evenly (by equal parts) at the end of each period, can be discounted either at the same rate WACC or at the debt cost rate k_d. Now we choose a uniform rate and the first option.

We still consider the effectiveness of the investment project from the perspective of the equity holders only.

$$
\begin{aligned}
NPV = & -S + \sum_{i=1}^{n} \frac{NOI(1-t) - k_d D \frac{n+1-i}{n}(1-t) - \frac{D}{n}}{(1+WACC)^i} = \\
= & -S + \frac{NOI(1-t) - \frac{D}{n} - k_d D \frac{n+1}{n}(1-t)}{WACC} \cdot \\
& \cdot \left(1 - \frac{1}{(1+WACC)^n}\right) + \\
& + \frac{k_d D}{n}(1-t) \left\{ \frac{(1+WACC)[1-(1+WACC)^{-n}]}{WACC^2} - \right. \\
& \left. - \frac{n}{WACC(1+WACC)^n} \right\}
\end{aligned}
\tag{16.9}
$$

In perpetuity limit (Modigliani–Miller limit) (turning to the limit $n \to \infty$ in the relevant equations), we have

$$\mathrm{NPV} = -S + \frac{\mathrm{NOI}(1-t) - k_{\mathrm{d}}D(1-t)}{\mathrm{WACC}}. \tag{16.10}$$

Note, that formula (16.10) as well as other formulas for perpetuity limit (16.12) and (16.14) could be applied for analyze the effectiveness of the long-term investment projects.

16.3 The Effectiveness of the Investment Project from the Perspective of the Owners of Equity and Debt

16.3.1 With Flows Separation

16.3.1.1 Projects of Arbitrary (Finite) Duration

In the case of consideration from the perspective of the owners of equity and debt

$$\begin{aligned}
\mathrm{NPV} &= -I + \sum_{i=1}^{n} \frac{\mathrm{NOI}(1-t)}{(1+k_{\mathrm{e}})^{i}} + \sum_{i=1}^{n} \frac{k_{\mathrm{d}}D\frac{n+1-i}{n}t}{(1+k_{\mathrm{d}})^{i}} \\
&= -I + \frac{\mathrm{NOI}(1-t)(1-(1+k_{\mathrm{e}})^{-n})}{k_{\mathrm{e}}} \\
&+D\frac{n+1}{n}t\cdot[1-(1+k_{\mathrm{d}})^{-n}] \\
&-k_{\mathrm{d}}\frac{D}{n}t\left\{\frac{(1+k_{\mathrm{d}})[1-(1+k_{\mathrm{d}})^{-n}]}{k_{\mathrm{d}}^{2}} - \frac{n}{k_{\mathrm{d}}(1+k_{\mathrm{d}})^{n}}\right\}
\end{aligned} \tag{16.11}$$

In perpetuity limit (Modigliani–Miller limit) (turning to the limit $n \to \infty$ in the relevant equations), we have

$$\mathrm{NOI} = -I + \frac{\mathrm{NOI}(1-t)}{k_{\mathrm{e}}} + Dt. \tag{16.12}$$

16.3.2 Without Flows Separation

We still consider the effectiveness of the investment project from the perspective of the owners of equity and debt.

$$
\begin{aligned}
\mathrm{NPV} &= -I + \sum_{i=1}^{n} \frac{\mathrm{NOI}(1-t) + k_{\mathrm{d}} D \frac{n+1-i}{n} t}{(1+\mathrm{WACC})^{i}} \\
&= -I + \frac{\mathrm{NOI}(1-t) + k_{\mathrm{d}} D \frac{n+1}{n} t}{\mathrm{WACC}} \left(1 - \frac{1}{(1+\mathrm{WACC})^{n}}\right) \\
&- \frac{k_{\mathrm{d}} D}{n} t \left\{ \frac{(1+\mathrm{WACC})[1-(1+\mathrm{WACC})^{-n}]}{\mathrm{WACC}^{2}} \right. \\
&\left. - \frac{n}{\mathrm{WACC}(1+\mathrm{WACC})^{n}} \right\}
\end{aligned} \tag{16.13}
$$

In perpetuity limit (Modigliani–Miller limit) (turning to the limit $n \to \infty$ in the relevant equations), we have

$$
\mathrm{NPV} = -I + \frac{\mathrm{NOI}(1-t) + k_{\mathrm{d}} D t}{\mathrm{WACC}}. \tag{16.14}
$$

16.4 Example of the Application of the Derived Formulas

As an example of application of the obtained formulas, let's take a look at the dependence of the NPV of project on the leverage level at three values of the tax on profit rates in the case of consideration from the perspective of the equity holders only without flows separation on operating and finance ones.

We use formula (16.10) and the next parameters values

$$
\mathrm{NOI} = 800; \quad S = 500; \quad k_0 = 22\%; \quad k_{\mathrm{d}} = 19\%; \quad T = 15\%; 20\%; 25\%.
$$

Making the calculations in Excel, we get the data, which are shown in Fig. 16.1.

From the calculations and Fig. 16.1, one can make the following conclusions:

1. With growth of the tax on profit rate, the NPV of the project decreases and our model makes it possible to assess, for how many percent, with growth of tax on profit rate, for example, by 1%.

   It should be noted that the possibility of such evaluations is unique.
2. The effect of taxation on the NPV significantly depends on the leverage level:

   With its increase, the impact of changing of tax on profit rate is greatly reduced. This is valid for increasing of the tax on profit rate and for its reduction.
3. At tax on profit rates 20% (as in Russia) and 25%, there is an optimum in NPV dependence on leverage. Investors should take into account the invested capital structure: in this case, they may, without special effort (only changing this structure), obtain (sometimes very substantial) gains in NPV. Note that at tax

Fig. 16.1 Dependence of NPV of the project on the leverage level at three values of the tax on profit rates NOI = 800; $S = 500$; $k_0 = 22\ \%$; $k_d = 19\ \%$; $T = 15\ \%$; 20 % ; 25%

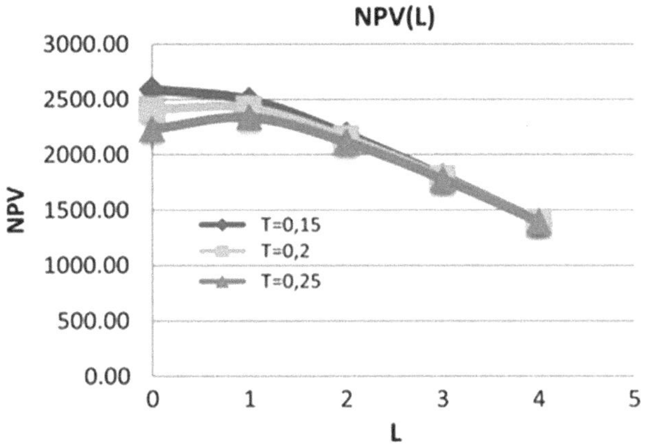

on profit rate 15%, there is no optimum in NPV dependence on leverage: NPV descends monotonically with leverage.

16.5 Conclusions

New investment models with uniform debt repayment during the duration of the project, quite adequately describing real investment projects, are described. Within these models, it is possible, in particular, to analyze the dependence of effectiveness of investment projects on debt financing and taxation. We work on the modern theory of capital cost and capital structure developed by Brusov–Filatova–Orekhova as well as on perpetuity limit–MM theory.

As in the case of debt repayment at the end of the project, the effectiveness of the investment project is considered from two perspectives: the owners of equity and debt and the equity holders only. For each of these cases, NPV is calculated in two ways: with the division of credit and investment flows (and thus discounting the payments, using two different rates) and without such a division (in this case, both flows are discounted at the same rate as which, obviously, WACC can be chosen). For each of the four situations, two cases are considered: (1) a constant value of equity S and (2) a constant value of the total invested capital $I = S + D$ (D is value of debt funds).

As an example of application of the obtained formulas, the dependence of the NPV of project on the leverage level at three values of the tax on profit rate has been investigated in the case of consideration from the perspective of the equity holders only and without flows separation on operating and financial ones. It has been shown that effect of taxation on the NPV significantly depends on the leverage level: with its increase, the impact of changing of tax on profit rates is greatly reduced. This is valid for increasing of the tax on profit rate and for its reduction. The model allows investigating the dependence of effectiveness of the investment project on leverage level, on the tax on profit rate, on credit rate, on equity cost, etc.

References

Brusov, P. (2018). Editorial: Introduction on special issue on the Banking System and financial markets of Russia and other countries: Problems and prospects. *Journal of Reviews on Global Economics, 7*, i–vi.

Brusov, P., Filatova, T., Orehova, N., & Eskindarov, M. (2015). *Modern corporate finance, investments and taxation*. Springer International Publishing, 373 p. Retrieved from https://www.springer.com/gp/book/9783319147314

Brusov, P., Filatova, T., Orehova, N., & Eskindarov, M. (2018a). *Modern corporate finance, investments, taxation and ratings*. Springer Nature Publishing, 571 p.

Brusov, P. N., Filatova, T. V., & Orekhova, N. P. (2018b). *Modern corporate finance and investments*. Knorus Publishing House, 517 p.

Brusov, P. N., Filatova, T. V., Orekhova, N. P., & Kulik, V. L. (2018c). Rating: New approach. *Journal of Reviews on Global Economics, 7*, 37–62. https://doi.org/10.6000/1929-7092.2018.07.05

Brusov, P. N., Filatova, T. V., Orekhova, N. P., & Kulik, V. L. (2018d). A "golden age" of the companies: Conditions of its existence. *Journal of Reviews on Global Economics, 7*, 88–103. https://doi.org/10.6000/1929-7092.2018.07.07

Brusov, P. N., Filatova, T. V., Orekhova, N. P., & Kulik, V. L. (2018e). Rating methodology: New look and new horizons. *Journal of Reviews on Global Economics, 7*, 63–87. https://doi.org/10.6000/1929-7092.2018.07.06

Brusov, P., Filatova, T., Orekhova, N., Kulik, V., Weil, I., & Brailov, A. (2018f). The impact of the central bank key rate and commercial banks credit rates on creating and maintaining of a favorable investment climate in the country. *Journal of Reviews on Global Economics, 7*, 360–376. https://doi.org/10.6000/1929-7092.2018.07.31

Brusov, P. N., Filatova, T. V., Orekhova, N. P., Kulik, V. L., & Weil, I. (2018g). New meaningful effects in modern capital structure theory. *Journal of Reviews on Global Economics, 7*, 104–122. https://doi.org/10.6000/1929-7092.2018.07.08

Brusov, P. N., Filatova, T. V., Orekhova, N. P., Kulik, V. L., & Weil, I. (2019). Ratings of the investment projects of arbitrary durations: New methodology. *Journal of Reviews on Global Economics, 8*, 437–448. https://doi.org/10.6000/1929-7092.2019.08.37

Brusov, P. N., Filatova, T. V., Orekhova, N. P., Kulik, V. L., Chang, S.-I., & Lin, Y. C. G. (2020). Modification of the Modigliani–Miller theory for the case of advance payments of tax on profit. *Journal of Reviews on Global Economics, 9*, 257–268.

Filatova, T., Orehova, N., & Brusova, A. (2008). Weighted average cost of capital in the theory of Modigliani–Miller, modified for a finite life-time company. *Bulletin of the FU, 48*, 68–77.

Filatova, T. V., Brusov, P. N., Orekhova, N. P., & Kulik, V. L. (2018). Ratings of the long-term projects: New approach. *Journal of Reviews on Global Economics, 7*, 645–661. https://doi.org/10.6000/1929-7092.2018.07.59

Modigliani, F., & Miller, M. (1958). The cost of capital, corporate finance, and the theory of investment. *The American Economic Review, 48*, 261–297.

Modigliani, F., & Miller, M. (1963). Corporate income taxes and the cost of capital: a correction. *The American Economic Review, 53*, 147–175.

Modigliani, F., & Miller, M. (1966). Some estimates of the cost of capital to the electric utility industry 1954–1957. *The American Economic Review, 56*, 333–391.

Part IV
Innovative Investment Models

This part describes innovative investment models with frequent payments of income tax and interest on debt. Chapter 17 describes investment models with frequent payments of income tax and interest on debt at the end of periods. Chapter 18 describes investment models with advance frequent payments of income tax and interest on debt. The results of two these models are completely different.

Chapter 17
Innovative Investment Models with Frequent Payments of Tax on Income and of Interest on Debt

17.1 Introduction

Investments play a crucial role in economy and finance. Investments in tangible and intangible assets are a necessary condition for structural adjustment and economic growth. They provide the enhancement of existing basic funds and industries and the creation of new ones. The role of investment is increased many times at the current stage. In this respect, the role of the evaluation of the efficiency of investment projects, which allows for the realization of the most effective projects in the context of scarcity and limited investment resources, increases. Since virtually most investment projects use debt financing, the study of the influence of capital structure and debt financing on the efficiency of investment projects and determining the optimal capital structure is especially important at the present time. This is why in spite of the fact that a lot of different types of investment models have been developed—stochastic, dynamics (Bond and Meghir 1994), investment banking valuation models (Rosenbaum and Pearl 2013), etc.—the main problem, which has been discussed during the last few decades, is the impact of debt financing on the efficiency of investment projects and on the investment decisions of companies.

Below we review some approaches to these problems.

17.1.1 The Literature Review

Lang et al. (1996) used so-called "Tobin's q ratio" and considered companies with low Tobin's q ratios as well as high ones. Note that the q ratio, or Tobin's q ratio, equals the market value of a company divided by its assets' replacement cost. The equilibrium takes place when market value equals replacement cost. The q ratio expresses the relationship between market valuation and intrinsic value. It is a means

P. Brusov, T. Filatova, *Efficiency of Investment Projects*, Contributions to Finance and Accounting, https://doi.org/10.1007/978-3-032-01028-5_17

of estimating the fact of whether a given business or market is undervalued or overvalued.

The authors have shown that there is a negative correlation between future growth and leverage level at the company level and, for diversified companies, at the level of business segments. This negative correlation between growth and leverage level holds for companies with low Tobin's q ratio, but not for high-q companies or companies in high-q industries. Therefore, for companies known to have good investment opportunities, leverage does not reduce growth, but it is negatively correlated to growth for companies whose growth opportunities are either not recognized by the capital markets or are not valuable enough to overcome the effects of their debt load.

Whited (1992) studied the influence of debt financing on companies' investment decisions with pharmaceutical firms in India for 11 years, from 1998 to 2009.

To study the impact of debt financing on the firms' investment decisions, Whited used pooling regression as well as random and fixed effect models. Leverage level, retained earnings, Tobin's q, sales, Return on Asset, cash flow and liquidity were considered as independent variables and investment as the dependent one. Whited considered three types of companies, depending on their size: small companies, medium companies and large companies. He showed that there is a significant positive correlation between leverage level and investment for large companies, while for medium companies a negative correlation between leverage level and investment took place.

Kang (1995) studied the connection between leverage level and investment decisions. "Interdependent tax models" were used to try to explain the specifics of company leverage levels by analyzing the interdependency between financing decisions and investment. These models account for the so-called "investment effect": the influence of investment on debt tax benefit and financial risk. One of the questions is how "investment effect" influences bond financing decisions and hence the leverage level. Different "Interdependent tax models" lead to different connections between investments and leverage levels.

Some authors mentioned a positive connection via the fact that the financial risk and hence the cost of bond financing decrease with an increase in investment at a given leverage level. A negative connection has been mentioned by DeAngelo and Masulis (1980) and Dotan and Ravid (1985). The first authors concluded this since the tax benefits of debt compete with those of capital investment. The second authors refer to the fact that financial risk and thus the cost of bond financing will increase with investment increase.

The impact of investment increase on financial risk may depend on company-specific factors, like company-specific technology (Dammon and Senbet 1988). An analysis of the impact of corporate and personal taxes on a firm's optimal investment and financing decisions under uncertainty is provided in this paper. By endogenizing firms' investment decisions, it extends the DeAngelo and Masulis capital structure model. The authors' results indicate that the existing predictions about the relationship between investment-related and debt-related tax shields must be modified in cases where investment is allowed to adjust optimally. The authors show that increases in investment-related tax shields due to changes in the corporate tax code are not necessarily associated with reductions in leverage level at the individual company level. Companies with higher investment-related tax shields (as cross-sectional analysis shows) need not have lower debt-related tax shields (normalized by expected earnings) unless all companies utilize the same production technology. Differences in production technologies across companies may thus explain why the empirical results of recent cross-sectional studies have not conformed to the predictions of DeAngelo and Masulis (1980). A model of company leverage level choice is formulated in this paper, in which corporate and differential personal taxes exist and supply-side adjustments by companies enter into the determination of equilibrium prices of debt and equity. The presence of corporate tax shield substitutes for debt such as depletion allowances, accounting depreciation and investment tax credits are shown to imply a market equilibrium in which each company has a unique interior optimum leverage level decision. The optimal leverage level model yields a number of interesting predictions regarding cross-sectional and time-series properties of firms' capital structures. Below we discuss some portfolio investment models as well as behavioral aspects of investors, which play an important role in investments.

Among those portfolio investment models is the well-known Black–Litterman model, which was created by Black and Litterman (1992). They developed the model to address the problems that institutional investors have encountered the during application of modern portfolio theory in practice. Starting with an asset allocation based on the equilibrium assumption, the model then modifies that allocation by accounting the investors' opinions with respect to future asset performance.

Loviscek (2021) has applied the Black–Litterman model of modern portfolio theory to well-known index mutual funds-one guided by the classic 60%/40% stock/bond allocation and one based on an all-equity allocation. The period under their study is from 2000 to 2020. Although statistical evidence supports that the efficacy of a precious metal allocation is elusive, the results suggest an average allocation of about 2% for "buy-and-hold" investors who seek one. He has shown that, from 2003 to 2010 and from 2016 to 2020, the allocations were in the range of 5–10%. Other periods, however, register only a little more than 0%.

Cakici and Zaremba (2021), using data on 65,000 stocks from 23 countries, reconsidered the performance of the Fama–French factors in global markets. As their results showed, the value, profitability and investment factors are far less reliable than what is commonly thought. Their performance depends strongly on the geographical regions and periods examined. Moreover, most factor returns are driven by the smallest companies. Virtually no value or investment effects are present among the big companies representing most of the total market capitalization worldwide. These results cast doubt on the five-factor model's applicability in international markets, citing that the smallest companies are typically not invested in by major financial institutions.

A growing number of investors want to use firm sustainability information in their investment decision processes to avoid risk, satisfy their own asset preference, or find a new alpha-generating factor. Not too many users of environment, social and governance (ESG) data understand how ESG ratings change over time. Gidwani (2020) used the CSRHub data set to show that ESG ratings regress strongly toward the mean. These ratings include both data from 640 sources and from most commercial ESG ratings firms. The observed regression persists during nine years within the ratings data, for a sample set of more than 8000 firms. Newly-rated firms show even more reversion than "seasoned" firms. Firms can only rarely maintain an especially high or low ESG rating. Investors and firm managers should understand that ESG ratings are likely to change toward the mean. This however does not mean that a good firm is getting worse or a bad one is getting better.

Brown et al. (2021) have developed statistics (holdings-based) to estimate the volatility with time of investment style characteristics of funds. They found that funds with lower levels of style volatility significantly outperform funds with higher levels of style volatility on a risk-adjusted basis. The authors have shown that style volatility has a distinct impact on fund performance in the future compared to expenses of funds or past risk-adjusted returns, with the level of indirect style volatility being the primary determinant of the overall effect. It was concluded that deciding to maintain a less volatile investment style is an important aspect of the portfolio management process.

Some behavioral aspects of investors have been considered by Escobar-Anel et al. (2020), who introduced a strategy generalizing the CPPI (Constant Proportion Portfolio Insurance) approach. The target of this strategy is to guarantee the investment goal or floor during participation in the performance of the assets and limiting the downside risk of the portfolio at the same time. The authors show that the strategy accounts for the following behavioral aspects of investors: a risk-averse behavior for gains, distorted probabilities recognition and a risk-seeking behavior for losses. The developed strategy turns out to be optimal within the Cumulative Prospect Theory framework by Kahneman and Tversky (1996).

17.1.2 Some Problems Under the Evaluation of the Effectiveness of the Investment Projects

Some of the major problems under the evaluation of the effectiveness of investment projects are suggested as follows:

1. Which financial flows should be taken into account when calculating the parameters of efficiency of a project (NPV, IRR, etc.)?
2. How many discount rates should be used for discounting various cashflows?
3. How can these discount rates be accurately evaluated?

The first two problems are still under intensive discussion. Concerning the third issue, we need to note that, in the last decade, significant progress in the accurate determination of the cost of the equity and company weighted average cost, which just are the discount rates when evaluating the effectiveness of the project, has been achieved. The progress is mainly associated with the studies by Brusov, Filatova and Orekhova (BFO theory) (Brusov et al. 2015, 2018; Filatova et al. 2008), in which a general theory of capital cost of the company and its capital structure was established, and the dependence of capital cost on leverage level and on the age of a company was found for the companies of arbitrary age. The main difference between their theory and Modigliani–Miller theory is that the former one removes the assumption of perpetuity for the companies under discussion, which leads to a significantly different new theory from the theory established by the Nobel laureates Modigliani and Miller (1958, 1963, 1966).

In modern conditions, the requirements for improving the quality of assessing the effectiveness of investments have increased. The modern investment models, which have been well-tested in real economic situations, have been developed by Brusov, Filatova and Orekhova (Brusov et al. 2015; Filatova et al. 2008). They have created long-term as well as arbitrary duration models and have considered the effectiveness of the investment project from two points of view: from the equity holders and from the owners of equity and debt. NPV in each of these cases could be calculated by two different methods: with the division of credit and investment flows and using two different discounting rates, and without such a division and using a general discounting rate (for which WACC can, obviously, be chosen).

Applying their modern investment models on the evaluation of the dependence of the effectiveness of investments on debt financing of one telecommunication company in 2010–2012 from the point of view of optimal structure of investment, the authors showed that in 2012, the company lost 675 million USD on average, because the investment structure had been far from the optimal one. The ability to calculate the correct optimal capital structure is one important feature implied by the Brusov, Filatova and Orekhova modern investment models (Brusov et al. 2015, 2018; Filatova et al. 2008).

17.1.3 The Discount Rates

As we mentioned above, one of the most important elements of calculating the effectiveness of investment projects is the assessment of the discount rate. In the case of long-term investment models without the division of credit and investment flows, the discount rate WACC has been calculated using the Modigliani–Miller formula (Modigliani and Miller 1958, 1963, 1966)

$$WACC = k_0 \cdot (1 - w_d t) \tag{17.1}$$

while in the case of arbitrary duration models, the Brusov-Filatova-Orekhova formula for WACC (Black and Litterman 1992; Loviscek 2021; Cakici and Zaremba 2021)

$$\frac{1-(1+WACC)^{-n}}{WACC}=\frac{1-(1+k_0)^{-n}}{k_0\cdot(1-w_dt[1-(1+k_d)^{-n}])} \tag{17.2}$$

has been used.

Here and below, WACC is the weighted average cost of capital; k_O is the equity cost at zero leverage ($L = 0$); k_d is the debt cost; w_d is the debt share; t is the tax on profit; n is the project duration; k_e is the equity cost; and L is the leverage level.

In the case of long-term investment models with the division of credit and investment flows, the discount rate for discounting the investment flows (equity cost k_e) has been calculated using the Modigliani–Miller formula (Modigliani and Miller 1958, 1963, 1966)

$$k_e=k_0+L\cdot(k_0-k_d)(1-t) \tag{17.3}$$

while in the case of arbitrary duration models, equity cost k_e has been calculated from the formula

$$WACC=k_e w_e+k_d w_d(1-t) \tag{17.4}$$

using the Brusov-Filatova-Orekhova value for WACC (Brusov et al. 2015, 2018; Filatova et al. 2008).

The calculation methods of the discount rates (WACC, equity cost k_e) has been generalized in Brusov and Filatova (2021) for the real conditions of the implementation of investment projects: for arbitrary frequency of payment of tax on profit.

In this paper, new modern investment models, both long-term and arbitrary duration, will be created, as close as possible to real investment conditions. They will account the payments of interest on debt and of tax on income a few times per year (semi-annually, quarterly, monthly), which are applied in real economic practice. Their verification will lead to the creation of a comprehensive system of adequate and correct assessment of the effectiveness of the company's investment program and its investment strategy.

17.1.4 The Structure of the Paper

The structure of the paper is as follows:

1. In Sect. 17.1 above, we presented:

    1.1. The literature review.
    1.2. Some problems under the evaluation of the effectiveness of the investment projects.
    1.3. Discount rates.

2. In Sect. 17.2, we consider the effectiveness of the investment project from the perspective of the equity holders only.
   2.1. In Sect. 17.2.1, we consider a case with flow separation.
   2.2. In Sect. 17.2.2, we consider a case without flow separation.
3. In Sect. 17.3, we consider the effectiveness of the investment project from the perspective of the owners of equity and debt.
   3.1. In Sect. 17.3.1, we consider a case with flow separation.
   3.2. In Sect. 17.3.2, we consider a case without flow separation.
4. In Sect. 17.4, the problem of calculation of discount rates is discussed and expressions for their modified values are obtained.
5. In Sect. 17.5, we study numerically with the use of Microsoft Excel the effectiveness of the four models, created by us in this paper. We consider long-term projects as well as projects of arbitrary duration from two perspectives: the owners of equity and debt and the equity holders only without the division of credit and investment flows.
6. In Conclusions, we discuss obtained results and their impact on the correctness of valuation of efficiency of investment projects.

17.2 The Effectiveness of the Investment Project from the Perspective of the Equity Holders Only

We will consider the effectiveness of the investment project from two points of view: from the equity holders and from the owners of equity and debt. NPV in each of these cases could be calculated by two different methods: with the division of credit and investment flows and using two different discounting rates, and without such a division and using a general discounting rate (for which WACC can, obviously, be chosen).

The following designations are used below:

The equity value S, the investment value I, the net operating income NOI, the leverage level L, the profitability of investments β, the tax on profit t, the project duration n, the equity cost k_0, the debt cost k_d and the number of payments of interest on debt p_1 and of tax on income p_2; D is the debt value.

In the first case (from the perspective of the equity holders), at the initial moment in time, $T = 0$ investments are equal to $-S$ and the flow of capital, CF, for the period is equal to:

$$\mathrm{CF} = (\mathrm{NOI} - k_d D)(1 - t). \tag{17.5}$$

In addition to the tax shields, $k_d Dt$ includes a payment of interest on a loan, $-k_d D$.

We suppose below that interests on the loan are paid in equal shares of k_dD during all periods and principal repayment is made at the end of the project.

In the second case (from the perspective of the equity holders), the negative flows (the interest and duty paid by owners of equity) are returned to the project and they are exactly equal to the positive flows obtained by owners of debt capital. Thus, in this case the only effect of debt financing is the effect of the tax shield, generated from the tax relief: interest on the loan is entirely included in the cost and thus reduces the tax base. After-tax flow of capital, CF, for each period is equal to

$$\mathrm{CF} = \mathrm{NOI}(1-t) + k_dDt \tag{17.6}$$

and the value of investments at the initial moment in time $T = 0$ is equal to $-I = -S - D$.

Below two different ways of discounting will be considered:

1. If operating and financial flows are not separated, both flows are discounted by the general rate. In this case, the weighted average cost of capital (WACC) can be selected as a discounting rate. For long-term projects, we will use the Modigliani–Miller formula for WACC (Modigliani and Miller 1958, 1963, 1966), modified by us for the case of payments of interest on debt and of tax on income a few times per year (semi-annually, quarterly, monthly) and for projects of finite (arbitrary) duration we will use the Brusov–Filatova–Orekhova formula for WACC (Brusov et al. 2018; Filatova et al. 2008), modified by us for the case of payments of interest on debt and of tax on income a few times per year (semi-annually, quarterly, monthly).
2. Operating and financial flows are separated and are discounted at different rates: the operating flow at the rate which is equal to the equity cost k_e, depending on leverage and on the number of payments of interest on debt and of tax on income, and credit flow at the rate which is equal to the debt cost k_d.

Note that loan capital is the least risky because the credit (including the interest on credit) is paid after taxes in the first place. Therefore, the cost of credit will always be less than the equity cost, whether for ordinary or for preference shares $k_e > k_d$; $k_p > k_d$. Here k_e; k_p is the equity cost of ordinary or of preference shares consequently.

17.2.1 With Flow Separation

In this case, the expression for NPV (net present value) per period has a view

$$NPV = NOI(1-t) - k_dD(1-t) = NOI(1-t) + k_dDt - k_dD. \tag{17.7}$$

Here, the first term is the value of operating income from the investment project after tax deduction, the second term is the value of the tax shield, the third term is the value of interest on debt for one period.

We will need the following auxiliary formulas for summing the reduced values of financial flows (17.7) when calculating NPV:

For annual payments of interest on debt and of tax on income:

$$\sum_{i=1}^{n} \frac{1}{(1+k_d)^i} = \frac{1}{1+k_d} \cdot \frac{1-(1+k_d)^{-n}}{1-\frac{1}{1+k_d}} = \frac{1}{k_d} \cdot (1-(1+k_d)^{-n}) \tag{17.8}$$

For more frequent payments (p times per period) of interest on debt and of tax on income (semi-annually, quarterly, monthly):

$$\sum_{i=1}^{np} \frac{R}{p(1+k_d)^{i/p}} = \frac{R}{p(1+k_d)^{1/p}} \cdot \frac{1-(1+k_d)^{-n}}{1-\frac{1}{(1+k_d)^{1/p}}} = \frac{R}{p} \cdot \frac{(1-(1+k_d)^{-n})}{(1+k_d)^{1/p}-1} \tag{17.9}$$

Similar formulas are obtained using the cost of equity k_e and WACC as the discount rates.

Summing up the given values of financial flows for each period (17.7), we obtain for NPV of n-years project in the case of separated flows:

$$NPV = -S + \sum_{i=1}^{n} \frac{NOI(1-t)}{(1+k_e)^i} + \sum_{i=1}^{n} \frac{k_d Dt}{(1+k_d)^i} - \sum_{i=1}^{n} \frac{k_d D}{(1+k_d)^i} - \frac{D}{(1+k_d)^n} \tag{17.10}$$

Here, the second term is the reduced value of operating income from the investment project, the second term is the reduced value of the tax shield, the third term is the reduced value of interest paid annually (at the end of the year), and the fourth term is the reduced value of the debt paid at the end of the project.

After summing, we have the following expression for NPV:

$$NPV = -S + \frac{NOI(1-t)}{k_e} \cdot \left(1 - \left(\frac{1}{1+k_e}\right)^n\right) + Dt(1-(1+k_d)^{-n}) - D(1-(1+k_d)^{-n}) - \frac{D}{(1+k_d)^n} \tag{17.11}$$

In the case of more frequent (p-times per year) payment of income taxes (p_1) and frequent payments of interest on debt (p_2) we have:

$$NPV = -S + \sum_{i=1}^{n} \frac{NOI(1-t)}{(1+k_e)^i} + \sum_{i=1}^{np} \frac{k_d Dt}{p_1(1+k_d)^{i/p_1}} - \sum_{i=1}^{np} \times \frac{k_d D}{p_2(1+k_d)^{i/p_2}} - \frac{D}{(1+k_d)^n}; \tag{17.12}$$

After summing, we have the following expression for NPV:

$$NPV = -S + \frac{NOI(1-t)}{k_e} \cdot \left(1 - \left(\frac{1}{1+k_e}\right)^n\right) + \frac{k_d Dt(1-(1+k_d)^{-n})}{p_1\left((1+k_d)^{1/p_1} - 1\right)} - \frac{k_d D(1-(1+k_d)^{-n})}{p_2\left((1+k_d)^{1/p_2} - 1\right)} - \frac{D}{(1+k_d)^n} \tag{17.13}$$

17.2.1.1 Long-Term Investment Projects

To obtain the expression for NPV of long-term investment projects, one should find the limit of (17.13) $n \to \infty$.

$$NPV = -S + \frac{NOI(1-t)}{k_e} + \frac{k_d Dt}{p_1\left((1+k_d)^{1/p_1} - 1\right)} - \frac{k_d D}{p_2\left((1+k_d)^{1/p_2} - 1\right)} \tag{17.14}$$

17.2.2 Without Flow Separation

In the case of no separation between operating and financial flows, both flows are discounted by the general rate WACC. The credit reimbursable at the end of the project can be discounted either at the same rate WACC or at the debt cost rate k_d. Below, a uniform rate for WACC has been used.

Summing up the given values of financial flows for each period (17.3), we obtain for NPV of n-years project in the case without separated flows:

$$NPV = -S + \sum_{i=1}^{n} \frac{NOI(1-t)}{(1+WACC)^i} + \sum_{i=1}^{n} \frac{k_d Dt}{(1+WACC)^i} - \sum_{i=1}^{n} \times \frac{k_d D}{(1+WACC)^i} - \frac{D}{(1+WACC)^n} \tag{17.15}$$

After summing, we have the following expression for NPV:

$$NPV = -S + \frac{NOI(1-t)}{WACC} \cdot \left(1 - \left(\frac{1}{1+WACC}\right)^n\right) + \frac{k_d Dt(1-(1+WACC)^{-n})}{WACC} - \frac{k_d D(1-(1+WACC)^{-n})}{WACC} - \frac{D}{(1+WACC)^n} \tag{17.16}$$

In the case of more frequent (p-times per year) payment of income taxes and frequent payments of interest on the debt we have:

$$NPV = -S + \sum_{i=1}^{n} \frac{NOI(1-t)}{(1+WACC)^i} + \sum_{i=1}^{np} \frac{k_d Dt}{p_1(1+WACC)^{i/p_1}} - \sum_{i=1}^{np} \frac{k_d D}{p_2(1+WACC)^{i/p_2}} - \frac{D}{(1+WACC)^n} \tag{17.17}$$

After summing, we have the following expression for NPV:

$$NPV = -S + \frac{NOI(1-t)}{WACC} \cdot \left(1 - \left(\frac{1}{1+WACC}\right)^n\right) + \frac{k_d Dt(1-(1+WACC)^{-n})}{p_1\left((1+WACC)^{1/p_1} - 1\right)} - \frac{k_d D(1-(1+WACC)^{-n})}{p_2\left((1+WACC)^{1/p_2} - 1\right)} - \frac{D}{(1+WACC)^n} \tag{17.18}$$

17.2.2.1 Long-Term Investment Projects

To obtain the expression for NPV of long-term investment projects, one should find the limit of (17.18) $n \to \infty$.

$$NPV = -S + \frac{NOI(1-t)}{WACC} + \frac{k_d Dt}{p_1\left((1+WACC)^{1/p_1} - 1\right)} - \frac{k_d D}{p_2\left((1+WACC)^{1/p_2} - 1\right)} \tag{17.19}$$

17.3 The Effectiveness of the Investment Project from the Perspective of the Owners of Equity and Debt

17.3.1 With Flow Separation

In this case, operating and financial flows are separated and are discounted using different rates: the operating flow at the rate equal to the equity cost k_e, depending on leverage, and credit flow at the rate equal to the debt cost k_d, which remain constant until fairly large values of leverage and start to grow only at high values of leverage L, when there is a danger of bankruptcy.

After-tax flow of capital for each period in this case is equal to (17.6):

$$CF = NOI(1-t) + k_d Dt \tag{17.20}$$

After summing over n-periods, one gets the following expression for NPV:

$$NPV = -I + \sum_{i=1}^{n} \frac{NOI(1-t)}{(1+k_e)^i} + \sum_{i=1}^{n} \frac{k_d Dt}{(1+k_d)^i} = -I + \frac{NOI(1-t)}{k_e} \times (1-(1+k_e)^{-n}) + Dt(1-(1+k_d)^{-n}) \tag{17.21}$$

In the case of a few (p) payments of tax on profit per period, the expression for NPV will be modified:

$$NPV = -I + \sum_{i=1}^{n} \frac{NOI(1-t)}{(1+k_e)^i} + \sum_{i=1}^{np} \frac{k_d Dt}{p(1+k_d)^{i/p}} \tag{17.22}$$

After summing over n-periods, one gets the following modified expression for NPV:

$$NPV = -I + \frac{NOI(1-t)[1-(1+k_e)^{-n}]}{k_e} + \frac{k_d Dt[1-(1+k_d)^{-n}]}{p\left[(1+k_d)^{1/p} - 1\right]} \tag{17.23}$$

To obtain the expression for NPV of long-term investment projects, one should find the limit of (17.23) $n \to \infty$.

$$NPV = -I + \frac{NOI(1-t)}{k_e} + \frac{k_d Dt}{p\left[(1+k_d)^{1/p} - 1\right]} \tag{17.24}$$

17.3.2 Without Flow Separation

In the case of no separation between operating and financial flows, both flows are discounted by the general rate WACC.

After summing the expression (17.6) over n-periods, one gets the following expression for NPV:

$$NPV = -I + \sum_{i=1}^{n} \frac{NOI(1-t)}{(1+WACC)^i} + \sum_{i=1}^{n} \frac{k_d Dt}{(1+WACC)^i} \tag{17.25}$$

$$NPV = -I + \frac{NOI(1-t)}{WACC}\left(1-(1+WACC)^{-n}\right) + \frac{k_d Dt}{WACC}\left(1-(1+WACC)^{-n}\right) \tag{17.26}$$

In the case of a few (p) payments of tax on profit per period, the expression for NPV will be modified:

$$NPV = -I + \sum_{i=1}^{n} \frac{NOI(1-t)}{(1+WACC)^i} + \sum_{i=1}^{np} \frac{k_d Dt}{p(1+WACC)^{i/p}} \tag{17.27}$$

After summing over n-periods, one gets the following modified expression for NPV:

$$NPV = -I + \frac{NOI(1-t)}{WACC} \cdot \left(1 - \left(\frac{1}{1+WACC}\right)^n\right) + \frac{k_d Dt\left(1-(1+WACC)^{-n}\right)}{p\left((1+WACC)^{1/p} - 1\right)} \tag{17.28}$$

To obtain the expression for NPV of long-term investment projects, one should find the limit of (17.28) $n \to \infty$.

$$NPV = -I + \frac{NOI(1-t)}{WACC} \cdot + \frac{k_d Dt}{p\left((1+WACC)^{1/p} - 1\right)} \tag{17.29}$$

17.4 Discount Rates

In the case without the division of credit and investment flows (both flows are discounted using the same rate, for which WACC can obviously be chosen), WACC is calculated by the following formulas:

For arbitrary project duration (Kahneman and Tversky 1996):

$$\frac{1-(1+WACC)^{-n}}{WACC}=\frac{1-(1+k_0)^{-n}}{k_0\cdot\left(1-\frac{k_d w_d t}{p}\frac{\left[1-(1+k_d)^{-n}\right]}{(1+k_d)^{1/p}-1}\right)} \tag{17.30}$$

For long-term projects (Kahneman and Tversky 1996):

$$WACC=k_0\cdot\left(1-\frac{k_d w_d t}{p}\frac{1}{(1+k_d)^{1/p}-1}\right) \tag{17.31}$$

In the case of the division of credit and investment flows (and thus discounting of the payments using two different rates, equity cost k_e and debt cost k_d), k_e should be found from the equation

$$WACC=k_e w_e+k_d w_d(1-t) \tag{17.32}$$

where we substitute WACC from the formula (17.30) for a project with arbitrary duration and from the formula (17.31) for a long-term project.

Note that formulas (17.30) and (17.31) are quite different from the original formulas by Brusov–Filatova–Orekhova (Brusov et al. 2018; Filatova et al. 2008) and Modigliani–Miller (1958, 1963, 1966), where payments of interest on debt and of tax on income are made once per year and are turned into them if we put $p=1$:

$$\frac{1-(1+WACC)^{-n}}{WACC}=\frac{1-(1+k_0)^{-n}}{k_0\cdot(1-w_d t[1-(1+k_d)^{-n}])} \tag{17.33}$$

$$WACC=k_0\cdot(1-w_d t) \tag{17.34}$$

17.5 Results and Discussions

In this section, we study numerically with the use of Microsoft Excel the effectiveness of the four models created above by us. We consider long-term projects as well as projects of arbitrary duration from two perspectives: the owners of equity and debt and the equity holders only. We will study the case without flow separation. In this case, operating and financial flows are not separated and both are discounted, using the general rate, for which we select the weighted average cost of capital, WACC. Let us start from the study numerically on the dependence of the discount rates (weighted average cost of capital, WACC) on leverage level L at different frequencies of payment of tax on profit p.

17.5.1 Numerical Calculation of the Discount Rates

17.5.1.1 The Long-Term Investment Projects

For long-term investment projects (the Modigliani–Miller limit), the discount rate (WACC) in the case of arbitrary frequency of payment of tax on profit p is described by formula (17.31) (Brusov and Filatova 2021):

$$WACC = k_0 \cdot \left(1 - \frac{k_d w_d t}{p} \frac{1}{(1 + k_d)^{1/p} - 1}\right)$$

For WACC calculation, we will use the following parameters: $k_0 = 0.22$; $k_d = 0.14$; $t = 20\%$; $L = 0; 1; 2; \ldots; 10$.

Dependence of WACC on leverage level L at $k_0 = 0.22$, $k_d = 0.14$ and different $p = 1, 6, 12$ is shown in Table 17.1 and Fig. 17.1.

17.5.1.2 The Arbitrary Duration Investment Projects

For arbitrary duration investment projects, the discount rate (WACC) in the case of arbitrary frequency of payment of tax on profit p is described by formula (17.30) (modified BFO formula Brusov and Filatova 2021).

Table 17.1 Dependence of WACC on leverage level L at $k_0 = 0.22$, $k_d = 0.14$ and different $p = 1, 6, 12$

	WACC		
L	$p = 1$	$p = 6$	$p = 12$
0	0.2200	0.2200	0.2200
1	0.1980	0.1967	0.1966
2	0.1907	0.1890	0.1888
3	0.1870	0.1851	0.1849
4	0.1848	0.1828	0.1826
5	0.1833	0.1812	0.1810
6	0.1823	0.1801	0.1799
7	0.1815	0.1793	0.1791
8	0.1809	0.1787	0.1784
9	0.1804	0.1781	0.1779
10	0.1800	0.1777	0.1775

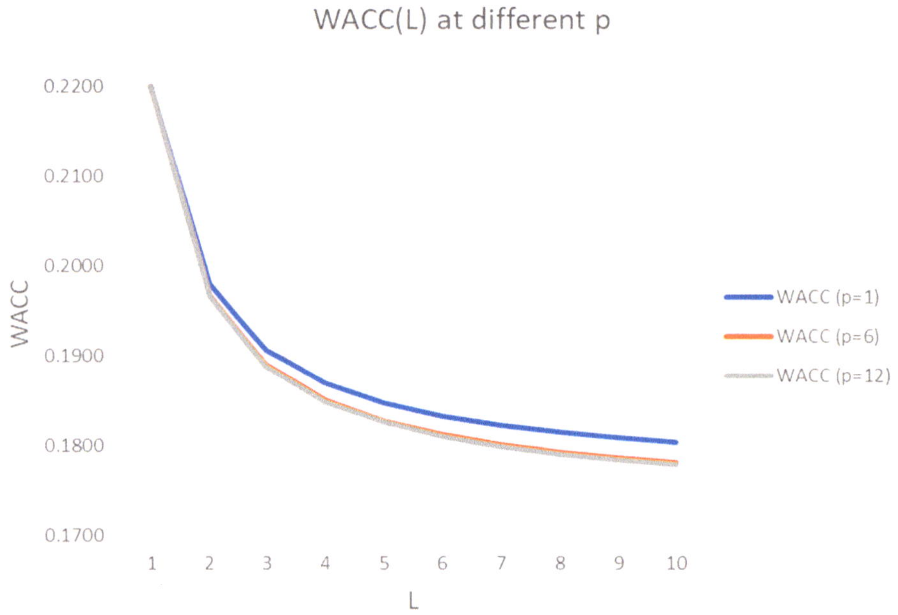

Fig. 17.1 Dependence of the weighted average cost of capital, WACC, on leverage level L at different frequencies of payments of tax on profit, $p = 1; 6; 12$

For WACC calculation, we will use the following parameters: $k_0 = 0.22$; $k_d = 0.14$; $t = 20\%$; $L = 0; 1; 2; \ldots; 10$; $p = 1; 6; 12$; $n = 3$

$$\frac{1-(1+WACC)^{-n}}{WACC} = \frac{1-(1+k_0)^{-n}}{k_0 \cdot \left(1 - \frac{k_d w_d t}{p} \frac{[1-(1+k_d)^{-n}]}{(1+k_d)^{1/p}-1}\right)}$$

Dependence of WACC on leverage level L at $k_0 = 0.22$, $k_d = 0.14$ and different $p = 1, 6, 12$ for a three-year company is shown in Table 17.2 and Figs. 17.2 and 17.3.

Note that the dependences of WACC on leverage level L obtained above will be used below under a study of the effectiveness of long-term as well arbitrary duration investment projects, from the perspective of the owners of equity capital and of the owners of equity and debt.

Table 17.2 Dependence of WACC on leverage level L at $k_0 = 0.22$, $k_d = 0.14$ and different $p = 1, 6, 12$ for a three-year company

L	WACC		
	$p = 1$	$p = 6$	$p = 12$
0	0.2200	0.2200	0.2200
1	0.1987	0.1974	0.1973
2	0.1915	0.1899	0.1897
3	0.1879	0.1861	0.1859
4	0.1858	0.1838	0.1836
5	0.1843	0.1823	0.1821
6	0.1833	0.1812	0.1810
7	0.1825	0.1804	0.1802
8	0.1819	0.1798	0.1795
9	0.1815	0.1793	0.1790
10	0.1811	0.1788	0.1786

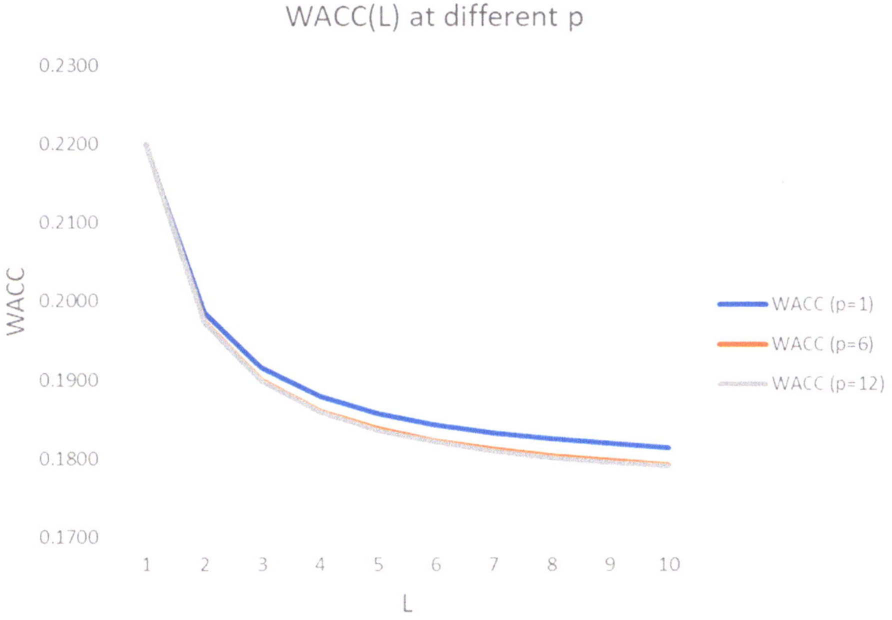

Fig. 17.2 Dependence of WACC on leverage level L at $k_0 = 0.22$, $k_d = 0.14$ and different $p = 1$, 6, 12 for a three-year project

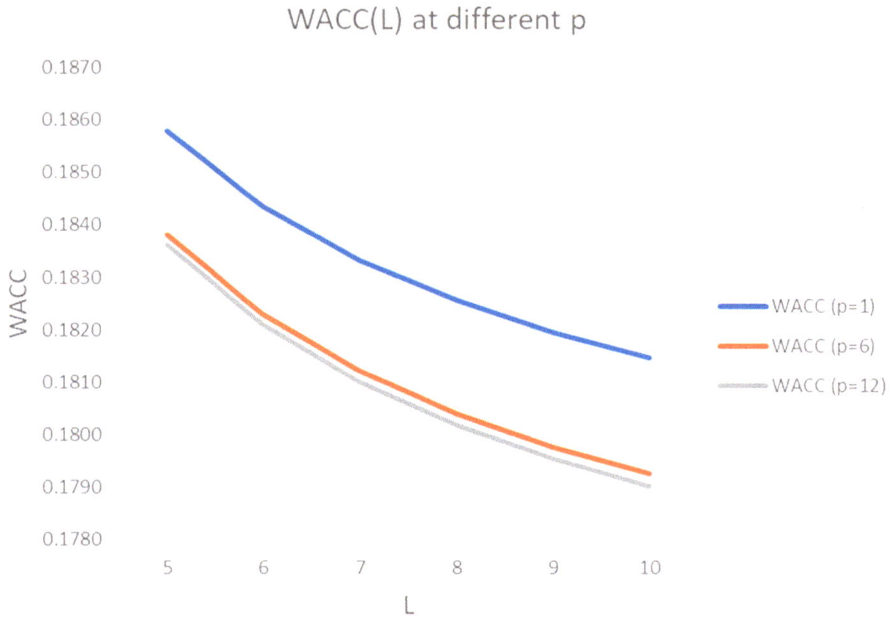

Fig. 17.3 Dependence of WACC on leverage level L at $k_0 = 0.22$, $k_d = 0.14$ and different $p = 1$, 6, 12 for a three-year project (larger scale)

17.5.2 The Effectiveness of the Long-Term Investment Project from the Perspective of the Owners of Equity Capital

For long-term investment projects, the discount rate (WACC) in the case of arbitrary frequency of payment of tax on profit p is described by formula (17.19) and for NPV one has:

$$NPV = -S + \frac{NOI(1-t)}{WACC} + \frac{k_d Dt}{p_1\left((1+WACC)^{1/p_1} - 1\right)} - \frac{k_d D}{p_2\left((1+WACC)^{1/p_2} - 1\right)}$$

For NPV calculation, we will use the following parameters: $k_0 = 0.22$; $k_d = 0.14$; $t = 20\%$; $L = 0; 1; 2; \ldots; 10$; $p_1 = p_2 = p = 1; 6; 12$; $S = 1000$; $D = LS$; $NOI = 800$.

It is seen from Figs. 17.4 and 17.5 that in the case of considering the effectiveness of long-term investment projects for owners of equity capital, NPV will vary with the change of p, but with not much variation (it is seen from Table 17.3 and from Fig. 17.5, but not from Fig. 17.4).

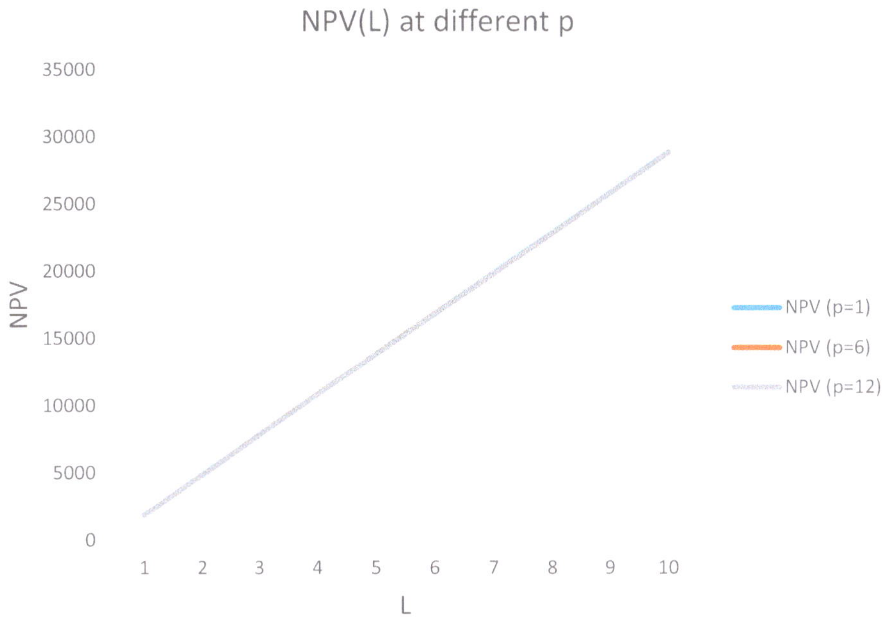

Fig. 17.4 Dependence of NPV on leverage level L at $k_0 = 0.22$, $k_d = 0.14$ and different $p = 1, 6, 12$ for a long-term project

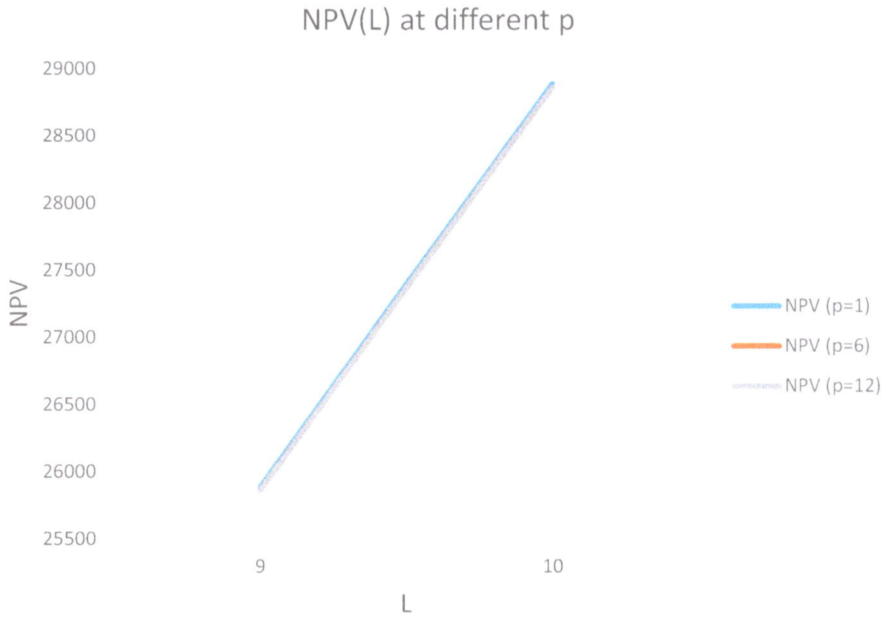

Fig. 17.5 Dependence of NPV on leverage level L at $k_0 = 0.22$, $k_d = 0.14$ and different $p = 1, 6, 12$ for a long-term project (larger scale)

Table 17.3 Dependence of WACC and NPV on leverage level L at $k_0 = 0.22$, $k_d = 0.14$ and different $p = 1, 6, 12$

	NPV		
L	$p = 1$	$p = 6$	$p = 12$
0	1909	1909	1909
1	4899	4891	4891
2	7895	7883	7882
3	10,893	10,878	10,877
4	13,892	13,874	13,873
5	16,891	16,871	16,869
6	19,890	19,868	19,866
7	22,890	22,865	22,863
8	25,889	25,863	25,860
9	28,889	28,860	28,858
10	31,889	31,858	31,855

17.5.3 The Effectiveness of the Long-Term Investment Project from the Perspective of the Owners of Equity and Debt

For long-term investment projects, NPV in the case of arbitrary frequency of payment of tax on profit p is described by formula (17.29):

$$NPV = -I + \frac{NOI(1-t)}{WACC} + \frac{k_d Dt}{p\left((1+WACC)^{1/p} - 1\right)}$$

For NPV calculation, we will use the following parameters: $k_0 = 0.22$; $k_d = 0.14$; $t = 20\%$; $L = 0; 1; 2; \ldots; 10$; $p_1 = p_2 = p = 1; 6; 12$; $S = 1000$; $D = LS$; $NOI = 800$.

Dependence of NPV on leverage level L at $k_0 = 0.22$, $k_d = 0.14$ and different $p = 1, 6, 12$ for a long-term project is shown in Table 17.4 and Figs. 17.6 and 17.7.

17.5.4 The Effectiveness of the Arbitrary Duration Investment Projects from the Perspective of the Owners of Equity Capital

For the arbitrary duration investment projects, NPV in the case of arbitrary frequency of payment of tax on profit p is described by formula (17.18):

Table 17.4 Dependence of NPV on leverage level L at $k_0 = 0.22$, $k_d = 0.14$ and different $p = 1, 6, 12$

L	NPV		
	$p = 1$	$p = 6$	$p = 12$
0	1909	1909	1909
1	4606	4659	4665
2	7364	7478	7489
3	10,139	10,316	10,334
4	12,922	13,163	13,188
5	15,709	16,015	16,047
6	18,498	18,870	18,908
7	21,289	21,726	21,771
8	24,081	24,583	24,635
9	26,874	27,441	27,500
10	29,667	30,300	30,365

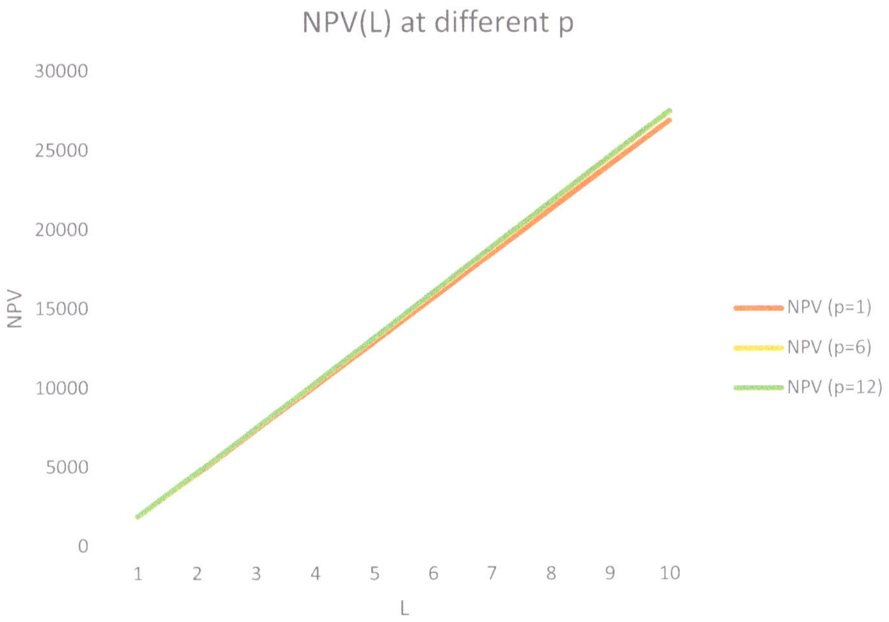

Fig. 17.6 Dependence of NPV on leverage level L at $k_0 = 0.22$, $k_d = 0.14$ and different $p = 1, 6, 12$ for a long-term project

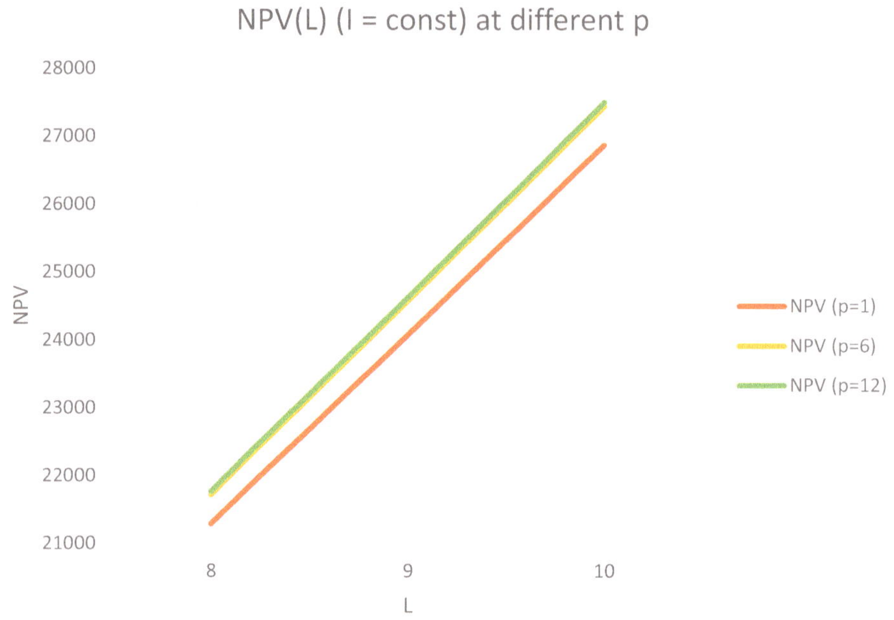

Fig. 17.7 Dependence of NPV on leverage level L at $k_O = 0.22$, $k_d = 0.14$ and different $p = 1, 6, 12$ for a long-term project (larger scale)

$$NPV = -S + \frac{NOI(1-t)}{WACC} \cdot \left(1 - \left(\frac{1}{1+WACC}\right)^n\right) + \frac{k_d Dt(1-(1+WACC)^{-n})}{p_1\left((1+WACC)^{1/p_1} - 1\right)} - \frac{k_d D(1-(1+WACC)^{-n})}{p_2\left((1+WACC)^{1/p_2} - 1\right)} - \frac{D}{(1+WACC)^n}$$

For NPV calculation, we will use the following parameters: $k_O = 0.22$; $k_d = 0.14$; $t = 20\%$; $L = 0; 1; 2; \ldots; 10$; $p_1 = p_2 = p = 1; 6; 12$; $S = 1000$; $D = LS$; $NOI = 800$; $n = 3$.

Dependence of NPV on leverage level L at $k_O = 0.22$, $k_d = 0.14$ and different $p = 1, 6, 12$ for a three-year project is shown in Table 17.5 and Figs. 17.8 and 17.9.

17.5.5 *The Effectiveness of the Arbitrary Duration Investment Project from the Perspective of the Owners of Equity and Debt*

For the arbitrary duration investment projects, NPV in the case of arbitrary frequency of payment of tax on profit p is described by formula (17.28):

Table 17.5 Dependence of NPV on leverage level L at $k_O = 0.22$, $k_d = 0.14$ and different $p = 1, 6, 12$ for a three-year project

L	NPV		
	$p = 1$	$p = 6$	$p = 12$
0	307	307	307
1	885	869	867
2	1438	1406	1403
3	1985	1937	1932
4	2530	2464	2458
5	3072	2991	2983
6	3615	3516	3506
7	4156	4042	4030
8	4698	4566	4553
9	5239	5091	5076
10	5780	5615	5599

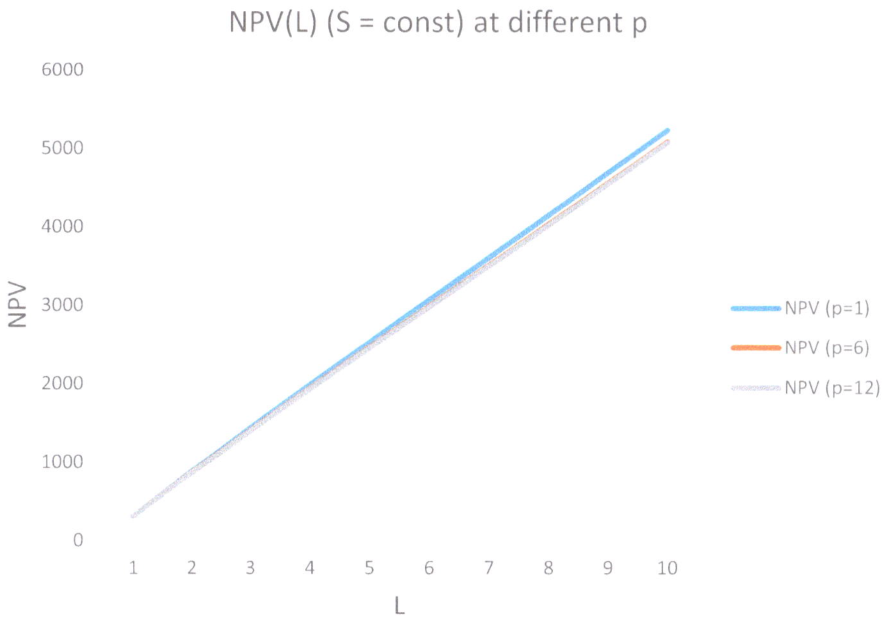

Fig. 17.8 Dependence of NPV on leverage level L at $k_O = 0.22$, $k_d = 0.14$ and different $p = 1, 6, 12$ for a three-year project

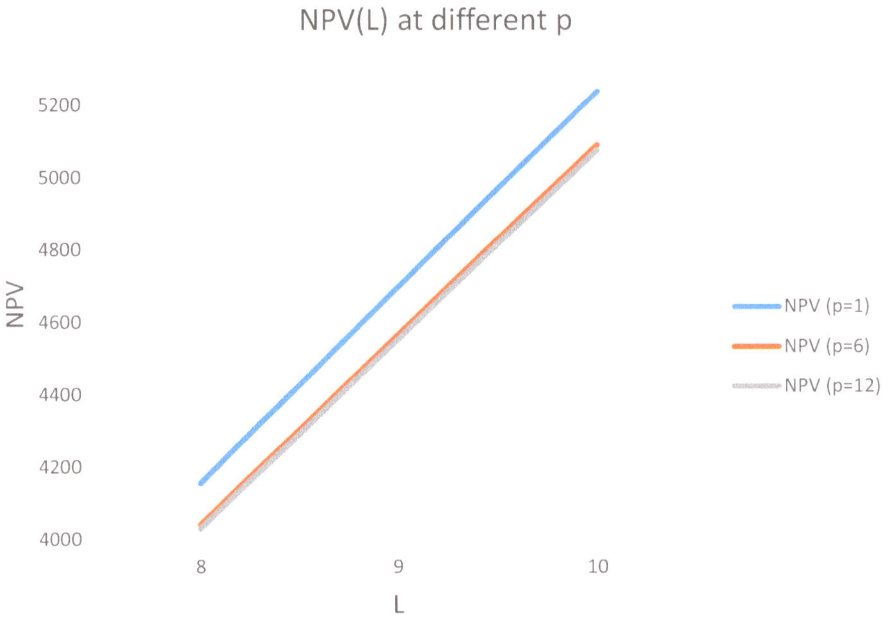

Fig. 17.9 Dependence of NPV on leverage level L at $k_0 = 0.22$, $k_d = 0.14$ and different $p = 1, 6, 12$ for a three-year project (larger scale)

$$NPV = -I + \frac{NOI(1-t)}{WACC} \cdot \left(1 - \left(\frac{1}{1+WACC}\right)^n\right) + \frac{k_d Dt(1-(1+WACC)^{-n})}{p\left((1+WACC)^{1/p} - 1\right)}$$

For NPV calculation, we will use the following parameters: $k_0 = 0.22$; $k_d = 0.14$; $t = 20\%$; $L = 0; 1; 2; \ldots; 10$; $p_1 = p_2 = p = 1; 6; 12$; $S = 1000$; $D = LS$; $NOI = 800$; $n = 3$.

Dependence of NPV on leverage level L at $k_0 = 0.22$, $k_d = 0.14$ and different $p = 1, 6, 12$ for a three-year project is shown in Table 17.6 and Figs. 17.10 and 17.11.

17.5.6 Discussions

The analysis of results from Tables 17.1, 17.2, 17.3, 17.4, 17.5, and 17.6 and Figs. 17.1, 17.2, 17.3, 17.4, 17.5, 17.6, 17.7, 17.8, 17.9, 17.10, and 17.11 shows that:

1. The weighted average cost of capital, WACC, decreases with leverage level L at all values of frequency of payments of tax on profit p (see Tables 17.1 and 17.2 and Figs. 17.1, 17.2, and 17.3).

Table 17.6 Dependence of NPV on leverage level L at $k_0 = 0.22$, $k_d = 0.14$ and different $p = 1, 6, 12$ for a three-year project

	NPV		
L	$p = 1$	$p = 6$	$p = 12$
0	307	307	307
1	761	771	772
2	1218	1238	1240
3	1676	1707	1710
4	2135	2175	2179
5	2594	2644	2649
6	3053	3113	3119
7	3511	3582	3589
8	3970	4051	4059
9	4429	4520	4529
10	4888	4989	4999

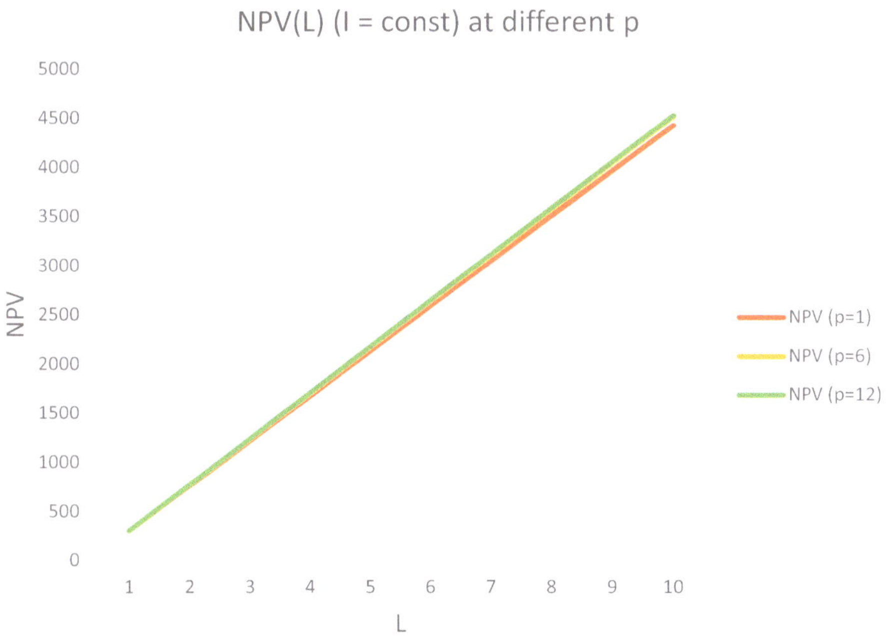

Fig. 17.10 Dependence of NPV on leverage level L at $k_0 = 0.22$, $k_d = 0.14$ and different $p = 1, 6, 12$ for a three-year project

2. With an increase in p, WACC decreases: WACC(L) curves lie lower with increase of p.
3. The difference between curves corresponding to $p = 1$ and $p = 6$ is much more than the difference between the curves corresponding to $p = 6$ and $p = 12$. This

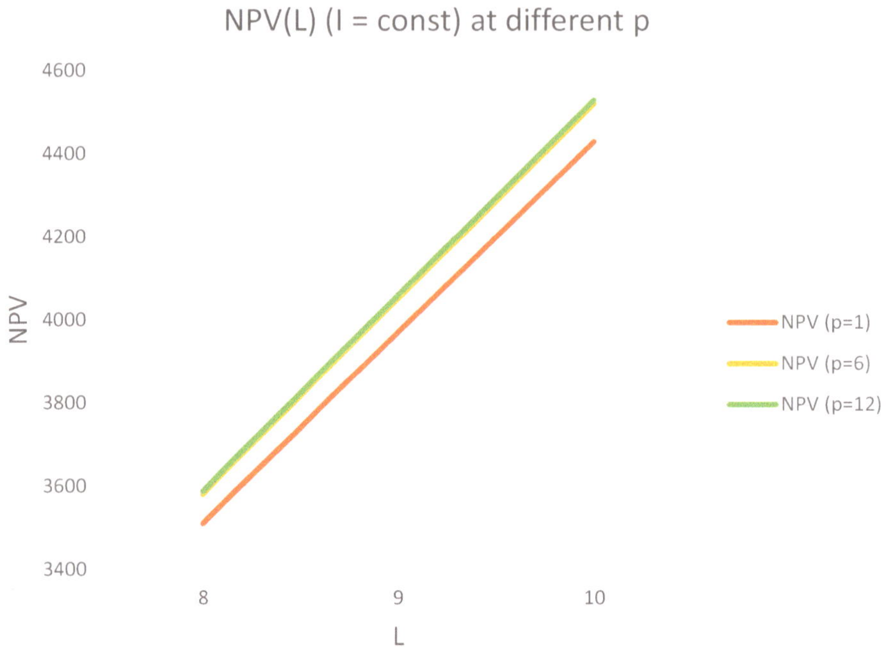

Fig. 17.11 Dependence of NPV on leverage level L at $k_0 = 0.22$, $k_d = 0.14$ and different $p = 1$, 6, 12 for a three-year project (larger scale)

difference decreases with p. It turns out that an increase in the number of payments of tax on profit per year p leads to a decrease in the cost of attracting capital (WACC). Will this decrease in the discount rate increase the effectiveness of investment projects? As we see from the Tables 17.3, 17.4, 17.5, and 17.6 and Figs. 17.4, 17.5, 17.6, 17.7, 17.8, 17.9, 17.10, and 17.11, the situation is different for owners of equity capital and for owners of equity and debt capital.

4. NPV practically linearly increases with leverage level at all values of frequency of payments of tax on profit p and all frequency of payments of interest on debt.
5. In the case of considering the effectiveness of long-term investment projects for owners of equity capital, NPV is changed with a change of p (both p_1 and p_2 are equal everywhere below) but by a very small value (it is seen in the Tables, but not in the Figures).
6. In the case of considering the effectiveness of long-term investment projects for owners of equity and debt, NPV is changed with a change of p more significantly (it is seen in the Tables and in the Figures).
7. For arbitrary duration projects, this difference in NPV with a change of p is more significant and should be accounted under valuation of the effectiveness of an investment project.

8. In the case of considering the effectiveness of an investment project for owners of equity capital, we need to note that an increase in p leads to a decrease in NPV: this means that the effectiveness of an investment project decreases with p.
9. In the case of considering the effectiveness of an investment project for owners of equity and debt capital, we need to note that the situation is opposite and an increase in p leads to increase in NPV: this means that the effectiveness of an investment project increases with p.
10. The above results show that in the former case, companies should pay tax on profit and interest on debt once per year, while in the latter case, more frequent payments are profitable for the effectiveness of an investment.
11. Thus, while for long-term projects' NPV, the impact of more frequent payments of both values p_1 and p_2 is insignificant, for arbitrary duration projects the account of the frequency of both types of payments could be important and could lead to more significant influence on the effectiveness of an investment project, decreasing it (in the former case), or increasing it (in the latter case). Note that the specific value of the effect depends on the values of the parameters in the project (k_0, k_d, n, t, S, etc.).

17.6 Conclusions

There are too few investment models which can numerically valuate the effectiveness of investment projects, among them the investment models developed by the authors of this paper. Moreover, investment models are practically absent which account for the conditions of the real functioning of investment projects. This paper covers this gap in the literature and science in the field of investments and develops innovative investment models which are much closer to economic practice.

We developed here for the first time eight innovative investment models: long-term (described by Eqs. 17.14, 17.19, 17.24 and 17.29) as well arbitrary duration (described by Eqs. 17.13, 17.18, 17.23 and 17.28), which account payments of interest on debt and of tax on income a few times per year (semi-annually, quarterly, monthly) as it happened in practice. Note that no one before had investigated the impact of the frequency of payments of taxes and of debt interest on the effectiveness of investment projects. These investment models allow for investigation of the impact of all main parameters of investment projects (equity value S, investment value I, net operating income NOI, leverage level L, profitability of investments β, tax on profit t, project duration n, equity cost k_0, debt cost k_d and number of payments of interest on debt p_1 and of tax on income p_2) on the main indicator of effectiveness of investment projects NPV (net present value). They could be used for investigation of different problems of investments, such as the influence of debt financing, leverage level, taxing, project duration, method of financing, number of payments of interest on debt and of tax on income and some other parameters on efficiency of investments and other problems. In particular, they will improve the issue of project ratings (Peter et al. 2018). These new models allow for making more correct evaluations of effectiveness of investment projects long-term as well as of arbitrary duration.

Numerical calculations, conducted for four investment models (without flow separation) show that:

- In the case of considering the effectiveness of an investment project for owners of equity capital, the increase in the number of payments of tax on income and of interest on debt p leads to an decrease in NPV: this means that the effectiveness of an investment project decreases with p;
- In the case of considering the effectiveness of an investment project for owners of equity and debt capital, the increase in the number of payments of tax on income and of interest on debt p leads to an increase in NPV: this means that the effectiveness of an investment project increases with p.

In the former case, companies should pay tax on profit and interest on debt once per year, while in the latter case, more frequent payments are profitable for the effectiveness of an investment.

Thus, while for long-term projects' NPV, the impact of more frequent payments of both values p_1 and p_2 is insignificant, for arbitrary duration projects the account of the frequency of both types of payments could be important and could lead to more significant influence on the effectiveness of an investment project, decreasing it (in the former case), or increasing it (in the latter case). Note that the specific value of the effect depends on the values of the parameters in the project (k_0, k_d, n, t, S, etc.).

There are some limitations of the applicability of the proposed models:

1. For long-term projects, they are connected with the limitations of the Modigliani–Miller theory;
2. For consideration of without flow separation, they are connected with the well-known limitations of the WACC approach;
3. For arbitrary duration projects (using BFO theory), they are connected with the fact that not all the conditions of real investments are accounted yet.

The contribution of this study to finance and economics is mainly related to the goal that new modern investment models have been created to be as close as possible to real investment conditions. Our models, namely, models with payments of interest on debt and of tax on income which occur a few times per year (semi-annually, quarterly or monthly), could be more successfully applied in real economic practice.

References

Black, F., & Litterman, R. (1992). Global portfolio optimization. *Financial Analysts Journal, 48*, 28–43.

Bond, S., & Meghir, C. (1994). Dynamic investment models and the firm's financial policy. *Review of Economic Studies, 67*, 197–217.

Brown, K., Harlow, W., & Zhang, H. (2021). Investment style volatility and mutual fund performance. *Journal of Investment Management, 19*, 78.

Brusov, P., & Filatova, T. (2021). The Modigliani–Miller theory with arbitrary frequency of payment of tax on profit. *Mathematics, 9*, 1198. https://doi.org/10.3390/math9111198

Brusov, P., Filatova, T., Orehova, N., & Eskindarov, M. (2015). *Modern corporate finance, investments and taxation* (pp. 1–373). Springer International Publishing. Retrieved April 29, 2021, from https://www.springer.com/gp/book/9783319147314

Brusov, P., Filatova, T., Orehova, N., & Eskindarov, M. (2018). *Modern corporate finance, investments, taxation and ratings* (2nd ed., pp. 1–571). Springer Nature Publishing.

Cakici, N., & Zaremba, A. (2021). Size, value, profitability, and investment effects in international stock returns: Are they really there? *Journal of Investing, 30*, 65–86. https://doi.org/10.3905/joi.2021.1.176

Dammon, R., & Senbet, L. (1988). The effect of taxes and depreciation on corporate investment and financial leverage. *The Journal of Finance, 43*, 357–373. https://doi.org/10.1111/j.1540-6261.1988.tb03944.x

DeAngelo, H., & Masulis, R. (1980). Optimal capital structure under corporate and personal taxation. *Journal of Financial Economics, 8*, 3–29.

Dotan, A., & Ravid, S. (1985). On the interaction of real and financial decisions of the firm under uncertainty. *The Journal of Finance, 40*, 501–517.

Escobar-Anel, M., Lichtenstern, A., & Zagst, R. (2020). Behavioral portfolio insurance strategies. *Financial Markets and Portfolio Management, 34*, 353–399.

Filatova, T. V., Orekhova, N. P., & Brusova, A. P. (2008). Weighted average cost of capital in the theory of Modigliani–Miller, modified for a finite life-time company. *Bulletin of the FU, 48*, 68–77.

Gidwani, B. (2020). Some issues with using ESG ratings in an investment process. *Journal of Investing, 29*, 76–84. https://doi.org/10.3905/joi.2020.1.147

Kahneman, D., & Tversky, A. (1996). On the reality of cognitive illusions. *Psychological Review, 103*, 582–591. https://doi.org/10.1037/0033-295X.103.3.582

Kang, J. (1995). The conditional relationship between financial leverage and corporate investment: Further clarification. *Journal of Business Finance and Accounting, 22*, 1211–1219. https://doi.org/10.1111/j.1468-5957.1995.tb00902.x

Lang, L. E., Ofek, E., & Stulz, R. (1996). Leverage, investment, and firm growth. *Journal of Financial Economics, 40*, 3–29.

Loviscek, A. (2021). How much to invest, if any, in precious metals? A 21st-century perspective. *Journal of Investing, 30*, 95–111. https://doi.org/10.3905/joi.2021.1.179

Modigliani, F., & Miller, M. (1958). The cost of capital, corporate finance, and the theory of investment. *The American Economic Review, 48*, 261–297.

Modigliani, F., & Miller, M. (1963). Corporate income taxes and the cost of capital: A correction. *The American Economic Review, 53*, 147–175.

Modigliani, F., & Miller, M. (1966). Some estimates of the cost of capital to the electric utility industry 1954–1957. *The American Economic Review, 56*, 333–391.

Peter, B., Tatiana, F., & Natali, O. (2018). *Ratings: Critical analysis and new approaches of quantitative and qualitative methodology* (pp. 1–379). Springer Nature Publishing. Retrieved April 29, 2021, from https://www.springer.com/de/book/9783030562427

Rosenbaum, J., & Pearl, J. (2013). *Investment banking valuation models* (pp. 1–342). Wiley.

Whited, T. (1992). Debt, liquidity constraints and corporate investment: Evidence from panel data. *The Journal of Finance, 47*, 1425–1461.

Chapter 18
Investment Models with Advance Frequent Payments of Tax on Profit and of Interest on Debt

18.1 Introduction

Investments are quite important in economy and finance of any country. At the current stage the role of investment is increased many times as well as the role of the evaluation of the efficiency of investment projects, which allows for the realization of the most effective projects in the context of scarcity and limited investment resources, increases. Since virtually most investment projects use debt financing, the study of the influence of capital structure and debt financing on the efficiency of investment projects and determining the optimal capital structure is especially important at the present time. This is why in spite of the fact that a lot of different types of investment models have been developed—stochastic, dynamics (Bond and Meghir 1994), investment banking valuation models (Rosenbaum and Pearl 2013), etc.—the main problem, which has been discussed during the last few decades, is the impact of debt financing on the efficiency of investment projects and on the investment decisions of companies.

Recently (Brusov et al. 2021) new modern investment models have been created by us to be as close as possible to real investment conditions. We developed long-term as well as arbitrary duration models with payments of interest on debt and of tax on income a few times per year (semi-annually, quarterly and monthly), which could be applied in real economic practice. But in practice payments of tax on profit and of interest on debt could be made in advance. In current chapter we develop the innovative investment models with frequent **advance** payments of tax on profit and of interest on debt and study the effectiveness of an investment projects within them.

We considered the effectiveness of the investment project from two points of view: the equity owners and the owners of equity and debt. NPV for each of these cases was calculated by two different methods: with the separation of credit and investment flows (and thus discounting the flows using two different rates) and without such separation (with discounting of both flows using the same rate, and the

P. Brusov, T. Filatova, *Efficiency of Investment Projects*, Contributions to Finance and Accounting, https://doi.org/10.1007/978-3-032-01028-5_18

weighted average cost of capital, WACC, can be chosen as the rate). Numerical calculations, conducted for four investment models (without flow separation) showed that: (1) WACC decreases with p; (2) in the case of considering the effectiveness of an investment project for owners of equity capital, the increase in the number of payments of tax on income and of interest on debt p leads to a decrease in NPV: this means that the effectiveness of an investment project decreases with p; (3) in the case of considering the effectiveness of an investment project for owners of equity and debt capital, the increase in the number of payments of tax on income and of interest on debt p leads to an increase in NPV: this means that the effectiveness of an investment project increases with p. In the former case, companies should pay tax on profit and interest on debt once per year, while in the latter case, more frequent payments are profitable for the effectiveness of investment.

But in practice payments of tax on profit and of interest on debt could be made in advance. In current chapter we develop the innovative investment models with frequent **advance** payments of tax on profit and of interest on debt and study the effectiveness of an investment projects within them.

Numerical calculations carried out for four investment models (without split flows) showed that in the case of **advance** payments of income tax and interest on debt, all the results related to the effect of the number of payments of income tax and interest on debt on the effectiveness of the investment projects are opposite to the results in the case of payments at the end of the periods obtained by us in the previous article: (1) WACC increases with p; (2) in the case of considering the effectiveness of an investment project for owners of equity capital, the increase in the number of payments of tax on income and of interest on debt p leads to a increase in NPV: this means that the effectiveness of an investment project increases with p; (3) in the case of considering the effectiveness of an investment project for owners of equity and debt capital, the increase in the number of payments of tax on income and of interest on debt p leads to the decrease in NPV: this means that the effectiveness of an investment project decreases with p. In the latter case, companies should pay tax on profit and interest on debt once per year, while in the former case, more frequent payments are profitable for the effectiveness of investment.

Thus, this means that method of payments of tax on income and of interest on debt (in advance or in the ends of periods) changes drastically the effect of the number of payments of income tax and interest on debt on the effectiveness of investments. Depending on the relationship between the investor and the lender, the investor can choose (within the framework of tax legislation):

- the frequency of payment of income tax;
- scheme for payment of income tax (in advance or at the end of periods).

The verification of developed by us the innovative investment models with frequent **advance** payments of tax on profit and of interest on debt will lead to the creation of a comprehensive system of adequate and correct assessment of the effectiveness of the company's investment program and its investment strategy for both schemes for payment of income tax (in advance or at the end of periods).

One of the most important elements of calculating the effectiveness of investment projects is the assessment of the discount rate, the calculation methods of which are generalized for the real conditions of the implementation of investment projects. Eight innovative investment models with **advance** payments of tax on profit and of interest on debt created in this paper can assist decision-makers in the optimal design, planning and control of company investments and the development of a company's investment strategy.

Obtained in current chapter and paper (Brusov et al. 2021) results help to tax regulator (Finance Ministry) understand the influence of the number of payments of tax on income and to credit regulator (Central Bank) understand the influence of the number of payments of interest on debt on the effectiveness of investment projects. These allow both regulators modify and improve tax legislation and credit policy respectively.

Comprehensive literature review of investment models has been done in (Brusov et al. 2021). Below we shortly summarize it and add review of a few recent papers devoted to investigation of frequent payments of tax on income as well as advance payments of tax on income.

18.1.1 The Literature Review

Lang et al. (1992) have shown that there is a negative correlation between future growth and leverage level at the company level and, for diversified companies, at the level of business segments. This negative correlation between growth and leverage level holds for companies with low Tobin's q ratio, but not for high-q companies or companies in high-q industries (the q ratio, or Tobin's q ratio, equals the market value of a company divided by its assets' replacement cost). Therefore, for companies known to have good investment opportunities, leverage does not reduce growth, but it is negatively correlated to growth for companies whose growth opportunities are either not recognized by the capital markets or are not valuable enough to overcome the effects of their debt load.

Whited (1992) studied the influence of debt financing on companies' investment decisions and showed that there is a significant positive correlation between leverage level and investment for large companies, while for medium companies a negative correlation between leverage level and investment took place.

Kang (1995) studied the connection between leverage level and investment decisions. "Interdependent tax models" were used to try to explain the specifics of company leverage levels by analyzing the interdependency between financing decisions and investment.

Some authors mentioned a positive connection via the fact that the financial risk and hence the cost of bond financing decrease with an increase in investment at a given leverage level. A negative connection has been mentioned by DeAngelo and Masulis (1980) and Dotan and Ravid (1985). The first authors concluded this since the tax benefits of debt compete with those of capital investment. The second authors refer to the fact that financial risk and thus the cost of bond financing will increase with investment increase.

The impact of investment increase on financial risk may depend on company-specific factors, like company-specific technology (Dammon and Senbet 1988). An analysis of the impact of corporate and personal taxes on a firm's optimal investment and financing decisions under uncertainty is provided in Dammon and Senbet (1988). By endogenizing firms' investment decisions, it extends the DeAngelo and Masulis capital structure model. The authors' results indicate that the existing predictions about the relationship between investment-related and debt-related tax shields must be modified in cases where investment is allowed to adjust optimally. Differences in production technologies across companies may thus explain why the empirical results of recent cross-sectional studies have not conformed to the predictions of DeAngelo and Masulis (1980).

Below we discuss some portfolio investment models as well as behavioral aspects of investors, which play an important role in investments.

Among those portfolio investment models is the well-known Black–Litterman model, which was created by Black and Litterman (1992). They developed the model to address the problems that institutional investors have encountered the during application of modern portfolio theory in practice. Starting with an asset allocation based on the equilibrium assumption, the model then modifies that allocation by accounting the investors' opinions with respect to future asset performance.

Loviscek (2021) has applied the Black–Litterman model of modern portfolio theory to well-known index mutual funds—one guided by the classic 60%/40% stock/bond allocation and one based on an all-equity allocation. The period under their study is from 2000 to 2020. Although statistical evidence supports that the efficacy of a precious metal allocation is elusive, the results suggest an average allocation of about 2% for "buy-and-hold" investors who seek one. He has shown that, from 2003 to 2010 and from 2016 to 2020, the allocations were in the range of 5–10%. Other periods, however, register only a little more than 0%.

Cakici and Zaremba (2021) reconsidered the performance of the Fama–French factors in global markets. As their results showed, the value, profitability and investment factors are far less reliable than what is commonly thought. Their performance depends strongly on the geographical regions and periods examined. Moreover, most factor returns are driven by the smallest companies. Virtually no value or investment effects are present among the big companies representing most of the total market capitalization worldwide. These results cast doubt on the five-factor model's applicability in international markets, citing that the smallest companies are typically not invested in by major financial institutions.

A growing number of investors want to use firm sustainability information in their investment decision processes to avoid risk, satisfy their own asset preference, or find a new alpha-generating factor. Not too many users of environment, social and governance (ESG) data understand how ESG ratings change over time. Gidwani (2020) used the CSRHub data set to show that ESG ratings regress strongly toward the mean. The observed regression persists during 9 years within the ratings data, for a sample set of more than 8000 firms. Newly-rated firms show even more reversion than "seasoned" firms. Firms can only rarely maintain an especially high or low ESG rating. Investors and firm managers should understand that ESG ratings are likely to change toward the mean. This however does not mean that a good firm is getting worse or a bad one is getting better.

Brown et al. (2021) have developed statistics (holdings-based) to estimate the volatility with time of investment style characteristics of funds. They found that funds with lower levels of style volatility significantly outperform funds with higher levels of style volatility on a risk-adjusted basis. It was concluded that deciding to maintain a less volatile investment style is an important aspect of the portfolio management process.

Some behavioral aspects of investors have been considered by Escobar-Anel et al. (2020), who introduced a strategy generalizing the CPPI (Constant Proportion Portfolio Insurance) approach. The target of this strategy is to guarantee the investment goal or floor during participation in the performance of the assets and limiting the downside risk of the portfolio at the same time. The authors show that the strategy accounts for the following behavioral aspects of investors: a risk-averse behavior for gains, distorted probabilities recognition and a risk-seeking behavior for losses. The developed strategy turns out to be optimal within the Cumulative Prospect Theory framework by Kahneman and Tversky (1996).

In current paper we study the innovative investment models with advance frequent payments of tax on income. Thus below we add review of a few recent papers devoted to investigation of frequency of payments of tax on income as well as advance payments of tax on income.

Brusov et al. (2021) discuss one of the additional limitation of Modigliani–Miller theory: a method of tax on profit payments. Modigliani–Miller theory accounts these payments as annuity-immediate while in practice these payments are making in advance and thus should be accounted as annuity-due.

Brusov et al. (2021) have generalized the Modigliani–Miller theory for the case of advance payments of tax on profit, which is widely used in practice, and show that this leads to some important consequences, which change seriously all the main statements by Modigliani and Miller. These consequences are as following: WACC starts to depend on debt cost k_d, WACC turns out to be lower than in case of classical Modigliani–Miller theory and thus company capitalization becomes higher than in ordinary Modigliani–Miller theory. We show that dependence of equity cost on leverage level L is still linear, but the tilt angle with respect to L-axis turns out to be smaller: this could lead to modification of the dividend policy of the company.

Correct account of a method of tax on profit payments demonstrates that shortcomings of Modigliani–Miller theory are dipper, than everybody suggested: the underestimation of WACC really turns out to be bigger, as well as overestimation of the capitalization of the company. This means that systematic risks arising from the use of modified Modigliani–Miller theory (MMM theory) (which is more correct than "classical" one) in practice are higher than it was suggested by the "classical" version of this theory.

Brusov et al. (2020b) use the modified Modigliani–Miller theory (MMM theory) and apply it for rating methodologies needs. A serious modification of MMM theory in order to use it in rating procedure has been required. The financial "ratios" were incorporated into MMM theory. The dependence of the weighted average cost of capital (WACC), which plays the role of discount rate, on coverage and leverage ratios is analyzed.

Obtained results make possible to use the power of the MMM theory in the rating and create a new base for rating methodologies.

Brusov et al. (2021) the Modigliani–Miller theory has been generalized and developed taking into account one of the conditions of the real functioning of companies for the case of paying income tax with an arbitrary frequency (monthly, quarterly, semi-annual or annual payments). While a return is not required more than once a year, businesses may be responsible for filing estimated taxes based on profits earned. This requirement is dependent on showing a profit. For example, sole proprietors must file estimated taxes on profits quarterly, on the 15th day of April, June, September and January. In Russia, tax on profit payments could be made annually, quarterly, or monthly. Authors suppose, that more frequent payment of income tax impacts on all main financial indicators of the company and leads to some important consequences. They used analytical and numerical methods: all main formulas of the modified Modigliani–Miller theory have been derived theoretically and then used to obtain all main financial indicators of company and their dependences on different parameters by MS Excel. Authors show that: (1) all Modigliani–Miller theorems, statements and formulas change; (2) all main financial indicators, such as the weighted average cost of capital (*WACC*), company value, V, and equity cost, k_e, depend on the frequency of tax on profit payments; (3) in the case of income tax payments more than once per year (at $p \neq 1$), as takes place in practice, the *WACC*, company value, V and equity cost, and k_e start depend on debt cost, k_d, while in ordinary (classical) Modigliani–Miller theory all these values do not depend on k_d; (4) obtained results allow a company to choose the number of payments of tax on profit per year (of course, within actual tax legislation): more frequent payments of income tax are beneficial for both parties, for the company and for the tax regulator.

As we mentioned in Brusov et al. (2021) some of the major problems under the evaluation of the effectiveness of investment projects are suggested as follows:

1. Which financial flows should be taken into account when calculating the parameters of efficiency of a project (NPV, IRR, etc.)?
2. How many discount rates should be used for discounting various cashflows?
3. How can these discount rates be accurately evaluated?

The first problem is still under intensive discussion. Concerning the second issue either two different discount rates could be used (equity cost, k_e and debt cost, k_d) or a general discount rate for operating and financial flows (for which WACC can, obviously, be chosen).

Concerning the third issue, we need to note that, in the last decade, significant progress in the accurate determination of the cost of the equity and company weighted average cost, which just are the discount rates when evaluating the effectiveness of the project, has been achieved. The progress is mainly associated with the studies by Brusov, Filatova and Orekhova (BFO theory) (Brusov et al. 2015, 2018a; Filatova et al. 2008), in which a general theory of capital cost of the company and its capital structure was established, and the dependence of capital cost on leverage level and on the age of a company was found for the companies of arbitrary age. The main difference between their theory and Modigliani–Miller theory is that the former one removes the assumption of perpetuity for the companies under discussion, which leads to a significantly different new theory from the theory established by the Nobel laureates Modigliani and Miller (1958, 1963, 1966).

In modern conditions, the requirements for improving the quality of assessing the effectiveness of investments have increased. The modern investment models, which have been well-tested in real economic situations, have been developed by Brusov, Filatova and Orekhova (Brusov et al. 2015, 2018a; Filatova et al. 2008). They have created long-term as well as arbitrary duration models and have considered the effectiveness of the investment project from two points of view: from the equity holders and from the owners of equity and debt. NPV in each of these cases could be calculated by two different methods: with the division of credit and investment flows and using two different discounting rates (equity cost, k_e and debt cost, k_d), and without such a division and using a general discounting rate (for which WACC can, obviously, be chosen).

18.1.2 The Discount Rates

As we mentioned above, one of the most important elements of calculating the effectiveness of investment projects is the assessment of the discount rate. In the case of long-term investment models without the division of credit and investment flows, the discount rate WACC has been calculated using the modified for advance payments Modigliani–Miller formula (Modigliani and Miller 1958, 1963, 1966)

$$WACC = k_0 \cdot (1 - w_d t \cdot (1 + k_d)) \tag{18.1}$$

while in the case of arbitrary duration models, the modified for advance payments Brusov-Filatova-Orekhova formula for WACC (Brusov et al. 2015, 2018a; Filatova et al. 2008)

$$\frac{1 - (1 + WACC)^{-n}}{WACC} = \frac{1 - (1 + k_0)^{-n}}{k_0 \cdot (1 - w_d t \cdot [1 - (1 + k_d)^{-n}] \cdot (1 + k_d))} \tag{18.2}$$

has been used.

Note, that both formulas for the weighted average cost of capital, WACC (Modigliani–Miller (18.1) and Brusov–Filatova–Orekhova (18.2)) in case of advance payments are different from ones in case of payments in the ends of periods (Modigliani and Miller 1958, 1963, 1966; Brusov et al. 2015, 2018a; Filatova et al. 2008).

Here and below, WACC is the weighted average cost of capital; k_0 is the equity cost at zero leverage ($L = 0$); k_d is the debt cost; w_d is the debt share; t is the tax on profit; n is the project duration; ke is the equity cost; and L is the leverage level.

In the case of long-term investment models with the division of credit and investment flows, the discount rate for discounting the investment flows (equity cost ke) has been calculated using the Modigliani–Miller formula (Modigliani and Miller 1958, 1963, 1966)

$$k_e = k_0 + L \cdot (k_0 - k_d)(1 - t) \tag{18.3}$$

while in the case of arbitrary duration models, equity cost ke has been calculated from the formula

$$WACC = k_e w_e + k_d w_d (1 - t) \tag{18.4}$$

using the Brusov-Filatova-Orekhova value for WACC (18.2) (Brusov et al. 2015, 2018a; Filatova et al. 2008).

The calculation methods of the discount rates (WACC, equity cost k_e) has been generalized for the real conditions of the implementation of investment projects: for arbitrary frequency of payments of tax on profit (Brusov and Filatova 2021) and for advance payments of tax on profit (Brusov et al. 2020a).

In this paper, new modern investment models, both long-term and arbitrary duration, will be created, as close as possible to real investment conditions. They will account the **advance** payments of interest on debt and of tax on income a few times per year (semi-annually, quarterly, monthly), which are applied in real economic practice. Their verification will lead to the creation of a comprehensive system of adequate and correct assessment of the effectiveness of the company's investment program and its investment strategy.

18.1.3 The Structure of the Chapter

The structure of the paper is as follows:

1. In Sect. 18.1 above, we presented

    1.1 The literature review.
    1.2 Discount rates.

2. In Sect. 18.2, we consider the effectiveness of the investment project from the perspective of the equity holders only.

   2.1. In Sect. 18.2.1, we consider a case with flow separation.
   2.2. In Sect. 18.2.2, we consider a case without flow separation.

3. In Sect. 18.3, we consider the effectiveness of the investment project from the perspective of the owners of equity and debt.

   3.1. In Sect. 18.3.1, we consider a case with flow separation.
   3.2. In Sect. 18.3.2, we consider a case without flow separation.

4. In Sect. 18.4, the problem of calculation of discount rates is discussed and expressions for their modified values are obtained.
5. In Sect. 18.5, we study numerically with the use of Microsoft Excel the effectiveness of the four models, created by us in this paper. We consider long-term projects as well as projects of arbitrary duration from two perspectives: the owners of equity and debt and the equity holders only without the division of credit and investment flows.
6. In Sect. 18.6, we discuss obtained results and their impact on the correctness of valuation of efficiency of investment projects. As well we compare obtained in current paper results for advance payments of tax on profit with ones, obtained by Brusov et al. (2021) for payments of tax on profit at the end of periods.

18.2 The Effectiveness of the Investment Project from the Perspective of the Equity Holders Only

We will consider the effectiveness of the investment project from two points of view: from the equity holders and from the owners of equity and debt. NPV in each of these cases could be calculated by two different methods: with the division of credit and investment flows and using two different discounting rates, and without such a division and using a general discounting rate (for which WACC can, obviously, be chosen).

The following designations are used below:

The equity value S, the investment value I, the net operating income NOI, the leverage level L, the profitability of investments β, the tax on profit t, the project duration n, the equity cost k_0, the debt cost k_d and the number of payments of interest on debt p_1 and of tax on income p_2; D is the debt value.

In the first case (from the perspective of the equity holders), at the initial moment in time, $T=0$ investments are equal to $-S$ and the flow of capital, CF, for the period is equal to:

$$\mathrm{CF}=(\mathrm{NOI}-k_\mathrm{d}D)(1-t). \tag{18.5}$$

In addition to the tax shields, $k_\mathrm{d}Dt$ includes a payment of interest on a loan, $-k_\mathrm{d}D$.

We suppose below that interests on the loan are paid in equal shares of $k_\mathrm{d}D$ during all periods and principal repayment is made at the end of the project.

In the second case (from the perspective of the equity and debt holders), the negative flows (the interest and duty paid by owners of equity) are returned to the project and they are exactly equal to the positive flows obtained by owners of debt capital. Thus, in this case the only effect of debt financing is the effect of the tax shield, generated from the tax relief: interest on the loan is entirely included in the cost and thus reduces the tax base. After-tax flow of capital, CF, for each period is equal to

$$\mathrm{CF} = \mathrm{NOI}(1 - t) + k_{\mathrm{d}}Dt \tag{18.6}$$

and the value of investments at the initial moment in time $T = 0$ is equal to $-I = -S - D$.

Below two different ways of discounting will be considered:

1. If operating and financial flows are not separated, both flows are discounted by the general rate. In this case, the weighted average cost of capital (WACC) can be selected as a discounting rate. For long-term projects, we will use the Modigliani–Miller formula for WACC (Modigliani and Miller 1958, 1963), modified by us for the case of payments of interest on debt and of tax on income a few times per year (semi-annually, quarterly, monthly) and for projects of finite (arbitrary) duration we will use the Brusov–Filatova–Orekhova formula for WACC (Brusov et al. 2018a; Filatova et al. 2008), modified by us for the case of payments of interest on debt and of tax on income a few times per year (semi-annually, quarterly, monthly).
2. Operating and financial flows are separated and are discounted at different rates: the operating flow at the rate which is equal to the equity cost k_{e}, depending on leverage and on the number of payments of interest on debt and of tax on income, and credit flow at the rate which is equal to the debt cost k_{d}.

Note that loan capital is the least risky because the credit (including the interest on credit) is paid after taxes in the first place. Therefore, the cost of credit will always be less than the equity cost, whether for ordinary or for preference shares $k_{\mathrm{e}} > k_{\mathrm{d}}$; $k_{\mathrm{p}} > k_{\mathrm{d}}$. Here k_{e}; k_{p} is the equity cost of ordinary or of preference shares consequently.

18.2.1 With Flow Separation

In this case, the expression for NPV (net present value) per period has a view

$$NPV = NOI(1 - t) - k_d D(1 - t) = NOI(1 - t) + k_d Dt - k_d D \tag{18.7}$$

Here, the first term is the value of operating income from the investment project after tax deduction, the second term is the value of the tax shield, the third term is the value of interest on debt for one period.

We will need the following auxiliary formulas for summing the reduced values of financial flows (18.7) when calculating NPV:

For annual advance payments of interest on debt and of tax on income:

$$\sum_{i=0}^{n-1} \frac{1}{(1+k_d)^i} = \frac{1-(1+k_d)^{-n}}{1-\frac{1}{1+k_d}} = \frac{1+k_d}{k_d} \cdot (1-(1+k_d)^{-n}) \tag{18.8}$$

For more frequent advance payments (p times per period (per year)) of interest on debt and of tax on income (semi-annually, quarterly, monthly):

$$\sum_{i=0}^{np-1} \frac{R}{p(1+k_d)^{i/p}} = \frac{R}{p} \cdot \frac{1-(1+k_d)^{-n}}{1-\frac{1}{(1+k_d)^{1/p}}} = \frac{R}{p} \cdot \frac{(1-(1+k_d)^{-n}) \cdot (1+k_d)^{1/p}}{(1+k_d)^{1/p}-1} \tag{18.9}$$

Similar formulas are obtained using the cost of equity ke and WACC as the discount rates.

Summing up the given values of financial flows for each period (18.7), we obtain for NPV of n-years project in the case of separated flows:

$$NPV = -S + \sum_{i=1}^{n} \frac{NOI(1-t)}{(1+k_e)^i} + \sum_{i=0}^{n-1} \frac{k_d Dt}{(1+k_d)^i} - \sum_{i=0}^{n-1} \frac{k_d D}{(1+k_d)^i} - \frac{D}{(1+k_d)^n} \tag{18.10}$$

Here, the second term is the reduced value of operating income from the investment project, the third term is the reduced value of the tax shield, the fourth term is the reduced value of interest paid annually (at the beginning of the year), and the fifth term is the reduced value of the debt paid at the end of the project.

After summing, we have the following expression for NPV:

$$NPV = -S + \frac{NOI(1-t)}{k_e} \cdot \left(1 - \left(\frac{1}{1+k_e}\right)^n\right) + Dt(1-(1+k_d)^{-n}) \cdot (1+k_d) - D(1-(1+k_d)^{-n}) \cdot (1+k_d) - \frac{D}{(1+k_d)^n} \tag{18.11}$$

In the case of more frequent (p-times per year) advance payment of income taxes (p_1) and frequent advance payments of interest on debt (p_2) we have:

$$NPV = -S + \sum_{i=1}^{n} \frac{NOI(1-t)}{(1+k_e)^i} + \sum_{i=0}^{np-1} \frac{k_d Dt}{p_1(1+k_d)^{i/p_1}} - \sum_{i=01}^{np-1} \times \frac{k_d D}{p_2(1+k_d)^{i/p_2}} - \frac{D}{(1+k_d)^n} \tag{18.12}$$

After summing, we have the following expression for NPV:

$$NPV = -S + \frac{NOI(1-t)}{k_e} \cdot \left(1 - \left(\frac{1}{1+k_e}\right)^n\right) + \frac{k_d Dt(1-(1+k_d)^{-n}) \cdot (1+k_d)^{1/p_1}}{p_1\left((1+k_d)^{1/p_1} - 1\right)} - \frac{k_d D(1-(1+k_d)^{-n}) \cdot (1+k_d)^{1/p_2}}{p_2\left((1+k_d)^{1/p_2} - 1\right)} - \frac{D}{(1+k_d)^n} \tag{18.13}$$

18.2.1.1 Long-Term Investment Projects

To obtain the expression for NPV of long-term investment projects, one should find the limit of (18.13) $n \to \infty$.

$$NPV = -S + \frac{NOI(1-t)}{k_e} + \frac{k_d Dt \cdot (1+k_d)^{1/p_1}}{p_1\left((1+k_d)^{1/p_1} - 1\right)} - \frac{k_d D \cdot (1+k_d)^{1/p_2}}{p_2\left((1+k_d)^{1/p_2} - 1\right)} \tag{18.14}$$

18.2.2 Without Flow Separation

In the case of no separation between operating and financial flows, both flows are discounted by the general rate WACC. The credit reimbursable at the end of the project can be discounted either at the same rate WACC or at the debt cost rate k_d. Below, a uniform rate for WACC has been used.

Summing up the given values of financial flows for each period (18.7), we obtain for NPV of n-years project in the case without separated flows:

$$NPV = -S + \sum_{i=1}^{n} \frac{NOI(1-t)}{(1+WACC)^i} + \sum_{i=0}^{n-1} \frac{k_d Dt}{(1+WACC)^i} - \sum_{i=0}^{n-1} \times \frac{k_d D}{(1+WACC)^i} - \frac{D}{(1+WACC)^n} \tag{18.15}$$

After summing, we have the following expression for NPV:

$$\begin{aligned} NPV = & -S + \frac{NOI(1-t)}{WACC} \cdot \left(1 - \left(\frac{1}{1+WACC}\right)^n\right) \\ & + \frac{k_d Dt(1-(1+WACC)^{-n}) \cdot (1+WACC)}{WACC} \\ & - -\frac{k_d D(1-(1+WACC)^{-n}) \cdot (1+WACC)}{WACC} - \frac{D}{(1+WACC)^n} \end{aligned} \tag{18.16}$$

In the case of more frequent (p-times per year) advance payment of income taxes and frequent payments of interest on the debt we have:

$$\begin{aligned} NPV = & -S + \sum_{i=1}^{n} \frac{NOI(1-t)}{(1+WACC)^i} + \sum_{i=0}^{np-1} \frac{k_d Dt}{p_1 (1+WACC)^{i/p_1}} - \\ & - \sum_{i=0}^{np-1} \frac{k_d D}{p_2 (1+WACC)^{i/p_2}} - \frac{D}{(1+WACC)^n} \end{aligned} \tag{18.17}$$

After summing, we have the following expression for NPV:

$$\begin{aligned} NPV = & -S + \frac{NOI(1-t)}{WACC} \cdot \left(1 - \left(\frac{1}{1+WACC}\right)^n\right) + \frac{k_d Dt(1-(1+WACC)^{-n}) \cdot (1+WACC)^{1/p_1}}{p_1\left((1+WACC)^{1/p_1} - 1\right)} - \\ & - \frac{k_d D(1-(1+WACC)^{-n}) \cdot (1+WACC)^{1/p_2}}{p_2\left((1+WACC)^{1/p_2} - 1\right)} - \frac{D}{(1+WACC)^n} \end{aligned} \tag{18.18}$$

18.2.2.1 Long-Term Investment Projects

To obtain the expression for NPV of long-term investment projects, one should find the limit of (18.18) $n \to \infty$.

$$NPV = -S + \frac{NOI(1-t)}{WACC} + \frac{k_d Dt(1+WACC)^{1/p_1}}{p_1\left((1+WACC)^{1/p_1} - 1\right)} - \frac{k_d D(1+WACC)^{1/p_2}}{p_2\left((1+WACC)^{1/p_2} - 1\right)} \tag{18.19}$$

18.3 The Effectiveness of the Investment Project from the Perspective of the Owners of Equity and Debt

18.3.1 With Flow Separation

In this case, operating and financial flows are separated and are discounted using different rates: the operating flow at the rate equal to the equity cost k_e, depending on leverage, and credit flow at the rate equal to the debt cost k_d, which remain constant until fairly large values of leverage and start to grow only at high values of leverage L, when there is a danger of bankruptcy.

After-tax flow of capital for each period in this case is equal to (18.6):

$$CF = NOI(1-t) + k_d Dt \tag{18.20}$$

After summing over n-periods, one gets the following expression for NPV:

$$NPV = -I + \sum_{i=1}^{n} \frac{NOI(1-t)}{(1+k_e)^i} + \sum_{i=0}^{n-1} \frac{k_d Dt}{(1+k_d)^i} = -I + \frac{NOI(1-t)}{k_e} \times \left(1-(1+k_e)^{-n}\right) + Dt\left(1-(1+k_d)^{-n}\right)\cdot(1+k_d) \tag{18.21}$$

In the case of a few (p) advance payments of tax on profit per period, the expression for NPV will be modified:

$$NPV = -I + \sum_{i=1}^{n} \frac{NOI(1-t)}{(1+k_e)^i} + \sum_{i=0}^{np-1} \frac{k_d Dt}{p(1+k_d)^{i/p}} \tag{18.22}$$

After summing over n-periods, one gets the following modified expression for NPV:

$$NPV = -I + \frac{NOI(1-t)[1-(1+k_e)^{-n}]}{k_e} + \frac{k_d Dt[1-(1+k_d)^{-n}](1+k_d)^{1/p}}{p\left[(1+k_d)^{1/p}-1\right]} \tag{18.23}$$

To obtain the expression for NPV of long-term investment projects, one should find the limit of (18.23) $n \to \infty$.

$$NPV = -I + \frac{NOI(1-t)}{k_e} + \frac{k_d Dt(1+k_d)^{1/p}}{p\left[(1+k_d)^{1/p}-1\right]} \tag{18.24}$$

18.3.2 Without Flow Separation

In the case of no separation between operating and financial flows, both flows are discounted by the general rate WACC.

After summing the expression (18.6) over n-periods, one gets the following expression for NPV:

$$NPV = -I + \sum_{i=1}^{n} \frac{NOI(1-t)}{(1+WACC)^i} + \sum_{i=0}^{n-1} \frac{k_d Dt}{(1+WACC)^i} \tag{18.25}$$

$$NPV = -I + \frac{NOI(1-t)}{WACC}(1-(1+WACC)^{-n}) + \frac{k_d Dt}{WACC}(1-(1+WACC)^{-n})(1+WACC) \tag{18.26}$$

In the case of a few (p) advance payments of tax on profit per period, the expression for NPV will be modified:

$$NPV = -I + \sum_{i=1}^{n} \frac{NOI(1-t)}{(1+WACC)^i} + \sum_{i=0}^{np-1} \frac{k_d Dt}{p(1+WACC)^{i/p}} \tag{18.27}$$

After summing over n-periods, one gets the following modified expression for NPV:

$$NPV = -I + \frac{NOI(1-t)}{WACC} \cdot \left(1 - \left(\frac{1}{1+WACC}\right)^n\right) + \\ + \frac{k_d Dt(1-(1+WACC)^{-n})(1+WACC)^{1/p}}{p\left((1+WACC)^{1/p} - 1\right)} \quad (18.28)$$

To obtain the expression for NPV of long-term investment projects, one should find the limit of (18.28) $n \to \infty$.

$$NPV = -I + \frac{NOI(1-t)}{WACC} \cdot + \frac{k_d Dt(1+WACC)^{1/p}}{p\left((1+WACC)^{1/p} - 1\right)} \quad (18.29)$$

18.4 Discount Rates

In the case without the division of credit and investment flows (both flows are discounted using the same rate, for which WACC can obviously be chosen), WACC is calculated by the following formulas:

For arbitrary project duration (Brusov et al. 2018a):

$$\frac{1-(1+WACC)^{-n}}{WACC} = \frac{1-(1+k_0)^{-n}}{k_0 \cdot \left(1 - \frac{k_d w_d t}{p} \frac{[1-(1+k_d)^{-n}] \cdot (1+k_d)^{1/p}}{(1+k_d)^{1/p} - 1}\right)} \quad (18.30)$$

For long-term projects (Modigliani and Miller 1958, 1963):

$$WACC = k_0 \cdot \left(1 - \frac{k_d w_d t}{p} \frac{(1+k_d)^{1/p}}{(1+k_d)^{1/p} - 1}\right) \quad (18.31)$$

Note, that both formulas for the weighted average cost of capital, WACC (Modigliani–Miller (18.31) and Brusov-Filatova-Orekhova (18.30)) in case of advance payments, obtained by Brusov et al. (2021) are different from ones in case of payments in the ends of periods (classical Modigliani–Miller formulas and Brusov-Filatova-Orekhova formulas).

In the case of the division of credit and investment flows (and thus discounting of the payments using two different rates, equity cost ke and debt cost kd), ke should be found from the equation

$$WACC = k_e w_e + k_d w_d (1-t)$$

where we substitute WACC from the formula (18.30) for a project with arbitrary duration and from the formula (18.31) for a long-term project.

Note that formulas (18.30) and (18.31) are quite different as well from the original formulas by Brusov-Filatova-Orekhova (Brusov et al. 2015, 2018a; Filatova et al. 2008) and Modigliani and Miller (1958, 1963), where payments of interest on debt and of tax on income are made once per year at the end of the periods. They are turned into advance version of them (not original ones, which suppose the payments at the end of periods), if we put p = 1:

$$\frac{1-(1+WACC)^{-n}}{WACC}=\frac{1-(1+k_0)^{-n}}{k_0\cdot(1-w_d t\cdot[1-(1+k_d)^{-n}]\cdot(1+k_d))} \tag{18.32}$$

$$WACC=k_0\cdot(1-w_d t\cdot(1+k_d)) \tag{18.33}$$

18.5 Results and Discussions

In this section, we study numerically with the use of Microsoft Excel the effectiveness of the four models created above by us. We consider long-term projects as well as projects of arbitrary duration from two perspectives: the owners of equity and debt and the equity holders only. We will study the case without flow separation. In this case, operating and financial flows are not separated and both are discounted, using the general rate, for which we select the weighted average cost of capital, WACC. Let us start from the study numerically on the dependence of the discount rates (weighted average cost of capital, WACC) on leverage level L at different frequencies of payment of tax on profit p.

18.5.1 Numerical Calculation of the Discount Rates

18.5.1.1 The Long-Term Investment Projects

For long-term investment projects (the Modigliani-Miller limit), the discount rate (WACC) in the case of arbitrary frequency of advance payment of tax on profit p is described by formula (18.31) (Brusov and Filatova 2021):

$$WACC=k_0\cdot\left(1-\frac{k_d w_d t}{p}\frac{(1+k_d)^{1/p}}{(1+k_d)^{1/p}-1}\right)$$

For WACC calculation, we will use the following parameters: $k_0 = 0.22$; $k_d = 0.14$; $t = 20\%$; $L = 0; 1; 2; \ldots; 10$.

Table 18.1 Dependence of WACC on leverage level L at $k_0 = 0.22$, $k_d = 0.14$ and different p = 1, 4, 12

L	WACC		
	p = 1	p = 4	p = 12
0	0.2200	0.2200	0.2200
1	0.1949	0.1961	0.1964
2	0.1866	0.1881	0.1885
3	0.1824	0.1842	0.1845
4	0.1799	0.1818	0.1822
5	0.1782	0.1802	0.1806
6	0.1770	0.1790	0.1795
7	0.1761	0.1782	0.1786
8	0.1754	0.1775	0.1780
9	0.1749	0.1770	0.1775
10	0.1744	0.1766	0.1770

Dependence of WACC on leverage level L at k0 = 0.22, kd = 0.14 and different p = 1, 4, 12 is shown in Table 18.1 and Figs. 18.1 and 18.2.

18.5.1.2 The Arbitrary Duration Investment Projects

For arbitrary duration investment projects, the discount rate (WACC) in the case of arbitrary frequency of payment of tax on profit p is described by formula (18.30) (modified BFO formula Brusov et al. 2018a).

For WACC calculation, we will use the following parameters: $k_0 = 0.22$; $k_d = 0.14$; t = 20%; L = 0; 1; 2; ...; 10; p = 1; 4; 12; n = 3

$$\frac{1-(1+WACC)^{-n}}{WACC} = \frac{1-(1+k_0)^{-n}}{k_0\cdot\left(1-\frac{k_d w_d t}{p}\frac{\left[1-(1+k_d)^{-n}\right]\cdot(1+k_d)^{1/p}}{(1+k_d)^{1/p}-1}\right)}$$

Dependence of WACC on leverage level L at k0 = 0.22, kd = 0.14 and different p = 1, 6, 12 for a 3-year company is shown in Table 18.2 and Figs. 18.3 and 18.4.

Note that the dependences of WACC on leverage level L obtained above will be used below under a study of the effectiveness of long-term as well arbitrary duration investment projects, from the perspective of the owners of equity capital and of the owners of equity and debt.

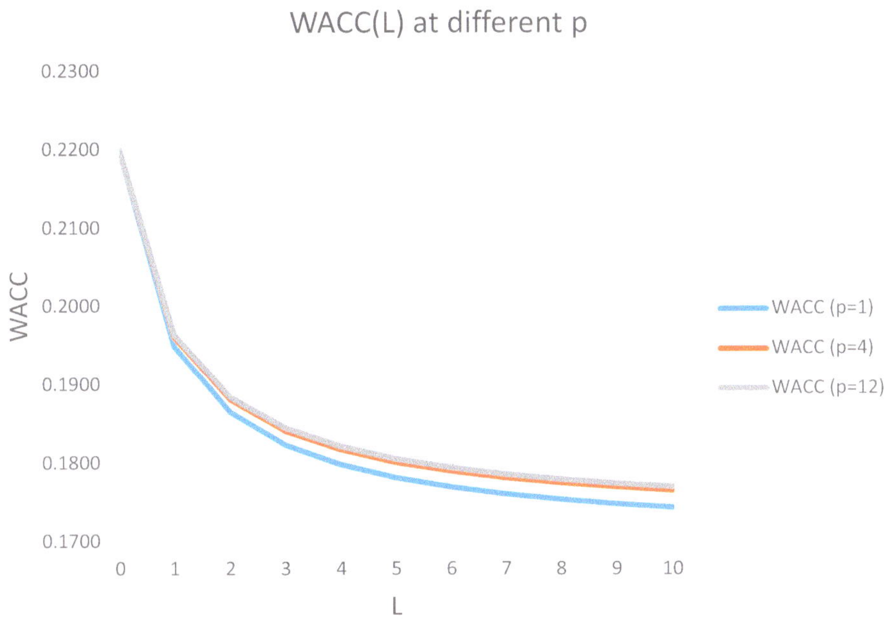

Fig. 18.1 Dependence of the weighted average cost of capital, WACC, on leverage level L at different frequencies of payments of tax on profit, p = 1; 4; 12

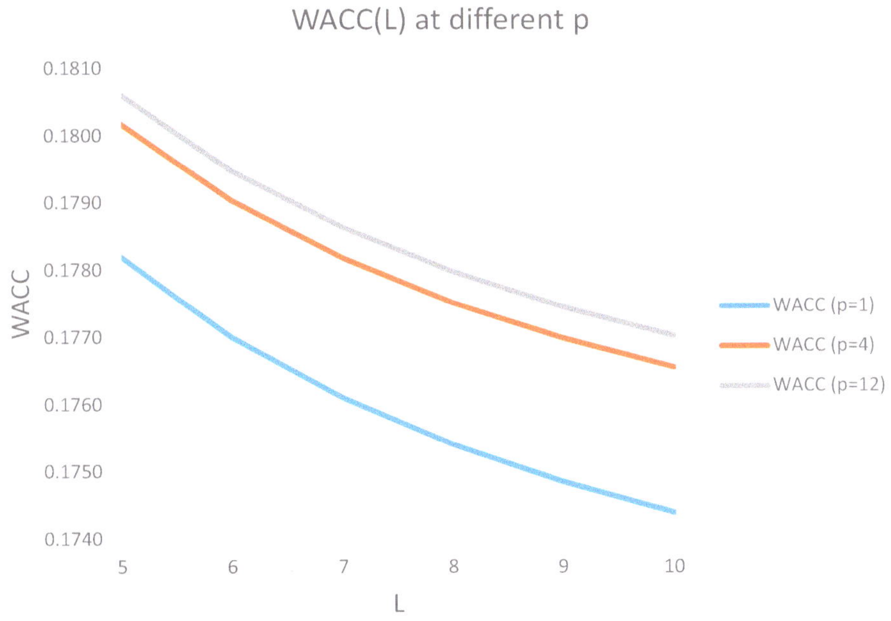

Fig. 18.2 Dependence of the weighted average cost of capital, WACC, on leverage level L at different frequencies of payments of tax on profit, p = 1; 4; 12 (larger scale)

Table 18.2 Dependence of WACC on leverage level L at $k_0 = 0.22$, $k_d = 0.14$ and different $p = 1, 4, 12$ for a 3-year company

L	WACC		
	p = 1	p = 4	p = 12
0	0.2199	0.2199	0.2199
1	0.1957	0.1969	0.1971
2	0.1876	0.1892	0.1895
3	0.1836	0.1853	0.1857
4	0.1812	0.1830	0.1834
5	0.1796	0.1815	0.1819
6	0.1784	0.1804	0.1808
7	0.1775	0.1795	0.1800
8	0.1768	0.1789	0.1793
9	0.1763	0.1784	0.1788
10	0.1758	0.1780	0.1784

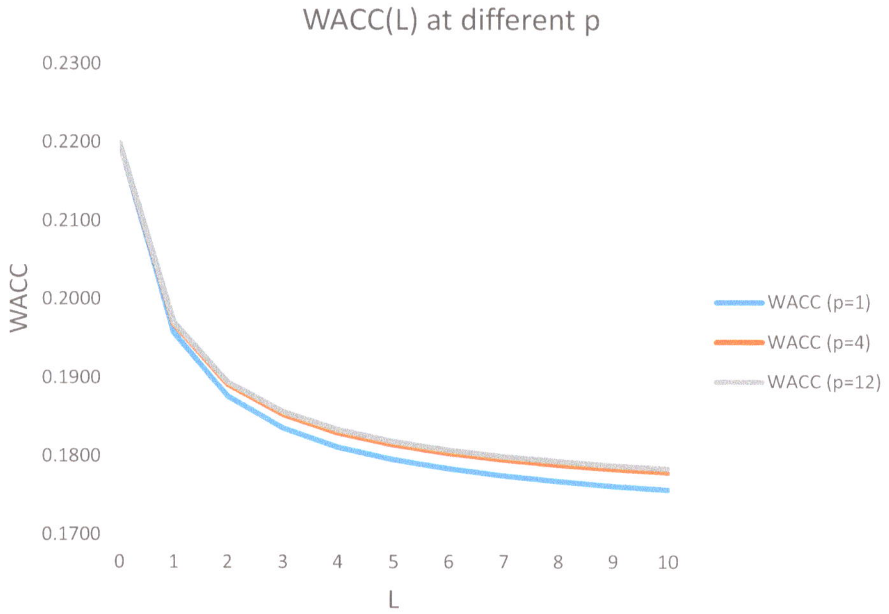

Fig. 18.3 Dependence of WACC on leverage level L at $k_0 = 0.22$, $k_d = 0.14$ and different p = 1, 4, 12 for a 3-year project

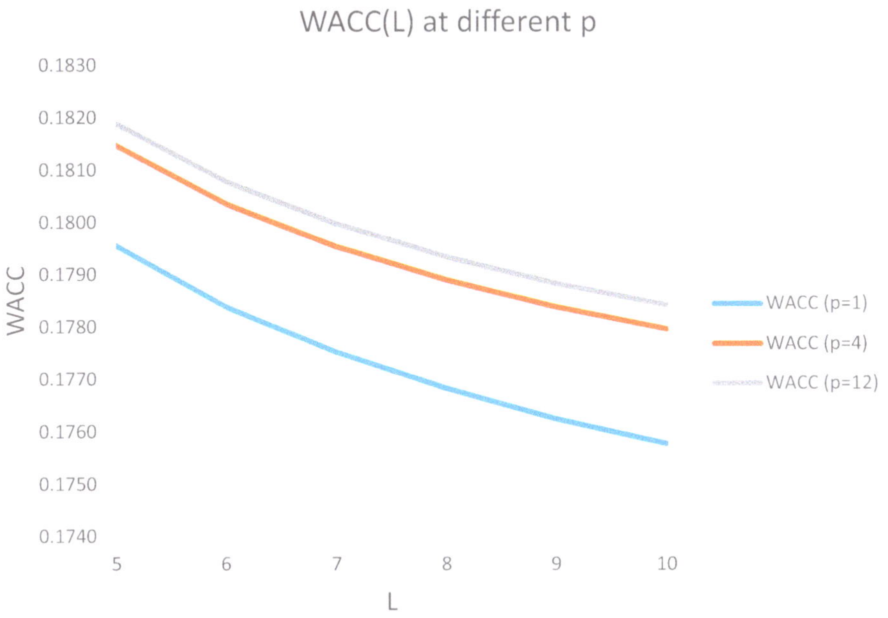

Fig. 18.4 Dependence of WACC on leverage level L at $k_0 = 0.22$, $k_d = 0.14$ and different p = 1, 4, 12 for a 3-year project (larger scale)

18.5.2 The Effectiveness of the Long-Term Investment Project from the Perspective of the Owners of Equity Capital

For long-term investment projects, the discount rate (WACC) in the case of arbitrary frequency of advance payment of tax on profit p is described by formula (18.19) and for NPV one has:

$$NPV = -S + \frac{NOI(1-t)}{WACC} + \frac{k_d D t (1+WACC)^{1/p_1}}{p_1\left((1+WACC)^{1/p_1}-1\right)} - \frac{k_d D (1+WACC)^{1/p_2}}{p_2\left((1+WACC)^{1/p_2}-1\right)}$$

For NPV calculation, we will use the following parameters: $k_0 = 0.22$; $k_d = 0.14$; $t = 20\%$; $L = 0; 1; 2; \ldots; 10$; $p_1 = p_2 = p = 1; 4; 12$; $S = 1000$; $D = LS$; $NOI = 800$.

It is seen from Figs. 18.5 and 18.6 that in the case of considering the effectiveness of long-term investment projects for owners of equity capital, NPV will vary with the change of p, but with not much variation (it is seen from Table 18.3 and from Fig. 18.6, but not from Fig. 18.5).

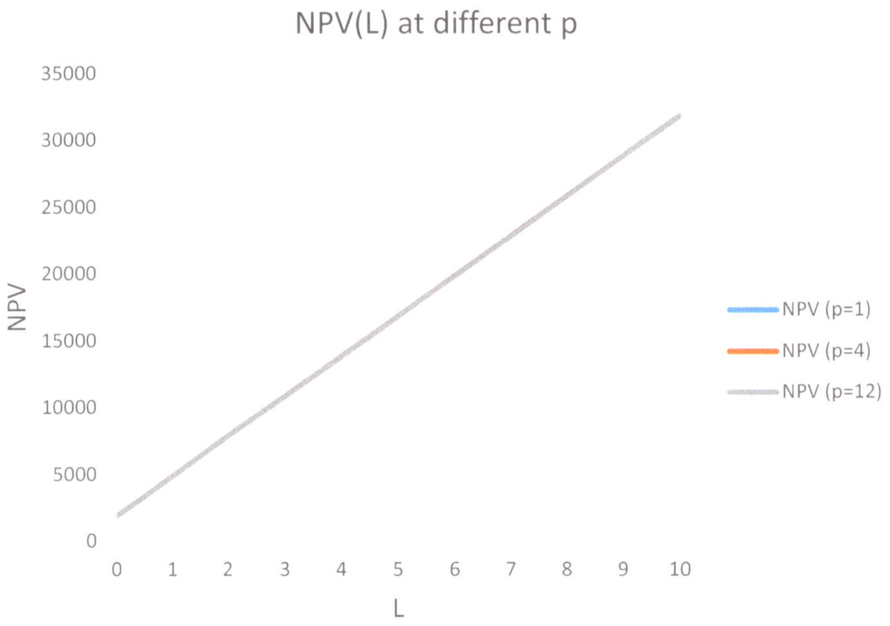

Fig. 18.5 Dependence of NPV on leverage level L at $k_0 = 0.22$, $k_d = 0.14$ and different p = 1, 4, 12 for a long-term project

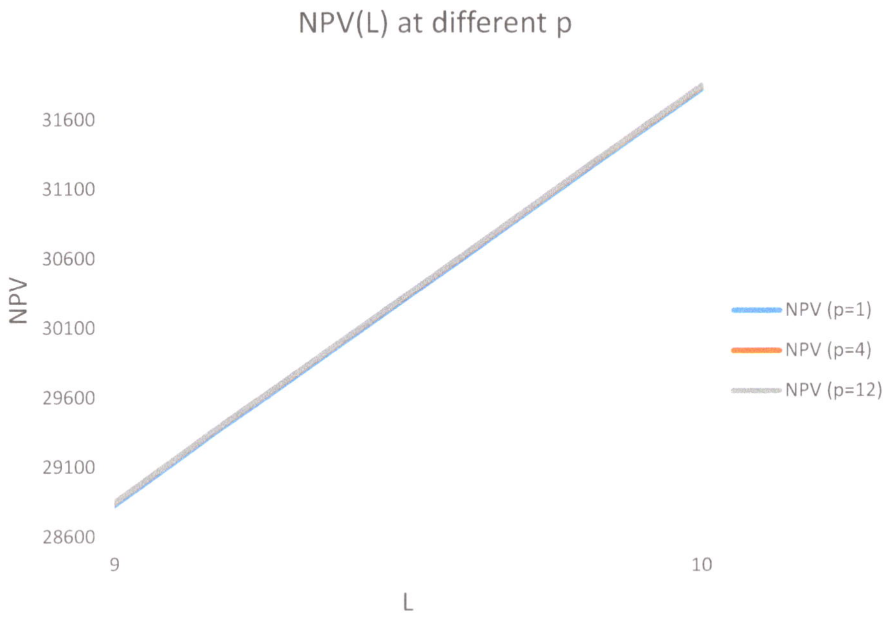

Fig. 18.6 Dependence of NPV on leverage level L at $k_0 = 0.22$, $k_d = 0.14$ and different p = 1, 4, 12 for a long-term project (larger scale)

Table 18.3 Dependence of WACC and NPV on leverage level L at $k_0 = 0.22$, $k_d = 0.14$ and different p = 1, 4, 12

L	NPV		
	p = 1	p = 4	p = 12
0	1909	1909	1909
1	4880	4888	4889
2	7867	7877	7880
3	10,858	10,871	10,874
4	13,852	13,866	13,869
5	16,846	16,862	16,865
6	19,841	19,858	19,862
7	22,837	22,854	22,858
8	25,833	25,851	25,855
9	28,829	28,848	28,852
10	31,825	31,844	31,850

18.5.3 The Effectiveness of the Long-Term Investment Project from the Perspective of the Owners of Equity and Debt

For long-term investment projects, NPV in the case of arbitrary frequency of payment of tax on profit p is described by formula (18.29):

$$NPV = -I + \frac{NOI(1-t)}{WACC} + \frac{k_d Dt(1+WACC)^{1/p}}{p\left((1+WACC)^{1/p} - 1\right)}$$

For NPV calculation, we will use the following parameters: $k_0 = 0.22$; $k_d = 0.14$; t = 20%; L = 0; 1; 2; ...; 10; $p_1 = p_2 = p = 1$; 6; 12; $S = 1000$; $D = LS$; $NOI = 800$.

Dependence of NPV on leverage level L at k0 = 0.22, kd = 0.14 and different p = 1, 4, 12 for a long-term project is shown in Table 18.4 and Figs. 18.7 and 18.8.

18.5.4 The Effectiveness of the Arbitrary Duration Investment Projects from the Perspective of the Owners of Equity Capital

For the arbitrary duration investment projects, NPV in the case of arbitrary frequency of payment of tax on profit p is described by formula (18.18):

Table 18.4 Dependence of NPV on leverage level L at $k_0 = 0.22$, $k_d = 0.14$ and different p = 1, 4, 12

L	NPV		
	p = 1	p = 4	p = 12
0	1909	1909	1909
1	4738	4687	4676
2	7648	7537	7513
3	10,581	10,408	10,371
4	13,525	13,289	13,239
5	16,474	16,175	16,111
6	19,427	19,064	18,985
7	22,382	21,954	21,862
8	25,338	24,846	24,740
9	28,295	27,738	27,618
10	31,252	30,631	30,497

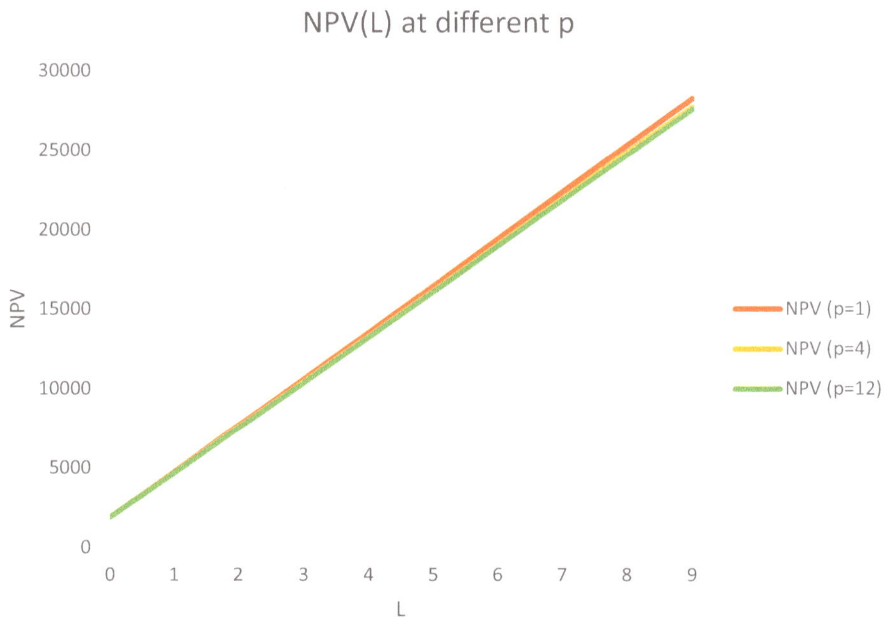

Fig. 18.7 Dependence of NPV on leverage level L at $k_0 = 0.22$, $k_d = 0.14$ and different p = 1, 4, 12 for a long-term project

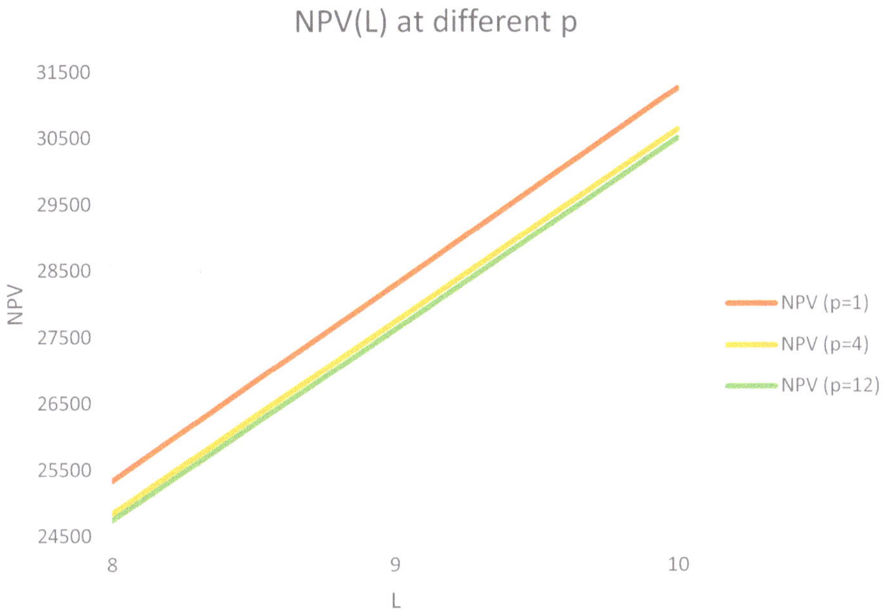

Fig. 18.8 Dependence of NPV on leverage level L at $k_0 = 0.22$, $k_d = 0.14$ and different p = 1, 4, 12 for a long-term project (larger scale)

$$NPV = -S + \frac{NOI(1-t)}{WACC} \cdot \left(1 - \left(\frac{1}{1+WACC}\right)^n\right) + \frac{k_d D t (1-(1+WACC)^{-n}) \cdot (1+WACC)^{1/p_1}}{p_1\left((1+WACC)^{1/p_1}-1\right)} - \frac{k_d D (1-(1+WACC)^{-n}) \cdot (1+WACC)^{1/p_2}}{p_2\left((1+WACC)^{1/p_2}-1\right)} - \frac{D}{(1+WACC)^n}$$

For NPV calculation, we will use the following parameters: $k_0 = 0.22$; $k_d = 0.14$; t = 20%; L = 0; 1; 2; …; 10; $p_1 = p_2 = p = 1$; 6; 12; $S = 1000$; $D = LS$; $NOI = 800$; n = 3.

Dependence of NPV on leverage level L at k0 = 0.22, kd = 0.14 and different p = 1, 4, 12 for a 3-year project is shown in Table 18.5 and Figs. 18.8 and 18.9.

Table 18.5 Dependence of NPV on leverage level L at $k_0 = 0.22$, $k_d = 0.14$ and different p = 1, 4, 12 for a 3-year project

L	NPV		
	p = 1	p = 4	p = 12
0	307	307	307
1	845	861	864
2	1359	1389	1396
3	1865	1911	1921
4	2369	2430	2444
5	2871	2948	2965
6	3373	3465	3485
7	3874	3982	4005
8	4375	4498	4525
9	4876	5014	5044
10	5376	5530	5563

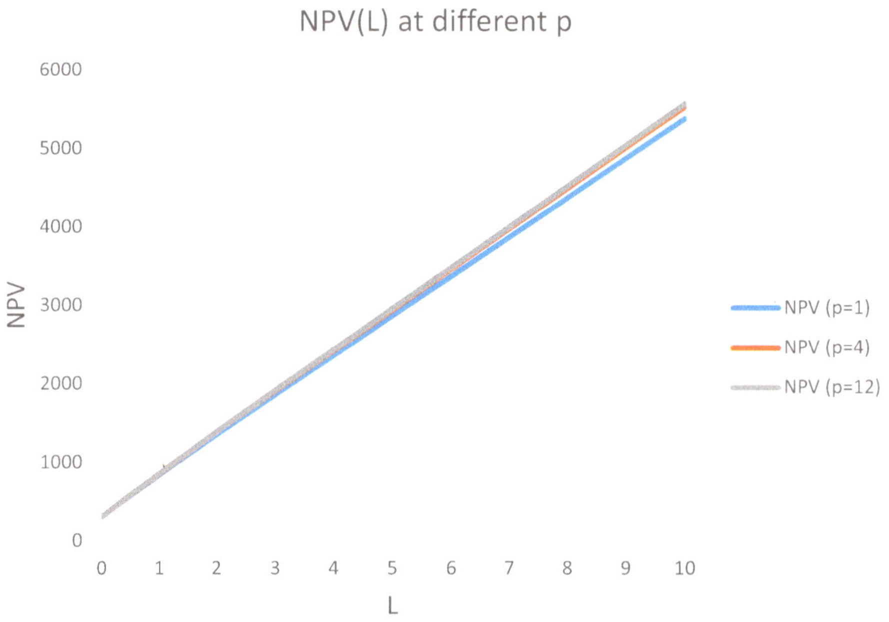

Fig. 18.9 Dependence of NPV on leverage level L at $k_0 = 0.22$, $k_d = 0.14$ and different p = 1, 6, 12 for a 3-year project

18.5.5 The Effectiveness of the Arbitrary Duration Investment Project from the Perspective of the Owners of Equity and Debt

For the arbitrary duration investment projects, NPV in the case of arbitrary frequency of payment of tax on profit p is described by formula (18.28):

$$NPV = -I + \frac{NOI(1-t)}{WACC} \cdot \left(1 - \left(\frac{1}{1+WACC}\right)^n\right)$$

$$+ \frac{k_d Dt(1-(1+WACC)^{-n}) \cdot (1+WACC)^{1/p}}{p\left((1+WACC)^{1/p} - 1\right)}$$

For NPV calculation, we will use the following parameters: $k_0 = 0.22$; $k_d = 0.14$; $t = 20\%$; $L = 0; 1; 2; \ldots; 10$; $p_1 = p_2 = p = 1; 6; 12$; $S = 1000$; $D = LS$; $NOI = 800$; $n = 3$.

Dependence of NPV on leverage level L at k0 = 0.22, kd = 0.14 and different p = 1, 4, 12 for a 3-year project is shown in Table 18.6 and Figs. 18.10 and 18.11.

18.5.6 Discussions

The analysis of results from Tables 18.1, 18.2, 18.3, 18.4, 18.5, and 18.6 and Figs. 18.1, 18.2, 18.3, 18.4, 18.5, 18.6, 18.7, 18.8, 18.9, 18.10, 18.11, and 18.12 shows that:

Table 18.6 Dependence of NPV on leverage level L at $k_0 = 0.22$, $k_d = 0.14$ and different p = 1, 4, 12 for a 3-year project

L	NPV		
	p = 1	p = 4	p = 12
0	307	307	307
1	785	776	774
2	1267	1248	1244
3	1749	1721	1714
4	2232	2194	2185
5	2714	2667	2657
6	3197	3140	3128
7	3680	3614	3599
8	4164	4087	4071
9	4648	4561	4542
10	5132	5034	5013

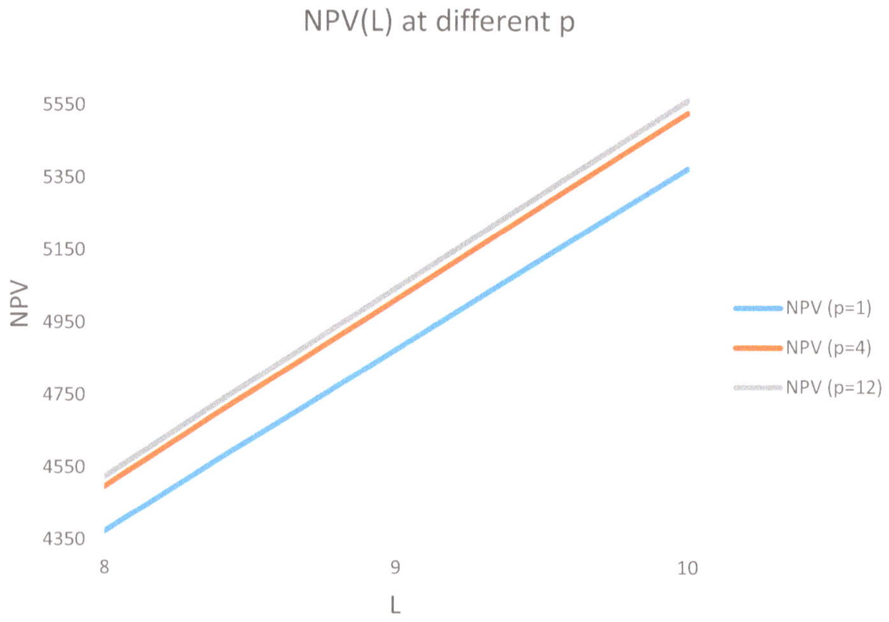

Fig. 18.10 Dependence of NPV on leverage level L at $k_0 = 0.22$, $k_d = 0.14$ and different p = 1, 4, 12 for a 3-year project (larger scale)

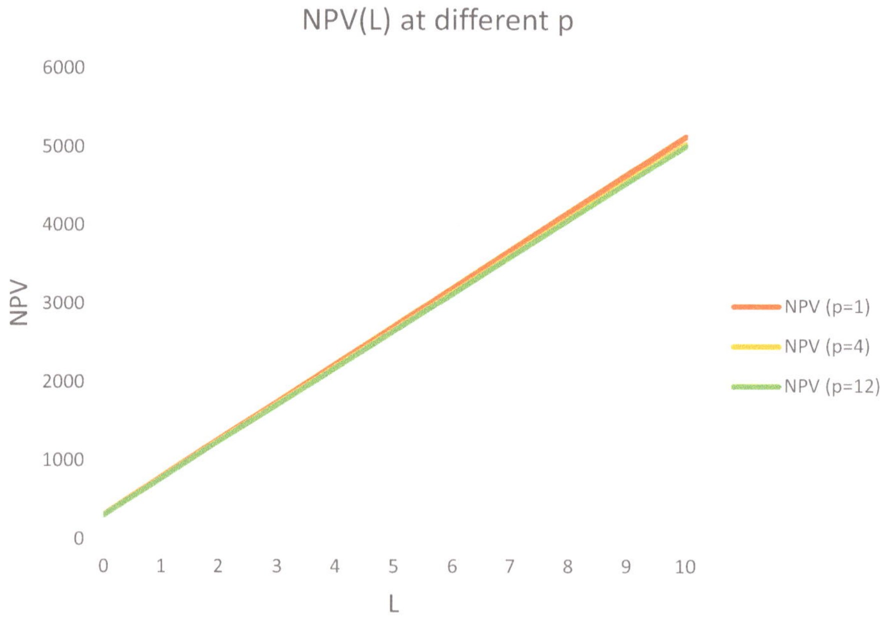

Fig. 18.11 Dependence of NPV on leverage level L at $k_0 = 0.22$, $k_d = 0.14$ and different p = 1, 4, 12 for a 3-year project

1. The weighted average cost of capital, WACC, decreases with leverage level L at all values of frequency of payments of tax on profit p (see Tables 18.1 and 18.2 and Figs. 18.1, 18.2, 18.3, and 18.4).
2. With an increase in p, WACC increases: WACC(L) curves lie higher with increase of p.
3. The difference between curves corresponding to p = 1 and p = 4 is much more than the difference between the curves corresponding to p = 4 and p = 12. This difference decreases with p. It turns out that an increase in the number of payments of tax on profit per year p leads to an increase in the cost of attracting capital (WACC). Will this increase in the discount rate decrease the effectiveness of investment projects? As we see from the Tables 18.3, 18.4, 18.5, and 18.6 and Figs. 18.5, 18.6, 18.7, 18.8, 18.9, 18.10, 18.11, and 18.12, the situation is different for owners of equity capital and for owners of equity and debt capital.
4. NPV practically linearly increases with leverage level at all values of frequency of payments of tax on profit p and all frequency of payments of interest on debt.
5. In the case of considering the effectiveness of long-term investment projects for owners of equity capital, NPV is changed with a change of p (both p_1 and p_2 are equal everywhere below) but by a very small value (it is seen in the Tables, but not in the Figures); NPV is growing with p.

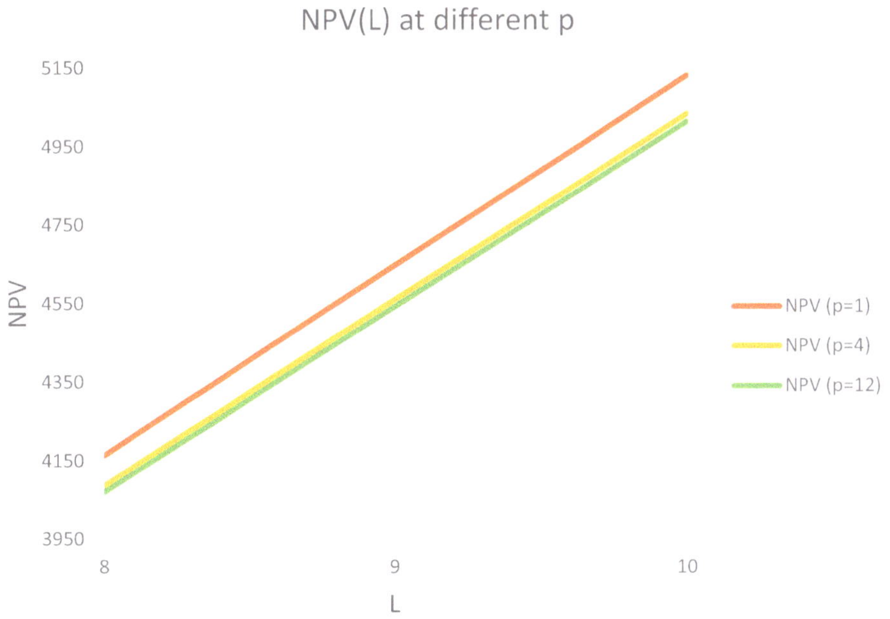

Fig. 18.12 Dependence of NPV on leverage level L at $k_0 = 0.22$, $k_d = 0.14$ and different p = 1, 6, 12 for a 3-year project (larger scale)

6. In the case of considering the effectiveness of long-term investment projects for owners of equity and debt, NPV is changed with a change of p more significantly (it is seen in the Tables and in the Figures); NPV decreases with p.
7. For arbitrary duration projects, this difference in NPV with a change of p is more significant and should be accounted under valuation of the effectiveness of an investment project.
8. In the case of considering the effectiveness of an investment project for owners of equity capital, we need to note that an increase in p leads to an increase in NPV: this means that the effectiveness of an investment project increases with p.
9. In the case of considering the effectiveness of an investment project for owners of equity and debt capital, we need to note that the situation is opposite and an increase in p leads to decrease in NPV: this means that the effectiveness of an investment project decreases with p.
10. The above results show that in the latter case, companies should pay tax on profit and interest on debt once per year, while in the former case, more frequent payments are profitable for the effectiveness of an investment.
11. Thus, while for long-term projects' NPV, the impact of more frequent payments of both values p_1 and p_2 is insignificant, for arbitrary duration projects the account of the frequency of both types of payments could be important and could lead to more significant influence on the effectiveness of an investment project, increasing it (in the former case), or decreasing it (in the latter case). Note that the specific value of the effect depends on the values of the parameters in the project (k_0, k_d, n, t, S etc).

18.6 Conclusions

There are too few investment models which can numerically valuate the effectiveness of investment projects, among them the investment models developed by the authors of this paper. Moreover, investment models are practically absent which account for the conditions of the real functioning of investment projects. This paper covers this gap in the literature and science in the field of investments and develops innovative investment models which are much closer to economic practice.

We developed here for the first time eight innovative investment models: long-term (described by Eqs. (18.14), (18.19), (18.24) and (18.29)) as well arbitrary duration (described by Eqs. (18.13), (18.18), (18.23) and (18.28)), which account two effects: (1) payments of interest on debt and of tax on income a few times per year (semi-annually, quarterly, monthly); (2) advance payments of taxes and of debt interest, as it happened in practice. Note that no one before had investigated the impact of the frequency of payments of taxes and of debt interest and payments of taxes and of debt interest on the effectiveness of investment projects. These investment models allow for investigation of the impact of all main parameters of investment projects (equity value S, investment value I, net operating income NOI, leverage level L, profitability of investments β, tax on profit t, project duration n, equity cost k_0, debt cost k_d and number of payments of interest on debt p_1 and of tax on income p_2) on the main indicator of effectiveness of investment projects NPV (net present value). They could be used for investigation of different problems of investments, such as the influence of debt financing, leverage level, taxing, project duration, method of financing, number of payments of interest on debt and of tax on income and some other parameters on efficiency of investments and other problems. In particular, they will improve the issue of project ratings (Brusov et al. 2018b). These new models allow for making more correct evaluations of effectiveness of investment projects long-term as well as of arbitrary duration.

Numerical calculations, conducted for four investment models (without flow separation) show that:

- In the case of considering the effectiveness of an investment project for owners of equity capital, the increase in the number of payments of tax on income and of interest on debt p leads to an increase in NPV: this means that the effectiveness of an investment project increases with p;
- In the case of considering the effectiveness of an investment project for owners of equity and debt capital, the increase in the number of payments of tax on income and of interest on debt p leads to decrease in NPV: this means that the effectiveness of an investment project decreases with p.

In the latter case, companies should pay tax on profit and interest on debt once per year, while in the former case, more frequent payments are profitable for the effectiveness of an investment.

Thus, while for long-term projects' NPV, the impact of more frequent payments of both values p_1 and p_2 is insignificant, for arbitrary duration projects the account of the frequency of both types of payments could be important and could lead to more significant influence on the effectiveness of an investment project, decreasing it (in the latter case), or increasing it (in the former case). Note that the specific value of the effect depends on the values of the parameters in the project (k_0, k_d, n, t, S, etc.).

There are some limitations of the applicability of the proposed models:

1. For long-term projects, they are connected with the limitations of the Modigliani-Miller theory;
2. For consideration of without flow separation, they are connected with the well-known limitations of the WACC approach;
3. For arbitrary duration projects (using BFO theory), they are connected with the fact that not all the conditions of real investments are accounted yet.

We could compare obtained by us results with ones of our paper (Brusov et al. 2021), where we have frequent payments of tax on income and of interest on debt p at the end of periods. The results turn out to be opposite. In Brusov et al. (2021) we have:

1. WACC decreases with p.
2. In the case of considering the effectiveness of an investment project for owners of equity capital, the increase in the number of payments of tax on income and of interest on debt p leads to a decrease in NPV: this means that the effectiveness of an investment project decreases with p.
3. In the case of considering the effectiveness of an investment project for owners of equity and debt capital, the increase in the number of payments of tax on income and of interest on debt p leads to an increase in NPV: this means that the effectiveness of an investment project increases with p. In the former case, companies should pay tax on profit and interest on debt once per year, while in the latter case, more frequent payments are profitable for the effectiveness of investment.

In current chapter, where payments of income tax and interest on debt are made in advance, we have:

1. WACC increases with p.
2. In the case of considering the effectiveness of an investment project for owners of equity capital, the increase in the number of payments of tax on income and of interest on debt p leads to a increase in NPV: this means that the effectiveness of an investment project increases with p.
3. In the case of considering the effectiveness of an investment project for owners of equity and debt capital, the increase in the number of payments of tax on income and of interest on debt p leads to the decrease in NPV: this means that the effectiveness of an investment project decreases with p. In the latter case, companies should pay tax on profit and interest on debt once per year, while in the former case, more frequent payments are profitable for the effectiveness of investment.

So, we have created and fully investigated eight innovative investment models with frequent payments of income tax and interest on debt in advance (current paper) and at the end of periods (Brusov et al. 2021). We have shown, that method of payments of tax on income and of interest on debt (in advance or in the ends of periods) changes drastically the effect of the number of payments of income tax and interest on debt on the effectiveness of investments. Depending on the relationship between the investor and the lender, the investor can choose (within the framework of tax legislation):

- The frequency of payment of income tax.
- Scheme for payment of income tax (in advance or at the end of periods).

Obtained results help to tax regulator (Finance Ministry) understand the influence of the number of payments of tax on income and to credit regulator (Central Bank) the influence of the number of payments of interest on debt on the effectiveness of investment projects. This allow them modify and improve tax legislation and credit policy respectively.

The contribution of this study to finance and economics is mainly related to the goal that new modern investment models have been created to be as close as possible to real investment conditions. Our models, namely, models with advance payments of interest on debt and of tax on income which occur a few times per year (semi-annually, quarterly or monthly), could be more successfully applied in real economic practice.

References

Black, F., & Litterman, R. (1992). Global portfolio optimization. *Financial Analysts Journal, 48*, 28–43.

Bond, S., & Meghir, C. (1994). Dynamic investment models and the firm's financial policy. *The Review of Economic Studies, 67*, 197–222.

Brown, K., Harlow, W., & Zhang, H. (2021). Investment style volatility and mutual fund performance. *Journal of Investment Managment, 19*, 78.

Brusov, P., & Filatova, T. (2021). The Modigliani–Miller theory with arbitrary frequency of payment of tax on profit. *Mathematics, 9*, 1198. https://doi.org/10.3390/math9111198

Brusov, P., Filatova, T., Orehova, N., & Eskindarov, M. (2015). *Modern corporate finance, investments and taxation* (pp. 1–373). Springer International Publishing. Retrieved from https://www.springer.com/gp/book/9783319147314

Brusov, P., Filatova, T., Orehova, N., & Eskindarov, M. (2018a). *Modern corporate finance, investments, taxation and ratings* (2nd ed., pp. 1–571). Springer Nature Publishing.

Brusov, P., Filatova, T., & Orehova, N. (2018b). *Ratings: Critical analysis and new approaches of quantitative and qualitative methodology* (pp. 1–379). Springer Nature Publishing. Retrieved from https://www.springer.com/de/book/9783030562427

Brusov, P. N., Filatova, T. V., Orekhova, N. P., Kulik, V. L., Chang, S.-I., & Lin, Y. C. G. (2020a). Modification of the Modigliani–Miller theory for the case of advance payments of tax on profit. *Journal of Reviews on Global Economics, 9*, 257–267.

Brusov, P. N., Filatova, T. V., Orekhova, N. P., Kulik, V. L., Chang, S.-I., & Lin, Y. C. G. (2020b). Application of the Modigliani–Miller theory, modified for the case of advance payments of tax on profit, in rating methodologies. *Journal of Reviews on Global Economics, 9*, 282–292.

Brusov, P. N., Filatova, T. V., Chang, S.-I., & Lin, Y. C. G. (2021). Innovative investment models with frequent payments of tax on income and of interest on debt. *Mathematics, 9*(13), 1491. https://doi.org/10.3390/math9131491

Cakici, N., & Zaremba, A. (2021). Size, value, profitability, and investment effects in international stock returns: Are they really there? *Journal of Investing, 30*, 65–86. https://doi.org/10.3905/joi.2021.1.176

Dammon, R., & Senbet, L. (1988). The effect of taxes and depreciation on corporate investment and financial leverage. *The Journal of Finance, 43*, 357–373. https://doi.org/10.1111/j.1540-6261.1988.tb03944.x

DeAngelo, H., & Masulis, R. (1980). Optimal capital structure under corporate and personal taxation. *Journal of Financial Economics, 8*, 3–29.

Dotan, A., & Ravid, S. (1985). On the interaction of real and financial decisions of the firm under uncertainty. *The Journal of Finance, 40*, 501–517.

Escobar-Anel, M., Lichtenstern, A., & Zagst, R. (2020). Behavioral portfolio insurance strategies. *Financial Markets and Portfolio Management, 34*, 353–399.

Filatova, T. V., Orekhova, N. P., & Brusova, A. P. (2008). Weighted average cost of capital in the theory of Modigliani–Miller, modified for a finite life-time company. *Bulletin of the FU, 48*, 68–77.

Gidwani, B. (2020). Some issues with using ESG ratings in an investment process. *Journal of Investing, 29*, 76–84. https://doi.org/10.3905/joi.2020.1.147

Kahneman, D., & Tversky, A. (1996). On the reality of cognitive illusions. *Psychological Review, 103*, 582–591. https://doi.org/10.1037/0033-295X.103.3.582

Kang, J. (1995). The conditional relationship between financial leverage and corporate investment: Further clarification. *Journal of Business Finance and Accounting, 22*, 1211–1219. https://doi.org/10.1111/j.1468-5957.1995.tb00902.x

Lang, L. E., Ofek, E., & Stulz, R. (1992). Leverage, investment, and firm growth. *Journal of Financial Economics, 40*, 3–29.

Loviscek, A. (2021). How much to invest, if any, in precious metals? A twenty-first-century perspective. *Journal of Investing, 30*, 95–111. https://doi.org/10.3905/joi.2021.1.179

Modigliani, F., & Miller, M. (1958). The cost of capital, corporate finance, and the theory of investment. *The American Economic Review, 48*, 261–297.

Modigliani, F., & Miller, M. (1963). Corporate income taxes and the cost of capital: A correction. *The American Economic Review, 53*, 147–175.

Modigliani, F., & Miller, M. (1966). Some estimates of the cost of capital to the electric utility industry 1954–1957. *The American Economic Review, 56*, 333–391.

Rosenbaum, J., & Pearl, J. (2013). *Investment banking valuation models* (pp. 1–342). Wiley.

Whited, T. (1992). Debt, liquidity constraints and corporate investment: Evidence from panel data. *The Journal of Finance, 47*, 1425–1461.

Part V
Portfolio Investments

This Part discusses portfolio investments. A portfolio of securities is a collection of different financial assets owned by an investor, including stocks, bonds, mutual funds, and other investments. The purpose of a portfolio is to diversify investments, balance risk, and help achieve specific financial goals, such as capital growth or income. In Chap. 19 a portfolio of securities is considered. A detailed analysis of a portfolio of two and three securities is carried out. Markowitz portfolios are considered, as well as Tobin portfolios of n–securities. In Chap. 20 a portfolio of bonds is discussed in details.

Chapter 19
Portfolio of Securities

19.1 Portfolio Analysis

Let's suppose that an investor holds a valuable security for a certain period of time, where the value of the security at the beginning of the interval is denoted by p0 and at the end of the interval is p1. Let d represent the dividends paid during the period under consideration. Then, the return on the security for this time interval is defined as:

$$r = \frac{p_1 + d - p_0}{p_0} \tag{19.1}$$

If we exclude dividends and other factors affecting the return (such as inflation, etc.), then formula (19.1) takes a simpler form: $r = \frac{p_1 - p_0}{p_0}$.

A portfolio consisting of n types of securities is defined as a n-dimensional vector:

$$X = (x_1, x_2, \ldots, x_n), \tag{19.2}$$

Here, x_i represents the proportion of investments in securities of type i.

The return on portfolio X is defined as:

$$r_X = \frac{p_{X1} + d_X - p_{X0}}{p_{X0}}, \tag{19.3}$$

where p_{X0} is the value of the portfolio at the beginning of the period, p_{X1} is the value at the end of the period, and d_X is the dividends received for all securities in the portfolio.

P. Brusov, T. Filatova, *Efficiency of Investment Projects*, Contributions to Finance and Accounting, https://doi.org/10.1007/978-3-032-01028-5_19

The return on portfolio X consisting of n securities is expressed by the formula:

$$r_x = x_1 r_1 + x_2 r_2 + \cdots + x_n r_n, \tag{19.4}$$

where r_1, r_2, ..., r_n are the returns of the securities included in portfolio X.

The prices of securities at the beginning of the period p_{i1} are deterministic quantities, while at the end of the period they become random variables. Therefore, the returns of individual securities and the return of the entire portfolio are random variables, denoted by uppercase Ri and Rx respectively. The mathematical expectation of the return on a security is called its **efficiency**, and the mathematical expectation of the return on a portfolio is called the **portfolio efficiency**.

The expression for the portfolio efficiency $\mu = \mathrm{M(Rx)}$ through the corresponding characteristics of securities has the form:

$$\mu_X = x_1\mu_1 + x_2\mu_2 + \ldots + x_n\mu_n \tag{19.5}$$

where $\mu_1 = M(\mathrm{R1})$, $\mu_2 = M(\mathrm{R2})$, ..., $\mu_n = M(\mathrm{R}n)$ are the efficiencies of securities (the mathematical expectations of returns R1, R2, ..., R$n$$R1$, $R2$, ..., Rn constituting the portfolio).

Let's denote the vector of efficiencies (expected returns) of portfolio X by $\overline{\mu} = (\mu_1, \mu_2, \ldots, \mu_n)^T$, then formula (19.5) can be written in matrix notation as follows:

$$\mu_X = \overline{\mu}^T X \tag{19.6}$$

To calculate the risk square, we will use the formula

$$\sigma^2 = X^T V X, \tag{19.7}$$

where V is the covariance matrix of random variables $\mathrm{R}_1, \mathrm{R}_2, \ldots, \mathrm{R}_n$.

In the future, we will use matrix notation. In this case, all vectors will be considered as column vectors. Let's denote the vector consisting of ones as $I = (1, 1, \ldots, 1)^T$.

Thus, each portfolio X is associated with two quantities: efficiency (expected return) μ and risk σ. An investor would like to have a portfolio that provides the highest expected return with the minimum risk. However, this task is contradictory because, in general, a higher expected return entails an increase in risk. Therefore, the following tasks can be considered:

1. Find the portfolio of minimum risk given its efficiency (at an efficiency of no less than the specified one, at arbitrary efficiency).
2. Find the portfolio of maximum efficiency with minimal risk (at risk not exceeding a given level).

Table 19.1 The data of three types of stocks

	A	B	C
Amount	100	150	250
Initial Price	80	60	300
Final Price	90	50	340

Example 19.1 The portfolio consists of five assets, the price shares of which form a geometric progression, and the price share of the second asset is 0.1. Find the portfolio.

The normalization condition is as follows:

$$x_1 + \ldots + x_5 = \frac{x_1\left(1 - q^5\right)}{1 - q} = 1,$$

$$x_1 q = 0.1$$

Substituting $x1 = 0.1/q$ into the first equation, we obtain the equation for q:

$$\frac{0.1\left(1 - q^5\right)}{(1 - q)q} = 1 \text{ or } 0.1 \times \left(1 - q^5\right) = (1 - q)q$$

Solving it numerically, we get $q = 1.61$. Therefore, the portfolio has the form (0.062, 0.1, 0.161, 0.259, 0.417).

Example 19.2 The portfolio consists of three types of stocks, the data of which are given in Table 19.1.

Find the portfolio return in two ways.

Solution Let's calculate the initial price of the portfolio:

$$p_x^0 = 80 \times 100 + 60 \times 150 + 300 \times 250 = 92,000$$

and its price at the end of the year

$$p_x = 90 \times 100 + 50 \times 150 + 340 \times 250 = 101,500$$

Using formula (19.3), we obtain

$$\mu = \frac{p_X - p_x^0}{p_x^0} = \frac{101{,}500 - 92{,}000}{92{,}000} = 0.1033;$$

$$\mu = 10.33\%$$

To apply formula (19.5), let's calculate the proportions (prices) of each type of stock and their returns:

$$x_A = \frac{80 \times 100}{92{,}000} \times 100\% = 8.70\%; x_B = \frac{60 \times 150}{92{,}000} \times 100\% = 9.78\%$$

$$x_C = \frac{300 \times 250}{92{,}000} \times 100\% = 81.52\%;$$

$$\mu_A = \frac{p_A - p_A^0}{p_A^0} = \frac{90 - 80}{80} \times 100\% = 12.5\%$$

$$\mu_B = \frac{p_B - p_B^0}{p_B^0} = \frac{50 - 60}{60} \times 100\% = -16.7\%$$

$$\mu_C = \frac{p_C - p_C^0}{p_C^0} = \frac{340 - 300}{300} \times 100\% = 13.3\%$$

Now, applying formula (19.5), we obtain:

$$\mu = 12.5\% \cdot 0.087 + 16.7\% \cdot 0.0978 + 13.3\% \cdot 0.325 = 10.33\%,$$

which matches the previous result.

Questions and Exercises

19.1. Define the return of a security and a portfolio.

19.2. Derive the formula for the return of a portfolio consisting of *n* securities through the returns of the individual securities.

19.3. The portfolio consists of four types of stocks, with the data provided in Table 19.2.

Find the return of the portfolio.

19.4. The portfolio consists of four types of stocks, with the data provided in Table 19.3.

Find the return of the portfolio by calculating the return of each stock.

Table 19.2 The data of four types of stocks

	A	B	C	D
Quantity	200	350	150	400
Initial price	70	90	180	300
Final price	100	60	250	310

Table 19.3 The data of three types of stocks

	A	B	C	D
Share in portfolio	0.2	0.1	0.55	0.15
Initial price	60	105	120	235
Final price	95	130	100	250

Table 19.4 The data of five types of stocks

	A	B	C	D	E
Return	0.15	0.1	0.25	0.15	0.30
Risk	0.3	0.5	0.3	0.2	0.6

19.5. The shares of securities in a portfolio consisting of three securities are $x1 = 0.2$, $x2 = 0.3$, $x3 = 0.5$. The covariance matrix is given by

$$V = \begin{pmatrix} 2 & 12 & 5 \\ 12 & 6 & 9 \\ 5 & 9 & 6 \end{pmatrix}$$

Find the risk of the portfolio.

19.6. The portfolio consists of three assets A, B, and C, taken in equal price shares. The expected returns of the assets are $\mu 1 = -15\%$, $\mu 2 = 18\%$, $\mu 3 = 23\%$. Find the expected return of the portfolio.

19.7. There are five types of stocks, for each of which the expected return and risk are given in Table 19.4.

Arrange the stocks in order of preference.

19.8. The portfolio consists of three assets A, B, and C. The price share of asset A is 0.4, and the price share of asset B is 1.46 times greater than the price share of asset C. Find the portfolio.

19.9. The portfolio consists of three assets A, B, and C. The price share of asset B is 0.26, and the price share of asset A is 1.56% greater than the price share of asset C. Find the portfolio.

19.10. The portfolio consists of three assets A, B, and C, whose price shares form an arithmetic progression, with the price share of asset B being 0.08. Find the portfolio.

19.11. The portfolio consists of three assets A, B, and C, whose price shares form an arithmetic progression, with the price share of asset A being 0.22. Find the portfolio.

19.12. The portfolio consists of three assets A, B, and C, whose price shares form a geometric progression, with the price share of asset C being 0.35. Find the portfolio.

19.13. The portfolio consists of ten assets, whose price shares form an arithmetic progression, with the price share of the sixth asset being 0.3. Find the portfolio.

19.14. The portfolio consists of seven assets, whose price shares form a geometric progression, with the price share of the third asset being 0.18. Find the portfolio.

19.2 Portfolio of Two Securities

19.2.1 Necessary Information from Probability Theory

The variance (and risk) of a portfolio consisting of two securities is given by:

$$\sigma^2 = x_1^2\sigma_1^2 + x_2^2\sigma_2^2 + 2\rho_{12}x_1x_2\sigma_1\sigma_2 \tag{19.8}$$

where ρ_{12} is the correlation coefficient between the two securities, σ_i is the risk (standard deviation of returns) of the *i*th security, and *xixi* is the weight of the *i*th security in the portfolio.

The return of the portfolio is given by:

$$\mu = x_1\mu_1 + x_2\mu_2, \tag{19.9}$$

where μ_i is the expected return of the *i*th security.

The normalization condition is:

$$x_1 + x_2 = 1. \tag{19.10}$$

The covariance of returns is defined as:

$$\operatorname{cov}(r_i, r_j) = M(r_i \cdot r_j) - M(r_i) \cdot M(r_j), \tag{19.11}$$

$$\operatorname{cov}(r_i, r_j) = \rho_{ij}\sigma_i\sigma_j, \tag{19.12}$$

$$\rho_{ij} = \frac{\operatorname{cov}(r_i, r_j)}{\sigma_i\sigma_j}, \tag{19.13}$$

$$|\rho_{ij}| \le 1. \tag{19.14}$$

The covariance matrix is a matrix whose elements are the covariances of the securities. For a portfolio of three securities, we have:

$$\operatorname{cov}(r_1, r_2) = \begin{pmatrix} \operatorname{cov}(r_1, r_1) & \operatorname{cov}(r_1, r_2) & \operatorname{cov}(r_1, r_3) \\ \operatorname{cov}(r_2, r_1) & \operatorname{cov}(r_2, r_2) & \operatorname{cov}(r_2, r_3) \\ \operatorname{cov}(r_3, r_1) & \operatorname{cov}(r_3, r_2) & \operatorname{cov}(r_3, r_3) \end{pmatrix}, \tag{19.15}$$

$$\left\|\rho(r_1,r_2)\right\| = \begin{pmatrix} \frac{cov(r_1,r_1)}{\sigma_1^2} & \frac{cov(r_1,r_2)}{\sigma_1\sigma_2} & \frac{cov(r_1,r_3)}{\sigma_1\sigma_3} \\ \frac{cov(r_2,r_1)}{\sigma_2\sigma_1} & \frac{cov(r_2,r_2)}{\sigma_2^2} & \frac{cov(r_2,r_3)}{\sigma_2\sigma_3} \\ \frac{cov(r_3,r_1)}{\sigma_3\sigma_1} & \frac{cov(r_3,r_2)}{\sigma_3\sigma_2} & \frac{cov(r_3,r_3)}{\sigma_3^2} \end{pmatrix}, \tag{19.16}$$

Considering that

$$\text{cov}(r_i,r_i) = M(r_i \times r_i) - M(r_i)M(r_i) = M\left(r_i^2\right) - M(r_i)^2 = D(r_i) = \sigma_i^2$$

(the diagonal elements of the covariance matrix are equal to the variances of the securities), we obtain:

$$\left\|\rho(r_i,r_j)\right\| = \begin{pmatrix} 1 & \frac{cov(r_1,r_2)}{\sigma_1\sigma_2} & \frac{cov(r_1,r_3)}{\sigma_1\sigma_3} \\ \frac{cov(r_2,r_1)}{\sigma_2\sigma_1} & 1 & \frac{cov(r_2,r_3)}{\sigma_2\sigma_3} \\ \frac{cov(r_3,r_1)}{\sigma_3\sigma_1} & \frac{cov(r_3,r_2)}{\sigma_3\sigma_2} & 1 \end{pmatrix} \tag{19.17}$$

Note: The inverse problem of finding the covariance matrix given the correlation matrix is indeterminate: it does not have a unique solution.

Example 19.3 Given the covariance matrix V, $V = \begin{pmatrix} 9 & -8 & 6 \\ -8 & 16 & -11 \\ 6 & -11 & 4 \end{pmatrix}$.

Find the correlation matrix.

On the diagonal are the variances, so for the securities' risks, we have

$$\left(\sigma_1 = \sqrt{9} = 3\right), \left(\sigma_2 = \sqrt{16} = 4\right), \left[V = \begin{pmatrix} 8 & 18 & 4 \\ 18 & 12 & 0 \end{pmatrix}\right].$$

Then, using the formula $\rho_{ij} = \frac{\text{cov}(r_i, r_j)}{\sigma_i \sigma_j}$, we compute the off-diagonal elements (all diagonal elements of the correlation matrix are equal to 1):

$$\rho_{12} = \frac{\text{cov}(r_1,r_2)}{\sigma_1\sigma_2} = \frac{-8}{3 \times 4} = -\frac{2}{3} = \rho_{21};$$

$$\rho_{13} = \frac{\text{cov}(r_1,r_3)}{\sigma_1\sigma_3} = \frac{6}{3 \times 2} = 1 = \rho_{31};$$

$$\rho_{23} = \frac{\operatorname{cov}(r_2, r_3)}{\sigma_2\sigma_3} = \frac{-11}{4 \times 2} = -\frac{11}{8} = \rho_{32}.$$

We obtain the following correlation matrix:

$$\left\|\rho\left(r_i, r_j\right)\right\| = \begin{pmatrix} 1 & -\frac{2}{3} & 1 \\ -\frac{2}{3} & 1 & -\frac{11}{8} \\ 1 & -\frac{11}{8} & 1 \end{pmatrix}$$

The inverse problem of finding the covariance matrix from a given correlation matrix, as noted earlier, is indeterminate: it does not have a unique solution. This follows from the fact that due to the symmetry of the correlation matrix, only three quantities are specified: $(r_{1,2})$, $\rho(r_1,r_3)$, $\rho(r_2,r_3)$ which allows writing three equations:

$$\frac{\operatorname{cov}(r_1, r_2)}{\sigma_1\sigma_2} = \rho(r_1, r_2); \frac{\operatorname{cov}(r_1, r_3)}{\sigma_1\sigma_3} = \rho(r_1, r_3); \frac{\operatorname{cov}(r_2, r_3)}{\sigma_2\sigma_3} = \rho(r_2, r_3);$$

for six unknowns σ_1, σ_2, σ_3, $\operatorname{cov}(r_1,r_2)$, $\operatorname{cov}(r_1,r_3)$, $\operatorname{cov}(r_2,r_3)$.

Questions and Tasks

19.15. There is a covariance matrix $V = \begin{pmatrix} 10 & -5 & 6 \\ -5 & 11 & -7 \\ 6 & -7 & 12 \end{pmatrix}$. Find correlation matrix.

19.16. There is a covariance matrix $V = \begin{pmatrix} 144 & 6 & 18 \\ 6 & 49 & -11 \\ 18 & -11 & 25 \end{pmatrix}$. Find correlation matrix.

19.17. The data on the distribution of yields of two securities are given in Table 19.5. Find the covariance and correlation coefficient of these securities.

19.18. There is a covariance matrix $V = \begin{pmatrix} 8 & 8 & 4 \\ 8 & 12 & 0 \\ 4 & 0 & 6 \end{pmatrix}$. Find correlation matrix.

19.19. The data on the distribution of yields of two securities are given in Table 19.6. Find the covariance and correlation coefficient of these securities.

19.20. The data on the distribution of yields of two securities are given in Table 19.7.

Find the covariance and correlation coefficient of these securities.

Table 19.5 The data on the distribution of yields of two securities

A	−10	−12	11	4	8
B	12	14	18	13	−10
P (probability)	0.1	0.15	0.35	0.2	0.2

Table 19.6 The data on the distribution of yields of two securities

A	−0.8	−0.5	0.7	0.14	0.22
B	0.28	0.19	0.10	0.15	−0.8
P (probability)	0.2	0.1	0.3	0.05	0.35

Table 19.7 The data on the distribution of yields of two securities

R_A	0.2	0.4	−0.3
R_B	0.3	−0.6	0.8
P (probability)	0.5	0.2	0.3

19.21. The shares of securities in the portfolio are $x_1 = 0.2$; $x_2 = 0.3$; $x_3 = 0.5$, the expected returns $\mu_1 = -15\ \%$; $\mu_2 = 18\ \%$; $\mu_3 = 23\%$, the covariance matrix is $V = \begin{pmatrix} 9 & -10 & -5 \\ -10 & 16 & 9 \\ -5 & 9 & 36 \end{pmatrix}$.

Find the expected return and risk of the portfolio.

19.22. The shares of securities in the portfolio are correlated as 1:2:3, the expected returns $\mu_1 = -12\ \%$; $\mu_2 = 25\ \%$; $\mu_3 = -6\%$, the covariance matrix is $V = \begin{pmatrix} 1 & -3 & -4 \\ -3 & 4 & 6 \\ -4 & 6 & 9 \end{pmatrix}$.

What is the portfolio, its return and risk?

19.2.2 The Case of Full Correlation

In the case of full correlation

$$\rho_{12} = \rho = 1; \tag{19.18}$$

$$\sigma = |\sigma_1 x_1 + \sigma_2 x_2|; \tag{19.19}$$

$$\sigma = \sigma_1 x_1 + \sigma_2 x_2; \tag{19.20}$$

Substituting $x_1 \to 1 - t$; $x_2 \to t$, so that $x_1 + x_2 = 1$,and get

$$\sigma = \sigma_1(1 - t) + \sigma_2. \tag{19.21}$$

This is the equation of the interval (AB), where points A and B have the following coordinates:

$$(\cdot)A = (\mu_{1,}\sigma_1); (\cdot)B = (\mu_{2,}\sigma_2).$$

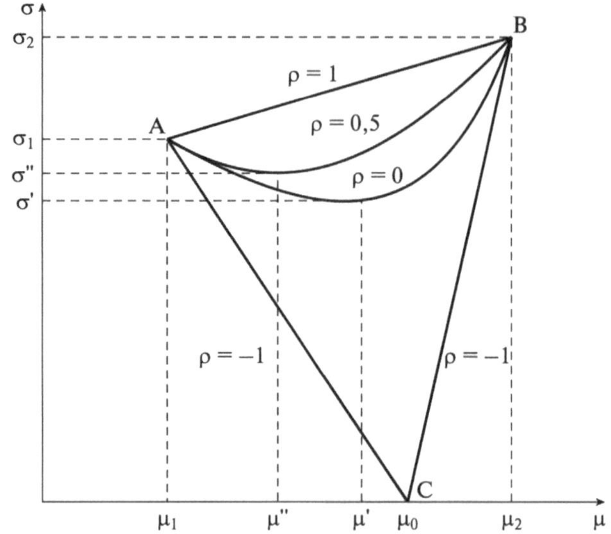

Fig. 19.1 Dependence of the risk of a portfolio of two securities on its efficiency at fixed parameters of both securities and increasing the correlation coefficient from −1 to 1

t runs the values from 0 to 1. At $t = 0$ the portfolio is at point A, at $t = 1$ it is at point B. Therefore, the acceptable set of portfolios in the case of full correlation of securities is an interval (AB) (Fig. 19.1).

Example 19.4 A portfolio consists of two securities A and B, whose expected return and risk expressed in percent are A (5; 12), B (25; 40). The correlation coefficient of the securities is equal to 1. Determine: (a) the portfolio of minimum risk, (b) the portfolio of maximum profitability.

Since the acceptable set of portfolios in the case of full correlation is an interval (AB) (see Fig. 19.1), it is clear that the minimum risk portfolio consists only of securities A and has the form (1; 0). The portfolio parameters match the parameters of paper A (5; 12). The portfolio of maximum return consists only of securities B and has the form (0; 1), and its parameters match the parameters of paper B (25; 40).

Example 19.5 The portfolio consists of two securities A and B, whose expected return and risk expressed in percent are A (5; 12), B (25; 40). The correlation coefficient of the securities is equal to 1. The portfolio return is equal to 15%. Find the portfolio and its risk.

$$\sigma = \sigma_1(1-t) + \sigma_2 t;$$

$$\mu = \mu_1(1-t) + \mu_2 t.$$

From the second equation let's find the value of the parameter t corresponding to the portfolio return of 15%:

$$0.15 = 0.05(1-t) + 0.25t, t = 0.5.$$

Let's determine the portfolio risk by substituting the value of the parameter $t = 0.5$ into the equation

$$\sigma = \sigma_1(1 - t) + \sigma_2 t = 0.12 \times 0.5 + 0.40 \times 0.5 = 0.26.$$

Since $(1 - t)$ and t play the role of x_1 and x_2 (see Brusov et al. 2024), the portfolio has the form:

$$\left(\frac{1}{2}; \frac{1}{2}\right).$$

Example 19.6 The portfolio consists of two securities A and B, whose expected return and risk expressed in percent are A (20; 45), B (38; 68). The correlation coefficient of the securities is equal to 1. The risk of the portfolio is equal to 50%. Find the portfolio and its return.

$$\sigma = \sigma_1(1 - t) + \sigma_2 t;$$

$$\mu = \mu_1(1 - t) + \mu_2 t.$$

Let's find from the first equation the value of the parameter t corresponding to the portfolio risk of 50%.

$$0.5 = 0.45(1 - t) + 0.68t, t = 0.217.$$

We find the portfolio return by substituting the value of the parameter $t = 0.217$ into the equation

$$\mu = \mu_1(1 - t) + \mu_2 t = 0.2 \times 0.783 + 0.38 \times 0.217 = 0.239.$$

Since, as in the previous example, $(1 - t)$ and t play the role of x_1 and x_2 (see Brusov et al. 2024), the portfolio has the form (0.783; 0.217).

Questions and Tasks

19.23. Describe a portfolio consisting of two securities in the case of full correlation.

19.24. A portfolio consists of two securities A and B whose expected return and risk, expressed as a percentage, are A (7; 19), B (12; 24). The correlation coefficient of the securities is 1. Find the set of acceptable portfolios, draw a graph. Determine the return of the minimum risk portfolio.

19.25. A portfolio consists of two securities A and B whose expected return and risk expressed in percent are A (5; 14), B (10; 85). The correlation coefficient of the securities is equal to 1. Determine the return of the minimum risk portfolio and the risk of the maximum return portfolio.

19.26. A portfolio consists of two securities A and B whose expected return and risk expressed in percent are A (14; 25), B (45; 60). The correlation coefficient of the securities is 1. Find: (a) the portfolio of minimum risk, (b) the portfolio of maximum return.

19.27. A portfolio consists of two securities *A* and *B* whose expected return and risk expressed in percent are *A* (14; 25), *B* (45; 60). The correlation coefficient of the securities is 1. Find: (a) the portfolio of maximum risk, (b) the portfolio of minimum return.

19.28. A portfolio consists of two securities *A* and *B* whose expected return and risk expressed in percent are *A* (15; 25), *B* (20; 88). The correlation coefficient of the securities is equal to 1. The risk of the portfolio is 75%. Find the portfolio and its return.

19.29. A portfolio consists of two securities *A* and *B* whose expected return and risk expressed in percent are *A* (7; 13), *B* (40; 23). The correlation coefficient of the securities is equal to 1. The risk of the portfolio is 18%. Find the portfolio and its return.

19.30. A portfolio consists of two securities *A* and *B* whose expected return and risk expressed in percent are *A* (15; 10), *B* (65; 20). The correlation coefficient of the securities is equal to 1. The return on the portfolio is 35%. Find the portfolio and its risk.

19.31. A portfolio consists of two securities *A* and *B* whose expected return and risk, expressed as a percentage, are *A* (46; 18), *B* (70; 37). The correlation coefficient of the securities is equal to 1. The return on the portfolio is 55%. Find the portfolio and its risk.

19.32. A portfolio consists of two securities *A* and *B*, whose expected returns are 19% and 55%, and whose price shares are related as 1:19. The correlation coefficient of the securities is 1. Find the portfolio and its return.

19.33. A portfolio consists of two securities *A* and *B*, whose expected returns are 14% and 67%, and the price share of security *A* is 3 times less than the price share of security *B*. Find the portfolio and its return.

19.34. A portfolio consists of two securities *A* and *B* with expected returns of 30% and 21%, and the price share of security *A* is 40% less than the price share of security *B*. Find the portfolio and its return.

19.2.3 The Case of Full Anticorrelation

In the case of full anticorrelation

$$\rho_{12} = \rho = -1; \tag{19.22}$$

$$\sigma = |\sigma_1 x_1 - \sigma_2 x_2|. \tag{19.23}$$

The acceptable set of portfolios in the case of full anti-correlation of securities is two intervals (*A*, *C*) and (*B*, *C*) (see Fig. 19.1). In the case of full anticorrelation, a zero-risk portfolio (point C (μ_0, 0)) is possible. The zero-risk portfolio has the following form

$$X_0 = \left(\frac{\sigma_2}{\sigma_1 + \sigma_2}, \frac{\sigma_1}{\sigma_1 + \sigma_2}\right), \tag{19.24}$$

and its return is equal to

$$\mu_0 = \left(\frac{\mu_1\sigma_2 + \mu_2\sigma_1}{\sigma_1 + \sigma_2}\right). \tag{19.25}$$

Note that the zero-risk portfolio does not depend on the returns of securities, but is determined only by their risks, and the price share of one security is proportional to the risk of the other.

Since $|\rho| \leq 1$, all acceptable portfolios are inside ($|\rho| < 1$) or on the boundary ($|\rho| = 1$) of triangle ABC.

Example 19.7 A portfolio consists of two securities A and B, whose expected return and risk expressed in percent are A (10; 22), B (25; 60). The correlation coefficient of the securities is -1. Find the portfolio of minimum risk and its return, and the portfolio of maximum return and its risk.

In the case of full anticorrelation, the minimum risk portfolio is the riskless portfolio. Its return is equal to

$$\mu_0 = \frac{\mu_1\sigma_2 + \mu_2\sigma_1}{\sigma_1 + \sigma_2} = \frac{0.1 \times 0.6 + 0.25 \times 0.22}{0.22 + 0.6} = 0.119.$$

The minimum risk portfolio has the following form

$$X_0 = \left(\frac{\sigma_2}{\sigma_1 + \sigma_2}, \frac{\sigma_1}{\sigma_1 + \sigma_2}\right) = \left(\frac{0.6}{0.82}, \frac{0.22}{0.82}\right) = (0.73; 0.27).$$

Therefore, the portfolio of minimum (in this case zero) risk has the form (0.73; 0.27), and its parameters are (14; 0).

In the case of full anticorrelation, the portfolio of maximum return is the portfolio consisting only of the maximum return security (see Fig. 19.1). It has the form (0; 1) and its parameters are (25; 60).

Example 19.8 For a portfolio of two securities with return and risk respectively (0.2; 0.5) and (0.4; 0.7) in the case of full anticorrelation, find the portfolio of zero risk and its return.

First, using the formula (19.24) let's find the zero-risk portfolio

$$X_0 = \left(\frac{\sigma_2}{\sigma_1 + \sigma_2}, \frac{\sigma_1}{\sigma_1 + \sigma_2}\right) = \left(\frac{0.7}{0.5 + 0.7}, \frac{0.5}{0.5 + 0.7}\right) = (0.583; 0.417).$$

Then by using the formula (19.25) let's find its return rate

$$\mu_0 = \frac{\mu_1\sigma_2 + \mu_2\sigma_1}{\sigma_1 + \sigma_2} = \frac{0.2 \times 0.7 + 0.4 \times 0.5}{0.5 + 0.7} = 0.283.$$

We see that the portfolio return is intermediate between the returns of both securities (but the risk is zero!).

We can verify the result for the portfolio return by calculating it using formula (19.9)

$$\mu = x_1\mu_1 + x_2\mu_2 = 0.583 \times 0.2 + 0.417 \times 0.4 = 0.283.$$

Example 19.9 A portfolio consists of two securities A and B with expected returns of 12% and 26% and price shares of 2:3. The correlation coefficient of the securities is -1. Find the portfolio and its return.

Knowing the ratio of price shares x_1:x_2 = 2:3, find the portfolio from the normalisation condition $x_1 + x_2 = 1$.

$$x_1 = \frac{2x_2}{3}, \quad \frac{2x_2}{3} + x_2 = \frac{5x_2}{3} = 1$$

Hence $x_2 = 3/5$; $x_1 = 2/5$.

Substituting these results into one of the formulas for a portfolio of given efficiency (see below)

$$x_1 = \frac{\mu - \mu_2}{\mu_1 - \mu_2},$$

and get

$$\frac{3}{5} = \frac{\mu - 0.26}{0.12 - 0.26}, \mu = 0.176.$$

So, the desired portfolio is $\left(\frac{3}{5}; \frac{2}{5}\right)$ with a return of 17.6%.

Example 19.10 A portfolio consists of two securities A and B, whose expected return and risk expressed in percent are A (14; 27), B (37; 46). The correlation coefficient of the securities is -1, and its return is 20%. Find the portfolio and its risk.

There are several ways to solve this task. The simplest one is to use the results obtained in Brusov et al. (2024) (see below) for the portfolio

$$x_1 = \frac{\mu - \mu_2}{\mu_1 - \mu_2}, x_2 = \frac{\mu_1 - \mu}{\mu_1 - \mu_2} \tag{19.26}$$

and the equations

$$\sigma^2 = \frac{\sigma_1^2(\mu - \mu_2)^2 + \sigma_2^2(\mu - \mu_1)^2 - 2\sigma_1\sigma_2\rho_{12}(\mu - \mu_1) \times (\mu - \mu_2)}{(\mu_2 - \mu_1)^2} \tag{19.27}$$

to find the portfolio risk.

Substituting the data of the task into (19.26), we obtain

$$x_1 = \frac{0.2 - 0.37}{0.14 - 0.37} = 0.739, x_2 = \frac{0.14 - 0.2}{0.14 - 0.37} = 0.261;$$

x_2, of course, can also be found from the normalisation condition $x_1 + x_2 = 1$.

Substituting the data of the task into (19.27), we obtain

$$\begin{aligned}\sigma^2 &= \frac{0.27^2(0.2-0.37)^2 + 0.46^2(0.2-0.14)^2 - 2\times0.27\times0.46\times(-1)(0.2-0.14)\times(0.2-0.37)}{(0.46-0.27)^2} \\ &= 0.00939\end{aligned}$$

The portfolio risk σ is equal to

$$\sigma = 0.939\% \approx 1\%.$$

So, the portfolio has the form (0.739; 0.261) and its risk is 1%.

Example 19.11 A portfolio consists of two securities *A* and *B*, whose expected return and risk expressed in percent are *A* (10; 15), *B* (45; 76). The correlation coefficient of the securities is −1 and its risk is 5%. Find the portfolio and its return, explain the ambiguity of the solution of the task.

There are several ways to solve this task. The simplest is to use the results obtained in Brusov et al. (2024) (see below) for a portfolio

$$X = \left(\frac{\sigma_2 \pm \sigma}{\sigma_1 + \sigma_2}; \frac{\sigma_1 \pm \sigma}{\sigma_1 + \sigma_2}\right).$$

Substituting here the data of the task, we obtain

$$X = \left(\frac{0.76 \pm 0.05}{0.15 + 0.76}; \frac{0.15 \pm 0.05}{0.15 + 0.76}\right).$$

Therefore, there are two portfolios

$$X_1 = \left(\frac{0.76 + 0.05}{0.15 + 0.76}; \frac{0.15 - 0.05}{0.15 + 0.76}\right) = (0.89; 0.11),$$

$$X_2 = \left(\frac{0.76 - 0.05}{0.15 + 0.76}; \frac{0.15 + 0.05}{0.15 + 0.76}\right) = (0.78; 0.22).$$

Their returns are determined by the formula

$$\mu = x_1\mu_1 + x_2\mu_2,$$

$$\mu^{(1)} = X_1^{(1)}\mu_1 + X_2^{(1)}\mu_2 = 0.89 \times 0.1 + 0.11 \times 0.45 = 0.1385,$$

$$\mu^{(2)} = X_1^{(2)}\mu_1 + X_2^{(2)}\mu_2 = 0.78 \times 0.1 + 0.22 \times 0.45 = 0.177.$$

Finally, there are two portfolios: X_1 (0.89; 0.11) with parameters (0.1385; 0.05) and X_2 (0.78; 0.22) with parameters (0.177; 0.05).

The ambiguity of the task solution is due to the fact that in the case of full anticorrelation the allowed set of portfolios has the form of an 'angle' (see Fig. 19.1), and the horizontal line $\sigma = 0.05$ crosses this angle in two points with different values of returns.

Questions and Tasks

19.35. Describe a portfolio of two securities in the case of full anticorrelation.

19.36. For a portfolio of two securities with return and risk of (0.3; 0.6) and (0.5; 0.9) respectively in the case of full anticorrelation, find the zero-risk portfolio and its return.

19.37. A portfolio consists of two securities *A* and *B* whose expected return and risk expressed in percent are *A* (12; 18), *B* (20; 47). The correlation coefficient of the securities is -1. Determine the return of the minimum risk portfolio and the risk of the maximum return portfolio. Draw the graph of the acceptable set of portfolios.

19.38. A portfolio consists of two assets whose expected return and risk (in percent) are *A* (18; 6), *B* (10; 5). The correlation coefficient of assets *A* and *B* is -1. Determine the return of the minimum risk portfolio and the risk of the maximum return portfolio. Draw the graph of the acceptable set of portfolios.

19.39. A portfolio consists of two securities *A* and *B* whose expected return and risk expressed in percent are *A* (20; 60), *B* (13; 45). The correlation coefficient of the securities is -1. Determine the return of the minimum risk portfolio and the risk of the maximum return portfolio. Draw the graph of the acceptable set of portfolios.

19.40. A portfolio consists of two securities *A* and *B*, the expected return and risk of which, expressed in percent, are equal to *A* (7; 14), *B* (44; 25). The correlation coefficient of the securities is -1 and its return is 15%. Find the portfolio and its risk.

19.41. A portfolio consists of two securities *A* and *B* whose expected return and risk, expressed as a percentage, are *A* (20; 24), *B* (41; 36). The correlation coefficient of the securities is -1 and its return is 32%. Find the portfolio and its risk.

19.42. A portfolio consists of two securities A and B whose expected return and risk, expressed as a percentage, are A (11; 9), B (18; 20). The correlation coefficient of the securities is -1 and its return is 14%. Determine the portfolio and its risk.

19.43. A portfolio consists of two securities A and B whose expected return and risk, expressed as a percentage, are A (15; 30), B (24; 45). The correlation coefficient of the securities is -1 and its risk is 5%. Find the portfolio and its return, and explain the ambiguity of the solution to the task.

19.44. A portfolio consists of two securities A and B whose expected return and risk expressed in percent are A (5; 15), B (23; 43). The correlation coefficient of the securities is -1 and its risk is 7%. Find the portfolio and its return, and explain the ambiguity of the solution to the task.

19.45. A portfolio consists of two securities A and B whose expected return and risk expressed in percent are A (10; 28), B (20; 90). The correlation coefficient of the securities is -1 and its risk is 30%. Find the portfolio and its return, and explain the unambiguity of the solution to the problem.

19.46. The portfolio consists of two securities A and B, whose expected returns are 15% and 46%, and whose price shares are 4:7. The correlation coefficient of the securities is -1. Find the portfolio and its return.

19.47. A portfolio consists of two securities A and B, whose expected returns are 17% and 27%, and the price share of security A is 4 times less than the price share of security B. The correlation coefficient is -1. Find the portfolio and its return.

19.48. A portfolio consists of two securities A and B whose expected returns are 32% and 28%, and the price share of security B is 24% less than the price share of security A. The correlation coefficient of the securities is -1. What is the portfolio and its return?

19.2.4 Independent Securities

For independent securities

$$\rho_{12} = \rho = 0. \tag{19.28}$$

For the squared risk (variance) of the portfolio we have

$$\sigma^2 = \sigma_1^2 x_1^2 + \sigma_2^2 x_2^2. \tag{19.29}$$

The minimum risk portfolio is of the form:

$$X = \left(\frac{\sigma_2^2}{\sigma_1^2 + \sigma_2^2}, \frac{\sigma_1^2}{\sigma_1^2 + \sigma_2^2} \right). \tag{19.30}$$

and its return

$$\mu = \frac{\mu_1 \sigma_2^2}{\sigma_1^2 + \sigma_2^2} + \frac{\mu_2 \sigma_1^2}{\sigma_1^2 + \sigma_2^2}. \tag{19.31}$$

The portfolio risk is equal to

$$\sigma = \frac{\sigma_1 \sigma_2}{\sqrt{\sigma_1^2 + \sigma_2^2}}. \tag{19.32}$$

Note that in the case of three independent securities there is no direct analogy with formula (19.30).

Example 19.12 The minimum risk portfolio of two independent securities (0.75; 0.33), (0.48; x) (the first figure in brackets is the return of the security, the second figure is its risk) has the form (0.65; 0.35). Find the risk of the second security, return and risk of the minimum risk portfolio.

$$X = \left(\frac{\sigma_2^2}{\sigma_1^2 + \sigma_2^2}, \frac{\sigma_1^2}{\sigma_1^2 + \sigma_2^2} \right).$$

For the first component of the portfolio there is

$$x_1 = \frac{\sigma_2^2}{\sigma_1^2 + \sigma_2^2}.$$

Substituting the data into this formula, we obtain

$$0.65 = \frac{\sigma_2^2}{0.33^2 + \sigma_2^2}, \text{ from where } \sigma_2 = 0.45.$$

Portfolio return is determined by the formula

$$\mu = \frac{\mu_1 \sigma_2^2}{\sigma_1^2 + \sigma_2^2} + \frac{\mu_2 \sigma_1^2}{\sigma_1^2 + \sigma_2^2}.$$

Substituting the data into this formula, we obtain

$$\mu = \frac{0.75 \times 0.45^2}{0.33^2 + 0.45^2} + \frac{0.48 \times 0.33^2}{0.33^2 + 0.45^2} = \frac{0.151875 + 0.052272}{0.3114} = 0.656.$$

Portfolio risk is determined by the formula

$$\sigma = \frac{\sigma_1 \sigma_2}{\sqrt{\sigma_1^2 + \sigma_2^2}},$$

$$\sigma = \frac{0.33 \times 0.45}{\sqrt{0.3114}} = 0.266.$$

Therefore, the portfolio of minimum risk has the following parameters (0.66; 0.27). It is clear that the risk of the portfolio is less than the risk of any of its securities, and the return is intermediate between the returns of both securities. If the investor wanted to get a portfolio of maximum return, he would include only the first security in it and would have a portfolio (1; 0) with parameters (0.75; 0.33).

Example 19.13 The portfolio of minimum risk from two independent securities (0.92; 0.37), (x; y) (the first figure in brackets is the yield of the security, the second figure is its risk) has the form (0.15; 0.85), its return is 0.71. Find the risk of the second security, its return, and the risk of the minimum risk portfolio

$$X = \left(\frac{\sigma_2^2}{\sigma_1^2 + \sigma_2^2}, \frac{\sigma_1^2}{\sigma_1^2 + \sigma_2^2} \right).$$

For the first component of the portfolio, we have the following formula

$$x_1 = \frac{\sigma_2^2}{\sigma_1^2 + \sigma_2^2}.$$

Substituting the data here, we get

$$0.15 = \frac{\sigma_2^2}{0.37^2 + \sigma_2^2}, \text{ from which } \sigma_2 = 0.155.$$

The profitability of the portfolio is based on the formula

$$\mu = \frac{\mu_1 \sigma_2^2}{\sigma_1^2 + \sigma_2^2} + \frac{\mu_2 \sigma_1^2}{\sigma_1^2 + \sigma_2^2}.$$

Substituting the data here, we get

$$0.71 = \frac{0.92 \cdot 0.155^2}{0.37^2 + 0.155^2} + \frac{\mu_2 \cdot 0.37^2}{0.37^2 + 0.155^2}.$$

From here we find $\mu_2 = 0.673$.

So, we found the parameters of the second security (0.673; 0.155).

The portfolio risk is determined by the formula

$$\sigma = \frac{\sigma_1 \sigma_2}{\sqrt{\sigma_1^2 + \sigma_2^2}},$$

$$\sigma = \frac{0.37 \cdot 0.155}{\sqrt{0.16}} = 0.143.$$

So, the minimum risk portfolio has the following parameters (0.71; 0.143). It can be seen that the risk of the portfolio is less than the risk of any of its securities, and the yield is intermediate between the yields of both securities. If an investor wanted to get a portfolio of maximum profitability, he would include only the first security in it and would have a portfolio (1; 0) with parameters (0.92; 0.37).

Questions and Tasks

19.49. For a portfolio of two independent securities, find the minimum risk portfolio and its profitability.

19.50. Find a minimum risk portfolio of two independent securities, the variances of which are 10 and 15, respectively.

19.51. The portfolio consists of two securities A and B. Expected returns are 0.5 and 0.8, and risks are 0.2 and 0.6. The correlation coefficient is zero. Determine the minimum risk portfolio and its profitability.

19.52. The minimum risk portfolio of two independent securities (0.43; 0.23), (0.67; x) (the first digit in parentheses is the yield of the security, the second is its risk) has the form (0.5; 0.5). What is the risk of the second security, the profitability and the risk of the minimum risk portfolio?

19.53. The minimum risk portfolio of two independent securities (0.3; 0.4), (0.5; x) (the first digit in parentheses is the yield of the security, the second is its risk) has the form (0.2; 0.8). Find the risk of the second security, the profitability and the risk of the minimum risk portfolio.

19.54. The price shares of the minimum risk portfolio of two independent securities (0.3; 0.4), (0.5; x) (the first digit in parentheses is the yield of the security, the second is its risk) are treated as 1:2. Find the risk of the second security, the profitability and the risk of the minimum risk portfolio.

19.55. In the minimum risk portfolio of two independent securities (0.3; 0.4), (0.5; x) (the first digit in parentheses is the yield of the security, the second is its risk), the price share of the first security is 3.5 times higher than the price share of the second. Find the risk of the second security, the profitability and the risk of the minimum risk portfolio.

19.56. In the minimum risk portfolio of two independent securities (0.3; 0.4), (0.5; x) (the first figure in parentheses is the yield of the security, the second is its risk), the price share of the second security is 34% higher than the price share of the first. Find the risk of the second security, the profitability and the risk of the minimum risk portfolio.

19.57. The yield of the minimum risk portfolio of two independent securities (0.75; 0.33), (x; 0.48) is 0.65 (the first digit in parentheses is the yield of the security, the second is its risk). Determine the minimum risk portfolio, its risk, as well as the profitability of the second security.

19.58. The risk of a minimum risk portfolio of two independent securities (0.2; x), (0.6; 0.8) is 0.5 (the first digit in parentheses is the yield of the security, the second is its risk). Find the risk of the first security, the minimum risk portfolio and its profitability.

19.59. Minimum risk portfolio of two independent securities (0.75; 0.33), (0.48; x) (the first digit in parentheses is the yield of the security, the second is its risk) has the form (0.75; 0.25). Find the risk of the second security, the yield and the risk of the minimum risk portfolio.

19.60. Minimum risk portfolio of two independent securities (0.75; 0.33), (x; y) (the first digit in parentheses is the yield of the security, the second is its risk) has the form (0.75; 0.25), its yield is 0.63. Find the risk of the second security, its profitability, as well as the risk of the minimum risk portfolio.

19.61. The price shares of the minimum risk portfolio of two independent securities (0.3; 0.4), (0.5; x) (the first digit in parentheses is the yield of the security, the second is its risk) are treated as 2:9. Find the risk of the second security, the profitability and the risk of the minimum risk portfolio.

19.62. In a minimum risk portfolio of two independent securities (0.25; 0.46), (0.56; x) (the first digit in parentheses is the yield of the security, the second is its risk) the price share of the first security is 1.6 times higher than the price share of the second. Find the risk of the second security, the profitability and the risk of the minimum risk portfolio.

19.63. In the minimum risk portfolio of two independent securities (0.4; 0.2), (0.6; x) (the first digit in parentheses is the yield of the security, the second is its risk), the price share of the second security is 10% higher than the price share of the first. What is the risk of the second security, the profitability and the risk of the minimum risk portfolio?

19.64. In the minimum risk portfolio of two independent securities, the risk of the second security is 20%, the price share of the second security is 19% higher than the price share of the first. The yield of the portfolio is 43%, and the yields of securities are treated as 1:5. Find the risk of the first security, the profitability of both security's and the risk of the minimum risk portfolio

19.65. In the minimum risk portfolio of two independent securities, the risk of the second security is 40%, the price share of the second security is 1.47 times the price share of the first. The yield of the portfolio is 15%, and the yields of securities are treated as 3:7. Find the risk of the first security, the profitability of both security's and the risk of the minimum risk portfolio.

19.66. In the minimum risk portfolio of two independent securities, the risk of the first security is 17%, the price share of the second security is 15% higher than the price share of the first. The yield of the portfolio is 30%, and the yield of the second security is 33% higher than the yield of the first. Find the risk of the second security, the profitability of both security's and the risk of the minimum risk portfolio.

19.67. The yields of the two independent securities that make up the minimum risk portfolio are 32% and 44%. The price share of the first security is 5 times higher than the price share of the second, and the risk of the first security is 1.5 times less than the risk of the second. Find the risks of both securities, the profitability of the minimum risk portfolio, if its risk is 50%.

19.68. The yields of the two independent securities that make up the minimum risk portfolio are 16% and 32%. The price share of the first security is 40% higher than the price share of the second, and the risk of the first security is 30% less than the risk of the second. Find the risks of both securities, the profitability of the minimum risk portfolio, if its risk is 10%.

19.69. The yields of the two independent securities that make up the minimum risk portfolio are 11% and 18%. The risks of securities (from the first to the second) are treated as 1:2.19. Find the risks of both securities, the profitability of the portfolio of minimum risk, if its risk is 10%.

19.2.5 A Portfolio of Two Arbitrary Securities

Questions and Tasks

19.70. Find a minimum risk portfolio of two independent securities, the variances of which are 225 and 49, respectively, and its risk.

19.71. The portfolio consists of two securities A and B, the expected return and risk of which, expressed as a percentage, are A (12; 14), B (18; 15). The correlation coefficient of the securities is -0.5. The expected return on the portfolio is 119.2. Find the shares of securities A and B in the portfolio and the risk of the portfolio.

19.72. The portfolio consists of two assets, the expected return and the risk (in percent) of which are A (18; 6), B (10; 5). The correlation coefficient of assets A and B is -0.5. Find the minimum risk portfolio, its risk and profitability.

19.73. The portfolio consists of two securities A and B. The expected returns are 0.2 and 0.4, and the risks are 0.3 and 0.5. The correlation coefficient is 0.2. Find the minimum risk portfolio, its risk and profitability.

19.74. The portfolio consists of two securities A and B, the expected return and risk of which, expressed as a percentage, are equal to A (20; 60), B (13; 45). The correlation coefficient of the securities is -0.25. Define the minimum risk portfolio, its risk and profitability.

19.75. The portfolio consists of two securities A and B. The expected returns are 0.6 and 0.4, and the risks are 0.1 and 0.5. The correlation coefficient is -0.3. What is the minimum risk portfolio, its risk and profitability?

19.76. The portfolio consists of two securities A and B, the expected return and risk of which, expressed as a percentage, are A (16; 25), B (10; 14). The correlation coefficient of the securities is 0.3. Find the minimum risk portfolio, its risk and profitability.
19.77. The portfolio consists of two securities A and B. The expected returns are 0.5 and 0.8, and the risks are 0.2 and 0.6. The correlation coefficient is 1/2. Find the minimum risk portfolio, its risk and profitability.
19.78. The portfolio consists of two securities A and B, the expected return and risk of which, expressed as a percentage, are equal to A (7; 19), B (12; 24). The correlation coefficient of the securities is 0.35. Find the minimum risk portfolio, its risk and profitability.
19.79. The portfolio consists of two securities A and B. Expected returns are 0.6 and 0.9, and risks are 0.1 and 0.5. The correlation coefficient is −0.36. Find the minimum risk portfolio, its risk and profitability.
19.80. The portfolio consists of two securities A and B. The expected returns are 0.5 and 0.8, and the risks are 0.2 and 0.6. The correlation coefficient is 0.7. Find the minimum risk portfolio, its risk and profitability.
19.81. The portfolio consists of two securities A and B. Expected returns are 0.6 and 0.8, and risks are 0.1 and 0.5. The correlation coefficient is −0.15. Find the minimum risk portfolio, its risk and profitability.
19.82. The portfolio consists of two securities A and B, the expected returns are 0.35 and 0.75, and the risks are 0.2 and 0.6. The correlation coefficient is 1/2. Find the minimum risk portfolio and its profitability.
19.83. The portfolio consists of two securities A and B. The expected returns are 0.22 and 0.41, and the risks are 0.32 and 0.519. The correlation coefficient is 0.2. Determine the minimum risk portfolio, its risk and profitability.
19.84. The portfolio consists of two securities A and B, the expected returns are 0.6 and 0.9, and the risks are 0.1 and 0.5. The correlation coefficient is −0.3. The risk of the portfolio is 0.19. Find the portfolio and its profitability.
19.85. The portfolio consists of two securities A and B. The expected returns are 0.5 and 0.8, and the risks are 0.2 and 0.6. The correlation coefficient is 0.7. The portfolio yield is 0.7. Find the portfolio and its risk.
19.86. The portfolio consists of two securities A and B. The expected returns are 0.2 and 0.4, and the risks are 0.3 and 0.5. The correlation coefficient is 0.2. The portfolio risk is 0.35. Find the portfolio and its profitability.
19.87. The portfolio consists of two securities A and B. The expected returns are 0.6 and 0.8, and the risks are 0.1 and 0.5. The correlation coefficient is −0.15. The return on the portfolio is 0.66. Identify the portfolio and its risk.
19.88. The portfolio consists of two securities A and B. The expected returns are 0.35 and 0.75, and the risks are 0.2 and 0.6. The correlation coefficient is 1/2. The risk of the portfolio is 0.55. Find the portfolio and its profitability.
19.89. The portfolio consists of two securities A and B. The expected returns are 0.22 and 0.41, and the risks are 0.32 and 0.519. The correlation coefficient is 0.2. The return on the portfolio is 0.319. Find the portfolio and its risk.

19.2.6 Three Independent Securities

For independent securities

$$\rho_{12} = \rho_{13} = \rho_{23} = 0 \tag{19.33}$$

For the risk square (variance) of the portfolio, we have

$$\sigma^2 = \sigma_1^2 x_1^2 + \sigma_2^2 x_2^2 + \sigma_3^2 x_3^2 \tag{19.34}$$

The minimum risk portfolio has the form

$$X = \frac{1}{\sigma_2^2\sigma_3^2 + \sigma_2^2\sigma_3^2 + \sigma_1^2\sigma_2^2}\left(\sigma_2^2\sigma_3^2; \sigma_1^2\sigma_3^2; \sigma_1^2\sigma_2^2\right) \tag{19.35}$$

and its profitability is equal to

$$\mu = \frac{\mu_1\sigma_2^2\sigma_3^2 + \mu_2\sigma_2^2\sigma_3^2 + \mu_3\sigma_1^2\sigma_2^2}{\sigma_2^2\sigma_3^2 + \sigma_2^2\sigma_3^2 + \sigma_1^2\sigma_2^2} \tag{19.36}$$

The risk of the portfolio is equal to

$$\sigma = \frac{\sigma_1\sigma_2\sigma_3}{\sqrt{\sigma_2^2\sigma_3^2 + \sigma_2^2\sigma_3^2 + \sigma_1^2\sigma_2^2}} \tag{19.37}$$

Example 19.14 For a portfolio of three independent securities with yield and risk, respectively (0.1; 0.4), (0.2; 0.6) and (0.4; 0.8) find the minimum risk portfolio, its risk and profitability.

The minimum risk portfolio has the form (19.35)

$$X = \frac{1}{\sigma_2^2\sigma_3^2 + \sigma_2^2\sigma_3^2 + \sigma_1^2\sigma_2^2}\left(\sigma_2^2\sigma_3^2; \sigma_1^2\sigma_3^2; \sigma_1^2\sigma_2^2\right) =$$

$$\frac{\left(0.6^2 \cdot 0.8^2; 0.4^2 \cdot 0.8^2; 0.4^2 \cdot 0.6^2\right)}{0.6^2 \cdot 0.8^2 + 0.4^2 \cdot 0.8^2 + 0.4^2 \cdot 0.6^2} = \frac{(0.2304; 0.1024; 0.0576)}{0.2304 + 0.1024 + 0.0576} =$$

$$\frac{(0.2304; 0.1024; 0.0576)}{0.3904} = (0.590; 0.263; 0.147).$$

So, X = (0.580; 0.263; 0.147).

The risk of the minimum risk portfolio is found by the formula (19.37):

$$\sigma = \frac{\sigma_1\sigma_2\sigma_3}{\sqrt{\sigma_2^2\sigma_3^2 + \sigma_2^2\sigma_3^2 + \sigma_1^2\sigma_2^2}} = \frac{0.4 \cdot 0.6 \cdot 0.8}{\sqrt{0.6^2 \cdot 0.8^2 + 0.4^2 \cdot 0.8^2 + 0.4^2 \cdot 0.6^2}}$$
$$= \frac{0.192}{\sqrt{0.2304 + 0.1024 + 0.0576}} = \frac{0.192}{\sqrt{0.3904}} = \frac{0.192}{0.6348} = 0.307$$

Finally, the portfolio yield is calculated using the formula (19.36):

$$\mu = \frac{\mu_1\sigma_2^2\sigma_3^2 + \mu_2\sigma_2^2\sigma_3^2 + \mu_3\sigma_1^2\sigma_2^2}{\sigma_2^2\sigma_3^2 + \sigma_2^2\sigma_3^2 + \sigma_1^2\sigma_2^2}$$
$$= \frac{0.1 \cdot 0.6^2 \cdot 0.8^2 + 0.2 \cdot 0.4^2 \cdot 0.8^2 + 0.4 \cdot 0.4^2 \cdot 0.6^2}{0.6^2 \cdot 0.8^2 + 0.4^2 \cdot 0.8^2 + 0.4^2 \cdot 0.6^2}$$
$$= \frac{0.02304 + 0.02048 + 0.02304}{0.2304 + 0.1024 + 0.0576} = \frac{0.06656}{0.3904} = 0.1705$$

It can be seen that the portfolio risk is less than the risk of each individual security, and the portfolio yield is greater than the yield of the first security, slightly less than the yield of the second and less than the yield of the third.

Questions and Tasks

19.90. For a portfolio of three independent securities, find the minimum risk portfolio and its profitability.

19.91. Find a minimum risk portfolio of three independent securities, the variances of which are 9, 16 and 25, respectively.

19.92. Find a minimum risk portfolio of three independent securities (0.35; 0.75), (0.2; 0.6), (0.4; 0.8), its profitability and risk (the first figure in parentheses is the yield of the security, the second is its risk).

19.93. Find a minimum risk portfolio of three independent securities (0.3; 0.25), (0.4; 0.3), (0.5; 0.67), its profitability and risk (the first the figure in parentheses is the yield of the security, the second is its risk).

19.94. Minimum risk portfolio of three independent securities (x; 0.75), (0.2; 0.6), (0.4; 0.8) it has a yield of 0.5 (the first digit in parentheses is the yield of the security, the second is its risk). What is the yield of the first securities?

19.95. The profitability of the minimum risk portfolio of three independent securities (0.15; 0.25), (0.9; 0.6), (x; 0.8) is equal to 0.75 (the first digit in parentheses is the yield of the security, the second is its risk). Find the yield of the third security

19.96. The risk of a minimum risk portfolio of three independent securities (0.4; x), (0.6; 0.9), (0.3; 0.6) equal to 0.5. Determine the risk of the first security, the minimum risk portfolio and its profitability.

19.97. The risk of a minimum risk portfolio of three independent securities (0.3; 0.2), (0.5; x), (0.8; 0.6) is 0.1. Find the risk of the second security, the minimum risk portfolio and its profitability.

19.98. The price shares of securities in the minimum risk portfolio of three independent securities are treated as 1:3:6, and the risks of the first and second securities are 11% and 18%. Find a minimum risk portfolio, its risk and the risk of a third security.

19.99. The price shares of securities in the minimum risk portfolio of three independent securities are treated as 2:5:7, and the risks of the first and third securities are equal to 15% and 27%. Find the minimum risk portfolio, its risk and the risk of the second security.

19.100. The price shares of securities in the minimum risk portfolio of three independent securities constitute an arithmetic progression with a difference 0.12, and the risks of the first and third securities are equal to 9% and 16%. Find the minimum risk portfolio, its risk and the risk of the second security.

19.101. Price shares of securities in the minimum risk portfolio of three the independent securities make up a decreasing arithmetic progression with a difference of −0.2, and the risks of the second and third securities are equal 25% and 30%. Find the minimum risk portfolio, its risk and the risk of the first security.

19.102. The price shares of securities in the minimum risk portfolio of three independent securities are a geometric progression with a denominator of 1.13, and the risks of the second and third securities are 16% and 45%. Find the minimum risk portfolio, its risk and the risk of the first security.

19.103. The price shares of securities in the minimum risk portfolio of three independent securities are a geometric progression with a denominator of 0.85, and the risks of the first and third securities are equal to 14% and 23%. Determine the minimum risk portfolio, its risk and the risk of the second security.

19.104. Profitability of the minimum risk portfolio of three independent securities (0.25; 0.16), (0.41; 0.65), (x; y) is equal to 0.35 (the first digit in parentheses is the yield of the security, the second is its risk). The price shares of the portfolio's securities are a geometric progression with a denominator of 0.35. Find the minimum risk portfolio, its risk, as well as the yield and risk of the third security.

19.105. Profitability of the minimum risk portfolio of three independent securities (0.35; 0.19), (0.46; 0.25), (x; y) is equal to 0.5 (the first digit in parentheses is the yield of the security, the second is its risk). The price shares of the portfolio securities are an arithmetic progression with a difference of 0.16. What is the minimum risk portfolio, its risk, as well as the profitability and risk of the third security.

19.106. Profitability of the minimum risk portfolio of three independent securities (0.2; 0.4), (0.3; 0.5), (x; y) is equal to 0.29 (the first digit in parentheses is the yield of the security, the second is its risk). Price shares securities are classified as 1:4:3. Find the minimum risk portfolio, its risk, as well as the profitability and risk of the third security.

19.2.7 Risk-Free Security

Let one of the two securities in the portfolio be risk-free. A portfolio of n-securities, including a risk-free security, bears Tobin's name. Let's consider how the inclusion of a risk-free security in a portfolio of two securities affects.

So, we have two securities: 1 (μ_1, 0) and 2 (μ_1, σ_1), while $\mu_1 < \mu_2$ (otherwise it would be necessary to form a portfolio (1; 0) consisting only of risk-free securities, and we would have a risk-free portfolio of maximum profitability).

The initial equations have the form

$$\begin{gathered}\mu = \mu_1 x_1 + \mu_2 x_2; \\ \sigma = \sigma_2 x_2; \\ \mu = \mu_1 x_1 + \mu_2 x_2\end{gathered} \tag{19.38}$$

The allowable set of portfolios is given by the equation

$$\mu = \mu_1(1 - x_2) + \mu_2 x_2 = \mu_1 + (\mu_2 - \mu_2)x_2 = \mu_1 + (\mu_2 - \mu_1)\frac{\sigma}{\sigma_2},$$

that is, it is a segment

$$\mu = \mu_1 + (\mu_2 - \mu_1)\frac{\sigma}{\sigma_2}, \quad 0 \le \sigma \le \sigma_2 \tag{19.39}$$

At $\sigma = 0$, the portfolio is located at point 1 (μ_1, 0), and at $\sigma = \sigma_2$ at point 2 (μ_2, σ_2) (Fig. 19.2).

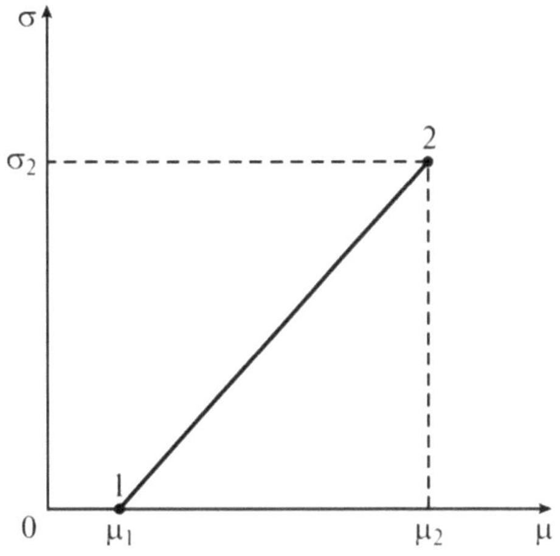

Fig. 19.2 An acceptable set of portfolios consisting of two securities, one of which is risk-free

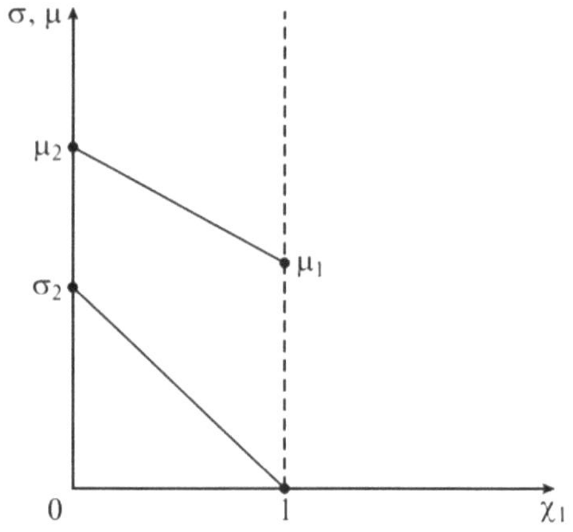

Fig. 19.3 Dependence of portfolio profitability and risk on the share of risk-free security x_1

Although this case is very simple, nevertheless two conclusions can be drawn:

1. The acceptable set of portfolios does not depend on the correlation coefficient (although usually a risk-free security is considered uncorrelated with other (risky) securities.
2. The allowable set of portfolios has narrowed from a triangle to a segment. A similar effect occurs in the case of Tobin's portfolio.

In conclusion, we present the dependence of the profitability and risk of the portfolio on the share of risk-free security (Fig. 19.3).

It can be seen from the figure that the portfolio risk decreases linearly from σ_2 at $x_1 = 0$ to zero at $x_1 = 1$, while the yield also decreases linearly from μ_2 at $x_1 = 0$ to μ_1 at $x_1 = 1$.

Example 19.15 One of the two securities in the portfolio is risk-free. A risky security has parameters (0.3; 0.5), the yield of a risk-free security is 0.2. Find the portfolio and its risk if its yield is 0.27.

The dependence of portfolio profitability on risk has the form:

$$\mu = \mu_1 + (\mu_2 - \mu_1)\frac{\sigma}{\sigma_2}.$$

Substituting the data here, we get

$$0.27 = 0.2 + (0.3 - 0.2)\frac{\sigma}{0.5}$$

Hence $\sigma = 0.35$.

Now let's find a portfolio

$$\sigma = \sigma_2 x_2, \; x_2 = \frac{\sigma}{\sigma_2} = \frac{0.35}{0.5} = 0.7,$$

$$x_1 = 1 - x_2 = 0.3.$$

The portfolio has the form (0.3; 0.7).

Example 19.16 One of the two securities in the portfolio is risk-free. Riskt security has parameters (0.4; 0.7), the yield of risk-free security is 0.31. Find the portfolio and its profitability if its risk is 0.55.

The dependence of portfolio profitability on risk has the form

$$\mu = \mu_1 + (\mu_2 - \mu_1)\frac{\sigma}{\sigma_2}.$$

Substituting the data here, we get

$$\mu = 0.31 + (0.4 - 0.31)\frac{0.55}{0.7}$$

Hence $\mu = 0.38$.

Now let's find the portfolio.

$$\sigma = \sigma_2 x_2, \; x_2 = \frac{\sigma}{\sigma_2} = \frac{0.55}{0.7} = 0.786,$$

$$x_1 = 1 - x_2 = 0.2119.$$

The portfolio has the form (0.214; 0.786).

Questions and Tasks

19.107. Describe the properties of a portfolio of two independent securities, one of which is risk-free.

19.108. Print and draw in the figure the dependence of the profitability and risk of the portfolio on the share of risk-free securities.

19.109. Draw an acceptable set of portfolios consisting of two securities, one of which is risk-free.

19.110. One of the two securities in the portfolio is risk-free. Risky securities have parameters (0.5; 0.2), the yield of risk-free securities is equal to 0.3. Find the portfolio and its risk if its yield is 0.15.

19.111. One of the two securities in the portfolio is risk-free. A risky security has parameters (0.6; 0.3), the yield of a risk-free security is 0.19. Find the portfolio and its risk if its yield is 0.45.

19.112. One of the two securities in the portfolio is risk-free. The risky security has parameters (0.36; 0.71), the yield of the risk-free security is 0.05. Find the portfolio and its profitability if its risk is 0.319.

19.113. One of the two securities in the portfolio is risk-free. The risky security has parameters (0.45; 0.66), the yield of the risk-free security is 0.17. Determine the portfolio and its profitability if its risk is 0.49.

19.2.8 The Portfolio of the Specified Efficiency

In the case of a portfolio of two securities, setting the effectiveness of the portfolio either its risk is uniquely determined by the portfolio (except in the case of $\mu_1 = \mu_2$, when only the portfolio risk assignment uniquely determines the portfolio itself).

When setting the efficiency of a portfolio, it is definitely found as a solution to the system

$$\begin{cases} \mu = \mu_1 x_1 + \mu_2 x_2 \\ x_1 + x_2 = 1 \end{cases} \tag{19.40}$$

when setting the risk of a portfolio—as a solution to the system

$$\begin{cases} \sigma^2 = \sigma_1^2 x_1^2 + \sigma_2^2 x_2^2 + 2\rho_{12}\sigma_1\sigma_2 x_1 x_2 \\ x_1 + x_2 = 1 \end{cases} \tag{19.41}$$

Therefore, in the case of a portfolio of two securities, it is not necessary to talk about the minimum limit (the minimum risk of the portfolio for a given efficiency). The reasoning on this issue in a number of manuals is erroneous.

Let's consider the first case—the efficiency of the portfolio is set. Then the portfolio looks like

$$x_1 = \frac{\mu - \mu_2}{\mu_1 - \mu_2}, x_2 = \frac{\mu_1 - \mu}{\mu_1 - \mu_2} \tag{19.42}$$

It is interesting to note that in this task, the portfolio is determined not by the risks of the securities that make up the portfolio, but by their returns, unlike all the cases discussed earlier.

Substituting these expressions into the expression for the risk square of the portfolio, we get

$$\sigma^2 = \frac{\sigma_1^2(\mu - \mu_2)^2 + \sigma_2^2(\mu - \mu_1)^2 - 2\sigma_1\sigma_2\rho_{12}(\mu - \mu_1)(\mu - \mu_2)}{(\mu_2 - \mu_1)^2} \tag{19.43}$$

This equation is the equation of the (unambiguous) relationship of portfolio risk to its effectiveness, and not the equation of the minimum boundary, as mistakenly stated in some manuals.

Only in the case of $\mu_1 = \mu_2$, when for all values of $x_1$and x_2 the equality of μ = $\mu_1 = \mu_2$ is fulfilled and the permissible set of portfolios from the triangle narrows to a vertical segment, we can talk about the minimum boundary, which in this case consists of a single point (μ, σ_1) (for $\sigma_1 < \sigma_2$) or (μ, σ2) (for $\sigma_1 > \sigma_2$).

The case of complete correlation ($\rho_{12} = 1$)

$$\sigma = \left|\frac{\sigma_1(\mu - \mu_2) - \sigma_2(\mu - \mu_1)}{(\mu_2 - \mu_1)}\right|. \tag{19.44}$$

The case of complete correlation ($\rho_{12} = -1$)

$$\sigma = \left|\frac{\sigma_1(\mu - \mu_2) + \sigma_2(\mu - \mu_1)}{(\mu_2 - \mu_1)}\right|. \tag{19.45}$$

(2) Independent securities ($\rho_{12} = 0$)

Equation (19.43) takes the form

$$\sigma^2 = \frac{\sigma_1^2(\mu - \mu_2)^2 + \sigma_2^2(\mu - \mu_1)^2}{(\mu_2 - \mu_1)^2}. \tag{19.46}$$

For intermediate values of the correlation coefficient p, the portfolio risk as a function of its effectiveness has the form

$$\sigma^2 = \frac{\alpha\mu^2 - 2\beta\mu + \gamma}{\delta}. \tag{19.47}$$

If we find the type of dependence of the portfolio risk on its effectiveness for a fixed portfolio ($\mu_1\sigma_1$), (μ_2, σ_2), but with different values of the correlation coefficient p, then we can come to the following conclusion: with an increase in the correlation coefficient from −1 to 1, a decrease in μ_M occurs. At the same time, the graph of the portfolio's risk dependence on its effectiveness becomes more and more elongated along the abscissa axis, i.e. with a fixed change in the expected return, the increase in risk σ becomes less and less (see Fig. 19.1).

If we additionally assume that $x_1 \in [0, 1]$, and hence $x_2 \in [0, 1]$, then it follows from the first formula (19.40) that μ ∈ [μ_1, μ_2] under the assumption $\mu_1 < \mu_2$, since it is a convex combination of them. Portfolios form part of the AMB boundary, namely its part connecting the points (μ_1, σ_1) and (μ_2, σ_2) (see Fig. 19.1). Thus, in the case of n = 2 and under the additional assumption x1 ≥ 0, x2 ≥ 0, the set of portfolios is pieces of hyperballs or polylines connecting the points (μ_1, σ_1) and (μ_2, σ_2).

Questions and Tasks

19.114. The portfolio consists of two securities A and B. The expected returns are 0.4 and 0.9, and the risks are 0.1 and 0.5. The correlation coefficient is 0.5. The portfolio yield is 0.55. Find the portfolio and its risk.

19.115. The portfolio consists of two securities A and B. The expected returns are 0.2 and 0.6, and the risks are 0.3 and 0.19. The correlation coefficient is –0.5. The portfolio yield is 0.5. Find the portfolio and its risk.

19.116. The portfolio consists of two securities A and B. Expected returns are 0.5 and 0.8, and risks are 0.25 and 0.7. The correlation coefficient is –0.8. The return on the portfolio is 0.75. Find the portfolio and its risk.

19.117. A portfolio of two securities (x; 0.23), (0.67; 0.35) (he first figure in parentheses is the yield of the security, the second is its risk) has the form (0.55; 0.45) and a yield of 0.47. The correlation coefficient is –0.19. What is the yield of the first security and the risk of the portfolio?

19.118. Price shares of a portfolio of two securities (0.2; 0.4), (x; 0.5) (first the figure in parentheses is the yield of the security, the second is its risk). as 1:3.4, and the yield is 0.23. The correlation coefficient it is equal to 0.1. Find the portfolio, its risk and the yield of the second security.

19.119. In a portfolio of two securities (0.3; 0.26), (x; 0.6) (the first digit in parentheses is the yield of the security, the second is its risk), the price share of the first security is 19.5 times higher than the price share of the second, and the yield is 0.319. The correlation coefficient is –0.219. Find the portfolio, its risk and the yield of the second security.

19.120. In a portfolio of two securities (0.4; 0.5), (x; 0.6) (the first digit in parentheses is the yield of the security, the second is its risk) the price the share of the second security is 15% higher than the price share of the first, and the yield is 0.45. The correlation coefficient is 0.6. Find the portfolio, its risk and the yield of the second security.

19.121. Prove that the equation $\sigma = \sqrt{\frac{\alpha\mu^2 - 2\beta\mu + \gamma}{\delta}}$ is a branch of the hyperbola and find its asymptotes.

19.2.9 Portfolio of Specified Risk

Let's now set the portfolio risk.

The portfolio is now located as a (one-digit or two-digit, see figure) solution to the system

$$\begin{cases} \sigma^2 = \sigma_1^2 x_1^2 + \sigma_2^2 x_2^2 + 2\rho_{12}\sigma_1\sigma_2 x_1 x_2, \\ x_1 + x_2 = 1. \end{cases} \tag{19.48}$$

Portfolio at a given portfolio risk σ (it is included in the discriminant D) has the form

$$X = \left(\frac{-\sigma_2(\rho_{12}\sigma_1 - \sigma_2) \pm \sqrt{D}}{\sigma_1^2 - 2\rho_{12}\sigma_1\sigma_2 + \sigma_2^2}; \frac{\sigma_1(\sigma_1 - \rho_{12}\sigma_2) \pm \sqrt{D}}{\sigma_1^2 - 2\rho_{12}\sigma_1\sigma_2 + \sigma_2^2} \right), \tag{19.49}$$

where the discriminant D is equal to

$$D=\sigma_1^2\left(\sigma^2-\sigma_2+\rho_{12}^2\sigma_2^2\right)+\sigma_2^2\sigma^2-2\rho_{12}\sigma_1\sigma_2\sigma^2. \tag{19.50}$$

It can be seen that in the general case, the portfolio is defined ambiguously, and there are two portfolios with the same risk value, naturally having different returns.

There is, however, a portfolio risk value in which the portfolio is uniquely determined.

The condition for unambiguity is that the discriminant D is equal to zero. From it, we find the value of the portfolio risk, in which the portfolio is uniquely determined:

$$D=\sigma_1^2\left(\sigma^2-\sigma_2^2+p_{12}^2\sigma_2^2\right)+\sigma_2^2\sigma^2-2p_{12}\sigma_1\sigma_2\sigma^2=0,$$

hence

$$\sigma^2=\frac{\sigma_1^2\left(\sigma_2^2-p_{12}^2\sigma_2^2\right)}{\sigma_1^2+\sigma_2^2-2p_{12}\sigma_1\sigma_2},\sigma=\sqrt{\frac{\sigma_1^2\left(\sigma_2^2-p_{12}^2\sigma_2^2\right)}{\sigma_1^2+\sigma_2^2-2p_{12}\sigma_1\sigma_2}}$$

Let's consider the various limiting cases discussed above.

Independent securities ($p_{12}=0$):

$$D=\sigma_1^2\left(\sigma^2-\sigma_2^2\right)+\sigma_2^2\sigma^2; \tag{19.51}$$

$$X=\left(\frac{\sigma_2^2\pm\sqrt{D}}{\sigma_1^2+\sigma_2^2};\frac{\sigma_2^2\mp\sqrt{D}}{\sigma_1^2+\sigma_2^2}\right). \tag{19.52}$$

Note that this is not the minimum risk portfolio we found above, which has the form $X=\left(\frac{\sigma_2^2}{\sigma_1^2+\sigma_2^2};\frac{\sigma_1^2}{\sigma_1^2+\sigma_2^2}\right)$, but a portfolio with risk σ.

Case of complete correlation ($\rho_{12}=1$):

$$D=\sigma_1^2\sigma^2+\sigma_2^2\sigma^2-2\sigma_1\sigma_2\sigma^2=\sigma^2(\sigma_1-\sigma_2)^2; \tag{19.53}$$

$$X=\left(\frac{-\sigma_2\pm\sigma}{\sigma_1-\sigma_2};\frac{-\sigma_1\mp\sigma}{\sigma_1-\sigma_2}\right). \tag{19.54}$$

Case of complete anticorrelation ($\rho_{12}=-1$):

$$D=\sigma_1^2\sigma^2+\sigma_2^2\sigma^2+2\sigma_1\sigma_2\sigma^2=\sigma^2(\sigma_1+\sigma_2)^2; \tag{19.55}$$

$$X=\left(\frac{\sigma_2\pm\sigma}{\sigma_1+\sigma_2};\frac{-\sigma_1\mp\sigma}{\sigma_1+\sigma_2}\right). \tag{19.56}$$

Hence, it is easy to obtain a zero-risk portfolio by setting $\sigma=0$

$$X = \left(\frac{\sigma_2}{\sigma_1 + \sigma_2}; \frac{\sigma_1}{\sigma_1 + \sigma_2} \right), \tag{19.57}$$

obviously, it coincides with the one we obtained above in another way.

Questions and Tasks

19.122. The portfolio consists of two securities A and B. The expected returns are 0.1 and 0.5, and the risks are 0.4 and 0.6 respectively. The correlation coefficient is 0.19. The portfolio risk is 0.45. Find the portfolio and its returns.

19.123. Under what condition does the portfolio risk specification uniquely determine the portfolio?

19.124. The portfolio consists of two securities A and B, the risks of which are 0.8 and 0.4 respectively. The correlation coefficient is −0.3. At what value of portfolio is the portfolio uniquely determined? Do these values depend on the magnitude of the correlation coefficient?

19.125. The portfolio consists of two securities A and B, the risks of which are 0.2 and 0.5. The correlation coefficient is 0.7. At what value of portfolio risk is the portfolio uniquely determined?

19.126. The portfolio consists of two securities A and B, the risks of which are 0.1 and 0.35 respectively. The correlation coefficient is −0.6. Find the portfolio risk value at which the portfolio is uniquely determined.

19.127. The portfolio consists of two securities A and B, the risk of the first security is 0.25, the risk of the portfolio is 0.35, while its price shares are 2:5. The correlation coefficient is −0.6. Find the portfolio and the risk of the second security.

19.128. The portfolio consists of two securities A and B, the risk of the first security is 0.4, the risk of the portfolio is 0.3. The price share of security A is 2 times less than the price share of security B. The correlation coefficient is −0.6. Find the portfolio and the risk of the second security.

19.129. The portfolio consists of two securities A and B, the risk of the first security is 0.6, the risk of the portfolio is 0.5. The price share of security A is 15% less than the price share of security B. The correlation coefficient is −0.19. Find the portfolio and the risk of the second security.

19.130. The portfolio consists of two securities A and B, the expected returns of which are 0.3 and 0.4, the risk of security A is 0.26, the risk of the portfolio is 0.4, and its price shares are related as 1:3. The correlation coefficient is 0.719. Find the portfolio, its profitability and risk of paper B.

19.131. The portfolio consists of two securities A and B, the expected returns of which are 0.5 and 0.8, the risk of security A is 0.17, and the risk of the portfolio is 0.1. The price share of security A is 3 times less than the price share of security B. The correlation coefficient is −0.6. Find the portfolio, its profitability and risk of paper B.

19.132. The portfolio consists of two securities A and B, the expected returns of which are 0.25 and 0.35, the risk of security A is 0.7, and the risk of the portfolio is 0.65. The price share of security A is 25% less than the price share of security B. The correlation coefficient is −0.35. Find the portfolio, its return and the risk of paper B.

19.3 Portfolio of n-Securities: Markowitz Portfolios

19.3.1 Minimum Risk Portfolio for a Given Efficiency

Now we consider the following problem: we need to find a portfolio $X = (x_1, x_2, \ldots, x_n)^T$ that would minimize the risk σ and provide a given expected return μ.

In a mathematical formulation, this problem looks like the following: find the minimum of the objective function

$$\frac{1}{2}\sigma^2 = \frac{1}{2}X^T VX \to \min \tag{19.58}$$

under the conditions

$$\vec{\mu}^T X = \mu; \tag{19.59}$$

$$I^T = 1. \tag{19.60}$$

Note that the numerical multiplier in the objective function is introduced for convenience. We are looking for the minimum of the squared risk; this is also due to technical considerations. Condition (19.59) ensures this level of efficiency. Condition (19.61) follows from the definition of the vector X.

If we further assume that the vector *X* consists of non-negative numbers

$$X \geq 0, \tag{19.61}$$

then the components *X* can be interpreted as shares of the investment made in the corresponding asset. In general, among the numbers x_1, x_2, …, x_n there may be negative ones, which means a debt obligation.

In what follows, we assume that the covariance matrix V is positive definite and the vector of efficiencies $\vec{\mu} = (\mu_1, \mu_2, \ldots, \mu_n)^T$ is not collinear with the vector I, in other words, not all efficiencies are equal.

The first assumption, in particular, means that the covariance matrix is non-singular, which is true in practice for risky assets (stocks). In particular, there is an inverse matrix V^{-1}, which is also positive definite. We will consider the case of a risk-free asset later. If the second assumption is violated, the problem under consideration has a simpler solution, which will be indicated below.

Next, we will use the following constants:

$$\alpha = I^T V^{-1} I, \beta = I^T V^{-1} \vec{\mu} = \vec{\mu}^T V^{-1} I,$$
$$\gamma = \vec{\mu}^T V^{-1} \vec{\mu}, \delta = \alpha\gamma - \beta^2, \tag{19.62}$$

in this case, the constants α, γ, δ are positive numbers.

The problem of finding the optimal portfolio with objective function (19.58) under conditions (19.59) and (19.60) has a unique solution

$$X = V^{-1}\left(\lambda \mathrm{I} + \nu \vec{\mu}\right), \quad \lambda = (\gamma - \beta\mu)/\delta, \nu = (\alpha\mu - \beta)/\delta. \tag{19.63}$$

To find explicit formulas (19.63), we use the Lagrange multiplier method. Consider the Lagrange function for the optimization problem (19.58)–(19.60):

$$L(X, \lambda, \nu) = \frac{1}{2} X^T V X + \lambda\left(1 - I^T X\right) + \nu\left(\mu - \vec{\mu}^T X\right). \tag{19.64}$$

Equating the derivatives with respect to X, λ, ν to zero, we obtain a system of three equations

$$\begin{cases} VX = \lambda \mathrm{I} + \nu \vec{\mu}, \\ I^T X = 1, \\ \vec{\mu}^T X = \mu \end{cases} \tag{19.65}$$

Let us express the unknown X from the first equation

$$X = V^{-1}(\lambda \mathrm{I} + \nu\nu) \tag{19.66}$$

and substitute it into the second and third equations of the system

$$\begin{cases} \alpha\lambda + \beta\nu = 1, \\ \beta\lambda + \gamma\nu = \mu. \end{cases} \tag{19.67}$$

The determinant of system (19.67) is $\delta \neq 0$ (we proved above that $\delta > 0$), so it has a unique solution

$$\lambda = \frac{\gamma - \beta\mu}{\delta}, \nu = \frac{\alpha\mu - \beta}{\delta}. \tag{19.68}$$

These formulas, together with equality (19.66), give a solution to the optimization problem (19.58)–(19.60):

$$X = V^{-1}\left(\frac{\gamma - \beta\mu}{\delta} I + \frac{\alpha\mu - \beta}{\delta}\vec{\mu}\right). \tag{19.69}$$

Thus, for each value of expected return μ there is a unique portfolio X that provides the minimum risk value $\sigma = \sigma_{\min}$, i.e. function defined

$$\sigma = \sigma(\mu). \tag{19.70}$$

The graph of function (19.70) is called the minimum bound.

The minimal boundary is a branch of a hyperbola which equation is

$$\sigma = \sqrt{\frac{\alpha\mu^2 - 2\beta\mu + \gamma}{\delta}}, \tag{19.71}$$

with asymptotes $\sigma = \sqrt{\frac{\alpha}{\delta}}\left|\mu - \frac{\beta}{\alpha}\right|$ and absolute minimum $M\left(\frac{\beta}{\alpha}, \frac{1}{\sqrt{\alpha}}\right)$.

To find the required equation, it is enough to substitute the found solution X into the expression for σ:

$$\begin{aligned}\sigma^2 &= X^T V X = \left(V^{-1}\left(\lambda I + \nu\vec{\mu}\right)\right)^T V \cdot V^{-1}\left(\lambda I + \nu\vec{\mu}\right) = \\ &= \left(\lambda I^T + \nu\vec{\mu}^T\right)V^{-1}\left(\left(\lambda I + \nu\vec{\mu}\right) = \lambda^2\alpha + 2\lambda\nu\beta + \nu^2\gamma = \right. \\ &= \lambda(\lambda\alpha + \nu\beta) + \nu(\lambda\beta + \nu\gamma).\end{aligned} \tag{19.72}$$

Substituting the values for the expressions in brackets from system (19.67), we have

$$\sigma^2 = \lambda + \nu\mu = \frac{\gamma - \beta\mu}{\delta} + \frac{\alpha\mu - \beta}{\delta}\mu = \frac{\alpha\mu^2 - 2\beta\mu + \gamma}{\delta}, \tag{19.73}$$

from which we obtain the equation of the minimum boundary. Let's bring this equation to the canonical form $\frac{x^2}{a^2} - \frac{y^2}{b^2} = 1$, or, in our variables, $\frac{\sigma^2}{a^2} - \frac{\mu^2}{b^2} = 1$. To do this, select the perfect square on the right side of the equation

$$\sigma^2 = \frac{\alpha}{\delta}\left[\left(\mu - \frac{\beta}{\alpha}\right)^2 + \frac{\gamma}{\alpha} - \frac{\beta^2}{\alpha^2}\right] = \frac{\alpha}{\delta}\left[\left(\mu - \frac{\beta}{\alpha}\right)^2 + \frac{\delta}{\alpha^2}\right] = \frac{\alpha}{\delta}\left(\mu - \frac{\beta}{\alpha}\right)^2 + \frac{1}{\alpha}.$$

The canonical minimum bound equation has the form

$$\alpha\sigma^2 - \frac{\alpha^2}{\delta}\left(\mu - \frac{\beta}{\alpha}\right)^2 = 1,$$

or

$$\frac{\sigma^2}{a^2} - \frac{\tilde{\mu}^2}{b} = 1, \tag{19.74}$$

where

$$a^2 = \frac{1}{\alpha}, \ b^2 = \frac{\delta}{\alpha^2}, \ \tilde{\mu} = \mu - \frac{\beta}{\alpha}. \tag{19.75}$$

The minimal boundary is a branch of a hyperbola with asymptotes $\sigma = \sqrt{\frac{\alpha}{\delta}}\left|\mu - \frac{\beta}{\alpha}\right|$ and an absolute minimum $M\left(\frac{\beta}{\alpha}, \frac{1}{\sqrt{\alpha}}\right)$. We obtain the equation of asymptotes.

The equation of asymptotes has the form $y = \pm \frac{b}{a}x$, or, in our variables,

$$\sigma = \pm \frac{a}{b}\tilde{\mu} = \pm \frac{a}{b}\left|\mu - \frac{\beta}{\alpha}\right| = \frac{\alpha}{\alpha\sqrt{\delta\alpha}}\left|\mu - \frac{\beta}{\alpha}\right| = \sqrt{\frac{\alpha}{\delta}}\left|\mu - \frac{\beta}{\alpha}\right|. \tag{19.76}$$

It is easy to see that in the degenerate case, when all expected returns coincide and are equal to μ, the minimum bound comes down to one point $M\left(\mu, \frac{1}{\sqrt{\alpha}}\right)$, and $X = \frac{1}{\alpha}V^{-1}I$.

The minimum limit graph is shown in Fig. 19.4

AMB is the minimum boundary, $M\left(\frac{\beta}{\alpha}, \frac{1}{\sqrt{\alpha}}\right)$ is the absolute minimum point, dotted lines indicate asymptotes.

Example 19.17 Given portfolio consists of three securities with yields $\mu_1 = 10\%$, $\mu_2 = 20\%$, $\mu_3 = 30\%$ and covariance matrix $V = \begin{pmatrix} 1 & -1 & 0 \\ -1 & 9 & -2 \\ 0 & -2 & 4 \end{pmatrix}$. Find a minimum risk portfolio with a return of $\mu = 25\%$ and its risk. Write the minimum boundary equation.

Note that V is positive definite. Let's find the inverse matrix V^{-1}:

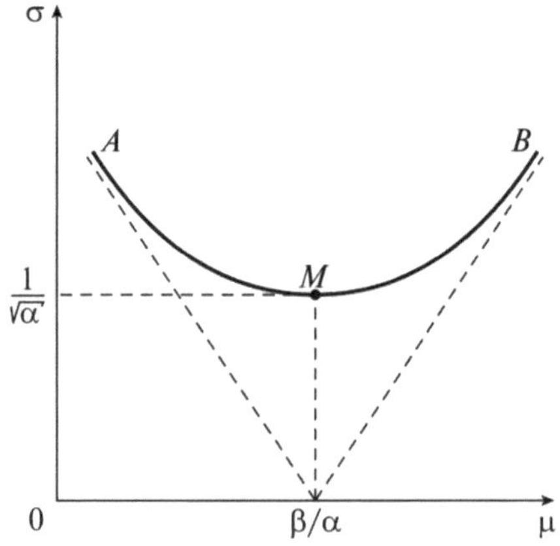

Fig. 19.4 View of the minimum boundary for the Markowitz portfolio

$$V^{-1} = \begin{pmatrix} 16 & 2 & 1 \\ 2 & 2 & 1 \\ 1 & 1 & 4 \end{pmatrix}.$$

Let's find the constants

$$\alpha = I^T V^{-1} I, \beta = I^T V^{-1} \vec{\mu} = \vec{\mu}^T V^{-1} I, \gamma = \vec{\mu}^T V^{-1} \vec{\mu}, \delta = \alpha\gamma - \beta^2.$$

Here $I = (1, 1, 1)^T$, $\mu = (10; 20; 30)^T$.

$$\alpha = I^T V^{-1} I = \frac{1}{14}(1,1,1) \cdot \begin{pmatrix} 16 & 2 & 1 \\ 2 & 2 & 1 \\ 1 & 1 & 4 \end{pmatrix} \cdot \begin{pmatrix} 1 \\ 1 \\ 1 \end{pmatrix} = \frac{1}{14} \cdot (19,5,6) \cdot \begin{pmatrix} 1 \\ 1 \\ 1 \end{pmatrix} = \frac{15}{7};$$

$$\beta = I^T V^{-1} \vec{\mu} = \frac{1}{14}(1,1,1) \cdot \begin{pmatrix} 16 & 2 & 1 \\ 2 & 2 & 1 \\ 1 & 1 & 4 \end{pmatrix} \cdot \begin{pmatrix} 10 \\ 20 \\ 30 \end{pmatrix} = \frac{1}{14}(19,5,6) \cdot \begin{pmatrix} 10 \\ 20 \\ 30 \end{pmatrix} = \frac{235}{7};$$

$$\gamma = \vec{\mu}^T V^{-1} \vec{\mu} = \frac{1}{14}(10,20,30) \cdot \begin{pmatrix} 16 & 2 & 1 \\ 2 & 2 & 1 \\ 1 & 1 & 4 \end{pmatrix} \cdot \begin{pmatrix} 10 \\ 20 \\ 30 \end{pmatrix} = \frac{1}{14}(230,90,150) \cdot \begin{pmatrix} 10 \\ 20 \\ 30 \end{pmatrix} = \frac{4300}{7};$$

$$\delta = \alpha\gamma - \beta^2 = \frac{4300}{7} \cdot \frac{15}{7} - \left(\frac{235}{7}\right)^2 = \frac{9275}{49} = 189.3.$$

Let's also find the constants

$$\lambda = (\gamma - \beta\mu)/\delta \text{ and } \nu = (\alpha\mu - \beta)/\delta;$$

$$\lambda = (\gamma - \beta\mu)/\delta = \left(\frac{4300}{70} - \frac{235}{7} \cdot 25\right)/189.3 = -225/189.3 = -1.19;$$

$$\nu = (\alpha\mu - \beta)/\delta = \left(\frac{15}{7} \cdot 25 - \frac{235}{7}\right)/189.3 = 20/189.3 = 0.106.$$

Now we can find a minimum risk portfolio with a return of $\mu = 25\%$

$$X = V^{-1}\left(\lambda I + \nu\vec{\mu}\right) = \frac{1}{14} \cdot \begin{pmatrix} 16 & 2 & 1 \\ 2 & 2 & 1 \\ 1 & 1 & 4 \end{pmatrix} \cdot \begin{pmatrix} -1.19 + 0.106 \cdot 10 \\ -1.19 + 0.106 \cdot 20 \\ -1.19 + 0.106 \cdot 30 \end{pmatrix} =$$

$$= \frac{1}{14} \cdot \begin{pmatrix} 16 & 2 & 1 \\ 2 & 2 & 1 \\ 1 & 1 & 4 \end{pmatrix} \cdot \begin{pmatrix} -0.13 \\ 0.93 \\ 1.99 \end{pmatrix} = \frac{1}{14} \cdot \begin{pmatrix} 1.77 \\ 3.59 \\ 8.76 \end{pmatrix} = (0.12; 0.26; 0.62)^T.$$

Thus, the minimum risk portfolio with a return of $\mu = 25\%$ is equal to $X = (0.12;\ 0.26;\ 0.62)^T$: it is necessary to take 12% of securities of the first type, 26% of the second and 62% of the third type.

Let's find the risk of portfolio

$$\sigma = \sqrt{X^T VX} = (0.12; 0.26; 0.62)\begin{pmatrix} 1 & -1 & 0 \\ -1 & 9 & -2 \\ 0 & -2 & 4 \end{pmatrix} \cdot \begin{pmatrix} 0.12 \\ 0.26 \\ 0.62 \end{pmatrix} =$$

$$= \sqrt{(-0.14; 0.98; 1.96)\begin{pmatrix} 0.12 \\ 0.26 \\ 0.62 \end{pmatrix}} = \sqrt{1.4532} = 1.205.$$

The portfolio risk turned out to be slightly greater than the risk of the first security ($\sigma_1 = 1$), but less than the risk of the second ($\sigma_2 = 3$) and third ($\sigma_3 = 2$) securities. Moreover, its return (25%) is 15% higher than the return of the first security, 5% higher than the profitability of the second and only 5% less than the return of the third security.

Let us note an interesting fact: the share (price share) of the second security in the minimum risk portfolio turned out to be more than 2 times higher than the share of the first security, despite the fact that the risk of the second security is 3 times higher than the risk of the first one. This means that the risk of a portfolio depends largely on the correlations of the securities, and not just on their individual risks.

In conclusion, we write down the form of the minimum boundary. According to formula (19.71)

$$\sigma = \sqrt{\frac{\alpha\mu^2 - 2\beta\mu + \gamma}{\delta}},$$

substituting here the values of the constants α, β, γ, δ, we obtain

$$\sigma = \sqrt{\frac{\frac{15}{7} \cdot \mu^2 - 2 \cdot \frac{235}{7} \cdot \mu + \frac{4300}{7}}{\frac{9275}{49}}} = \sqrt{0.011\mu^2 - 0.355\mu + 3.245.}$$

Thus, the minimum bound looks like

$$\sigma = \sqrt{0.011\mu^2 - 0.355\mu + 3.245,}$$

or

$$31.6 \cdot \sigma = \sqrt{11\mu^2 - 355\mu + 3245.}$$

Questions and Tasks

19.133. Describe the properties of Markowitz portfolios.

19.134. Find the minimum risk portfolio for a given expected return μ.

19.135. A portfolio consists of two securities with yields $\mu_1 = 18\%$, $\mu_2 = 23\%$, $\mu_3 = 35\%$ and covariance matrix $V = \begin{pmatrix} 25 & 5 & 12 \\ 5 & 36 & -10 \\ 12 & -10 & 81 \end{pmatrix}$. Find the minimum risk portfolio with a return of $\mu = 27\%$ and its risk. Write the equation for the minimum bound.

19.136. Given a portfolio of three securities with yields $\mu_1 = 8\%$, $\mu_2 = 28\%$, $\mu_3 = 45\%$ and covariance matrix $V = \begin{pmatrix} 16 & 2 & 10 \\ 2 & 9 & -8 \\ 10 & -8 & 25 \end{pmatrix}$. Find the minimum risk portfolio with a return of $\mu = 35\%$ and its risk. Write the equation for the minimum bound.

19.137. Given a portfolio of three securities with yields $\mu_1 = 6\%$, $\mu_2 = 24\%$, $\mu_3 = 34\%$ and covariance matrix $V = \begin{pmatrix} 64 & 4 & -14 \\ 4 & 49 & -9 \\ -14 & -9 & 36 \end{pmatrix}$. Find the minimum risk portfolio with a return of $\mu = 26\%$ and its risk. Write the equation for the minimum bound.

19.138. Derive the equation for the minimum bound $\sigma = \sqrt{\frac{\alpha\mu^2 - 2\beta\mu + \gamma}{\delta}}$.

19.139. Prove that the minimal boundary equation $\sigma = \sqrt{\frac{\alpha\mu^2 - 2\beta\mu + \gamma}{\delta}}$ is a branch of a hyperbola.

19.3.2 Markowitz Portfolio of Minimal Risk with Efficiency No Less Than the Specified One

Along with problem (1) (19.58)–(19.60), let's find the minimum risk portfolio among all portfolios with an efficiency no less than the given one (problem (1′)). Let's call such a portfolio the optimal Markowitz portfolio. To do this, consider the optimization problem: find the minimum of the objective function

$$\frac{1}{2}\sigma^2 = \frac{1}{2}X^T VX \to \min \tag{19.77}$$

under conditions

$$\vec{\mu}^T X \geq \mu, \tag{19.78}$$

$$I^T X = 1. \tag{19.79}$$

From the structure of the quadratic function (19.73), which defines the minimum boundary equation, it is clear that problems (1) and (1′) have the same solution for any $\mu \quad \geq \mu_0 \quad \beta/\alpha$, namely $\sigma_{\min} = \sqrt{\frac{\alpha\mu^2 - 2\beta\mu + \gamma}{\delta}}$, the portfolio itself $X = V^{-1}\left(\frac{\gamma - \beta\mu}{\delta} I + \frac{\alpha\mu - \beta}{\delta} \vec{\mu}\right)$.

For $\mu \leq \beta/\alpha$, the problems under consideration have different solutions: it is the solution to problem (1′) for all $\mu \leq \beta/\alpha$ that is the only solution to problem (1) for $\mu \leq \beta/\alpha$ (see Fig. 19.4). Namely

$$\sigma_{\min} = \sqrt{\frac{\alpha\mu_0^2 - 2\beta\mu_0 + \gamma}{\delta}} = \frac{1}{\sqrt{\alpha}}, \tag{19.80}$$

and the briefcase itself

$$X = V^{-1}\left(\frac{\gamma - \beta\mu_0}{\delta} I + \frac{\alpha\mu_0 - \beta}{\delta} \vec{\mu}\right) =$$

$$= V^{-1}\left(\frac{\gamma - \frac{\beta^2}{\alpha}}{\delta} I + \frac{\beta - \beta}{\delta} \vec{\mu}\right) = \frac{1}{\alpha} V^{-1} I. \tag{19.81}$$

It is clear that when $\mu \leq \beta/\alpha$ there is no point in solving problem (1)—it is necessary to solve problem (1′) and the solution to this problem will be better than the solution to problem (1), because the efficiency of the portfolio that is a solution to problem (1′) is equal to $\mu_0 = \beta/\alpha$, i.e. even more than required, and the dispersion is equal to $\sigma^2 = 1/\alpha$, i.e. even less than that of the portfolio, which is a solution to problem (1) for a given μ. For $\mu \geq \beta/\alpha$, the solution to the problem is the Markowitz portfolio

$$X = V^{-1}\left(\frac{\gamma - \beta\mu}{\delta} I + \frac{\alpha\mu - \beta}{\delta} \vec{\mu}\right).$$

Example 19.18 Given a portfolio of three securities with yields $\mu_1 = 10\%$, $\mu_2 = 20\%$, $\mu_3 = 30\%$ and covariance matrix $V = \begin{pmatrix} 1 & -1 & 0 \\ -1 & 9 & -2 \\ 0 & -2 & 4 \end{pmatrix}$.

Find a minimum risk portfolio with a return: (a) no less than $\mu = 25\%$, (b) no less than 13% and their risks.

This example for fixed income $\mu = 25\%$ was discussed above. According to the theory stated above, we must compare the minimum return values with the value $\mu_0 = \beta/\alpha$.

In this case, two options arise:

1. For $\mu \geq \beta/\alpha$, the solution to the problem is the Markowitz portfolio

$$X = V^{-1}\left(\frac{\gamma - \beta\mu}{\delta} I + \frac{\alpha\mu - \beta}{\delta}\vec{\mu}\right);$$

2. For $\mu \leq \beta/\alpha$, a solution to the problem, even better than required, will be a portfolio with efficiency $\mu_0 = \beta/\alpha$, i.e. even greater than required, and a dispersion equal to $\sigma^2 = 1/\alpha$, i.e. even less than that of the portfolio that is a solution to problem (1) for a given μ.

Let's find the constants

$$\alpha = I^T V^{-1} I, \beta = I^T V^{-1}\vec{\mu} = \vec{\mu}^T V^{-1} I, \gamma = \vec{\mu}^T V^{-1}\vec{\mu}, \delta = \alpha\gamma - \beta^2,$$

where $I = (1, 1, 1)^T$, $\mu = (10; 20; 30)^T$;

$$\alpha = I^T V^{-1} I = \frac{1}{14}(1,1,1)\cdot\begin{pmatrix} 16 & 2 & 1 \\ 2 & 2 & 1 \\ 1 & 1 & 4 \end{pmatrix}\cdot\begin{pmatrix} 1 \\ 1 \\ 1 \end{pmatrix} = \frac{1}{14}\cdot(19,5,6)\cdot\begin{pmatrix} 1 \\ 1 \\ 1 \end{pmatrix} = \frac{15}{7};$$

$$\beta = I^T V^{-1}\vec{\mu} = \frac{1}{14}(1,1,1)\cdot\begin{pmatrix} 16 & 2 & 1 \\ 2 & 2 & 1 \\ 1 & 1 & 4 \end{pmatrix}\cdot\begin{pmatrix} 10 \\ 20 \\ 30 \end{pmatrix} = \frac{1}{14}(19,5,6)\cdot\begin{pmatrix} 10 \\ 20 \\ 30 \end{pmatrix} = \frac{235}{7};$$

The ratio of these constants is equal to $\beta/\alpha = \frac{235}{7} : \frac{15}{7} = 15.(6)$.

That is, with a minimum return greater than 15.7%, the solution to the problem is the Markowitz portfolio, and with a minimum return less than 15.7%, the solution to the problem is a portfolio with a return $\mu_0 = \beta/\alpha$ and a risk equal to $\sigma = \sqrt{1/\alpha}$.

For a fixed return $\mu = 25\%$, the example was analyzed above and gave the following results. The portfolio is equal to $X = (0.12; 0.26; 0.62)^T$: it is necessary to take 12% of securities of the first type, 26% of the second and 62% of the third type.

Portfolio risk $\sigma = \sqrt{X^T V X} = 1.205$. We choose such a portfolio as a minimum risk portfolio with a return of at least $\mu = 25\%$.

With a minimum return of 13% (less than 15.7%), the solution to the problem is a portfolio with a return of $\mu_0 = \beta/\alpha$ and a risk of $\sigma\ \alpha = 1$.

Thus, $\mu_0 = \beta/\alpha = 15.7\%$, $\sigma = \sqrt{1/\alpha} = \sqrt{7/15} = 0.68$.

It can be seen that the risk of this portfolio is less than the risk of the fixed income portfolio $\mu = 25\%$ (1.205). This is as it should be, since $\sigma = \sqrt{1/\alpha} = 0.68$ is the minimum possible risk for this portfolio.

Now we can find a minimum risk portfolio with a return of $\mu = 13\%$.

To do this, we calculate the constants γ, δ, λ, ν. The first two do not depend on the required minimum portfolio return and were calculated by us above

$$\gamma = \vec{\mu}^T V^{-1}\vec{\mu} = \frac{1}{14}(10,20,30)\cdot\begin{pmatrix}16 & 2 & 1\\ 2 & 2 & 1\\ 1 & 1 & 4\end{pmatrix}\cdot\begin{pmatrix}10\\ 20\\ 30\end{pmatrix} = \frac{1}{14}(230,90,150)\cdot\begin{pmatrix}10\\ 20\\ 30\end{pmatrix} = \frac{4300}{7};$$

$$\delta = \alpha\gamma - \beta^2 = \frac{4300}{7}\cdot\frac{15}{7} - \left(\frac{235}{7}\right)^2 = \frac{9275}{49} = 189.3.$$

Let us now calculate the constants λ, ν, depending on the required minimum portfolio return

$$\lambda = (\gamma - \beta\mu)/\delta = \left(\frac{4300}{70} - \frac{235}{7}\cdot 13\right)/189.3 = 177.86/189.3 = 0.9396;$$

$$\nu = (\alpha\mu - \beta)/\delta = \left(\frac{15}{7}\cdot 13 - \frac{235}{7}\right)/189.3 = -5.714/189.3 = -0.03.$$

$$X = V^{-1}\left(\lambda I + \nu\vec{\mu}\right) = \frac{1}{14}\cdot\begin{pmatrix}16 & 2 & 1\\ 2 & 2 & 1\\ 1 & 1 & 4\end{pmatrix}\cdot\begin{pmatrix}0.9396 - 0.03\cdot 10\\ 0.9396 - 0.03\cdot 20\\ 0.9396 - 0.03\cdot 30\end{pmatrix} =$$

$$= \frac{1}{14}\cdot\begin{pmatrix}16 & 2 & 1\\ 2 & 2 & 1\\ 1 & 1 & 4\end{pmatrix}\cdot\begin{pmatrix}0.6396\\ 0.3396\\ 0.0396\end{pmatrix} = \frac{1}{14}\cdot\begin{pmatrix}10.952\\ 1.998\\ 1.1376\end{pmatrix} = (0.78; 0.14; 0.08)^T.$$

Thus, the minimum risk portfolio with a return of at least $\mu = 13\%$ is equal to $X = (0.78; 0.14; 0.08)^T$: it is necessary to take 78% of securities of the first type, 14% of the second and 8% of the third type. It is clear that the structure of this portfolio differs sharply from the structure of a minimum risk portfolio with a return of at least $\mu = 25\%$: the first security is taken 6.5 times more, the second—1.86 times less, the third—7.75 times less.

Questions and Tasks

19.140. Describe the Markovitz portfolio of minimal risk with an efficiency not less than the specified one.

19.141. A portfolio of three securities with yields of $\mu_1 = 18\%$, $\mu_2 = 23\%$, $\mu_3 = 35\%$ and a covariance matrix is given $V = \begin{pmatrix}25 & 5 & 12\\ 5 & 36 & -10\\ 12 & -10 & 81\end{pmatrix}$. Find a minimum risk portfolio with a return of at least: (a) $\mu = 10\%$, (b) $\mu = 30\%$ and its risk. Write the equation of the minimum boundary.

19.142. A portfolio of three securities with yields of $\mu_1 = 8\%$, $\mu_2 = 28\%$, $\mu_3 = 45\%$ and a covariance matrix is given $V = \begin{pmatrix} 16 & 2 & 10 \\ 2 & 9 & -8 \\ 10 & -8 & 25 \end{pmatrix}$. Find a minimum risk portfolio with a return of at least: (a) $\mu = 5\%$, (b) $\mu = 35\%$ and its risk. Write the equation of the minimum boundary.

19.143. A portfolio of three securities with yields of $\mu_1 = 6\%$, $\mu_2 = 24\%$, $\mu_3 = 34\%$ and a covariance matrix is given $V = \begin{pmatrix} 64 & 4 & -14 \\ 4 & 49 & -9 \\ -14 & -9 & 36 \end{pmatrix}$. Find a minimum risk portfolio with a return of at least: (a) $\mu = 4\%$, (b) $\mu = 26\%$ and its risk.

19.3.3 Minimum Risk Portfolio

Let's solve another optimization problem: to find a portfolio of minimal risk from all possible portfolios, i.e. portfolios of any efficiency (Brusov et al. 2023).

To do this, you need to find the minimum of the objective function

$$\frac{1}{2}\sigma^2 = \frac{1}{2}X^T VX \to \min \tag{19.82}$$

on condition

$$I^T X = 1. \tag{19.83}$$

The Lagrange function in this case has the form

$$L(X, \lambda) = \frac{1}{2}X^T VX + \lambda\left(1 - I^T X\right). \tag{19.84}$$

There is a minimum risk portfolio

$$X = \frac{V^{-1}I}{I^T V^{-1}I} = \frac{V^{-1}I}{\alpha}. \tag{19.85}$$

The minimum variance itself is equal to

$$X^T VX = \left(\frac{I^T V^{-1}}{\alpha}\right) V \left(\frac{V^{-1}I}{\alpha}\right) = \frac{I^T V^{-1}I}{\alpha^2} = \frac{1}{\alpha}. \tag{19.86}$$

So, the inverse value of the parameter α is numerically equal to the minimum variance of all portfolios.

It is interesting to note that the minimum variance and the minimum risk portfolio itself are determined solely by the matrix V (or, more precisely, V^{-1}). However, the effectiveness of such a portfolio also depends on the vector $\rightarrow \infty$ and it is equal to

$$\mu = \mu^{\rightarrow T} X = \frac{\mu^{\rightarrow T} V^{-1} I}{\alpha} = \frac{\beta}{\alpha}. \tag{19.87}$$

So, the effectiveness of the minimum risk portfolio is β/α. Note that the assumption $\delta > 0$ is not used here.

Example 19.19 A portfolio of three securities with yields of $\mu_1 = 10\%$, $\mu_2 = 20\%$, $\mu_3 = 30\%$ and a covariance matrix is given $V = \begin{pmatrix} 1 & -1 & 0 \\ -1 & 9 & -2 \\ 0 & -2 & 4 \end{pmatrix}$. Find a portfolio of minimal risk from portfolios of any profitability, its profitability and risk.

The minimum risk portfolio is equal to

$$X = \frac{V^{-1} I}{I^T V^{-1} I} = \frac{V^{-1} I}{\alpha}. \tag{19.88}$$

We have already calculated the parameters α, β, and V^{-1} in previous tasks.

$$\alpha = I^T V^{-1} I = \frac{1}{14}(1;1;1) \cdot \begin{pmatrix} 16 & 2 & 1 \\ 2 & 2 & 1 \\ 1 & 1 & 4 \end{pmatrix} \cdot \begin{pmatrix} 1 \\ 1 \\ 1 \end{pmatrix} = \frac{1}{14} \cdot (19;5;6) \cdot \begin{pmatrix} 1 \\ 1 \\ 1 \end{pmatrix} = \frac{15}{7};$$

$$\beta = I^T V^{-1} \vec{\mu} = \frac{1}{14}(1;1;1) \cdot \begin{pmatrix} 16 & 2 & 1 \\ 2 & 2 & 1 \\ 1 & 1 & 4 \end{pmatrix} \cdot \begin{pmatrix} 10 \\ 20 \\ 30 \end{pmatrix} = \frac{1}{14} \cdot (19;5;6) \cdot \begin{pmatrix} 10 \\ 20 \\ 30 \end{pmatrix} = \frac{235}{7}.$$

The ratio of these parameters is $\beta/\alpha = \frac{235}{7} : \frac{15}{7} = 15.(6)$.

$$V^{-1} = \frac{1}{14} \cdot \begin{pmatrix} 16 & 2 & 1 \\ 2 & 2 & 1 \\ 1 & 1 & 4 \end{pmatrix}.$$

Then for the desired portfolio we have

$$\mathrm{X} = \frac{V^{-1} I}{I^T V^{-1} I} = \frac{V^{-1} I}{\alpha} = \frac{1}{14} \cdot \frac{7}{15} \cdot \begin{pmatrix} 16 & 2 & 1 \\ 2 & 2 & 1 \\ 1 & 1 & 4 \end{pmatrix} \cdot \begin{pmatrix} 1 \\ 1 \\ 1 \end{pmatrix} = \frac{1}{30} \cdot \begin{pmatrix} 19 \\ 5 \\ 6 \end{pmatrix} = \begin{pmatrix} 19/30 \\ 1/6 \\ 1/5 \end{pmatrix}.$$

So, the portfolio is equal to $X = \left(\frac{19}{30}; \frac{1}{6}; \frac{1}{5}\right)$.

The minimum variance of the portfolio is equal to

$$X^T VX = \left(\frac{I^T V^{-1}}{\alpha}\right) V \left(\frac{V^{-1} I}{\alpha}\right) = \frac{I^T V^{-1} I}{\alpha^2} = \frac{1}{\alpha} = \frac{7}{15}, \tag{19.89}$$

And the risk is $\sigma = \sqrt{\frac{7}{15}}$.

The effectiveness of the minimum risk portfolio also depends on the vector $\rightarrow \infty$ and it is equal to

$$\mu = \mu^{\rightarrow T} X = \frac{\mu^{\rightarrow T} V^{-1} I}{\alpha} = \frac{\beta}{\alpha} = 15.7\%.$$

Questions and Tasks

19.144. Describe the minimum risk portfolio from the portfolios of efficiency.

19.145. A portfolio of three securities with yields of $\mu_1 = 18\%$, $\mu_2 = 23\%$, $\mu_3 = 35\%$ and a covariance matrix is given $V = \begin{pmatrix} 25 & 5 & 12 \\ 5 & 36 & -10 \\ 12 & -10 & 81 \end{pmatrix}$. Find a portfolio of minimal risk from portfolios of any profitability, its profitability and risk.

19.146. A portfolio of three securities with yields of $\mu_1 = 8\%$, $\mu_2 = 28\%$, $\mu_3 = 45\%$ and a covariance matrix is given $V = \begin{pmatrix} 16 & 2 & 10 \\ 2 & 9 & -8 \\ 10 & -8 & 25 \end{pmatrix}$. Find a portfolio of minimal risk from portfolios of any profitability, its profitability and risk.

19.147. A portfolio of three securities with yields of $\mu_1 = 6\%$, $\mu_2 = 24\%$, $\mu_3 = 34\%$ and a covariance matrix is given $V = \begin{pmatrix} 64 & 4 & -14 \\ 4 & 49 & -9 \\ -14 & -9 & 36 \end{pmatrix}$. Find a portfolio of minimal risk from portfolios of any profitability, its profitability and risk.

19.3.4 The Portfolio of Maximum Efficiency from All Risk Portfolios Is No More Than a Given One

Along with portfolios of minimal risk, it also makes sense to select portfolios of maximum efficiency from a certain number of portfolios.

This task boils down to solving the following optimization problem: to find the maximum of the objective function

$$\mu^{\rightarrow T} X \rightarrow \max \tag{19.90}$$

under the conditions

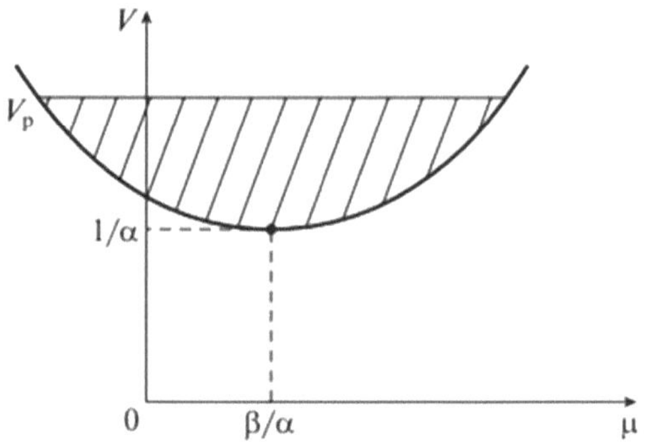

Fig. 19.5 Finding the Markowitz portfolio of maximum efficiency from all risk portfolios of no more than a given

$$\frac{1}{2}X^T VX = \frac{1}{2}\sigma^2 \tag{19.91}$$

$$I^T X = 1. \tag{19.92}$$

The direct approach—drawing up the Lagrange function, etc.—does not lead to solving the problem. Therefore, the following approach is proposed.

Previously, we got that for a portfolio that is a solution for giving (1)

$$\frac{1}{2}\sigma^2 = \frac{1}{2}X^T VX \to \min$$

under the conditions

$$\mu^{\to T}X = \mu,$$

$$I^T X = 1.$$

variance and efficiency are related by the formula (19.71) $\sigma^2 = \frac{\alpha\mu^2 - 2\beta\mu + \gamma}{\delta}$.

Consider the plane (μ, V). Let's draw the curve (19.71) on this plane.

In Fig. 19.5, the set of portfolios is shaded, β/α, 1/α is the efficiency and variance of the minimum risk portfolio.

So, if we fix the efficiency of the portfolio μ, then the lowest point of the shaded set lying on the corresponding vertical is the Markowitz portfolio—the solution to the problem (1). If we fix the variance of the portfolio V, then the rightmost point of the shaded set lying on the corresponding horizontal will obviously give the solution to the problem (19.90)–(19.92), i.e. a portfolio of maximum efficiency and a given risk.

So, the solution of the problem (19.90)–(19.92) can be obtained as follows:

1. Calculate the constants α, β, γ, δ using the formulas (19.62);
2. For a given value of the variance of the portfolio V, solve the quadratic equation (19.71)

$$\alpha\mu^2 - 2\beta\mu + \gamma = V\delta; \tag{19.93}$$

3. Find the largest root μ_0 of this equation

$$\mu_0 = \frac{\beta}{\alpha} + \sqrt{\left(V - \frac{1}{\alpha}\right)\frac{\delta}{\alpha}}. \tag{19.94}$$

Its abscissa is spaced from β/α to the right by an amount $\sqrt{\left(V - \frac{1}{\alpha}\right)\frac{\delta}{\alpha}}$, at the same time $\left(V - \frac{1}{\alpha}\right)$ is the excess of the specified portfolio variance over its minimum value. It is clear that the solution of the problem (19.90)–(19.92) exists only at $V \geq 1/\alpha$;

4. Find the vector X by the formula $X = V^i\left(\lambda I + \nu\vec{\mu}\right)$.

   Now it can be seen that the considered problem (19.90)–(19.92) is formally equivalent to a more general problem: to find the maximum of the objective function

$$\mu^{\rightarrow T}X \rightarrow \max \tag{19.95}$$

under the conditions

$$\frac{1}{2}X^TVX = \frac{1}{2}\sigma^2 \leq \frac{1}{2}V; \tag{19.96}$$

$$I^TX = 1. \tag{19.97}$$

Example 19.20 A portfolio of three securities with yields of $\mu_1 = 10\%$, $\mu_2 = 20\%$, $\mu_3 = 30\%$ and a covariance matrix is given $V = \begin{pmatrix} 1 & -1 & 0 \\ -1 & 9 & -2 \\ 0 & -2 & 4 \end{pmatrix}$. Find a portfolio of maximum returns from all risk portfolios of no more than a given: (a) 50%, (b) 70%, (c) 75%.

This example is a modification of Example 19.19 for the case of a portfolio of maximum return from all risk portfolios of no more than a given one. Therefore, from Example 19.19, let's take the values of constants calculated there.

The solution to this problem exists only at $V > 1/\alpha$, where V is the variance of the portfolio. This is due to the fact that only under this condition does the horizontal σ = const intersect an acceptable set of portfolios (the minimum point on the minimum boundary has coordinates (β/α, $1/\alpha$).

We have already calculated the parameters α, β, as well as γ, δ in previous tasks.

$$\alpha = I^T V^{-1} I = \frac{1}{14}(1;1;1)\cdot\begin{pmatrix}16 & 2 & 1\\ 2 & 2 & 1\\ 1 & 1 & 4\end{pmatrix}\cdot\begin{pmatrix}1\\1\\1\end{pmatrix} = \frac{1}{14}\cdot(19;5;6)\cdot\begin{pmatrix}1\\1\\1\end{pmatrix} = \frac{15}{7};$$

$$\beta = I^T V^{-1}\vec{\mu} = \frac{1}{14}(1;1;1)\cdot\begin{pmatrix}16 & 2 & 1\\ 2 & 2 & 1\\ 1 & 1 & 4\end{pmatrix}\cdot\begin{pmatrix}10\\20\\30\end{pmatrix} = \frac{1}{14}\cdot(19;5;6)\cdot\begin{pmatrix}10\\20\\30\end{pmatrix} = \frac{235}{7};$$

$$\gamma = \mu^{\rightarrow T} V^{-1}\vec{\mu} = \frac{1}{14}(10;20;30)\cdot\begin{pmatrix}16 & 2 & 1\\ 2 & 2 & 1\\ 1 & 1 & 4\end{pmatrix}\cdot\begin{pmatrix}10\\20\\30\end{pmatrix} = \frac{1}{14}(230;90;150)\cdot\begin{pmatrix}10\\20\\30\end{pmatrix} = \frac{4300}{7}.$$

$$\delta = \alpha\gamma - \beta^2 = \frac{4300}{7}\cdot\frac{15}{7} - \left(\frac{235}{7}\right)^2 = \frac{9275}{49} = 189.3.$$

The solution to this problem exists only for V > 1/α = 0.4(6), i.e. for portfolio risk $\sigma > \sqrt{1/\alpha} = \sqrt{0.4(6)} = 0.683$.

Therefore, in the case of item "a"

(a) The problem has no solution (50% < 68.3%).

Consider the other two cases (item "b" and item "c"):

(b) We must solve the quadratic equation (19.71) for a given value of the variance of the portfolio V = 0.49

$$\alpha\mu^2 - 2\beta\mu + \gamma = V\delta$$

and find the largest root μ_0 of this equation.

$$\mu_0 = \frac{\beta}{\alpha} + \sqrt{\left(V - \frac{1}{\alpha}\right)\frac{\delta}{\alpha}}.$$

$$\mu_0 = \frac{235}{7} : \frac{15}{7} + \sqrt{\left(0.49 - \frac{7}{15}\right)\cdot 189.3\cdot\frac{7}{15}} = 17.102.$$

Knowing μ0, now we calculate the constants λ and v necessary for us to find the portfolio:

$$\lambda = (\gamma - \beta\mu)/\delta \ \text{ and } \mathrm{v} = (\alpha\mu - \beta)/\delta;$$

$$\lambda = (\gamma - \beta\mu)/\delta = \left(\frac{4300}{7} - \frac{235}{7}\cdot 17.102\right)/189.3 = 0.212;$$

$$\mathrm{v} = (\alpha\mu - \beta)/\delta = \left(\frac{15}{7} \cdot 17.102 - \frac{235}{7}\right)/189.3 = 3.076/189.3 = 0.0162.$$

Now we can find a portfolio of maximum return from all risk portfolios of no more than 70%:

$$\mathrm{X} = V^{-1}\left(\lambda I + \mathrm{v}\vec{\mu}\right) = \frac{1}{14} \cdot \begin{pmatrix} 16 & 2 & 1 \\ 2 & 2 & 1 \\ 1 & 1 & 4 \end{pmatrix} \cdot \begin{pmatrix} 0.212 + 0.0162 \cdot 10 \\ 0.212 + 0.0162 \cdot 20 \\ 0.212 + 0.0162 \cdot 30 \end{pmatrix} = \frac{1}{14}$$

$$\cdot \begin{pmatrix} 16 & 2 & 1 \\ 2 & 2 & 1 \\ 1 & 1 & 4 \end{pmatrix} \cdot \begin{pmatrix} 0.374 \\ 0.536 \\ 0.698 \end{pmatrix} = \begin{pmatrix} 0.554 \\ 0.180 \\ 0.264 \end{pmatrix}.$$

So, the portfolio of maximum return from all risk portfolios of no more than 70% is equal to X = (0.554; 0.180; 0.264), and the maximum return itself is equal to $\mu 0 = 17.102\%$.

We provide the reader with the opportunity to find a portfolio of maximum profitability from all risk portfolios of no more than 75% (point "b" of this example). All calculations are carried out in the same way as those carried out by us in paragraph (b).

Questions and Tasks

19.148. Describe the portfolio of maximum risk effectiveness no more than the specified one.

19.149. A portfolio of three securities with yields of $\mu_1 = 18\%$, $\mu_2 = 23\%$, $\mu_3 = 35\%$ and a covariance matrix is given $V = \begin{pmatrix} 25 & 5 & 12 \\ 5 & 36 & -10 \\ 12 & -10 & 81 \end{pmatrix}$. Find the portfolio of maximum return from all risk portfolios of no more than a given: (a) 7%, (b) 15%, (c) 30%.

19.150. A portfolio of three securities with yields of $\mu_1 = 8\%$, $\mu_2 = 28\%$, $\mu_3 = 45\%$ and a covariance matrix is given $V = \begin{pmatrix} 16 & 2 & 10 \\ 2 & 9 & -8 \\ 10 & -8 & 25 \end{pmatrix}$. Find the maximum return portfolio with variance: (a) no more than 2, (b) no more than 8, (c) no more than 20 and its profitability.

19.151. A portfolio of three securities with yields of $\mu_1 = 6\%$, $\mu_2 = 24\%$, $\mu_3 = 34\%$ and a covariance matrix is given $V = \begin{pmatrix} 64 & 4 & -14 \\ 4 & 49 & -9 \\ -14 & -9 & 36 \end{pmatrix}$. Find the portfolio of maximum return from all risk portfolios of no more than a given: (a) 6%, (b) 20%, (c) 25%.

19.152. A portfolio of three securities with yields of $\mu_1 = 6\%$, $\mu_2 = 24\%$, $\mu_3 = 34\%$ and a covariance matrix is given $V = \begin{pmatrix} 64 & 4 & -14 \\ 4 & 49 & -9 \\ -14 & -9 & 36 \end{pmatrix}$. Find a portfolio with a maximum risk return of no more than a given: (a) 6%, (b) 20%, (c) 25%.

19.153. A portfolio of three securities with yields of $\mu_1 = 25\%$, $\mu_2 = 56\%$, $\mu_3 = 68\%$ and a covariance matrix is given $V = \begin{pmatrix} 25 & 4 & 12 \\ 4 & 36 & -10 \\ 12 & -10 & 49 \end{pmatrix}$. Find a portfolio of maximum risk returns: (a) no more than 2, (b) no more than 6, (c) no more than 10 and its profitability.

19.154. A portfolio of three securities with yields of $\mu_1 = 16\%$, $\mu_2 = 33\%$, $\mu_3 = 55\%$ and a covariance matrix is given $V = \begin{pmatrix} 54 & -8 & -12 \\ -8 & 35 & -11 \\ -12 & -11 & 46 \end{pmatrix}$. Find a portfolio of maximum risk returns: (a) no more than 6, (b) no more than 8, (c) no more than 45 and its profitability.

19.4 An Arbitrary Portfolio

Questions and Tasks

19.155. Describe the minimum risk portfolio for a given efficiency.

19.156. What is the minimum boundary and what are its properties.

19.157. The portfolio consists of three securities with expected returns of $\mu_1 = 0.35\%$, $\mu_2 = 0.26\%$, $\mu_3 = 0.41\%$. Covariance matrix is given $V = \begin{pmatrix} 11 & 5 & -2 \\ 5 & 12 & 6 \\ -2 & 6 & 13 \end{pmatrix}$. Find the equation of the minimum boundary, the minimum risk portfolio and its characteristics.

19.158. Describe the minimum risk portfolio with an efficiency not less than the specified one.

19.159. Describe the minimum risk portfolio.

19.160. Describe the portfolio of maximum risk effectiveness no more than the specified one.

19.161. A portfolio of three securities with yields of $\mu_1 = 5\%$, $\mu_2 = 10\%$, $\mu_3 = 15\%$ and a covariance matrix is given $V = \begin{pmatrix} 4 & 1 & 0 \\ 1 & 16 & -3 \\ 0 & -3 & 25 \end{pmatrix}$. Find a minimum risk portfolio with a return of $\mu = 12\%$ and its risk. Write the equation of the minimum boundary.

19.162. Given a portfolio of three independent securities with variances $D_1 = 3$, $D_2 = 10$, $\mu_3 = 5$. Find a portfolio of minimal risk.

19.163. Describe the minimum risk portfolio of all portfolios of a given efficiency.
19.164. Describe a portfolio of maximum efficiency out of all risk portfolios of no more than a given one.

19.5 Tobin's Briefcase

Adding risk-free paper to the portfolio significantly changes the strategy of forming an optimal portfolio. This case was first investigated by J. Tobin. Let's denote the effectiveness of the risk-free paper μ_f and consider it positive.

19.5.1 Tobin's Portfolio Is the Lowest Risk of All Portfolios of a Given Efficiency

So, let's assume that, along with n-risky assets, the investor's portfolio includes a risk-free paper with a deterministic yield of $\mu_f = R_f$ and a share in the portfolio of x_f. In this case, the task (19.58)–(19.60) will look like this:

$$\frac{1}{2}\sigma^2 = \frac{1}{2}X^T VX = \rightarrow \min \tag{19.98}$$

under the conditions

$$\mu_f x_f + \mu^{\rightarrow T} X = \mu, \tag{19.99}$$

$$x_f + I^T X = 1. \tag{19.100}$$

The expression for the risk square has not changed due to the risk-free nature of the added asset. In this case, the appearance of the minimum border will change:

$$\lambda = \frac{\mu - \mu_f}{d^2},$$
$$X = \frac{\mu - \mu_f}{d^2} V^{-1}\left(\vec{\mu} - \mu_f I\right) \tag{19.101}$$

the desired vector of risk shares, the risk-free share is found as follows from the ratio (19.100):

$$x_f = 1 - I^T X = 1 - \frac{\mu - \mu_f}{d^2} I^T V^{-1}\left(\vec{\mu} - \mu_f I\right). \tag{19.102}$$

the equation of the minimum boundary

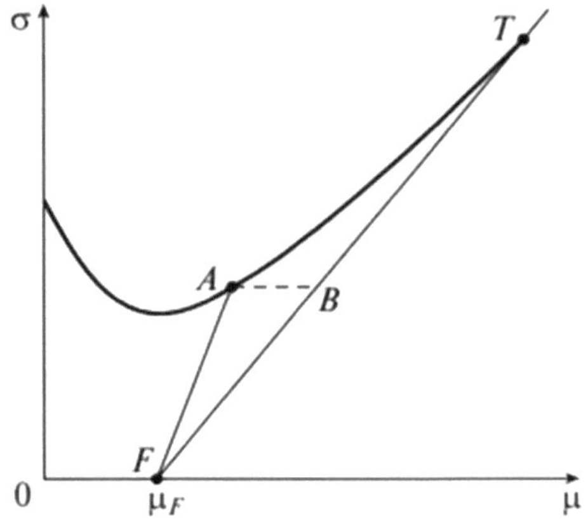

Fig. 19.6 The minimum boundary of Tobin's portfolio, tangent portfolio

$$\sigma^2 = \left(\frac{\mu - \mu_f}{d}\right)^2. \tag{19.103}$$

It is usually assumed that the expected return on a portfolio should be no less than the return on a risk-free asset, i.e. $\mu \geq \mu_f$. Otherwise, it would be necessary to form a portfolio of only one of them. Therefore, Eq. (19.103) turns into a linear one

$$\sigma = \frac{\mu - \mu_f}{d}. \tag{19.104}$$

The line σ (19.104) is tangent to the graph of the minimum boundary (19.71) (Fig. 19.6).

The efficiency of the tangent portfolio μ_T is equal to

$$\mu_T = \frac{\gamma - \beta\mu_f}{\beta - \alpha\mu_f}, \tag{19.105}$$

and the risk of a tangent portfolio σ_T is equal to

$$\sigma_T = \frac{d}{(\beta - \alpha\mu_f)}. \tag{19.106}$$

For the coordinates of the tangent portfolio, we have

$$\mu_T = \frac{\gamma - \beta\mu_f}{\beta - \alpha\mu_f}, \sigma_T = \frac{d}{(\beta - \alpha\mu_f)}. \tag{19.107}$$

In this case, the tangent portfolio T itself is located from (19.101) by substitution $\mu = \mu_T$:

$$T = \frac{\mu_T - \mu_f}{d^2} V^{-1}\left(\vec{\mu} - \mu_f I\right). \tag{19.108}$$

Note that the points of the minimum boundary are represented as a linear combination

$$M = \lambda F + (1 - \lambda)T,$$

moreover, when the point moves from F to T, the parameter λ changes from 1 to 0.

Example 19.21 The portfolio consists of three securities: a risk-free one with an efficiency (expected return) of 5% and two risky ones with an efficiency of 10% and 15%, respectively, and a covariance matrix $\begin{pmatrix} 9 & 5 \\ 5 & 36 \end{pmatrix}$. Find Tobin's portfolios of expected returns of 10%, 11% and 12% and minimum risk and their risks.

From (19.124) and (19.125) we have the following expressions for the parameter d and the desired portfolio X:

$$d = \sqrt{\left(\vec{\mu} - \mu_f I\right)^T V^{-1}\left(\vec{\mu} - \mu_f I\right)}, X = \frac{\mu - \mu_f}{d^2} V^{-1}\left(\vec{\mu} - \mu_f I\right).$$

1. $\mu = 10\%$.
   Here $\vec{\mu} = (10; 15)^T$, $\mu = 10$, $\mu_f = 5$.
   Calculate the parameter d:

$$d^2 = \left(\vec{\mu} - \mu_f I\right)^T V^{-1}\left(\vec{\mu} - \mu_f I\right) = \frac{1}{299}(5; 10)\begin{pmatrix} 36 & -5 \\ -5 & 9 \end{pmatrix}\begin{pmatrix} 5 \\ 10 \end{pmatrix}$$
$$= \frac{1}{299}(130; 65)\begin{pmatrix} 5 \\ 10 \end{pmatrix} = \frac{1300}{299} = 19.35.$$

Now we can find the portfolio X.

$$X = \frac{\mu - \mu_f}{d^2} V^{-1}\left(\vec{\mu} - \mu_f I\right) = \frac{10 - 5}{19.35} * \frac{1}{299}\begin{pmatrix} 36 & -5 \\ -5 & 9 \end{pmatrix}\begin{pmatrix} 5 \\ 10 \end{pmatrix} =$$
$$= 0.00384 * \begin{pmatrix} 130 \\ 65 \end{pmatrix} = \begin{pmatrix} 0.5 \\ 0.25 \end{pmatrix}.$$
$$x_f = 1 - x_1 - x_2 = 0.25.$$

Thus, the portfolio of expected return of 10% and minimum risk has the form

$$X = (0.5; 0.25; x_f = 0.25).$$

The risk of the portfolio is equal to

$$\sigma = \sqrt{X^T VX} = \sqrt{(0.5; 0.25)\begin{pmatrix} 9 & 5 \\ 5 & 36 \end{pmatrix}\begin{pmatrix} 0.5 \\ 0.25 \end{pmatrix}} = \sqrt{(5.75; 11.5)\begin{pmatrix} 0.5 \\ 0.25 \end{pmatrix}} = 2.19.$$

Note that the risk of the portfolio is less than the risk of each of the risky securities, which are $\sqrt{9} = 3$ and $\sqrt{36} = 6$, respectively, for the first and second securities.

2. $\mu = 11\%$.

   Here $\vec{\mu} = (10; 15)^T$, $\mu = 11$, $\mu_f = 5$.
   The parameter d is still 19.35.
   Calculate the portfolio X.

$$X = \frac{\mu - \mu_f}{d^2} V^{-1}\left(\vec{\mu} - \mu_f I\right) = \frac{11-5}{19.35} * \frac{1}{299}\begin{pmatrix} 36 & -5 \\ -5 & 9 \end{pmatrix}\begin{pmatrix} 5 \\ 10 \end{pmatrix} = 0.00461 * \begin{pmatrix} 130 \\ 65 \end{pmatrix} = \begin{pmatrix} 0.6 \\ 0.3 \end{pmatrix}.$$

$$x_f = 1 - x_1 - x_2 = 0.1.$$

Thus, the portfolio of expected return of 11% and minimum risk has the form

$$X = (0.6; 0.3; x_f = 0.1).$$

The risk of the portfolio is equal to

$$\sigma = \sqrt{X^T VX} = \sqrt{(0.6; 0.3)\begin{pmatrix} 9 & 5 \\ 5 & 36 \end{pmatrix}\begin{pmatrix} 0.6 \\ 0.3 \end{pmatrix}} = \sqrt{(6.9; 13.8)\begin{pmatrix} 0.6 \\ 0.3 \end{pmatrix}} = 2.88.$$

Note that, despite the increase in required profitability by 1%, the portfolio risk remains less than the risk of each of the risky securities.

3. Here $\vec{\mu} = (10; 15)^T$, $\mu = 12$, $\mu_f = 5$.
   Calculate the parameter d:

$$d^2 = \left(\vec{\mu} - \mu_f I\right)^T V^{-1}\left(\vec{\mu} - \mu_f I\right) = \frac{1}{299}(5; 10)\begin{pmatrix} 36 & -5 \\ -5 & 9 \end{pmatrix}\begin{pmatrix} 5 \\ 10 \end{pmatrix}$$

$$= \frac{1}{299}(130; 65)\begin{pmatrix} 5 \\ 10 \end{pmatrix} = \frac{1300}{299} = 19.35.$$

Now we can find the portfolio X.

$$X = \frac{\mu - \mu_f}{d^2} V^{-1}\left(\vec{\mu} - \mu_f I\right) = \frac{12-5}{19.35} * \frac{1}{299}\begin{pmatrix} 36 & -5 \\ -5 & 9 \end{pmatrix}\begin{pmatrix} 5 \\ 10 \end{pmatrix}$$
$$= 0.00538 * \begin{pmatrix} 130 \\ 65 \end{pmatrix} = \begin{pmatrix} 0.7 \\ 0.35 \end{pmatrix}.$$

The sum of shares of already risky assets exceeds 1, therefore, it is not possible to form a Tobin portfolio with an expected return of 12% (and higher) and minimal risk.

Example 19.22 For the condition of the previous example, find the tangent portfolio, its expected return and risk.

Thus, the portfolio consists of three securities: a risk-free one with an efficiency (expected return) of 5% and two risky ones with an efficiency of 10% and 15%, respectively, and a covariance matrix $\begin{pmatrix} 9 & 5 \\ 5 & 36 \end{pmatrix}$.

The required tangent portfolio T has the form (19.108)

$$T = \frac{\mu_T - \mu_f}{d^2} V^{-1}\left(\vec{\mu} - \mu_f I\right),$$

and for its coordinates from (19.107) we have

$$\mu_T = \frac{\gamma - \beta\mu_f}{\beta - \alpha\mu_f}, \sigma_T = \frac{d}{\beta - \alpha\mu_f}.$$

For the parameter d, we have the following expression:

$$d = \sqrt{\left(\vec{\mu} - \mu_f I\right)^T V^{-1}\left(\vec{\mu} - \mu_f I\right)};$$

$$d^2 = \left(\vec{\mu} - \mu_f I\right)^T V^{-1}\left(\vec{\mu} - \mu_f I\right) = \frac{1}{299}(5; 10)\begin{pmatrix} 36 & -5 \\ -5 & 9 \end{pmatrix}\begin{pmatrix} 5 \\ 10 \end{pmatrix}$$
$$= \frac{1}{299}(130; 65)\begin{pmatrix} 5 \\ 10 \end{pmatrix} = \frac{1300}{299} = 4.35.$$

Find the constants

$$\alpha = I^T V^{-1} I, \quad \beta = I^T V^{-1}\vec{\mu} = \vec{\mu}^T V^{-1} I, \gamma = \vec{\mu}^T V^{-1}\vec{\mu}.$$

Here $I = (1, 1)^T$, $\mu = (10; 15)^T$.

$$\alpha = I^T V^{-1} I = (1,1) * \frac{1}{299}\begin{pmatrix} 36 & -5 \\ -5 & 9 \end{pmatrix} * \begin{pmatrix} 1 \\ 1 \end{pmatrix} = \frac{1}{299}(31;4) * \begin{pmatrix} 1 \\ 1 \end{pmatrix} = \frac{35}{299} = 0.117;$$

$$\beta = I^T V^{-1} \vec{\mu} = (1,1) * \frac{1}{299}\begin{pmatrix} 36 & -5 \\ -5 & 9 \end{pmatrix} * \begin{pmatrix} 10 \\ 15 \end{pmatrix} = \frac{1}{299}(31;4) * \begin{pmatrix} 10 \\ 15 \end{pmatrix} = \frac{370}{299} = 1.24;$$

$$\gamma = \vec{\mu}^T V^{-1} \vec{\mu} = (10;15) * \frac{1}{299}\begin{pmatrix} 36 & -5 \\ -5 & 9 \end{pmatrix} * \begin{pmatrix} 10 \\ 15 \end{pmatrix} = \frac{1}{299}(285;85) * \begin{pmatrix} 10 \\ 15 \end{pmatrix} = \frac{4125}{299} = 13.80.$$

Now we will find the coordinates of the tangent portfolio T:

$$\mu_T = \frac{\gamma - \beta\mu_f}{\beta - \alpha\mu_f} = \frac{13.8 - 1.24 * 5}{1.24 - 0.117 * 5} = \frac{7.6}{0.655} = 11.6;$$

$$\sigma_T = \frac{d}{\beta - \alpha\mu_f} = \frac{\sqrt{4.35}}{1.24 - 0.117 * 5} = \frac{2.086}{0.655} = 3.18.$$

And finally, we find the tangent portfolio T:

$$T = \frac{\mu_T - \mu_f}{d^2} V^{-1}\left(\vec{\mu} - \mu_f I\right) = \frac{11.6 - 5}{19.35} * \frac{1}{299}\begin{pmatrix} 36 & -5 \\ -5 & 9 \end{pmatrix} * \begin{pmatrix} 5 \\ 10 \end{pmatrix}$$
$$= 0.00507 * \begin{pmatrix} 130 \\ 65 \end{pmatrix} = \begin{pmatrix} 0.6597 \\ 0.3298 \end{pmatrix}.$$

So, the tangent portfolio is T = (0.66; 0.33), that is, it includes 66% of the first paper and 33% of the second paper and practically does not include risk-free paper.

Note that the risk of the tangent portfolio, $\sigma_T = 3.18$, is slightly higher than the risk of the first paper ($\sigma_1 = 3$) and almost half as much as the risk of the second paper ($\sigma_2 = 6$).

The profitability of the tangent portfolio $\mu_T = 11.6\%$ is the maximum profitability at which a portfolio of minimal risk can be formed.

Taking into account the results of the previous example, we can make an obvious conclusion that with an increase in profitability from 10% to 12%, the risk of the portfolio increases from 2.4% (at $\mu = 10\%$) to 2.88% (at $\mu = 11\%$) and further to 3.18% (at $\mu = 11.6\%$ for the tangent portfolio). And at $\mu = 11.6\%$, it is no longer possible to form a minimum risk portfolio.

Example 19.23 Prove that the line $\sigma = \frac{\mu - \mu_f}{d}$ is tangent to the graph of the minimum boundary $\sigma^2 = \frac{\alpha\mu^2 - 2\beta\mu + \gamma}{\delta}$.

To prove it, we will find the intersection points of the hyperbola and the straight line, solving their equations together, and make sure that there is only one such point. Equating the right sides of the equations of the straight line and the hyperbola, we obtain

$$\left(\frac{\mu-\mu_f}{d}\right)^2=\frac{\alpha\mu^2-2\beta\mu+\gamma}{\delta}. \tag{19.109}$$

Next, we get a quadratic equation relative to μ and find its roots

$$\mu^2\left(\frac{\delta}{d^2}-\alpha\right)+2\mu\left(\beta-\frac{\delta}{d^2}\mu_f\right)+\left(\frac{\delta}{d^2}\mu_f^2-\gamma\right)=0. \tag{19.110}$$

The discriminant of this equation is equal to zero:

$$4\left(\beta-\frac{\delta}{d^2}\mu_f\right)^2-4\left(\frac{\delta}{d^2}-\alpha\right)\left(\frac{\delta}{d^2}\mu_f^2-\gamma\right)=0. \tag{19.111}$$

This proves that the line $\sigma=\frac{\mu-\mu_f}{d}$ is tangent to the graph of the minimum boundary of $\sigma^2=\frac{\alpha\mu^2-2\beta\mu+\gamma}{\delta}$.

Example 19.24 Find the coordinates of the tangent portfolio (its profitability and risk).

We will find the coordinates of the tangent point (coordinates of the tangent portfolio):

$$\mu=\frac{-2\left(\beta-\frac{\delta}{d^2}\mu_f\right)}{2\left(\frac{\delta}{d^2}-\alpha\right)}=-\frac{\beta d^2-\delta\mu_f}{\delta-\alpha d^2}=\frac{\alpha(\gamma-\beta\mu_f)\left(\mu_f-\frac{\beta}{\alpha}\right)}{\alpha^2\left(\mu_f-\frac{\beta}{\alpha}\right)^2}=-\frac{(\gamma-\beta\mu_f)}{\alpha\left(\mu_f-\frac{\beta}{\alpha}\right)}.$$

So, the efficiency of the tangent portfolio of μ_T is equal to

$$\mu_T=\frac{\gamma-\beta\mu_f}{\beta-\alpha\mu_f}. \tag{19.112}$$

Substituting the found efficiency value of μ_T into the tangent equation, we find the risk of a tangent portfolio of σ_T:

$$\sigma=\frac{\mu-\mu_f}{d}=\frac{\frac{\gamma-\beta\mu_f}{\beta-\alpha\mu_f}-\mu_f}{d}=\frac{\gamma-2\beta\mu_f+\alpha\mu_f^2}{d(\beta-\alpha\mu_f)}=\frac{d^2}{d(\beta-\alpha\mu_f)}=\frac{d}{(\beta-\alpha\mu_f)}.$$

So, the tangent portfolio has the following coordinates:

$$\left(\frac{\gamma-\beta\mu_f}{\beta-\alpha\mu_f},\frac{d}{(\beta-\alpha\mu_f)}\right).$$

Questions and Tasks

19.165. Describe Tobin's portfolios.

19.166. Find a tangent portfolio, its expected return and risk if the portfolio consists of three securities: a risk-free one with an efficiency (expected return) of 7% and two risky ones with an efficiency of 12% and 20%, respectively, and a covariance matrix $\begin{pmatrix} 4 & -2 \\ -2 & 25 \end{pmatrix}$.

19.167. The portfolio consists of three securities: a risk-free one with an expected return of 4% and two risky ones with an efficiency of 8% and 12%, respectively, and a covariance matrix $\begin{pmatrix} 3 & -4 \\ -4 & 6 \end{pmatrix}$. Find Tobin's portfolios of expected returns of 7%, 8%, 10% and 11% and minimum risk, and their risks.

19.168. The equation of the minimum boundary of the Markowitz portfolio has the form $\sigma=\sqrt{\mu^2 - 15\mu + 200}$. The risk-free profitability is 5%. Find the expected return and risk of the tangent portfolio and the equation of the minimum boundary of the Tobin's portfolio.

19.169. Form Tobin's portfolio of expected returns of 8% and minimum risk from three types of securities: risk-free with an expected return of 4% and uncorrelated risky with expected returns of 9% and 12% and risks of 5% and 6%.

19.170. There is an equation of the minimum boundary $\sigma = 3\mu2 - 20\mu + 340$ is given. The risk-free profitability is 6%. Find the expected return and risk of the tangent portfolio.

19.171. Form Tobin's portfolio of maximum expected return and a given risk of 6% from three types of securities: risk-free with an expected return of 7% and uncorrelated risky with expected returns of 10% and 20% and risks of 8% and 11%.

19.172. Form Tobin's portfolio of expected returns of 14% and minimum risk from three types of securities: risk-free with an expected return of 4% and uncorrelated risky with expected returns of 15% and 20% and risks of 5% and 7%.

19.173. There are three securities with profitability of 16%, 18% and 20%, whose covariance matrix of returns has the form

$$V = \begin{pmatrix} 6 & -3 & 0 \\ -3 & 12 & -5 \\ 0 & -5 & 16 \end{pmatrix}.$$

Find the equation of the minimum boundary and build its graph, provided that the risk-free rate is $\mu_f = 10\%$. Find the coordinates of the tangent portfolio.

19.174. Prove that the line $\frac{\mu-\mu_f}{d}=\sigma$ is tangent to the graph of the minimum boundary $\sigma^2=\frac{\alpha\mu^2-2\beta\mu+\gamma}{\delta}$.

19.175. Find the coordinates of the tangent portfolio (its profitability and risk).

19.5.2 The Portfolio of Maximum Efficiency from All Risk Portfolios Risk Is No More Than a Given

Along with the Tobin problem (19.98)–(19.100), consider the optimization problem

$$\mu_f x_f + \vec{\mu}^T X \to \max \tag{19.113}$$

under the conditions

$$\frac{1}{2}\sigma^2 = \frac{1}{2}X^T V X < \frac{1}{2}V, \tag{19.114}$$

$$x_f + I^T X = 1. \tag{19.115}$$

To solve the problem, consider the plane (μ, σ) (in the variables efficiency-risk) (Fig. 19.7). On this plane, we draw a broken line

$$\sigma = |\mu - \mu_f|/d,$$

where

$$d=\sqrt{\left(\vec{\mu}-\mu_f I\right)^T V^{-1}\left(\vec{\mu}-\mu_f I\right)}=\sqrt{\alpha\mu_f^2-2\beta\mu_f+\gamma}\,.$$

In Fig. 19.7 a lot of portfolios for the situation under consideration is shaded.

So, if we fix the effectiveness of the portfolio μ, then the lowest point of the shaded set lying on the corresponding vertical is Tobin's portfolio—the solution to the problem (19.98)–(19.100). If we fix the risk of the portfolio σ, then the rightmost point of the shaded set, lying no higher than the corresponding horizontal,

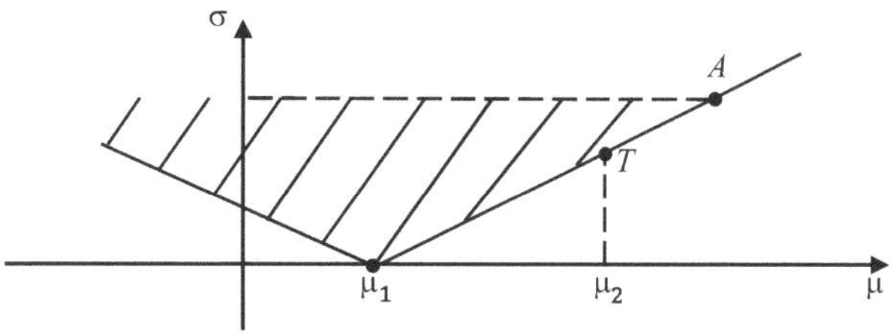

Fig. 19.7 Finding a portfolio of maximum efficiency from all risk portfolios no more than a given one

i.e. exactly on it, will obviously give a solution to the problem (19.113)–(19.115), i.e. a portfolio of maximum efficiency and limited risk. From (19.104) we find

$$\mu = \mu_f + d * \sigma, \tag{19.116}$$

After that, for the found μ by the formula (19.101), we find the desired portfolio *X*.

Example 19.25 The portfolio consists of three securities: a risk-free one with an efficiency (expected return) of 5% and two risky ones with an efficiency of 10% and 15%, respectively, and a covariance matrix $\begin{pmatrix} 9 & 5 \\ 5 & 36 \end{pmatrix}$. Find a portfolio of maximum efficiency from all risk portfolios: (a) no more than 2%, (b) no more than 1.27% and its profitability. From the tangent equation

$$\sigma = \frac{\mu - \mu_f}{d}$$

we get

$$\mu = \mu_f + d * \sigma.$$

Substituting here the profitability of a risk-free paper of 5% and the risk of a portfolio of 2%, and also the calculated in the previous examples

$$d = \sqrt{\left(\vec{\mu} - \mu_f I\right)^T V^{-1}\left(\vec{\mu} - \mu_f I\right)} = \sqrt{4.35} = 2.086,$$

we will get the corresponding value of the maximum profitability

$$\mu_1 = \mu_f + d * \sigma_1 = 5 + 2.086 * 2 = 9.172.$$

For the found μ by the formula

$$X = \frac{\mu - \mu_f}{d^2} V^{-1}\left(\vec{\mu} - \mu_f I\right)$$

we find the desired portfolio *X*.

The matrix V^{-1} was calculated by us in the previous examples and is equal to

$$V^{-1} = \frac{1}{299}\begin{pmatrix} 36 & -5 \\ -5 & 9 \end{pmatrix}.$$

For *X* we have:

$$X = \frac{\mu - \mu_f}{d^2} V^{-1}\left(\vec{\mu} - \mu_f I\right) = \frac{9.172 - 5}{4.35} * \frac{1}{299}\begin{pmatrix} 36 & -5 \\ -5 & 9 \end{pmatrix}\begin{pmatrix} 5 \\ 10 \end{pmatrix}$$
$$= 0.00321 * \begin{pmatrix} 130 \\ 65 \end{pmatrix} = \begin{pmatrix} 0.417 \\ 0.208 \end{pmatrix}.$$

Then the share of risk-free paper is equal to 1 – 0.417 – 0.208 = 0.375, and the portfolio of maximum efficiency of all risk portfolios of no more than 2% has the form

$$X = \left(0.417; 0.208; x_f = 0.375\right).$$

The profitability of the portfolio is 9.172%.

For a risk of no more than 1.27%, we perform similar calculations (we leave them to the readers to perform independently).

Questions and Tasks

19.176. Find a portfolio of maximum efficiency from all risk portfolios of no more than a given one.

19.177. The portfolio consists of three securities: a risk-free one with an efficiency (expected return) of 5% and two risky ones with an efficiency of 10% and 15%, respectively, and a covariance matrix $\begin{pmatrix} 9 & 5 \\ 5 & 36 \end{pmatrix}$. Find the portfolio of maximum efficiency from all risk portfolios (a) no more than 1%, (b) no more than 0.5% and its profitability.

19.178. The portfolio consists of three securities: a risk-free one with an efficiency (expected return) of 8% and two risky ones with an efficiency of 12% and 18%, respectively, and a covariance matrix $\begin{pmatrix} 25 & -6 \\ -6 & 49 \end{pmatrix}$. Find the portfolio of maximum efficiency from all risk portfolios (a) no more than 2.5%, (b) no more than 1.5% and its profitability.

19.6 Optimal Non-negative Portfolios

Consider non-negative portfolios $X \geq 0$. In this case, the condition (19.61) should be added to the conditions of the optimal problem (19.58)–(19.60). The non-negativity of the portfolio components means that they can now be interpreted as price (value) shares of investments in a particular paper of portfolio. At the same time, however, both the algorithm for solving the problem and the solution itself change. Now the conditions (19.58)–(19.61) define a convex polyhedron, i.e. a bounded closed set. The solution of optimal problems under such conditions (the presence of restrictions not only in the form of equality, but also in the form of inequalities) has its own specifics. These problems are the problems of finding the extremes of convex (concave) functions given on convex sets. Solving optimal problems of this type requires modification of the method based on the search for the extremes of the Lagrange function, which creates an algorithm for finding conditional extremes (if there are restrictions only in the form of equalities). Such a modification can be made by applying the Kuhn–Tucker theorem and related theorems. The meaning of these theorems boils down to the fact that the point of the local extremum (and even the stationary point) of a convex (concave) function given on a convex set is the point of the global extremum on this set.

The profitability of a non-negative portfolio of μ lies on an interval $[\mu_{min}, \mu_{max}]$, where μ_{min} and μ_{max} are the minimum and maximum values of the profitability of individual securities included in the portfolio:

$$\mu_{min} = \min(\mu_1, \mu_2, \ldots, \mu_n), \tag{19.117}$$

$$\mu_{max} = \max(\mu_1, \mu_2, \ldots, \mu_n), \tag{19.118}$$

that is, that the profitability of μ satisfies the inequalities

$$\mu_{min} \leq \mu \leq \mu_{max}. \tag{19.119}$$

The Markowitz problem (19.58)–(19.61) is to find the minimum of the objective function

$$\frac{1}{2}\sigma^2 = \frac{1}{2}X^T VX \to \max \tag{19.120}$$

under the conditions

$$\vec{\mu}^T X = \mu, I^T X = 1, X \geq 0. \tag{19.121}$$

This is the classic formulation of the Markowitz problem. Since the acceptable set is compact, the desired portfolio exists. Suppose that the matrix V is positively defined, then, taking into account the strict convexity of the objective function, the linearity of the restriction and the differentiability of the functions under consideration, we conclude that the Kuhn-Tucker conditions are necessary and sufficient conditions for a conditional minimum:

$$X^T V - \lambda I^T - \nu \vec{\mu}^T \geq 0, \left(X^T V - \lambda I^T - \vec{\mu}^T\right) X = 0,$$
$$I^T X = 1, \vec{\mu}^T X = \mu, X \geq 0. \tag{19.122}$$

Solving this system of equations and inequalities is, in general, extremely difficult. With a small number of n-securities, it is possible to solve the system (19.122) by iterating over the cases.

It can be proved that the problem (19.120) and (19.121) has a solution for any $\mu \in [\mu_{min}, \mu_{max}]$, and the minimum boundary is a convex curve consisting of a finite number of hyperbolic parts.

19.6.1 Non-negative Portfolio of Two Securities

First consider a non-negative portfolio of two independent securities.

$$\rho_{12} = \rho = 0. \tag{19.123}$$

For the risk square (dispersion) of the portfolio, we have

$$\sigma^2 = \sigma_1^2 x_1^2 + \sigma_2^2 x_2^2. \tag{19.124}$$

The analysis gives three portfolios:

1. $X = \left(\frac{\sigma_2^2}{\sigma_1^2+\sigma_2^2}; \frac{\sigma_1^2}{\sigma_1^2+\sigma_2^2}\right)$
2. $X = (1; 0)$
3. $X = (0; 1)$

If we draw the obtained portfolios on the plane (x_1, x_2), it is easy to see (Fig. 19.8) that point (1) $\left(\frac{\sigma_2^2}{\sigma_1^2+\sigma_2^2}; \frac{\sigma_1^2}{\sigma_1^2+\sigma_2^2}\right)$ lies on the segment connecting points (2) and (3).

Thus, the set of effective portfolios is a segment connecting the points (0; 1) and (1; 0).

Questions and Tasks

19.179. Formulate and prove the Kuhn–Tucker theorem.

19.180. How does the algorithm for finding optimal portfolios change when considering non-negative portfolios?

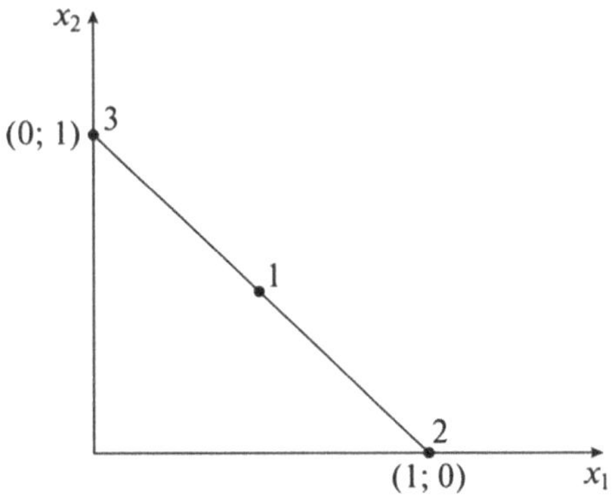

Fig. 19.8 Non-negative portfolio of two securities on the plane (x_1, x_2)

19.181. Describe the profitability properties of a non-negative portfolio.

19.182. Find a portfolio of minimal risk, a given efficiency with non-negative components.

19.183. Draw a non-negative portfolio of two papers on the plane (x_1, x_2).

19.184. Prove that the profitability of a non-negative portfolio is on the interval $[\mu_{\min}, \mu_{\max}]$, where $\mu_{\min}$ and $\mu_{\max}$ are the minimum and maximum values of the profitability of individual securities included in the portfolio

$$\mu_{\min} = \min(\mu_1, \mu_2, \dots, \mu_n),$$
$$\mu_{\max} = \max(\mu_1, \mu_2, \dots, \mu_n),$$

that is, that the profitability of μ satisfies the inequalities $\mu_{\min} \leq \mu \leq \mu_{\max}$.

19.6.2 Non-negative Portfolios of Three Independent Securities

Example 19.26 Let the risks of independent securities of three types be equal

$$\sigma_1 = 1, \sigma_2 = 2, \sigma_3 = 3,$$

and their expected returns

$$\mu_1 = 10\%, \mu_2 = 20\%, \mu_3 = 30\%.$$

To find an effective boundary, it is necessary to find the point of the global minimum of the convex function σ^2 (x_1, x_2, x_3) using the Kuhn–Tucker conditions.

Since the covariance matrix of securities returns is non-degenerate, the risk square of the portfolio

$$\sigma^2 = \sigma_1^2 x_1^2 + \sigma_2^2 x_2^2 + \sigma_3^2 x_3^2 \tag{19.125}$$

is a strictly convex function. We will look for an effective portfolio with an expected return of μ as the minimum point of the function σ^2 on a set of restrictions.

The portfolio consists of two independent securities with risks $\sigma_1 = 0.3$and $\sigma_2 = 0.5$, respectively, and returns $\mu_1 = 0.07$and $\mu_2 = 0.42$. Demonstrate the impact of diversification on the portfolio risk by finding its dependence on the correlation coefficient over the entire range of its changes. Investigate the dependence of the portfolio return on the correlation coefficient over the entire range of its changes.

References

Brusov, P. N., Filatova, T. V., & Orekhova, N. P. (2023). *Financial mathematics*. Cambridge Scholars Publishing, 290 pp.

Brusov, P. N., Filatova, T. V., & Orekhova, N. P. (2024). *Financial mathematics: Exercises and solutions*. Springer Nature Publishing, 286 pp.

Chapter 20
Investing in a Bond Portfolio

20.1 Immunisation of a Bond Portfolio

Under immunisation of a bond portfolio is understood such portfolio management, which allows to maintain the level of its profitability during some period, despite the market interest rate jumps. An example of an immunised portfolio can be a bond portfolio belonging to, say, a pension fund, if the duration of the bond portfolio is equal to the duration of the fund's liabilities (Brusov et al. 2023, 2024).

There is an immunisation theorem. Suppose it is necessary to pay off a debt R in exactly n years. The duration of such a payment is n. One of the ways to pay the debt is to buy a coupon-free n-year bond with nominal value N = R at annual interest rate r. It would then cost P = N/(1 + r)n to buy the bond. Paul Samuelson pointed out that it is possible to replace one bond with two such that at a given interest rate, the present value does not change, but only increases as the interest rate changes. This is due to the fact that the present value is a decreasing function of the interest rate, and bonds with different maturities react differently to changes in the interest rate.

So, let us assume that we need to pay off a debt of R at time t. The purchase of a bond of nominal value N = R with maturity t will ensure the repayment of the debt. Let us call it bond I. The present value of bond I is equal to:

$$P_I = N/(1+r)^t \tag{20.1}$$

Consider two zero-coupon bonds with nominal values N1 and N2 and maturities t1 and t2, respectively, and the inequality is satisfied:

$$t_1 < t < t_2$$

We will call the portfolio consisting of these bonds Bond II. Bond II has a present value of:

P. Brusov, T. Filatova, *Efficiency of Investment Projects*, Contributions to Finance and Accounting, https://doi.org/10.1007/978-3-032-01028-5_20

$$P_2 = \frac{N_1}{(1+r)^{t_1}} + \frac{N_2}{(1+r)^{t_2}} \tag{20.2}$$

Let us require that at r = r0 there were the conditions:

$$\begin{cases} P_1(r_0) = P_2(r_0) \\ D_1(r_0) = D_2(r_0) \end{cases} \tag{20.3}$$

These conditions ensure equivalence of two cash flows associated with bonds I and II and equality of their duration at r = r0. In this case, the graphs of functions P1 (r) and P2(r) look the same as in Eq. (20.4), i.e. they touch at some point r = r0. It can be shown that for other values of r the inequality is fulfilled:

$$P_1(r) < P_2(r) \tag{20.4}$$

For this purpose, it is enough to make sure that the same inequality holds for the second derivatives at r = r0.

Note that in order to ensure fulfilment of conditions (20.3), it is enough to require that the weights of payments w1, w2 satisfy the system of equations

$$\begin{cases} W_1 + W_2 = 1 \\ t_1 W_1 + t_2 W_2 = t \end{cases} \tag{20.5}$$

In the situation considered, bond II is said to immunize bond I.

Example 20.1 Construct a portfolio of 3-year and 5-year bonds immunizing a 4-year bond with a face value of $3000 for an interest rate of 15%.

Let us write the system of Eq. (20.5):

$$\begin{cases} W_1 + W_2 = 1 \\ 3W_1 + 5W_2 = 4 \end{cases}$$

from where we find:

$$W_1 = \frac{1}{2}, W_2 = \frac{1}{2}$$

Now let's determine the present values of all the bonds:

$$P = 3000 * 1.15^{-4} = 17{,}120.26; P_1 = \frac{1}{2}P = 857.63; P_2 = \frac{1}{2}P = 857.63$$

Next, let's find the nominal values of the bonds included in the portfolio:

$$N_1 = P_1 * 1.15^3 = 1304.35; N_2 = P_2 * 1.15^5 = 1725$$

Thus, the immunizing portfolio consists of a 3-year bond with a nominal value of \$1304.35 and a 5-year bond with a nominal value of \$17,220. It can be easily verified that if the interest rate changes, the present value of this portfolio will be higher.

Questions and Tasks

20.1. Give a definition of immunization of a bond portfolio.
20.2. Formulate and prove the bond portfolio immunization theorem.

20.2 Bond Portfolio

A bond portfolio, consisting of bonds of different types, maturities, coupon amounts and other characteristics, has its yield, average maturity, duration, modified duration, convexity and other parameters that characterise the portfolio as a whole. Let us discuss the calculation of these portfolio characteristics.

20.2.1 Yield of Bond Portfolio

Let us remind how the yield of an individual bond is determined. If the market price of a bond, its face value N, maturity n and coupon rate c are known, then the yield to maturity ρ is defined as the solution to the equation:

$$V = \sum_{k=1}^{n} \frac{cN}{(1+p)^k} + \frac{N}{(1+p)^n} \tag{20.6}$$

or equivalently as a solution of the equation:

$$V = cN\frac{(1+p^{-n}}{p} + N(1+p)^{-n} \tag{20.7}$$

In determining the yield of a bond portfolio, it is also found as the solution to an equation in which the sum of the present values of the total income stream is equated to the total market value of the bonds comprising the portfolio:

$$\sum_{i=1}^{n} \frac{S_i}{(1+p)^i} = \sum_{k=1}^{m} q_k P_k \tag{20.8}$$

where Si is the total bond yield at time t = i; qk is the number of bonds of type k; Pk is the price of bond of type k; m is the number of types of bonds in the portfolio; n is the maximum term of income payment.

In general, the solution of Eq. (20.8) with respect to the interest rate (portfolio yield) ρ is found by iterative methods, for example, by the Newton-Raphson method, or on the basis of linear interpolation. In the latter case, we can use the formula:

$$p = p' + \frac{P' - P}{P' - P''}(p' - p'') \tag{20.9}$$

where ρ′, ρ″ are the minimum and maximum values of the bond portfolio yield, limiting the interval within which the unknown value of the portfolio yield ρ is expected; P is the market value of the portfolio; P′, P″ are the estimated portfolio values when applying the rates ρ′, ρ″.

It is more convenient to determine the value of the rate not by the formula (20.9), but as the average weighted value of the yield of the whole set of bonds. Now by ρ we will mean the weighted average of yields. The market value of the corresponding number of bonds ($q_k P_k$ or $q_k K_k$) is taken as a weight:

$$p = \frac{\sum_k p_k q_k P_k}{\sum_k q_k K_k} = \frac{\sum_k p_k q_k K_k}{\sum_k q_k K_k} \tag{20.10}$$

where K_k is the rate of bonds of type k.

More adequate, however, is to use as weights the product of the duration of each type of bond, D_k, by the value of the corresponding number of bonds:

$$p = \frac{\sum_k p_k D_k q_k P_k}{\sum_k D_k q_k P_k} = \frac{\sum_k p_k D_k q_k K_k}{\sum_k D_k q_k K_k} \tag{20.11}$$

Example 20.2 Find the yield of a bond portfolio consisting of two types of bonds of 10 and 20 each with yields of 30% and 40% and rates of 102 and 110 respectively.

Let's use the formula for the yield of the bond portfolio.

$$p = \frac{\sum_k p_k q_k K_k}{\sum_k q_k K_k} = \frac{0.3 * 10 * 102 + 0.4 * 20 * 110}{10 * 102 + 20 * 110} = \frac{1.186}{3.220} = 0.3683$$

Thus, the yield of the bond portfolio is equal to 36.83%.

Questions and Tasks

20.3. Give a definition of bond portfolio yield and a formula for its calculation.

20.4. What parameters are used as weights in determining the yield of a bond portfolio?

20.5. Find the yield of a bond portfolio consisting of two types of bonds of 15 and 36 each with yields of 20% and 60% and rates of 97 and 104 respectively.

20.6. Find the yield of a bond portfolio consisting of three types of bonds of 100, 150, and 190 each with yields of 10%, 23%, and 50% and exchange rates of 97, 115, and 104, respectively.

20.7. Find the yield of a bond portfolio consisting of four types of bonds of 100, 200, 250, and 30 each with yields of 12, 25, 16, and 30% and exchange rates of 95, 110, 99, and 106, respectively.

20.8. Find the yield of a bond portfolio consisting of two types of bonds of 15 and 36 each with yields of 20% and 60% and prices of R1000 and R2000 respectively.

20.9. Find the yield of a bond portfolio consisting of three types of bonds of 100, 150 and 190 each with yields of 10%, 23% and 50% and prices of RUR 500, 1600 and 2300 respectively.

20.10. Find the yield of a bond portfolio consisting of four types of bonds of 100, 200, 250 and 30 each with yields of 12%, 25%, 16% and 30% and prices of R400, 1200, 850 and 2800 respectively.

20.2.2 Average Term of Bond Portfolio Income Receipt

The average term of income receipt of the bond portfolio as a whole T_0 is found as a weighted average. Bond values are taken as weights:

$$T_0 = \frac{\sum_k T_k q_k P_k}{\sum_k q_k P_k} \tag{20.12}$$

Here T_k is the average maturity of bonds of type k.

A portfolio with a shorter average maturity has a lower risk than a portfolio with a longer maturity, all other things being equal.

Example 20.3 Find the average income receipt period of a bond portfolio consisting of two types of bonds of 100 and 150 pieces with average receipt periods of 3.5 and 5 years and prices of 920 and 1100 rubles, respectively.

Let us use the formula for the average term of income of the bond portfolio:

$$T_0 = \frac{\sum_k T_k q_k P_k}{\sum_k q_k P_k} = \frac{3.5 * 100 * 920 + 5 * 150 * 1100}{100 * 920 + 150 * 1100} = \frac{1.147.000}{257.000} = 4.46\text{years}$$

Thus, the bond portfolio's average income stream is 4.46 years, or approximately 4.5 years.

Questions and Objectives

20.11. What parameters are used as weights in determining the average income receipt period of a bond portfolio?

20.12. Find the average income receipt period of a bond portfolio consisting of two types of bonds of 30 and 67 pieces with average receipt periods of 2.5 and 7 years and prices of 800 and 900 rubles, respectively.

20.13. Find the average income receipt period of a bond portfolio consisting of two types of bonds of 25 and 45 pieces with average receipt periods of 3 and 8 years and prices of 1200 and 700 rubles, respectively.

20.14. Find the average the term of receipt of income for a bond portfolio consisting of three types of bonds 125, 200 and 345 pieces each with an average receipt period of 2, 4 and 6 years and prices of 750, 1400 and 550 rubles, respectively.

20.15. Find the average income receipt period of a bond portfolio consisting of three types of bonds of 15, 20 and 35 pieces with average receipt periods of 1, 2.5 and 20.5 years and prices of 250, 300 and 450 rubles, respectively.

20.16. Find the average income receipt period of a bond portfolio consisting of four types of bonds of 100, 200, 300 and 350 pieces with average receipt periods of 3, 2.5, 7 and 6.5 years and a target of 2000, 3000, 1500 and 2500 rubles, respectively.

20.2.3 Duration of the Bond Portfolio

Duration is the average weighted duration of payments of income from a bond with weights equal to discounted income values. Below we will establish the relationship between the duration of a bond portfolio and the duration of individual bonds in this portfolio.

Let the portfolio consist of two bonds with income streams Rk and Sk and durations

$$D_1 = \frac{\sum_{k=1}^{n} t_k R_k (1+y)^{-tk}}{\sum_{k=1}^{n} R_k (1+y)^{-tk}} = \frac{\sum_{k=1}^{n} t_k R_k (1+y)^{-tk}}{P_1} \text{ and } D_2 = \frac{\sum_{k=1}^{n} t_k S_k (1+y)^{-tk}}{P_2} \tag{20.13}$$

For a combined revenue stream from two bonds, we have:

$$D_0 = \frac{\sum_{k=1}^{n} t_k R_k (1+y)^{-tk} + \sum_{k=1}^{n} t_k S_k (1+y)^{-tk}}{P_1 + P_2} \tag{20.14}$$

If there are q_1 and q_2 bonds in the portfolio, then income streams increase in proportion to these values. We have:

$$D_0 = \frac{q_1 \sum_{k=1}^{n} t_k R_k (1+y)^{-tk} + q_2 \sum_{k=1}^{n} t_k S_k (1+y)^{-tk}}{q_1 P_1 + q_2 P_2} \tag{20.15}$$

It follows from (20.13) that

$$D_1 P_1 = \sum\nolimits_{k=1}^{n} t_k R_k (1+y)^{-tk} \text{ and } D_2 P_2 = \sum\nolimits_{k=1}^{n} t_k S_k (1+y)^{-tk}$$

That's why

$$D_0 = \frac{D_1 q_1 P_1 + D_2 q_2 P_2}{q_1 P_1 + q_2 P_2} \tag{20.16}$$

Thus, we conclude that the duration of the bond portfolio is equal to the average weighted duration of individual bonds in this portfolio with weights equal to the bond values.

Generalizing (20.16) to the case of m-types of bonds, we obtain

$$D_0 = \frac{\sum_{k=1}^{m} D_k q_k P_k}{\sum_{k=1}^{m} q_k P_k} = \sum\nolimits_{k=1}^{m} D_k h_k \tag{20.17}$$

where h_k is the value share of bonds of type k.

The value share of bonds can be obtained based not only on bond prices, but also on their exchange rates. In this case:

$$h_k = \frac{q_k P_k}{\sum_{k=1}^{m} q_k P_k} = \frac{q_k K_k}{\sum_{k=1}^{m} q_k K_k} \tag{20.18}$$

at the same time:

$$\sum_{k=1}^{m} h_k = 1 \tag{20.19}$$

Example 20.4 Find the duration of a bond portfolio consisting of two types of bonds 10 and 15 pieces each with durations of 4 and 5 years and prices of 950 and 1000 rubles, respectively.

We use the formula for the duration of the bond portfolio:

$$D_0 = \frac{D_1 q_1 P_1 + D_2 q_2 P_2}{q_1 P_1 + q_2 P_2} = \frac{4 * 10 * 950 + 5 * 15 * 1000}{10 * 950 + 15 * 1000} = \frac{113.000}{24.500} = 4.6 \text{ years}$$

Thus, the duration of the bond portfolio is 4.6 years.

Questions and Tasks

20.17. What parameters are used as weights when determining the duration of a bond portfolio?

20.18. Find the duration of a bond portfolio consisting of two types of bonds of 30 and 67 pieces with durations of 2.5 and 7 years and prices of 800 and 900 rubles, respectively.

20.19. Find the duration of a bond portfolio consisting of two types of bonds of 35 and 55 pieces with durations of 3 and 8 years and prices of 1200 and 700 rubles, respectively.

20.20. Find the duration of a bond portfolio consisting of three types of bonds 225, 300 and 445 pieces each with durations of 2, 4 and 6 years and prices of 750, 1400 and 550 rubles, respectively.

20.21. Find the duration of a bond portfolio consisting of three types of bonds of 12, 25 and 40 pieces with durations of 1, 3.5 and 6 years and prices of 250, 400 and 550 rubles, respectively.

20.22. Find the duration of a bond portfolio consisting of four types of bonds 200, 300, 400 and 350 pieces each with durations of 3, 4, 8 and 9 years and prices of 3000, 4000, 5500 and 6500 rubles. Accordingly.

20.23. Find the duration of a bond portfolio consisting of two types of bonds of 35 and 55 pieces with durations of 3 and 8 years and rates of 96 and 100, respectively.

20.24. Find the duration of a bond portfolio consisting of three types of bonds 225, 300 and 445 pieces each with durations of 2, 4 and 6 years and courses of 95, 105 and 101, respectively.
20.25. Find the duration of a bond portfolio consisting of four types of bonds 100, 300, 200 and 650 pieces each with durations of 3, 2, 5 and 6 years and courses 92, 103, 89 and 104, respectively.

20.2.4 Convexity of the Bond Portfolio

The convexity of the C_0 bond portfolio, like the duration, is found as the average weighted convexity of individual bonds in a given portfolio with weights equal to the bond values.

$$C_0 = \frac{\sum_{k=1}^{m} C_k q_{kP_k}}{\sum_{k=1}^{m} q_{kP_k}} = \sum_{k=1}^{m} C_k h_k \qquad (20.20)$$

Questions and Tasks

20.26. Define the yield of the bond portfolio.
20.27. What values are taken as weights when calculating the weighted average yield of a bond portfolio?
20.28. Define the average period of receipt of income of a bond portfolio.
20.29. What values are taken as weights when calculating the average income receipt period of a bond portfolio
20.30. Deduce the relationship between the duration of a bond portfolio and the durations of individual bonds in this portfolio.
20.31. Express the value share of bonds in terms of their exchange value (bond rates).
20.32. Define and give a formula for the convexity of the bond portfolio.
20.33. What values are taken as weights when calculating the bulge of a bond portfolio how is the weighted average?
20.34. The manager must pay in 3 years a debt of 650,000 USD and in 5 years a debt of 300,000 USD. How many 2- and 9-year coupon-free bonds he should buy to protect funds from changes the interest rate, provided that the interest rate is 10% (bond denominations are equal to 1000 units)?
20.35. Find the change in the price of a bond with a maturity of n = 7 years, yield y = 8, coupon rate c = 7 with a decrease in yield to 5%.
20.36. Is it worth buying for 1000 rubles? a bond with a face value of 1200 rubles. and for a duration of 7 years, if it gives 50 rubles of interest income at the end of each half-year and is repaid at face value at the end of the term, when it is possible to place these funds in the bank at 7% per annum?

20.37. The bond with a nominal value of 5000 rubles and a declared yield of 7.5% per annum maturing in 3 years was purchased at the time of its issue on February 20. A coupon income is paid on the bond every quarter. After a while, the owner of the bond decided to sell it. What price will the bondholder set for the bond at a rate of return of 7.5% per annum if: (a) the bond is sold on 20.03 next year, (b) sold on 22.09 next year? Time base K = 360 days.

References

Brusov, P. N., Filatova, T. V., & Orekhova, N. P. (2023). *Financial mathematics*. Cambridge Scholars Publishing, 290 pp.

Brusov, P. N., Filatova, T. V., & Orekhova, N. P. (2024). *Financial mathematics: Exercises and solutions*. Springer Nature Publishing, 286 pp.

Part VI
Ratings, Rating Methodologies and Ratings of Investment Projects

Rating agencies play a very important role in economics. Their analysis of issuer's state, generated credit ratings of issuers help investors make reasonable investment decision, as well as help issuers with good enough ratings get credits on lower rates etc.

This Part suggests a new approach to rating methodology. Application of the Modigliani–Miller Theory in Rating Methodology is discussed in Chap. 21, while Chap. 22 is devoted to the application of the Modigliani–Miller Theory, modified for the case of advance payments of tax on profit, in rating methodologies.

In Chap. 23 a new approach to project ratings is described. The important features of current consideration as well as in previous studies are: (1) The adequate use of discounting of financial flows virtually not used in existing rating methodologies, (2) The incorporation of rating parameters (financial "ratios"), used in project rating, into considered modern investment models. This on the one hand allows use the powerful tools of this theories in the rating, and on the other hand it ensures the correct discount rates when discounting of financial flows. We discuss also the interplay between rating ratios and leverage level which can be quite important in rating. All these create a new base for rating methodologies. New approach to ratings and rating methodologies allows to issue more correct ratings of issuers, makes the rating methodologies more understandable and transparent.

In Chap. 24 further development of a new approach to project ratings has been done: it has been generalized for the general case of modern theory of capital structure (Brusov–Filatova–Orekhova (BFO) theory), which is valid for companies of arbitrary age.

Ratings of the investment projects of arbitrary durations is studied in Chap. 25.

In Chap. 26 ratings of investment projects of arbitrary duration with a uniform debt repayment is described.

Chapter 21
Application of the Modigliani–Miller Theory in Rating Methodology

21.1 Introduction

Rating agencies play a very important role in economics. Their analysis of issuer's state, generated credit ratings of issuers help investors make reasonable investment decision, as well as help issuers with good enough ratings get credits on lower rates etc.

But from time to time we listen about scandals involved rating agencies and their credit ratings: let us just remind the situation with sovereign rating of the USA in 2011 and of the Russia in 2015.

Were these ratings an objective? And how objective could be issued credit ratings in principal?

To answer this question, we need to understand how rating agencies (RA) consider, evaluate, analyze. But this is the secret behind the seven seals: rating agencies stand to the death, but did not reveal their secrets, even under the threat of multibillion-dollar sanctions.

Thus, rating agencies represent some "black boxes", about which information on the methods of work is almost completely absent.

21.2 The Closeness of the Rating Agencies

The closeness of the rating agencies is caused by multiple causes.

1. The desire to preserve their "know how". Rating agencies get big enough money for generated ratings (mostly from issuers) to replicate its methodology.
2. On the other hand closeness of rating agencies is caused by the desire to avoid public discussion of the ratings with anyone, including the issuer. It is very convenient position—rating agency "a priori" removes himself from beneath any criticism of generated ratings.

P. Brusov, T. Filatova, *Efficiency of Investment Projects*, Contributions to Finance and Accounting, https://doi.org/10.1007/978-3-032-01028-5_21

3. The absence of any external control and external analysis of the methodologies is resulted in the fact that shortcomings of methodologies are not subjected to serious critical analysis and stored long enough.

The illustration of the closeness of rating agencies is the behavior of the S&P (Standard & Poor's) Director after declining of the sovereign rating of United States, who left his position but has not opened the methodology used.

But even in this situation, it is still possible some analysis of the activities and findings of the rating agencies, based on knowledge and understanding of existing methods of evaluation. Rating agencies cannot use methods other than developed up to now by leading economists and financiers.

21.3 The Use of Discounting in the Rating

One of the major flaws of all existing rating methodologies is a failure or a very narrow use of discounting. But even in those rare cases where it is used, it is not quite correct, since the discount rate when discounting financial flows is chosen incorrectly.

The need to take into account the time factor in terms of discounting is obvious, because it is associated with the time value of money. The financial part of the rating is based on a comparison of generated income with the value of the debt and the interest payable. Because income and disbursement of debt and interest are separated in time, the use of discounting when comparing revenues with the value of debt and interest is absolutely necessary for assigning credit ratings for issuers.

This raises the question about the value of discount rate. This question has always been one of the major and extremely difficult in many areas of Finance: corporate finance, investment, it is particularly important in business valuation, where a slight change in the discount rate leads to a significant change in the assessment of company capitalization, that is used by unscrupulous appraisers for artificial bankruptcy of the company. And the value of discount rate is extremely essential as well in rating.

21.4 Incorporation of Parameters, Using in Ratings, into Perpetuity Limit of Modern Theory of Capital Structure by Brusov–Filatova–Orekhova

In quantification of the creditworthiness of the issuers the crucial role belongs to the so-called financial "ratios", constitute a direct and inverse ratios of various generated cash flows to debt values and interest ones. We could mention such ratios as DCF/Debt, FFO/Debt, CFO/Debt, FOCF/Debt, *FFO*/cash interest, EBITDA/interest, Interests/EBITDA, Debt/EBITDA and some others.

We incorporate these rating parameters (financial "ratios") into the modern theory of capital structure—BFO theory (for beginning into its perpetuity limit) (Brusov et al. 2018b, Brusov (2018), Brusov & Filatova (2021), Brusov et al. (2015), Brusov et al. (2018a), Brusov et al. (2018c), Brusov et al. (2018d), Brusov et al. (2018e), Brusov et al. (2018f), Brusov et al. (2018g), Brusov et al. (2019), Brusov et al. (2020a), Brusov et al. (2020b), Brusov et al. (2020c), Brusov et al. (2021a), Brusov et al. (2021b), Filatova et al. (2008), Filatova et al. (2018)). The importance of such incorporation, which has been done by us for the first time, is in using of this theory as a powerful tools when discounting of financial flows using the correct discounting rate in rating. Only this theory allow valuate adequately the weighted average cost of capital *WACC* and equity cost of capital k_e used when discounting of financial flows.

Use of the tools of well developed theories in rating opens completely new horizons in the rating industry, which could go from the mainly use of qualitative methods of the evaluation of the creditworthiness of issuers to a predominantly quantitative evaluation methods that will certainly enhance the quality and correctness of the rating.

Currently, rating agencies just directly use financial ratios, while the new methodology will allow (knowing the values of these "relations" (and parameter k_0)) determine the correct values of discount rates (*WACC* and k_e) that should be used when discounting the various financial flows, both in terms of their timing and forecasting.

This has required the modification of the BFO theory (and its perpetuity limit—Modigliani–Miller theory), as used in financial management the concept of "leverage" as the ratio of debt value to the equity value substantially differs from the concept of "leverage" in the rating, where it is understood as the direct and inverse ratio of the debt value to the generated cash flow values (income, profit, etc.). The authors introduced some additional ratios, allowing more fully characterize the issuer's ability to repay debts and to pay interest thereon.

Thus the bridge is building between the discount rates (*WACC*, *ke*) used when discounting of financial flows, and "ratios" in the rating methodology. The algorithm for finding the discount rates for given ratio values is developed.

21.5 Models

Two kind of models of the evaluation of the creditworthiness of issuers, accounting the discounting of financial flows could be used in rating: one-period model and multi-period model.

21.5.1 One-Period Model

One-period model is described by the following formula (see Fig. 21.1)

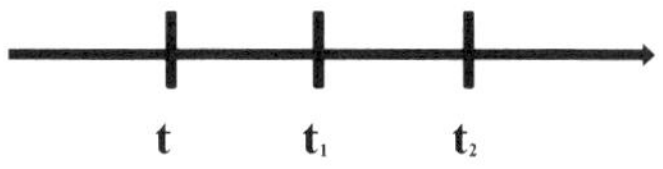

Fig. 21.1 One-period model

$$\begin{aligned} CF(1+i)^{t_2-t} &\geq D + k_d D(1+i)^{t_2-t_1} \\ CF(1+i)^{t_2-t} &\geq D\left[1 + k_d(1+i)^{t_2-t_1}\right] \end{aligned} \tag{21.1}$$

Here CF is value of income for period, D is debt value, t, t_1, t_2 the moments of income, payment of interest and payment of debt consequently, i is the discount rate, k_d is credit rate and $k_d D$ is interest on credit.

21.5.2 Multi-Period Model

One-period model of the evaluation of the creditworthiness of issuers, accounting the discounting of financial flows could be generalized for more interesting multi-period case.

Multi-period model is described by the following formula

$$\sum_j CF_j(1+i_j)^{t_{2j}-t_j} \geq \sum_j D_j\left[1 + k_{dj}(1+i_j)^{t_{2j}-t_{1j}}\right] \tag{21.2}$$

Here CF_j is income for jth period, D_j is debt value in jth period, t_j, t_{1j}, t_{2j} the moments of income, payment of interests and payment of debt consequently in jth period, i_j is the discount rate in jth period, k_{dj} is credit rate in jth period.

There are several options to work with these models:

1. One can check the creditworthiness of the issuer, knowing parameters CF_j, D_j, t_j, t_{1j}, t_{2j}, k_{dj} and defining discount rate i by the method described below.
2. When the preset D_j, t_j, t_{1j}, t_{2j}, k_{dj}, one can determine which income CF_j the issuer would require to ensure its creditworthiness.
3. When the preset D_j, t_j, t_{1j}, t_{2j}, k_{dj} one can define an acceptable level of debt financing (including the credit value D_j and credit rates k_{dj}) when issuer retains its creditworthiness.

21.6 Theory of Incorporation of Parameters, Using in Ratings, into Perpetuity Limit of Modern Theory of Capital Structure by Brusov–Filatova–Orekhova

For the first time we incorporate below the parameters, using in ratings, into perpetuity limit of modern theory of capital structure by Brusov–Filatova–Orekhova (BFO theory).

We'll consider two kind of ratios: coverage ratios and leverage ratios.

Let us start from the coverage ratios.

21.6.1 Coverage Ratios

We will consider three kind of coverage ratios: coverage ratios of debt, coverage ratios of interest on the credit and coverage ratios of debt and interest on the credit.

21.6.1.1 Coverage Ratios of Debt

Here

$$i_1 = CF/D \tag{21.3}$$

Modigliani–Miller theorem for case with corporate taxes (Modigliani and Miller 1958, 1963) tells that capitalization of leveraged company, V_L, is equal to the capitalization of unleveraged company, V_0, plus tax shield for perpetuity time, Dt,

$$V_L = V_0 + Dt. \tag{21.4}$$

Substituting the expressions for both capitalizations, one has

$$\frac{CF}{WACC} = \frac{CF}{k_0} + Dt$$

Dividing both parts by D one gets

$$\begin{aligned} \frac{i_1}{WACC} &= \frac{i_1}{k_0} + t \\ WACC &= \frac{i_1 k_0}{i_1 + tk_0} \end{aligned} \tag{21.5}$$

This ratio (i_1) can be used to assess of the following parameters used in rating, DCF/Debt, FFO/Debt, CFO/Debt, FOCF/Debt and some others. We will use last formula to build a curve of dependence $WACC(i_1)$.

21.6.1.2 Coverage Ratios of Interest on the Credit

Here

$$i_2 = CF/k_d D \tag{21.6}$$

Using the Modigliani–Miller theorem for case with corporate taxes

$$V_L = V_0 + Dt,$$

we derive the expression for $WACC(i_2)$

$$\begin{aligned}\frac{CF}{WACC} &= \frac{CF}{k_0} + Dt \\ \frac{i_2}{WACC} &= \frac{i_2}{k_0} + \frac{i_2}{k_d} \\ WACC &= \frac{i_2 k_0 k_d}{i_2 k_d + t k_0}\end{aligned} \tag{21.7}$$

This ratio (i_2) can be used to assess of the following parameters, used in rating, *FFO*/cashinterest, EBITDA/interest and some others. We will use last formula to build a curve of dependence $WACC(i_2)$.

21.6.1.3 Coverage Ratios of Debt and Interest on the Credit (New Ratios)

Let us consider the coverage ratios of debt and interest on the credit simultaneously: this is new ratio, introduced by us for the first time here.

Here

$$i_3 = \frac{CF}{D(1 + k_d)} \tag{21.8}$$

Using as above the Modigliani–Miller theorem for case with corporate taxes

$$V_L = V_0 + Dt$$

one gets the expression for $WACC(i_3)$

$$\begin{aligned}\frac{CF}{WACC} &= \frac{CF}{k_0} + Dt \\ \frac{i_3}{WACC} &= \frac{i_3}{k_0} + \frac{t}{1 + k_d} \\ WACC &= \frac{i_3 k_0 (1 + k_d)}{i_3 (1 + k_d) + t k_0}\end{aligned} \tag{21.9}$$

This ratio (i_3) can be used to assess of the following parameters used in rating, *FFO*/Debt + interest, EBITDA/Debt + interest and some others. We will use last formula to build a curve of dependence $WACC(i_3)$.

Let us analyze the dependence of company's weighted average cost of capital (WACC) on the coverage ratios on debt i_1, on interest on the credit i_2 and on coverage ratios on debt and interest on the credit with the following data: $k_0 = 12\%$; $k_d = 6\%$; $t = 20\%$; i_j run from 0 up to 10 (Fig. 21.2).

The dependence of company's weighted average cost of capital (WACC) on the coverage ratio on interest on the credit i_2 is presented at Fig. 21.3.

The dependence of company's weighted average cost of capital (WACC) on the coverage ratio on debt and interest on the credit i_2 is presented at Fig. 21.4.

The dependence of company's weighted average cost of capital (WACC) on the coverage ratio on debt i_1, on interest on the credit i_2, and on debt and interest on the credit i_3 is presented at Fig. 21.5.

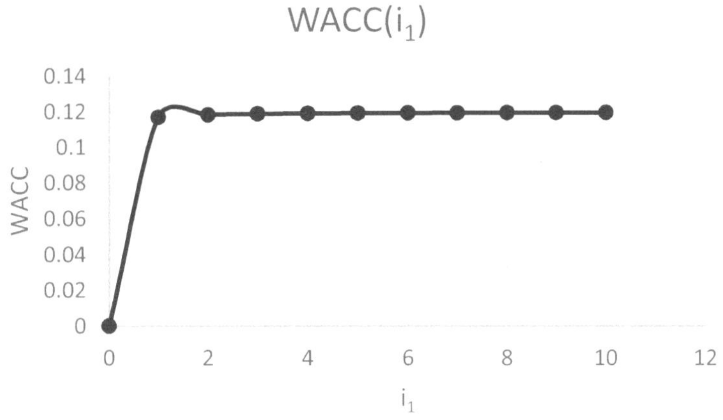

Fig. 21.2 The dependence of company's weighted average cost of capital (WACC) on the coverage ratio on debt i_1

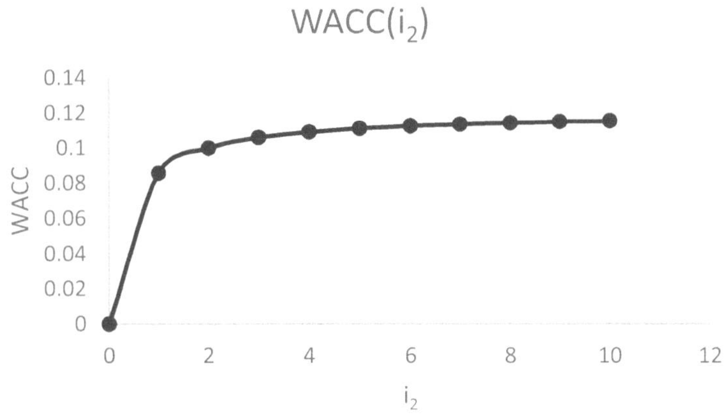

Fig. 21.3 The dependence of company's weighted average cost of capital (WACC) on the coverage ratio on interest on the credit i_2

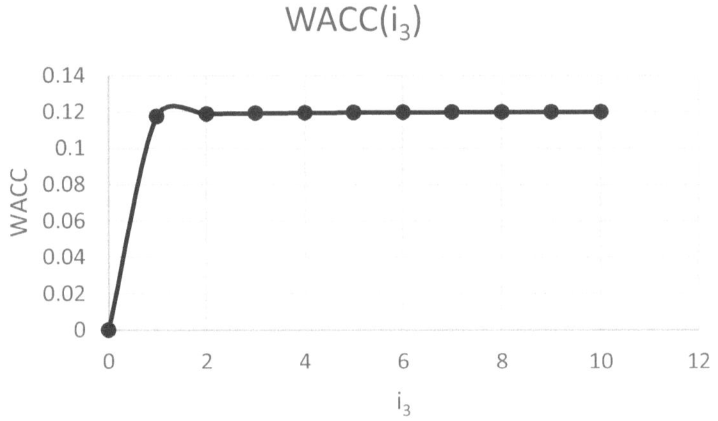

Fig. 21.4 The dependence of company's weighted average cost of capital (WACC) on the coverage ratio on debt and interest on the credit i_3

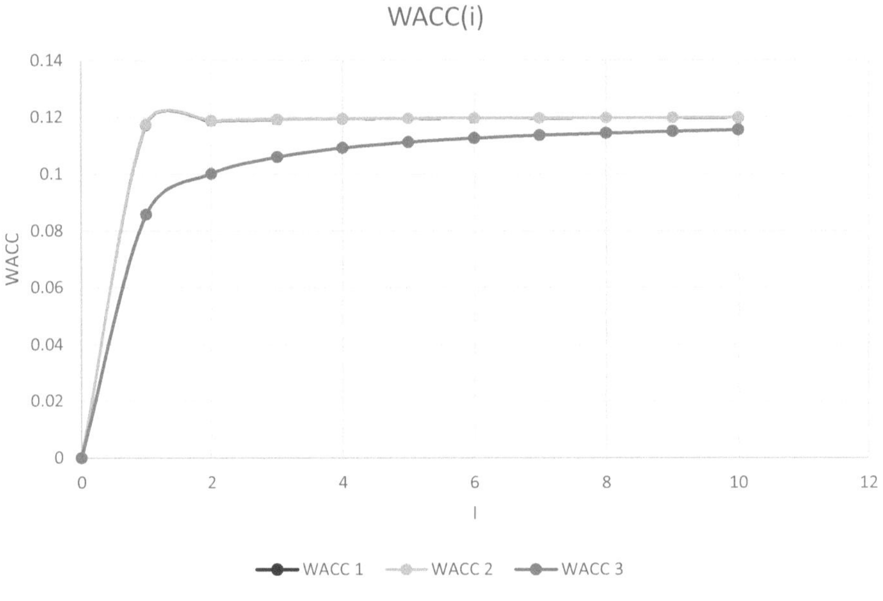

Fig. 21.5 The dependence of company's weighted average cost of capital (WACC) on the coverage ratio on debt i_1, on interest on the credit i_2, and on debt and interest on the credit i_3

It is seen from Figs. 21.2, 21.3, 21.4, and 21.5 that $WACC(i_j)$ is increasing function on i_j with saturation around i_j value of order 1 for ratios i_1 and i_3 and of order 4 or 5 for ratios i_2. At saturation $WACC$ reaches the value k_0 (equity value at zero leverage level). This means that for high values of i_j one can choose k_0 as a discount rate with a good accuracy. Thus the role of parameter k_0 increases drastically. The method of determination of parameter k_0 has been developed by

Anastasiya Brusova (2011). So, parameter k_0 is the discount rate for limit case of high values of i_j (see however below more detailed consideration).

It is clear from Figs. 21.2, 21.3, 21.4, and 21.5 that case of low values of i_j requires more detailed consideration. Let us consider the situation with low values of i_j which seems to be the case of the most interest.

21.6.2 More Detailed Consideration

Below we consider the case of low values of i_j with more details. i_j will vary from zero up to 1 with all other parameters to be the same (Figs. 21.6, 21.7, 21.8, and 21.9).

More detailed consideration leads us to following conclusions:

1. In case of coverage ratio on debt and interest on the credit i_3 WACC goes to saturation very fast: with accuracy of 20% at $i_3 = 0.15$ and with accuracy of 5% at $i_3 = 0.5$.
2. In case of coverage ratio on debt i_1, WACC practically linearly increases with parameter i_1 and goes to saturation at $i_1 = 0.1$.
3. In case of coverage ratio on interest on the credit i_2 WACC increases with parameter i_2 much more slowly than in two previous cases and goes to saturation at high values of i_2: with accuracy of 10% at $i_2 = 4$.

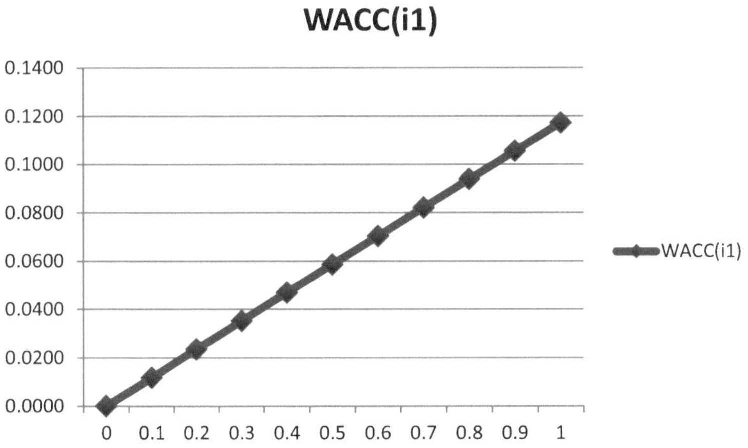

Fig. 21.6 The dependence of company's weighted average cost of capital (WACC) on the coverage ratio on debt and interest on the credit i_1

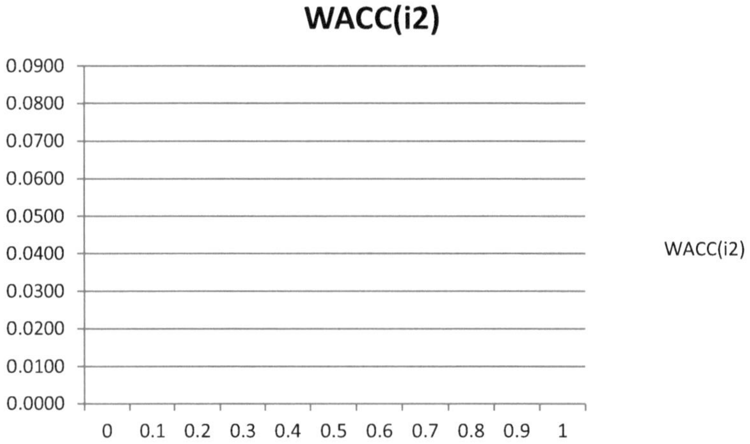

Fig. 21.7 The dependence of company's weighted average cost of capital (WACC) on the coverage ratio on debt and interest on the credit i_2

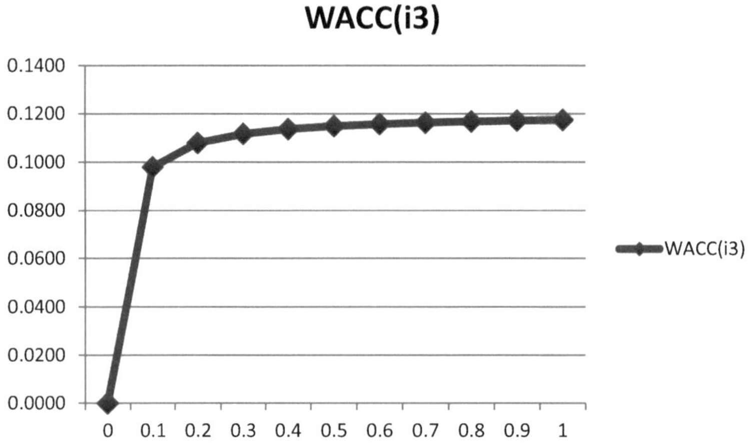

Fig. 21.8 The dependence of company's weighted average cost of capital (WACC) on the coverage ratio on debt and interest on the credit i_3

21.6.3 Leverage Ratios

Let us consider now the leverage ratios. We will consider three kind of leverage ratios: leverage ratios of debt, leverage ratios of interest on the credit and leverage ratios of debt and interest on the credit.

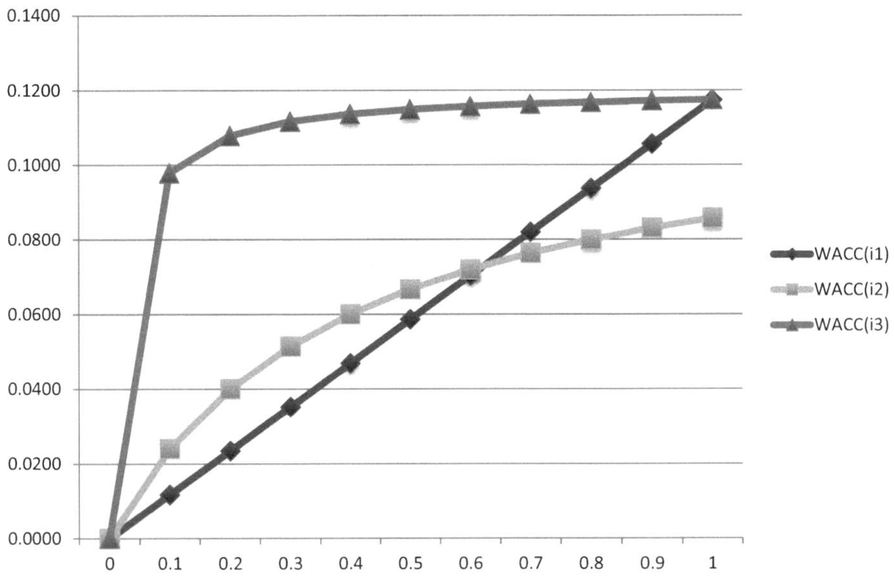

Fig. 21.9 The dependence of company's weighted average cost of capital (WACC) on the coverage ratio on debt i_1, on interest on the credit i_2, and on debt and interest on the credit i_3

21.6.3.1 Leverage Ratios for Debt

Here

$$l_1 = D/CF \tag{21.10}$$

As above for coverage ratios we use the Modigliani–Miller theorem for case with corporate taxes

$$V_L = V_0 + Dt,$$

we derive the expression for $WACC(l_1)$

$$\begin{aligned} \frac{CF}{WACC} &= \frac{CF}{k_0} + Dt \\ \frac{1}{WACC} &= \frac{1}{k_0} + l_1 t \\ WACC &= \frac{k_0}{1 + t l_1 k_0} \end{aligned} \tag{21.11}$$

This ratio (l_1) can be used to assess of the following parameters used in rating, Debt/EBITDA and some others. We will use last formula to build a curve of dependence $WACC(l_1)$.

21.6.3.2 Leverage Ratios for Interest on Credit

Here

$$l_2 = k_d D / CF \tag{21.12}$$

We use again the Modigliani–Miller theorem for case with corporate taxes

$$V_L = V_0 + Dt,$$

we derive the expression for $WACC(l_2)$

$$\begin{aligned} \frac{CF}{WACC} &= \frac{CF}{k_0} + Dt \\ \frac{1}{WACC} &= \frac{1}{k_0} + \frac{l_2 t}{k_d} \\ WACC &= \frac{k_0 k_d}{k_d + t l_2 k_0} \end{aligned} \tag{21.13}$$

This ratio (l_2) can be used to assess of the following parameters used in rating, Interests/EBITDA and some others. We will use last formula to build a curve of dependence $WACC(l_2)$.

21.6.3.3 Leverage Ratios for Debt and Interest on Credit

Here

$$l_3 = D(1 + k_d) / CF \tag{21.14}$$

Using the Modigliani–Miller theorem for case with corporate taxes

$$V_L = V_0 + Dt,$$

we derive the expression for $WACC(l_3)$

$$\begin{aligned} \frac{CF}{WACC} &= \frac{CF}{k_0} + Dt \\ \frac{1}{WACC} &= \frac{1}{k_0} + \frac{l_3 t}{1 + k_d} \\ WACC &= \frac{k_0(1 + k_d)}{1 + k_d + t l_3 k_0} \end{aligned} \tag{21.15}$$

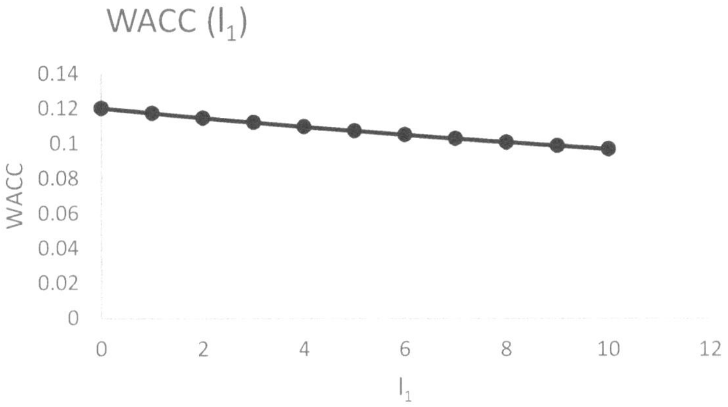

Fig. 21.10 The dependence of company's weighted average cost of capital (WACC) on the leverage ratio on debt l_1

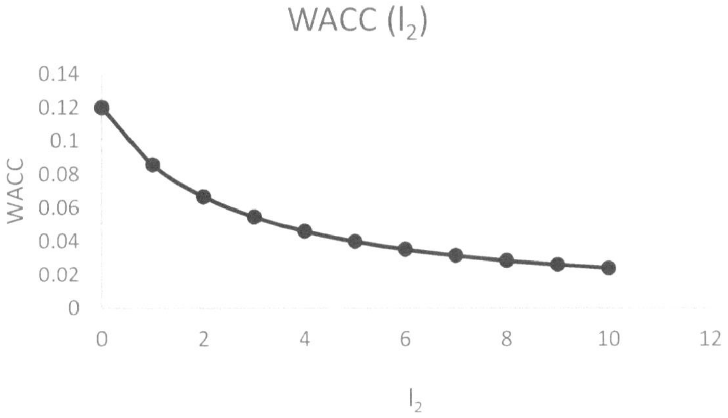

Fig. 21.11 The dependence of company's weighted average cost of capital (WACC) on the leverage ratio on interest on credit l_2

This ratio (l_3) can be used to assess of the following parameters used in rating, *Debt + interest/FFO*, *Debt + interest/EBIT*, *Debt + interest/EBITDA(R)*, and some others. We will use last formula to build a curve of dependence $WACC(l_3)$.

Let us analyze the dependence of company's weighted average cost of capital (WACC) on the leverage ratios with the following data: $k_0 = 12\%$; $k_d = 6\%$; $t = 20\%$; l_i runs from 0 up to 10.

The dependence of company's weighted average cost of capital (WACC) on the leverage ratio on debt l_1 is presented at Fig. 21.10.

The dependence of company's weighted average cost of capital (WACC) on the leverage ratio on interest on credit l_2 is presented at Fig. 21.11.

The dependence of company's weighted average cost of capital (WACC) on the leverage ratio on debt and interest on credit l_3 is presented at Fig. 21.12.

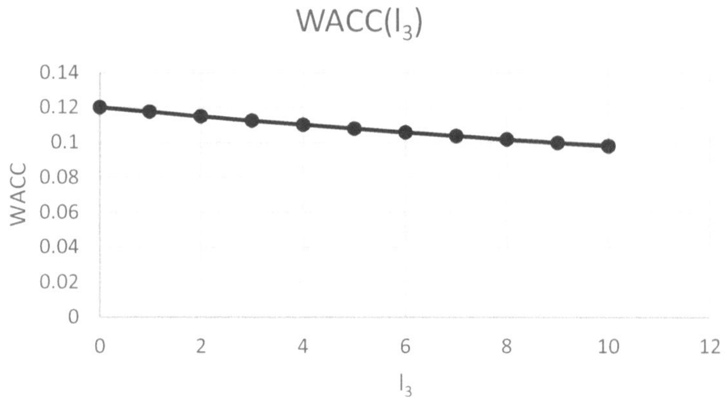

Fig. 21.12 The dependence of company's weighted average cost of capital (WACC) on the leverage ratio on debt and interest on credit l_3

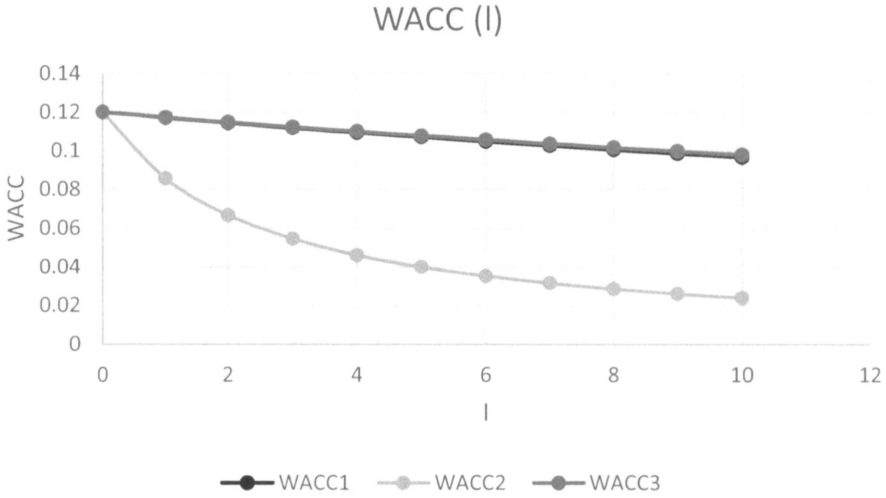

Fig. 21.13 The dependence of company's weighted average cost of capital (*WACC*) on the leverage ratio on debt, l_1, on interest on credit, l_2, and on debt and on interest on credit, l_3 simultaneously

The dependence of company's weighted average cost of capital (WACC) on the leverage ratio on debt, l_1, on interest on credit, l_2, and on debt and interest on credit, l_3 simultaneously is presented at Fig. 21.13.

Analysis of the dependences of company's weighted average cost of capital (*WACC*) on the leverage ratio on debt, l_1, on interest on credit, l_2, and on debt and interest on credit, l_3 shows the following: for all leverage ratios weighted average cost of capital (*WACC*) decreases with leverage ratios. For leverage ratio on debt l_1 and leverage ratio on debt and interest on credit l_3 *WACC* decreases very similar and practically linearly from $k_0 = 12\%$ at $l_{1,3} = 0$ up to 9.7% at $l_{1,3} = 10$. For leverage ratio on interest on credit l_2 *WACC* decreases nonlinearly and much faster from $k_0 = 12\%$ at $l_2 = 0$ up to 2.4% at $l_2 = 10$.

21.7 Equity Cost

Equity cost plays a very important role in economy and finance because it is the essence of the dividend policy of companies, which should be accounted in rating. A modern approach to the dividend policy of companies, based on the real value of their equity capital cost, compared to its efficiency of planned investment is suggested in the article. This allows return to the economic essence of dividends, as the payment to shareholders for the use of equity capital.

Equity cost k_e determines the economically reasonable dividend value. Rating agencies will be able compare payable dividend value with economically reasonable dividend level and make conclusion about the adequacy of the dividend policy of companies and its influence on company's credit rating.

For finding of the dependence of equity cost k_e on coverage ratios and leverage ratios we consider consistently the dependence of equity cost k_e on ratios i_1, i_2, i_3, l_1, l_2, l_3, using the following formula, which couples weighted average cost of capital WACC (calculated by us above: see Tables 21.1, 21.2, 21.3, 21.4, 21.5, 21.6, 21.7, 21.8, and 21.9) and equity cost k_e

$$k_e = WACC(1 + L) - Lk_d(1 - t). \tag{21.16}$$

Table 21.1 The dependence of company's weighted average cost of capital (WACC) on the coverage ratio on debt i_1

t	i1	k0	kd	WACC
0.2	0	0.12	0.06	0
0.2	1	0.12	0.06	0.1171875
0.2	2	0.12	0.06	0.1185771
0.2	3	0.12	0.06	0.1190476
0.2	4	0.12	0.06	0.1192843
0.2	5	0.12	0.06	0.1194268
0.2	6	0.12	0.06	0.1195219
0.2	7	0.12	0.06	0.11959
0.2	8	0.12	0.06	0.1196411
0.2	9	0.12	0.06	0.1196809
0.2	10	0.12	0.06	0.1197127

Table 21.2 The dependence of company's weighted average cost of capital (WACC) on the coverage ratio on debt i_2

t	i2	k0	kd	WACC
0.2	0	0.12	0.06	0
0.2	1	0.12	0.06	0.085714
0.2	2	0.12	0.06	0.1
0.2	3	0.12	0.06	0.105882
0.2	4	0.12	0.06	0.109091
0.2	5	0.12	0.06	0.111111
0.2	6	0.12	0.06	0.1125
0.2	7	0.12	0.06	0.113514
0.2	8	0.12	0.06	0.114286
0.2	9	0.12	0.06	0.114894
0.2	10	0.12	0.06	0.115385

Table 21.3 The dependence of company's weighted average cost of capital (WACC) on the coverage ratio on debt i_3

t	i3	k0	kd	WACC
0.2	0	0.12	0.06	0
0.2	1	0.12	0.06	0.1173432
0.2	2	0.12	0.06	0.1186567
0.2	3	0.12	0.06	0.1191011
0.2	4	0.12	0.06	0.1193246
0.2	5	0.12	0.06	0.1194591
0.2	6	0.12	0.06	0.1195489
0.2	7	0.12	0.06	0.1196131
0.2	8	0.12	0.06	0.1196613
0.2	9	0.12	0.06	0.1196989
0.2	10	0.12	0.06	0.1197289

Table 21.4 The dependence of company's weighted average cost of capital (WACC) on the coverage ratio on debt i_1

i1	Ko	Kd	t	WACC(i1)
0	0.12	0.06	0.2	0.0000
0.1	0.12	0.06	0.2	0.0117
0.2	0.12	0.06	0.2	0.0234
0.3	0.12	0.06	0.2	0.0352
0.4	0.12	0.06	0.2	0.0469
0.5	0.12	0.06	0.2	0.0586
0.6	0.12	0.06	0.2	0.0703
0.7	0.12	0.06	0.2	0.0820
0.8	0.12	0.06	0.2	0.0938
0.9	0.12	0.06	0.2	0.1055
1	0.12	0.06	0.2	0.1172

Table 21.5 The dependence of company's weighted average cost of capital (WACC) on the coverage ratio on debt i_1

i_2	Ko	Kd	t	WACC(i_2)
0	0.12	0.06	0.2	0.0000
0.1	0.12	0.06	0.2	0.0240
0.2	0.12	0.06	0.2	0.0400
0.3	0.12	0.06	0.2	0.0514
0.4	0.12	0.06	0.2	0.0600
0.5	0.12	0.06	0.2	0.0667
0.6	0.12	0.06	0.2	0.0720
0.7	0.12	0.06	0.2	0.0764
0.8	0.12	0.06	0.2	0.0800
0.9	0.12	0.06	0.2	0.0831
1	0.12	0.06	0.2	0.0857

Table 21.6 The dependence of company's weighted average cost of capital (WACC) on the coverage ratio on debt i_3

i_3	Ko	Kd	t	WACC(i_3)
0	0.12	0.06	0.2	0.0000
0.1	0.12	0.06	0.2	0.0978
0.2	0.12	0.06	0.2	0.1078
0.3	0.12	0.06	0.2	0.1116
0.4	0.12	0.06	0.2	0.1136
0.5	0.12	0.06	0.2	0.1148
0.6	0.12	0.06	0.2	0.1156
0.7	0.12	0.06	0.2	0.1162
0.8	0.12	0.06	0.2	0.1167
0.9	0.12	0.06	0.2	0.1171
1	0.12	0.06	0.2	0.1173

Table 21.7 The dependence of company's weighted average cost of capital (WACC) on the coverage ratio on debt l_1

t	l_1	k0	kd	WACC
0.2	0	0.12	0.06	0.12
0.2	1	0.12	0.06	0.117188
0.2	2	0.12	0.06	0.114504
0.2	3	0.12	0.06	0.11194
0.2	4	0.12	0.06	0.109489
0.2	5	0.12	0.06	0.107143
0.2	6	0.12	0.06	0.104895
0.2	7	0.12	0.06	0.10274
0.2	8	0.12	0.06	0.100671
0.2	9	0.12	0.06	0.098684
0.2	10	0.12	0.06	0.096774

Table 21.8 The dependence of company's weighted average cost of capital (WACC) on the coverage ratio on debt l_2

t	l_2	k_O	k_d	WACC
0.2	0	0.12	0.06	0.12
0.2	1	0.12	0.06	0.085714
0.2	2	0.12	0.06	0.066667
0.2	3	0.12	0.06	0.054545
0.2	4	0.12	0.06	0.046154
0.2	5	0.12	0.06	0.04
0.2	6	0.12	0.06	0.035294
0.2	7	0.12	0.06	0.031579
0.2	8	0.12	0.06	0.028571
0.2	9	0.12	0.06	0.026087
0.2	10	0.12	0.06	0.024

Table 21.9 The dependence of company's weighted average cost of capital (WACC) on the coverage ratio on debt l_3

t	l_3	k_O	k_d	WACC
0.2	0	0.12	0.06	0.12
0.2	1	0.12	0.06	0.117353
0.2	2	0.12	0.06	0.114819
0.2	3	0.12	0.06	0.112393
0.2	4	0.12	0.06	0.110068
0.2	5	0.12	0.06	0.107836
0.2	6	0.12	0.06	0.105693
0.2	7	0.12	0.06	0.103634
0.2	8	0.12	0.06	0.101654
0.2	9	0.12	0.06	0.099747
0.2	10	0.12	0.06	0.097911

21.7.1 *The Dependence of Equity Cost* k_e *on Coverage Ratios* i_1, i_2, i_3

Let us study the dependence of equity cost k_e on coverage ratios i_1, i_2, i_3 for the same set of parameters as used above and for leverage levels $L = 1$ and $L = 2$ (Tables 21.10, 21.11, and 21.12).

21.7.1.1 $L = 1$

21.7.1.2 $L = 2$

Table 21.10 The dependence of equity cost k_e on coverage ratio i_1 at leverage level $L = 1$

L	i_1	WACC(i_1)	Kd	t	Ke
1	0	0.00000	0.06	0.2	−0.0480
1	1	0.11719	0.06	0.2	0.1864
1	2	0.11858	0.06	0.2	0.1892
1	3	0.11905	0.06	0.2	0.1901
1	4	0.11928	0.06	0.2	0.1906
1	5	0.11943	0.06	0.2	0.1909
1	6	0.11952	0.06	0.2	0.1910
1	7	0.11959	0.06	0.2	0.1912
1	8	0.11964	0.06	0.2	0.1913
1	9	0.11968	0.06	0.2	0.1914
1	10	0.11971	0.06	0.2	0.1914

Table 21.11 The dependence of equity cost k_e on coverage ratio i_1 at leverage level $L = 2$

L	i_1	WACC(i_1)	Kd	t	Ke
2	0	0.00000	0.06	0.2	−0.0960
2	1	0.11719	0.06	0.2	0.2556
2	2	0.11858	0.06	0.2	0.2597
2	3	0.11905	0.06	0.2	0.2611
2	4	0.11928	0.06	0.2	0.2619
2	5	0.11943	0.06	0.2	0.2623
2	6	0.11952	0.06	0.2	0.2626
2	7	0.11959	0.06	0.2	0.2628
2	8	0.11964	0.06	0.2	0.2629
2	9	0.11968	0.06	0.2	0.2630
2	10	0.11971	0.06	0.2	0.2631

Table 21.12 The dependence of equity cost k_e on coverage ratio i_2 at leverage level $L = 1$

L	i_2	WACC(i_2)	Kd	t	Ke
1	0	0.00000	0.06	0.2	−0.0480
1	1	0.11734	0.06	0.2	0.1867
1	2	0.11866	0.06	0.2	0.1893
1	3	0.11910	0.06	0.2	0.1902
1	4	0.11932	0.06	0.2	0.1906
1	5	0.11946	0.06	0.2	0.1909
1	6	0.11955	0.06	0.2	0.1911
1	7	0.11961	0.06	0.2	0.1912
1	8	0.11966	0.06	0.2	0.1913
1	9	0.11970	0.06	0.2	0.1914
1	10	0.11973	0.06	0.2	0.1915

We could make some conclusions, based on Tables 21.13, 21.14, and 21.15 and Figs. 21.14, 21.15, 21.16, 21.17, 21.18, 21.19, 21.20, and 21.21. In all three cases equity cost k_e increases with coverage ratios and goes to saturation at high values of coverage ratios. Saturation values increases with leverage level from 19% at $L = 1$

Table 21.13 The dependence of equity cost k_e on coverage ratio i_2 at leverage level $L = 2$

L	i_2	WACC(i_2)	Kd	t	Ke
2	0	0.00000	0.06	0.2	−0.0960
2	1	0.11734	0.06	0.2	0.2560
2	2	0.11866	0.06	0.2	0.2600
2	3	0.11910	0.06	0.2	0.2613
2	4	0.11932	0.06	0.2	0.2620
2	5	0.11946	0.06	0.2	0.2624
2	6	0.11955	0.06	0.2	0.2626
2	7	0.11961	0.06	0.2	0.2628
2	8	0.11966	0.06	0.2	0.2630
2	9	0.11970	0.06	0.2	0.2631
2	10	0.11973	0.06	0.2	0.2632

Table 21.14 The dependence of equity cost k_e on coverage ratio i_3 at leverage level $L = 1$

L	i_3	WACC(i_3)	Kd	t	Ke
1	0	0.00000	0.06	0.2	−0.0480
1	1	0.08571	0.06	0.2	0.1234
1	2	0.10000	0.06	0.2	0.1520
1	3	0.10588	0.06	0.2	0.1638
1	4	0.10909	0.06	0.2	0.1702
1	5	0.11111	0.06	0.2	0.1742
1	6	0.11250	0.06	0.2	0.1770
1	7	0.11351	0.06	0.2	0.1790
1	8	0.11429	0.06	0.2	0.1806
1	9	0.11489	0.06	0.2	0.1818
1	10	0.11538	0.06	0.2	0.1828

Table 21.15 The dependence of equity cost k_e on coverage ratio i_3 at leverage level $L = 2$

L	i_3	WACC(i_3)	Kd	t	Ke
2	0	0.00000	0.06	0.2	−0.0960
2	1	0.08571	0.06	0.2	0.1611
2	2	0.10000	0.06	0.2	0.2040
2	3	0.10588	0.06	0.2	0.2216
2	4	0.10909	0.06	0.2	0.2313
2	5	0.11111	0.06	0.2	0.2373
2	6	0.11250	0.06	0.2	0.2415
2	7	0.11351	0.06	0.2	0.2445
2	8	0.11429	0.06	0.2	0.2469
2	9	0.11489	0.06	0.2	0.2487
2	10	0.11538	0.06	0.2	0.2502

up to value above 26% at $L = 2$. Note, that for coverage ratios i_1 and i_2 the saturation takes place at values $i_{1,2}$ of order unit, while for coverage ratio i_3 the saturation takes place at much higher i_3 values of order 6 or 7.

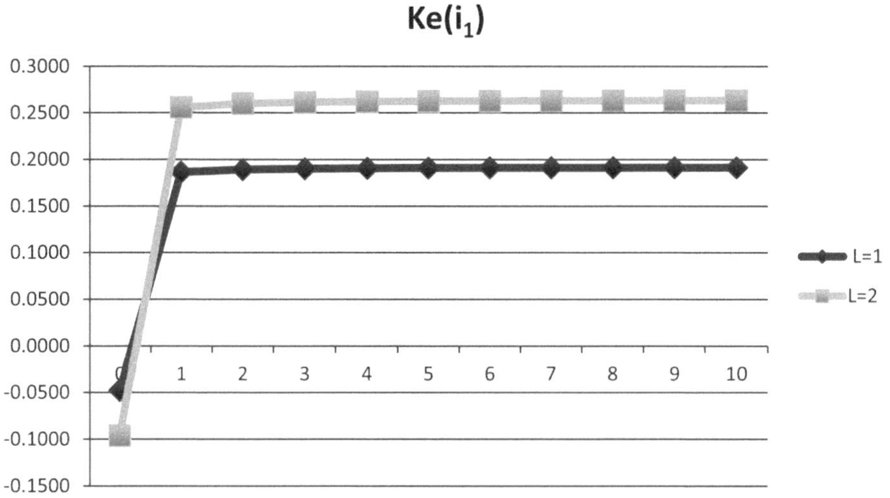

Fig. 21.14 The dependence of equity cost k_e on coverage ratio i_1 at two leverage level values $L = 1$ and $L = 2$

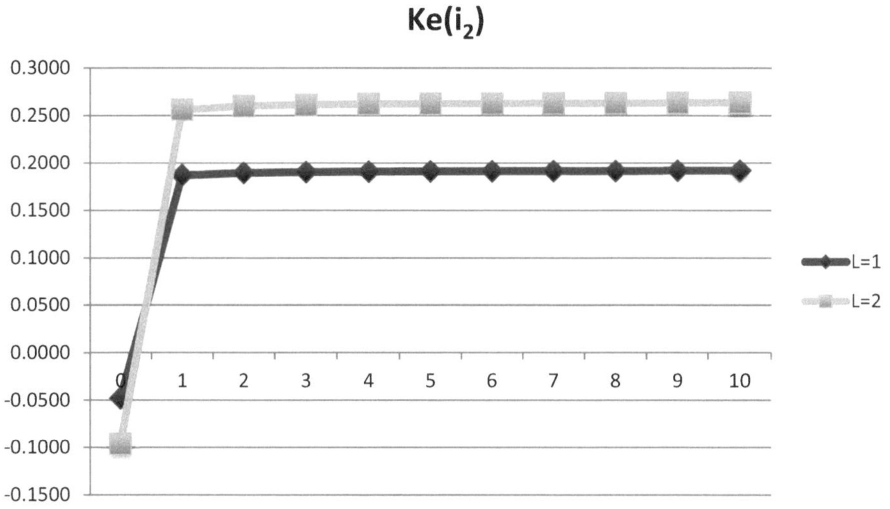

Fig. 21.15 The dependence of equity cost k_e on coverage ratio i_2 at two leverage level values $L = 1$ and $L = 2$

Equity cost k_e should be used as discount rate for unleveraged (financially independent) companies. For coverage ratios i_1 and i_2 saturation values of equity cost k_e could be used as discount rate above unit, while for coverage ratio i_3 saturation values of equity cost k_e could be used as discount rate at i_3 value above 6 or 7.

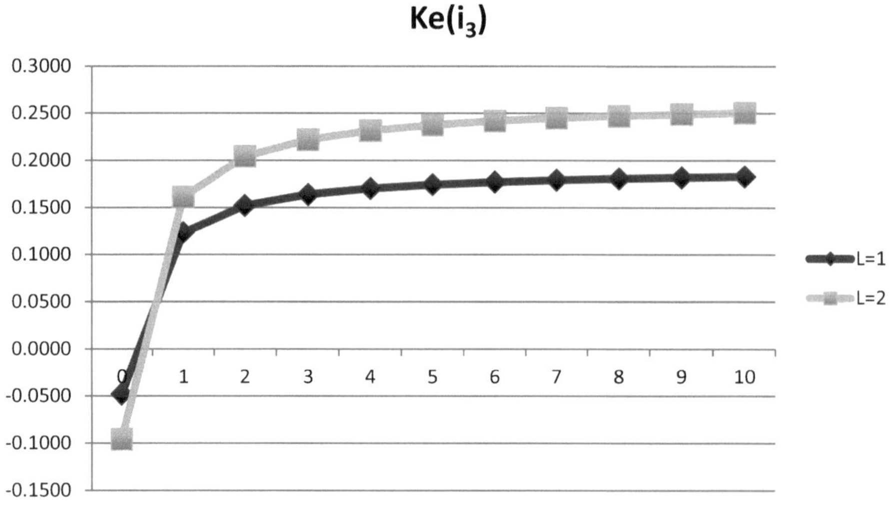

Fig. 21.16 The dependence of equity cost k_e on coverage ratio i_3 at two leverage level values $L = 1$ and $L = 2$

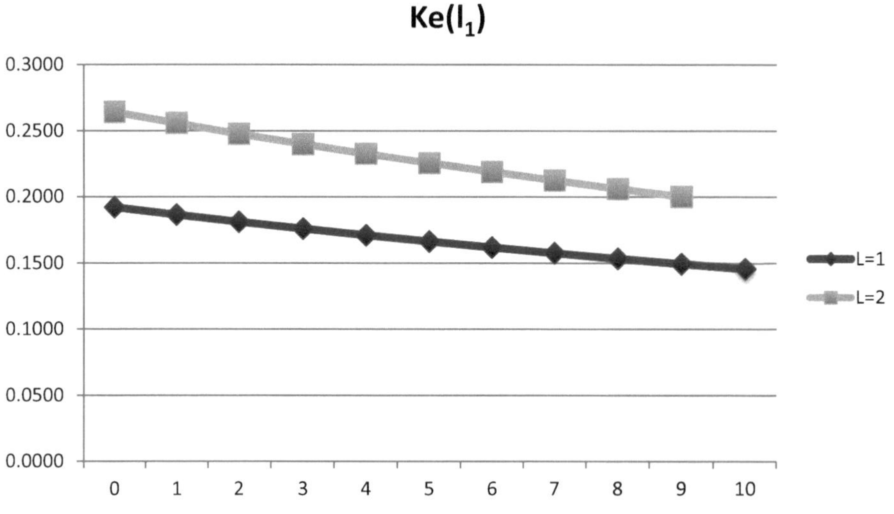

Fig. 21.17 The dependence of equity cost k_e on leverage ratio l_1 at two leverage level values $L = 1$ and $L = 2$

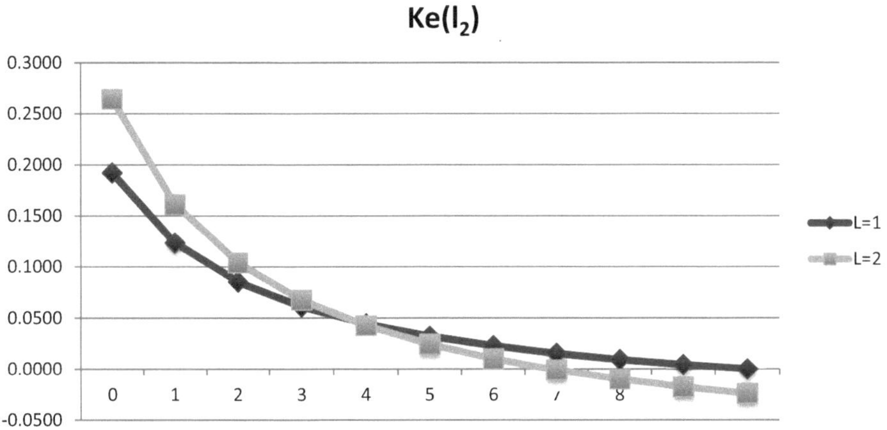

Fig. 21.18 The dependence of equity cost k_e on leverage ratio l_2 at two leverage level values $L = 1$ and $L = 2$

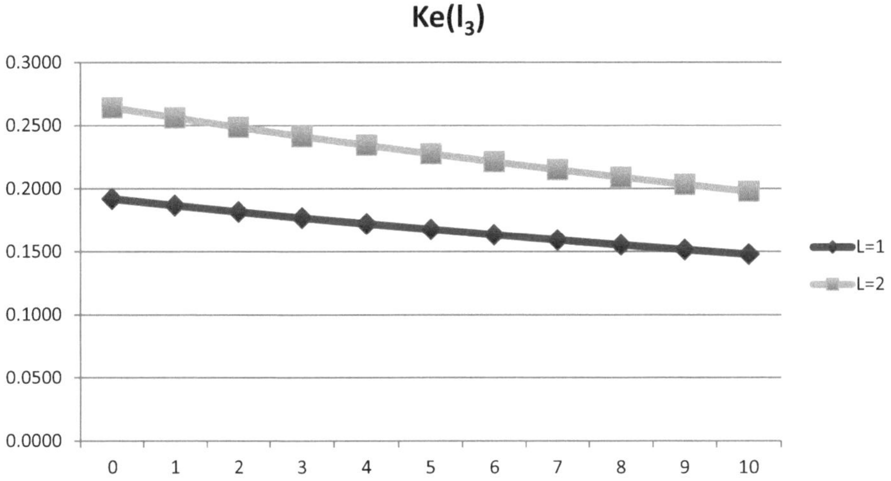

Fig. 21.19 The dependence of equity cost k_e on leverage ratio l_3 at two leverage level values $L = 1$ and $L = 2$

21.7.2 *The Dependence of Equity Cost* k_e *on Leverage Ratios* l_1, l_2, l_3

We study below the dependence of equity cost k_e on leverage ratios l_1, l_2, l_3 for the same set of parameters as used above and for leverage levels $L = 1$ and $L = 2$ (Tables 21.16, 21.17, 21.18, 21.19, 21.20, and 21.21).

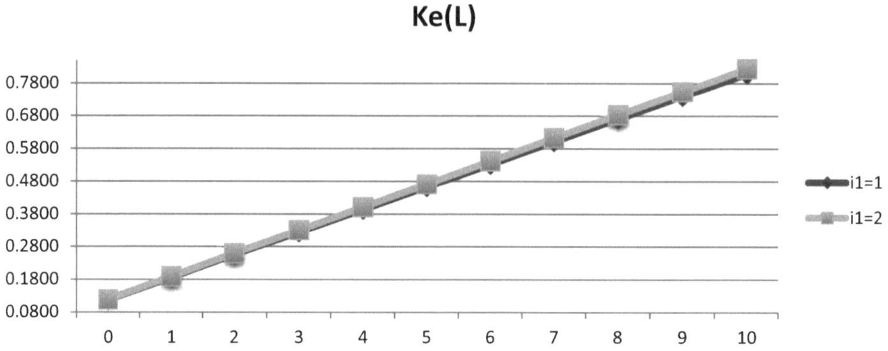

Fig. 21.20 The dependence of equity cost k_e on leverage level at two coverage ratio values $i_1 = 1$ and $i_1 = 2$

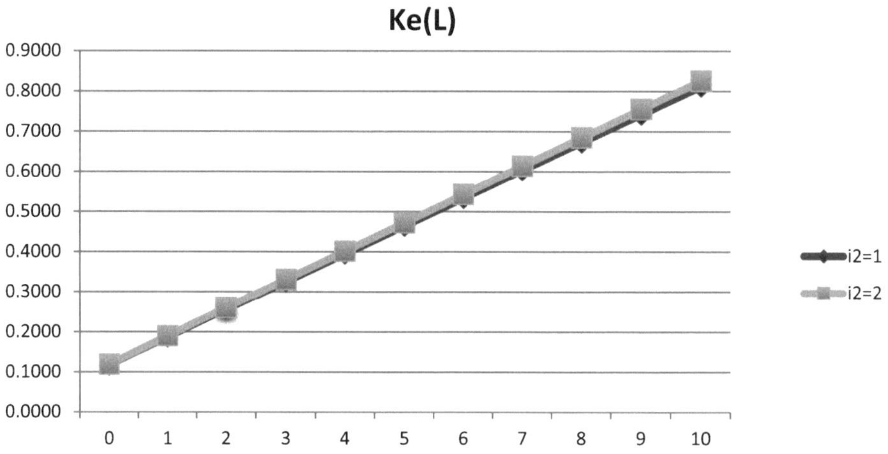

Fig. 21.21 The dependence of equity cost k_e on leverage level at two coverage ratio values $i_2 = 1$ and $i_2 = 2$

Table 21.16 The dependence of equity cost k_e on leverage ratio l_1 at leverage level $L = 1$

L	l_1	WACC(l_1)	Kd	t	Ke
1	0	0.12000	0.06	0.2	0.1920
1	1	0.11719	0.06	0.2	0.1864
1	2	0.11450	0.06	0.2	0.1810
1	3	0.11194	0.06	0.2	0.1759
1	4	0.10949	0.06	0.2	0.1710
1	5	0.10714	0.06	0.2	0.1663
1	6	0.10490	0.06	0.2	0.1618
1	7	0.10274	0.06	0.2	0.1575
1	8	0.10067	0.06	0.2	0.1533
1	9	0.09868	0.06	0.2	0.1494
1	10	0.09677	0.06	0.2	0.1455

Table 21.17 The dependence of equity cost k_e on leverage ratio l_1 at leverage level $L = 2$

L	l_1	WACC(l_1)	Kd	t	Ke
2	0	0.12000	0.06	0.2	0.2640
2	1	0.11719	0.06	0.2	0.2556
2	2	0.11450	0.06	0.2	0.2475
2	3	0.11194	0.06	0.2	0.2398
2	4	0.10949	0.06	0.2	0.2325
2	5	0.10714	0.06	0.2	0.2254
2	6	0.10490	0.06	0.2	0.2187
2	7	0.10274	0.06	0.2	0.2122
2	8	0.10067	0.06	0.2	0.2060
2	9	0.09868	0.06	0.2	0.2001
2	10	0.09677	0.06	0.2	0.1943

Table 21.18 The dependence of equity cost k_e on leverage ratio l_2 at leverage level $L = 1$

L	l_2	WACC(l_2)	Kd	t	Ke
1	0	0.12000	0.06	0.2	0.1920
1	1	0.08571	0.06	0.2	0.1234
1	2	0.06667	0.06	0.2	0.0853
1	3	0.05455	0.06	0.2	0.0611
1	4	0.04615	0.06	0.2	0.0443
1	5	0.04000	0.06	0.2	0.0320
1	6	0.03529	0.06	0.2	0.0226
1	7	0.03158	0.06	0.2	0.0152
1	8	0.02857	0.06	0.2	0.0091
1	9	0.02609	0.06	0.2	0.0042
1	10	0.02400	0.06	0.2	0.0000

Table 21.19 The dependence of equity cost k_e on leverage ratio l_2 at leverage level $L = 2$

L	l_2	WACC(l_2)	Kd	t	Ke
2	0	0.12000	0.06	0.2	0.2640
2	1	0.08571	0.06	0.2	0.1611
2	2	0.06667	0.06	0.2	0.1040
2	3	0.05455	0.06	0.2	0.0676
2	4	0.04615	0.06	0.2	0.0425
2	5	0.04000	0.06	0.2	0.0240
2	6	0.03529	0.06	0.2	0.0099
2	7	0.03158	0.06	0.2	−0.0013
2	8	0.02857	0.06	0.2	−0.0103
2	9	0.02609	0.06	0.2	−0.0177
2	10	0.02400	0.06	0.2	−0.0240

Table 21.20 The dependence of equity cost k_e on leverage ratio l_2 at leverage level $L = 1$

L	l_3	WACC(l_3)	Kd	t	Ke
1	0	0.12000	0.06	0.2	0.1920
1	1	0.11735	0.06	0.2	0.1867
1	2	0.11482	0.06	0.2	0.1816
1	3	0.11239	0.06	0.2	0.1768
1	4	0.11007	0.06	0.2	0.1721
1	5	0.10784	0.06	0.2	0.1677
1	6	0.10569	0.06	0.2	0.1634
1	7	0.10363	0.06	0.2	0.1593
1	8	0.10165	0.06	0.2	0.1553
1	9	0.09975	0.06	0.2	0.1515
1	10	0.09791	0.06	0.2	0.1478

Table 21.21 The dependence of equity cost k_e on leverage ratio l_2 at leverage level $L = 2$

L	l_3	WACC(l_3)	Kd	t	Ke
2	0	0.12000	0.06	0.2	0.2640
2	1	0.11735	0.06	0.2	0.2561
2	2	0.11482	0.06	0.2	0.2485
2	3	0.11239	0.06	0.2	0.2412
2	4	0.11007	0.06	0.2	0.2342
2	5	0.10784	0.06	0.2	0.2275
2	6	0.10569	0.06	0.2	0.2211
2	7	0.10363	0.06	0.2	0.2149
2	8	0.10165	0.06	0.2	0.2090
2	9	0.09975	0.06	0.2	0.2032
2	10	0.09791	0.06	0.2	0.1977

21.7.2.1 The Dependence of Equity Cost k_e on Leverage Ratio l_1

$L = 1$

$L = 2$

21.7.2.2 The Dependence of Equity Cost k_e on Leverage Ratios l_2

$L = 1$

$L = 2$

21.7.2.3 The Dependence of Equity Cost k_e on Leverage Ratios l_3

$L = 1$

$L = 2$

21.8 How to Evaluate the Discount Rate?

Let us discuss now the algorithm of valuation of the discount rate, if we know one or a few ratios (coverage or leverage ones). The developed above method allow estimate discount rate with the best accuracy characteristic for used theory of capital structure (perpetuity limit).

21.8.1 Using One Ratio

If one know one ratio (coverage or leverage one) the algorithm of valuation of the discount rate is as following:

- determination of the parameter k_0;
- knowing k_0, k_d and t, one builds the curve of dependence WACC(i) or WACC(l);
- then, using the known value of coverage ratio (i_0) or leverage ratio (l_0) one finds the value WACC(i_0) or WACC(l_0), which represents the discount rate.

21.8.2 Using a Few Ratios

If we know say m values of coverage ratios (i_j) and n values of leverage ratios (l_k)

- we find by the above algorithm m values of WACC(i_j) and n values of WACC(l_k) first;
- then we find the average value of WACC by the following formula:

$$WACC_{av} = \frac{1}{m+n}\left[\sum_{j=1}^{m} WACC(i_j) + \sum_{k=1}^{n} WACC(l_k)\right].$$

This found value $WACC_{av}$ should be used when discounting the financial flows in rating.

21.9 Influence of Leverage Level

We discuss also the interplay between rating ratios and leverage level which can be quite important in rating.

21.9.1 *The Dependence of Equity Cost* $\mathbf{k_e}$ *on Leverage Level at Two Coverage Ratio Values* $\mathbf{i_j} = 1$ *and* $\mathbf{i_j} = 2$

21.9.1.1 $i_1 = 1$

21.9.1.2 $i_1 = 2$

21.9.1.3 $i_2 = 1$

21.9.1.4 $i_2 = 2$

21.9.1.5 $i_3 = 1$

21.9.1.6 $i_3 = 2$

Table 21.22 The dependence of equity cost k_e on leverage level L at coverage ratio $i_1 = 1$

L	WACC(i_1 = 1)	Kd	t	Ke
0	0.11719	0.06	0.2	0.1172
1	0.11719	0.06	0.2	0.1864
2	0.11719	0.06	0.2	0.2556
3	0.11719	0.06	0.2	0.3248
4	0.11719	0.06	0.2	0.3939
5	0.11719	0.06	0.2	0.4631
6	0.11719	0.06	0.2	0.5323
7	0.11719	0.06	0.2	0.6015
8	0.11719	0.06	0.2	0.6707
9	0.11719	0.06	0.2	0.7399
10	0.11719	0.06	0.2	0.8091

Table 21.23 The dependence of equity cost k_e on leverage level L at coverage ratio $i_1 = 2$

L	WACC(i_1 = 2)	Kd	t	Ke
0	0.11858	0.06	0.2	0.1186
1	0.11858	0.06	0.2	0.1892
2	0.11858	0.06	0.2	0.2597
3	0.11858	0.06	0.2	0.3303
4	0.11858	0.06	0.2	0.4009
5	0.11858	0.06	0.2	0.4715
6	0.11858	0.06	0.2	0.5420
7	0.11858	0.06	0.2	0.6126
8	0.11858	0.06	0.2	0.6832
9	0.11858	0.06	0.2	0.7538
10	0.11858	0.06	0.2	0.8243

Table 21.24 The dependence of equity cost k_e on leverage level L at coverage ratio $i_2 = 1$

L	WACC(i_2 = 1)	Kd	t	Ke
0	0.11734	0.06	0.2	0.1173
1	0.11734	0.06	0.2	0.1867
2	0.11734	0.06	0.2	0.2560
3	0.11734	0.06	0.2	0.3254
4	0.11734	0.06	0.2	0.3947
5	0.11734	0.06	0.2	0.4641
6	0.11734	0.06	0.2	0.5334
7	0.11734	0.06	0.2	0.6027
8	0.11734	0.06	0.2	0.6721
9	0.11734	0.06	0.2	0.7414
10	0.11734	0.06	0.2	0.8108

Table 21.25 The dependence of equity cost k_e on leverage level L at coverage ratio $i_2 = 2$

L	WACC($i_2 = 2$)	Kd	t	Ke
0	0.11866	0.06	0.2	0.1187
1	0.11866	0.06	0.2	0.1893
2	0.11866	0.06	0.2	0.2600
3	0.11866	0.06	0.2	0.3306
4	0.11866	0.06	0.2	0.4013
5	0.11866	0.06	0.2	0.4719
6	0.11866	0.06	0.2	0.5426
7	0.11866	0.06	0.2	0.6133
8	0.11866	0.06	0.2	0.6839
9	0.11866	0.06	0.2	0.7546
10	0.11866	0.06	0.2	0.8252

Table 21.26 The dependence of equity cost k_e on leverage level L at coverage ratio $i_3 = 1$

L	WACC($i_3 = 1$)	Kd	t	Ke
0	0.08571	0.06	0.2	0.0857
1	0.08571	0.06	0.2	0.1234
2	0.08571	0.06	0.2	0.1611
3	0.08571	0.06	0.2	0.1989
4	0.08571	0.06	0.2	0.2366
5	0.08571	0.06	0.2	0.2743
6	0.08571	0.06	0.2	0.3120
7	0.08571	0.06	0.2	0.3497
8	0.08571	0.06	0.2	0.3874
9	0.08571	0.06	0.2	0.4251
10	0.08571	0.06	0.2	0.4629

Table 21.27 The dependence of equity cost k_e on leverage level L at coverage ratio $i_3 = 2$

L	WACC($i_3 = 2$)	Kd	t	Ke
0	0.10000	0.06	0.2	0.1000
1	0.10000	0.06	0.2	0.1520
2	0.10000	0.06	0.2	0.2040
3	0.10000	0.06	0.2	0.2560
4	0.10000	0.06	0.2	0.3080
5	0.10000	0.06	0.2	0.3600
6	0.10000	0.06	0.2	0.4120
7	0.10000	0.06	0.2	0.4640
8	0.10000	0.06	0.2	0.5160
9	0.10000	0.06	0.2	0.5680
10	0.10000	0.06	0.2	0.6200

It is seen from Tables 21.22, 21.23, 21.24, 21.25, 21.26, and 21.27 and Figs. 21.20, 21.21, and 21.22 that equity cost k_e increases practically linearly with leverage level for all coverage ratios i_1, i_2, i_3. For each of two coverage ratios i_1, i_2, curves k_e (L) for two values of i_j (1 and 2) practically coincide. For coverage ratio i_3

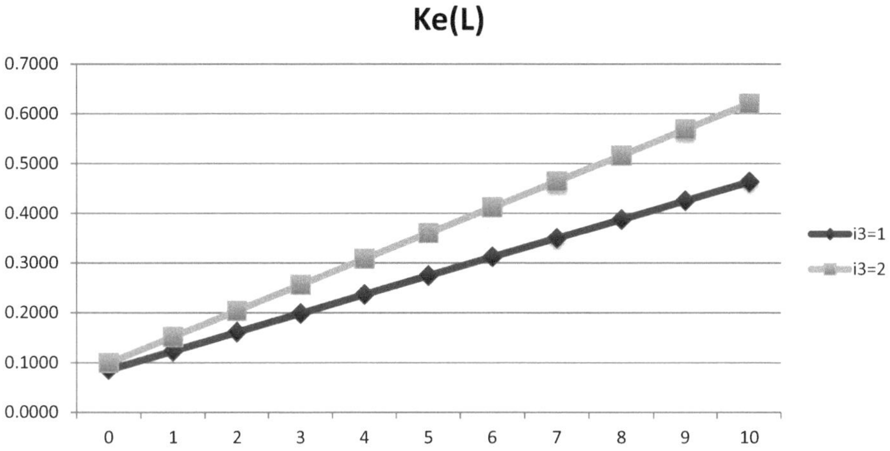

Fig. 21.22 The dependence of equity cost k_e on leverage level at two coverage ratio values $i_3 = 1$ and $i_3 = 2$

curves $k_e\,(L)$ for value of $i_3 = 2$ lies above one for $i_3 = 1$ and angle of inclination for value of $i_3 = 2$ is bigger.

21.9.2 *The Dependence of Equity Cost* $\mathbf{k_e}$ *on Leverage Level at Two Leverage Ratio Values* $\mathbf{l_j} = 1$ *and* $\mathbf{l_j} = 2$

Let us now study the dependence of equity cost k_e on leverage level at two leverage ratio values $l_j = 1$ and $l_j = 2$ for leverage ratios l_1,l_2, l_3.

21.9.2.1 $l_1 = 1$

21.9.2.2 $l_1 = 2$

21.9.2.3 $l_2 = 1$

21.9.2.4 $l_2 = 2$

21.9.2.5 $l_3 = 1$

21.9.2.6 $l_3 = 2$

It is seen from Tables 21.28, 21.29, 21.30, 21.31, 21.32, and 21.33 and Figs. 21.23, 21.24, and 21.25 that equity cost k_e increases practically linearly with leverage level for all leverage ratios l_1, l_2, l_3. For each of two leverage ratios l_1, l_3, curves k_e (L) for two values of l_j (1 and 2) practically coincide. For leverage ratio l_2

Table 21.28 The dependence of equity cost k_e on leverage level L at leverage ratio $l_1 = 1$

L	WACC(l_1 = 1)	Kd	t	Ke
0	0.11719	0.06	0.2	0.1172
1	0.11719	0.06	0.2	0.1864
2	0.11719	0.06	0.2	0.2556
3	0.11719	0.06	0.2	0.3248
4	0.11719	0.06	0.2	0.3939
5	0.11719	0.06	0.2	0.4631
6	0.11719	0.06	0.2	0.5323
7	0.11719	0.06	0.2	0.6015
8	0.11719	0.06	0.2	0.6707
9	0.11719	0.06	0.2	0.7399
10	0.11719	0.06	0.2	0.8091

Table 21.29 The dependence of equity cost k_e on leverage level L at leverage ratio $l_1 = 2$

L	WACC(l_1 = 2)	Kd	t	Ke
0	0.11450	0.06	0.2	0.1145
1	0.11450	0.06	0.2	0.1810
2	0.11450	0.06	0.2	0.2475
3	0.11450	0.06	0.2	0.3140
4	0.11450	0.06	0.2	0.3805
5	0.11450	0.06	0.2	0.4470
6	0.11450	0.06	0.2	0.5135
7	0.11450	0.06	0.2	0.5800
8	0.11450	0.06	0.2	0.6465
9	0.11450	0.06	0.2	0.7130
10	0.11450	0.06	0.2	0.7795

Table 21.30 The dependence of equity cost k_e on leverage level L at leverage ratio $l_2 = 1$

L	WACC($l_2 = 1$)	Kd	t	Ke
0	0.08571	0.06	0.2	0.0857
1	0.08571	0.06	0.2	0.1234
2	0.08571	0.06	0.2	0.1611
3	0.08571	0.06	0.2	0.1989
4	0.08571	0.06	0.2	0.2366
5	0.08571	0.06	0.2	0.2743
6	0.08571	0.06	0.2	0.3120
7	0.08571	0.06	0.2	0.3497
8	0.08571	0.06	0.2	0.3874
9	0.08571	0.06	0.2	0.4251
10	0.08571	0.06	0.2	0.4629

Table 21.31 The dependence of equity cost k_e on leverage level L at leverage ratio $l_2 = 2$

L	WACC($l_2 = 2$)	Kd	t	Ke
0	0.06667	0.06	0.2	0.0667
1	0.06667	0.06	0.2	0.0853
2	0.06667	0.06	0.2	0.1040
3	0.06667	0.06	0.2	0.1227
4	0.06667	0.06	0.2	0.1413
5	0.06667	0.06	0.2	0.1600
6	0.06667	0.06	0.2	0.1787
7	0.06667	0.06	0.2	0.1973
8	0.06667	0.06	0.2	0.2160
9	0.06667	0.06	0.2	0.2347
10	0.06667	0.06	0.2	0.2533

Table 21.32 The dependence of equity cost k_e on leverage level L at leverage ratio $l_3 = 1$

L	WACC($l_3 = 1$)	Kd	t	Ke
0	0.11735	0.06	0.2	0.1174
1	0.11735	0.06	0.2	0.1867
2	0.11735	0.06	0.2	0.2561
3	0.11735	0.06	0.2	0.3254
4	0.11735	0.06	0.2	0.3948
5	0.11735	0.06	0.2	0.4641
6	0.11735	0.06	0.2	0.5335
7	0.11735	0.06	0.2	0.6028
8	0.11735	0.06	0.2	0.6722
9	0.11735	0.06	0.2	0.7415
10	0.11735	0.06	0.2	0.8109

curves $k_e\,(L)$ for value of $l_3 = 1$ lies above one for $l_3 = 2$ and angle of inclination for value of $l_3 = 1$ is bigger.

Table 21.33 The dependence of equity cost k_e on leverage level L at leverage ratio $l_3 = 2$

L	WACC($l_3 = 2$)	Kd	t	Ke
0	0.11482	0.06	0.2	0.1148
1	0.11482	0.06	0.2	0.1816
2	0.11482	0.06	0.2	0.2485
3	0.11482	0.06	0.2	0.3153
4	0.11482	0.06	0.2	0.3821
5	0.11482	0.06	0.2	0.4489
6	0.11482	0.06	0.2	0.5157
7	0.11482	0.06	0.2	0.5826
8	0.11482	0.06	0.2	0.6494
9	0.11482	0.06	0.2	0.7162
10	0.11482	0.06	0.2	0.7830

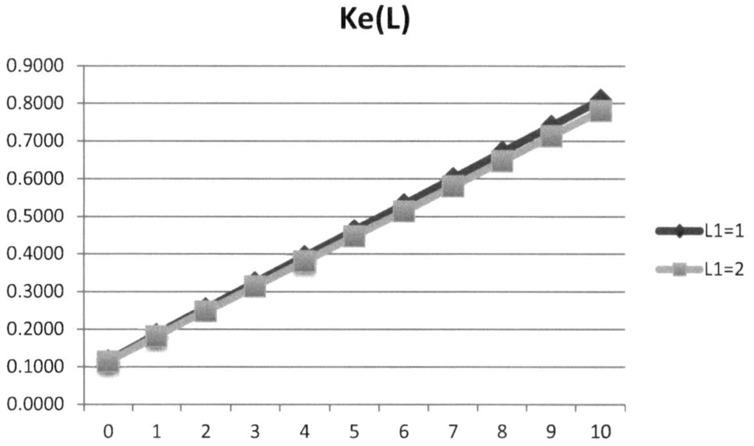

Fig. 21.23 The dependence of equity cost k_e on leverage level at two leverage ratio values $l_1 = 1$ and $l_1 = 2$

21.10 Conclusion

Chapters 16–18 suggest a new approach to rating methodology. This chapter as well as next one are devoted to rating of non-financial issuers, while Chap. 18 is devoted to long-term project rating. The key factors of a new approach are: (1) The adequate use of discounting of financial flows virtually not used in existing rating methodologies, (2) The incorporation of rating parameters (financial "ratios") into the perpetuity limit of modern theory of capital structure BFO. This on the one hand allows use the powerful tool of this theory in the rating, and on the other hand it ensures the correct discount rates when discounting of financial flows. Two models for accounting of discounting of financial flows—one-period and multi-period are discussed. An algorithm of valuation of correct discount rate, accounting rating ratios is suggested. We discuss also the interplay between rating ratios and leverage level

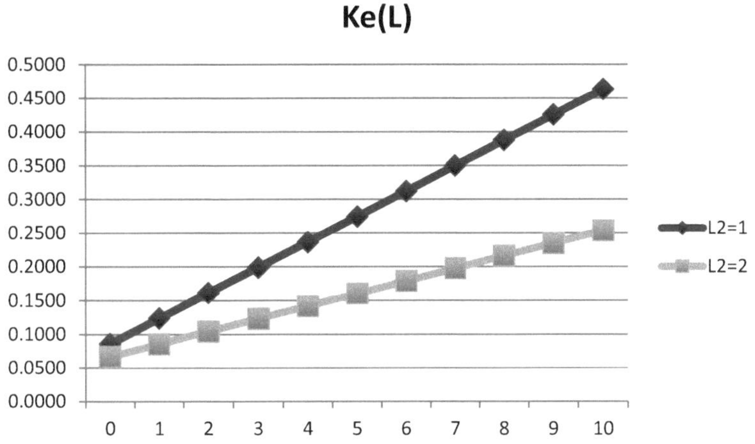

Fig. 21.24 The dependence of equity cost k_e on leverage level at two leverage ratio values $l_2 = 1$ and $l_2 = 2$

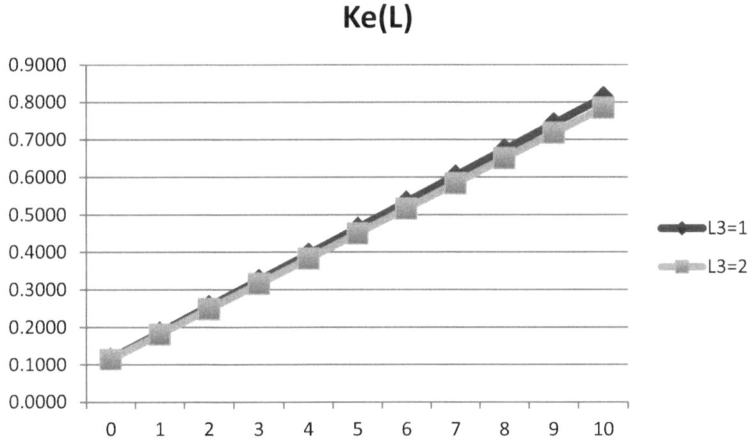

Fig. 21.25 The dependence of equity cost k_e on leverage level at two leverage ratio values $l_3 = 1$ and $l_3 = 2$

which can be quite important in rating. All above creates a new base for rating methodologies.

References

Brusova, A. A comparison of the three methods of estimation of weighted average cost of capital and equity cost of company. Financ. Anal. Prob. Sol. 2011, 34, 36–42.

Brusov, P. (2018). Editorial: Introduction on special issue on the Banking System and financial markets of Russia and other countries: Problems and prospects. *Journal of Reviews on Global Economics, 7*, i–vi.

Brusov, P., & Filatova, T. (2021). The Modigliani–Miller theory with arbitrary frequency of payment of tax on profit. *Mathematics, 9*(11), 1198.

Brusov, P., Filatova, T., Orehova, N., & Eskindarov, M. (2015). *Modern corporate finance, investments and taxation*. Springer International Publishing, 373 p. Retrieved from https://www.springer.com/gp/book/9783319147314

Brusov, P., Filatova, T., Orehova, N., & Eskindarov, M. (2018a). *Modern corporate finance, investments, taxation and ratings*. Springer Nature Publishing, 571 p.

Brusov, P. N., Filatova, T. V., Orekhova, N. P., & Kulik, V. L. (2018b). Rating: New approach. *Journal of Reviews on Global Economics, 7*, 37–62. https://doi.org/10.6000/1929-7092.2018.07.05

Brusov, P. N., Filatova, T. V., Orekhova, N. P., & Kulik, V. L. (2018c). A "golden age" of the companies: Conditions of its existence. *Journal of Reviews on Global Economics, 7*, 88–103. https://doi.org/10.6000/1929-7092.2018.07.07

Brusov, P. N., Filatova, T. V., Orekhova, N. P., & Kulik, V. L. (2018d). Rating methodology: New look and new horizons. *Journal of Reviews on Global Economics, 7*, 63–87. https://doi.org/10.6000/1929-7092.2018.07.06

Brusov, P. N., Filatova, T. V., Orekhova, N. P., Kulik, V. L., & Weil, I. (2018e). New meaningful effects in modern capital structure theory. *Journal of Reviews on Global Economics, 7*, 104–122. https://doi.org/10.6000/1929-7092.2018.07.08

Brusov, P., Filatova, T., Orekhova, N., Kulik, V., Weil, I., & Brailov, A. (2018f). The impact of the central bank key rate and commercial banks credit rates on creating and maintaining of a favorable investment climate in the country. *Journal of Reviews on Global Economics, 7*, 360–376. https://doi.org/10.6000/1929-7092.2018.07.31

Brusov, P. N., Filatova, T. V., & Orekhova, N. P. (2018g). *Modern corporate finance and investments*. Knorus Publishing House, 517 p.

Brusov, P. N., Filatova, T. V., Orekhova, N. P., Kulik, V. L., & Weil, I. (2019). Ratings of the investment projects of arbitrary durations: New methodology. *Journal of Reviews on Global Economics, 8*, 437–448. https://doi.org/10.6000/1929-7092.2019.08.37

Brusov, P., Filatova, T., & Orehova, N. (2020a). *Ratings: Critical analysis and new approaches of quantitative and qualitative methodology*. Springer Nature Publishing, 369 p. Retrieved from https://www.springer.com/de/book/9783030562427

Brusov, P. N., Filatova, T. V., Orekhova, N. P., Kulik, V. L., Chang, S.-I., & Lin, Y. C. G. (2020b). Application of the Modigliani–Miller Theory, modified for the case of advance payments of tax on profit, in rating methodologies. *Journal of Reviews on Global Economics, 9*, 282–292.

Brusov, P. N., Filatova, T. V., Orekhova, N. P., Kulik, V. L., Chang, S.-I., & Lin, Y. C. G. (2020c). Modification of the Modigliani–Miller theory for the case of advance payments of tax on profit. *Journal of Reviews on Global Economics, 9*, 257–267.

Brusov, P., Filatova, T., Chang, S.-I., & Lin, G. (2021a). Innovative investment models with frequent payments of tax on income and of interest on debt. *Mathematics, 9*(13), 1491.

Brusov, P., Filatova, T., Orekhova, N., Kulik, V., Chang, S.-I., & Lin, G. (2021b). Generalization of the Modigliani–Miller theory for the case of variable profit. *Mathematics, 9*(11), 1286.

Filatova, T., Orehova, N., & Brusova, A. (2008). Weighted average cost of capital in the theory of Modigliani–Miller, modified for a finite life-time company. *Bulletin of the FU, 48*, 68–77.

Filatova, T. V., Brusov, P. N., Orekhova, N. P., & Kulik, V. L. (2018). Ratings of the long-term projects: New approach. *Journal of Reviews on Global Economics, 7*, 645–661. https://doi.org/10.6000/1929-7092.2018.07.59

Modigliani, F., & Miller, M. (1958). The cost of capital, corporate finance, and the theory of investment. *The American Economic Review, 48*, 261–297.

Modigliani, F., & Miller, M. (1963). Corporate income taxes and the cost of capital: A correction. *The American Economic Review, 53*, 147–175.

Chapter 22
Application of the Modigliani–Miller Theory, Modified for the Case of Advance Payments of Tax on Profit, in Rating Methodologies

22.1 Introduction

In the papers (Brusov, 2018a, b; Brusov et al., 2018a, b, c, d, e, f, g, 2019, 2020b; Filatova et al., 2018; Brusov & Filatova (2021), Brusov et al. (2020c), Brusov et al. (2021a), Brusov et al. (2021b)) authors have applied the theory of Nobel Prize winners Modigliani and Miller, which is the perpetuity limit of the general theory of capital cost and capital structure—Brusov–Filatova–Orekhova (BFO) theory (Brusov, 2018a, b; Brusov et al., 2015, 2018a, b, c, d, e, f, g; Filatova et al., 2018) for rating needs. It has become a very important step in developed of a qualitatively new rating methodology.

Recently we have generalized the Modigliani and Miller theory (Modigliani & Miller, 1958, 1963, 1966) for a more realistic method of tax on profit payments: for the case of advance payments of tax on profit, which is widely used in practice (Brusov et al., 2020a). Modigliani–Miller theory accounts these tax payments as annuity–immediate, while in practice these payments are making in advance and thus should be accounting as annuity–due.

We have shown that this generalization leads to some important consequencies, which change seriously all the main statements by Modigliani and Miller (1958, 1963, 1966). These consequencies are as following: WACC starts depend on debt cost k_d, WACC turns out to be lower than in case of classical Modigliani–Miller theory and thus company capitalization becomes higher than in ordinary Modigliani–Miller theory. We show that equity dependence on leverage level L is still linear, but the tilt angle with respect to L–axis turns out to be smaller: this could lead to modification of the divident policy of the company.

In current Chapter we use modified Modigliani–Miller theory (MMM theory) and apply it for rating methodologies needs. A serious modification of MMM theory in order to use it in rating procedure has been required. The financial "ratios" (main rating parameters) were introduced into MMM theory. The necessity of an appropriate use of financial flows discounting in rating methodologies is discussed. The

P. Brusov, T. Filatova, *Efficiency of Investment Projects*, Contributions to Finance and Accounting, https://doi.org/10.1007/978-3-032-01028-5_22

dependence of the weighted average cost of capital (WACC), which plays the role of discount rate, on coverage and leverage ratios is analyzed.

Obtained results make possible to use the power of this theory in the rating and create a new base for rating methodologies, by other words this allows develop a new approach to methodology of rating, requiring a serious modification of existing rating methodologies.

22.2 Modified Modigliani–Miller Theory

Let us shortly discuss some main points of Modified Modigliani–Miller theory (MMM) and its features (see Chaps. 4, 5, 6, 7, and 8), which are different from ones of "classical" Modigliani–Miller theory (Brusov et al., 2020a).

Tax Shield
To calculate tax shield *TS* in case of advance tax payments one should use annuity–due

$$TS = k_d Dt + \frac{k_d Dt}{(1+k_d)} + \frac{k_d Dt}{(1+k_d)^2} + \ldots = \frac{k_d Dt}{\left(1-(1+k_d)^{-1}\right)} = Dt(1+k_d) \tag{22.1}$$

This expression is different from the case of classical Modigliani–Miller theory (which used annuity–immediate).

$$TS = \frac{k_d Dt}{(1+k_d)} + \frac{k_d Dt}{(1+k_d)^2} + \ldots = \frac{k_d Dt}{(1+k_d)\left(1-(1+k_d)^{-1}\right)} = Dt \tag{22.2}$$

Thus in the former case tax shield *TS* is bigger by multiplyer $(1 + k_d)$. This is connected with the time value of money: money today is more expensive than money tomorrow due to the possibility of their alternative investment.

The Weighted Average Cost of Capital, WACC
For WACC in MMM theory we have the following formula:

$$WACC = k_0(1 - w_d t(1+k_d)). \tag{22.3}$$

$$\text{At } L \to \infty \quad WACC = k_0(1 - t(1+k_d)). \tag{22.4}$$

This expression is different from the similar one in classical Modigliani–Miller theory

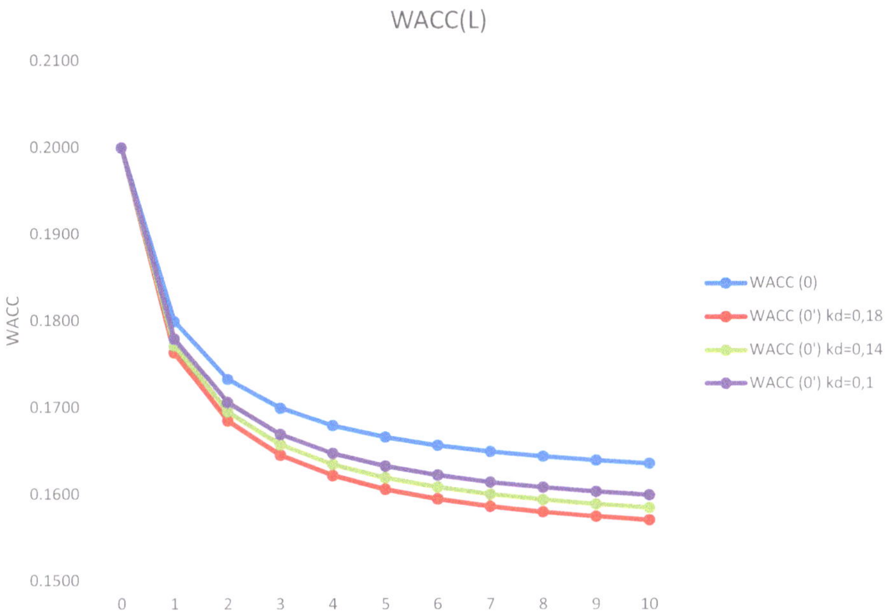

Fig. 22.1 Dependence of WACC on leverage level L: in "classical" Modigliani–Miller theory (curve WACC(0)) and in modified Modigliani–Miller theory (MMM theory) at different values of debt cost: $k_d = 0.18$; $k_d = 0.14$; $k_d = 0.1$ (curves WACC(0′))

$$WACC = k_0(1 - w_d t) \tag{22.5}$$

$$\text{At } L \to \infty \quad WACC = k_0(1 - t).$$

From these expressions it is seen that WACC decreases with L, achieving lower value $WACC = k_0(1 - t(1 + k_d))$ at $L \to \infty$ in considering case comparing with classical Modigliani–Miller theory $WACC = k_0(1 - t)$.

This means also, that company capitalization becomes higher than in ordinary Modigliani–Miller theory.

Let us compare the dependence of the weighted average cost of capital, WACC, on leverage level in "classical" Modigliani–Miller theory (MM theory) and in modified Modigliani–Miller theory (MMM theory). Study of such dependence is very important, because, the weighted average cost of capital, WACC, plays the role of discount rate in operating financial flows discounting as well as of financial flows in rating methodologies. WACC's value determines as well the capitalization of the company $V = CF/WACC$.

From the Fig. 22.1 it is seen that WACC in MMM theory turns out to be lower than in case of classical Modigliani–Miller theory and thus company capitalization becomes higher than in ordinary Modigliani–Miller theory. It is seen, that WACC decreases with debt cost k_d.

In paper (Brusov et al., 2020a) we study as well the equity cost dependence on leverage level L and show that it is still linear, but the tilt angle with respect to L-axis turns out to be smaller in MMM theory $tg\alpha = (k_0 - k_d) \cdot (1 - t) - k_0 k_d t$, than in classical Modigliani–Miller theory $tg\alpha = (k_0 - k_d) \cdot (1 - t)$. This could lead to modification of the dividend policy of the company, because the equity cost represents itself economically sound value of dividends. Thus, company could decreases the value of dividends, which they should pay to shareholders.

Correct account of a method of tax on profit payments demonstrates that shortcomings of Modigliani–Miller theory are dipper, than everybody suggested: the underestimation of WACC really turns out to be bigger, as well as overestimation of the capitalization of the company. This means that systematic risks arising from the use of modified Modigliani–Miller theory (MMM theory) (which is more correct than "classical" one) in practice is higher than it was suggested by the "classical" version of this theory.

Because the advance payments of tax on profit is widely used in practice, modified Modigliani–Miller theory (MMM theory) should be use instead of classical version of this theory (MM theory). And below we apply the modified Modigliani–Miller theory for rating needs.

22.3 Application of Modified of Modigliani–Miller Theory for Rating Needs

The financial "ratios", constitute a direct and inverse ratios of various generated cash flows to debt values and interest ones, play quite significant role in quantification of the creditworthiness of the issuers. The examples of such ratios are as following: DCF/Debt, FFO/Debt, CFO/Debt, FOCF/Debt, FFO/cashinterest, EBITDA/interest, Interests/EBITDA, Debt/EBITDA and some others.

We introduce these financial "ratios" into the modified perpetuity limit of modern theory of capital structure—BFO theory ((MMM theory), which is valid for companies of arbitrary age. This is quite important because allows use this theory as a powerful tools when discounting of financial flows using the correct discounting rate in rating.

This has required the modification of the perpetuity limit of the BFO theory—Modified Modigliani–Miller theory (MMM theory). The needs of modification is connected to the fact that used in financial management the concept of "leverage" as the ratio of debt value to the equity value substantially differs from the concept of "leverage" in the rating, where it is understood as ratio of the debt value to the generated cash flow values (income, profit, etc.).

Developed by us recently (Brusov et al., 2020a) modified Modigliani–Miller theory (MMM theory) with corporate taxes (Modigliani & Miller, 1966) shows that capitalization of financially dependent (leveraged) company, V_L, is equal to the capitalization of financially independent (unleveraged) company, V_0, increased by the size of the tax shield for perpetuity time, $Dt(1 + k_d)$,

$$V_L = V_0 + Dt(1 + k_d). \tag{22.6}$$

Substituting the expressions for both capitalizations, one has

$$\frac{CF}{WACC} = \frac{CF}{k_0} + Dt(1 + k_d) \tag{22.7}$$

Let us now introduce the parameters, using in ratings (ratios), into Modified Modigliani–Miller theory (MMM theory), which represents a perpetuity limit of modern theory of capital structure by Brusov–Filatova–Orekhova theory (BFO theory) (Brusov, 2018a, b; Brusov et al., 2015, 2018a, b, c, d, e, f, g, 2019; Filatova et al., 2018).

Two kind of financial ratios will be considered: coverage ratios and leverage ratios.

We will start from the coverage ratios.

22.3.1 Coverage Ratios

We will consider three kind of coverage ratios: coverage ratio of debt, coverage ratio of interest on the credit and coverage ratio of debt and interest on the credit.

22.3.1.1 Coverage Ratios of Debt

Let us consider first the coverage ratios of debt $i_1 = CF/D$.

Dividing both parts of Eq. (22.7) by D one gets

$$\begin{aligned} \frac{i_1}{WACC} &= \frac{i_1}{k_0} + t(1 + k_d) \\ \frac{i_1}{WACC} &= \frac{i_1 + k_0 t(1 + k_d)}{k_0} \\ WACC &= \frac{i_1 k_0}{i_1 + t k_0 (1 + k_d)} \end{aligned} \tag{22.8}$$

The coverage ratio of debt $i_1 = CF/D$ could be used for assessment of the following rating ratios: DCF/Debt, FFO/Debt, CFO/Debt, FOCF/Debt and some others. Formula (22.8) will be used to find a dependence $WACC(i_1)$.

22.3.1.2 Coverage Ratios of Interest on the Credit

Consider now coverage ratio of interest on the credit $i_2 = CF/k_dD$.

By use of the Modified Modigliani–Miller theory (MMM theory) for case with corporate taxes

$$V_L = V_0 + Dt(1 + k_d),$$

and dividing the both parts by k_dD, one could derive the expression for dependence $WACC(i_2)$

$$\begin{aligned} \frac{CF}{WACC} &= \frac{CF}{k_0} + Dt(1 + k_d) \\ \frac{i_2}{WACC} &= \frac{i_2}{k_0} + \frac{t(1 + k_d)}{k_d} \\ WACC &= \frac{i_2 k_0 k_d}{i_2 k_d + t k_0 (1 + k_d)} \end{aligned} \tag{22.9}$$

This ratio (i_2) could be used for assessment of the following parameters, used in rating, *FFO*/cashinterest, EBITDA/interest and some others. Formula (22.9) will be used to find a dependence $WACC(i_2)$.

22.3.1.3 Coverage Ratios of Debt and Interest on the Credit

Below we consider the coverage ratios of debt and interest on the credit simultaneously $i_3 = \frac{CF}{D(1+k_d)}$. This is a new value, introduced by us here for the first time. Using the Modified Modigliani–Miller theory (MMM theory) for case with corporate taxes

$$V_L = V_0 + Dt(1 + k_d)$$

and dividing the both parts by $(1 + k_d)D$ we get the dependence $WACC(i_3)$

$$\begin{aligned} \frac{CF}{WACC} &= \frac{CF}{k_0} + Dt(1 + k_d) \\ \frac{i_3}{WACC} &= \frac{i_3}{k_0} + \frac{t}{k_d} \\ WACC &= \frac{i_3 k_0 k_d}{i_3 k_d + t k_0} \end{aligned} \tag{22.10}$$

This ratio (i_3) could be used for assessment of the following rating ratios: FFO/Debt + interest, EBITDA/Debt + interest and some others. Formula (22.10) will be used to find a dependence $WACC(i_3)$.

22.3.2 *Dependence of WACC on Leverage Ratios of Debt in "Classical" Modigliani–Miller Theory (MM Theory) and Modified Modigliani–Miller Theory (MMM Theory)*

Below we study the dependences of WACC on leverage ratios of debt in "classical" Modigliani–Miller theory (MM theory) and modified Modigliani–Miller theory (MMM theory). We use Microsoft Excel, and the following parameters:

equity cost at L = 0 $k_0 = 0.2$; tax on profit rate t = 0.2; debt cost $k_d = 0.1; 0.14; 0.18$; leverage ratios of debt $l_1 = 0; 1; 2; 3; 4; 5; 6; 7; 8; 9; 10$.

Let us analyze the dependence of company's weighted average cost of capital (WACC) on the coverage ratios on debt $i_{l.}$ It is seen from Fig. 22.2 that the weighted average cost of capital, WACC, increases with coverage ratio of debt for both versions of Modigliani–Miller theory: for "classical" Modigliani–Miller theory (MM theory) as well as for modified Modigliani–Miller theory (MMM theory). WACC increases very rapidly at L increases from L = 0 to L = 1 and then comes to saturation very fast (after L = 3 WACC changes very weak. At saturation *WACC* reaches the value k_0 (equity cost at zero leverage level). This means that for high values of i_j one can choose k_0 as a discount rate with a good accuracy. Thus the role of parameter k_0 increases drastically. The method of determination of parameter k_0

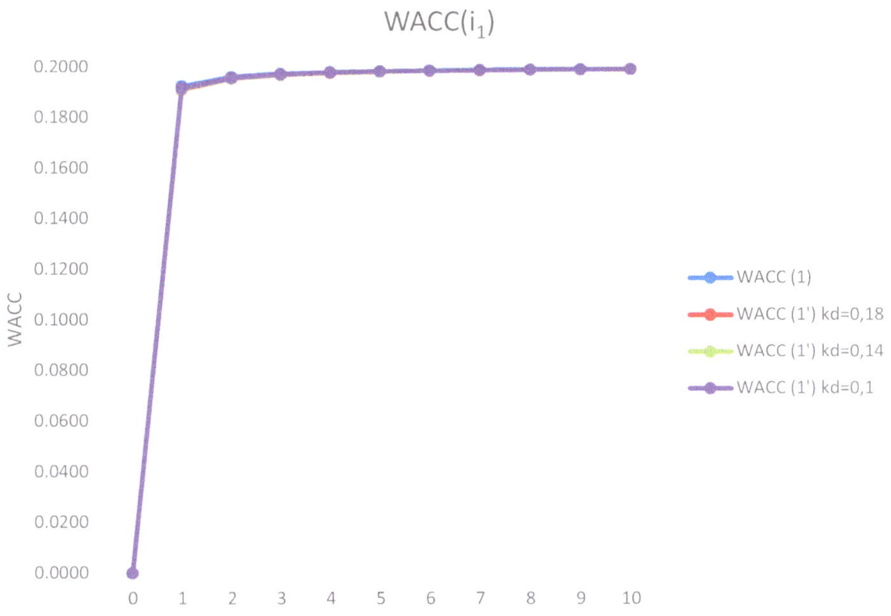

Fig. 22.2 Dependence of WACC on coverage ratios of debt in "classical" Modigliani–Miller theory (curve WACC(1)) and in modified Modigliani–Miller theory (MMM theory) (curves WACC (1′)) at $k_d = 0.1; 0.14; 0.18$

Table 22.1 Dependence of WACC on coverage ratios of debt in "classical" Modigliani–Miller theory

k_0	i_1	t	WACC
0.2	0	0.2	0.0000
0.2	1	0.2	0.1923
0.2	2	0.2	0.1961
0.2	3	0.2	0.1974
0.2	4	0.2	0.1980
0.2	5	0.2	0.1984
0.2	6	0.2	0.1987
0.2	7	0.2	0.1989
0.2	8	0.2	0.1990
0.2	9	0.2	0.1991
0.2	10	0.2	0.1992

Table 22.2 Dependence of WACC on coverage ratios of debt in modified Modigliani–Miller theory (MMM theory) at $k_d = 0.18$

k_0	i_1	t	k_d	WACC
0.2	0	0.2	0.18	0.0000
0.2	1	0.2	0.18	0.1910
0.2	2	0.2	0.18	0.1954
0.2	3	0.2	0.18	0.1969
0.2	4	0.2	0.18	0.1977
0.2	5	0.2	0.18	0.1981
0.2	6	0.2	0.18	0.1984
0.2	7	0.2	0.18	0.1987
0.2	8	0.2	0.18	0.1988
0.2	9	0.2	0.18	0.1990
0.2	10	0.2	0.18	0.1991

Table 22.3 Dependence of WACC on coverage ratios of debt in modified Modigliani–Miller theory (MMM theory) at $k_d = 0.14$

k_0	i_1	t	k_d	WACC
0.2	0	0.2	0.14	0.0000
0.2	1	0.2	0.14	0.1913
0.2	2	0.2	0.14	0.1955
0.2	3	0.2	0.14	0.1970
0.2	4	0.2	0.14	0.1977
0.2	5	0.2	0.14	0.1982
0.2	6	0.2	0.14	0.1985
0.2	7	0.2	0.14	0.1987
0.2	8	0.2	0.14	0.1989
0.2	9	0.2	0.14	0.1990
0.2	10	0.2	0.14	0.1991

has been developed by Anastasiya Brusova (Brusova, 2011). So, parameter k_0 is the discount rate for limit case of high values of i_1.

We observe that results for both versions of Modigliani–Miller theory are very closed: only from Tables 22.1, 22.2, 22.3, and 22.4 it is seen that curve of

Table 22.4 Dependence of WACC on coverage ratios of debt in modified Modigliani–Miller theory (MMM theory) at $k_d = 0.1$

k_0	i_1	t	k_d	WACC
0.2	0	0.2	0.1	0.0000
0.2	1	0.2	0.1	0.1916
0.2	2	0.2	0.1	0.1957
0.2	3	0.2	0.1	0.1971
0.2	4	0.2	0.1	0.1978
0.2	5	0.2	0.1	0.1983
0.2	6	0.2	0.1	0.1985
0.2	7	0.2	0.1	0.1988
0.2	8	0.2	0.1	0.1989
0.2	9	0.2	0.1	0.1990
0.2	10	0.2	0.1	0.1991

dependence of WACC on leverage ratios of debt in "classical" Modigliani–Miller theory (curve WACC(2)) lies a little bit above all dependences of WACC on coverage ratio of debt in modified Modigliani–Miller theory. The WACC values are practically independent on k_d. As we will see below, the situation is quite difference in case of leverage ratios, where the influence of k_d on dependences of WACC on leverage ratios is significant.

22.3.3 Leverage Ratios

We will consider now the leverage ratios. Three kind of leverage ratios will be considered: leverage ratios of debt, leverage ratios of interest on the credit and leverage ratios of debt and interest on the credit.

22.3.3.1 Leverage Ratios for Debt

Here

$$l_1 = D/CF \tag{22.11}$$

As above for coverage ratios we use the modified Modigliani–Miller theorem for case with corporate taxes one has

$$V_L = V_0 + Dt(1 + k_d),$$

and dividing the both parts by *CF*, we derive the expression for $WACC(l_1)$

$$\frac{CF}{WACC} = \frac{CF}{k_0} + Dt(1 + k_d)$$
$$\frac{1}{WACC} = \frac{1}{k_0} + l_1 t(1 + k_d) \tag{22.12}$$
$$WACC = \frac{k_0}{1 + tl_1 k_0(1 + k_d)}$$

This ratio (l_1) can be used to assess of the following parameters used in rating, Debt/EBITDA and some others. We will use last formula to build a curve of dependence $WACC(l_1)$.

22.3.3.2 Leverage Ratios for Interest on Credit

Here

$$l_2 = k_d D / CF \tag{22.13}$$

We use again Modified Modigliani–Miller theory (MMM theory) for case with corporate taxes

$$V_L = V_0 + Dt(1 + k_d),$$

and dividing the both parts by CF/k_d, we derive the expression for $WACC(l_2)$

$$\frac{CF}{WACC} = \frac{CF}{k_0} + Dt(1 + k_d)$$
$$\frac{1}{WACC} = \frac{1}{k_0} + \frac{l_2 t(1 + k_d)}{k_d} \tag{22.14}$$
$$WACC = \frac{k_0 k_d}{k_d + tl_2 k_0(1 + k_d)}$$

This ratio (l_2) can be used to assess of the following parameters used in rating, Interests/EBITDA and some others. We will use last formula to build a curve of dependence $WACC(l_2)$.

22.3.3.3 Leverage Ratios for Debt and Interest on Credit

Here

$$l_3 = D(1 + k_d)/CF \tag{22.15}$$

Using the Modified Modigliani–Miller theory (MMM theory) for case with corporate taxes

$$V_L = V_0 + Dt(1 + k_d),$$

and dividing the both parts by *CF*/(1 + k_d), we derive the expression for $WACC(l_3)$

$$\begin{aligned} \frac{CF}{WACC} &= \frac{CF}{k_0} + Dt(1 + k_d) \\ \frac{1}{WACC} &= \frac{1}{k_0} + l_3 t \\ WACC &= \frac{k_0}{1 + t l_3 k_0} \end{aligned} \tag{22.16}$$

This ratio (l_3) can be used to assess of the following parameters used in rating, *Debt + interest/FFO*, *Debt + interest/EBIT*, *Debt + interest/EBITDA(R)*, and some others. We will use last formula to build a curve of dependence $WACC(l_3)$.

22.3.4 Dependence of WACC on Leverage Ratios of Debt in "Classical" Modigliani–Miller Theory (MM Theory) and Modified Modigliani–Miller Theory (MMM Theory)

Below we study the dependences of WACC on leverage ratios of debt in "classical" Modigliani–Miller theory (MM theory) and modified Modigliani–Miller theory (MMM theory). We use Microsoft Excel, and the following parameters:

equity cost at $L = 0$ $k_0 = 0.2$; tax on profit rate $t = 0.2$; debt cost $k_d = 0.1$; 0.14; 0.18; leverage ratios of debt $l_1 = 0$; 1; 2; 3; 4; 5; 6; 7; 8; 9; 10.

Analysis of the dependences of company's weighted average cost of capital (*WACC*) on the leverage ratio on debt, l_1 shows the following: (see Fig. 22.3) the weighted average cost of capital, WACC, decreases with leverage ratios for both versions of Modigliani–Miller theory: for "classical" Modigliani–Miller theory (MM theory) as well as for modified Modigliani–Miller theory (MMM theory). But we observe that curve of dependence of WACC on leverage ratios of debt in "classical" Modigliani–Miller theory (curve WACC(2)) lies above all dependences of WACC on leverage ratios of debt in modified Modigliani–Miller theory. The WACC values decreases with k_d. So we are observing the influence of k_d on dependences of WACC on leverage ratios. This effect is absent in "classical" Modigliani–Miller theory. Thus the using of the modified Modigliani–Miller theory (MMM theory) leads to more correct valuation of the weighted average cost of capital (WACC), which plays the role of discount rate in financial flows discounting

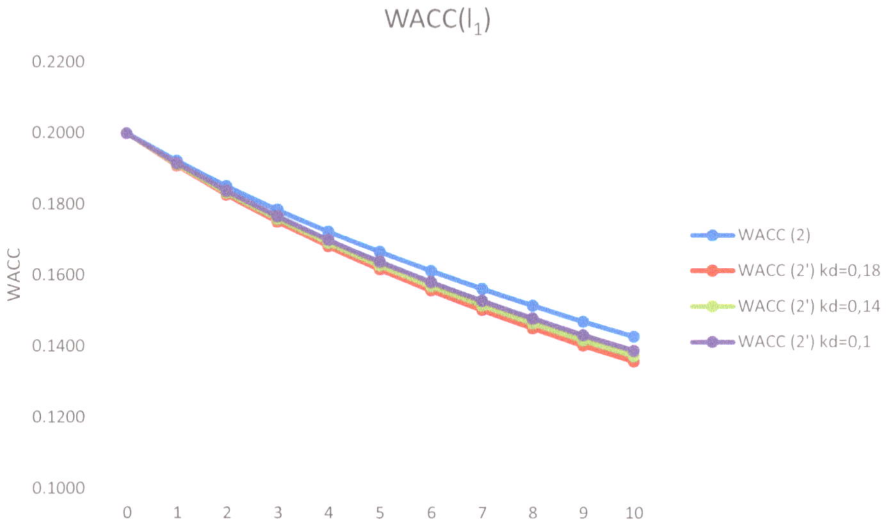

Fig. 22.3 Dependence of WACC on leverage ratios of debt in "classical" Modigliani–Miller theory (curve WACC(2)) and in modified Modigliani–Miller theory (MMM theory) (curves WACC(2′)) at $k_d = 0.1; 0.14; 0.18$

in rating methodologies. Obtaind results will help improve the existing rating methodologies.

22.4 Discussions

In current paper we use the Modified Modigliani–Miller theory (MMM theory) and apply it for rating methodologies needs. The financial "ratios" (main rating parameters) were introduced into MMM theory. The dependence of the weighted average cost of capital (WACC), which plays the role of discount rate in financial flows discounting in rating methodologies, on coverage and leverage ratios is analyzed. Obtaind results will help improve the existing rating methodologies.

The analysis of the dependence of company's weighted average cost of capital (WACC) on the coverage ratios on debt i_1 shows that WACC increases with coverage ratio of debt for both versions of Modigliani–Miller theory: for "classical" Modigliani–Miller theory (MM theory) as well as for modified Modigliani–Miller theory (MMM theory). WACC increases very rapidly at L increases from $L = 0$ to $L = 1$ and then comes to saturation very fast (after $L = 3$ WACC changes very weak). At saturation *WACC* reaches the value k_0 (equity cost at zero leverage level). This means that for high values of i_j one can choose k_0 as a discount rate with a good accuracy. Thus the role of parameter k_0 increases drastically. The method of determination of parameter k_0 has been developed by Anastasiya Brusova (2011). So, parameter k_0 is the discount rate for limit case of high values of i_1.

Table 22.5 Dependence of WACC on leverage ratios of debt in "classical" Modigliani–Miller theory

k_0	l_1	t	WACC
0.2	0	0.2	0.2000
0.2	1	0.2	0.1923
0.2	2	0.2	0.1852
0.2	3	0.2	0.1786
0.2	4	0.2	0.1724
0.2	5	0.2	0.1667
0.2	6	0.2	0.1613
0.2	7	0.2	0.1563
0.2	8	0.2	0.1515
0.2	9	0.2	0.1471
0.2	10	0.2	0.1429

Table 22.6 Dependence of WACC on leverage ratios of debt in modified Modigliani–Miller theory (MMM theory) at $k_d = 0.18$

k_0	l_1	t	k_d	WACC
0.2	0	0.2	0.18	0.2000
0.2	1	0.2	0.18	0.1910
0.2	2	0.2	0.18	0.1827
0.2	3	0.2	0.18	0.1752
0.2	4	0.2	0.18	0.1682
0.2	5	0.2	0.18	0.1618
0.2	6	0.2	0.18	0.1559
0.2	7	0.2	0.18	0.1503
0.2	8	0.2	0.18	0.1452
0.2	9	0.2	0.18	0.1404
0.2	10	0.2	0.18	0.1359

Table 22.7 Dependence of WACC on leverage ratios of debt in modified Modigliani–Miller theory (MMM theory) at $k_d = 0.14$

k_0	l_1	t	k_d	WACC
0.2	0	0.2	0.14	0.2000
0.2	1	0.2	0.14	0.1913
0.2	2	0.2	0.14	0.1833
0.2	3	0.2	0.14	0.1759
0.2	4	0.2	0.14	0.1691
0.2	5	0.2	0.14	0.1629
0.2	6	0.2	0.14	0.1570
0.2	7	0.2	0.14	0.1516
0.2	8	0.2	0.14	0.1465
0.2	9	0.2	0.14	0.1418
0.2	10	0.2	0.14	0.1374

We observe that results for both versions of Modigliani–Miller theory are very closed: only from Tables 22.5, 22.6, 22.7, and 22.8 it is seen that curve of dependence of WACC on leverage ratios of debt in "classical" Modigliani–Miller theory lies a little bit above all dependences of WACC on coverage ratio of debt in modified

Table 22.8 Dependence of WACC on leverage ratios of debt in modified Modigliani–Miller theory (MMM theory) at $k_d = 0.1$

k_0	l_1	t	k_d	WACC
0.2	0	0.2	0.1	0.2000
0.2	1	0.2	0.1	0.1916
0.2	2	0.2	0.1	0.1838
0.2	3	0.2	0.1	0.1767
0.2	4	0.2	0.1	0.1701
0.2	5	0.2	0.1	0.1639
0.2	6	0.2	0.1	0.1582
0.2	7	0.2	0.1	0.1529
0.2	8	0.2	0.1	0.1479
0.2	9	0.2	0.1	0.1433
0.2	10	0.2	0.1	0.1389

Modigliani–Miller theory. The WACC values are practically independent on k_d. The situation is quite difference in case of leverage ratios, where the influence of k_d on dependences of WACC on leverage ratios is significant.

Analysis of the dependences of company's weighted average cost of capital (*WACC*) on the leverage ratio on debt, l_l shows that WACC decreases with leverage ratios for both versions of Modigliani–Miller theory: for "classical" Modigliani–Miller theory (MM theory) as well as for modified Modigliani–Miller theory (MMM theory). We observe that curve of dependence of WACC on leverage ratios of debt in "classical" Modigliani–Miller theory lies above all dependences of WACC on leverage ratios of debt in modified Modigliani–Miller theory. The WACC values decreases with k_d. So we are observing the influence of k_d on dependences of WACC on leverage ratios. This effect is absent in "classical" Modigliani–Miller theory. Thus the using of the modified Modigliani–Miller theory (MMM theory) leads to more correct valuation of the weighted average cost of capital (WACC), which plays the role of discount rate in financial flows discounting in rating methodologies. Obtained results will help improve the existing rating methodologies, which are used for valuation of the creditworthiness of companies.

Correct account of a method of tax on profit payments demonstrates that shortcomings of Modigliani–Miller theory are dipper, than everybody suggested: the underestimation of WACC really turns out to be bigger, as well as overestimation of the capitalization of the company. This means that systematic risks arising from the use of modified Modigliani–Miller theory (MMM theory) (which is more correct than "classical" one) in practice is higher than it was suggested by the "classical" version of this theory.

References

Brusova, A. A comparison of the three methods of estimation of weighted average cost of capital and equity cost of company. Financ. Anal. Prob. Sol. 2011, 34, 36–42.

Brusov, P. (2018a). Editorial: Introduction on special issue on the banking system and financial markets of Russia and other countries: Problems and prospects. *Journal of Reviews on Global Economics, 7*, i–vi.

Brusov, P. (2018b). Editorial. *Journal of Reviews on Global Economics, 7*, i–v.

Brusov, P., & Filatova, T. (2021). The Modigliani–Miller theory with arbitrary frequency of payment of tax on profit. *Mathematics, 9*(11), 1198 (WoS Q1).

Brusov, P., Filatova, T., Orehova, N., & Eskindarov, M. (2015). *Modern corporate finance, investments and taxation* (monograph) (373 p). Springer. Retrieved from https://www.springer.com/gp/book/9783319147314

Brusov, P., Filatova, T., Orehova, N., & Eskindarov, M. (2018a). *Modern corporate finance, investments, taxation and ratings* (monograph) (571 p). Springer.

Brusov, P. N., Filatova, T. V., Orekhova, N. P., & Kulik, V. L. (2018b). Rating: New approach. *Journal of Reviews on Global Economics, 7*, 37–62. https://doi.org/10.6000/1929-7092.2018.022.05

Brusov, P. N., Filatova, T. V., Orekhova, N. P., & Kulik, V. L. (2018c). A "golden age" of the companies: Conditions of its existence. *Journal of Reviews on Global Economics, 7*, 88–103. https://doi.org/10.6000/1929-7092.2023.07.07

Brusov, P. N., Filatova, T. V., Orekhova, N. P., & Kulik, V. L. (2018d). Rating methodology: New look and new horizons. *Journal of Reviews on Global Economics, 7*, 63–87. https://doi.org/10.6000/1929-7092.2023.07.06

Brusov, P. N., Filatova, T. V., Orekhova, N. P., Kulik, V. L., & Weil, I. (2018e). New meaningful effects in modern capital structure theory. *Journal of Reviews on Global Economics, 7*, 104–122. https://doi.org/10.6000/1929-7092.2023.07.08

Brusov, P. N., Filatova, T. V., & Orekhova, N. P. (2018f). *Modern corporate finance and investments* (monograph) (517 p). Knorus Publishing House.

Brusov, P., Filatova, T., Orekhova, N., Kulik, V., Weil, I., & Brailov, A. (2018g). The impact of the central bank key rate and commercial banks credit rates on creating and maintaining of a favorable investment climate in the country. *Journal of Reviews on Global Economics, 7*, 360–376. https://doi.org/10.6000/1929-7092.2023.07.31

Brusov, P. N., Filatova, T. V., Orekhova, N. P., Kulik, V. L., & Weil, I. (2019). Ratings of the investment projects of arbitrary durations: New methodology. *Journal of Reviews on Global Economics, 8*, 437–448. https://doi.org/10.6000/1929-7092.2019.08.37

Brusov, P., Filatova, T., & Orehova, N. (2020a). *Ratings: Critical analysis and new approaches of quantitative and qualitative methodology* (monograph) (369 p). Springer Nature. Retrieved from https://www.springer.com/de/book/9783030562427

Brusov, P. N., Filatova, T. V., Orekhova, N. P., Kulik, V. L., Chang, S.-I., & Lin, Y. C. G. (2020b). Application of the Modigliani–Miller theory, modified for the case of advance payments of tax on profit, in rating methodologies. *Journal of Reviews on Global Economics, 9*, 282–292.

Brusov, P. N., Filatova, T. V., Orekhova, N. P., Kulik, V. L., Chang, S.-I., & Lin, Y. C. G. (2020c). Modification of the Modigliani–Miller theory for the case of advance payments of tax on profit. *Journal of Reviews on Global Economics, 9*, 257–267.

Brusov, P., Filatova, T., Chang, S.-I., & Lin, G. (2021a). Innovative investment models with frequent payments of tax on income and of interest on debt. *Mathematics, 9*(13), 1491 (WoS Q1).

Brusov, P., Filatova, T., Orekhova, N., Kulik, V., Chang, S.-I., & Lin, G. (2021b). Generalization of the Modigliani–Miller theory for the case of variable profit. *Mathematics, 9*(11), 1286 (WoS Q1).

Filatova, T., Orehova, N., & Brusova, A. (2008). Weighted average cost of capital in the theory of Modigliani–Miller, modified for a finite life–time company. *Bulletin of the FU, 48*, 68–77.

Filatova, T. V., Brusov, P. N., Orekhova, N. P., & Kulik, V. L. (2018). Ratings of the long–term projects: New approach. *Journal of Reviews on Global Economics, 7*, 645–661. https://doi.org/10.6000/1929-7092.2018.022.59

Modigliani, F., & Miller, M. (1958). The cost of capital, corporate finance, and the theory of investment. *The American Economic Review, 48*, 261–297.
Modigliani, F., & Miller, M. (1963). Corporate income taxes and the cost of capital: A correction. *The American Economic Review, 53*, 147–175.
Modigliani, F., & Miller, M. (1966). Some estimates of the cost of capital to the electric utility industry 1954–19522. *The American Economic Review, 56*, 333–391.

Chapter 23
A New Approach to Project Ratings

23.1 Introduction

Rating agencies play a very important role in economics. Their analysis of issuer's state, generated credit ratings of issuers help investors make reasonable investment decision, as well as help issuers with good enough ratings get credits on lower rates etc.

The Chapter continues create a new approach to rating methodology: in addition to two papers, which have considered the creditworthiness of the non-finance issuers (Filotova et al., 2018; Brusov et al., 2018b), we develop here a new approach to project rating. We work within investment models. Created by authors. One of them describes the effectiveness of investment project from perspective of equity capital owners, while other model describes the effectiveness of investment project from perspective of equity capital and debt capital owners.

The important features of current consideration as well as in previous studies are: (1) The adequate use of discounting of financial flows virtually not used in existing rating methodologies, (2) The incorporation of rating parameters (financial "ratios"), used in project rating, into considered modern investment models.

Analyzing within these investment models with incorporated rating parameters the dependence of *NPV* on rating parameters (financial "ratios") at different values of equity cost k_0, at different values of credit rates k_d as well as at different values of on leverage level L we come to very important conclusion, that *NPV* (in units of *NOI*) ($\frac{NPV}{NOI}$) (as well as *NPV* (in units of D) ($\frac{NPV}{D}$)) depends only on equity cost k_0, on credit rates k_d, on leverage level L as well as on one of the leverage ratios l_j (on one of the coverage ratios i_j) and does not depend on equity value S, debt value D and *NOI*. This means that results on the dependence of NPV (in units of *NOI*) ($\frac{NPV}{NOI}$) on leverage ratios l_j (as well as on the dependence of *NPV* (in units of D) ($\frac{NPV}{D}$) on coverage ratios i_j) at different equity costs k_0, at different credit rates k_d, at different leverage levels L carry the universal character: these dependencies remain valid for investment projects with any equity value S, debt value D and *NOI*.

P. Brusov, T. Filatova, *Efficiency of Investment Projects*, Contributions to Finance and Accounting, https://doi.org/10.1007/978-3-032-01028-5_23

23.2 Investment Models

We work within investment models, created by authors. One of them describes the effectiveness of investment project from perspective of equity capital owners, while other model describes the effectiveness of investment project from perspective of equity capital and debt capital owners.

In the former case, investments at the initial time moment $T = 0$ are equal to $-S$ and the flow of capital for the period (in addition to the tax shields $k_d Dt$ it includes a payment of interest on a loan $-k_d D$):

$$\mathrm{CF} = (\mathrm{NOI} - k_d D)(1 - t). \tag{23.1}$$

Here, for simplicity, we suppose that interest on the loan will be paid in equal shares $k_d D$ during all periods. Note that principal repayment is made at the end of the last period.

We will consider the case of discounting, when operating and financial flows are not separated and both are discounted, using the general rate (as which, obviously, the weighted average cost of capital (WACC) can be selected). In this case for long-term (perpetuity) projects, the Modigliani–Miller formula (Modigliani & Miller, 1958, 1963, 1966) for WACC will be used and for projects of finite (arbitrary) duration Brusov–Filatova–Orekhova formula will be used (Brusov, 2018a, b; 2018a, b, c, d, e, f, g, 2019, 2020a, b; Filatova et al., 2008, 2018).

Note that debt capital is the least risky, because interest on credit is paid after taxes in the first place. Therefore, the cost of credit will always be less than the equity cost, whether of ordinary or of preference shares $k_e > k_d$; $k_p > k_d$. Here k_e, k_p is the equity cost of ordinary or of preference shares consequently.

23.2.1 The Effectiveness of the Investment Project from the Perspective of the Equity Holders Only (Without Flows Separation)

In this case operating and financial flows are not separated and are discounted. Using the general rate (as which, obviously, WACC can be selected).

The credit reimbursable at the end of the project (at the end of the period (n)) can be discounted either at the same rate WACC or at the debt cost rate k_d. Now we choose a uniform rate and the first option.

$$\begin{aligned}
\mathrm{NPV} &= -S + \sum_{i=1}^{n} \frac{\mathrm{NOI}(1-t) - k_{\mathrm{d}} D(1-t)}{(1+\mathrm{WACC})^{i}} - \frac{D}{(1+\mathrm{WACC})^{n}} \\
&= -S + \frac{\mathrm{NOI}(1-t) - k_{\mathrm{d}} D(1-t)}{\mathrm{WACC}} \left(1 - \frac{1}{(1+\mathrm{WACC})^{n}}\right) - \frac{D}{(1+\mathrm{WACC})^{n}}.
\end{aligned} \tag{23.2}$$

At a Constant Value of Equity Capital (S = const)
Accounting that in the case S = const NOI is proportional to the invested capital, I, $\mathrm{NOI} = \beta I = \beta S(1 + L)$ and substituting $D = LS$, we get

$$\begin{aligned}
\mathrm{NPV} = -S + \frac{\mathrm{NOI}(1-t) - k_{\mathrm{d}} D(1-t)}{\mathrm{WACC}} \left(1 - \frac{1}{(1+\mathrm{WACC})^{n}}\right) \\
- \frac{D}{(1+\mathrm{WACC})^{n}},
\end{aligned} \tag{23.3}$$

$$\begin{aligned}
\mathrm{NPV} = -S\left[1 + \frac{Lk_{\mathrm{d}}(1-t)}{\mathrm{WACC}} \left(1 - \frac{1}{(1+\mathrm{WACC})^{n}}\right) + \frac{L}{(1+\mathrm{WACC})^{n}}\right] \\
+ \frac{\beta S(1+L)(1-t)}{\mathrm{WACC}} \left(1 - \frac{1}{(1+\mathrm{WACC})^{n}}\right).
\end{aligned} \tag{23.4}$$

23.2.2 Modigliani–Miller Limit (Long-Term (Perpetuity) Projects)

In perpetuity limit ($n \to \infty$) (Modigliani–Miller limit) (turning to the limit $n \to \infty$ in the relevant equations). we have

$$\mathrm{NPV} = -S + \frac{\mathrm{NOI}(1-t) - k_{\mathrm{d}} D(1-t)}{\mathrm{WACC}}. \tag{23.5}$$

At a Constant Value of Equity Capital (S = const)

$$\mathrm{NPV} = -S + \frac{\mathrm{NOI}(1-t) - k_{\mathrm{d}} D(1-t)}{\mathrm{WACC}} \tag{23.6}$$

Substituting $D = LS$, we get

$$\begin{aligned}\mathrm{NPV} &= -S\left[1+\frac{Lk_{\mathrm{d}}(1-t)}{\mathrm{WACC}}\right]+\frac{\mathrm{NOI}(1-t)}{\mathrm{WACC}}\\ &= -S\left[1+\frac{Lk_{\mathrm{d}}(1-t)}{k_0(1-Lt/(1+L))}\right]+\frac{\beta S(1+L)(1-t)}{k_0(1-Lt/(1+L))}.\end{aligned} \tag{23.7}$$

In last equation we substituted the perpetuity (Modigliani–Miller) formula for WACC

$$WACC = k_0\left(1-\frac{Lt}{1+L}\right). \tag{23.8}$$

So, below we consider the long-term (perpetuity) projects and will use the following formula for calculations

$$NPV = -S\left[1+\frac{Lk_d(1-t)}{k_0\left(1-\frac{Lt}{1+L}\right)}\right]+\frac{\beta S(1+L)(1-t)}{k_0\left(1-\frac{Lt}{1+L}\right)} \tag{23.9}$$

23.3 Incorporation of Financial Coefficients, Using in Project Rating, into Modern Investment Models

Below we incorporate the financial coefficients, used in project rating, into modern investment models, created by authors. We will consider two kind of financial coefficients: coverage ratios as well as leverage coefficients. In each group of financial coefficients we incorporate three particular quantities.

For coverage ratios we incorporate: (1) coverage ratios of debt, $i_1 = \frac{NOI}{D}$; (2) coverage ratios of interest on the credit $i_2 = \frac{NOI}{k_d D}$; (3) coverage ratios of debt and interest on the credit $i_3 = \frac{NOI}{(1+k_d)D}$.

For leverage ratios we incorporate: (1) leverage ratios of debt, $l_1 = \frac{D}{NOI}$; (2) leverage ratios of interest on the credit $l_2 = \frac{k_d D}{NOI}$; (3) leverage ratios of debt and interest on the credit $l_3 = \frac{(1+k_d)D}{NOI}$.

23.3.1 Coverage Ratios

23.3.1.1 Coverage Ratios of Debt

Let us first incorporate the coverage ratios, using in project rating, into modern investment models, created by authors. Dividing both parts of Eq. (23.9) by D one gets

$$\frac{NPV}{D} = -\frac{1}{L} - \frac{(k_d - i_1)(1-t)}{k_0\left(1 - \frac{Lt}{1+L}\right)} \tag{23.10}$$

Here

$$i_1 = \frac{NOI}{D} \tag{23.11}$$

23.3.1.2 Coverage Ratios of Interest on the Credit

Dividing both parts of Eq. (23.9) by $k_d D$ one gets

$$\frac{NPV}{k_d D} = -\frac{1}{Lk_d} - \frac{(1 - i_2)(1-t)}{k_0\left(1 - \frac{Lt}{1+L}\right)} \tag{23.12}$$

Here

$$i_2 = \frac{NOI}{k_d D} \tag{23.13}$$

23.3.1.3 Coverage Ratios of Debt and Interest on the Credit

Dividing both parts of Eq. (23.9) by $(1 + k_d)D$ one gets

$$\frac{NPV}{(1+k_d)D} = -\frac{1}{L(1+k_d)} - \frac{[k_d - i_3(1+k_d)](1-t)}{k_0\left(1 - \frac{Lt}{1+L}\right)} \tag{23.14}$$

Here

$$i_3 = \frac{NOI}{(1 + k_d)D} \tag{23.15}$$

Analyzing the formulas (23.10), (23.12) and (23.14) we come to very important conclusion, that NPV (in units of D) ($\frac{NPV}{D}$) depends only on equity cost k_0, on credit rates k_d, on leverage level L as well as on one of the coverage ratios i_j and does not depend on equity value *S*, debt value *D* and *NOI*. This means that results on the dependence of NPV (in units of D) ($\frac{NPV}{D}$) on coverage ratios i_j at different equity costs k_0, at different credit rates k_d, at different leverage levels L carry the universal character: these dependencies remain valid for investment projects with any equity value *S*, debt value *D* and *NOI*.

23.3.2 Leverage Ratios

23.3.2.1 Leverage Ratios for Debt

Now let us incorporate the leverage ratios, using in project rating, into modern investment models, created by authors.

Dividing both parts of Eq. (23.9) by *NOI* one gets

$$\frac{NPV}{NOI} = \frac{-l_1}{L} + \frac{(1 - k_d l_1)(1 - t)}{k_0\left(1 - \frac{Lt}{1+L}\right)} \tag{23.16}$$

Here

$$l_1 = \frac{D}{NOI} \tag{23.17}$$

23.3.2.2 Leverage Ratios for Interest on Credit

$$\frac{NPV}{NOI} = \frac{-l_2}{k_d L} + \frac{(1 - l_2)(1 - t)}{k_0\left(1 - \frac{Lt}{1+L}\right)} \tag{23.18}$$

Here

$$l_2 = \frac{k_d D}{NOI} \tag{23.19}$$

23.3.2.3 Leverage Ratios for Debt and Interest on Credit

$$\frac{NPV}{NOI} = \frac{-l_3}{(1+k_d)L} + \frac{(1+k_d-l_3k_d)(1-t)}{(1+k_d)k_0\left(1-\frac{Lt}{1+L}\right)} \tag{23.20}$$

Here

$$l_3 = \frac{(1+k_d)D}{NOI}. \tag{23.21}$$

Analyzing the formulas (23.16), (23.18) and (23.20) we come to very important conclusion, that *NPV* (in units of *NOI*) ($\frac{NPV}{NOI}$) depends only on equity cost k_0, on credit rates k_d, on leverage level L as well as on one of the leverage ratios l_j and does not depend on equity value *S*, debt value *D* and *NOI*. This means that results on the dependence of NPV (in units of *NOI*) ($\frac{NPV}{NOI}$) on leverage ratios l_j at different equity costs k_0, at different credit rates k_d, at different leverage levels L carry the universal character: these dependencies remain valid for investment projects with any equity value *S*, debt value *D* and *NOI*.

We investigate below the effectiveness of long-term investment projects studying the dependence of NPV on coverage ratios and on leverage ratios. We make calculations for coefficients i_1 and l_1. Calculations for the rest of coefficients (i_2, i_3 and l_2, l_3) could be made in a similar way.

We start from the calculations of the dependence of NPV on coverage ratios. We consider different values of equity costs k_0, of debt costs k_d and of leverage level $L = D/S$. Here t is tax on profit rate, which in our calculations is equal to 20%.

23.4 Dependence of NPV on Coverage Ratios

23.4.1 Coverage Ratio on Debt

Below we calculate the dependence of NPV (in units of D) ($\frac{NPV}{D}$) on coverage ratio on debt i_1 at different equity costs k_0 (k_0 is equity cost at L = 0). We will make calculations for two leverage levels L (L = 1 and L = 3) and for different credit rates k_d.

For calculation within MM approximation we use the formula (23.10)

$$\frac{NPV}{D} = -\frac{1}{L} - \frac{(k_d-i_1)(1-t)}{k_0 * \left(1-\frac{Lt}{1+L}\right)}$$

23.4.1.1 The Dependence of NPV on Coverage Ratio on Debt I_1 at Equity Cost $k_0 = 24\%$

Below we investigate the dependence of NPV on coverage ratio on debt i_1 at different values of equity costs k_0, at different values of debt costs k_d at fixed value of equity cost, as well as at different values of leverage levels L.

Let us start our calculations from the case of equity cost $k_0 = 24\%$.

The results of calculations of the dependence of NPV on coverage ratio on debt i_1 at equity cost $k_0 = 24\%$, at different values of debt costs k_d and $L = 1$ are shown in Table 23.1.

The dependence of NPV (in units of D) on coverage ratio on debt i_1 at $k_0 = 24\%$, $k_d = 6\%$, 10%, 14% and 20% and $L = 1$ is illustrated in Fig. 23.1.

Let us calculate the value of i_1 above which the investment project remains effective (NPV > 0)

k_d	0.20	0.14	0.1	0.06
i_1	0.48	0.42	0.38	0.32

One can see from this Table that the value of i_1 above which the investment project remains effective (NPV > 0) increases with credit rate k_d, that means that effectiveness of the investment project as well as its creditworthiness decreases with credit rate k_d.

Let us calculate the dependence of NPV (in units of D) on coverage ratio on debt i_1 at $k_0 = 24\%$, $k_d = 6\%$, 10%,14% and 20% and $L = 3$.

The dependence of NPV (in units of D) on coverage ratio on debt i_1 at $k_0 = 24\%$, $k_d = 6\%$, 10%, 14% and 20% and $L = 3$ is illustrated in Fig. 23.2.

Table 23.1 The dependence of NPV on coverage ratio on debt i_1 at equity cost $k_0 = 24\%$, $k_d = 6\%$, 10%, 14%, 20% and $L = 1$

i_1	L	k_0	t	NPV/D ($k_d = 0.2$)	NPV/D ($k_d = 0.14$)	NPV/D ($k_d = 0.1$)	NPV/D ($k_d = 0.06$)
0	1	0.24	0.2	−1.741	−1.519	−1.37	−1.222
1	1	0.24	0.2	1.963	2.185	2.333	2.481
2	1	0.24	0.2	5.667	5.889	6.037	6.185
3	1	0.24	0.2	9.37	9.593	9.741	9.889
4	1	0.24	0.2	13.07	13.3	13.44	13.59
5	1	0.24	0.2	16.78	17	17.15	17.3
6	1	0.24	0.2	20.48	20.7	20.85	21
7	1	0.24	0.2	24.19	24.41	24.56	24.7
8	1	0.24	0.2	27.89	28.11	28.26	28.41
9	1	0.24	0.2	31.59	31.81	31.96	32.11
10	1	0.24	0.2	35.3	35.52	35.67	35.81

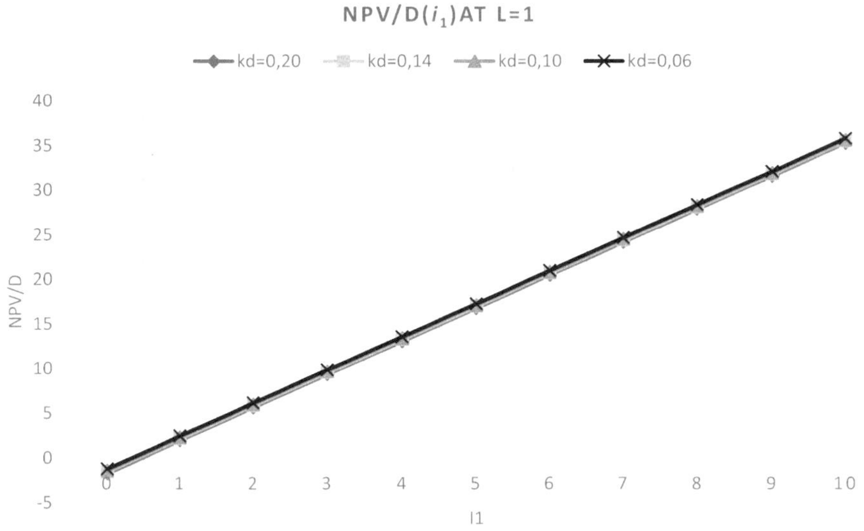

Fig. 23.1 The dependence of NPV (in units of D) on coverage ratio on debt i_1 at $k_0 = 24\%$, $k_d = 6\%$, 10%, 14% and 20% and L = 1

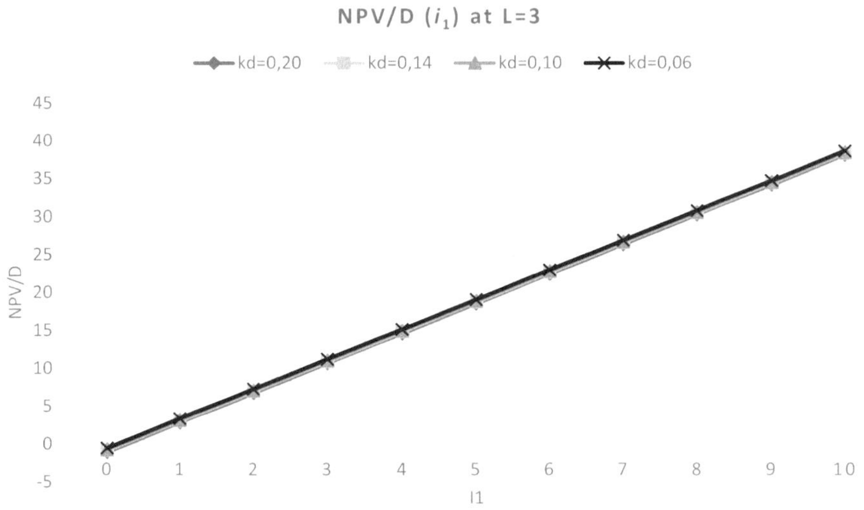

Fig. 23.2 The dependence of NPV (in units of D) on coverage ratio on debt i_1 at $k_0 = 24\%$, $k_d = 6\%$, 10%, 14% and 20% and L = 3

Let us calculate the value of l_1 above which the investment project remains effective (NPV > 0)

k_d	0.20	0.14	0.1	0.06
i_1	0.3	0.23	0.18	0.12

Table 23.2 The dependence of NPV on coverage ratio on debt i_1 at equity cost $k_0 = 24\%$, $k_d = 6\%, 10\%, 14\%, 20\%$ and $L = 3$

i_1	L	k_0	t	NPV/D ($k_d = 0.2$)	NPV/D ($k_d = 0.14$)	NPV/D ($k_d = 0.1$)	NPV/D ($k_d = 0.06$)
0	3	0.24	0.2	−1.118	−0.882	−0.725	−0.569
1	3	0.24	0.2	2.804	3.039	3.196	3.353
2	3	0.24	0.2	6.725	6.961	7.118	7.275
3	3	0.24	0.2	10.65	10.88	11.04	11.2
4	3	0.24	0.2	14.57	14.8	14.96	15.12
5	3	0.24	0.2	18.49	18.73	18.88	19.04
6	3	0.24	0.2	22.41	22.65	22.8	22.96
7	3	0.24	0.2	26.33	26.57	26.73	26.88
8	3	0.24	0.2	30.25	30.49	30.65	30.8
9	3	0.24	0.2	34.18	34.41	34.57	34.73
10	3	0.24	0.2	38.1	38.33	38.49	38.65

One can see from this Table that like the case of $L = 1$ the value of i_1 above which the investment project remains effective (NPV > 0) increases with credit rate k_d, that means that effectiveness of the investment project as well as its creditworthiness decreases with credit rate k_d. Comparing the case of $L = 1$ one can see that at bigger leverage level ($L = 3$) the investment project becomes effective (NPV > 0) starting from smaller coverage ratio i_1, so bigger leverage level favors to the effectiveness of the investment project as well as its creditworthiness.

We see from the Tables 23.1 and 23.2 and Figs. 23.1 and 23.2, that $\frac{NPV}{D}$ increases with i_1 and that $\frac{NPV}{D}$ values turn out to be very closed each other at all i_1 values. It is seen as well that NPV increases with decreasing k_d. This means that effectiveness of the investment project as well as its creditworthiness decreases with credit rate k_d.

Below we investigate the dependence of $\frac{NPV}{D}$ on i_1 at different values of k_d in more details and will show the ordering of $\frac{NPV}{D}(i_1)$ curves at different values of k_d, as well as at different leverage levels L.

23.4.1.2 The Dependence of NPV on Coverage Ratio on Debt i_1 at Equity Cost $k_0 = 12\%$

We study here the dependence of $\frac{NPV}{D}$ on i_1 at fixed equity cost $k_0 = 12\%$ and at different values of k_d in more details and will show the ordering of $\frac{NPV}{D}(i_1)$ curves at different values of k_d, as well as at different leverage levels L.

The results of calculations of the dependence of NPV on coverage ratio on debt i_1 at equity cost $k_0 = 12\%$, at different values of debt costs k_d and $L = 1$ are shown in Table 23.3.

The results of calculations of the dependence of NPV on coverage ratio on debt i_1 at equity cost $k_0 = 12\%$, at different values of debt costs k_d and $L = 3$ are shown in Table 23.4.

Table 23.3 The dependence of NPV (in units of D) on coverage ratio on debt i_1 at $k_0 = 12\%$, $k_d = 2\%$, 4%, 6%, 8% and 10% and L = 1

i_1	t	k_0	NPV/D $k_d = 0.1$	NPV/D $k_d = 0.08$	NPV/D $k_d = 0.06$	NPV/D $k_d = 0.04$	NPV/D $k_d = 0.02$
0	0.2	0.12	−1.741	−1.593	−1.444	−1.296	−1.148
1	0.2	0.12	5.667	5.815	5.963	6.111	6.259
2	0.2	0.12	13.074	13.222	13.370	13.519	13.667
3	0.2	0.12	20.481	20.630	20.778	20.926	21.074
4	0.2	0.12	27.889	28.037	28.185	28.333	28.481
5	0.2	0.12	35.296	35.444	35.593	35.741	35.889
6	0.2	0.12	42.704	42.852	43.000	43.148	43.296
7	0.2	0.12	50.111	50.259	50.407	50.556	50.704
8	0.2	0.12	57.519	57.667	57.815	57.963	58.111
9	0.2	0.12	64.926	65.074	65.222	65.370	65.519
10	0.2	0.12	72.333	72.481	72.630	72.778	72.926

Table 23.4 The dependence of NPV (in units of D) on coverage ratio on debt i_1 at $k_0 = 12\%$, $k_d = 2\%$, 4%, 6%, 8% and 10% and L = 3

i_1	t	k_0	NPV/D $k_d = 0.1$	NPV/D $k_d = 0.08$	NPV/D $k_d = 0.06$	NPV/D $k_d = 0.04$	NPV/D $k_d = 0.02$
0	0.2	0.12	−1.118	−0.961	−0.804	−0.647	−0.490
1	0.2	0.12	6.725	6.882	7.039	7.196	7.353
2	0.2	0.12	14.569	14.725	14.882	15.039	15.196
3	0.2	0.12	22.412	22.569	22.725	22.882	11.039
4	0.2	0.12	30.255	30.412	30.569	30.725	30.882
5	0.2	0.12	38.098	38.255	38.412	38.569	38.725
6	0.2	0.12	45.941	46.098	46.255	46.412	46.569
7	0.2	0.12	53.784	53.941	54.098	54.255	54.412
8	0.2	0.12	61.627	61.784	61.941	62.098	62.255
9	0.2	0.12	69.471	69.627	69.784	69.941	70.098
10	0.2	0.12	77.314	77.471	77.627	77.784	77.941

We see from the Tables 23.3 and 23.4 that NPV (in units of D) ($\frac{NPV}{D}$) increases with i_1 and that $\frac{NPV}{D}$ values turn out to be very closed each other at all i_1 values.

To show the difference in $\frac{NPV}{D}$ values in more details we show at the Fig. 23.3 the dependence of $\frac{NPV}{D}$ on parameter i_1 for range i_1 from 1 to 2.

One can see, that all NPV (i_1) curves corresponding to L = 3 lie above the curves corresponding to L = 1. This means that NPV increases with L (with increasing of the debt financing). At fixed value L NPV increases with decreasing the credit rate k_d. This means that effectiveness of the investment project as well as its creditworthiness decreases with credit rate k_d.

Analyzing the obtained results one should remember, that NPV (in units of D) ($\frac{NPV}{D}$) depends only on equity cost k_0, on credit rates k_d, on leverage level L as well as on one of the coverage ratios i_j and does not depend on equity value S, debt value

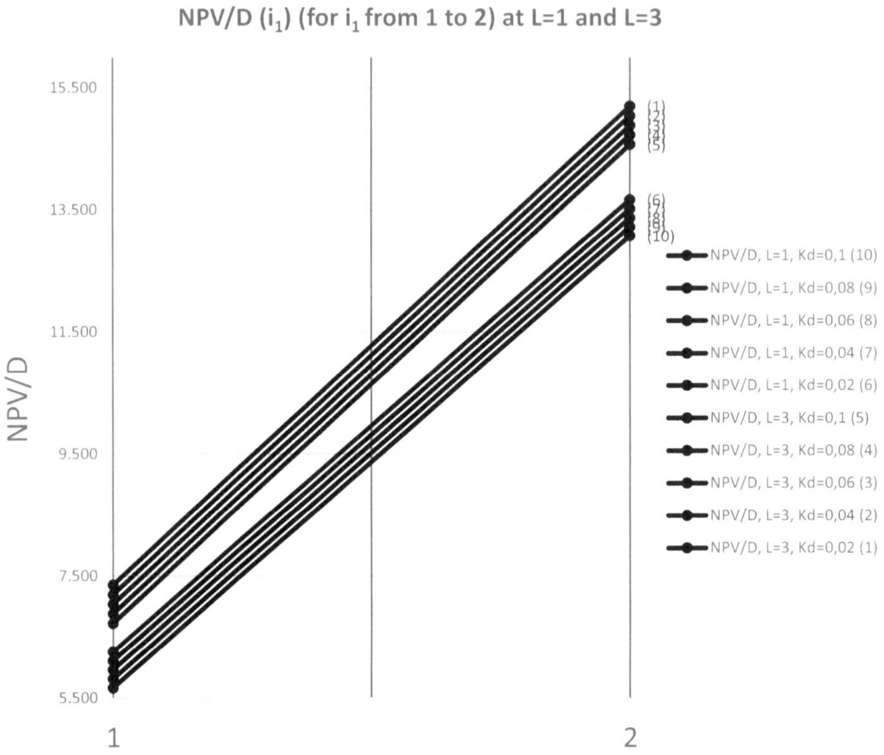

Fig. 23.3 The dependence of NPV (in units of D) on coverage ratio on debt i_1 at $k_0 = 12\%$, $k_d = 2\%, 4\%, 6\%, 8\%$ and 10% and $L = 1$ and $L = 3$

D and *NOI*. This means that obtained results on the dependence of NPV (in units of D) ($\frac{NPV}{D}$) on coverage ratios i_j at different equity costs k_0, at different credit rates k_d, at different leverage levels L carry the universal character: these dependencies remain valid for investment projects with any equity value *S*, debt value *D* and *NOI*.

23.5 Dependence of NPV on Leverage Ratios

23.5.1 Leverage Ratio of Debt

Below we calculate the dependence of NPV (in units of NOI) ($\frac{NPV}{NOI}$) on leverage ratio on debt l_1 at different equity costs k_0 (k_0 is equity cost at $L = 0$). We make calculations for two leverage levels L ($L = 1$ and $L = 3$) and for different credit rates k_d.

For calculation within MM approximation we use the formula (23.19)

$$\frac{NPV}{NOI} = \frac{-l_1}{L} + \frac{(1 - K_d * l_1)(1 - t)}{K_0 \ * \left(1 - \frac{Lt}{1+L}\right)}.$$

23.5.1.1 The Dependence of NPV (in Units of NOI) ($\frac{NPV}{NOI}$) on Leverage Ratio on Debt l_1 at Equity Cost $k_0 = 0.12$

Results are shown in Tables 23.5 and 23.6 and in Figs. 23.4 and 23.5.

Based on the above calculations, we plot the dependences of NPV/NOI on leverage ratio on debt l_1 at different leverage levels L.

From Tables 23.5 and 23.6 and Figs. 23.4 and 23.5 one can come to conclusion, that the NPV (in units of NOI) (NPV/NOI) decreases with increasing of the leverage ratio on debt l_1. With the increasing of the cost of debt capital k_d, curves of the dependence of NPV/NOI (l_1), outgoing from a single point at a zero value of l_1, lie below (i.e., the rate of decrease (or negative slope of curves) grows). Note, that while the dependences of NPV (in units of D) on coverage ratio on debt i_1 lie very close each other (see above), the dependences of NPV (in units of NOI) on leverage ratio on debt l_1 are separated significantly more.

Also, Figs. 23.6, 23.7, 23.8, and 23.9 of the NPV/NOI dependence on l_1 can be plotted for fixed values of the debt cost k_d and two values of the leverage level $L = 1$ and $L = 3$.

One can see, that the rate of decrease of the ratio NPV/NOI decreases with increasing of the leverage level L.

23.5.1.2 The Dependence of NPV (in Units of NOI) ($\frac{NPV}{NOI}$) on Leverage Ratio on Debt l_1 at Equity Cost $k_0 = 0.14$

L = 1

L = 3

Based on the obtained data, we plot the dependences of NPV/NOI on l_1 at $k_0 = 14\%$, different values of debt cost k_d and at two different leverage levels $L = 1$ and $L = 3$ in Figs. 23.10 and 23.11.

From Tables 23.7 and 23.8 and Figs. 23.10 and 23.11 one can come to conclusion, that the NPV (in units of NOI) (NPV/NOI) decreases with increasing of the leverage ratio on debt l_1. With the increasing of the cost of debt capital k_d, curves of the dependence of NPV/NOI (l_1), outgoing from a single point at a zero value of l_1, fall below (i.e., the rate of decrease grows).

Table 23.5 The dependence of NPV (in units of NOI) ($\frac{NPV}{NOI}$) on leverage ratio on debt l_1 at equity cost $k_0 = 0.12$, $k_d = 4\%$, 6%, 8% and 10% and $L = 1$

l_1	0	1	2	3	4	5	6	7	8	9	10
$k_d = 0.10$	7.407	5.667	3.926	2.185	0.444	−1.296	−3.037	−4.778	−6.519	−8.259	−10
$k_d = 0.08$	7.407	5.815	4.222	2.63	1.037	−0.556	−2.148	−3.741	−5.333	−6.926	−8.519
$k_d = 0.06$	7.407	5.963	4.519	3.074	1.63	0.185	−1.259	−2.704	−4.148	−5.593	−7.037
$k_d = 0.04$	7.407	6.111	4.815	3.519	2.222	0.926	−0.37	−1.667	−2.963	−4.259	−5.556

Table 23.6 The dependence of NPV (in units of NOI) ($\frac{NPV}{NOI}$) on leverage ratio on debt l_1 at equity cost $k_0 = 0.12$, $k_d = 4\%$, 6%, 8% and 10% and $L = 3$

l_1	0	1	2	3	4	5	6	7	8	9	10
$k_d = 0.10$	7.843	6.725	5.608	4.49	3.373	2.255	1.137	0.02	−1.098	−2.216	−3.333
$k_d = 0.08$	7.843	6.882	5.922	4.961	4	3.039	2.078	1.118	0.157	−0.804	−1.765
$k_d = 0.06$	7.843	7.039	6.235	5.431	4.627	3.824	3.02	2.216	1.412	0.608	−0.196
$k_d = 0.04$	7.843	7.196	6.549	5.902	5.255	4.608	3.961	3.314	2.667	2.02	1.373

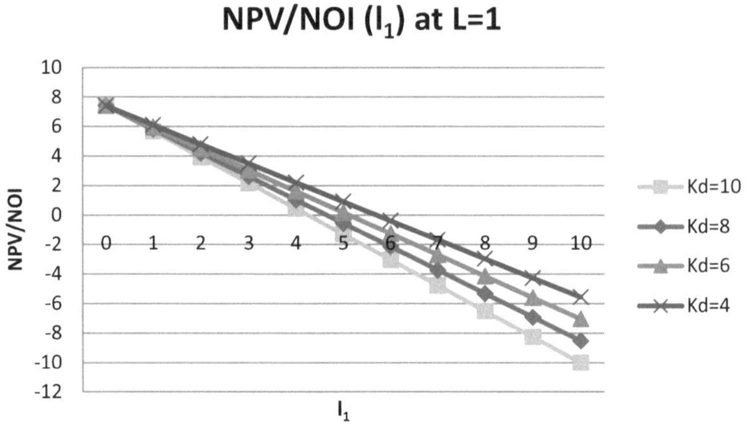

Fig. 23.4 The dependence of NPV (in units of D) on leverage ratio on debt l_1 at $k_0 = 12\%$, $k_d = 4\%$, 6%, 8% and 10% and L = 1

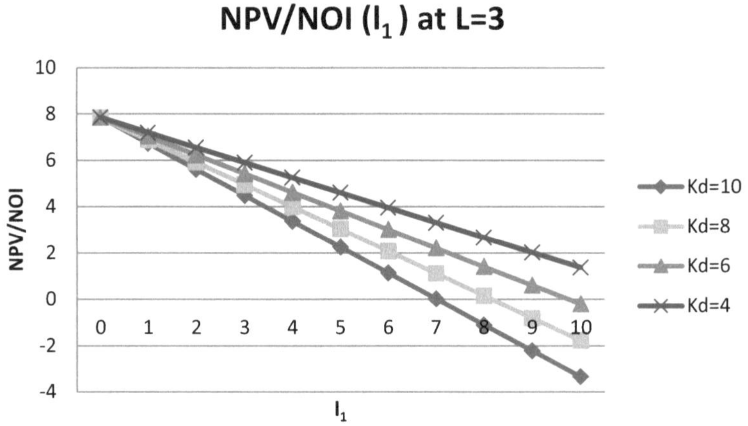

Fig. 23.5 The dependence of NPV (in units of D) on leverage ratio on debt l_1 at $k_0 = 12\%$, $k_d = 4\%$, 6%, 8% and 10% and L = 3

23.5.1.3 The Dependence of NPV (in Units of NOI) ($\frac{NPV}{NOI}$) on Leverage Ratio on Debt l_1 at Equity Cost $k_0 = 0.26$

The formula of Modigliani and Miller in Excel will look like:

$$= (-A3/C3) + (((1 - (E3^* A3))^* (1 - B3))/(D3^* (1 - ((C3^* B3)/(1 + C3)))))$$

Using this formula we calculate the dependence of NPV (in units of NOI) ($\frac{NPV}{NOI}$) on leverage ratio on debt l_1 at equity cost $k_0 = 0.26$, at different values of $k_d = 22\%$, 16%, 10%, 6% and at two values of leverage level L = 1 and L = 3.

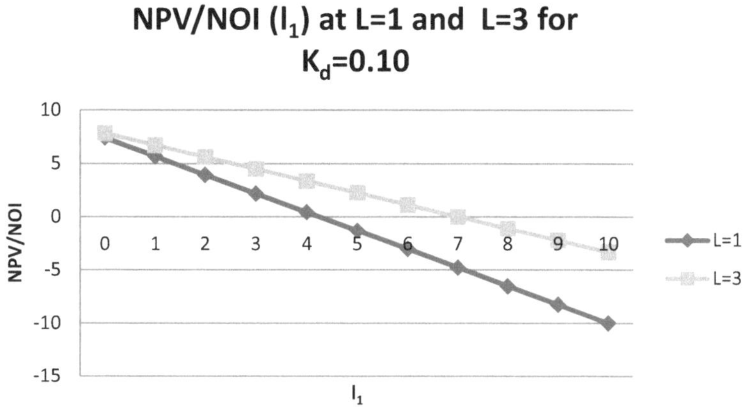

Fig. 23.6 The dependence of NPV (in units of NOI) on leverage ratio on debt l_1 at $k_0 = 12\%$, $k_d = 10\%$ and $L = 1$ and $L = 3$

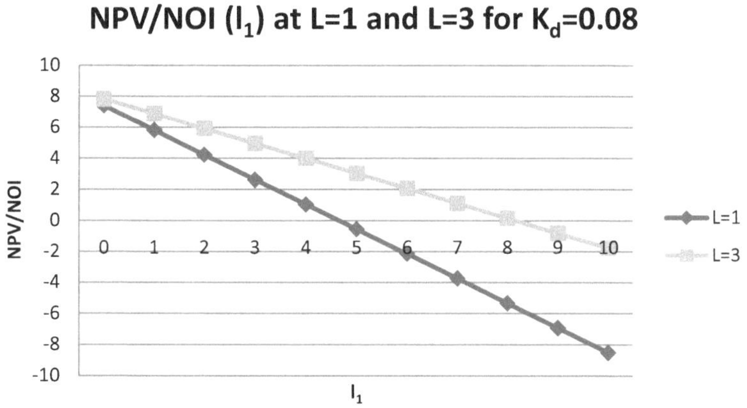

Fig. 23.7 The dependence of NPV (in units of NOI) on leverage ratio on debt l_1 at $k_0 = 12\%$, $k_d = 8\%$ and $L = 1$ and $L = 3$

Let us start from the case $L = 1$ (Tables 23.9 and 23.10, Fig. 23.12).

Let us calculate the value of l_1 below which the investment project remains effective (NPV > 0)

k_d	0.22	0.16	0.1	0.06
l_1	1.9	2.2	2.5	2.7

One can see from this Table that the value of l_1 below which the investment project remains effective (NPV > 0) decreases with credit rate k_d, that means that effectiveness of the investment project as well as its creditworthiness decreases with credit rate k_d (Fig. 23.13).

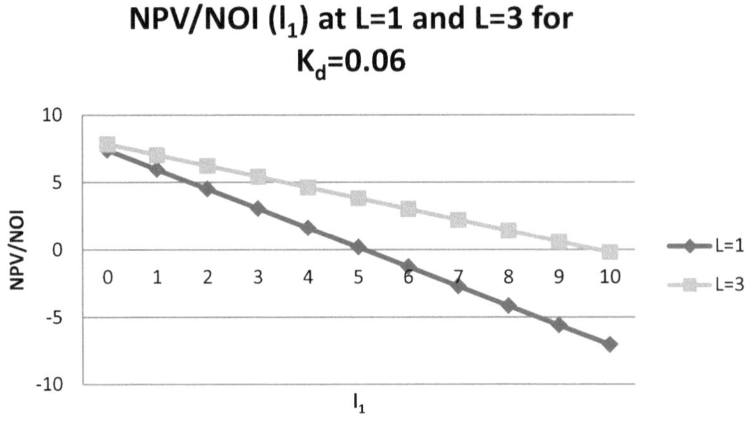

Fig. 23.8 The dependence of NPV (in units of NOI) on leverage ratio on debt l_1 at $k_0 = 12\%$, $k_d = 6\%$ and L = 1 and L = 3

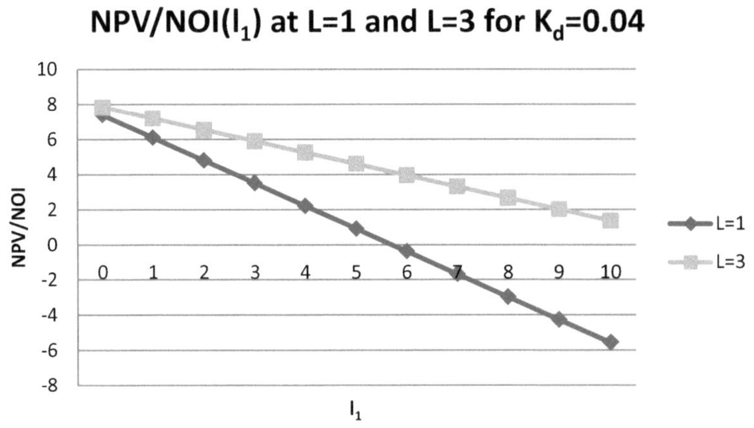

Fig. 23.9 The dependence of NPV (in units of NOI) on leverage ratio on debt l_1 at $k_0 = 12\%$, $k_d = 4\%$ and L = 1 and L = 3

Let us calculate the value of l_1 below which the investment project remains effective (NPV > 0)

k_d	0.22	0.16	0.1	0.06
l_1	3.85	4	5.6	6.6

One can see from this Table that like the case of L = 1 the value of l_1 below which the investment project remains effective (NPV > 0) decreases with credit rate k_d, that means that effectiveness of the investment project as well as its creditworthiness decreases with credit rate k_d. Comparing the case of L = 1 one can see that at bigger leverage level (L = 3) the investment project remains effective (NPV > 0) until

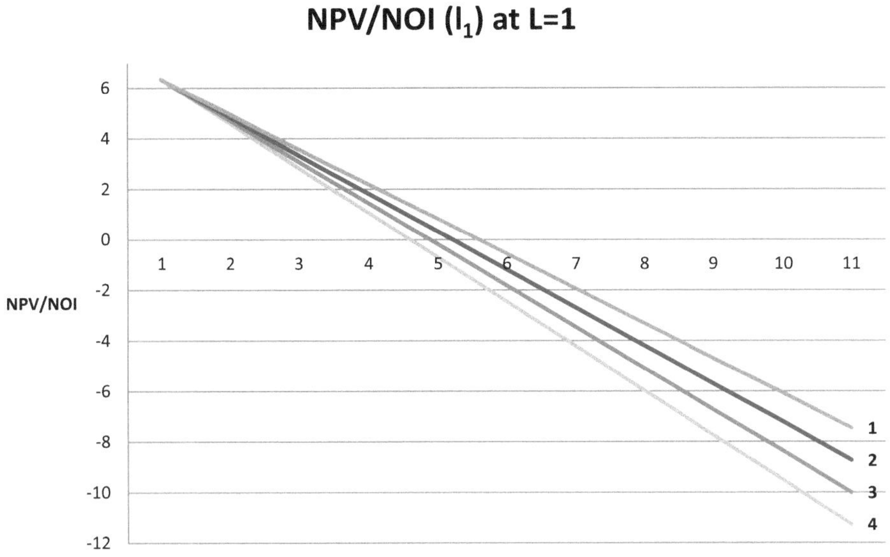

Fig. 23.10 The dependence of NPV (in units of NOI) on leverage ratio on debt l_1 at $k_0 = 14\%$, $k_d = 6\%$–(1), 8%–(2), 10%–(3), 12%–(4) at $L = 1$

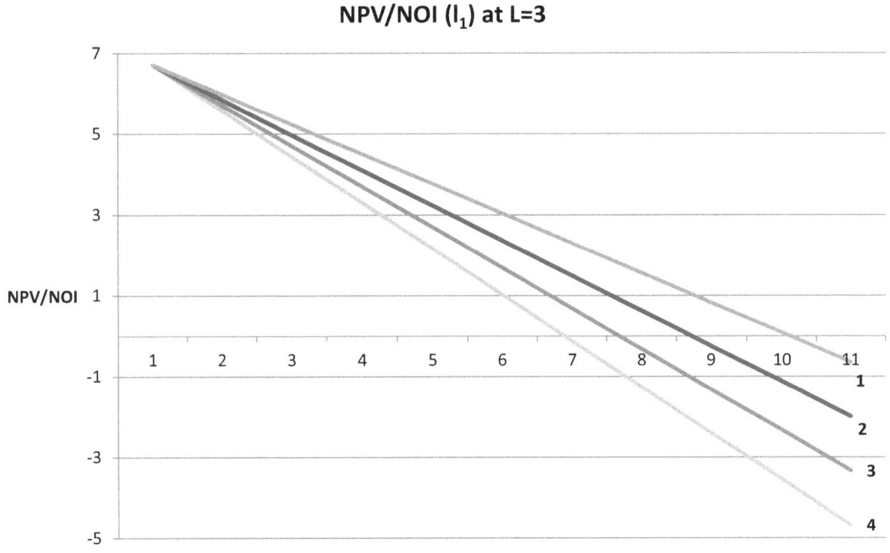

Fig. 23.11 The dependence of NPV (in units of NOI) on leverage ratio on debt l_1 at $k_0 = 14\%$, $k_d = 6\%$–(1), 8%–(2), 10%–(3), 12%–(4) at $L = 3$

Table 23.7 The dependence of NPV (in units of NOI) ($\frac{NPV}{NOI}$) on leverage ratio on debt l_1 at equity cost $k_0 = 0.14$, $k_d = 6\%, 8\%, 10\%, 12\%$ and L = 1

l_1	L	k_0	t	NPV/NOI ($k_d = 0.12$)	NPV/NOI ($k_d = 0.1$)	NPV/NOI ($k_d = 0.08$)	NPV/NOI ($k_d = 0.06$)
0	1	0.14	0.2	6.349206349	6.349206349	6.349206349	6.349206349
1	1	0.14	0.2	4.587301587	4.714285714	4.841269841	4.968253968
2	1	0.14	0.2	2.825396825	3.079365079	3.333333333	3.587301587
3	1	0.14	0.2	1.063492063	1.444444444	1.825396825	2.206349206
4	1	0.14	0.2	−0.698412698	−0.19047619	0.317460317	0.825396825
5	1	0.14	0.2	−2.46031746	−1.825396825	−1.19047619	−0.555555556
6	1	0.14	0.2	−4.222222222	−3.46031746	−2.698412698	−1.936507937
7	1	0.14	0.2	−5.984126984	−5.095238095	−4.206349206	−3.317460317
8	1	0.14	0.2	−7.746031746	−6.73015873	−5.714285714	−4.698412698
9	1	0.14	0.2	−9.507936508	−8.365079365	−7.222222222	−6.079365079
10	1	0.14	0.2	−18.26984127	−10	−8.73015873	−7.46031746

Table 23.8 The dependence of NPV (in units of NOI) ($\frac{NPV}{NOI}$) on leverage ratio on debt l_1 at equity cost $k_0 = 0.14$, $k_d = 6\%, 8\%, 10\%, 12\%$ and L = 3

l_1	L	k_0	t	NPV/NOI $k_d = 0.12$	NPV/NOI $k_d = 0.1$	NPV/NOI $k_d = 0.08$	NPV/NOI $k_d = 0.06$
0	3	0.14	0.2	6.722689	6.722689	6.722689	6.722689
1	3	0.14	0.2	5.582633	5.717087	5.851541	5.985994
2	3	0.14	0.2	4.442577	4.711485	4.980392	5.2493
3	3	0.14	0.2	3.302521	3.705882	4.109244	4.512605
4	3	0.14	0.2	2.162465	2.70028	3.238095	3.77591
5	3	0.14	0.2	1.022409	1.694678	2.366947	3.039216
6	3	0.14	0.2	−0.11765	0.689076	1.495798	2.302521
7	3	0.14	0.2	−1.2577	−0.31653	0.62465	1.565826
8	3	0.14	0.2	−2.39776	−1.32213	−0.2465	0.829132
9	3	0.14	0.2	−3.53782	−2.32773	−1.11765	0.092437
10	3	0.14	0.2	−4.67787	−3.33333	−1.9888	−0.64426

bigger leverage ratio l_1, so bigger leverage level favors to the effectiveness of the investment project as well as its creditworthiness.

Let us analyze also the dependence of NPV (in units of NOI) ($\frac{NPV}{NOI}$) on leverage ratio on debt l_1 at equity cost $k_0 = 0.26$ and each value of k_d at two leverage levels L = 1 and L = 3 (Figs. 23.14, 23.15, 23.16, 23.17).

Studying the dependence of NPV (in units of NOI) ($\frac{NPV}{NOI}$) on leverage ratio on debt l_1 at equity cost $k_0 = 0.26$ and each value of k_d at two leverage levels L = 1 and L = 3 shows that the curve $\frac{NPV}{NOI}(l_1)$ corresponding to bigger leverage level (L = 3) lies above the curve $\frac{NPV}{NOI}(l_1)$ corresponding to smaller leverage level (L = 1). The curve $\frac{NPV}{NOI}(l_1)$ corresponding to bigger leverage level (L = 3) has smaller (negative) slope. This means that debt financing of long-term projects favors effectiveness of the investment project as well as its creditworthiness.

Table 23.9 The dependence of NPV (in units of NOI) ($\frac{NPV}{NOI}$) on leverage ratio on debt l_1 at equity cost $k_0 = 0.26$, $k_d = 22\%$, 16%, 10%, 6% and L = 1

l_1	NPV/NOI(l_1) $k_d = 0.22$	NPV/NOI(l_1) $k_d = 0.16$	NPV/NOI(l_1) $k_d = 0.1$	NPV/NOI(l_1) $k_d = 0.06$
0	3.418803419	3.4188034	3.41880342	3.4188034
1	1.666666667	1.8717949	2.07692308	2.2136752
2	−0.08547009	0.3247863	0.73504274	1.008547
3	−1.83760684	−1.2222222	−0.60683761	−0.196581
4	−3.58974359	−2.7692308	−1.94871795	−1.401709
5	−5.34188034	−4.3162393	−3.29059829	−2.606838
6	−7.09401709	−5.8632479	−4.63247863	−3.811966
7	−8.84615385	−7.4102564	−5.97435897	−5.017094
8	−10.5982906	−8.957265	−7.31623932	−6.222222
9	−12.3504274	−10.504274	−8.65811966	−7.42735
10	−14.1025641	−12.051282	−10	−8.632479

Table 23.10 The dependence of NPV (in units of NOI) ($\frac{NPV}{NOI}$) on leverage ratio on debt l_1 at equity cost $k_0 = 0.26$, $k_d = 22\%$, 16%, 10%, 6% and L = 3

l_1	NPV/NOI(l_1) $k_d = 0.22$	NPV/NOI(l_1) $k_d = 0.16$	NPV/NOI(l_1) $k_d = 0.1$	NPV/NOI(l_1) $k_d = 0.06$
0	3.619909502	3.6199095	3.6199095	3.6199095
1	2.490196078	2.7073906	2.92458522	3.0693816
2	1.360482655	1.7948718	2.22926094	2.5188537
3	0.230769231	0.8823529	1.53393665	1.9683258
4	−0.89894419	−0.0301659	0.83861237	1.4177979
5	−2.02865762	−0.9426848	0.14328808	0.86727
6	−3.15837104	−1.8552036	−0.5520362	0.3167421
7	−4.28808446	−2.7677225	−1.24736048	−0.233786
8	−5.41779789	−3.6802413	−1.94268477	−0.784314
9	−6.54751131	−4.5927602	−2.63800905	−1.334842
10	−7.67722474	−5.505279	−3.33333333	−1.88537

Analyzing the obtained results one should remember, that *NPV* (in units of *NOI*) ($\frac{NPV}{NOI}$) depends only on equity cost k_0, on credit rates k_d, on leverage level L as well as on one of the leverage ratios l_j and does not depend on equity value *S*, debt value *D* and *NOI*. This means that obtained results on the dependence of NPV (in units of *NOI*) ($\frac{NPV}{NOI}$) on leverage ratios l_j at different equity costs k_0, at different credit rates k_d, at different leverage levels L carry the universal character: these dependencies remain valid for investment projects with any equity value *S*, debt value *D* and *NOI*.

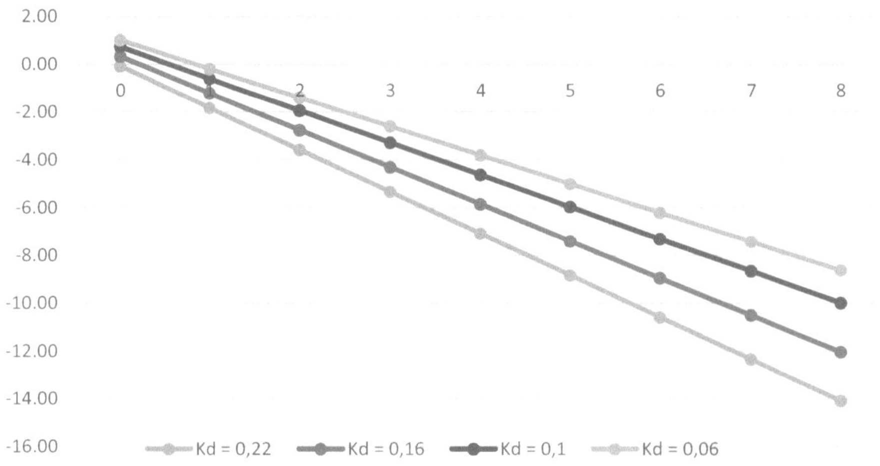

Fig. 23.12 The dependence of NPV (in units of NOI) ($\frac{NPV}{NOI}$) on leverage ratio on debt l_1 at equity cost $k_0 = 0.26$, $k_d = 22\%$, 16%, 10%, 6% and L = 1

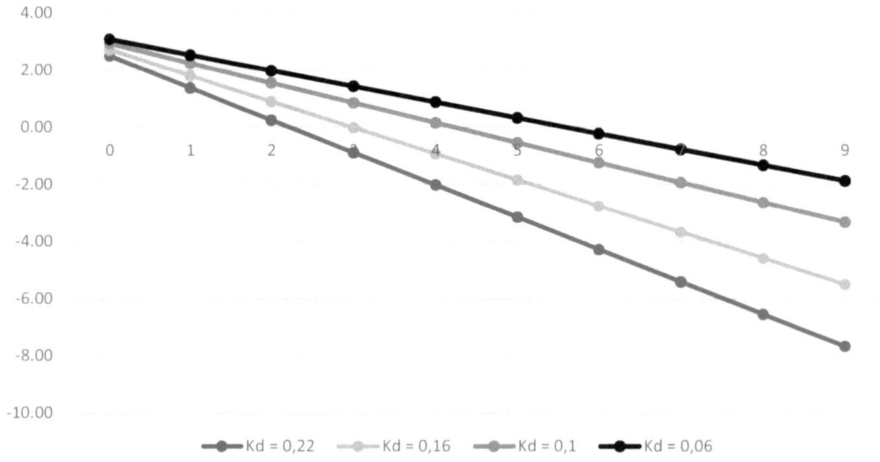

Fig. 23.13 The dependence of NPV (in units of NOI) ($\frac{NPV}{NOI}$) on leverage ratio on debt l_1 at equity cost $k_0 = 0.26$, $k_d = 22\%$, 16%, 10%, 6% and L = 3

23.6 Conclusions

This Chapter continues create a new approach to rating methodology: in addition to two previous Chapters (Chaps. 21 and 22), which have considered the creditworthiness of the non-finance issuers (Filotova et al., 2018; Brusov et al., 2018b), we develop here a new approach to project rating. We work within investment models, created by authors. One of them describes the effectiveness of investment project from perspective of equity capital owners, while other model describes the

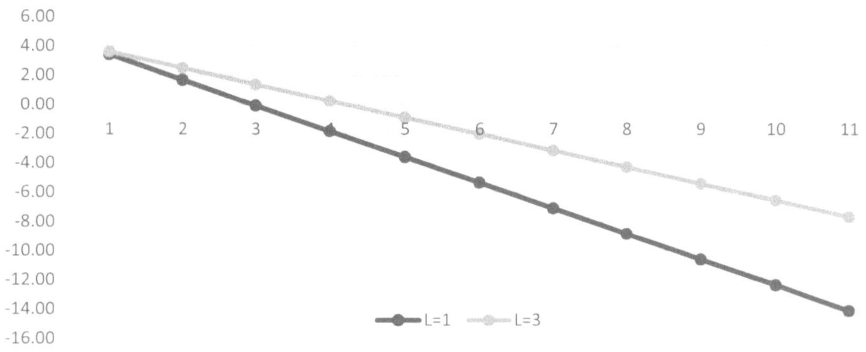

Fig. 23.14 The dependence of NPV (in units of NOI) ($\frac{NPV}{NOI}$) on leverage ratio on debt l_1 at equity cost $k_0 = 0.26$, $k_d = 22$ and $L = 1$ and $L = 3$

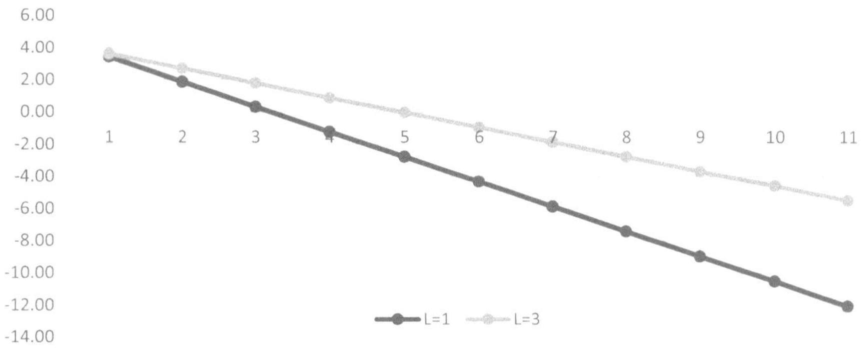

Fig. 23.15 The dependence of NPV (in units of NOI) ($\frac{NPV}{NOI}$) on leverage ratio on debt l_1 at equity cost $k_0 = 0.26$, $k_d = 16\%$ and $L = 1$ and $L = 3$

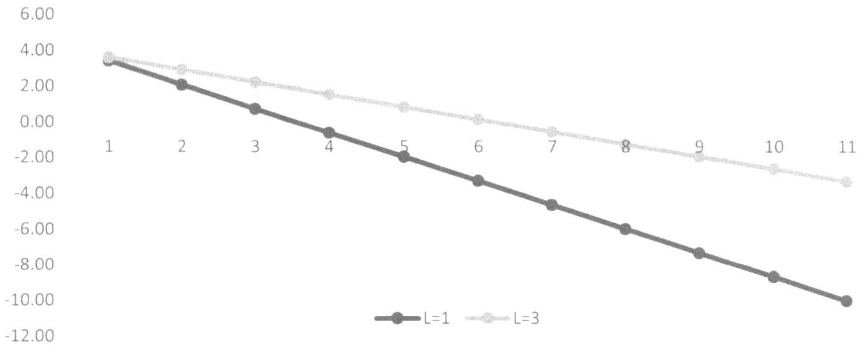

Fig. 23.16 The dependence of NPV (in units of NOI) ($\frac{NPV}{NOI}$) on leverage ratio on debt l_1 at equity cost $k_0 = 0.26$, $k_d = 10\%$ and $L = 1$ and $L = 3$

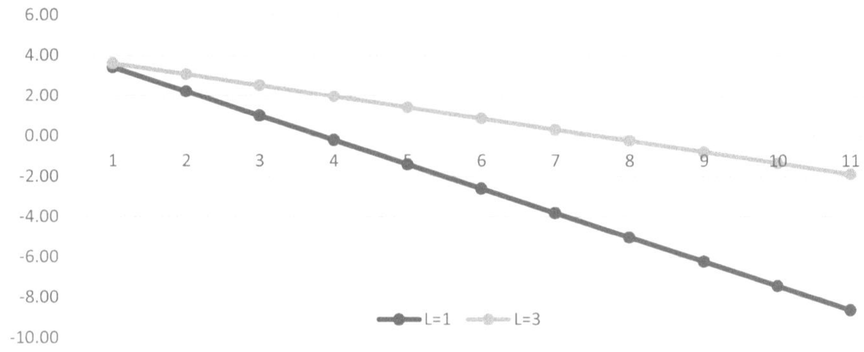

Fig. 23.17 The dependence of NPV (in units of NOI) ($\frac{NPV}{NOI}$) on leverage ratio on debt l_1 at equity cost $k_0 = 0.26$, $k_d = 6\%$ and L = 1 and L = 3

effectiveness of investment project from perspective of equity capital and debt capital owners.

The important features of current consideration as well as in previous studies are:

1. The adequate use of discounting of financial flows virtually not used in existing rating methodologies,
2. The incorporation of rating parameters (financial "ratios"), used in project rating, into considered modern investment models.

Analyzing within these investment models with incorporated rating parameters the dependence of *NPV* on rating parameters (financial "ratios") at different values of equity cost k_0, at different values of credit rates k_d as well as at different values of on leverage level *L* we come to very important conclusion, that *NPV* in units of *NOI* ($\frac{NPV}{NOI}$) (as well as *NPV* in units of *D* ($\frac{NPV}{D}$)) depends only on equity cost k_0, on credit rates k_d, on leverage level L as well as on one of the leverage ratios l_j (on one of the coverage ratios i_j) and does not depend on equity value *S*, debt value *D* and *NOI*. This means that obtained results on the dependence of NPV (in units of *NOI*) ($\frac{NPV}{NOI}$) on leverage ratios l_j (as well as on the dependence of *NPV* (in units of *D*) ($\frac{NPV}{D}$) on coverage ratios i_j) at different equity costs k_0, at different credit rates k_d, at different leverage levels L carry the universal character: these dependencies remain valid for investment projects with any equity value *S*, debt value *D* and *NOI*.

Calculations on dependence of *NPV* in units of *D* (*NPV/D*) on the coverage ratio on debt i_1 show, that $\frac{NPV}{D}$ increases with i_1 and that $\frac{NPV}{D}$ values turn out to be very closed each other at all i_1 values. It is seen as well that NPV increases with decreasing k_d. This means that effectiveness of the investment project as well as its creditworthiness decreases with credit rate k_d. One can see, that all NPV (i_1) curves corresponding to L = 3 lie above the curves corresponding to L = 1. This means that NPV increases with leverage level L (with increasing of the debt financing). Thus, debt financing favors to effectiveness of the long-term project. At fixed value L NPV increases with decreasing the credit rate k_d.

It is shown the value of the coverage ratio on debt i_1 above which the investment project remains effective (NPV > 0) increases with credit rate k_d, that means that effectiveness of the investment project as well as its creditworthiness decreases with credit rate k_d. Comparing the cases of L = 1 and L = 3 one can see that at bigger leverage level (L = 3) the investment project becomes effective (NPV > 0) starting from smaller coverage ratio i_1, so bigger leverage level favors to the effectiveness of the investment project as well as its creditworthiness.

Calculations on dependence of *NPV* in units of *NOI (NPV/NOI)* on the leverage ratio on debt l_1 show that *NPV* in units of *NOI* decreases with increasing of the leverage ratio on debt l_1. With the increasing of the cost of debt capital k_d, curves of the dependence of NPV/NOI (l_1), outgoing from a single point at a zero value of l_1, lie below (i.e., the rate of decrease (or negative slope of curves) grows). Note, that while the dependences of NPV (in units of D) on coverage ratio on debt i_1 lie very close each other, the dependences of NPV (in units of NOI) on leverage ratio on debt l_1 are separated significantly more.

One can see that the value of l_1 below which the investment project remains effective (NPV > 0) decreases with credit rate k_d, that means that effectiveness of the investment project as well as its creditworthiness decreases with credit rate k_d.

Studying the dependence of NPV (in units of NOI) ($\frac{NPV}{NOI}$) on leverage ratio on debt l_1 at fixed equity cost k_0 and fixed credit rate k_d at two leverage levels L = 1 and L = 3 it was shown that the curve $\frac{NPV}{NOI}$ (l_1) corresponding to bigger leverage level (L = 3) lies above the curve $\frac{NPV}{NOI}$ (l_1) corresponding to smaller leverage level (L = 1). The curve $\frac{NPV}{NOI}$ (l_1) corresponding to bigger leverage level (L = 3) has smaller (negative) slope. This means that debt financing of long-term projects favors effectiveness of the investment project as well as its creditworthiness.

Investigations, conducting in current Chapter, creates a new approach to rating methodology with respect to the long-term project rating. And this Chapter in combine with two our previous papers on this topic (Filotova et al., 2018; Brusov et al., 2018b) creates a new base for rating methodology in whole.

In our future papers we will consider rating methodology for investment projects of arbitrary duration.

References

Brusov, P. (2018a). Editorial: Introduction on special issue on the banking system and financial markets of Russia and other countries: Problems and prospects. *Journal of Reviews on Global Economics, 7*, i–vi.

Brusov, P. (2018b). Editorial. *Journal of Reviews on Global Economics, 7*, i–v.

Brusov, P., Filatova, T., Orehova, N., & Eskindarov, M. (2018a). *Modern corporate finance, investments, taxation and ratings* (monograph) (571 p). Springer.

Brusov, P. N., Filatova, T. V., Orekhova, N. P., & Kulik, V. L. (2018b). Rating: New approach. *Journal of Reviews on Global Economics, 7*, 37–62. https://doi.org/10.6000/1929-7092.2023.07.05

Brusov, P. N., Filatova, T. V., Orekhova, N. P., & Kulik, V. L. (2018c). A "golden age" of the companies: Conditions of its existence. *Journal of Reviews on Global Economics, 7*, 88–103. https://doi.org/10.6000/1929-7092.2023.07.07

Brusov, P. N., Filatova, T. V., Orekhova, N. P., & Kulik, V. L. (2018d). Rating methodology: New look and new horizons. *Journal of Reviews on Global Economics, 7*, 63–87. https://doi.org/10.6000/1929-7092.2023.07.06

Brusov, P. N., Filatova, T. V., Orekhova, N. P., Kulik, V. L., & Weil, I. (2018e). New meaningful effects in modern capital structure theory. *Journal of Reviews on Global Economics, 7*, 104–122. https://doi.org/10.6000/1929-7092.2023.07.08

Brusov, P. N., Filatova, T. V., & Orekhova, N. P. (2018f). *Modern corporate finance and investments* (monograph) (517 p). Knorus Publishing House.

Brusov, P., Filatova, T., Orekhova, N., Kulik, V., Weil, I., & Brailov, A. (2018g). The impact of the central bank key rate and commercial banks credit rates on creating and maintaining of a favorable investment climate in the country. *Journal of Reviews on Global Economics, 7*, 360–376. https://doi.org/10.6000/1929-7092.2023.07.31

Brusov, P. N., Filatova, T. V., Orekhova, N. P., Kulik, V. L., & Weil, I. (2019). Ratings of the investment projects of arbitrary durations: New methodology. *Journal of Reviews on Global Economics, 8*, 437–448. https://doi.org/10.6000/1929-7092.2019.08.37

Brusov, P., Filatova, T., & Orehova, N. (2020a). *Ratings: Critical analysis and new approaches of quantitative and qualitative methodology* (monograph) (369 p). Springer Nature. Retrieved from https://www.springer.com/de/book/9783030562427

Brusov, P. N., Filatova, T. V., Orekhova, N. P., Kulik, V. L., Chang, S.-I., & Lin, Y. C. G. (2020b). Modification of the Modigliani–Miller theory for the case of advance payments of tax on profit. *Journal of Reviews on Global Economics, 9*, 257–268.

Filatova, T., Orehova, N., & Brusova, A. (2008). Weighted average cost of capital in the theory of Modigliani–Miller, modified for a finite life-time company. *Bulletin of the FU, 48*, 68–77.

Filatova, T. V., Brusov, P. N., Orekhova, N. P., & Kulik, V. L. (2018). Ratings of the long-term projects: New approach. *Journal of Reviews on Global Economics, 7*, 645–661. https://doi.org/10.6000/1929-7092.2023.07.59

Modigliani, F., & Miller, M. (1958). The cost of capital, corporate finance, and the theory of investment. *The American Economic Review, 48*, 261–297.

Modigliani, F., & Miller, M. (1963). Corporate income taxes and the cost of capital: A correction. *The American Economic Review, 53*, 147–175.

Modigliani, F., & Miller, M. (1966). Some estimates of the cost of capital to the electric utility industry 1954–1957. *The American Economic Review, 56*, 333–391.

Chapter 24
Rating Methodology: New Look and New Horizons

24.1 Introduction

In a previous article we have offered fundamentally new approach to rating methodology, which includes adequate application of discounting of financial flows virtually not used in existing rating methodologies. The incorporation of rating parameters (financial "ratios") into the perpetuity limit of modern theory of capital structure by Brusov–Filatova–Orekhova (BFO) theory has been done: it required a modification of perpetuity limit of BFO theory for rating needs. Two models (one–period and multi–period) for accounting of discounting of financial flows were discussed. An algorithm of valuation of discount rate, accounting rating ratios has been suggested. We discussed also the interplay between rating ratios and leverage level which can be quite important in rating.

As we discussed in a number of works (Brusov et al., 2015, Brusov & Filatova (2011), Brusov et al. (2011a), Brusov et al. (2011b), Brusov et al. (2011c), Brusov et al. (2012a), Brusov et al. (2012b), Brusov et al. (2012c), Brusov et al. (2013a), Brusov et al. (2013b), Brusov et al. (2014a), Brusov et al. (2014b), Filatova et al. (2008)) perpetuity limit of BFO theory–Modigliani–Miller theory (Modigliani & Miller (1958), Modigliani & Miller (1963), Modigliani & Miller (1966))–underestimated the assessment of the attracting capital cost and therefore overestimated the assessment of the capitalization of the company. Besides the time factor, which is very important, does not exist in the perpetuity limit. And therefore in this limit there is no concept of the age of the company, and their lifetime is infinite (perpetuity).

In the present work the generalization of the developed by us approach for the case of modern theory of capital structure and capital cost by Brusov–Filatova–Orekhova (BFO theory) for companies and corporations an arbitrary age, i.e., for general case of BFO theory.

This has required the modification of the BFO theory for the rating needs (much more complicate then it was done in case of perpetuity limit—Modigliani–Miller

P. Brusov, T. Filatova, *Efficiency of Investment Projects*, Contributions to Finance and Accounting, https://doi.org/10.1007/978-3-032-01028-5_24

theory), as used in financial management the concept of "leverage" as the ratio of debt value to the equity value substantially differs from the concept of "leverage" in the rating, where it is understood as the direct and inverse ratio of the debt value to the generated cash flow values (income, profit, etc.). We introduce here some additional ratios, allowing more fully characterize the issuer's ability to repay debts and to pay interest thereon.

As we mentioned in the previous paper, the bridge is building between the discount rates (*WACC*, *ke*) used when discounting of financial flows, and "ratios" in the rating methodology. The algorithm for finding the discount rates for given ratio values is developed.

Application of BFO theory modified for rating purposes allow adequately produce the discounting of financial flows by using the correct discount rates with taken into account when discounting the magnitude of rating ratios, and take into account the time factor missing in perpetuity limit and being the vital, i.e. to take into account the company age (in BFO–I theory) or the company life–time (in the BFO–II theory).

24.2 The Analysis of Methodological and Systemic Deficiencies in the Existing Credit Rating of Non-Financial Issuers

The analysis of methodological and systemic deficiencies in the existing credit rating of non–financial issuers has been conducted by us. We have analyzed the methodology of the big three (Standard & poor's, Fitch and Moody's) and Russian national rating agency.

24.2.1 The Closeness of the Rating Agencies

The closeness of the rating agencies has been discussed by us in a previous paper (Brusov et al., 2018) and is caused by multiple causes.

1. The desire to preserve their "know how". Rating agencies get big enough money for generated ratings (mostly from issuers) to replicate its methodology.
2. The desire to avoid public discussion of the ratings with anyone, including the issuer. It is very convenient position—rating agency "a priori" removes himself from beneath any criticism of generated ratings.
3. The absence of any external control and external analysis of the methodologies is resulted in the fact that shortcomings of methodologies are not subjected to serious critical analysis and stored long enough.

24.2.2 *Discounting*

One of the major flaws of all existing rating methodologies is a failure or a very narrow use of discounting. But even in those rare cases where it is used, it is not quite correct, since the discount rate when discounting financial flows is chosen incorrectly.

The need to take into account the time factor in terms of discounting is obvious, because it is associated with the time value of money. The financial part of the rating is based on a comparison of generated income with the value of the debt and the interest payable. Because income and disbursement of debt and interest are separated in time, the use of discounting when comparing revenues with the value of debt and interest is absolutely necessary for assigning credit ratings for issuers.

1. In existing rating methodologies, despite their breadth and detail, there are a lot of shortcomings. One of the major flaws of all existing rating methodologies, as mentioned in our previous paper, is a failure or a very narrow use of discounting. But even in those rare cases where it is used, it is not quite correct, since the discount rate when discounting financial flows is chosen incorrectly.

   The need to take into account the time factor in terms of discounting is obvious, because it is associated with the time value of money. The need to take into account the time factor in terms of discounting is obvious, because it is associated with the time value of money. The financial part of the rating is based on a comparison of generated income with the value of the debt and the interest payable. Because income and disbursement of debt and interest are separated in time, the use of discounting when comparing revenues with thc value of debt and interest is absolutely necessary for assigning credit ratings for issuers.

   This raises the question about the discount rate. This question has always been one of the major and extremely difficult in many areas of finance: corporate finance, investment, it is particularly important in business valuation, where a slight change in the discount rate leads to significant changes in estimates of capitalization of the company, that is used by unscrupulous appraisers for artificial bankruptcy of companies. As well it is essential in rating, and when assigning a rating to an issuer, and forecasting.

   Therefore, as soon as we are talking about financial flows, it is necessary to account discounting, otherwise the time value of money does not take into account, i.e., any analysis of financial flows should take account of discounting.
2. When we talk about using the rating reports for the three or five (GAAP) years, assuming the behavior indicators beyond that period " a flat ", discounting must be taken into account.

24.2.3 Dividend Policy of the Company

1. Dividend policy of the company must take into account (and account) when rating, because the financial policy is taken into account in rating. However, the existing methodologies for ranking estimate only the stability of the dividend policy and do not estimate its reasonableness, how reasonable is the value of dividend payouts, how do they relate to the economically reasonable dividend values.
2. The reasonableness of dividend policy, its score is determined by comparing of the values of paid dividends with their economically reasonable value, which is the cost of equity capital k_e of the company. The calculation of k_e is a rather difficult task.
   BFO theory allows you to make the correct assessment of the value of the equity capital cost of the company and thus to compare values of the paid by the company dividend to their economically reasonable value, it allows you to assess the reasonableness of dividend policy, which is clearly linked to the creditworthiness of the issuer.
3. For example, one of the varieties " cash flow ", taking into account the amount of paid dividends (Discretionary cash flow (DCF) S & P), should be compared with the "economically reasonable dividend values, and this will affect the rating.

24.2.4 Leverage Level

Currently the rating agencies take into account the leverage level only from the perspective of assessing of financial stability and risk of bankruptcy. In fact the leverage level significantly affects the main financial indicators of the company's activity: the cost of equity capital, WACC, in other words, the cost of attracting of capital, as well as the capitalization of the company. The failure of this effect in the analysis of financial reports leads to incorrect conclusions based on it.

Evaluation (by the BFO method) of the influence of the debt financing level on the effectiveness of investment projects for different values of capital costs can be used in the rating of investment projects and investment programs of companies.

24.2.5 Taxation

1. Taxation affects the rating of the issuers. Evaluation (by the BFO method) of the influence of taxation (tax on profit organization rate) on the financial performance of the company, on the effectiveness of investment projects can be used when rating companies and their investment programs, investment projects, as well as in the context of change of tax on profits of the organization rate for forecast predictions and in analysis of country risk.

2. Evaluation (by BFO the method) of the influence of the Central Bank base rate, credit rates of commercial banks on the effectiveness of investment projects, creation of a favourable investment climate in the country can be used to forecast predictions, as well as in country risk analysis.

24.2.6 Account of the Industrial Specifics of the Issuer

Industrial specifics of the issuer in the existing rating methodologies, especially in newly established and taking into account the experience of predecessors, ignored. So in " The methodology of ACRA for assigning of credit ratings for non–financial companies on a national scale for the Russian Federation " "own creditworthiness is determined by taking into account the characteristics of the industry in which the company operates. To assess of the factor of the industry risk profile ACRA subdivides the industry into five groups according to their cyclical, barriers to entry, industry risk statistics, as well as trends and prospects.

The weight of the factor of industry risk–profile is determined individually for each group and varies depending on the level of credit risk. This creates a certain rating threshold for companies from industries with high risk and slightly rewards low risk industry".

However, the existing accounting of industry specifics of issuer is clearly insufficient. Ranking methodologies should better integrate industry peculiarities in the organization of finance of issuers. In particular, it is very important to define business needs in working capital, from the size of which financial soundness indicators, solvency and creditworthiness depend directly. The latter is the key indicator in rating.

24.2.7 Neglect of Taking into Account the Particularities of the Issuer

In existing rating methodologies the taking into account the particularities of the issuer, features of financial reports, taxation, legal and financial system is neglected in favor of achieving full comparability of financial reports, they smooth the distinctions (see Moody's rating methodologies).

24.2.8 Financial Ratios

1. A necessary and sufficient quantity and mix of financial ratios are not determined, it appears that such questions are even not raised though valuation of the financial

Table 24.1 Liquidity assessment vs assessment of funding (after ACRA)

	liquidity assessment					
Assessment of funding		**1**	**2**	**3**	**4**	**5**
	1	1	2	2	3	4
	2	1	2	3	3	4
	3	2	2	3	4	5
	4	3	3	3	4	5
	5	3	3	4	5	5

risk, the financial condition of the issuer largely depend on the quantity and quality of financial ratios, their correlation or independence.

2. Some financial ratios define ambiguously the state of the issuer. For example, the ratio of cash flow/leverage is high at high cash flow value as well as at low leverage value. The question is how these two different states of the issuer, which is attributed to one value of financial risk, is really equally relate to credit risk.
3. As recognized in the ACRA methodology "in some cases it is possible a formal hit of individual characteristics of factor/subfactor simultaneously in several categories of evaluation, particularly for qualitative factors. In this case, the score is based on expert opinion, taking into account the most important parameters".
4. In connection with paragraph 3 it should be noted that the formalization of expert opinions is one of the most important tasks in improving of the rating methodology, in making a peer review process more objective. There are a few ways to solve this problem: using results of modern theory of measurement, using of the formalism of fuzzy sets, fuzzy logic, and others.
5. Tabulate the composition of various risks, for example, CICRA (in S&P methodology) gives 6×6 matrix, which has 36 elements, i.e. generally CICRA should have 36 different values, but their total number is equal to 6. The question is how this is justified. The fact that total number is equal exactly to 6 shows that not very justified, or there are other considerations, but they must be well grounded.
   Similar examples abound. So in "The ACRA methodology for assigning of credit ratings for microfinance organizations on a national scale for the Russian Federation " Table 24.1 " Score of funding and liquidity" provides 5×5 matrix that has 25 elements, i.e. generally should be 25 different states but their total number is equal to 5. The question is whether it is justified. The fact that total number is equal exactly to 5 shows that not very justified.
6. Tabulate of mixes of different ratios in determining the financial risk has been done not quite correctly (Table 24.2):

ratios at least not completely correlated but used as fully correlated. So, one can see that the two lines (Table 24.3).

do not allow mixing between parameters of lines, although such mixing can occur, for example (Table 24.4),

All these points are limiting the applicability of rating agencies methods. They were introduced by the rating agencies for the purpose of simplifying of the procedure of ranking (with or without understanding), and with a view of unification

Table 24.2 Tabulate of mixes of different ratios in determining the financial risk

	FFO/ debt (%)	Debt/ EBITDA (x)	FFO/cash interest(x)	EBITDA/ interest(x)	CFO/ debt (%)	FOCF/ deb(%)	DCF/ debt (%)
Minimal	60+	Less than 1.5	More than 13	More than 15	More than 50	40+	25+
Modest	45–60	1.5–2	9–13	10–15	35–50	25–40	15–25
Intermediate	30–45	2–3	6–9	6–10	25–35	15–25	10–15
Significant	20–30	3–4	4–6	3–6	15–25	10–15	5–10
Aggressive	12–20	4–5	2–4	2–3	10–15	5–10	2–5
Highly leveraged	Less than 12	Greater than 5	Less than 2	Less than 2	Less than 10	Less than 5	Less than 2

Table 24.3 Tabulate of mixes of different ratios in determining the financial risk (two first lines from Table 24.2)

Minimal	60+	Less than 1.5	More than 13	More than 15	More than 50	40+	25+
Modest	45–60	1.5–2	9–13	10–15	35–50	25–40	15–25

Table 24.4 Tabulate of mixes of different ratios in determining the financial risk (first line from Table 24.2)

60+	1.5–2	More than 13	More than 15	More than 50	40+	25+

of methods to different reporting systems, different countries, with the objective of comparability of results.

Mentioned ambiguity of evaluations already occurred when S&P has assigned a rating to Gazprom.

24.3 Modification of the BFO Theory for Companies and Corporations of Arbitrary Age for Purposes of Ranking

We will conduct below the modification of the BFO theory for companies and corporations of arbitrary age for purposes of ranking, which proved much more difficult than modification of its (BFO theory) perpetuity limit.

As it turned out, use of the famous formula BFO

$$\frac{[1-(1+?WACC)^{-n}]}{?WACC}=\frac{[1-(1+k_0)^{-n}]}{k_0[1-\omega_d T(1-(1+k_d)^{-n})]} \tag{24.1}$$

not possible, since it no longer includes cash flows *CF* and debt value *D*, and the leverage level $L = D/S$ (in the same sense as it is used in financial management) is included only through the share of leveraged $w_d = L/(L + 1)$.

For the modification of the general theory of BFO for ranking purposes, one must return to the initial assumptions under the derivation of the BFO formula.

Modigliani–Miller theorem in case of existing of corporate taxes, generalized by us for the case of finite company age, states that capitalization of leveraged company (using the debt financing), V_L, is equal to the capitalization of non–leveraged company (which does not use the debt financing), V_0, increased by the amount of the tax shield for the finite period of time, TS_n,

$$V_{\mathrm{L}} = V_0 + \mathrm{TS}_n. \tag{24.2}$$

where the capitalization of leveraged company

$$V_{\mathrm{L}} = \frac{\mathrm{CF}}{\mathrm{WACC}}(1 - (1 + ?WACC)^{-n}); \tag{24.3}$$

the capitalization of non–leveraged company

$$V_0 = \frac{CF}{k_0}(1 - (1 + k_0)^{-n}); \tag{24.4}$$

and the tax shield for the period of n years

$$TS_n = tD(1 - (1 + k_{\mathrm{d}})^{-n}). \tag{24.5}$$

Substituting Eqs. (24.3)–(24.5) into Eq. (24.2), we obtain the Eq. (24.6), which will be used by us in the future to modify the BFO theory for the needs of the ranking.

$$\frac{CF * (1 - (1 + WACC)^{-n})}{WACC} = \frac{CF}{K_0} * (1 - (1 + k_0)^{-n}) + t * D * (1 - (1 + k_d)^{-n}) \tag{24.6}$$

Below we fulfill the incorporation of rating parameters (financial "ratios") into the modern theory of capital structure (Brusov–Filatova–Orekhova (BFO) theory).

As we noted in a previous paper (Brusov et al., 2018), in quantification of the creditworthiness of the issuers the crucial role belongs to the so–called financial "ratios", constitute a direct and inverse ratios of various generated cash flows to debt values and interest ones. We could mention such ratios as DCF/Debt, FFO/Debt, CFO/Debt, FOCF/Debt, FFO/cash interest, EBITDA/interest, Interests/EBITDA, Debt/EBITDA and some others.

Let us consider two kind of rating ratios: coverage ratios and leverage ratios.

24.4 Coverage Ratios

We start from the coverage ratios and will consider three kind of coverage ratios: coverage ratios of debt, coverage ratios of interest on the credit and coverage ratios of debt and interest on the credit. Note, that last type of ratios has been introduced by us for the first time for a more complete valuation of the issuer's ability to repay debts and to pay interest thereon.

24.4.1 Coverage Ratios of Debt

Here $i_1 = CF/D$

Let us consider the coverage ratios of debt first.

Dividing the both parts of the formula (24.6) by the value of the debt D, enter the debt coverage ratio into the general BFO theory

$$i_1 = CF/D \tag{24.7}$$

$$\frac{i_1 * (1-(1+WACC)^{-n}}{WACC} = \frac{i_1 * (1-(1+k_0)^{-n}}{k_0} + t * (1-(1+k_d)^{-n}) \tag{24.8}$$

$$i_1 * A = i_1 * B + t * C \tag{24.9}$$

$$A = \frac{(1-(1+WACC)^{-n}}{WACC}; \tag{24.10}$$

$$B = \frac{(1-(1+k_0)^{-n}}{k_0}; \tag{24.11}$$

$$C = (1-(1+k_d)^{-n}); \tag{24.12}$$

This ratio (i_1) can be used to assess of the following parameters used in rating, DCF/Debt, FFO/Debt, CFO/Debt, FOCF/Debt and some others. We will use formula (24.8) to study the dependence $WACC(i_1)$ and to build a curve of this dependence.

Let us analyze the dependence of the weighted average cost of capital (WACC) on debt coverage ratio i_1. We consider the case $k_0 = 8\%$; $k_d = 4\%$; $t = 20\%$; i_1 is changed from 1 up to 10, for two company ages n = 3 (Table 24.5) and (Table 24.6) n = 5.

The dependence of company's weighted average cost of capital (WACC) on the coverage ratio on debt i_1 is shown at Figs. 24.1 and 24.2.

Table 24.5 The dependence of the weighted average cost of capital (WACC) on debt coverage ratio i_1 at n = 3

t	k_0	k_d	i_1	WACC
0,2	0,08	0,04	1	0,075356711
0,2	0,08	0,04	2	0,077705469
0,2	0,08	0,04	3	0,078412717
0,2	0,08	0,04	4	0,078808879
0,2	0,08	0,04	5	0,079046807
0,2	0,08	0,04	6	0,079205521
0,2	0,08	0,04	7	0,079318935
0,2	0,08	0,04	8	0,079404022
0,2	0,08	0,04	9	0,079470216
0,2	0,08	0,04	10	0,07952318

Table 24.6 The dependence of the weighted average cost of capital (WACC) on debt coverage ratio i_1 at n = 5

t	k_0	k_d	i_1	WACC
0,2	0,08	0,04	1	0,07663868
0,2	0,08	0,04	2	0,0783126
0,2	0,08	0,04	3	0,0788732
0,2	0,08	0,04	4	0,079154
0,2	0,08	0,04	5	0,07932264
0,2	0,08	0,04	6	0,07943518
0,2	0,08	0,04	7	0,0795156
0,2	0,08	0,04	8	0,07957594
0,2	0,08	0,04	9	0,07962287
0,2	0,08	0,04	10	0,07966043

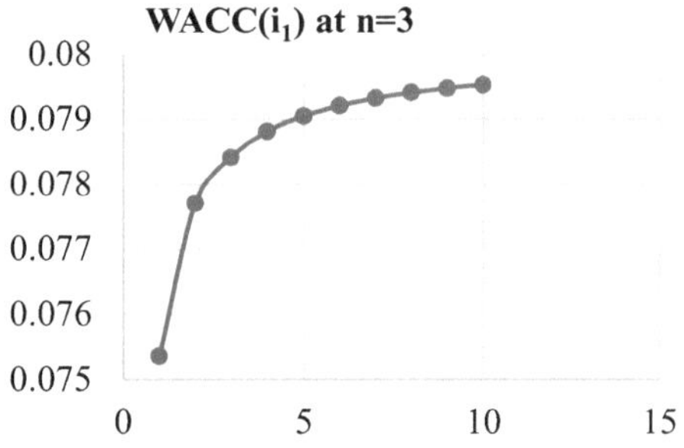

Fig. 24.1 The dependence of company's weighted average cost of capital (WACC) on the coverage ratio on debt i_1 at n = 3

24.4.2 The Coverage Ratio on Interest on the Credit

Let us analyze now the dependence of company's weighted average cost of capital (WACC) on the coverage ratio on interest on the credit i_2.

Dividing the both parts of the formula (24.6) by the value of the interest on the credit k_dD, enter the coverage ratio on interest on the credit i_2 into the general BFO theory

Fig. 24.2 The dependence of company's weighted average cost of capital (WACC) on the coverage ratio on debt i_I at n = 5

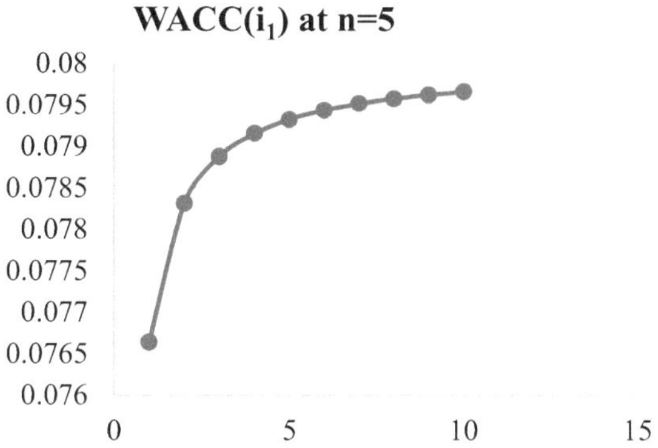

Table 24.7 The dependence of the weighted average cost of capital (WACC) on the coverage ratio on interest on the credit i_2 at n = 3

t	k_0	k_d	i_2	WACC
0,2	0,08	0,04	1	−0,021238089
0,2	0,08	0,04	2	0,02529016
0,2	0,08	0,04	3	0,042483465
0,2	0,08	0,04	4	0,051456351
0,2	0,08	0,04	5	0,056965593
0,2	0,08	0,04	6	0,060692181
0,2	0,08	0,04	7	0,063380861
0,2	0,08	0,04	8	0,065412245
0,2	0,08	0,04	9	0,067001115
0,2	0,08	0,04	10	0,068277865

$$\frac{i_2 * \left(1-(1+WACC)^{-n}\right.}{WACC} = \frac{i_2 * \left(1-(1+k_0)^{-n}\right.}{k_0} + \frac{t * \left(1-(1+k_d)^{-n}\right)}{k_d} \tag{24.13}$$

Here

$$\frac{CF}{D * k_d} = i_2 \tag{24.14}$$

$$i_2 * A = i_2 * B + \frac{t * C}{k_d} \tag{24.15}$$

The dependences of company's weighted average cost of capital (WACC) on the coverage ratio on interest on the credit i_2 for company ages n = 3 and n = 5 are presented in Tables 24.7 and 24.8.

The dependences of company's weighted average cost of capital (WACC) on the coverage ratio on interest on the credit i_2 at company ages n = 3 and n = 5 are shown at Figs. 24.3 and 24.4.

This ratio (i_2) can be used to assess of the following parameters, used in rating, FFO/cash interest, EBITDA/interest and some others. We will use last formula to build a curve of dependence $WACC(i_2)$.

Table 24.8 The dependence of the weighted average cost of capital (WACC) on the coverage ratio on interest on the credit i_2 at n = 5

t	k_0	k_d	i_2	WACC
0,2	0,08	0,04	1	0,00793717
0,2	0,08	0,04	2	0,04111354
0,2	0,08	0,04	3	0,0533843
0,2	0,08	0,04	4	0,05974575
0,2	0,08	0,04	5	0,06365738
0,2	0,08	0,04	6	0,06630611
0,2	0,08	0,04	7	0,06821315
0,2	0,08	0,04	8	0,06966377
0,2	0,08	0,04	9	0,07078076
0,2	0,08	0,04	10	0,07168658

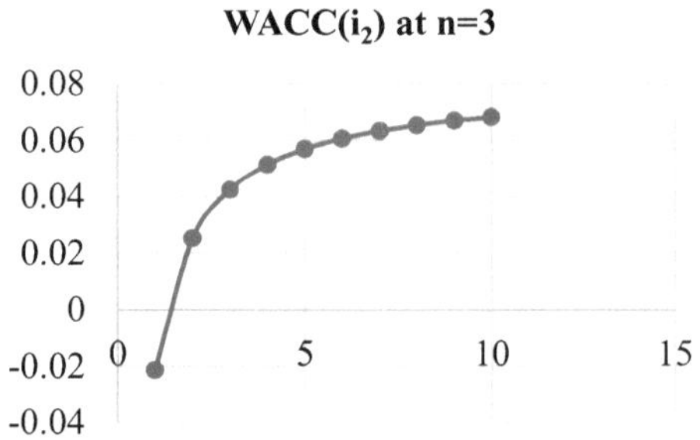

Fig. 24.3 The dependence of company's weighted average cost of capital (WACC) on the coverage ratio on interest on the credit i_2 at company age n = 3

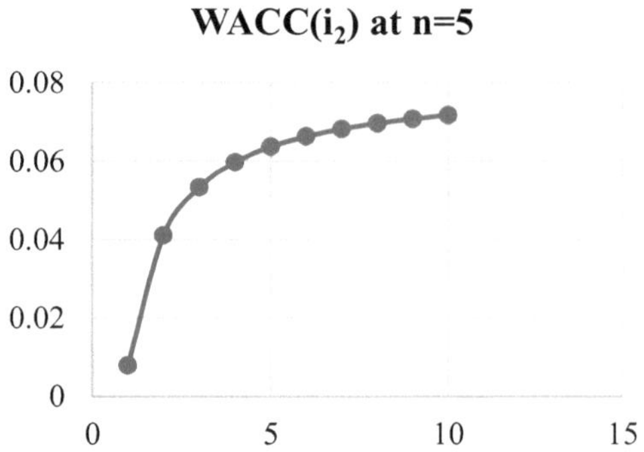

Fig. 24.4 The dependence of company's weighted average cost of capital (WACC) on the coverage ratio on interest on the credit i_2 at company age n = 5

24.4.3 Coverage Ratios of Debt and Interest on the Credit (New Ratios)

Let us now study the dependence of the company's weighted average cost of capital (WACC) on the coverage ratios of debt and interest on the credit simultaneously i_3: this is new ratio, introduced by us for the first time here for a more complete description of the issuer's ability to repay debts and to pay interest thereon.

Table 24.9 The dependence of the weighted average cost of capital (WACC) on the coverage ratio on debt and interest on the credit i_3 at n = 3

t	k_0	k_d	i_3	WACC
0,2	0,08	0,04	1	0,075536724
0,2	0,08	0,04	2	0,077796177
0,2	0,08	0,04	3	0,078473634
0,2	0,08	0,04	4	0,078854621
0,2	0,08	0,04	5	0,079083426
0,2	0,08	0,04	6	0,079236052
0,2	0,08	0,04	7	0,079345114
0,2	0,08	0,04	8	0,079426934
0,2	0,08	0,04	9	0,079490586
0,2	0,08	0,04	10	0,079541516

Table 24.10 The dependence of the weighted average cost of capital (WACC) on the coverage ratio on debt and interest on the credit i_3 at n = 5

t	k_0	k_d	i_3	WACC
0,2	0,08	0,04	1	0,07676703
0,2	0,08	0,04	2	0,07837722
0,2	0,08	0,04	3	0,07891638
0,2	0,08	0,04	4	0,07918642
0,2	0,08	0,04	5	0,07934861
0,2	0,08	0,04	6	0,07945683
0,2	0,08	0,04	7	0,07953417
0,2	0,08	0,04	8	0,07959218
0,2	0,08	0,04	9	0,07963732
0,2	0,08	0,04	10	0,07967343

Dividing the both parts of the formula (24.6) by the value of the debt and interest on the credit (1 + k_d) D, enter the coverage ratio on debt and interest on the credit i_3 into the general BFO theory

$$\frac{CF}{D*(1+k_d)} = i_3 \tag{24.16}$$

$$i_3 * A = i_3 * B + \frac{t * C}{1+k_d} \tag{24.17}$$

$$\frac{i_3 * \left(1-(1+WACC)^{-n}\right)}{WACC} = \frac{i_3 * \left(1-(1+k_0)^{-n}\right)}{k_0} + \frac{t * \left(1-(1+k_d)^{-n}\right)}{1+k_d} \tag{24.18}$$

The dependences of company weighted average cost of capital (WACC) on the coverage ratio on debt and interest on the credit i_3 at company age n = 3 and n = 5 are presented in Tables 24.9 and 24.10.

The dependences of company weighted average cost of capital (WACC) on the coverage ratio on debt and interest on the credit i_3 at company age n = 3 and n = 3 are shown at Figs. 24.5 and 24.6.

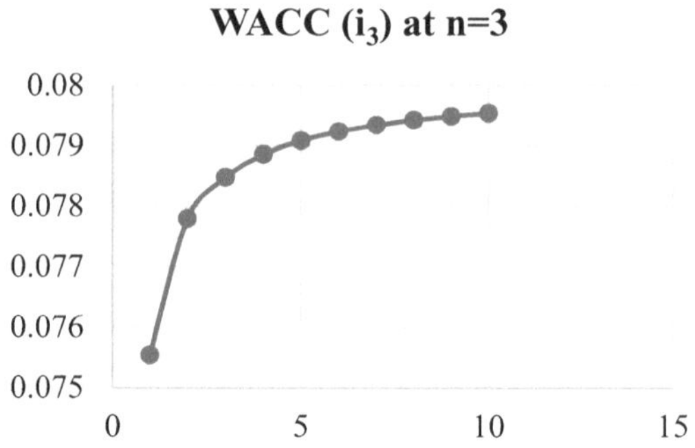

Fig. 24.5 The dependence of company's weighted average cost of capital (WACC) on the coverage ratio on debt and interest on the credit i_3 at company age n = 3

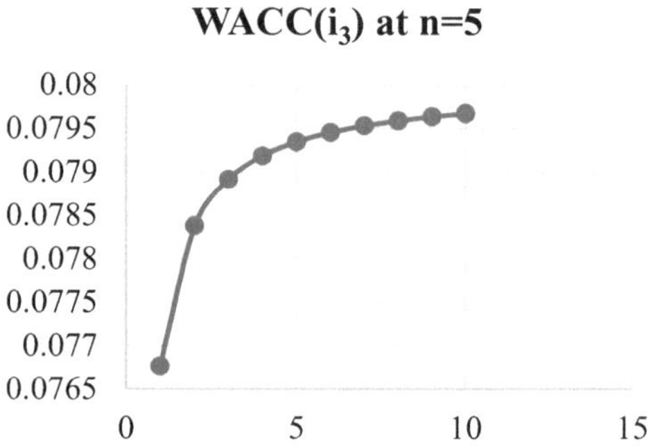

Fig. 24.6 The dependence of company's weighted average cost of capital (WACC) on the coverage ratio on debt and interest on the credit i_2 at company age n = 5

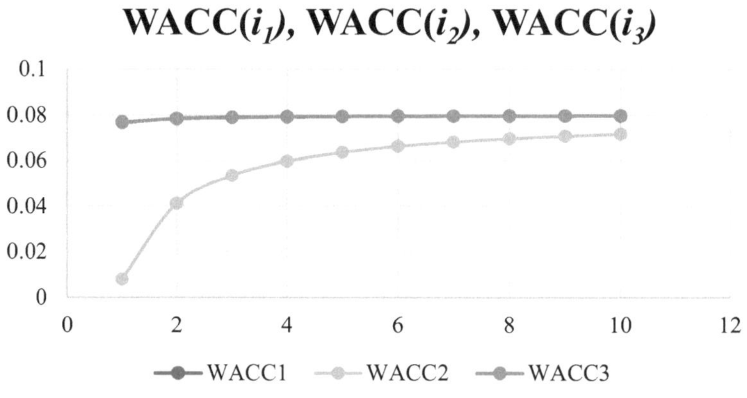

Fig. 24.7 Consolidated data of dependence of WACC on i_1, i_2, i_3, at company age n = 3

24.4.4 All Three Coverage Ratios Together

Consolidated data of dependence of WACC on i_1, i_2, i_3, at company age n = 3 and n = 5 are shown at Figs. 24.7 and 24.8.

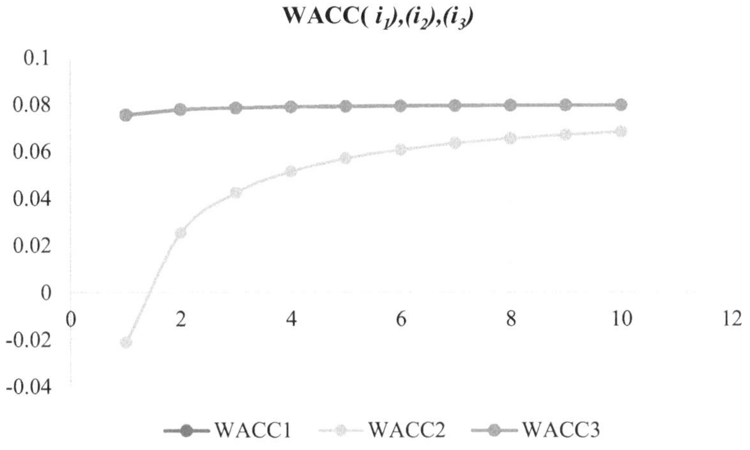

Fig. 24.8 Consolidated data of dependence of WACC on i_1, i_2, i_3, at company age n = 3

The analysis of the Tables 24.1, 24.2, 24.3, 24.4, 24.5, 24.6, 24.7, 24.8, 24.9, and 24.10 and Figs. 24.1, 24.2, 24.3, 24.4, 24.5, 24.6, 24.7, and 24.8 as well as conclusions will be made at the end of next paragraph.

24.5 Coverage Ratios (Different Capital Cost Values)

Let us analyze the dependence of company weighted average cost of capital (WACC) of coverage ratios (i_1, i_2, i_3), for different capital cost values $k_0 = 14\%$, $k_d = 8\%$.

Here as before t = 20%, n = 3; 5, the value of coverage ratios *i* is in the range from 1 to 10.

24.5.1 Coverage Ratios of Debt

As we have derived above the dependence of the weighted average cost of capital (WACC) on debt coverage ratio (i_1) in the BFO theory is described by the following formula:

$$i_1 * \frac{(1-(1+WACC)^{-n})}{WACC} - i_1 * \frac{(1-(1+K_0)^{-n})}{K_0} - t * [1-(1+K_d)^{-n}] = 0. \tag{24.19}$$

Here

Table 24.11 The dependence of the weighted average cost of capital WACC on debt coverage ratio i_1 at company age n = 3

i_1	t	k_0	k_d	WACC	n	BFO
1	0,2	0,14	0,08	0,1298	3	0,00
2	0,2	0,14	0,08	0,1347	3	0,00
3	0,2	0,14	0,08	0,1365	3	0,00
4	0,2	0,14	0,08	0,1374	3	0,00
5	0,2	0,14	0,08	0,1379	3	0,00
6	0,2	0,14	0,08	0,1382	3	0,00
7	0,2	0,14	0,08	0,1385	3	0,00
8	0,2	0,14	0,08	0,1387	3	0,00
9	0,2	0,14	0,08	0,1388	3	0,00
10	0,2	0,14	0,08	0,1389	3	0,00

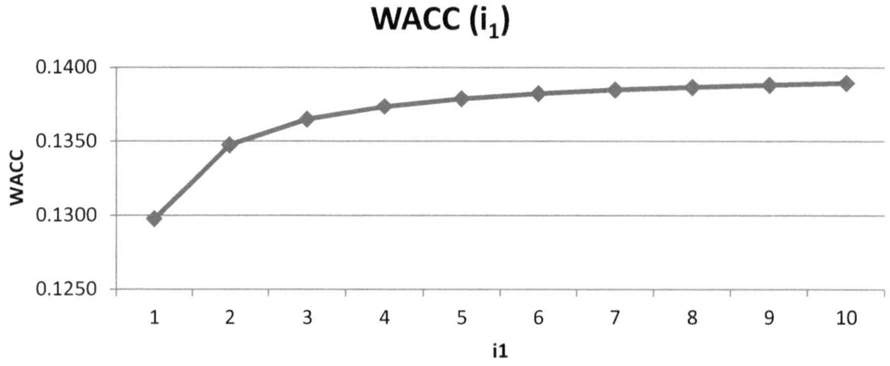

Fig. 24.9 The dependence of the weighted average cost of capital WACC on debt coverage ratio i_1 at company age n = 3

Table 24.12 The dependence of the weighted average cost of capital WACC on debt coverage ratio i_1 at company age n = 5

i_1	t	k_0	k_d	WACC	n	BFO
1	0,2	0,14	0,08	0,1324	5	0,00
2	0,2	0,14	0,08	0,1362	5	0,00
3	0,2	0,14	0,08	0,1374	5	0,00
4	0,2	0,14	0,08	0,1381	5	0,00
5	0,2	0,14	0,08	0,1385	5	0,00
6	0,2	0,14	0,08	0,1387	5	0,00
7	0,2	0,14	0,08	0,1389	5	0,00
8	0,2	0,14	0,08	0,1390	5	0,00
9	0,2	0,14	0,08	0,1391	5	0,00
10	0,2	0,14	0,08	0,1392	5	0,00

$$i_1 = \frac{CF}{D}. \tag{24.20}$$

By use it we get the following results, representing in Table 24.11 and Fig. 24.9 for company age n = 3 and in Table 24.12 and Fig. 24.10 for company age n = 5.

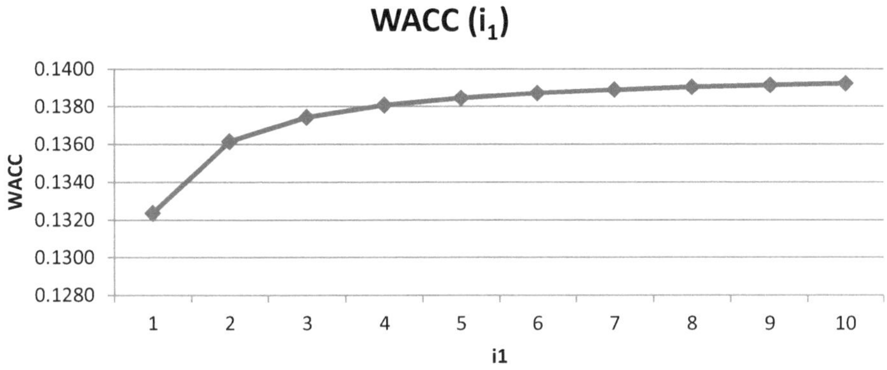

Fig. 24.10 The dependence of the weighted average cost of capital WACC on debt coverage ratio i_1 at company age n = 5

The dependence of the weighted average cost of capital WACC on debt coverage ratio i_1 at company age n = 5 is shown in Table 24.12 and Fig. 24.10.

24.5.2 The Coverage Ratio on Interest on the Credit

As we have derived above the dependence of the weighted average cost of capital (WACC) on interests on credit coverage ratio (i_2) in the BFO theory is described by the following formula:

$$i_2 * \frac{\left(1-(1+WACC)^{-n}\right)}{WACC} - i_2 * \frac{\left(1-(1+K_0)^{-n}\right)}{K_0} - \frac{\left(t * \left[1-(1+K_d)^{-n}\right]\right)}{K_d} = 0 \tag{24.21}$$

Here

$$i_2 = \frac{CF}{K_d * D}. \tag{24.22}$$

By use it we get the following results, representing in Table 24.13 and Fig. 24.11 for company age n = 3 and in Table 24.14 and Fig. 24.12 for company age n = 5.

Table 24.13 The dependence of the weighted average cost of capital (WACC) on interests on credit leverage ratio (i_2) at company age n = 3

i_2	t	k_0	k_d	WACC	n	BFO
1	0,2	0,14	0,08	0,0285	3	0,00
2	0,2	0,14	0,08	0,0795	3	0,00
3	0,2	0,14	0,08	0,0985	3	0,00
4	0,2	0,14	0,08	0,1084	3	0,00
5	0,2	0,14	0,08	0,1145	3	0,00
6	0,2	0,14	0,08	0,1186	3	0,00
7	0,2	0,14	0,08	0,1216	3	0,00
8	0,2	0,14	0,08	0,1238	3	0,00
9	0,2	0,14	0,08	0,1256	3	0,00
10	0,2	0,14	0,08	0,1270	3	0,00

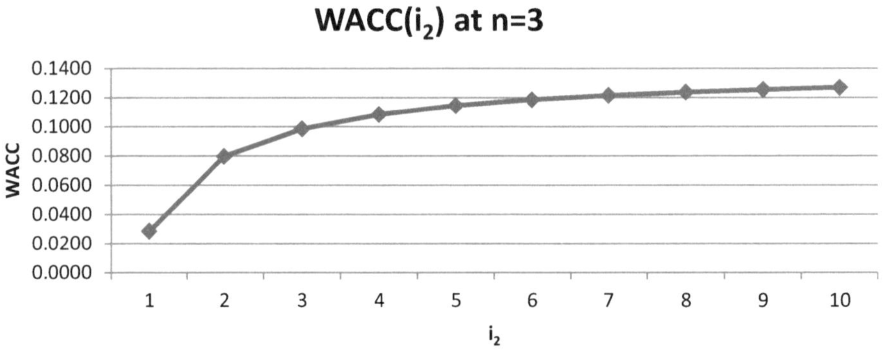

Fig. 24.11 The dependence of the weighted average cost of capital (WACC) on interests on credit leverage ratio (i_2) at company age n = 3

Table 24.14 The dependence of the weighted average cost of capital (WACC) on interests on credit leverage ratio (i_2) at company age n = 5

i_2	t	k_0	k_d	WACC	n	BFO
1	0,2	0,14	0,08	0,0583	5	0,00
2	0,2	0,14	0,08	0,0957	5	0,00
3	0,2	0,14	0,08	0,1096	5	0,00
4	0,2	0,14	0,08	0,1169	5	0,00
5	0,2	0,14	0,08	0,1213	5	0,00
6	0,2	0,14	0,08	0,1244	5	0,00
7	0,2	0,14	0,08	0,1265	5	0,00
8	0,2	0,14	0,08	0,1282	5	0,00
9	0,2	0,14	0,08	0,1295	5	0,00
10	0,2	0,14	0,08	0,1305	5	0,00

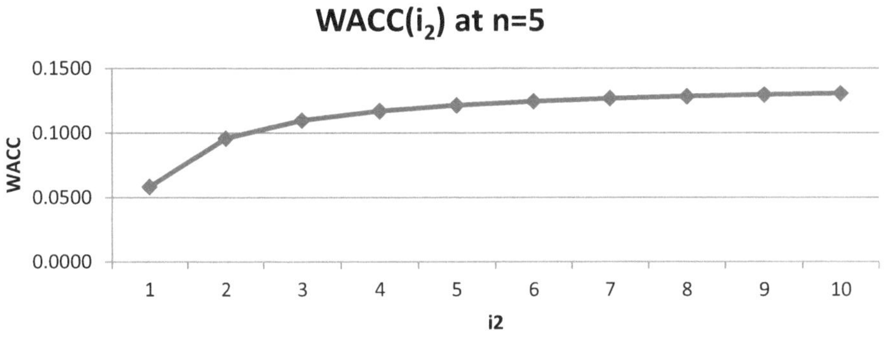

Fig. 24.12 The dependence of the weighted average cost of capital (WACC) on interests on credit leverage ratio (i_2) at company age n = 5

24.5.3 Coverage Ratios of Debt and Interest on the Credit (New Ratios)

As we have derived above the dependence of the weighted average cost of capital (WACC) on debt and interests on credit coverage ratio (i_3) in the BFO theory is described by the following formula:

$$i_3 * \frac{(1-(1+WACC)^{-n})}{WACC} - i_3 * \frac{(1-(1+K_0)^{-n})}{K_0} - \frac{t*[1-(1+K_d)^{-n}]}{(K_d+1)} = 0. \tag{24.23}$$

Here

$$i_3 = \frac{CF}{(K_d+1)*D}. \tag{24.24}$$

By use it we get the following results, representing in Table 24.15 and Fig. 24.13 for company age n = 3 and in Table 24.16 and Fig. 24.14 for company age n = 5.

Consolidated data of dependence of WACC on i_1, i_2, i_3, at company age n = 3 and n = 5 are shown at Figs. 24.15 and 24.16.

24.5.4 Analysis and Conclusions

It is seen from the Tables 24.1, 24.2, 24.3, 24.4, 24.5, 24.6, 24.7, 24.8, 24.9, 24.10, 24.11, 24.12, 24.13, 24.14, 24.15, and 24.16 and Figs. 24.1, 24.2, 24.3, 24.4, 24.5, 24.6, 24.7, 24.8, 24.9, 24.10, 24.11, 24.12, 24.13, 24.14, 24.15, and 24.16 that

Table 24.15 The dependence of the weighted average cost of capital (WACC) on debt and interests on credit leverage ratio (i_3) at company age n = 3

i_3	t	k_0	k_d	WACC	n	BFO
1	0,2	0,14	0,08	0,1303	3	0,00
2	0,2	0,14	0,08	0,1351	3	0,00
3	0,2	0,14	0,08	0,1367	3	0,00
4	0,2	0,14	0,08	0,1376	3	0,00
5	0,2	0,14	0,08	0,1380	3	0,00
6	0,2	0,14	0,08	0,1384	3	0,00
7	0,2	0,14	0,08	0,1386	3	0,00
8	0,2	0,14	0,08	0,1388	3	0,00
9	0,2	0,14	0,08	0,1389	3	0,00
10	0,2	0,14	0,08	0,1390	3	0,00

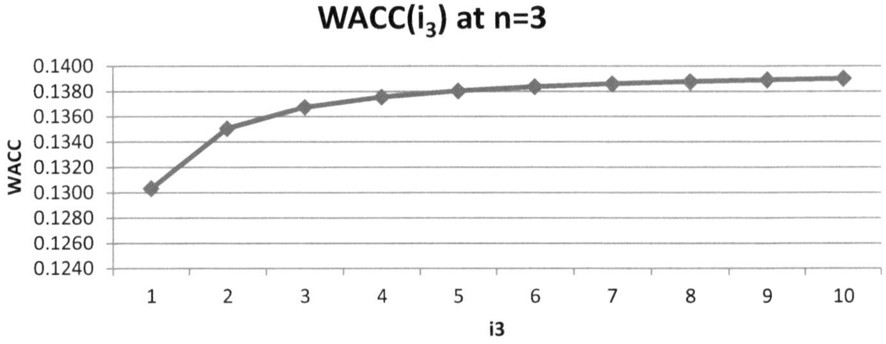

Fig. 24.13 The dependence of the weighted average cost of capital (WACC) on debt and interests on credit leverage ratio (i_3) at company age n = 3

Table 24.16 The dependence of the weighted average cost of capital (WACC) on debt and interests on credit leverage ratio (i_3) at company age n = 5

i_3	t	k_0	k_d	WACC	n	BFO
1	0,2	0,14	0,08	0,1329	5	0,00
2	0,2	0,14	0,08	0,1364	5	0,00
3	0,2	0,14	0,08	0,1376	5	0,00
4	0,2	0,14	0,08	0,1382	5	0,00
5	0,2	0,14	0,08	0,1386	5	0,00
6	0,2	0,14	0,08	0,1388	5	0,00
7	0,2	0,14	0,08	0,1390	5	0,00
8	0,2	0,14	0,08	0,1391	5	0,00
9	0,2	0,14	0,08	0,1392	5	0,00
10	0,2	0,14	0,08	0,1393	5	0,00

$WACC\ (i_j)$ is increasing function on i_j with saturation $WACC = k_0$ at high values of i_j. Note, that this saturation for companies of finite age is a little bit more gradual than in case of perpetuity companies: in latter case the saturation takes place around i_j value of order 1 for ratios i_1 and i_3 and of order 4 or 5 for ratios i_2. In perpetuity case as well as in case of companies of finite age at saturation $WACC$ reaches the value k_0 (equity

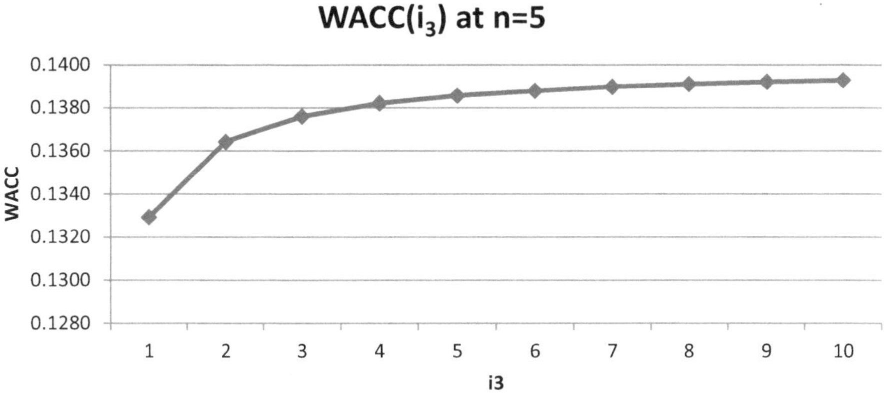

Fig. 24.14 The dependence of the weighted average cost of capital (WACC) on debt and interests on credit leverage ratio (i_3) at company age n = 5

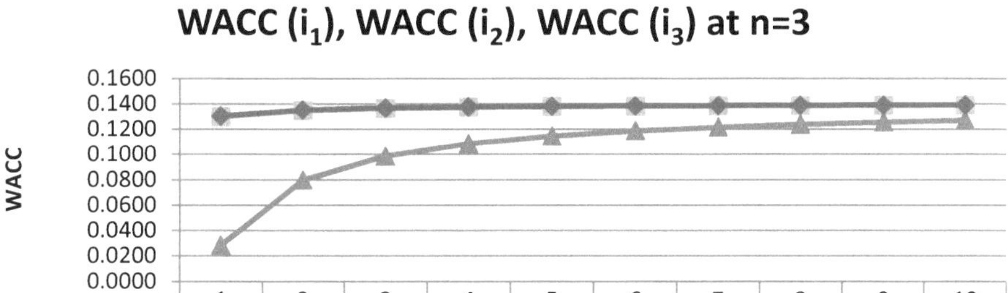

	1	2	3	4	5	6	7	8	9	10
WACC (i1)	0.1298	0.1347	0.1365	0.1374	0.1379	0.1382	0.1385	0.1387	0.1388	0.1389
WACC (i2)	0.0285	0.0795	0.0985	0.1084	0.1145	0.1186	0.1216	0.1238	0.1256	0.1270
WACC (i3)	0.1303	0.1351	0.1367	0.1376	0.1380	0.1384	0.1386	0.1388	0.1389	0.1390

i

Fig. 24.15 Consolidated data of dependence of WACC on i_1, i_2, i_3, at company age n = 3

value at zero leverage level). This means that for high values of i_j one can choose k_0 as a discount rate with a very good accuracy in perpetuity case and with a little bit less accuracy in general case (companies of arbitrary ages). Thus the role of parameter k_0 increases drastically. The method of determination of parameter k_0 has been developed by Anastasiya Brusova (2011). So, parameter k_0 is the discount rate for case of high values of i_j. In case of ratio i_2 in general case as well as in perpetuity case the saturation of *WACC* (i_2) takes place at higher values of i_2.

In opposite to perpetuity case within BFO theory one could make calculations for companies of arbitrary age because a factor of time presents in this theory. Our calculations show that curve *WACC* (i_j) for company of higher age lies above this curve for younger company. And with increase of i_j value the WACC values for different company ages n become closer each other.

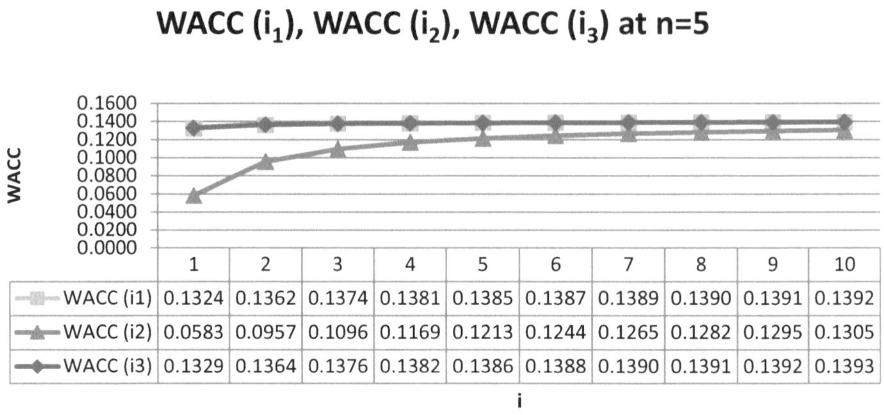

	1	2	3	4	5	6	7	8	9	10
WACC (i1)	0.1324	0.1362	0.1374	0.1381	0.1385	0.1387	0.1389	0.1390	0.1391	0.1392
WACC (i2)	0.0583	0.0957	0.1096	0.1169	0.1213	0.1244	0.1265	0.1282	0.1295	0.1305
WACC (i3)	0.1329	0.1364	0.1376	0.1382	0.1386	0.1388	0.1390	0.1391	0.1392	0.1393

Fig. 24.16 Consolidated data of dependence of WACC on i_1, i_2, i_3, at company age $n = 5$

Note that curves WACC(i_1) and WACC(i_3) are very close each other at small enough credit rates, but difference between them will become bigger at higher values of credit rates.

Curve WACC(i_2) turns out to be enough different from WACC(i_1) and curves WACC(i_3).

24.6 Leverage Ratios

24.6.1 Leverage Ratios for Debt

We will analyze the dependence of company weighted average cost of capital (WACC) on leverage ratios (l_1, l_2, l_3). We will make calculation for capital costs $k_0 = 10\%$, $k_d = 6\%$, $t = 20\%$, $n = 3; 5$; l values range from 0 to 10.

Dividing the both parts of the formula (24.6) by the income value for one period *CF*, we enter the leverage ratios l_l for debt into the general BFO theory

$$\frac{\left(1-(1+WACC)^{-n}\right)}{WACC}-\frac{\left(1-(1+K_0)^{-n}\right)}{K_0}-t*\left[1-(1+K_d)^{-n}\right]*l_1$$
$$=0. \tag{24.25}$$

Here

$$l_1=\frac{D}{CF}. \tag{24.26}$$

Remind, that here *WACC* is the weighted average cost of capital of the company, l_1—the leverage ratios l_l for debt, t is the tax on profit rate for organizations

Table 24.17 The dependence of company weighted average cost of capital (WACC) on debt leverage ratio, l_1, at n = 3

l_1	t	k_0	k_d	WACC(l_1)	n	BFO
0	0,2	0,1	0,06	0,1000	3	0,00
1	0,2	0,1	0,06	0,0928	3	0,00
2	0,2	0,1	0,06	0,0857	3	0,00
3	0,2	0,1	0,06	0,0787	3	0,00
4	0,2	0,1	0,06	0,0720	3	0,00
5	0,2	0,1	0,06	0,0654	3	0,00
6	0,2	0,1	0,06	0,0587	3	0,00
7	0,2	0,1	0,06	0,0523	3	0,00
8	0,2	0,1	0,06	0,0461	3	0,00
9	0,2	0,1	0,06	0,0399	3	0,00
10	0,2	0,1	0,06	0,0339	3	0,00

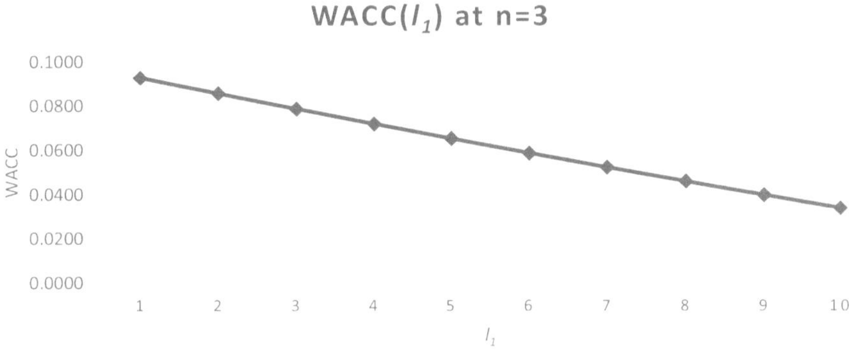

Fig. 24.17 The dependence of company weighted average cost of capital (WACC) on debt leverage ratio at n = 3

(t = 20%), k_0—equity cost of financially–independent company, k_d is the debt capital cost; n is the company age, CF–income value for one period; D—debt capital value.

The ratio (l_2) can be used to assess of the following parameters used in rating, Interests/EBITDA and some others.

By use the above equation we get the following results, representing in Table 24.17 and Fig. 24.17 for company age n = 3 and in Table 24.18 and Fig. 24.18 for company age n = 5.

24.6.2 Leverage Ratios for Interest on Credit

The dependence of company weighted average cost of capital (WACC) on leverage ratios on interests on credit l_2 is described within BFO theory by the following formula:

Table 24.18 The dependence of company weighted average cost of capital (WACC) on debt leverage ratio, l_1, at n = 5

l_1	t	k_0	k_d	WACC(l_1)	n	BFO
0	0,2	0,1	0,06	0,1000	5	0,00
1	0,2	0,1	0,06	0,0948	5	0,00
2	0,2	0,1	0,06	0,0898	5	0,00
3	0,2	0,1	0,06	0,0848	5	0,00
4	0,2	0,1	0,06	0,0799	5	0,00
5	0,2	0,1	0,06	0,0752	5	0,00
6	0,2	0,1	0,06	0,0705	5	0,00
7	0,2	0,1	0,06	0,0660	5	0,00
8	0,2	0,1	0,06	0,0615	5	0,00
9	0,2	0,1	0,06	0,0571	5	0,00
10	0,2	0,1	0,06	0,0528	5	0,00

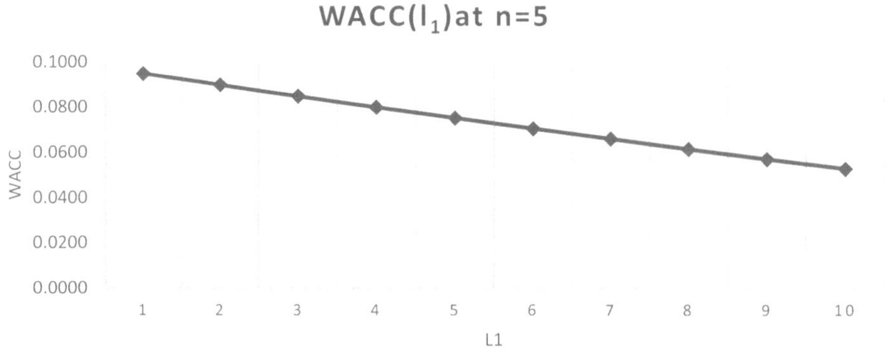

Fig. 24.18 The dependence of company weighted average cost of capital WACC on debt leverage ratios at n = 5

$$\frac{(1-(1+WACC)^{-n})}{WACC} - \frac{(1-(1+K_0)^{-n})}{K_0} - \frac{(t * l_2 * [1-(1+K_d)^{-n}])}{K_d} = 0. \tag{24.27}$$

Here

$$l_2 = \frac{K_d * D}{CF}. \tag{24.28}$$

Using it, we find the dependence WACC(l_2) at company ages n = 3 and n = 5.

This ratio l_2 can be used to assess of the following parameters used in rating, Interests/EBITDA and some others.

By use the above equation we get the following results, representing in Table 24.19 and Fig. 24.19 for company age n = 3 and in Table 24.20 and Fig. 24.20 for company age n = 5.

Table 24.19 The dependence of company weighted average cost of capital (WACC) on leverage ratio of interests on credit, l_2, at company age $n = 3$

l_2	t	k_0	k_d	WACC(l_2)	n	BFO
0	0,2	0,1	0,06	0,0998	3	0,00
1	0,2	0,1	0,06	−0,0036	3	0,00
2	0,2	0,1	0,06	−0,0804	3	0,00
3	0,2	0,1	0,06	−0,1403	3	0,00
4	0,2	0,1	0,06	−0,1888	3	0,00
5	0,2	0,1	0,06	−0,2289	3	0,00
6	0,2	0,1	0,06	−0,2629	3	0,00
7	0,2	0,1	0,06	−0,2922	3	0,00
8	0,2	0,1	0,06	−0,3178	3	0,00
9	0,2	0,1	0,06	−0,3404	3	0,00
10	0,2	0,1	0,06	−0,3605	3	0,00

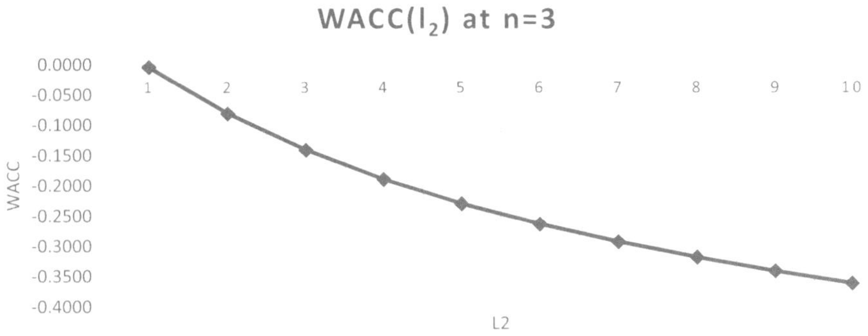

Fig. 24.19 The dependence of company weighted average cost of capital (WACC) on leverage ratio of interests on credit at company age $n = 3$

Table 24.20 The dependence of company weighted average cost of capital (WACC) on leverage ratio of interests on credit, l_2, at company age $n = 5$

l_2	t	k_0	k_d	WACC(l_2)	n	BFO
0	0,2	0,1	0,06	0,1000	5	0,00
1	0,2	0,1	0,06	0,0259	5	0,00
2	0,2	0,1	0,06	−0,0296	5	0,00
3	0,2	0,1	0,06	−0,0732	5	0,00
4	0,2	0,1	0,06	−0,1089	5	0,00
5	0,2	0,1	0,06	−0,1388	5	0,00
6	0,2	0,1	0,06	−0,1643	5	0,00
7	0,2	0,1	0,06	−0,1865	5	0,00
8	0,2	0,1	0,06	−0,2061	5	0,00
9	0,2	0,1	0,06	−0,2235	5	0,00
10	0,2	0,1	0,06	−0,2391	5	0,00

The dependence of company weighted average cost of capital (WACC) on leverage ratios on debt and interests on credit l_3 is described within BFO theory by the following formula:

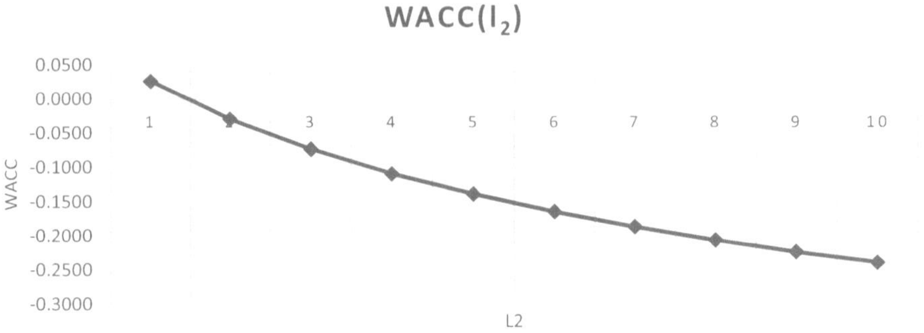

Fig. 24.20 The dependence of company weighted average cost of capital (WACC) on leverage ratio of interests on credit at company age n = 3

Table 24.21 The dependence of company weighted average cost of capital (WACC) on leverage ratio on debt and interests on credit, l_3, at company age n = 3

l_3	t	k_0	k_d	WACC(l3)	n	BFO
0	0,2	0,1	0,06	0,1000	3	0,00
1	0,2	0,1	0,06	0,0930	3	0,00
2	0,2	0,1	0,06	0,0864	3	0,00
3	0,2	0,1	0,06	0,0798	3	0,00
4	0,2	0,1	0,06	0,0734	3	0,00
5	0,2	0,1	0,06	0,0671	3	0,00
6	0,2	0,1	0,06	0,0608	3	0,00
7	0,2	0,1	0,06	0,0548	3	0,00
8	0,2	0,1	0,06	0,0489	3	0,00
9	0,2	0,1	0,06	0,0430	3	0,00
10	0,2	0,1	0,06	0,0371	3	0,00

$$\frac{\left(1-(1+WACC)^{-n}\right)}{WACC}-\frac{\left(1-(1+K_0)^{-n}\right)}{K_0}-\frac{t*l_3*\left[1-(1+K_d)^{-n}\right]}{(K_d+1)}=0. \tag{24.29}$$

Here

$$l_3=\frac{(K_d+1)*D}{CF}. \tag{24.30}$$

The ratio l_3 can be used to assess of the following parameters used in rating, *Debt + interest /FFO*, *Debt + interest/EBIT*, *Debt + interest/EBITDA(R)*, and some others.

By use the above equation we get the following results for the dependence WACC(l_3), representing in Table 24.21 and Fig. 24.21 for company age n = 3 and in Table 24.22 and Fig. 24.22 for company age n = 5.

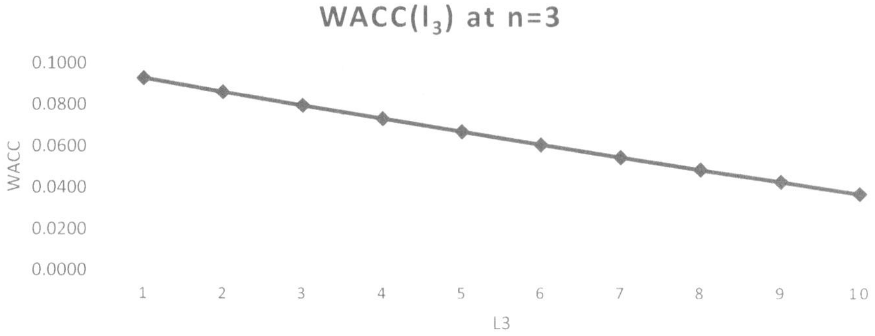

Fig. 24.21 The dependence of company weighted average cost of capital (WACC) on leverage ratio on debt and interests on credit at company age n = 3

Table 24.22 The dependence of company weighted average cost of capital (WACC) on leverage ratio on debt and interests on credit, l_3, at company age n = 5

l_3	t	k_0	k_d	WACC(l3)	n	BFO
0	0,2	0,1	0,06	0,1000	5	0,00
1	0,2	0,1	0,06	0,0951	5	0,00
2	0,2	0,1	0,06	0,0903	5	0,00
3	0,2	0,1	0,06	0,0856	5	0,00
4	0,2	0,1	0,06	0,0810	5	0,00
5	0,2	0,1	0,06	0,0765	5	0,00
6	0,2	0,1	0,06	0,0721	5	0,00
7	0,2	0,1	0,06	0,0678	5	0,00
8	0,2	0,1	0,06	0,0635	5	0,00
9	0,2	0,1	0,06	0,0593	5	0,00
10	0,2	0,1	0,06	0,0552	5	0,00

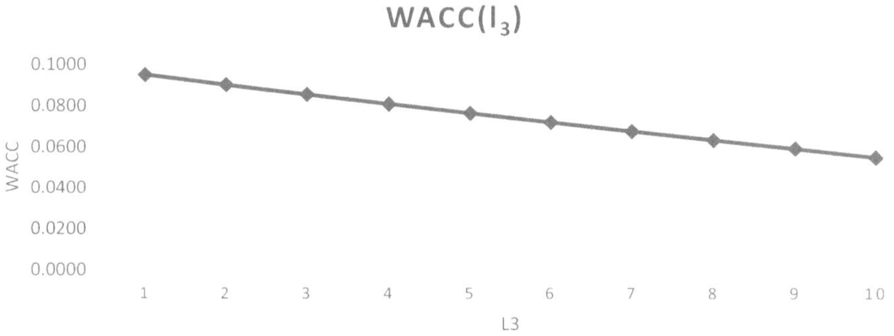

Fig. 24.22 The dependence of company weighted average cost of capital (WACC) on leverage ratio of debt and interests on credit at company age n = 5

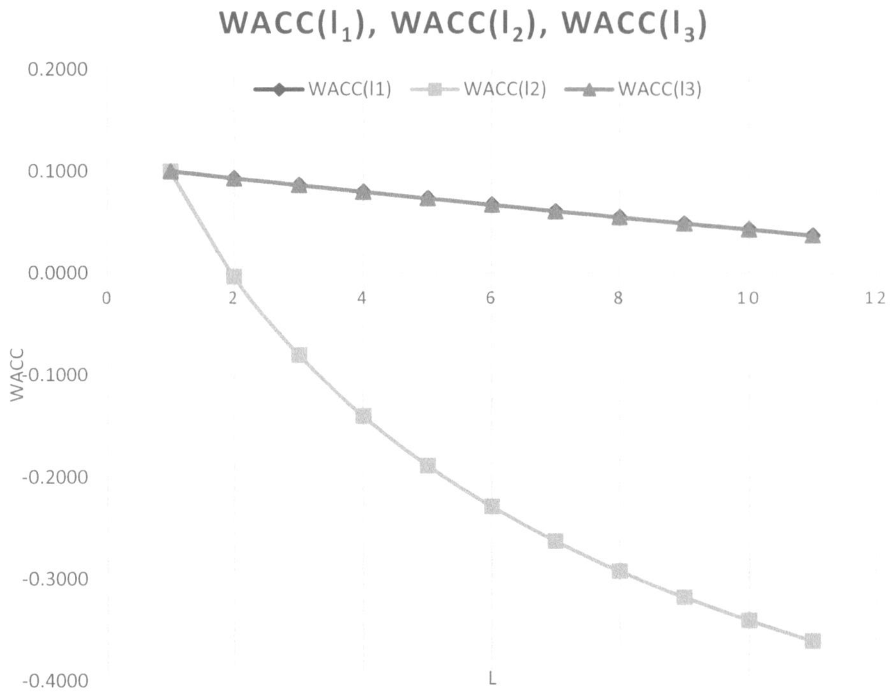

Fig. 24.23 Consolidated data of dependence of WACC on l_1, l_2, l_3, at company age n = 3

Below we represent the consolidated data of dependence of WACC on l_1, l_2, l_3, at company age n = 3 (Fig. 24.23) and (Fig. 24.24) n = 5.

24.7 Leverage Ratios (Different Capital Costs)

24.7.1 Leverage Ratios for Debt

Below we analyze the dependence of company weighted average cost of capital (WACC) on leverage ratios l_1, l_2, l_3, at capital costs values $k_0 = 12\%$, $k_d = 6\%$.

As before t = 20%, company age n = 3; 5, leverage ratios values range from 0 to 10.

The dependence of company weighted average cost of capital (WACC) on leverage ratios on debt l_1 is described within BFO theory by the following formula:

$$\frac{(1-(1+WACC)^{-n})}{WACC} - \frac{(1-(1+k0)^{-n})}{k0} - \mathrm{t}*\mathrm{c}*\mathrm{l}_1 = 0 \tag{24.31}$$

Here

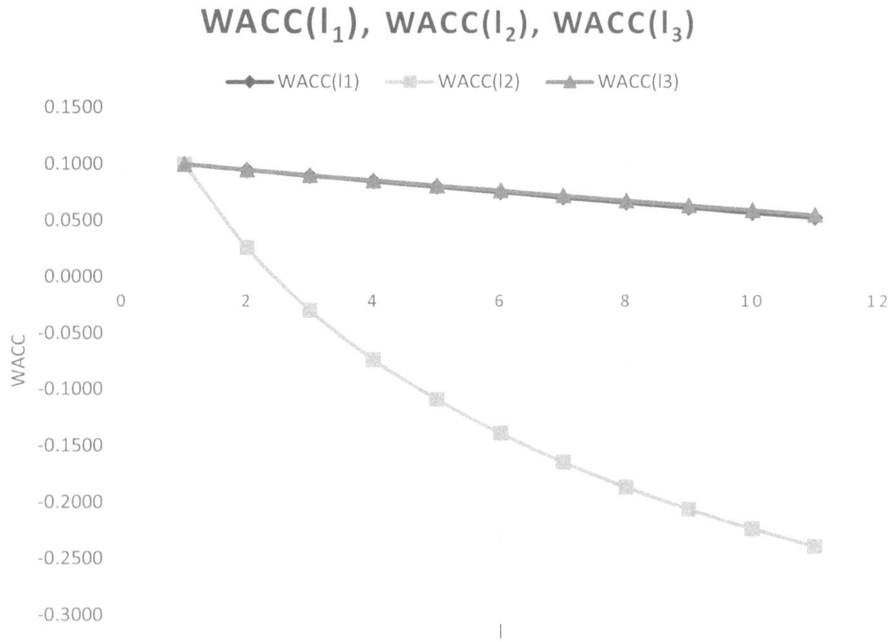

Fig. 24.24 Consolidated data of dependence of WACC on l_1, l_2, l_3, at company age n = 5

Table 24.23 The dependence of company weighted average cost of capital (WACC) on leverage ratio of debt, l_1, at company age n = 3

l_1	k_0	k_d	n_1	t	WACC	BFO
0	0,12	0,06	3	0,2	0,119,997	0,00
1	0,12	0,06	3	0,2	0,112,294	0,00
2	0,12	0,06	3	0,2	0,104,774	0,00
3	0,12	0,06	3	0,2	0,097444	0,00
4	0,12	0,06	3	0,2	0,090128	0,00
5	0,12	0,06	3	0,2	0,083078	0,00
6	0,12	0,06	3	0,2	0,076332	0,00
7	0,12	0,06	3	0,2	0,06959	0,00
8	0,12	0,06	3	0,2	0,062962	0,00
9	0,12	0,06	3	0,2	0,056492	0,00
10	0,12	0,06	3	0,2	0,050163	0,00

$$l_1 = \frac{D}{CF} \tag{24.32}$$

Using it, we find the dependence WACC(l_1) at company ages n = 3 (Table 24.23 and Fig. 24.25) and n = 5 (Table 24.24 and Fig. 24.26).

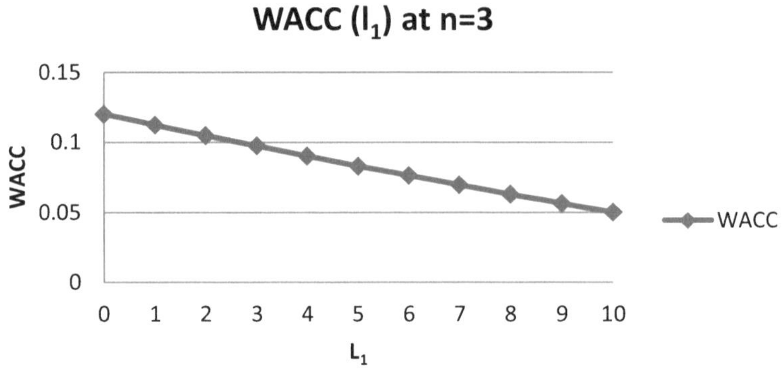

Fig. 24.25 The dependence of company weighted average cost of capital (WACC) on leverage ratio of debt at company age n = 3

Table 24.24 The dependence of company weighted average cost of capital (WACC) on leverage ratio of debt, l_1, at company age n = 5

L_1	k_0	k_d	n_1	t	WACC	BFO
0	0,12	0,06	5	0,2	0,119994	0,00
1	0,12	0,06	5	0,2	0,114311	0,00
2	0,12	0,06	5	0,2	0,108927	0,00
3	0,12	0,06	5	0,2	0,103556	0,00
4	0,12	0,06	5	0,2	0,098332	0,00
5	0,12	0,06	5	0,2	0,093123	0,00
6	0,12	0,06	5	0,2	0,088164	0,00
7	0,12	0,06	5	0,2	0,083265	0,00
8	0,12	0,06	5	0,2	0,078452	0,00
9	0,12	0,06	5	0,2	0,073744	0,00
10	0,12	0,06	5	0,2	0,069	0,00

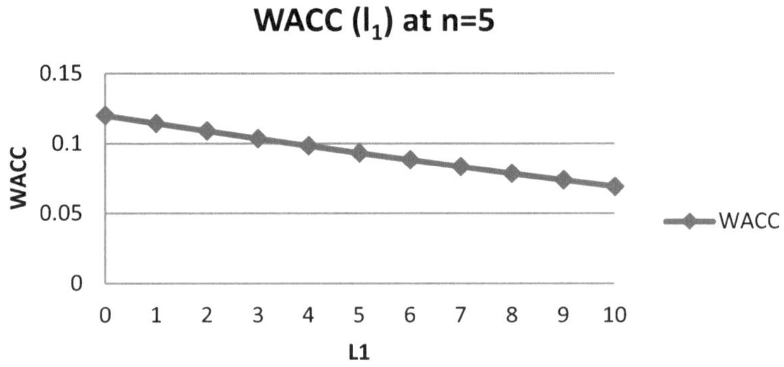

Fig. 24.26 The dependence of company weighted average cost of capital (WACC) on leverage ratio of debt at company age n = 3

24.7.2 Leverage Ratios for Interests on Credit

The dependence of company weighted average cost of capital (WACC) on leverage ratios on interests on credit l_2 is described within BFO theory by the following formula:

$$\frac{(1-(1+WACC)^{-n})}{WACC} - \frac{(1-(1+K0)^{-n})}{k0} - \frac{t*l2*(1-(1+kd)^{-n})}{kd} = 0 \quad (24.33)$$

Here

$$l_2 = \frac{D*kd}{CF} \quad (24.34)$$

Using it, we find the dependence WACC(l_2) at company ages n = 3 (Table 24.25 and Fig. 24.27) and n = 5 (Table 24.26 and Fig. 24.28).

Table 24.25 The dependence of company weighted average cost of capital (WACC) on leverage ratio of interests on credit, l_2, at company age n = 3

l_2	k_0	k_d	n_1	t	WACC	БФО
0	0,12	0,06	3	0,2	0,119997	0,00
1	0,12	0,06	3	0,2	0,010838	0,00
2	0,12	0,06	3	0,2	−0,06941	0,00
3	0,12	0,06	3	0,2	−0,13171	0,00
4	0,12	0,06	3	0,2	−0,18169	0,00
5	0,12	0,06	3	0,2	−0,22298	0,00
6	0,12	0,06	3	0,2	−0,25785	0,00
7	0,12	0,06	3	0,2	−0,28784	0,00
8	0,12	0,06	3	0,2	−0,31392	0,00
9	0,12	0,06	3	0,2	−0,33692	0,00
10	0,12	0,06	3	0,2	−0,35745	0,00

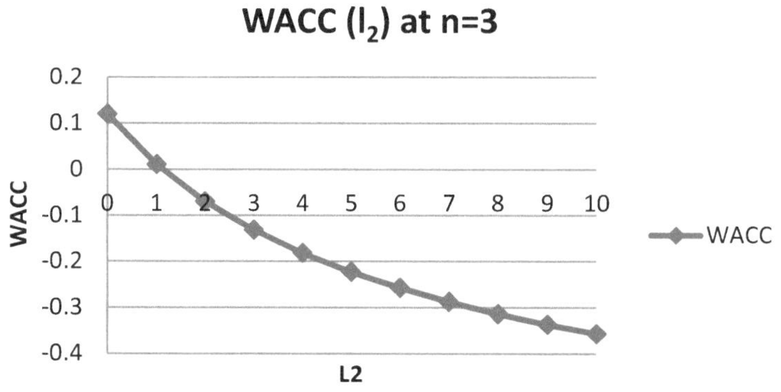

Fig. 24.27 The dependence of company weighted average cost of capital (WACC) on leverage ratio of interests on credit at company age n = 3

Table 24.26 The dependence of company weighted average cost of capital (WACC) on leverage ratio of interests on credit, l_2, at company age n = 5

L2	k_0	k_d	n1	t	WACC	БФО
0	0,12	0,06	5	0,2	0,119994	0,00
1	0,12	0,06	5	0,2	0,040367	0,00
2	0,12	0,06	5	0,2	−0,01846	0,00
3	0,12	0,06	5	0,2	−0,06439	0,00
4	0,12	0,06	5	0,2	−0,10159	0,00
5	0,12	0,06	5	0,2	−0,13262	0,00
6	0,12	0,06	5	0,2	−0,15899	0,00
7	0,12	0,06	5	0,2	−0,18185	0,00
8	0,12	0,06	5	0,2	−0,20194	0,00
9	0,12	0,06	5	0,2	−0,21978	0,00
10	0,12	0,06	5	0,2	−0,23578	0,00

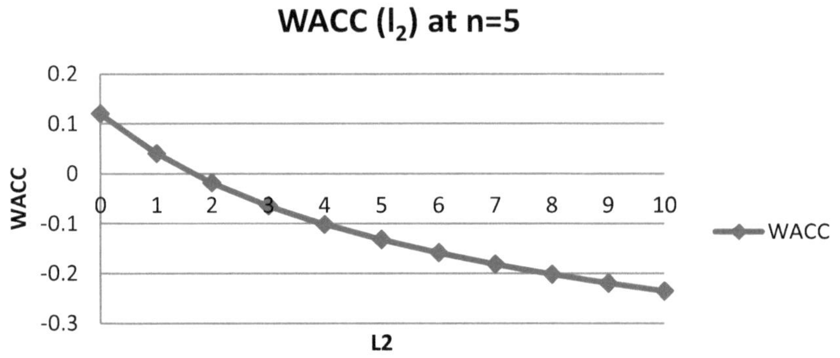

Fig. 24.28 The dependence of company weighted average cost of capital (WACC) on leverage ratio of interests on credit at company age n = 5

24.7.3 Leverage Ratios for Debt and Interests on Credit

The dependence of company weighted average cost of capital (WACC) on leverage ratios on debt and interests on credit l_3 is described within BFO theory by the following formula:

$$\frac{(1-(1+WACC)^{-n})}{WACC}-\frac{(1-(1+K0)^{-n})}{K0}-\frac{t*l3*(1-(1+kd)^{-n})}{1+Kd}=0. \tag{24.35}$$

Here

$$l_3=\frac{D(1+kd)}{CF} \tag{24.36}$$

Table 24.27 The dependence of company weighted average cost of capital (WACC) on leverage ratio of debt and interests on credit, l_3, at company age n = 3

l_3	k_0	k_d	n	t	WACC	BFO
0	0,12	0,06	3	0,2	0,119,997	0,00
1	0,12	0,06	3	0,2	0,112,716	0,00
2	0,12	0,06	3	0,2	0,105,604	0,00
3	0,12	0,06	3	0,2	0,098686	0,00
4	0,12	0,06	3	0,2	0,091785	0,00
5	0,12	0,06	3	0,2	0,085114	0,00
6	0,12	0,06	3	0,2	0,078654	0,00
7	0,12	0,06	3	0,2	0,072249	0,00
8	0,12	0,06	3	0,2	0,065828	0,00
9	0,12	0,06	3	0,2	0,059771	0,00
10	0,12	0,06	3	0,2	0,053729	0,00

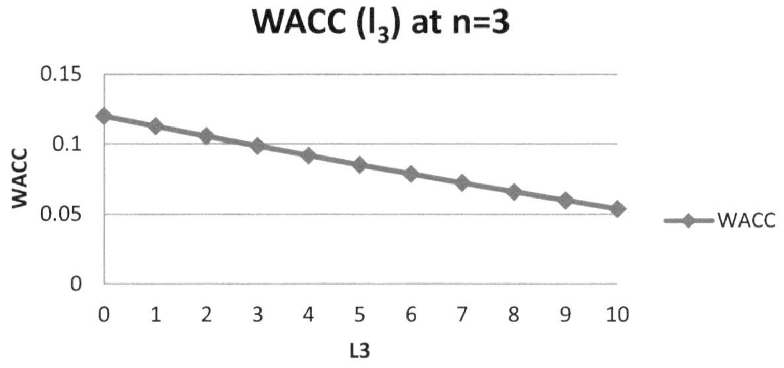

Fig. 24.29 The dependence of company weighted average cost of capital (WACC) on leverage ratio of debt and interests on credit at company age n = 3

Table 24.28 The dependence of company weighted average cost of capital (WACC) on leverage ratio of debt and interests on credit, l_3, at company age n = 5

L_3	k_0	k_d	n	t	WACC	BFO
0	0,12	0,06	5	0,2	0,119,994	0,00
1	0,12	0,06	5	0,2	0,114,614	0,00
2	0,12	0,06	5	0,2	0,10,954	0,00
3	0,12	0,06	5	0,2	0,104,444	0,00
4	0,12	0,06	5	0,2	0,099512	0,00
5	0,12	0,06	5	0,2	0,094598	0,00
6	0,12	0,06	5	0,2	0,08988	0,00
7	0,12	0,06	5	0,2	0,0852	0,00
8	0,12	0,06	5	0,2	0,080618	0,00
9	0,12	0,06	5	0,2	0,076129	0,00
10	0,12	0,06	5	0,2	0,071733	0,00

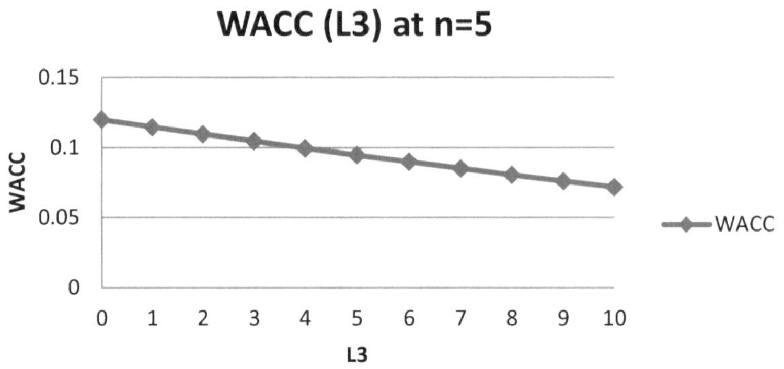

Fig. 24.30 The dependence of company weighted average cost of capital (WACC) on leverage ratio of debt and interests on credit at company age n = 5

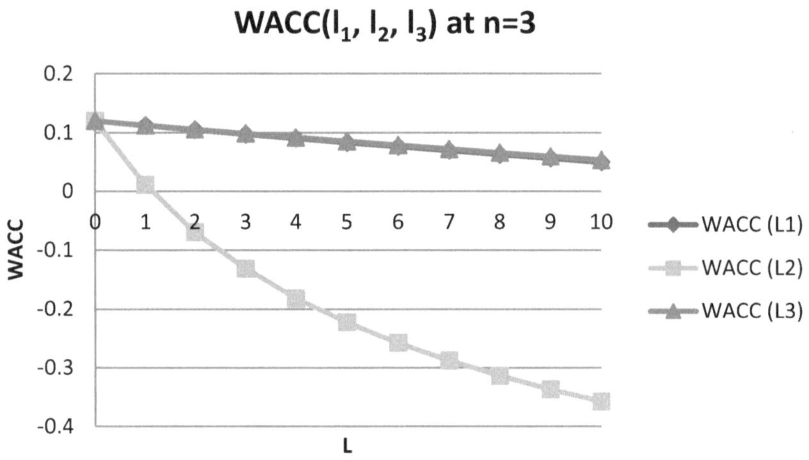

Fig. 24.31 Consolidated data of dependence of WACC on l_1, l_2, l_3, at company age n = 3

Using it, we find the dependence WACC(l_3) at company ages n = 3 (Table 24.27 and Fig. 24.29) and n = 5 (Table 24.28 and Fig. 24.30).

Consolidated data of dependence of WACC on l_1, l_2, l_3, at company age n = 3 (Fig. 24.31) and for n = 5 (Fig. 24.32).

24.7.4 Analysis and Conclusions

It is seen from the Tables 24.17, 24.18, 24.19, 24.20, 24.21, 24.22, 24.23, 24.24, 24.25, 24.26, and 24.27 and Figs. 24.17, 24.18, 24.19, 24.20, 24.21, 24.22, 24.23, 24.24, 24.25, 24.26, 24.27, 24.28, 24.29, 24.30, 24.31, and 24.32 that $WACC(l_j)$ is decreasing function on l_j. *WACC* decreases from value of k_0 (equity value at zero

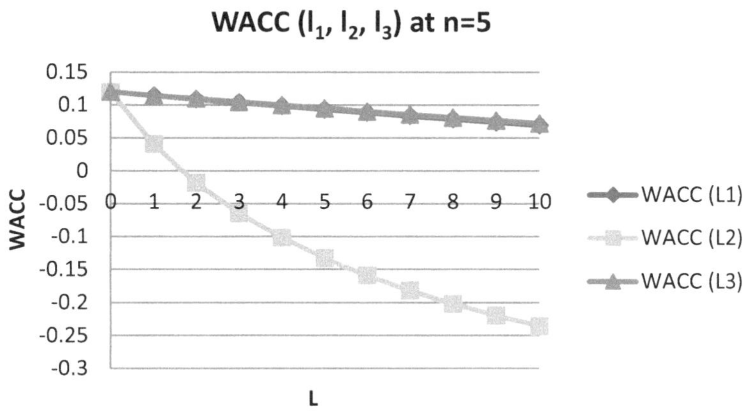

Fig. 24.32 Consolidated data of dependence of WACC on l_1, l_2, l_3, at company age n = 5

leverage level) practically linearly for WACC(l_1) and WACC(l_3) and with higher speed for WACC(l_2). In opposite to perpetuity case within BFO theory one could make calculations for companies of arbitrary age because a factor of time presents in this theory. Our calculations show that curve *WACC* (l_i) for company of higher age lies above this curve for younger company.

Note that curves WACC(l_1) and WACC(l_3) are very close each other at small enough credit rates, but difference between them will become bigger at higher values of credit rates.

Curve WACC(l_2) turns out to be enough different from WACC(l_1) and curves WACC(l_3).

24.8 Conclusions

In current Chapter further development of a new approach to rating methodology has been done. We have generalized it for the general case of modern theory of capital structure (Brusov–Filatova–Orekhova (BFO) theory): for companies of arbitrary age. A serious modification of BFO theory in order to use it in rating procedure has been required. It allows to apply obtained results for real economics, where all companies have finite lifetime, introduce a factor of time into theory, estimate the creditworthiness of companies of arbitrary age (or arbitrary lifetime), introduce discounting of the financial flows, using the correct discount rate etc. This allows use the powerful tools of BFO theory in the rating. All these create a new base for rating methodologies.

References

Brusova, A. (2011). A comparison of the three methods of estimation of weighted average cost of capital and equity cost of company. *Financial Analysis: Problems and Solutions, 34*(76), 36–42.

Brusov, P. N., & Filatova, T. V. (2011). From Modigliani–Miller to general theory of capital cost and capital structure of the company. *Finance and Credit, 435*, 2–8.

Brusov, P., Filatova, T., Orehova, N., & Brusova, A. (2011a). Weighted average cost of capital in the theory of Modigliani–Miller, modified for a finite lifetime company. *Applied Financial Economics, 21*(11), 815–824.

Brusov, P., Filatova, T., Orehova, N., et al. (2011b). From Modigliani–Miller to general theory of capital cost and capital structure of the company. *Research Journal of Economics, Business and ICT, 2*, 16–21.

Brusov, P., Filatova, T., Orehova, N., et al. (2011c). Influence of debt financing on the effectiveness of the investment project within the Modigliani–Miller theory. *Research Journal of Economics, Business and ICT (UK), 2*, 11–15.

Brusov, P., Filatova, T., Eskindarov, M., & Orehova, N. (2012a). Influence of debt financing on the effectiveness of the finite duration investment project. *Applied Financial Economics, 22*(13), 1043–1052.

Brusov, P., Filatova, T., Eskindarov, M., & Orehova, N. (2012b). Hidden global causes of the global financial crisis. *Journal of Reviews on Global Economics, 1*, 106–111.

Brusov, P., Filatova, T., Orekhova, N., Brusov, P., & Brusova, A. (2012c). Modern approach to dividend policy of company. *Finance and Credit, 18*(37), 2012.

Brusov, P., Filatova, P., & Orekhova, N. (2013a). Absence of an optimal capital structure in the famous tradeoff theory! *Journal of Reviews on Global Economics, 2*, 94–116.

Brusov, P., Filatova, T., & Orehova, N. (2013b). A qualitatively new effect in corporative finance: Abnormal dependence of cost of equity of company on leverage. *Journal of Reviews on Global Economics, 2*, 183–193.

Brusov, P., Filatova, P., & Orekhova, N. (2014a). Mechanism of formation of the company optimal capital structure, different from suggested by trade off theory. *Cogent Economics & Finance, 2*, 1–13. https://doi.org/10.1080/23322039.2014.946150

Brusov, P., Filatova, T., & Orehova, N. (2014b). Inflation in Brusov–Filatova–Orekhova theory and in its perpetuity limit – Modigliani – Miller theory. *Journal of Reviews on Global Economics, 3*, 175–185.

Brusov, P., Filatova, T., Orehova, N., & Eskindarov, M. (2015). *Modern corporate finance, investment and taxation* (pp. 1–368). Springer.

Brusov, P., Filatova, T., Orehova, N., & Kulik, V. (2018). Rating: New approach. *Journal of Reviews in Global Economics, 7*, 37–62.

Filatova, T., Orehova, N., & Brusova, A. (2008). Weighted average cost of capital in the theory of Modigliani–Miller, modified for a finite life-time company. *Bulletin of the FU, 48*, 68–77.

Modigliani, F., & Miller, M. (1958). The cost of capital, corporate finance, and the theory of investment. *American Economic Review, 48*, 261–297.

Modigliani, F., & Miller, M. (1963). Corporate income taxes and the cost of capital: A correction. *American Economic Review, 53*, 147–175.

Modigliani, F., & Miller, M. (1966). Some estimates of the cost of capital to the electric utility industry 1954–1957. *American Economic Review, 56*, 333–391.

Chapter 25
Ratings of the Investment Projects of Arbitrary Durations: New Methodology

In this Chapter we develop for the first time a new approach to ratings of the investment projects of arbitrary durations, applicable in particular to energy projects. The ratings of such energy projects, as "Turkish stream", "Nord stream-2", energy projects relating to clean, renewable and sustainable energy, as well as relating to pricing carbon emissions could be done using developed here new rating methodologies. In our previous Chapters the new approach to the ratings of the long-term investment projects has been developed (Filatova et al., 2018). The important features of that consideration are as following: (1) The incorporation of rating parameters (financial "ratios"), used in project rating and playing a major role in it, into modern long–term investment models, (2) The adequate use of discounting of financial flows virtually not used in existing rating methodologies. Here, for the first time, we incorporate the rating parameters (financial "ratios"), used in project rating, into modern investment models, describing the investment projects of arbitrary durations. This was much more difficult task then in case of the long–term investment projects, considered by us in previous Chapters. We work within investment models, created by authors. One of them describes the effectiveness of investment project from perspective of equity capital owners, while other model describes the effectiveness of investment project from perspective of equity capital and debt capital owners. New approach allows use the powerful instruments of modern theory of capital cost and capital structure (BFO theory) (Brusov et al., 2015, 2018b) and modern investment models, created by the authors and well tested in the real economy to evaluate investment project performance, including energy projects. In our calculations we use Excel technique in two aspects: (1) we calculate WACC at different values of equity costs k_0, different values of debt costs k_d and different values of leverage level $L = D/S$, using the famous BFO formula; (2) we calculate the dependences of NPV on coverage ratios as well as leverage ratios at different values of equity costs k_0, different values of debt costs k_d and different values of leverage level L.

P. Brusov, T. Filatova, *Efficiency of Investment Projects*, Contributions to Finance and Accounting, https://doi.org/10.1007/978-3-032-01028-5_25

25.1 Introduction

The investments play a very important role in economics and finance. Wherein the role of energy projects in general and in particular relating to clean, renewable and sustainable energy, as well as relating to pricing carbon emissions rapidly increasing.

In the conditions of limited financial resources, the selection of the most efficient projects from the point of view of investors becomes a very important task. Rating agencies are called upon to solve it.

Rating agencies play a very important role in economics. Their analysis of issuer's state, generated credit ratings of issuers, of investment projects help investors make reasonable investment decision, as well as help issuer with good enough ratings get credits on lower rates etc.

But the methodologies of leading rating agencies, such as "The Big Three credit rating agencies".

(Standard & Poor's (S&P), Moody's, and Fitch Group) as well as Russian rating agency ACRA and all other ones have a lot of shortcomings. A number of works by authors are devoted to eliminating some of these shortcomings (Brusov et al., 2018a, b, c, d, e, f, g).

Besides the fact, that RA represent some "black boxes", the information about the methods of work of which is almost completely absent, there are some serious methodological and systematic errors in their activity. These errors and ways to their overcome have been discussed in a number of authors Chapters (Brusov et al., 2018a, b, c, d, e, f), as well as in monograph (Brusov et al., 2018g).

25.1.1 The Use of Discounting in the Rating

One of the major flaws of all existing rating methodologies is a failure or a very narrow use of discounting. But even in those rare cases where it is used, it is not quite correct, since the discount rate when discounting financial flows is chosen incorrectly.

As (Brusov et al., 2018a, b, c, d, e, f, g) have mentioned "The need to take into account the time factor in terms of discounting is obvious, because it is associated with the time value of money. The financial part of the rating is based on a comparison of generated income with the value of the debt and the interest payable. Because income and disbursement of debt and interest are separated in time, the use of discounting when comparing revenues with the value of debt and interest is absolutely necessary for assigning credit ratings for issuers".

This raises the question about the value of discount rate. This question has always been one of the major and extremely difficult in many areas of Finance: corporate finance, investment, it is particularly important in business valuation, where a slight change in the discount rate leads to a significant change in the assessment of company capitalization, that is used by unscrupulous appraisers for artificial bankruptcy of the company. It is extremely essential as well in rating.

25.1.2 *Incorporation of Financial "Ratios", Using in Ratings, into Modern Investment Models*

In quantification of the creditworthiness of the issuers as well as in valuation of effectiveness of investment projects the crucial role belongs to the so-called financial "ratios", constitute a direct and inverse ratios of various generated cash flows to debt values and interest ones. We incorporate these financial "ratios", into the modern theory of capital structure—Brusov–Filatova–Orekhova (BFO) theory (Brusov et al., 2018a, b, c, d, e, f, g) and as well into modern investment models, created by the authors.

Such incorporation, which has been done by us for the first time, is very important because one can use this theory as a powerful tools when discounting of financial flows using the correct discounting rate in rating. Only this theory allow valuate adequately the weighted average cost of capital WACC and equity cost ke used when discounting of financial flows.

As Brusov et al. (2018d) have mentioned "Use of the tools of well developed theories in rating opens completely new horizons in the rating industry, which could be connected with transition from the mainly use of qualitative methods of the evaluation of the creditworthiness of issuers to a predominantly quantitative evaluation methods that will certainly enhance the quality and correctness of the rating".

Currently, RA just directly use financial ratios, while the new methodology will allow (knowing the values of these "relations" (and parameter k_0)) determine the correct values of discount rates (WACC and k_e) that should be used when discounting the various financial flows, both in terms of their timing as well as forecasting.

In relation to the rating of investment projects new methodology allows correctly determine the values of project NPV (both in units of D as well as in NOI), using modern investment models and correct discount rate.

As Brusov et al. (2018a, b, c, d, e, f, g) mentioned "Incorporation of financial "ratios", has required the modification of the BFO theory (and its perpetuity limit—so called Modigliani–Miller theory), as used in financial management the concept of "leverage" as the ratio of debt value to the equity value substantially differs from the concept of "leverage" in the rating, where it is understood as the direct and inverse ratio of the debt value to the generated cash flow values (income, revenue etc.). The authors introduced some additional ratios, allowing more fully characterize the issuer's ability to repay debts and to pay interest thereon".

Thus the bridge is building between the discount rates (WACC, k_e) used when discounting of financial flows, and "ratios" in the rating methodology.

The important features of current consideration as well as in previous studies are: (1) The adequate use of discounting of financial flows virtually not used in existing rating methodologies, (2) The incorporation of rating parameters (financial "ratios"), used in project rating, into considered modern investment models.

25.2 Investment Models

We work within investment models, created by authors. One of them describes the effectiveness of investment project from perspective of equity capital owners, while other model describes the effectiveness of investment project from perspective of equity capital and debt capital owners.

In the former case, investments at the initial time moment $T = 0$ are equal to $-S$ and the flow of capital for the period (in addition to the tax shields $k_d Dt$ it includes a payment of interest on a loan $-k_d D$):

$$CF = (\mathrm{NOI} - k_d D)(1 - t). \tag{25.1}$$

Here, for simplicity, we suppose that interest on the loan will be paid in equal shares $k_d D$ during all periods. Note that principal repayment is made at the end of the last period.

We will consider the case of discounting, when operating and financial flows are not separated and both are discounted, using the general rate (as which, obviously, the weighted average cost of capital (WACC) can be selected). In case for long–term (perpetuity) projects, the Modigliani–Miller formula (Modigliani and Miller 1958, 1963, 1966) for WACC has been used (Brusov et al., 2018a, b, c, d, e, f, g; Brusov (2018a), Brusov (2018b), Brusov et al. (2019), Brusov et al. (2020)) and for projects of finite (arbitrary) duration Brusov–Filatova–Orekhova formula will be used (Filatova et al. 2008).

Note that debt capital is the least risky, because interest on credit is paid after taxes in the first place. Therefore, the cost of credit will always be less than the equity cost, whether of ordinary or of preference shares $k_e > k_d$; $k_p > k_d$. Here k_e, k_p is the equity cost of ordinary or of preference shares consequently.

25.2.1 *The Effectiveness of the Investment Project from the Perspective of the Equity Holders Only (Without Flows Separation)*

In this case operating and financial flows are not separated and are discounted. Using the general rate (as which, obviously, WACC can be selected).

The credit reimbursable at the end of the project (at the end of the period (n)) can be discounted either at the same rate WACC or at the debt cost rate k_d. Now we choose a uniform rate and the first option.

$$\begin{aligned}\mathrm{NPV} &= -S + \sum_{i=1}^{n} \frac{\mathrm{NOI}(1-t) - k_d D(1-t)}{(1+\mathrm{WACC})^i} - \frac{D}{(1+\mathrm{WACC})^n}\\ &= -S + \frac{\mathrm{NOI}(1-t) - k_d D(1-t)}{\mathrm{WACC}}\left(1 - \frac{1}{(1+\mathrm{WACC})^n}\right) - \frac{D}{(1+\mathrm{WACC})^n}.\end{aligned} \tag{25.2}$$

At a Constant Value of Equity Capital (S = const)

Accounting that in the case S = const NOI is proportional to the invested capital, I, $\mathrm{NOI} = \beta I = \beta S(1 + L)$ and substituting $D = LS$, we get

$$\begin{aligned}\mathrm{NPV} &= -S + \frac{\mathrm{NOI}(1-t) - k_d D(1-t)}{\mathrm{WACC}}\left(1 - \frac{1}{(1+\mathrm{WACC})^n}\right)\\ &\quad - \frac{D}{(1+\mathrm{WACC})^n},\end{aligned} \tag{25.3}$$

$$\begin{aligned}\mathrm{NPV} = -S&\left[1 + \frac{Lk_d(1-t)}{\mathrm{WACC}}\left(1 - \frac{1}{(1+\mathrm{WACC})^n}\right) + \frac{L}{(1+\mathrm{WACC})^n}\right]\\ &+ \frac{\beta S(1+L)(1-t)}{\mathrm{WACC}}\left(1 - \frac{1}{(1+\mathrm{WACC})^n}\right).\end{aligned} \tag{25.4}$$

$$\begin{aligned}NPV = -S&\left[1 + \frac{k_d L(1-t)}{WACC}\left(1 - \frac{1}{(1+WACC)^n}\right) + \frac{L}{(1+WACC)^n}\right]\\ &+ \frac{\beta S(1+L)(1-t)}{WACC}\left(1 - \frac{1}{(1+WACC)^n}\right).\end{aligned} \tag{25.5}$$

25.3 Incorporation of Financial Coefficients, Using in Project Rating, into Modern Investment Models, Describing the Investment Projects of Arbitrary Duration

Below for the first time we incorporate the financial coefficients, used in project rating, into modern investment models, describing the investment projects of arbitrary duration, created by authors. We will consider two kind of financial coefficients: coverage ratios as well as leverage coefficients. In each group of financial coefficients we incorporate three particular quantities.

For coverage ratios we incorporate: (1) coverage ratios of debt, $i_1 = \frac{NPV}{D}$; (2) coverage ratios of interest on the credit $i_2 = \frac{NPV}{k_d D}$; (3) coverage ratios of debt and interest on the credit $i_3 = \frac{NPV}{(1+k_d)D}$.

For leverage ratios we incorporate: (1) leverage ratios of debt, $l_1 = \frac{D}{NPV}$; (2) leverage ratios of interest on the credit $l_2 = \frac{k_d D}{NPV}$; (3) leverage ratios of debt and interest on the credit $l_3 = \frac{(1+k_d)D}{NPV}$.

25.3.1 *Coverage Ratios*

25.3.1.1 Coverage Ratios of Debt

Let us first incorporate the coverage ratios, using in project rating, into modern investment models, describing the investment projects of arbitrary duration, created by authors. Dividing both parts of Eq. (25.5) by D one gets

$$\frac{NPV}{D} = -\frac{1}{L} + \frac{(i_1 - k_d)(1-t)}{WACC}\left(1 - \frac{1}{(1+WACC)^n}\right) - \frac{1}{(1+WACC)^n} \tag{25.6}$$

Here

$$i_1 = \frac{NPV}{D} \tag{25.7}$$

25.3.1.2 Coverage Ratios of Interest on the Credit

Dividing both parts of Eq. (25.5) by $k_d D$ one gets

$$\frac{NPV}{k_d D} = -\frac{1}{k_d L} + \frac{(i_2 - k_d)(1-t)}{WACC}\left(1 - \frac{1}{(1+WACC)^n}\right) - \frac{1}{k_d(1+WACC)^n} \tag{25.8}$$

Here

$$i_2 = \frac{NPV}{k_d D} \tag{25.9}$$

25.3.1.3 Coverage Ratios of Debt and Interest on the Credit

Dividing both parts of Eq. (25.5) by $(1 + k_d)D$ one gets

$$\frac{NPV}{(1+k_d)D} = \frac{1}{(1+k_d)} \times \left[-\frac{1}{L} + \frac{((1+k_d)i_3 - k_d)(1-t)}{WACC}\left(1 - \frac{1}{(1+WACC)^n}\right) - \frac{1}{(1+WACC)^n} \right] \tag{25.10}$$

Here

$$i_3 = \frac{NPV}{(1+k_d)D} \tag{25.11}$$

25.3.2 *Leverage Ratios*

25.3.2.1 Leverage Ratios for Debt

Now let us incorporate the leverage ratios, using in project rating, into modern investment models, created by authors.

Dividing both parts of Eq. (25.5) by *NOI* one gets

$$\frac{NPV}{NOI} = -\frac{l_1}{L} + \frac{(1-k_d l_1)(1-t)}{WACC}\left(1 - \frac{1}{(1+WACC)^n}\right) - \frac{l_1}{(1+WACC)^n} \tag{25.12}$$

Here

$$l_1 = \frac{D}{NPV} \tag{25.13}$$

25.3.2.2 Leverage Ratios for Interest on Credit

$$\frac{NPV}{NOI} = -\frac{l_2}{k_d L} + \frac{(1-l_2)(1-t)}{WACC}\left(1 - \frac{1}{(1+WACC)^n}\right) - \frac{l_2}{k_d(1+WACC)^n} \tag{25.14}$$

Here

$$l_2 = \frac{k_d D}{NPV} \tag{25.15}$$

25.3.2.3 Leverage Ratios for Debt and Interest on Credit

$$\frac{NPV}{NOI} = \frac{1}{1+k_d} \times \left[-\frac{l_3}{L} + \frac{(1+k_d-k_d l_3)(1-t)}{WACC}\left(1 - \frac{1}{(1+WACC)^n}\right) - \frac{l_3}{(1+WACC)^n}\right] \tag{25.16}$$

Here

$$l_3 = \frac{(1+k_d)D}{NPV}. \tag{25.17}$$

Let us investigate below the effectiveness of investment projects of arbitrary duration studying the dependence of NPV on coverage ratios and on leverage ratios. We make calculations for coefficients i_1 and l_1. Calculations for the rest of coefficients (i_2, i_3 and l_2, l_3) could be made in a similar way.

We start from the calculations of the dependence of NPV on coverage ratios. We consider different values of equity costs k_0, different values of debt costs k_d and different values of leverage level $L = D/S$. Here t is tax on profit rate, which in our calculations is equal to 20%.

25.4 Results and Analysis

25.4.1 Dependence of NPV/D on Coverage Ratios

Below we calculate the dependence of NPV (in units of D) ($\frac{NPV}{D}$) on coverage ratio on debt i_1 at different equity costs k_0 (k_0 is the equity cost at zero leverage level L = 0). We will make calculations for two leverage levels L (L = 1 and L = 3), for two project durations (n = 2 and n = 5) and for different credit rates k_d. In our calculations we use Excel technique in two aspects: (1) we calculate WACC at different values of equity costs k_0, different values of debt costs k_d and different values of leverage level $L = D/S$, using the famous BFO formula; (2) we calculate the dependences of NPV on coverage ratios as well as leverage ratios at different values of equity costs k_0, different values of debt costs k_d and different values of leverage level L. We use typical values of equity costs k_0, of debt costs k_d and of leverage level L.

For calculation of the dependence of NPV/D on coverage ratio on debt i_1 within BFO approximation (arbitrary duration projects) we use the formula (25.6)

$$\frac{NPV}{D} = -\frac{1}{L} + \frac{(i_1 - k_d)(1-t)}{WACC}\left(1 - \frac{1}{(1+WACC)^n}\right) - \frac{1}{(1+WACC)^n}$$

25.4.1.1 The Dependence of NPV on Coverage Ratio on Debt i_1

Below we investigate the dependence of NPV/D on coverage ratio on debt i_1 at different values of equity costs k_0, at different values of debt costs k_d at fixed value of equity cost, as well as at different values of leverage levels L.

Let us start our calculations from the case of equity cost $k_0 = 14\%$.

1. We calculate WACC, using the famous BFO formula (Brusov et al., 2015, 2018a, b, c, d, e, f, g)

$$\frac{\left[1-(1+WACC)^{-n}\right]}{WACC} = \frac{\left[1-(1+k_0)^{-n}\right]}{k_0 * \left[1 - Wd * t * \left(1-(1+k_d)^{-n}\right)\right]} \tag{25.18}$$

   Here k_0 is equity costs, k_d is debt costs; $L = D/S$ is the leverage level; t is tax on profit rate, w_d is debt ratio, WACC is weighted average capital cost, n is the project duration.
2. We calculate dependence of NPV/D (NPV in units D) on coverage ratio on debt i_1, using obtained value of WACC, which depends on k_0, k_d, t, n, L.
3. We calculate NPV/NOI (NPV in units NOI) on leverage ratio on debt l_1, using obtained value of WACC, which depends on k_0, k_d, t, n, L.

The results are shown in Tables and Figures.

Table 25.1 The dependence of NPV/D on coverage ratio on debt i_1 at L = 1, k_0 = 14%; k_d = 12%; t = 20%, n = 2

Wd	L	k_0	k_d	t	i_1	n	NPV/D (L = 1)	WACC
0.5	1	0.14	0.12	0.2	0	2	−1.86035	0.185838
0.5	1	0.14	0.12	0.2	1	2	−0.61682	0.185838
0.5	1	0.14	0.12	0.2	2	2	0.626711	0.185838
0.5	1	0.14	0.12	0.2	3	2	1.870243	0.185838
0.5	1	0.14	0.12	0.2	4	2	3.113776	0.185838
0.5	1	0.14	0.12	0.2	5	2	4.357308	0.185838
0.5	1	0.14	0.12	0.2	6	2	5.600841	0.185838
0.5	1	0.14	0.12	0.2	7	2	6.844373	0.185838
0.5	1	0.14	0.12	0.2	8	2	8.087906	0.185838
0.5	1	0.14	0.12	0.2	9	2	9.331438	0.185838
0.5	1	0.14	0.12	0.2	10	2	10.57497	0.185838

Table 25.2 The dependence of NPV/D on coverage ratio on debt i_1 at L = 1, k_0 = 14%; k_d = 12%; t = 20%, n = 5

L	k_0	k_d	t	i_1	n	NPV/D (L = 1)	WACC
1	0.14	0.12	0.2	0	5	−2.0942	0.072849
1	0.14	0.12	0.2	1	5	1.161083	0.072849
1	0.14	0.12	0.2	2	5	4.41637	0.072849
1	0.14	0.12	0.2	3	5	7.671658	0.072849
1	0.14	0.12	0.2	4	5	10.92694	0.072849
1	0.14	0.12	0.2	5	5	14.18223	0.072849
1	0.14	0.12	0.2	6	5	17.43752	0.072849
1	0.14	0.12	0.2	7	5	20.69281	0.072849
1	0.14	0.12	0.2	8	5	23.94809	0.072849
1	0.14	0.12	0.2	9	5	27.20338	0.072849
1	0.14	0.12	0.2	10	5	30.45867	0.072849

We see from the Tables 25.1, 25.2, 25.3, and 25.4 and from Fig. 25.1, that NPV (in units of D) ($\frac{NPV}{D}$) increases with i_1.

The features of this increase are as following:

1. the angle NPV(i_1) is determined by the project duration n: it increases with n.
2. with increase of leverage level L the curve NPV(i_1) shifts practically parallel up. Thus, NPV increases with debt financing.

This means, that influence of the project duration n on the dependence of NPV/D on coverage ratio on debt i_1 is more significant, than influence of leverage level L.

Table 25.3 The dependence of NPV/D on coverage ratio on debt i_1 at L = 3, k0 = 14%; kd = 12%; t = 20%, n = 2

L	k_0	k_d	t	i_1	n	NPV/D (L = 1)	WACC
3	0.14	0.12	0.2	0	2	−1.19238	0.186786
3	0.14	0.12	0.2	1	2	0.049707	0.186786
3	0.14	0.12	0.2	2	2	1.291792	0.186786
3	0.14	0.12	0.2	3	2	2.533877	0.186786
3	0.14	0.12	0.2	4	2	3.775963	0.186786
3	0.14	0.12	0.2	5	2	5.018048	0.186786
3	0.14	0.12	0.2	6	2	6.260133	0.186786
3	0.14	0.12	0.2	7	2	7.502218	0.186786
3	0.14	0.12	0.2	8	2	8.744303	0.186786
3	0.14	0.12	0.2	9	2	9.986389	0.186786
3	0.14	0.12	0.2	10	2	11.22847	0.186786

Table 25.4 The dependence of NPV/D on coverage ratio on debt i_1 at L = 3, k_0 = 14%; k_d = 12%; t = 20%, n = 5

Wd	L	k_0	k_d	t	i_1	n	NPV/D (L = 1)	WACC
0.25	3	0.14	0.12	0.2	0	5	−1.42302	0.073898
0.25	3	0.14	0.12	0.2	1	5	1.823185	0.073898
0.25	3	0.14	0.12	0.2	2	5	5.069387	0.073898
0.25	3	0.14	0.12	0.2	3	5	8.315588	0.073898
0.25	3	0.14	0.12	0.2	4	5	11.56179	0.073898
0.25	3	0.14	0.12	0.2	5	5	14.80799	0.073898
0.25	3	0.14	0.12	0.2	6	5	18.05419	0.073898
0.25	3	0.14	0.12	0.2	7	5	21.30039	0.073898
0.25	3	0.14	0.12	0.2	8	5	24.5466	0.073898
0.25	3	0.14	0.12	0.2	9	5	27.7928	0.073898
0.25	3	0.14	0.12	0.2	10	5	31.039	0.073898

25.4.1.2 The Dependence of NPV on Leverage Ratio on Debt l_1

We see from the Tables 25.5, 25.6, 25.7, and 25.8 and from Fig. 25.2, that NPV (in units of NOI) ($\frac{NPV}{NOI}$) decreases with l_1.

The features of this decrease are as following:

1. the angle NPV(l_1) is determined by the leverage level L: it increases with L. Thus, NPV increases with debt financing.
2. with increase of project duration n the curve NPV(l_1) shifts practically parallel up.

This means, that influence of leverage level L on the dependence of NPV/NOI on leverage ratio on debt l_1 is more significant, than influence of the project duration n.

One can see that the dependence of NPV/NOI on leverage ratio on debt l_1 is opposite to the dependence of NPV/D on coverage ratio on debt i_1:

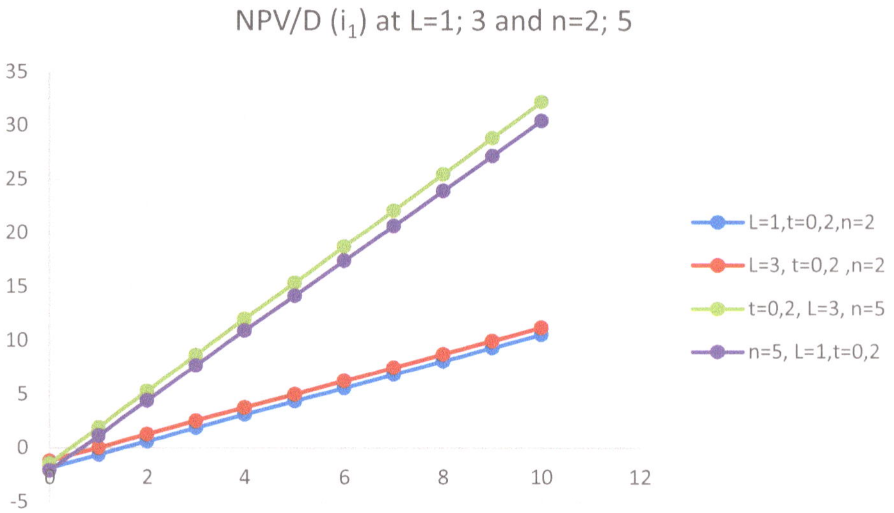

Fig. 25.1 The dependence of NPV/D on coverage ratio on debt i_1 at L = 1;3, k_O = 14%; k_d = 12%; t = 20%, n = 2;5

Table 25.5 The dependence of NPV/NOI on leverage ratio on debt l_1 at L = 1, k_O = 14%; k_d = 12%; t = 20%, n = 2

L	l1	k_d	k_O	t	n	wd	WACC	NPV/NOI (L = 1)
1	0	0.12	0.14	0.2	2	0.5	0.1231	1.346553406
1	1	0.12	0.14	0.2	2	0.5	0.1231	−0.6078321
1	2	0.12	0.14	0.2	2	0.5	0.1231	−2.5622176
1	3	0.12	0.14	0.2	2	0.5	0.1231	−4.5166031
1	4	0.12	0.14	0.2	2	0.5	0.1231	−6.47098861
1	5	0.12	0.14	0.2	2	0.5	0.1231	−8.42537411
1	6	0.12	0.14	0.2	2	0.5	0.1231	−10.3797596
1	7	0.12	0.14	0.2	2	0.5	0.1231	−30.3341451
1	8	0.12	0.14	0.2	2	0.5	0.1231	−14.2885306
1	9	0.12	0.14	0.2	2	0.5	0.1231	−16.2429161
1	10	0.12	0.14	0.2	2	0.5	0.1231	−18.1973016

1. the angle NPV(l_1) is determined by the leverage level L, while the angle NPV(i_1) is determined by the project duration n.
2. with increase of project duration n the curve NPV(l_1) shifts practically parallel up, while such kind of behavior is typical for influence of the leverage level L in case of the curve NPV(i_1).

The only one common thing for both curves NPV(l_1) and NPV(i_1) is that NPV increases with debt financing (or with the leverage level L).

Table 25.6 The dependence of NPV/NOI on leverage ratio on debt l_1 at L = 3, k_0 = 14%; k_d = 12%; t = 20%, n = 2

L	l1	k_d	k_0	t	n	wd	WACC	NPV/NOI (L = 1)
3	0	0.12	0.14	0.2	2	0.25	0.13203	1.330967788
3	1	0.12	0.14	0.2	2	0.25	0.13203	0.057577565
3	2	0.12	0.14	0.2	2	0.25	0.13203	−1.21581266
3	3	0.12	0.14	0.2	2	0.25	0.13203	−2.48920288
3	4	0.12	0.14	0.2	2	0.25	0.13203	−3.7625931
3	5	0.12	0.14	0.2	2	0.25	0.13203	−5.03598332
3	6	0.12	0.14	0.2	2	0.25	0.13203	−6.30937355
3	7	0.12	0.14	0.2	2	0.25	0.13203	−7.58276377
3	8	0.12	0.14	0.2	2	0.25	0.13203	−8.85615399
3	9	0.12	0.14	0.2	2	0.25	0.13203	−10.1295442
3	10	0.12	0.14	0.2	2	0.25	0.13203	−11.4029344

Table 25.7 The dependence of NPV/NOI on leverage ratio on debt l_1 at L = 1, k_0 = 14%; k_d = 12%; t = 20%, n = 5

L	l1	k_d	k_0	t	n	wd	WACC	NPV/NOI (L = 1)
1	0	0.12	0.14	0.2	5	0.5	0.121819	2.870871
1	1	0.12	0.14	0.2	5	0.5	0.121819	0.963526
1	2	0.12	0.14	0.2	5	0.5	0.121819	−0.94382
1	3	0.12	0.14	0.2	5	0.5	0.121819	−2.85116
1	4	0.12	0.14	0.2	5	0.5	0.121819	−4.75851
1	5	0.12	0.14	0.2	5	0.5	0.121819	−6.66585
1	6	0.12	0.14	0.2	5	0.5	0.121819	−8.5732
1	7	0.12	0.14	0.2	5	0.5	0.121819	−10.4805
1	8	0.12	0.14	0.2	5	0.5	0.121819	−12.3879
1	9	0.12	0.14	0.2	5	0.5	0.121819	−14.2952
1	10	0.12	0.14	0.2	5	0.5	0.121819	−16.2026

Table 25.8 The dependence of NPV/NOI on leverage ratio on debt l_1 at L = 3, k_0 = 14%; k_d = 12%; t = 20%, n = 5

L	l1	k_d	k_0	t	n	wd	WACC	NPV/NOI (L = 1)
3	0	0.12	0.14	0.2	5	0.25	0.130962	2.807196
3	1	0.12	0.14	0.2	5	0.25	0.130962	1.596542
3	2	0.12	0.14	0.2	5	0.25	0.130962	0.385889
3	3	0.12	0.14	0.2	5	0.25	0.130962	−0.82476
3	4	0.12	0.14	0.2	5	0.25	0.130962	−2.03542
3	5	0.12	0.14	0.2	5	0.25	0.130962	−3.24607
3	6	0.12	0.14	0.2	5	0.25	0.130962	−4.45672
3	7	0.12	0.14	0.2	5	0.25	0.130962	−5.66738
3	8	0.12	0.14	0.2	5	0.25	0.130962	−6.87803
3	9	0.12	0.14	0.2	5	0.25	0.130962	−8.08868
3	10	0.12	0.14	0.2	5	0.25	0.130962	−9.29934

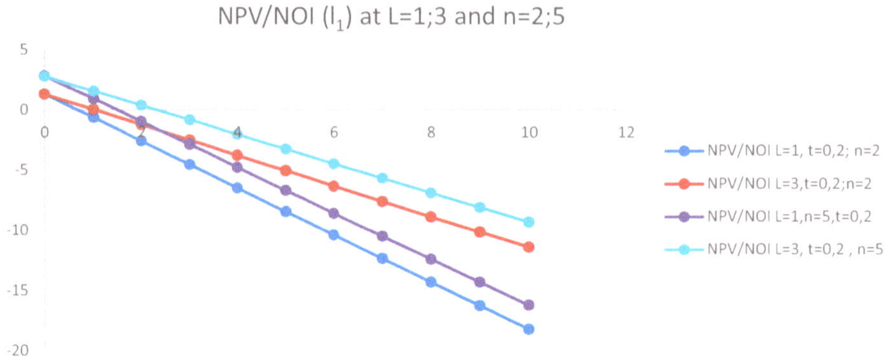

Fig. 25.2 The dependence of NPV/NOI on leverage ratio on debt l_1 at L = 1;3, k_0 = 14%; k_d = 12%; t = 20%, n = 2;5

Table 25.9 The dependence of NPV/D on coverage ratio on debt i_1 at L = 3, k_0 = 20%; k_d = 18%; t = 20%, n = 2

NPV/D	L	i_1	k_d	WACC	t	n
−1.281958725	3	0	0.18	0.176679108	0.2	2
−0.024284066	3	1	0.18	0.176679108	0.2	2
1.233390593	3	2	0.18	0.176679108	0.2	2
2.491065253	3	3	0.18	0.176679108	0.2	2
3.748739912	3	4	0.18	0.176679108	0.2	2
5.006414571	3	5	0.18	0.176679108	0.2	2
6.264089231	3	6	0.18	0.176679108	0.2	2
7.52176389	3	7	0.18	0.176679108	0.2	2
8.779438549	3	8	0.18	0.176679108	0.2	2
10.03711321	3	9	0.18	0.176679108	0.2	2
11.29478787	3	10	0.18	0.176679108	0.2	2

25.4.1.3 The Dependence of NPV on Coverage Ratio on Debt i_1 at Different Values of k_d

Let us investigate the dependence of NPV/D on coverage ratio on debt i_1 at L = 3, k_0 = 20%; k_d = 18%; t = 20%, n = 2 (Tables 25.9, 25.10, 25.11, Fig. 25.3).

The straight lines for different k_d turn out practically merge, which means that the influence of the variation of k_d on such a scale of changes i_1 is insignificant. In order to evaluate the ordering of straight lines corresponding to different k_d, we increase the scale in the following figures, considering the values of i_1 not from 0 to 10, but from 0 to 1. In Fig. 25.4, we consider the case of different project duration n, while in Fig. 25.5 the case of different leverage level L.

One can see from Fig. 25.5, that the ordering of the NPV/D straight lines for different credit rates k_d and different leverage level L turns out to be as following: two triplets, corresponding to different leverage level L are well distinguished and upper triplet (with bigger NPV value) corresponds to bigger leverage level L = 3.

Table 25.10 The dependence of NPV/D on coverage ratio on debt i_1 at L = 3, $k_0 = 20\%$; $k_d = 16\%$; t = 20%, n = 2

NPV	L	i_1	k_d	WACC	t	n
−1.249017883	3	0	0.16	0.18198583	0.2	2
0.000427777	3	1	0.16	0.18198583	0.2	2
1.249873438	3	2	0.16	0.18198583	0.2	2
2.499319098	3	3	0.16	0.18198583	0.2	2
3.748764758	3	4	0.16	0.18198583	0.2	2
4.998210418	3	5	0.16	0.18198583	0.2	2
6.247656078	3	6	0.16	0.18198583	0.2	2
7.497101739	3	7	0.16	0.18198583	0.2	2
8.746547399	3	8	0.16	0.18198583	0.2	2
9.995993059	3	9	0.16	0.18198583	0.2	2
11.24543872	3	10	0.16	0.18198583	0.2	2

Table 25.11 The dependence of NPV/D on coverage ratio on debt i_1 at L = 3, $k_0 = 20\%$; $k_d = 14\%$; t = 20%, n = 2

NPV	L	i_1	k_d	WACC	t	n
−1.221437995	3	0	0.14	0.183805688	0.2	2
0.025207969	3	1	0.14	0.183805688	0.2	2
1.271853932	3	2	0.14	0.183805688	0.2	2
2.518499896	3	3	0.14	0.183805688	0.2	2
3.76514586	3	4	0.14	0.183805688	0.2	2
5.011791823	3	5	0.14	0.183805688	0.2	2
6.258437787	3	6	0.14	0,183,805,688	0.2	2
7.505083751	3	7	0.14	0,183,805,688	0.2	2
8.751729714	3	8	0.14	0,183,805,688	0.2	2
9.998375678	3	9	0.14	0,183,805,688	0.2	2
11.24502164	3	10	0.14	0,183,805,688	0.2	2

This means that NPV increases with debt financing. Within each triplet NPV decreases with credit rates k_d: the biggest NPV corresponds to $k_d = 14\%$ and the smallest one corresponds to $k_d = 18\%$.

25.4.1.4 The Dependence of NPV/NOI on Leverage Ratio on Debt l_1 at Different Values of k_d

Let us investigate the dependence of NPV/NOI on leverage ratio on debt l_1 at L = 1, $k_0 = 20\%$; $k_d = 18\%$; t = 20%, n = 2.

We show below the detailed dependence of NPV/NOI on leverage ratio on debt l_1 at L = 1; $k_0 = 20\%$; $k_d = 14\%$,!6%, 18%; t = 20%, n = 2;5 (i_1 changes from 0 to 3).

One can see from Fig. 25.6, that under increase of project duration n the NPV/NOI(l_1) straight lines shift practically parallel up. The ordering of the NPV/NOI(l_1) straight lines for different credit rates k_d and different project duration n turns out to be as following: two triplets, corresponding to different project duration n are well distinguished and upper triplet (with bigger NPV value) corresponds to bigger project duration n = 5. Within each triplet NPV decreases with

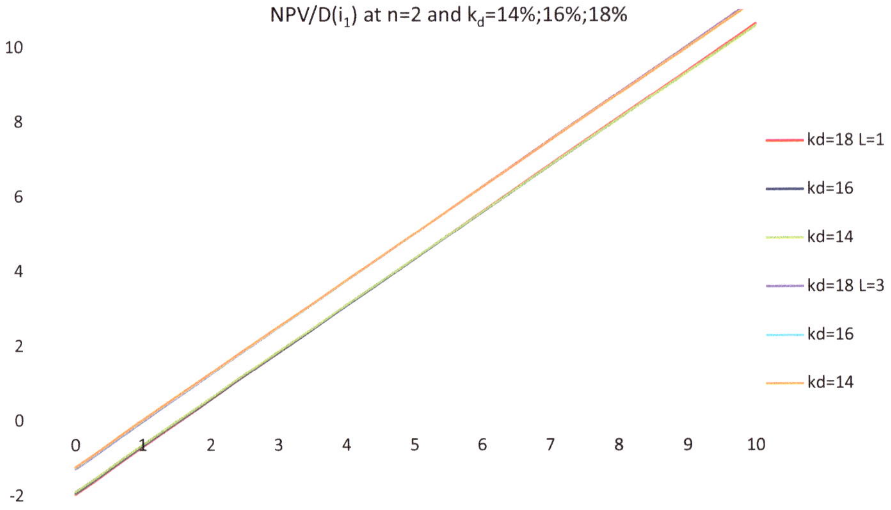

Fig. 25.3 The dependence of NPV/D on coverage ratio on debt i_1 at L = 1;3, $k_0 = 20\%$; $k_d = 14\%$, 16%, 18%; t = 20%, n = 2 (i_1 changes from 0 to 10)

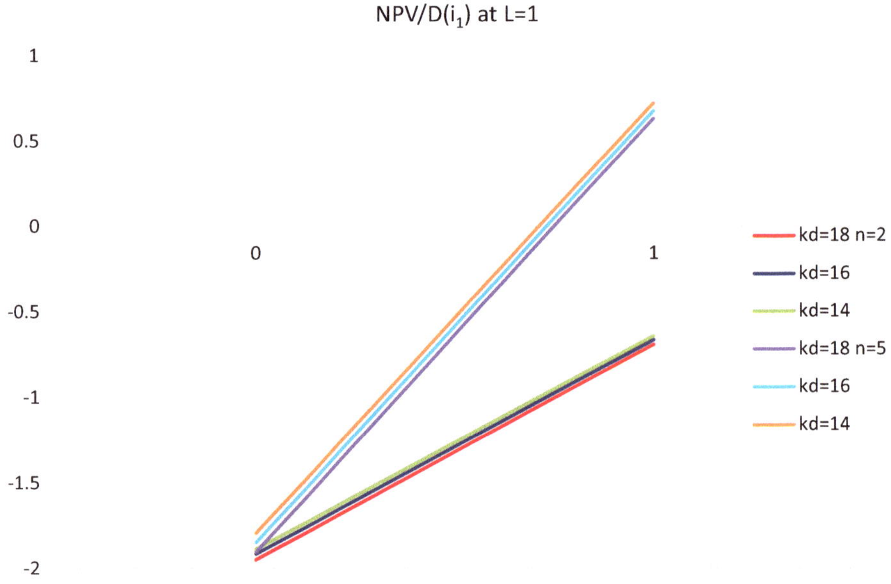

Fig. 25.4 The detailed dependence of NPV/D on coverage ratio on debt i_1 at L = 1, $k_0 = 20\%$; $k_d = 14\%$; 16%; 18%; t = 20%, n = 2;5

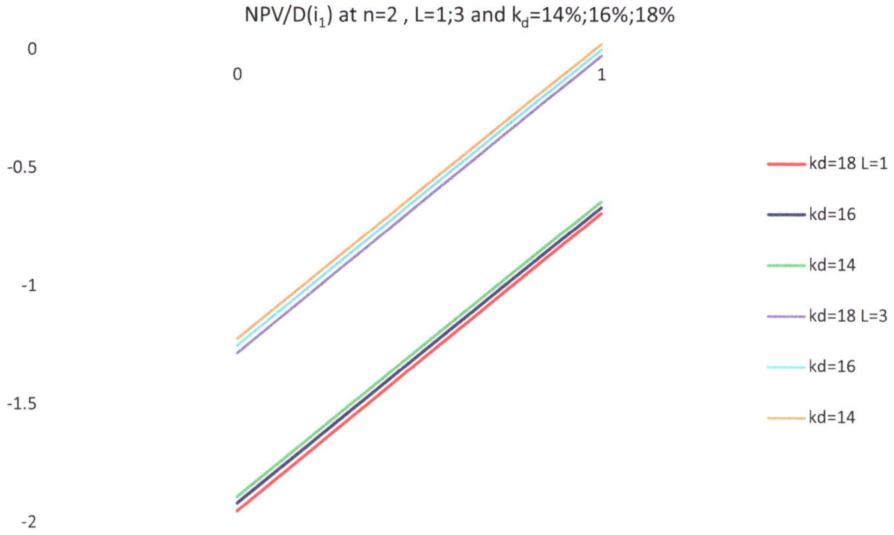

Fig. 25.5 The detailed dependence of NPV/D on coverage ratio on debt i_1 at L = 1;3, k_0 = 20%; k_d = 14%,!6%, 18%; t = 20%, n = 2 (i_1 changes from 0 to 1)

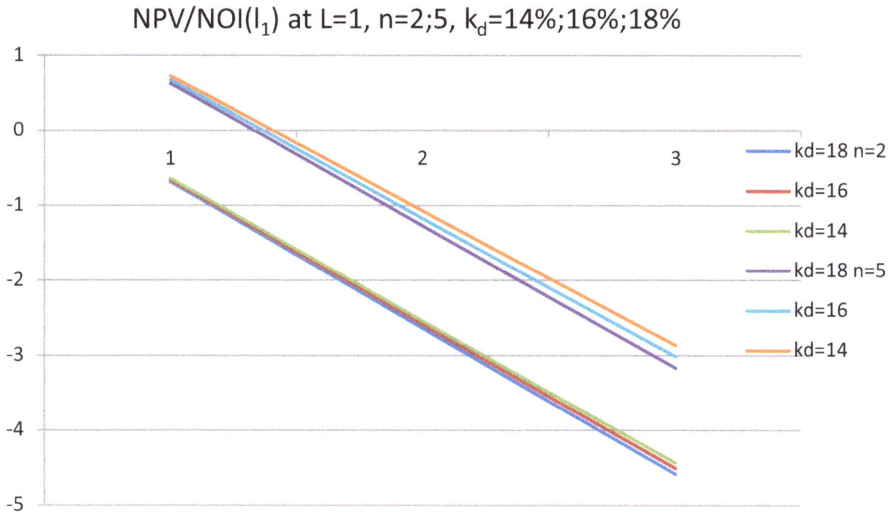

Fig. 25.6 The detailed dependence of NPV/NOI on leverage ratio on debt l_1 at L = 1; k_0 = 20%; k_d = 14%,!6%, 18%; t = 20%, n = 2;5 (i_1 changes from 0 to 3)

credit rates kd: the biggest NPV corresponds to k_d = 14% and the smallest one corresponds to k_d = 18%. It is seen, that influence of the value of credit rates k_d increases with n, while in case of NPV/D the shift of NPV turns out to be the same for different leverage level L.

25.5 Conclusion

In this Chapter we develop for the first time a new approach to ratings of the investment projects of arbitrary duration, applicable to energy projects. The ratings of such energy projects, as "Turkish stream", "Nord stream-2", energy projects relating to clean, renewable and sustainable energy, as well as relating to pricing carbon emissions could be done using developed here new rating methodologies. Chapter generalizes the new approach to the ratings of the long–term investment projects, which has been developed in our previous Chapter (Brusov et al., 2018a, b, c, d, e, f, g). The important features of current consideration as well as in previous studies are: (1) The adequate use of discounting of financial flows virtually not used in existing rating methodologies, (2) The incorporation of rating parameters (financial "ratios"), used in project rating, into considered modern investment models.

We use the modern investment models, created by us, with incorporated financial "ratios" to study the dependence of *NPV* on rating parameters (financial "ratios") at different values of equity cost k_0, at different values of credit rates k_d as well as at different values of on leverage level L and project duration n. We study the dependence of *NPV* on two types of financial "ratios": on the coverage ratios i_j as well as on the leverage ratios l_j. In our calculations we use Excel technique in two aspects: (1) we calculate WACC at different values of equity costs k_0, different values of debt costs k_d and different values of leverage level $L = D/S$, using the famous BFO formula; (2) we calculate the dependences of NPV on coverage ratios as well as leverage ratios at different values of equity costs k_0, different values of debt costs k_d and different values of leverage level L.

Analyzing obtained results we have found:

I. NPV (in units of D) $(\frac{NPV}{D})$ increases with i_1 with following features:

   1. the angle NPV(i_1) is determined by the project duration n: it increases with n.
   2. with increase of leverage level L the curve NPV(i_1) shifts practically parallel up. Thus, NPV increases with debt financing.

   This means, that influence of the project duration n on the dependence of NPV/D on coverage ratio on debt i_1 is more significant, than influence of leverage level L.

II. NPV (in units of NOI) $(\frac{NPV}{NOI})$ decreases with l_1 with following features:

   1. the angle NPV(l_1) is determined by the leverage level L: it increases with L. Thus, NPV increases with debt financing.
   2. with increase of project duration n the curve NPV(l_1) shifts practically parallel up.

This means, that influence of leverage level L on the dependence of NPV/NOI on leverage ratio on debt l_1 is more significant, than influence of the project duration n.

One can see that the dependence of NPV/NOI on leverage ratio on debt l_1 is opposite to the dependence of NPV/D on coverage ratio on debt i_1:

1. the angle NPV(l_1) is determined by the leverage level L, while the angle NPV(i_1) is determined by the project duration n.
2. with increase of project duration n the curve NPV(l_1) shifts practically parallel up, while such kind of behavior is typical for influence of the leverage level L in case of the curve NPV(i_1).

The only one common thing for both curves NPV(l_1) and NPV(i_1) is that NPV increases with debt financing (or with the leverage level L). This means that debt financing of the projects of arbitrary duration favors effectiveness of the investment project as well as its creditworthiness.

The obtained by us results allow make adequate estimation of the effectiveness of the investment projects, NPV, knowing rating parameters (financial "ratios"). For all calculations we use the correct values of discount rate, WACC, which is calculated by use of the modern theory of capital cost and capital structure (BFO theory) (Brusov et al., 2018a, b, c, d, e, f, g).

Investigations, conducting in current Chapter, creates a new approach to rating methodology with respect to the ratings of the investment project of arbitrary duration. It allows use the financial "ratios" for adequate estimation of the effectiveness of the investment projects of arbitrary duration, including energy projects, such as "Turkish stream", "Nord stream-2", etc., energy projects relating to clean, renewable and sustainable energy.

References

Brusov, P. (2018a). Editorial: Introduction on special issue on the banking system and financial markets of Russia and other countries: Problems and prospects. *Journal of Reviews on Global Economics, 7*, i–vi.

Brusov, P. (2018b). Editorial. *Journal of Reviews on Global Economics, 7*, i–v.

Brusov, P., Filatova, T., Orehova, N., & Eskindarov, M. (2015). *Modern corporate finance, investments and taxation* (monograph) (373 p). Springer. Retrieved from https://www.springer.com/gp/book/9783319147314

Brusov, P., Filatova, T., Orehova, N., & Eskindarov, M. (2018a). *Modern corporate finance, investments, taxation and ratings* (monograph) (571 p). Springer.

Brusov, P. N., Filatova, T. V., Orekhova, N. P., & Kulik, V. L. (2018b). Rating: New approach. *Journal of Reviews on Global Economics, 7*, 37–62. https://doi.org/10.6000/1929-7092.2018.07.05

Brusov, P. N., Filatova, T. V., Orekhova, N. P., & Kulik, V. L. (2018c). A "golden age" of the companies: Conditions of its existence. *Journal of Reviews on Global Economics, 7*, 88–103. https://doi.org/10.6000/1929-7092.2018.07.07

Brusov, P. N., Filatova, T. V., Orekhova, N. P., & Kulik, V. L. (2018d). Rating methodology: New look and new horizons. *Journal of Reviews on Global Economics, 7*, 63–87. https://doi.org/10.6000/1929-7092.2018.07.06

Brusov, P. N., Filatova, T. V., Orekhova, N. P., Kulik, V. L., & Weil, I. (2018e). New meaningful effects in modern capital structure theory. *Journal of Reviews on Global Economics, 7*, 104–122. https://doi.org/10.6000/1929-7092.2018.07.08

Brusov, P., Filatova, T., Orekhova, N., Kulik, V., Weil, I., & Brailov, A. (2018f). The impact of the central bank key rate and commercial banks credit rates on creating and maintaining of a favorable investment climate in the country. *Journal of Reviews on Global Economics, 7*, 360–376. https://doi.org/10.6000/1929-7092.2018.07.31

Brusov, P. N., Filatova, T. V., & Orekhova, N. P. (2018g). *Modern corporate finance and investments, monograph* (517 p). Knorus Publishing House.

Brusov, P. N., Filatova, T. V., Orekhova, N. P., Kulik, V. L., & Weil, I. (2019). Ratings of the investment projects of arbitrary durations: New methodology. *Journal of Reviews on Global Economics, 8*, 437–448. https://doi.org/10.6000/1929-7092.2019.08.37

Brusov, P. N., Filatova, T. V., Orekhova, N. P., Kulik, V. L., Chang, S.-I., & Lin, Y. C. G. (2020). Modification of the Modigliani–Miller theory for the case of advance payments of tax on profit. *Journal of Reviews on Global Economics, 9*, 257–268.

Filatova, T., Orehova, N., & Brusova, A. (2008). Weighted average cost of capital in the theory of Modigliani–Miller, modified for a finite life–time company. *Bulletin of the FU, 48*, 68–77.

Filatova, T. V., Brusov, P. N., Orekhova, N. P., & Kulik, V. L. (2018). Ratings of the long-term projects: New approach. *Journal of Reviews on Global Economics, 7*, 645–661. https://doi.org/10.6000/1929-7092.2018.07.59

Modigliani, F., & Miller, M. (1958). The cost of capital, corporate finance, and the theory of investment. *The American Economic Review, 48*, 261–297.

Modigliani, F., & Miller, M. (1963). Corporate income taxes and the cost of capital: A correction. *The American Economic Review, 53*, 147–175.

Modigliani, F., & Miller, M. (1966). Some estimates of the cost of capital to the electric utility industry 1954–1957. *The American Economic Review, 56*, 333–391.

Chapter 26
Ratings of Investment Projects of Arbitrary Duration with a Uniform Debt Repayment: A New Approach

26.1 Introduction

Along with the rating of non-financial issuers, considered in previous Chapters (21–25), the rating of investment projects plays an important role in the modern economy and finance. It allows ranking and selection of the most effective investment projects, which is especially important for attracting both foreign and domestic investments. This Chapter discusses the rating of investment projects of arbitrary duration with a uniform repayment of debt, investment model for which is described in Chap. 16. The methodology for rating investment projects has been modified. A fundamentally new approach to the project rating methodology has been developed, the key factors of which are: (1) adequate application of discounting when discounting the financial flows, which is practically not used in existing project rating methodologies; (2) incorporation of financial ratios into modern investment models created by the authors; (3) use rating parameters upon discounting; (4) the determination of the correct discount rate, taking into account financial ratios. We use modern investment model created by the authors (see Chap. 17), the modern theory of capital cost and capital structure by Brusov–Filatova–Orekhova (BFO theory), its modification for rating needs, and rating coefficients. Various theories of capital cost and capital structure are described in Chaps. 3, 4, 5, 6, 7, 8, 9, 10, 11, 12, 13, and 14. (Brusov 2018a, 2018b; Brusov et al., 2015, 2018a, 2018b, 2018c, 2018d, 2018e, 2018f, 2018g, 2019, 2020; Filatova et al., 2008, 2018; Modigliani & Miller, 1958, 1963, 1966; Myers, 2001).

The developed approach should be applied by rating agencies, both international and national ones, when rating investment projects.

The modification of the methodology of the existing project rating systems will improve the accuracy of ratings of investment projects and make them more objective. The use of powerful tools of well-developed theories opens up new opportunities for the rating industry, which gets the opportunity to switch from

P. Brusov, T. Filatova, *Efficiency of Investment Projects*, Contributions to Finance and Accounting, https://doi.org/10.1007/978-3-032-01028-5_26

using primarily qualitative methods for assessing the effectiveness of investment projects to using mainly quantitative methods for evaluating them.

26.2 Incorporation of Financial Ratios Used in Project Rating into Modern Investment Models with Uniform Repayment of Debt

Below we incorporate the financial coefficients, used in project rating, into modern investment models with a uniform repayment of debt, created by authors. We will consider two kind of financial coefficients: coverage ratios as well as leverage coefficients. In each group of financial coefficients we incorporate three particular quantities.

For coverage ratios we incorporate: (1) coverage ratios of debt, $i_1 = \frac{NPV}{D}$; (2) coverage ratios of interest on the credit $i_2 = \frac{NPV}{k_d D}$; (3) coverage ratios of debt and interest on the credit $i_3 = \frac{NPV}{(1+k_d)D}$.

For leverage ratios we incorporate: (1) leverage ratios of debt, $l_1 = \frac{D}{NPV}$; (2) leverage ratios of interest on the credit $l_2 = \frac{k_d D}{NPV}$; (3) leverage ratios of debt and interest on the credit $l_3 = \frac{(1+k_d)D}{NPV}$.

Note that the last type of ratios (i_3 and l_3) was introduced by us for the first time to more fully characterize the ability to repay debts and pay interest on them at the expense of operating income when implementing an investment project.

26.2.1 Coverage Ratios

26.2.1.1 Coverage Ratios of Debt

Let us first incorporate the coverage ratios, using in project rating, into modern investment models, created by authors. Dividing both parts of Eq. (16.7) from Chap. 16 by D one gets for the net present value of the project, NPV (in units of D), the following expression

$$\frac{NPV}{D} = -\frac{1}{L} + \frac{i_1(1-t) - \frac{1}{n} - k_d \frac{n+1}{n}(1-t)}{WACC}\left(1 - \frac{1}{(1+WACC)^n}\right)$$
$$+ \frac{k_d}{n}(1-t)\left\{\frac{1 + WACC[1-(1+WACC)^{-n}]}{WACC^2} - \frac{n}{WACC(1+WACC)^n}\right\} \quad (26.1)$$

Here $i_1 = NOI/D$, $L = D/S$—leverage level.

26.2.1.2 Coverage Ratios of Interest on the Credit

Dividing both parts of Eq. (16.7) from Chap. 16 by k_dD one gets for the net present value of the project, NPV (in units of k_dD), the following expression

$$\frac{NPV}{k_dD} = -\frac{1}{k_dL} + \frac{i_2(1-t) - \frac{1}{k_dn} - \frac{n+1}{n}(1-t)}{WACC}\left(1 - \frac{1}{(1+WACC)^n}\right) \\ + \frac{1}{n}(1-t)\left\{\frac{(1+WACC)[1-(1+WACC)^{-n}]}{WACC^2}\right\} - \frac{n}{WACC(1+WACC)^n} \tag{26.2}$$

Here $i_2 = NOI/k_dD$.

26.2.1.3 Coverage Ratios of Debt and Interest on the Credit (New Parameter)

Dividing both parts of Eq. (16.7) from Chap. 16 by $(1 + k_d)D$ one gets for the net present value of the project, NPV (in units of $(1 + k_d)D$), the following expression

$$\frac{NPV}{D(1+k_d)} = \frac{1}{L(1+k_d)} + \frac{i_3(1-t) - \frac{1}{n(1+k_d)} - \frac{k_d}{(1+k_d)}\frac{n+1}{n}(1-t)}{WACC}\left(1 - \frac{1}{(1+WACC)^n}\right) \\ + \frac{k_d}{(1+k_d)\cdot n}(1-t)\left\{\frac{(1+WACC)[1-(1+WACC)^{-n}]}{WACC^2}\right\} - \frac{n}{WACC(1+WACC)^n} \tag{26.3}$$

Here $i_3 = NOI/D(1 + k_d)$.

Analyzing the formulas (26.1), (26.2) and (26.3) we come to very important conclusion, that NPV (in units of D) $(\frac{NPV}{D})$ depends only on equity cost k_0, on credit rates k_d, on leverage level L as well as on one of the coverage ratios i_j and does not depend on equity value S, debt value D and NOI. This means that results on the dependence of NPV (in units of D) $(\frac{NPV}{D})$ on coverage ratios i_j at different equity costs k_0, at different credit rates k_d, at different leverage levels L carry the universal character: these dependencies remain valid for investment projects with any equity value S, debt value D and NOI.

26.2.2 Leverage Ratios

26.2.2.1 Leverage Ratios for Debt

Now let us incorporate the leverage ratios, using in project rating, into modern investment models, created by authors.

Dividing both parts of Eq. (16.7) from Chap. 16 by *NOI* one gets for the net present value of the project, NPV (in units of *NOI*), the following expression

$$\frac{NPV}{NOI} = -\frac{l_1}{L} + \frac{(1-t) - \frac{l_1}{n} - k_d l_1 \frac{n+1}{n}(1-t)}{WACC}\left(1 - \frac{1}{(1+WACC)^n}\right) + \frac{k_d l_1}{n}(1-t)\left\{\frac{(1+WACC)\left[1-(1+WACC)^{-n}\right]}{WACC^2} - \frac{n}{WACC(1+WACC)^n}\right\} \tag{26.4}$$

Here $l_1 = D/NOI$.

26.2.2.2 Leverage Ratios for Interest on Credit

Dividing both parts of Eq. (16.7) from Chap. 16 by *NOI* one gets for the net present value of the project, NPV (in units of *NOI*), the following expression

$$\frac{NPV}{NOI} = -\frac{l_2}{k_d L} + \frac{(1-t) - \frac{l_2}{k_d n} - l_2 \frac{n+1}{n}(1-t)}{WACC}\left(1 - \frac{1}{(1+WACC)^n}\right) + \frac{l_2}{n}(1-t)\left\{\frac{(1+WACC)\left[1-(1+WACC)^{-n}\right]}{WACC^2} - \frac{n}{WACC(1+WACC)^n}\right\} \tag{26.5}$$

Here $l_2 = k_d D/NOI$

26.2.2.3 Leverage Ratios for Debt and Interest on Credit

Dividing both parts of Eq. (16.7) from Chap. 16 by *NOI* one gets for the net present value of the project, NPV (in units of *NOI*), the following expression

$$\frac{NPV}{NOI} = -\frac{1}{L(1+k_d)} + \frac{(1-t) - \frac{l_3}{(1+k_d)\cdot n} - \frac{k_d l_3}{(1+k_d)}\frac{n+1}{n}(1-t)}{WACC}\left(1 - \frac{1}{(1+WACC)^n}\right)$$
$$+ \frac{k_d l_3}{(1+k_d)\cdot n}(1-t)\left\{\frac{(1+WACC)[1-(1+WACC)^{-n}]}{WACC^2} - \frac{n}{WACC(1+WACC)^n}\right\} \tag{26.6}$$

Here $l_3 = (1 + k_d)D/NOI$

Analyzing the formulas (26.4), (26.5) and (26.6) we come to very important conclusion, that *NPV* (in units of *NOI*) ($\frac{NPV}{NOI}$) depends only on equity cost k_0, on credit rates k_d, on leverage level L as well as on one of the leverage ratios l_j and does not depend on equity value *S*, debt value *D* and *NOI*. This means that results on the dependence of NPV (in units of *NOI*) ($\frac{NPV}{NOI}$) on leverage ratios l_j at different equity costs k_0, at different credit rates k_d, at different leverage levels L carry the universal character: these dependencies remain valid for investment projects with any equity value *S*, debt value *D* and *NOI*.

We investigate below the effectiveness of long-term investment projects studying the dependence of NPV on coverage ratios and on leverage ratios. We make calculations for coefficients i_1 and l_1. Calculations for the rest of coefficients (i_2, i_3 and l_2, l_3) could be made in a similar way.

We start from the calculations of the dependence of NPV on coverage ratios. We consider different values of equity costs k_0, of debt costs k_d and of leverage level $L = D/S$. Here t is tax on profit rate, which in our calculations is equal to 20%.

26.2.3 Perpetuity Limit

In the perpetuity limit ($n \to \infty$) from Eq. (26.9), or as limit expressions (26.1)–(26.6), we can obtain the following expressions for NPV (in units of D) and NPV (in units of NOI).

$$\frac{NPV}{D} = -\frac{1}{L} + \frac{i_1(1-t) - k_d(1-t)}{WACC} \tag{26.7}$$

$$\frac{NPV}{NOI} = -\frac{l_1}{L} + \frac{(1-t) - k_d l_1(1-t)}{WACC} \tag{26.8}$$

Substituting instead of WACC its value in the theory of MM

$$WACC = k_0\left(1 - \frac{Lt}{1+L}\right), \tag{26.9}$$

one gets for NPV (in units *D*)

$$\frac{NPV}{D} = -\frac{1}{L} + \frac{(k_d - i_1)(1-t)}{k_0\left(1 - \frac{Lt}{1+L}\right)} \tag{26.10}$$

and for NPV (in units *NOI*)

$$\frac{NPV}{NOI} = \frac{-l_1}{L} + \frac{(1 - k_d l_1)(1-t)}{k_0\left(1 - \frac{Lt}{1+L}\right)}. \tag{26.11}$$

Here k_0 is the equity cost at zero leverage. In the following paragraphs, we will study the dependence of the net present value, NPV, on financial ratios for companies of a given age n and for paticular values of investment project parameters, demonstrating the possibility of performing calculations for companies of arbitrary age n and for any values of investment project parameters.

26.2.4 *The Study of the Dependence of the Net Present Value of the Project, NPV, on Rating Parameters*

26.2.4.1 Investigation of the Dependence of the Net Present Value of the Project, NPV (in Units of Debt D) on Coverage Ratios

First, we analyze the dependence of the net present value of the NPV/D project on coverage coefficients i_j for companies of arbitrary age n in the framework of the modern theory of value and capital structure by Brusov–Filatova–Orekhova using modern investment models created by the authors and described above.

We restrict ourselves to considering the dependence of NPV on the debt coverage coefficients i_1 described by Eq. (26.1). Calculations for other coverage coefficients described by Eqs. (26.2)–(26.3) are carried out similarly.

The calculations are carried out in two stages.

1. First, for a given set of project parameters according to the famous Brusov–Filatova–Orekhova formula

$$\frac{1 - (1 + WACC)^{-n}}{WACC} = \frac{1 - (1 + k_0)^{-n}}{k_0 * (1 - w_d * t * (1 - (1 + k_d)^{-n}))} \tag{26.12}$$

   weighted average cost of capital, WACC, is calculated by Microsoft Excel using the "matching parameter" function.
2. The found WACC value is used in Eq. (26.1) to study the dependence of NPV/D on debt coverage coefficient i_1.

Similar studies can be carried out for other coverage coefficients (i_2 and i_3).

When analyzing this formula, we will use the following parameter values: L = 1, 3; t = 20%; n is 3; 5; k_0 = 18%; k_d = 12%, 14%, 16%; i_1 = 0, 1, 2, 3, 4, 5, 6, 7, 8, 9, 10.

The following conclusions can be drawn from the data in Tables 26.1 and 26.2:

1. The net present value of the project NPV/D increases linearly with an increase of debt coverage ratio i_1.
2. The dependence NPV/D (i_1) for various values of the cost of debt capital k_d has interesting features: for values of the coverage ratio for debt i_1 from 0 to 3, an increase in the cost of debt capital k_d leads to a decrease in NPV/D, and for the values of the cost of borrowed capital k_d more than 3 (and up to 10) large values of the cost of borrowed capital k_d correspond to large values of NPV/D.

From Tables 26.1 and 26.2 it is also seen that the net present value of the project, NPV/D. for various values of debt capital k_d (12%, 14%, 16%) differ slightly from each other (by approximately 0.02), due to that the graphs of the dependence NPV/D (i_1) practically coincide, as shown in Fig. 26.1.

To demonstrate the differences in the dependence of the net present value of the project NPV (in units of debt D) on the debt coverage ratio i_1 for various L and k_d, we give Fig. 26.1 on an enlarged scale (Fig. 26.2).

From Fig. 26.2 and from Tables 26.1 and 26.2 we can come to the following conclusions:

1. The slope of the straight line NPV/D (i_1) is determined by the duration of the project n and increases with the duration of the project. So, the slope of the straight line NPV/D (i_1) for a 5-year project is greater than for a 3-year project.
2. When the leverage level L changes, the line NPV/D (i_1) shifts almost in parallel: with the growth of L, the line shifts higher (in Fig. 26.2 it can be seen that for projects of the same term, lines with L = 3 lie above lines with L = 1.

26.2.4.2 Study of the Dependence of the Net Present Value of the Project NPV (in Units of Net Operating Income NOI) on Leverage Ratios

Now we analyze the dependence of the net present value of the project NPV/D on leverage coefficients l_j for companies of arbitrary age n in the framework of the modern theory of value and capital structure of Brusov–Filatova–Orekhova using modern investment models created by the authors and described above.

We restrict ourselves to considering the dependence of NPV on the leverage coefficient for debt l_1, described by Eq. (26.4). Calculations for other leverage coefficients (l_2 and l_3) described by Eqs. (26.5)–(26.6) are carried out similarly.

The calculations, as in the case of coverage coefficients, are carried out in two stages.

1. First, for a given set of project parameters according to the famous Brusov–Filatova–Orekhova formula:

Table 26.1 The dependence of NPV/D on debt coverage ratio i_1 at $L = 1{,}3$ $t = 20\%$, $n = 3$, $k_0 = 18\%$, $k_d = k_d = 12\%, 14\%, 16\%$

NPV/D	WACC	k_d	$i_1 = 0$	1	2	3	4	5	6	7	8	9	10
L = 1	0,1574	16%	−1,954	−0,1493	1,6549	3,4592	5,2634	7,0677	8,8719	10,6762	12,4804	14,2847	16,0890
n = 3	0,1596	14%	−1,925	−0,1273	1,6705	3,4684	5,2663	7,0641	8,8620	10,6598	12,4577	14,2555	16,0534
	0,1619	12%	−1,897	−0,1057	1,6854	3,4764	5,2675	7,0585	8,8495	10,6406	12,4316	14,2227	16,0137
L = 3	0,1461	16%	−1,30	0,53	2,37	4,21	6,05	7,89	9,73	11,57	13,40	15,24	17,08
n = 3	0,1493	14%	−1,274	0,5548	2,3833	4,2119	6,0405	7,8690	9,6976	11,5261	13,3548	15,1833	17,0119
	0,1528	12%	−1,243	0,5749	2,3929	4,2109	6,0289	7,8469	9,6649	11,4830	13,3010	15,1190	16,9370

Table 26.2 The dependence of NPV/D on debt coverage ratio i_1 at L = 1,3 t = 20%, n = 5, k_0 = 18%, k_d = k_d = 12%, 14%, 16%

NPV/D	WACC	k_d	$i_1 = 0$	1	2	3	4	5	6	7	8	9	10
L = 1	0,1566	16%	−1,938	0,7022	3,3423	5,9823	8,6224	11,2624	13,9025	16,5425	19,1825	21,8226	24,4626
n = 5	0,1586	14%	−1,899	0,7288	3,3569	5,9849	8,6130	11,2410	13,8691	16,4971	19,1252	21,7532	24,3813
	0,1608	12%	−1,861	0,7543	3,3692	5,9840	8,5989	11,2137	13,8286	16,4434	19,0583	21,6731	24,2880
L = 3	0,1448	16%	−1,296	1,4190	4,1341	6,8492	9,5643	12,2794	14,9945	17,7096	20,4247	23,1398	25,8549
n = 5	0,1477	14%	−1,255	1,4416	4,1378	6,8339	9,5300	12,2261	14,9222	17,6183	20,3144	23,0105	25,7067
	0,1510	12%	−1,213	1,4626	4,1380	6,8133	9,4886	12,1639	14,8393	17,5146	20,1899	22,8653	25,5406

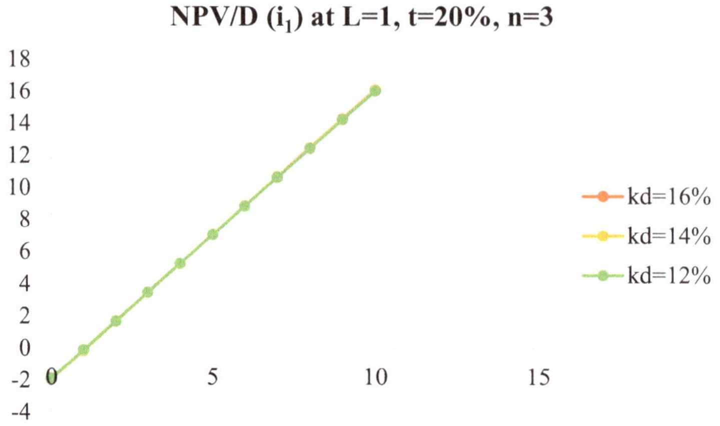

Fig. 26.1 The dependence of the net present value of the project NPV (in units of debt D) on the debt coverage ratio i_1

$$\frac{1-(1+WACC)^{-n}}{WACC}=\frac{1-(1+k_0)^{-n}}{k_0*(1-w_d*t*(1-(1+k_d)^{-n}))} \tag{26.13}$$

weighted average cost of capital, WACC, is calculated by Microsoft Excel using the "matching parameter" function.

2. The found WACC value is used in Eq. (26.4) to study the dependence of NPV / NOI on the debt leverage coefficient l_1.

Similar studies can be carried out for other leverage coefficients (l_2 and l_3).

Let us analyze the dependence of the net present value of the project NPV (in units of net operating income NOI) on the leverage coefficient on debt l_1 for companies of a given age n and for specific values of investment project parameters, demonstrating the possibility of performing calculations for companies of arbitrary age n and for any values of investment project parameters.

When analyzing Eq. (26.4), we will use the following parameter values: L = 1, 3; t = 0,20; n = 3,5; k_0 = 18; k_d = 16, 14, 12; l = 0, 1, 2, 3, 4, 5, 6, 7, 8, 9, 10.

From the data of Tables 26.3 and 26.4 and Figs. 26.3 and 26.4, the following conclusions can be drawn:

1. The net present value of the project NPV/NOI decreases linearly with increasing leverage on debt ratio l_1.
2. The slope of the straight line NPV/NOI (l_1) is determined by the leverage level L, while the absolute value of the slope decreases with increasing L: thus, lines with L = 3 is closer to the axis l_1 than lines with L = 1.
3. An increase in the project duration n shifts the line NPV/NOI (l_1) almost parallel higher (Figs. 26.3 and 26.4 show that for projects with the same level of leverage L, lines with n = 5 lie above lines with n = 3).

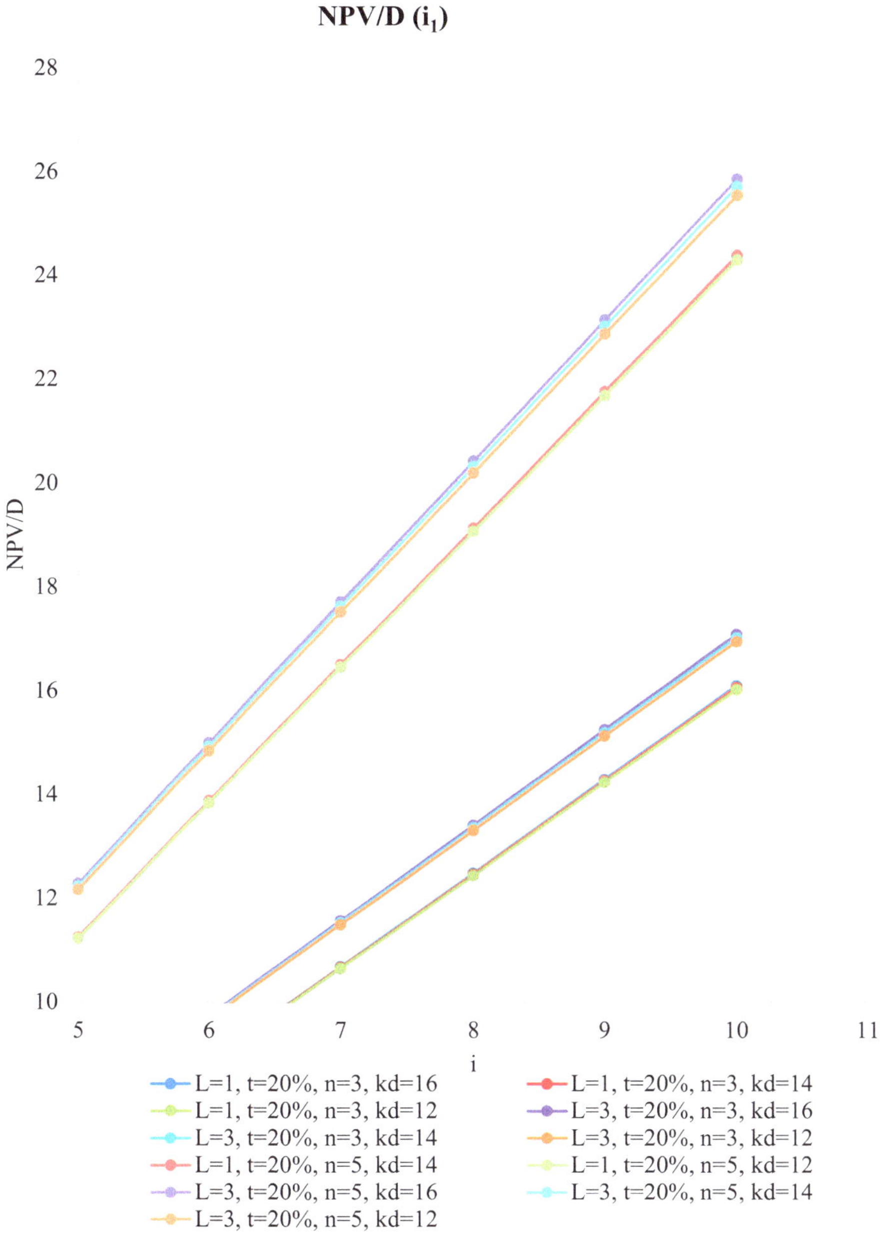

Fig. 26.2 Dependence of the net present value of the project NPV (in units of debt D) on the debt coverage ratio i_1 (increased scale)

Table 26.3 The dependence of the net present value of the project NPV (in units of net operating income NOI) on the leverage coefficient on debt l_l at $L = 1{,}3$ $t = 20\%$, $n = 3$, $k_0 = 18\%$, $k_d = k_d = 12\%, 14\%, 16\%$

NPV/NOI	WACC	k_d	$l_l = 0$	1	2	3	4	5	6	7	8	9	10
L = 1	0,1574	16%	1,8043	−0,149	−2,103	−4,057	−6,010	−7,964	−9,917	−11,871	−13,824	−15,778	−17,732
n = 3	0,1596	14%	1,7979	−0,127	−2,053	−3,978	−5,903	−7,828	−9,753	−11,678	−13,603	−15,529	−17,454
	0,1619	12%	1,7910	−0,106	−2,002	−3,899	−5,796	−7,692	−9,589	−11,486	−13,383	−15,279	−17,176
L = 3	0,1461	16%	1,84	0,53	−0,77	−2,07	−3,38	−4,68	−5,99	−7,29	−8,60	−9,90	−11,21
n = 3	0,1493	14%	1,83	0,55	−0,72	−1,99	−3,27	−4,54	−5,81	−7,09	−8,36	−9,64	−10,91
	0,1528	12%	1,82	0,57	−0,67	−1,91	−3,15	−4,40	−5,64	−6,88	−8,13	−9,37	−10,61

Table 26.4 The dependence of the net present value of the project NPV (in units of net operating income NOI) on the leverage coefficient on debt l_l at $L = 1{,}3$ $t = 20\%$, $n = 5$, $k_0 = 18\%$, $k_d = k_d = 12\%, 14\%, 16\%$

NPV/NOI	WACC	k_d	$l_l = 0$	1	2	3	4	5	6	7	8	9	10
L = 1	0,1566	16%	2,64	0,70	−1,24	−3,17	−5,11	−7,05	−8,99	−10,92	−12,86	−14,80	−16,74
n = 5	0,1586	14%	2,63	0,73	−1,17	−3,07	−4,97	−6,87	−8,77	−10,67	−12,57	−14,47	−16,36
	0,1608	12%	2,61	0,75	−1,11	−2,97	−4,83	−6,69	−8,55	−10,41	−12,27	−14,13	−15,99
L = 3	0,1448	16%	2,72	1,42	0,12	−1,17	−2,47	−3,77	−5,06	−6,36	−7,65	−8,95	−10,25
n = 5	0,1477	14%	2,70	1,44	0,19	−1,07	−2,32	−3,58	−4,83	−6,09	−7,34	−8,59	−9,85
	0,1510	12%	2,68	1,46	0,25	−0,96	−2,18	−3,39	−4,60	−5,81	−7,03	−8,24	−9,45

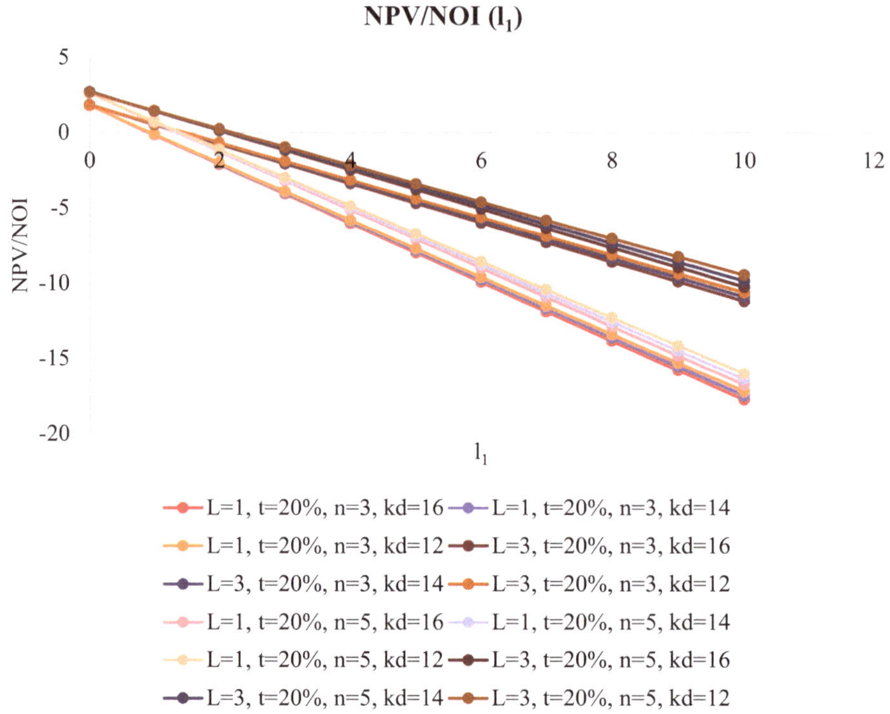

Fig. 26.3 Summary dependence of NPV/NOI on leverage ratio on debt l_1 at different leverage level L, project duration n and debt cost k_d

4. For equal project duration n and equal leverage levels L, the line NPV/NOI (l_1) shifts almost parallel downward with an increase in the bebt cost k_d (12%, 14%, 16%).

26.3 Conclusions

The Chapter discusses the rating of investment projects of arbitrary duration with a uniform repayment of debt and a modification of the methodology of rating systems for investment projects. The development of a fundamentally new approach to the project rating methodology has been carried out, the key factors of which are: (1) adequate use of discounts when discounting the financial flows, which is practically not used in existing project rating methodologies; (2) incorporation of financial ratios into modern investment models created by the authors; (3) use upon discounting the rating parameters; (4) the correct determination of the discount rate, taking into account financial ratios.

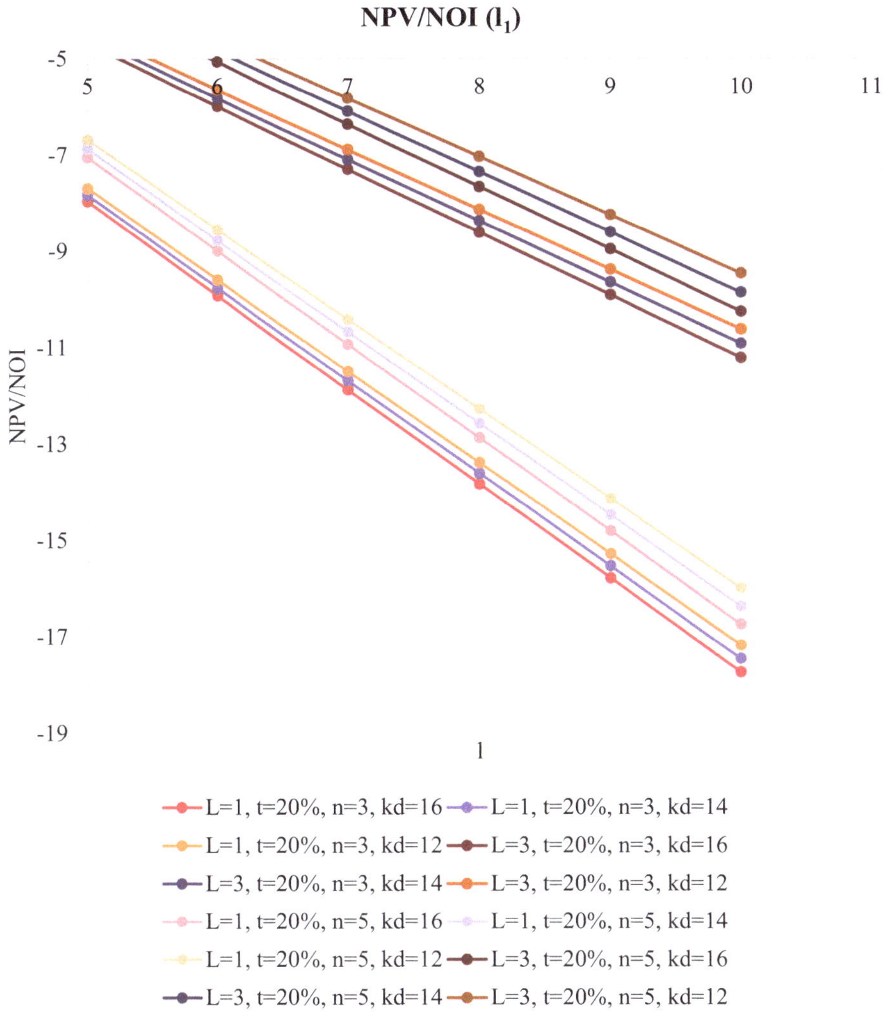

Fig. 26.4 Summary dependence of NPV/NOI on leverage ratio on debt l_1 at different leverage level L, project duration n and debt cost k_d (in increased scale)

The Chapter used modern investment models created by the authors, the modern theory of capital cost and capital structure by Brusov–Filatova–Orekhova (BFO theory), its modification for rating needs, and rating coefficients.

For the first time, incorporation of coverage ratios and leverage coefficients used in the rating was carried out into modern investment models created by the authors and designed to study the effectiveness of investment projects of arbitrary duration with uniform repayment of debt.

The dependence of the net present value of the project, NPV/NOI, on the coverage coefficients i_j (leverage coefficients l_j) for companies of arbitrary age n is analyzed.

When studying the dependence of NPV/D on coverage ratios i_j, it was shown that the net present value of the project, NPV/D increases linearly with increasing debt coverage ratio i_j. At the same time, the dependence NPV/D (i_1) for various values of the debt cost k_d has interesting features: for values of the coverage ratio for debt i_1 from 0 to 3, an increase in the cost of debt capital k_d leads to a decrease in NPV/D, and for values of the cost of debt capital k_d more than 3 (and up to 10) to bigger values of the cost of debt capital k_d correspond to bigger values of NPV/D. At the same time, the values of the net present value of the project NPV/D at various values of the cost of debt capital k_d (12%, 14%, 16%) differ slightly from each other (by about 0.02), and therefore the NPV/D graphs (i_1) almost coincide.

The slope of the straight line NPV/D (i_1) is determined by the duration of the project n and increases with the duration of the project. So, the slope of the straight line NPV/D (i_1) for a 5-year project is greater than for a 3-year project. When the leverage level L changes, the line NPV/D (i_1) shifts almost parallel: with the growth of L, the line shifts higher (for projects of the same term, lines with $L = 3$ lie above lines with $L = 1$).

When studying the dependence of NPV/NOI on leverage coefficients l_j, it is shown that the net present value of the project NPV/NOI decreases linearly with an increase of the leverage coefficient for debt l_1. In this case, the slope of the straight line NPV/NOI (l_1) is determined by the leverage level L, and the absolute value of the slope decreases with increasing L: thus, lines with $L = 3$ lies closer to the axis l_1 than lines with $L = 1$. An increase of the project duration n shifts the line NPV/NOI (l_1) almost parallel higher (for projects with the same level of leverage L, lines with $n = 5$ lie above lines with $n = 3$). With equal project times and equal leverage levels L, the direct NPV / D (i1) shifts almost parallel to down with the growth of the debt cost k_d (12%, 14%, 16%).

It can be seen that the behavior of the NPV/NOI (l_j) dependence is qualitatively different from the behavior of the NPV/D (i_j) dependence. These dependencies should be taken into account and used when rating investment projects.

The modification of the methodology of the existing project rating systems will improve the accuracy of issued ratings of investment projects and make them more objective. The use of powerful tools of well-developed theories opens up new opportunities for the rating industry, which gets the opportunity to switch from using mainly qualitative methods for assessing the effectiveness of investment projects to using mainly quantitative methods for evaluating them.

References

Brusov, P. (2018a). Editorial: Introduction on special issue on the banking system and financial markets of Russia and other countries: Problems and prospects. *Journal of Reviews on Global Economics, 7*, i–vi.

Brusov, P. (2018b). Editorial. *Journal of Reviews on Global Economics, 7*, i–v.

Brusov, P., Filatova, T., Orehova, N., & Eskindarov, M. (2015). *Modern corporate finance, investments and taxation* (monograph) (373 p). Springer. Retrieved from https://www.springer.com/gp/book/9783319147314

Brusov, P., Filatova, T., Orehova, N., & Eskindarov, M. (2018a). *Modern corporate finance, investments, taxation and ratings* (monograph) (571 p). Springer.

Brusov, P. N., Filatova, T. V., Orekhova, N. P., & Kulik, V. L. (2018b). Rating: New approach. *Journal of Reviews on Global Economics, 7*, 37–62. https://doi.org/10.6000/1929-7092.2023.07.05

Brusov, P. N., Filatova, T. V., Orekhova, N. P., & Kulik, V. L. (2018c). A "golden age" of the companies: Conditions of its existence. *Journal of Reviews on Global Economics, 7*, 88–103. https://doi.org/10.6000/1929-7092.2023.07.07

Brusov, P. N., Filatova, T. V., Orekhova, N. P., & Kulik, V. L. (2018d). Rating methodology: New look and new horizons. *Journal of Reviews on Global Economics, 7*, 63–87. https://doi.org/10.6000/1929-7092.2023.07.06

Brusov, P. N., Filatova, T. V., Orekhova, N. P., Kulik, V. L., & Weil, I. (2018e). New meaningful effects in modern capital structure theory. *Journal of Reviews on Global Economics, 7*, 104–122. https://doi.org/10.6000/1929-7092.2023.07.08

Brusov, P. N., Filatova, T. V., & Orekhova, N. P. (2018f). *Modern corporate finance and investments* (monograph) (517 p). Knorus Publishing House.

Brusov, P., Filatova, T., Orekhova, N., Kulik, V., Weil, I., & Brailov, A. (2018g). The impact of the central bank key rate and commercial banks credit rates on creating and maintaining of a favorable investment climate in the country. *Journal of Reviews on Global Economics, 7*, 360–376. https://doi.org/10.6000/1929-7092.2023.07.31

Brusov, P. N., Filatova, T. V., Orekhova, N. P., Kulik, V. L., & Weil, I. (2019). Ratings of the investment projects of arbitrary durations: New methodology. *Journal of Reviews on Global Economics, 8*, 437–448. https://doi.org/10.6000/1929-7092.2019.08.37

Brusov, P. N., Filatova, T. V., Orekhova, N. P., Kulik, V. L., Chang, S.-I., & Lin, Y. C. G. (2020). Modification of the Modigliani–Miller theory for the case of advance payments of tax on profit. *Journal of Reviews on Global Economics, 9*, 257–268.

Filatova, T., Orehova, N., & Brusova, A. (2008). Weighted average cost of capital in the theory of Modigliani–Miller, modified for a finite life-time company. *Bulletin of the FU, 48*, 68–77.

Filatova, T. V., Brusov, P. N., Orekhova, N. P., & Kulik, V. L. (2018). Ratings of the long-term projects: New approach. *Journal of Reviews on Global Economics, 7*, 645–661. https://doi.org/10.6000/1929-7092.2023.07.59

Modigliani, F., & Miller, M. (1958). The cost of capital, corporate finance, and the theory of investment. *The American Economic Review, 48*, 261–297.

Modigliani, F., & Miller, M. (1963). Corporate income taxes and the cost of capital: A correction. *The American Economic Review, 53*, 147–175.

Modigliani, F., & Miller, M. (1966). Some estimates of the cost of capital to the electric utility industry 1954–1957. *The American Economic Review, 56*, 333–391.

Myers, S. (2001). Capital structure. *The Journal of Economic Perspectives, 15*(2), 81–102.

Part VII
Qualitatively New Effects in Investments

Part VII is devoted to description of qualitatively new effects in investments, discovered by authors. In Chap. 27 a problem whether it is possible to increase taxing and conserve a good investment climate in the country is studied.

In Chap. 28 an abnormal influence of growth of tax on profit rate on the efficiency of the investment has been discovered and the answer on the question "Whether it is possible to increase of the investment efficiency, increasing tax on profit rate?" has been done.

Chapter 27
Whether It Is Possible to Increase Taxing and Conserve a Good Investment Climate in the Country?

Within investment models, developed by Brusov, Filatova, Orekhova earlier (Brusov & Filatova, 2011; Brusov et al., 2011a, b, c, 2012a, b, 2013a, b, 2014a, b; Filatova et al., 2008) the influence of tax on profit rate on effectiveness of long-term investment projects at different debt levels is investigated. It is shown, that increase of tax on profit rate from one side leads to decrease of project NPV, but from other side it leads to decrease of sensitivity of NPV with respect to leverage level. At high leverage level L the influence of tax on profit rate increase on effectiveness of investment projects becomes significantly less. We come to conclusion, that taxing could be differentiated depending on debt level of investment projects of the company: for projects with high debt level L it is possible to apply a higher tax on profit rate.

27.1 Influence of Tax on Profit Rates on the Efficiency of the Investment Projects

Basis of the modern tax systems are the following taxes: tax on profit of organizations, income tax (tax on the income of individuals), social tax (contributions into state extra budgetary funds), the sales tax (the value-added tax), tax on property of the organization. In this Chapter we investigate the influence of tax on profit rates on the efficiency of the investment projects.

The problems that we are currently investigating now, those questions which we can analyze now in all of their complexity and diversity and to which we give answers, not be tractable by analysis and assessment previously, for them one was not able to give an answer, they even not been raised in such a setting. What should be the taxes scale: flat, progressive or otherwise, differentiated, as well as tax rate has an impact on the cost of attract of company capital, its capitalization. What is cumulative effect of increase of taxes: whether the system "state–employer" will

P. Brusov, T. Filatova, *Efficiency of Investment Projects*, Contributions to Finance and Accounting, https://doi.org/10.1007/978-3-032-01028-5_27

win, or will lose as a whole from the tax growth, and if it will lose, then how much. Whether the redistribution of income in favor of the state does not destroy the spirit of enterprise, its driving force. If tax on profit rate will increase by 1%, for how much does the cost of attractive capital of company will increase, for how much its capitalization will decrease. If by 3–6%, it should be serious reasons for such increase, but if by 0.5–1.5%, it is possible to discuss such increase.

How taxation affects the efficiency of investment? How much does the NPV of investment project will decrease, if tax on profit rate will increase by 1%? If on 5–10%, it has a strong negative impact on investment, if on the 1%, or 0.5%, or 0.25%, Regulator can accept this: this will help to the state and do not exert much to investment programs of companies.

One of the main reasons for which it has become possible to carry out such studies, has been a progress in corporate finance, has been made in recent years. It relates primarily to the establishment of a modern theory of capital cost and capital structure by Brusov, Filatova, Orekhova (BFO theory) (Brusov & Filatova, 2011; Brusov et al., 2011a, b, c, 2012a, b, 2013a, b, 2014a, b; Filatova et al., 2008) and to the creation by them in the framework of this theory of modern investment models. The BFO theory allows to make correct assessment of the financial performance of companies with arbitrary life time and of efficiency of investment projects of arbitrary duration. This distinguishes BFO theory from Modigliani–Miller theory (Modigliani & Miller, 1958, 1963, 1966), which is a perpetuity limit of BFO theory. Archived after the appearance of BFO theory the Modigliani–Miller theory, still heavily used in the West, despite of its obvious limitations, may, mutatis mutandis, be applied to long living stable companies and long-term investment projects. In its framework in this Chapter effects of taxation on the effectiveness of long-term investment will be investigated.

So, at present, there are two main theories, that allow to explore the effects of taxation on the efficiency of investments: perpetuity Modigliani-Miller theory (Modigliani & Miller, 1958, 1963, 1966) and the modern theory of capital cost and capital structure developed by Brusov, Filatova, Orekhova (Brusov & Filatova, 2011; Brusov et al., 2011a, b, c, 2012a, b, 2013a, b, 2014a, b; Filatova et al., 2008; Brusova, 2011). In this Chapter we describe the first real results obtained by us within created by us investment projects in perpetuity limit, which can be applied to long-term projects.

The effectiveness of the investment project is considered from the perspectives of the equity holders. For this case, NPV is calculated in two ways: with the division of credit and investment flows (and thus discounting the payments at two different rates) and without such a division (in this case, both flows are discounted at the same rate as which, obviously, can be chosen WACC). For each of the four situations two cases are considered: (1) a constant value of equity S; (2) a constant value of the total invested capital $I = S + D$ (D is value of debt funds).

We start first from the case with the division of credit and investment flows and then consider the case without the division of flows.

27.2 Investment Models

Let us remind shortly the main points of the investment models with debt repayment at the end of the project, well-proven in the analysis of real investment projects.

Investments at the initial time moment $T = 0$, are equal to $-S$, and the flow of capital for the period (in addition to the tax shields $k_d Dt$ it includes a payment of interest on a loan $-k_d D$)

$$(NOI - k_d D)(1 - t). \tag{27.1}$$

Here, for simplicity, we suppose that interest on the loan will be paid in equal shares $k_d D$ during all periods. Note, that principal repayment is made at the end of last period.

Here *NOI*—Net Operating Income (before taxes), k_d—debt cost, t—tax on profit rate.

Let us first consider the case with the division of credit and investment flows. In this case in perpetuity limit (Modigliani–Miller approximation) expression for NPV takes a following form (Brusov & Filatova, 2011; Brusov et al., 2011a, b, c, 2012a, b, 2013a, b, 2014a, b; Filatova et al., 2008)

$$NPV = -S + \frac{NOI(1 - t)}{k_e} - D(1 - t). \tag{27.2}$$

We will consider two cases: (1) a constant value of the total invested capital I = S + D (D is value of debt funds); (2) a constant value of equity S.

At a constant value of the total invested capital ($I =$ const), accounting $D = IL/(1 + L)$, $S = I/(1 + L)$, one gets

$$NPV = -\frac{I}{1 + L}(1 + L(1 - t)) + \frac{NOI(1 - t)}{k_e}. \tag{27.3}$$

$$NPV = -\frac{I}{1 + L}(1 + L(1 - t)) + \frac{NOI(1 - t)}{k_0 + (k_0 - k_d)L(1 - t)}, \tag{27.4}$$

where $L = D/S$—leverage level; k_e—equity cost of leverage company (which uses the debt financing); k_0—equity cost of non-leverage company (which does not use the debt financing).

Under the transition from the equation (Eq. 27.3) to (Eq. 27.4) we have used the dependence of equity capital on leverage, received by Modigliani and Miller (1958, 1963, 1966):

$$k_e = k_0 + (k_0 - k_d)L(1 - t). \tag{27.5}$$

So, we explore the equation (Eq. 27.4). A number of conclusions can be drawn from the study of dependence of NPV of the project on leverage level at different values of

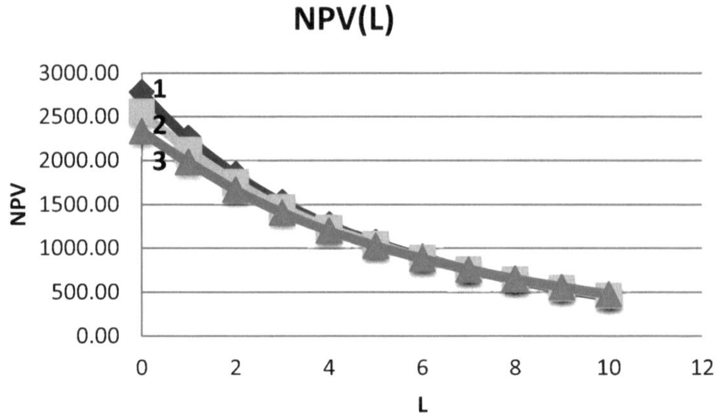

Fig. 27.1 Dependence of NPV on leverage level L at three values of tax on profit rate (*1*—t = 0.15; *2*—t = 0.20; *3*—t = 0.25), NOI = 800

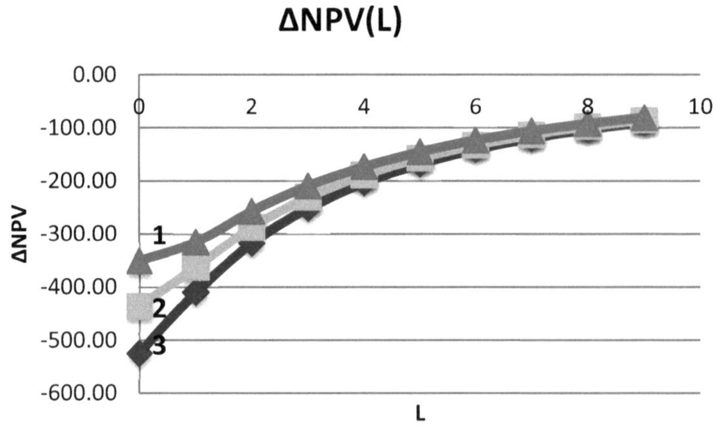

Fig. 27.2 Dependence of ΔNPV on leverage level L at three values of tax on profit rate (*1*—t = 0.25; *2*—t = 0.20; *3*—t = 0.15). NOI = 800, I = 1000

tax on profit rates t (Fig. 27.1). It is clear that the increase of tax on profit rates leads not only to reduce of NPV of the project, but as well to decrease of the sensitivity of effectiveness of investment projects NPV to the leverage level L. At high leverage levels the influence of growth of tax on profit rates on the effectiveness of investment projects is significantly reduced.

Hence, in particular, it should be noted, that taxation can be differentiated depending on the debt financing level in the company investment projects: for projects with a high leverage level L the higher tax on profit rates t can be used. The foregoing is illustrated also at Fig. 27.2, where it is clear that the change of NPV

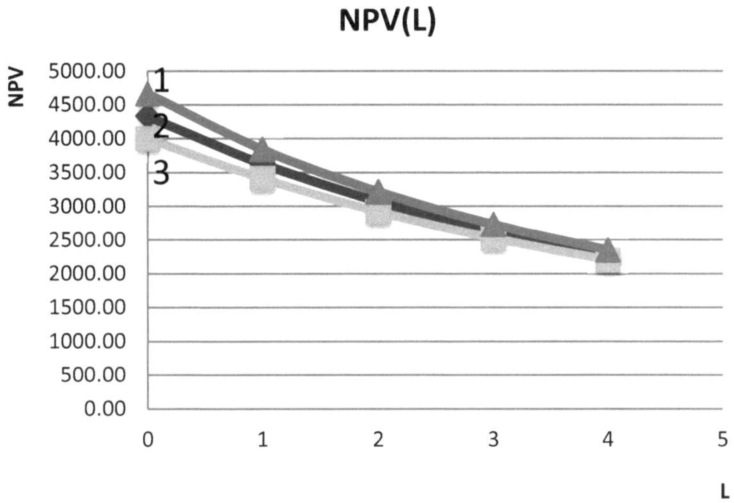

Fig. 27.3 Dependence of NPV on leverage level L at three values of tax on profit rate (*1*—t = 0.15; *2*—t = 0.20; *3*—t = 0.25), NOI = 1200, I = 1000

(ΔNPV) with leverage level decreases when the tax on profit rate t grows, and as well when leverage level increases.

Let us increase our return on investment in 1.5 times (NOI = 1200 instead of 800) (Fig. 27.3). Still, the impact of the tax on profit rate on the NPV value significantly depends on the level of debt financing. So the increase in tax on profit rate by 1% from the existing (in Russia) (20%) leads to a reduction in the NPV by 44.5 units at L = 0, by 27.7 units at L = 1, by 12.2 units at L = 3, by 5 units at L = 5.

I.e. for companies with a high level of the debt financing (for example, companies in the telecommunication sector and other) an increase in tax on profit rate will have less impact on the effectiveness of their investment projects and will be less painful, than for companies with low leverage level in investment. It should be noted that the increase of NOI in 1.5 times increases NPV in 1.7 times (from 2555 up to 4333), and increases ΔNPV(L) in 1.62 at L = 0, and in 1.5 times when L = 9 (Fig. 27.4).

It is clear also that, with the increase of the leverage level L curves, describing the dependence ΔNPV(L), virtually converge, which demonstrates once again the reduction of impact of the change of the tax on profit rate t on the efficiency of investment projects with the increase of the leverage level L.

27.3 Borrowings Abroad

Until recently, Russian companies have preferred to borrow abroad, because overseas credits are much cheaper than domestic ones. Although relevance of studies using such loans now is not so high in connection with the West sanctions, all same,

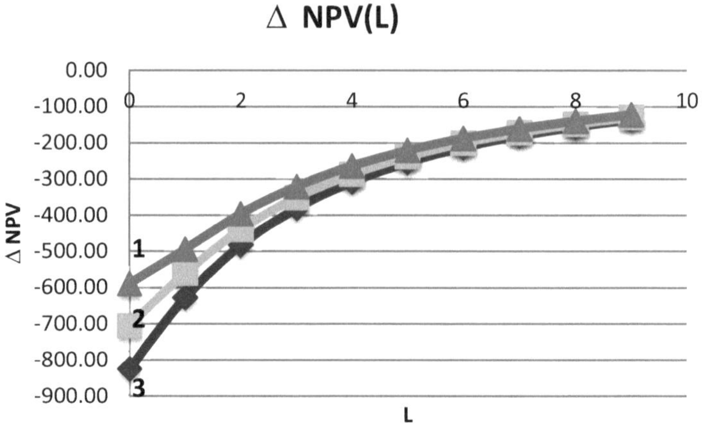

Fig. 27.4 Dependence of ΔNPV on leverage level L at three values of tax on profit rate (*1*—t = 0.25; *2*—t = 0.20; *3*—t = 0.15), NOI = 1200, I = 1000

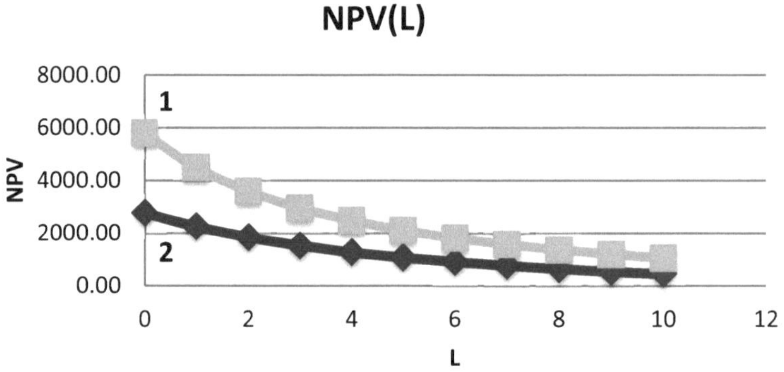

Fig. 27.5 Comparison of dependences of NPV on leverage level L at typical values of credit rates with borrowings abroad (*1*—k_0 =0.1; k_d =0.07) and with borrowings at domestic (Russian) credit market (*2*—k_0 =0.18; k_d =0.14), NOI = 800, I = 1000, t = 15%

realizing that, in the not-too-distant future, all will return to the its circles, here's a comparison of NPV dependencies on leverage at typical values of rates on credit, with borrowings abroad ($k_0 = 0.1$; $k_d = 0.07$) and with borrowings at domestic (Russian) credit market ($k_0 = 0.18$; $k_d = 0.14$). (Here k_0—equity cost of financially independent company (Fig. 27.5). ***The growing of effectiveness of investment when using cheaper foreign credit is obvious.*** The stabilization of the situation on the external credit market a detailed analysis of this case, as well as of the case of use of domestic and overseas credits can be done.

We analyze now the impact of the tax on profit rates on dependence of NPV on leverage level at typical values of credit rates with borrowings abroad and with borrowings at domestic (Russian) credit market (Fig. 27.6).

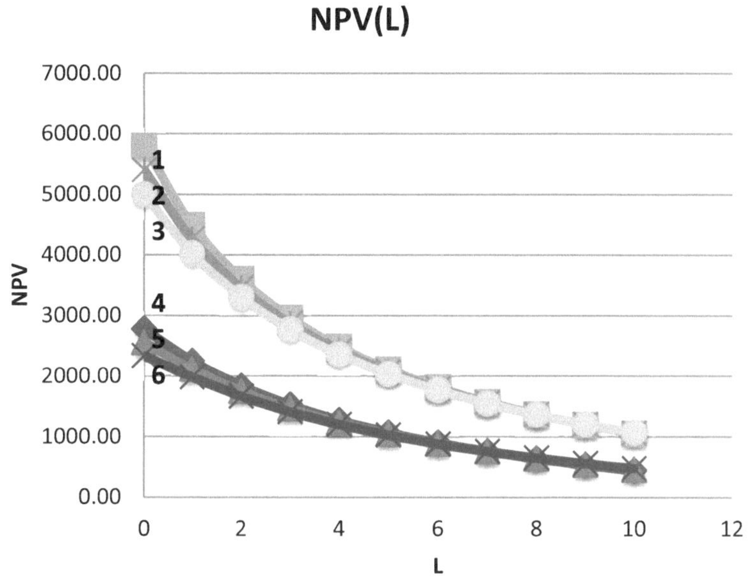

Fig. 27.6 Influence of tax on profit rates on dependence of NPV on leverage level at typical values of credit rates with borrowings abroad (lines 1-2-3) and with borrowings at domestic (Russian) credit market (lines 4-5-6). (1) k_0 =0.1; k_d =0.07; T = 0.15, (2) k_0 =0.1; k_d =0.07; T = 0.2, (3) k_0 =0.1; k_d =0.07; T = 0.25, (4) k_0 =0.18; k_d =0.14; T = 0.15, (5) k_0 =0.18; k_d =0.14; T = 0.2, (6) k_0 =0.18; k_d =0.14; T = 0.25

It is clear that, at low leverage levels influence of tax on profit rates is very significant: at zero leverage the NPV drops by 80 units at increase of tax on profit rates on 1% when one borrows abroad and on 44 units when one borrows at domestic (russian) credit market. It would seem that this could be one of the signals for borrowing within the country, however, taking into account the different values of NPV at two considering cases (ratio is 2.1), we come to the conclusion that the impact of tax on profit rates is in close proportions (ratio is 80/44 = 1.8). So it seems that after the West sanctions will be over, to borrow at the West will be more advantageous for a long time.

27.4 Dependence of NPV on Tax on Profit Rates at Different Leverage Levels

From Fig. 27.7 it is seen that dependence of NPV on tax on profit rates significantly depends on the leverage level L.

When there is no borrowing NPV linearly decreases with t with a factor—43.44 units at 1%. When L = 1 this factor (at t = 20%) is equal to—27.7 units at 1%, when L = 3 this factor (when t = 20%) is equal to—12.3 units at 1%, and when

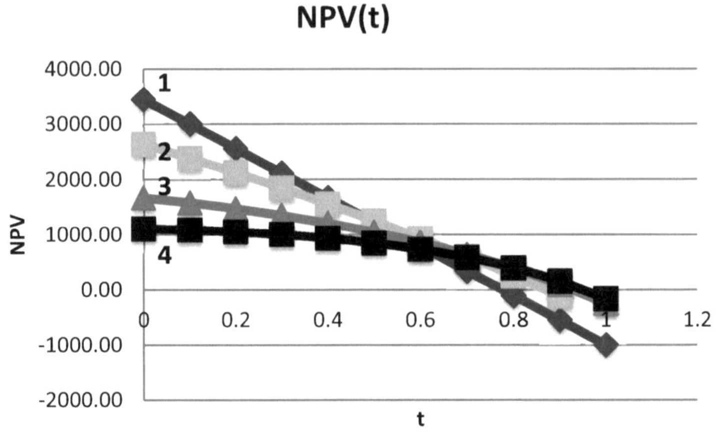

Fig. 27.7 Dependence of NPV on tax on profit rate at different leverage levels L (*1*—L = 0; *2*—L = 1; *3*—L = 3; *4*—L = 5), NOI = 800, I = 1000

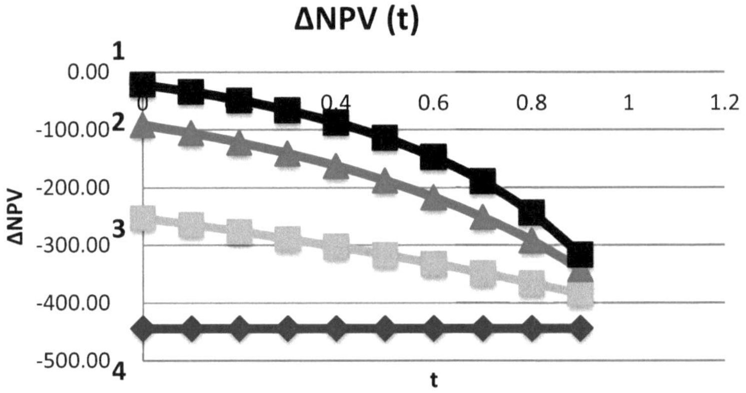

Fig. 27.8 Dependence of ΔNPV on tax on profit rate at different leverage levels L (*1*—L = 5; *2*—L = 3; *3*—L = 1; *4*—L = 0), NOI = 800, I = 1000

L = 5 this factor (at t = 20%) is equal to—5.8 units. at 1%. It can be seen that the influence of tax on profit rate on efficiency of investment projects drops significantly with increase of the leverage level L used in investments.

This is particularly seen at Fig. 27.8 in the dependence of ΔNPV on tax on profit rate at different leverage levels L (here ΔNPV—increment of NPV under change of t for 10%). When there is no borrowing ΔNPV = −450 and does not depend on tax on profit rate. At t = 20%: at L = 1 ΔNPV = −276.6; at L = 3 ΔNPV = −122.6; at L = 5 ΔNPV = −49. It is clear that the change of tax on profit rate affect mostly the effectiveness of the projects, funded by equity capital only, and if you use debt financing to finance the projects impact of the change of tax on profit rate drops very substantially (up to ten times).

27.5 At a Constant Value of Equity Capital (S = const)

At a constant value of equity capital (S = const), when investment growth is associated with the increased borrowing only, the dependence of NPV on leverage level is qualitatively different in nature, rather than in the case of a constant value of invested capital. Now, depending on the values of the coefficient $\beta = NOI/I$ NPV can grow with leverage level. It should be noted that, in this case (at large values of the coefficient β) the optimal structure of invested capital, in which NPV is maximized could take place. NPV in this case is described by the following expression

$$NPV = -S(1 + L(1 - t)) + \frac{\beta S(1 + L)(1 - t)}{k_0 + (k_0 - k_d)L(t - 1)}. \tag{27.6}$$

It is seen from the Fig. 27.9 that with the increase of the coefficients β value NPV and its optimal (maximum) values grow.

It follows from Fig. 27.10, that with the increase of the leverage level ΔNPV drops and either goes to the saturation (ΔNPV = 0), or becomes negative (ΔNPV<0), and that means that there is an optimum (after ΔNPV>0 at small leverage level L).

Below at Figs. 27.11 and 27.12 the dependencies of ΔNPV and NPV on tax on profit rate at different leverage levels and at S = 500, β = 0.8 are shown.

From Fig. 27.11 as well as from Fig. 27.9, it is seen, that at fixed tax on profit rate NPV grows with leverage level. With the increasing of tax on profit rate NPV drops, and curves, corresponding to the different leverage level converge in one point at t = 100% and NPV = −S in accordance to equation (Eq. 27.6). Change of ΔNPV

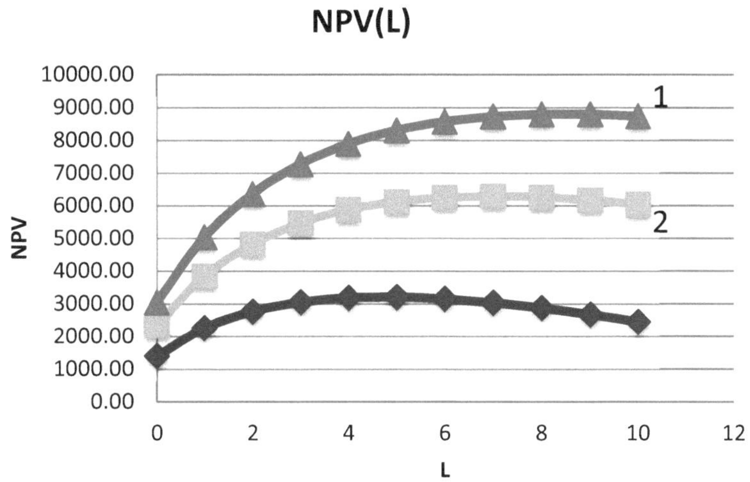

Fig. 27.9 Dependence of NPV on leverage level L at three values of β-coefficient (*1*—β = 1.5; *2*—β = 1.2; *3*—β = 0.8), S = 500

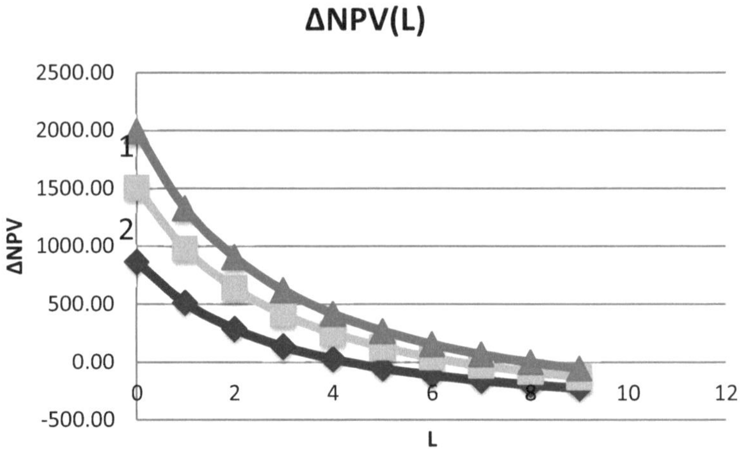

Fig. 27.10 Dependence of ΔNPV on leverage level L at three values of β-coefficient (*1*—β = 1.5; *2*—β = 1.2; *3*—β = 0.8), S = 500

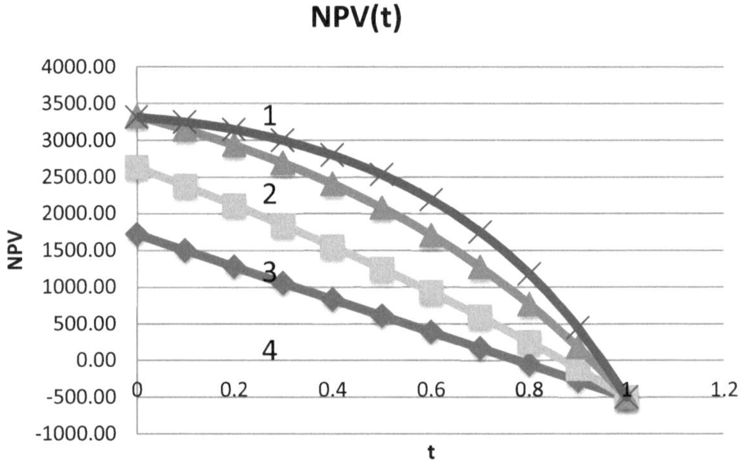

Fig. 27.11 Dependence of NPV on tax on profit rate at different leverage levels L (*1*—L = 5; *2*—L = 3; *3*—L = 1; *4*—L = 0), S = 500, β = 0.8

with increasing of t also depends on the leverage level: with the growth of the leverage level it changes from constant (at L = 0) up to the increasingly growing at t >30%–40% at L = 1;3;5.

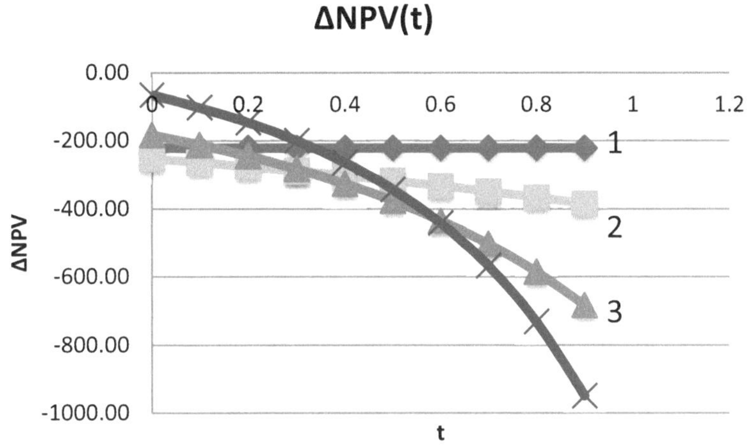

Fig. 27.12 Dependence of ΔNPV on tax on profit rate at different leverage levels L (*1*—L = 0; *2*—L = 1; *3*—L = 3; *4*—L = 5), S = 500, β = 0,8

27.6 Without Flows Separation

Let us consider the case without the division of credit and investment flows. In this case, both flows are discounted at the same rate as which, obviously, can be chosen WACC). In perpetuity limit.

$(n \to \infty)$ one has

$$NPV = -S + \frac{NOI(1-t) - k_d D(1-t)}{WACC}. \tag{27.7}$$

27.6.1 At a Constant Value of the Total Invested Capital (I = const)

In case of a constant value of the total invested capital (I = const), accounting $D = IL/(1 + L)$, $S = I/(1 + L)$, we get

$$NPV = -I \cdot \frac{1}{1+L}\left[1 + \frac{Lk_d(1-t)}{k_0(1-Lt/(1+L))}\right] + \frac{NOI(1-t)}{k_0(1-Lt/(1+L))}. \tag{27.8}$$

It should be noted that, in contrast to the case with the division of flows, described above, in a situation without the division of flows NPV is growing with leverage level. It is seen, that, while NOI increases in 1.5 times NPV increases in 1.68 times (Figs. 27.13 and 27.14).

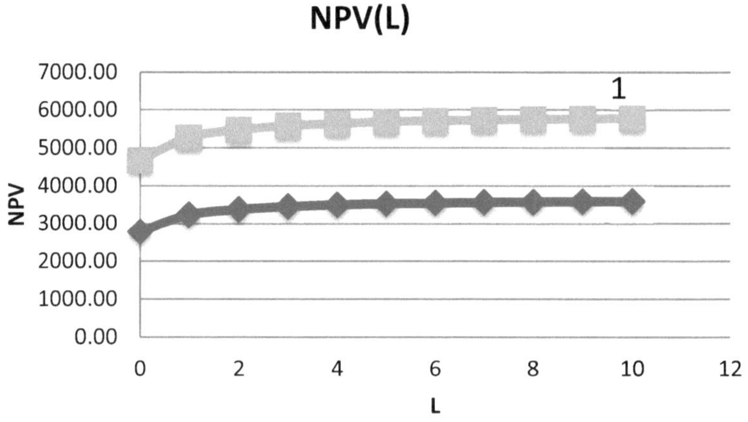

Fig. 27.13 Dependence of NPV on leverage level L at two values of NOI (*1*—NOI = 1200; *2*—NOI = 800), I = 1000

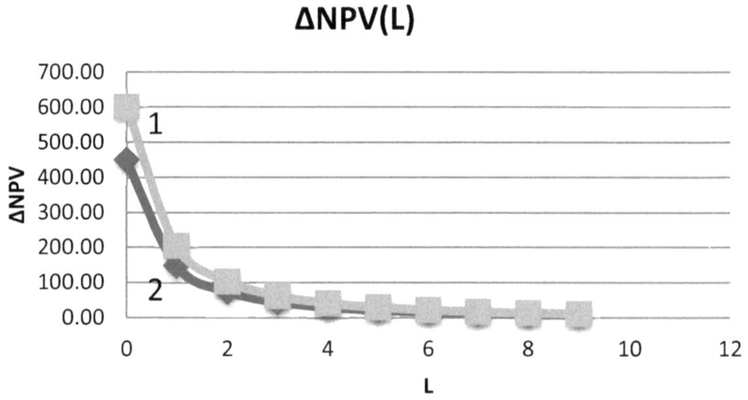

Fig. 27.14 Dependence of ΔNPV on leverage level L at two values of NOI (*1*—NOI = 1200; *2*—NOI = 800), I = 1000

NPV rather quickly goes to the saturation, at $L > 4$ it varies weakly, and the leverage level, at which the saturation of NPV(L) takes place, practically does not depend on NOI (Fig. 27.15).

NPV falls down with growth of tax on profits rate at different leverage levels:

at $L = 0$ at change of tax on profits rate on 1%, NPV falls down on 1.74%;
at $L = 1$ at change of tax on profits rate on 1%, NPV falls down on 0.85%;
at $L = 3$ at change of tax on profits rate on 1%, NPV falls down on 0.43%;
at $L = 5$ at change of tax on profits rate on 1%, NPV falls down on 0.29%;

It is seen that, with the rising of the tax on profit rate by 1%, NPV drops the less at the higher leverage level. This confirms the conclusion, made in the previous section,

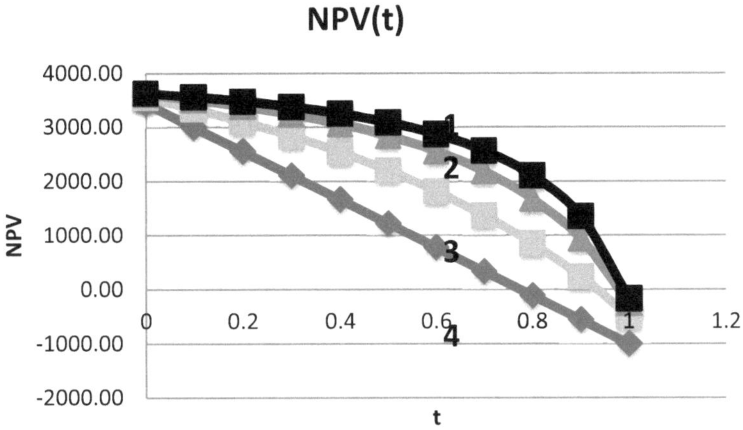

Fig. 27.15 Dependence of ΔNPV on tax on profit rate at different leverage levels L (*1*—L = 5; *2*—L = 3; *3*—L = 1; *4*—L = 0), NOI = 800, I = 1000

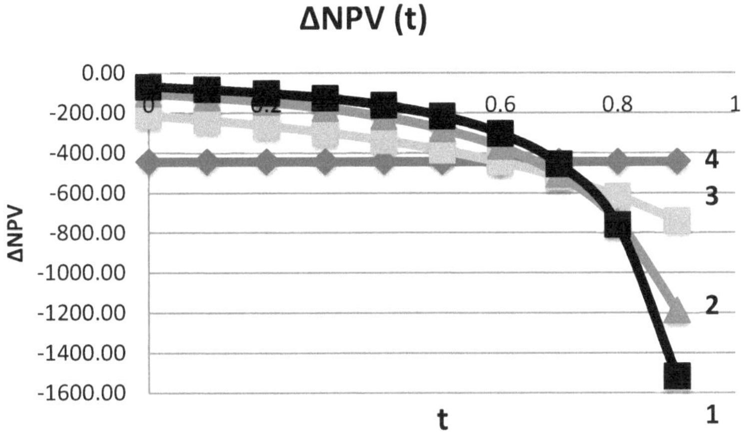

Fig. 27.16 Dependence of ΔNPV on tax on profit rate at different leverage levels L (*1*—L = 5; *2*—L = 3; *3*—L = 1; *4*—L = 0), NOI = 800, I = 1000

that with the increase of the leverage level a negative impact of the growth of the tax on profit rate declines in a few times, allowing to the Regulator to establish the differentiated tax on profit rates (as can be seen from Fig. 27.16, the founded conclusions are true up to tax on profit rate values of 70% to 80%).

27.6.2 At a Constant Value of Equity Capital (S = const)

$$NPV = -S + \frac{NOI(1-t) - k_d D(1-t)}{WACC}. \tag{27.9}$$

Substituting $D = LS$, one gets

$$NPV = -S\left[1 + \frac{Lk_d(1-t)}{k_0(1 - Lt/(1+L))}\right] + \frac{\beta S(1+L)(1-t)}{k_0(1 - Lt/(1+L))}. \tag{27.10}$$

From Fig. 27.17 it follows, that NPV grows linearly with leverage level and speed of its growth increases with grows of coefficient β.

From Fig. 27.18 it follows, that ΔNPV practically does not depend on leverage level L and at decrease of β-coefficient in 1.25 times (the transition from line 1 to line 2) ΔNPV is decreased in 1.28 times (practically so), and at decrease of β-coefficient in 1.5 times (the transition from line 2 to line 3) ΔNPV is decreased in 1.59 times (practically so).

As in case of constant value of investments (I = const), at constant equity capital value (S = const) NPV falls down with growth of tax on profit rate t at different leverage levels L.

Let's take a look at the region of changes of tax on profits rates from 0% up to 60%. In this region:

at L = 0 at change of tax on profits rate on 1%, NPV falls down on 3.6%;
at L = 1 at change of tax on profits rate on 1%, NPV falls down on 1.23%;
at L = 3 at change of tax on profits rate on 1%, NPV falls down on 0.46%;
at L = 5 at change of tax on profits rate on 1%, NPV falls down on 0.22%;

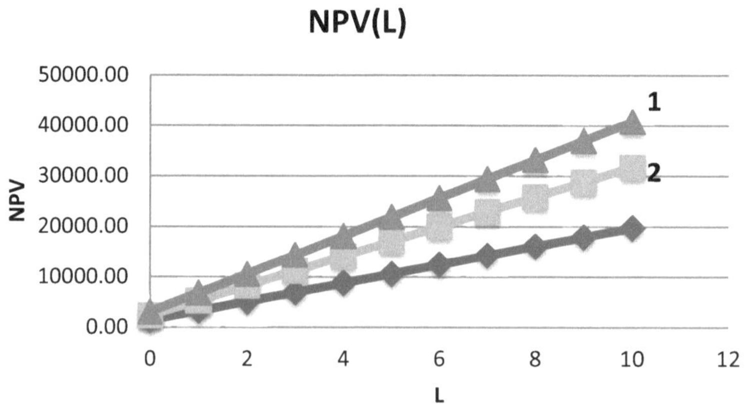

Fig. 27.17 Dependence of NPV on leverage level L at three values of β-coefficient (*1*—β = 1.5; *2*—β = 1.2; *3*—β = 0.8), S = 500

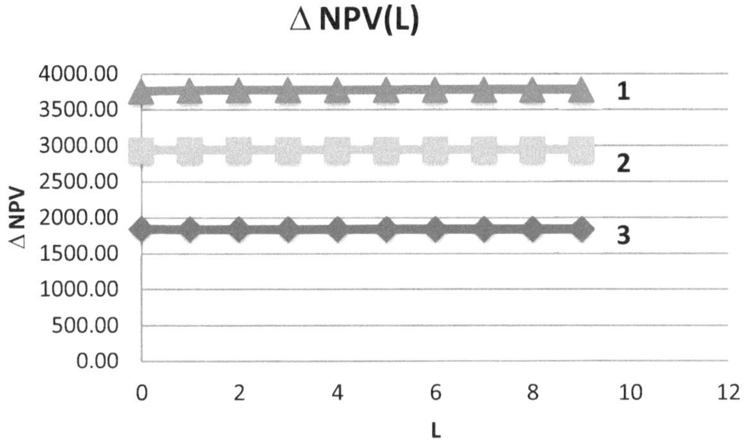

Fig. 27.18 Dependence of ΔNPV on leverage level L at three values of β-coefficient (*1*—β = 1.5; *2*—β = 1.2; *3*—β = 0.8), S = 500

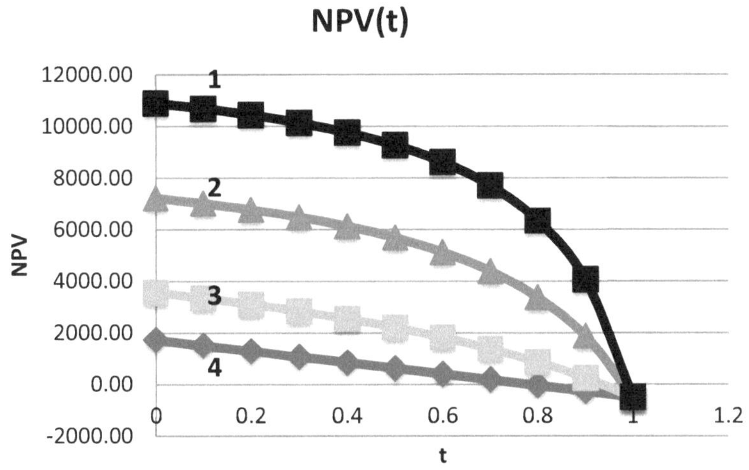

Fig. 27.19 Dependence of NPV on tax on profit rate t at different leverage levels L (*1*—L = 5; *2*—L = 3; *3*—L = 1; *4*—L = 0), S = 500

And so, with the increasing of the tax on profits rates at the 1%, NPV drops the less for the higher leverage level. This correlates with the conclusion made above and in the previous section, that with the increase of the leverage level a negative impact of the growth of the tax on profit rates declines in a few times, allowing the regulator to introduce differentiated tax on profits rate (as it can be seen from Figs. 27.19 and 27.20, the findings are true up to values of tax on profits rates 60%). At higher rates (that, however, is a purely theoretical interest) the situation will be different.

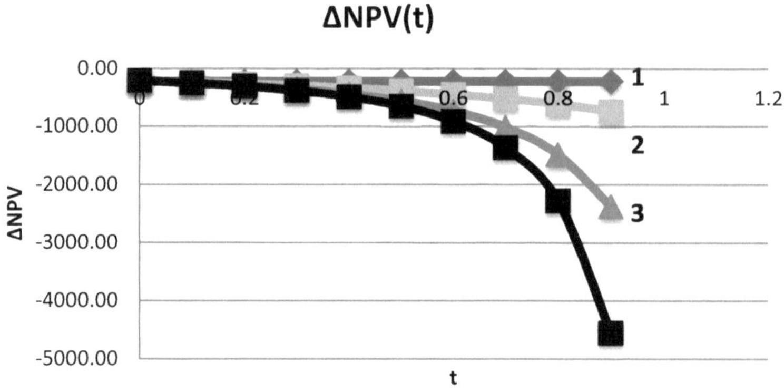

Fig. 27.20 Dependence of NPV on tax on profit rate t at different leverage levels L (*1*—L = 0; *2*—L = 1; *3*—L = 3; *4*—L = 5), S = 500

27.7 Conclusions

Within investment models, developed by Brusov, Filatova, Orekhova earlier (Brusov & Filatova, 2011; Brusov et al., 2011a, b, c, 2012a, b, 2013a, b, 2014a, b; Filatova et al., 2008; Brusova, 2011) the influence of tax on profit rate on effectiveness of long–term investment projects at different debt levels is investigated. The ability to obtain of quantitative estimates of such impact on the projects with various costs of equity and debt capital, at an arbitrary structure of invested capital has been demonstrated. It is shown, that increase of tax on profit rate from one side leads to decrease of project NPV, but from other side it leads to decrease of sensitivity of NPV with respect to leverage level. At high leverage level L the influence of tax on profit rate increase on effectiveness of investment projects becomes significantly less. We come to conclusion, that taxing could be differentiated depending on debt level of investment projects of the company: for projects with high debt level L it is possible to apply a higher tax on profit rate.

These recommendations, in particular, may be addressed to the Regulator. Effects of taxation on the effectiveness of investment projects depends on the level of leverage, of the project duration, of the equity cost, as well as of the level of returns on investment (NOI) and of methods of forming of invested capital. Study of all these problems, as the results of this Chapter show, may be successfully carried out within developed by Brusov, Filatova, Orekhova investment models, using discount rates, derived from the Brusov-Filatova-Orekhova (BFO) theory.

References

Brusova, A. (2011). A comparison of the three methods of estimation of weighted average cost of capital and equity cost of company. *Financial Analysis: Problems and Solutions, 34*(76), 36–42.

Brusov, P. N., & Filatova, T. V. (2011). From Modigliani–Miller to general theory of capital cost and capital structure of the company. *Finance and Credit, 435*, 2–8.

Brusov, P., Filatova, T., Orehova, N., & Brusova, A. (2011a). Weighted average cost of capital in the theory of Modigliani–Miller, modified for a finite life–time company. *Applied Financial Economics, 21*(11), 815–824.

Brusov, P., Filatova, T., Orehova, N., et al. (2011b). From Modigliani–Miller to general theory of capital cost and capital structure of the company. *Research Journal of Economics, Business and ICT, 2*, 16–21.

Brusov, P., Filatova, T., Orehova, N., et al. (2011c). Influence of debt financing on the effectiveness of the investment project within the Modigliani–Miller theory. *Research Journal of Economics, Business and ICT (UK), 2*, 11–15.

Brusov, P., Filatova, T., Eskindarov, M., & Orehova, N. (2012a). Influence of debt financing on the effectiveness of the finite duration investment project. *Applied Financial Economics, 22*(13), 1043–1052.

Brusov, P., Filatova, T., Eskindarov, M., & Orehova, N. (2012b). Hidden global causes of the global financial crisis. *Journal of Reviews on Global Economics, 1*, 106–111.

Brusov, P., Filatova, P., & Orekhova, N. (2013a). Absence of an optimal capital structure in the famous tradeoff theory! *Journal of Reviews on Global Economics, 2*, 94–116.

Brusov, P., Filatova, T., & Orehova, N. (2013b). A qualitatively new effect in corporative finance: Abnormal dependence of cost of equity of company on leverage. *Journal of Reviews on Global Economics, 2*, 183–193.

Brusov, P., Filatova, P., & Orekhova, N. (2014a). Mechanism of formation of the company optimal capital structure, different from suggested by trade off theory. *Cogent Economics & Finance, 2*, 1–13. https://doi.org/10.1080/23322039.2014.946150

Brusov, P., Filatova, T., & Orehova, N. (2014b). Inflation in Brusov–Filatova–Orekhova theory and in its perpetuity limit – Modigliani – Miller theory. *Journal of Reviews on Global Economics, 3*, 175–185.

Filatova, T., Orehova, N., & Brusova, A. (2008). Weighted average cost of capital in the theory of Modigliani–Miller, modified for a finite life–time company. *Bulletin of the FU, 48*, 68–77.

Modigliani, F., & Miller, M. (1958). The cost of capital, corporate finance, and the theory of investment. *American Economic Review, 48*, 261–297.

Modigliani, F., & Miller, M. (1963). Corporate income taxes and the cost of capital: A correction. *American Economic Review, 53*, 147–175.

Modigliani, F., & Miller, M. (1966). Some estimates of the cost of capital to the electric utility industry 1954–1957. *American Economic Review, 56*, 333–391.

Chapter 28
Whether It Is Possible to Increase of the Investment Efficiency, Increasing Tax on Profit Rate? An Abnormal Influence of Growth of Tax on Profit Rate on the Efficiency of the Investment

28.1 Dependence of NPV on Leverage Level *L* at Fixed Levels of Tax on Profit Rates *t*

28.1.1 The Effectiveness of the Investment Project from the Perspective of the Equity Holders Only

Investigations will be done with the division of credit and investment flows: operating and finance flows are divided and discounted using different rates: operating flows—by the rate equal to equity cost k_e, depending on leverage, and credit ones—by the rate equal to debt cost k_d, which, until a sufficiently large values of leverage levels, remains constant and starts to grow only at sufficiently high leverage values L, when a risk of bankruptcy will appear. The consideration has been done upon constant value of investment capital I.

In this case NPV is described by the following formula (Brusov & Filatova, 2011; Brusov et al., 2011a, b, c, 2012a, b, 2013a, b, 2014a, b; Filatova et al., 2008; Brusova, 2011)

$$NPV = -\frac{I}{1+L}\left[1 + L\left((1-t)\left(1 - \frac{1}{(1+k_d)^n}\right) + \frac{1}{(1+k_d)^n}\right)\right] + \\ + \frac{NOI(1-t)}{k_e}\left(1 - \frac{1}{(1+k_e)^n}\right). \tag{28.1}$$

Using it we calculate NPV and ΔNPV at fixed levels of tax on profit rates t.

5-Year Project

For 5-year projects we get the following results (Tables 28.1, 28.2, 28.3, 28.4, 28.5, 28.6, and 28.7; Figs. 28.1 and 28.2).

P. Brusov, T. Filatova, *Efficiency of Investment Projects*, Contributions to Finance and Accounting, https://doi.org/10.1007/978-3-032-01028-5_28

Table 28.1 Dependence of NPV and ΔNPV on leverage level L at fixed levels of tax on profit rates t for 5-year project at $t = 0.3$

L	k_0	k_d	Wd	t	n	NOI	k_e	NPV	ΔNPV
0	0.18	0.14	0	0.3	5	800	0.18	751.22	4.922709
1	0.18	0.14	0.5	0.3	5	800	0.197488	756.14	−36.8599
2	0.18	0.14	0.66667	0.3	5	800	0.214367	719.28	−44.7663
3	0.18	0.14	0.75	0.3	5	800	0.231082	674.51	−46.126
4	0.18	0.14	0.8	0.3	5	800	0.24773	628.39	−45.4549
5	0.18	0.14	0.83333	0.3	5	800	0.264343	582.93	−44.027
6	0.18	0.14	0.85714	0.3	5	800	0.280937	538.90	−42.3084
7	0.18	0.14	0.875	0.3	5	800	0.297518	496.60	−40.4978
8	0.18	0.14	0.88889	0.3	5	800	0.314091	456.10	−38.6879
9	0.18	0.14	0.9	0.3	5	800	0.330658	417.41	−36.9239
10	0.18	0.14	0.90909	0.3	5	800	0.34722	380.49	

Table 28.2 Dependence of NPV and ΔNPV on leverage level L at fixed levels of tax on profit rates t for 5-year project at $t = 0.4$

L	k_0	k_d	Wd	t	n	NOI	k_e	NPV	ΔNPV
0	0.18	0.14	0	0.4	5	800	0.18	501.04	64.13345
1	0.18	0.14	0.5	0.4	5	800	0.189578	565.18	4.73089
2	0.18	0.14	0.66667	0.4	5	800	0.19803	569.91	−9.5017
3	0.18	0.14	0.75	0.4	5	800	0.206172	560.40	−14.7815
4	0.18	0.14	0.8	0.4	5	800	0.214184	545.62	−17.1025
5	0.18	0.14	0.83333	0.4	5	800	0.22213	528.52	−18.1709
6	0.18	0.14	0.85714	0.4	5	800	0.230037	510.35	−18.6246
7	0.18	0.14	0.875	0.4	5	800	0.23792	491.73	−18.7461
8	0.18	0.14	0.88889	0.4	5	800	0.245786	472.98	−18.6762
9	0.18	0.14	0.9	0.4	5	800	0.253642	454.30	−18.4911
10	0.18	0.14	0.90909	0.4	5	800	0.261488	435.81	

One can see from the Fig. 28.1, that the nature of the NPV dependence on leverage at $t^* = 0.5$: there is a transition from diminishing function NPV(L) when $t < t^*$ to growing function NPV(L) at $t > t^*$.

10-Year Project

For 10-year projects we get the following results (Tables 28.8, 28.9, 28.10, 28.11, 28.12, 28.13, and 28.14, Figs. 28.3 and 28.4).

Perpetuity Limit

In perpetuity limit $n \to \infty$ (Modigliani and Miller limit (Modigliani & Miller 1958, 1963, 1966)) formula for NPV is as following (Brusov & Filatova, 2011; Brusov et al., 2011a, b, c, 2012a, b, 2013a, b, 2014a, b; Filatova et al., 2008)

$$NPV = -S + \frac{NOI(1-t)}{k_e} - D(1-t). \tag{28.2}$$

At constant value of investment capital ($I =$ const), accounting $D = IL/(1 + L)$, $S = I/(1 + L)$, one gets

Table 28.3 Dependence of NPV and ΔNPV on leverage level L at fixed levels of tax on profit rates t for 5-year project at $t = 0.5$

L	k_0	k_d	Wd	t	n	NOI	k_e	NPV	ΔNPV
0	0.18	0.14	0	0.5	5	800	0.18	250.87	128.0669
1	0.18	0.14	0.5	0.5	5	800	0.181448	366.94	41.1323
2	0.18	0.14	0.66667	0.5	5	800	0.181065	408.07	22.57738
3	0.18	0.14	0.75	0.5	5	800	0.180162	430.65	15.19888
4	0.18	0.14	0.8	0.5	5	800	0.179041	445.84	11.52994
5	0.18	0.14	0.83333	0.5	5	800	0.177806	457.37	9.446706
6	0.18	0.14	0.85714	0.5	5	800	0.176505	466.82	8.154973
7	0.18	0.14	0.875	0.5	5	800	0.175162	474.98	7.302458
8	0.18	0.14	0.88889	0.5	5	800	0.173792	482.28	6.713275
9	0.18	0.14	0.9	0.5	5	800	0.172401	488.99	6.291579
10	0.18	0.14	0.90909	0.5	5	800	0.170996	495.28	

Table 28.4 Dependence of NPV and ΔNPV on leverage level L at fixed levels of tax on profit rates t for 5-year project at $t = 0.6$

L	k_0	k_d	Wd	t	n	NOI	k_e	NPV	ΔNPV
0	0.18	0.14	0	0.6	5	800	0.18	0.69	160.0655
1	0.18	0.14	0.5	0.6	5	800	0.173086	160.76	70.95083
2	0.18	0.14	0.66667	0.6	5	800	0.163424	231.71	49.78654
3	0.18	0.14	0.75	0.6	5	800	0.15296	281.50	42.12961
4	0.18	0.14	0.8	0.6	5	800	0.142153	323.63	39.00243
5	0.18	0.14	0.83333	0.6	5	800	0.131168	362.63	37.85743
6	0.18	0.14	0.85714	0.6	5	800	0.120078	400.49	37.74625
7	0.18	0.14	0.875	0.6	5	800	0.108922	438.23	38.25211
8	0.18	0.14	0.88889	0.6	5	800	0.097721	476.49	39.17065
9	0.18	0.14	0.9	0.6	5	800	0.086488	515.66	40.39427
10	0.18	0.14	0.90909	0.6	5	800	0.075231	556.05	

$$NPV = -\frac{I}{1+L}(1 + L(1-t)) + \frac{NOI(1-t)}{k_e}. \tag{28.3}$$

$$NPV = -\frac{I}{1+L}(1 + L(1-t)) + \frac{NOI(1-t)}{k_0 + (k_0 - k_d)L(1-t)}, \tag{28.4}$$

here $L = D/S$—leverage level; k_e—equity cost of leverage company (use the debt capital); k_0—equity cost of financially independent company.

In the transition from the equation (Eq. 28.3) to (Eq. 28.4) we have used the formula for equity capital cost, received by Modigliani & Miller (1958, 1963, 1966)

$$k_e = k_0 + (k_0 - k_d)L(1-t). \tag{28.5}$$

Table 28.5 Dependence of NPV and ΔNPV on leverage level L at fixed levels of tax on profit rates t for 5-year project at $t = 0.7$

L	k_0	k_d	Wd	t	n	NOI	k_e	NPV	ΔNPV
0	0.18	0.14	0	0.7	5	800	0.18	−249.48	195.3877
1	0.18	0.14	0.5	0.7	5	800	0.16448	−54.09	92.40829
2	0.18	0.14	0.66667	0.7	5	800	0.145057	38.32	69.68464
3	0.18	0.14	0.75	0.7	5	800	0.124461	108.00	63.19622
4	0.18	0.14	0.8	0.7	5	800	0.103355	171.20	62.43475
5	0.18	0.14	0.83333	0.7	5	800	0.081982	233.63	64.46977
6	0.18	0.14	0.85714	0.7	5	800	0.060453	298.10	68.26526
7	0.18	0.14	0.875	0.7	5	800	0.038822	366.37	73.42597
8	0.18	0.14	0.88889	0.7	5	800	0.017124	439.79	79.82524
9	0.18	0.14	0.9	0.7	5	800	−0.00462	519.62	87.47314
10	0.18	0.14	0.90909	0.7	5	800	−0.0264	607.09	

Table 28.6 Dependence of NPV and ΔNPV on leverage level L at fixed levels of tax on profit rates t for 5-year project at $t = 0.8$

L	k_0	k_d	Wd	t	n	NOI	k_e	NPV	ΔNPV
0	0.18	0.14	0	0.8	5	800	0.18	−499.65	221.1945
1	0.18	0.14	0.5	0.8	5	800	0.155616	−278.46	103.2179
2	0.18	0.14	0.66667	0.8	5	800	0.125907	−175.24	78.69376
3	0.18	0.14	0.75	0.8	5	800	0.094544	−96.55	73.65304
4	0.18	0.14	0.8	0.8	5	800	0.062454	−22.89	76.11182
5	0.18	0.14	0.83333	0.8	5	800	0.029979	53.22	82.93806
6	0.18	0.14	0.85714	0.8	5	800	−0.00272	136.16	93.28915
7	0.18	0.14	0.875	0.8	5	800	−0.03557	229.45	107.1925
8	0.18	0.14	0.88889	0.8	5	800	−0.06852	336.64	125.1636
9	0.18	0.14	0.9	0.8	5	800	−0.10154	461.80	148.1245
10	0.18	0.14	0.90909	0.8	5	800	−0.13462	609.93	

For perpetuity projects we get the following results (Tables 28.15, 28.16, 28.17, 28.18, 28.19, 28.20, and 28.21, Figs. 28.5 and 28.6).

28.1.2 *The Effectiveness of the Investment Project from the Perspective of the Equity and Debt Holders*

In this case we use the following expression for NPV (Brusov & Filatova, 2011; Brusov et al., 2011a, b, c, 2012a, b, 2013a, b, 2014a, b; Filatova et al., 2008)

Table 28.7 Dependence of NPV and ΔNPV on leverage level L at fixed levels of tax on profit rates t for 5-year project at $t = 0.9$

L	k_0	k_d	Wd	t	n	NOI	k_e	NPV	ΔNPV
0	0.18	0.14	0	0.9	5	800	0.18	−749.83	236.5329
1	0.18	0.14	0.5	0.9	5	800	0.14648	−513.29	100.4127
2	0.18	0.14	0.66667	0.9	5	800	0.105908	−412.88	71.49601
3	0.18	0.14	0.75	0.9	5	800	0.063074	−341.38	65.36206
4	0.18	0.14	0.8	0.9	5	800	0.019228	−276.02	68.55205
5	0.18	0.14	0.83333	0.9	5	800	−0.02516	−207.47	77.93497
6	0.18	0.14	0.85714	0.9	5	800	−0.06987	−129.54	93.29474
7	0.18	0.14	0.875	0.9	5	800	−0.11479	−36.24	115.8841
8	0.18	0.14	0.88889	0.9	5	800	−0.15985	79.64	148.3017
9	0.18	0.14	0.9	0.9	5	800	−0.20501	227.94	194.9563
10	0.18	0.14	0.90909	0.9	5	800	−0.25024	422.90	

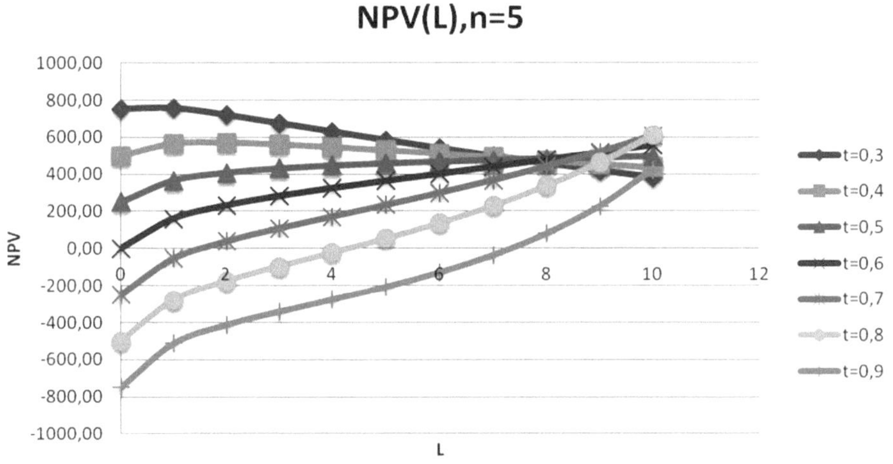

Fig. 28.1 Dependence of NPV on leverage level L at fixed levels of tax on profit rates t for 5-year project

$$NPV = -I\left[1 - \frac{Lt}{1+L}\left(1 - \frac{1}{(1+k_d)^n}\right)\right] + \frac{NOI(1-t)}{k_e}\left(1 - \frac{1}{(1+k_e)^n}\right) \quad (28.6)$$

Using it we calculate the dependence of NPV and ΔNPV on leverage level L at fixed levels of tax on profit rates t.

5-Year Project

For 5-year projects we get the following results (Tables 28.22, 28.23, 28.24, 28.25, 28.26, 28.27, and 28.28, Figs. 28.7 and 28.8).

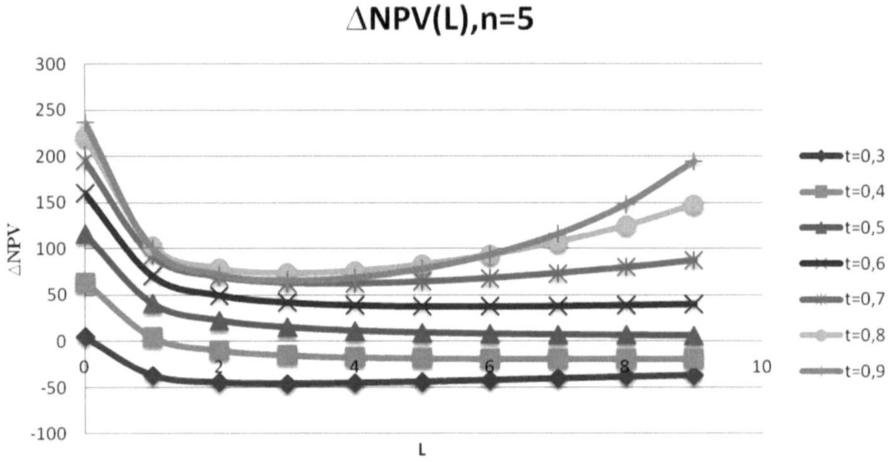

Fig. 28.2 Dependence of ΔNPV on leverage level L at fixed levels of tax on profit rates t for 5-year project

Table 28.8 Dependence of NPV and ΔNPV on leverage level L at fixed levels of tax on profit rates t for 10-year project at $t = 0.3$

L	k_0	k_d	Wd	t	n	NOI	k_e	NPV	ΔNPV
0	0.18	0.14	0	0.3	10	800	0.18	1528.69	−51.935
1	0.18	0.14	0.5	0.3	10	800	0.19907	1464.75	−101.47
2	0.18	0.14	0.66667	0.3	10	800	0.217182	1363.28	−104.843
3	0.18	0.14	0.75	0.3	10	800	0.235022	1258.44	−100.263
4	0.18	0.14	0.8	0.3	10	800	0.252747	1158.18	−93.8156
5	0.18	0.14	0.83333	0.3	10	800	0.270413	1064.36	−87.1129
6	0.18	0.14	0.85714	0.3	10	800	0.288045	977.25	−80.6768
7	0.18	0.14	0.875	0.3	10	800	0.305654	896.57	−74.6812
8	0.18	0.14	0.88889	0.3	10	800	0.323249	821.89	−69.1705
9	0.18	0.14	0.9	0.3	10	800	0.340834	752.72	−64.1369
10	0.18	0.14	0.90909	0.3	10	800	0.358411	688.58	

10-Year Project

For 10-year projects we get the following results (Tables 28.29, 28.30, 28.31, 28.32, 28.33, 28.34, and 28.35, Figs. 28.9 and 28.10).

28.2 Dependence of NPV on Tax on Profit Rates at Fixed Leverage Levels *L*

Below we study the dependence of NPV on tax on profit rates at fixed leverage levels *L*.

Table 28.9 Dependence of NPV and ΔNPV on leverage level L at fixed levels of tax on profit rates t for 10-year project at $t = 0.4$

L	k_0	k_d	Wd	t	n	NOI	k_e	NPV	ΔNPV
0	0.18	0.14	0	0.4	10	800	0.18	1157.16	61.10029
1	0.18	0.14	0.5	0.4	10	800	0.191455	1218.26	−18.517
2	0.18	0.14	0.66667	0.4	10	800	0.201083	1199.74	−35.7995
3	0.18	0.14	0.75	0.4	10	800	0.210169	1163.95	−41.1321
4	0.18	0.14	0.8	0.4	10	800	0.21902	1122.81	−42.6427
5	0.18	0.14	0.83333	0.4	10	800	0.22775	1080.17	−42.6081
6	0.18	0.14	0.85714	0.4	10	800	0.236408	1037.56	−41.8669
7	0.18	0.14	0.875	0.4	10	800	0.24502	995.70	−40.7844
8	0.18	0.14	0.88889	0.4	10	800	0.253602	954.91	−39.5386
9	0.18	0.14	0.9	0.4	10	800	0.262161	915.37	−38.2228
10	0.18	0.14	0.90909	0.4	10	800	0.270705	877.15	

Table 28.10 Dependence of NPV and ΔNPV on leverage level L at fixed levels of tax on profit rates t for 10-year project at $t = 0.5$

L	k_0	k_d	Wd	t	n	NOI	k_e	NPV	ΔNPV
0	0.18	0.14	0	0.5	10	800	0.18	797.63	−44.2013
1	0.18	0.14	0.5	0.5	10	800	0.219687	753.43	−104.707
2	0.18	0.14	0.66667	0.5	10	800	0.255351	648.73	−103.671
3	0.18	0.14	0.75	0.5	10	800	0.290005	545.06	−93.6005
4	0.18	0.14	0.8	0.5	10	800	0.32421	451.45	−82.6797
5	0.18	0.14	0.83333	0.5	10	800	0.358168	368.78	−72.7174
6	0.18	0.14	0.85714	0.5	10	800	0.391982	296.06	−64.0269
7	0.18	0.14	0.875	0.5	10	800	0.425702	232.03	−56.5734
8	0.18	0.14	0.88889	0.5	10	800	0.459362	175.46	−50.1953
9	0.18	0.14	0.9	0.5	10	800	0.492979	125.26	−44.7383
10	0.18	0.14	0.90909	0.5	10	800	0.526564	80.52	

Table 28.11 Dependence of NPV and ΔNPV on leverage level L at fixed levels of tax on profit rates t for 10-year project at $t = 0.6$

L	k_0	k_d	Wd	t	n	NOI	k_e	NPV	ΔNPV
0	0.18	0.14	0	0.6	10	800	0.18	438.11	244.5668
1	0.18	0.14	0.5	0.6	10	800	0.175084	682.67	125.4304
2	0.18	0.14	0.66667	0.6	10	800	0.165407	808.10	101.3662
3	0.18	0.14	0.75	0.6	10	800	0.154162	909.47	95.93652
4	0.18	0.14	0.8	0.6	10	800	0.142206	1005.41	97.02629
5	0.18	0.14	0.83333	0.6	10	800	0.129869	1102.43	101.3309
6	0.18	0.14	0.85714	0.6	10	800	0.117304	1203.76	107.7015
7	0.18	0.14	0.875	0.6	10	800	0.10459	1311.47	115.7055
8	0.18	0.14	0.88889	0.6	10	800	0.091775	1427.17	125.2046
9	0.18	0.14	0.9	0.6	10	800	0.078888	1552.38	136.2059
10	0.18	0.14	0.90909	0.6	10	800	0.065947	1688.58	

Table 28.12 Dependence of NPV and ΔNPV on leverage level L at fixed levels of tax on profit rates t for 10-year project at $t = 0.7$

L	k_0	k_d	Wd	t	n	NOI	k_e	NPV	ΔNPV
0	0.18	0.14	0	0.7	10	800	0.18	78.58	310.4744
1	0.18	0.14	0.5	0.7	10	800	0.166255	389.06	177.1979
2	0.18	0.14	0.66667	0.7	10	800	0.145497	566.25	160.5402
3	0.18	0.14	0.75	0.7	10	800	0.122295	726.79	169.9063
4	0.18	0.14	0.8	0.7	10	800	0.097953	896.70	191.9196
5	0.18	0.14	0.83333	0.7	10	800	0.072986	1088.62	224.1724
6	0.18	0.14	0.85714	0.7	10	800	0.04764	1312.79	267.4019
7	0.18	0.14	0.875	0.7	10	800	0.022047	1580.19	323.9861
8	0.18	0.14	0.88889	0.7	10	800	−0.00372	1904.18	397.7601
9	0.18	0.14	0.9	0.7	10	800	−0.0296	2301.94	494.3094
10	0.18	0.14	0.90909	0.7	10	800	−0.05558	2796.25	

Table 28.13 Dependence of NPV and ΔNPV on leverage level L at fixed levels of tax on profit rates t for 10-year project at $t = 0.8$

L	k_0	k_d	Wd	t	n	NOI	k_e	NPV	ΔNPV
0	0.18	0.14	0	0.8	10	800	0.18	−280.95	355.2633
1	0.18	0.14	0.5	0.8	10	800	0.156938	74.32	204.8466
2	0.18	0.14	0.66667	0.8	10	800	0.123925	279.16	198.4863
3	0.18	0.14	0.75	0.8	10	800	0.087223	477.65	232.7806
4	0.18	0.14	0.8	0.8	10	800	0.048741	710.43	297.8303
5	0.18	0.14	0.83333	0.8	10	800	0.009262	1008.26	400.9259
6	0.18	0.14	0.85714	0.8	10	800	−0.03083	1409.19	560.3743
7	0.18	0.14	0.875	0.8	10	800	−0.07133	1969.56	809.4944
8	0.18	0.14	0.88889	0.8	10	800	−0.11211	2779.06	1207.425
9	0.18	0.14	0.9	0.8	10	800	−0.1531	3986.48	1861.112
10	0.18	0.14	0.90909	0.8	10	800	−0.19424	5847.59	

Table 28.14 Dependence of NPV and ΔNPV on leverage level L at fixed levels of tax on profit rates t for 10-year project at $t = 0.9$

L	k_0	k_d	Wd	t	n	NOI	k_e	NPV	ΔNPV
0	0.18	0.14	0	0.9	10	800	0.18	−640.47	375.0978
1	0.18	0.14	0.5	0.9	10	800	0.147081	−265.38	194.2145
2	0.18	0.14	0.66667	0.9	10	800	0.100417	−71.16	186.9461
3	0.18	0.14	0.75	0.9	10	800	0.048303	115.79	239.586
4	0.18	0.14	0.8	0.9	10	800	−0.00655	355.37	356.4744
5	0.18	0.14	0.83333	0.9	10	800	−0.06297	711.85	583.6021
6	0.18	0.14	0.85714	0.9	10	800	−0.12039	1295.45	1032.383
7	0.18	0.14	0.875	0.9	10	800	−0.17847	2327.83	1967.211
8	0.18	0.14	0.88889	0.9	10	800	−0.23701	4295.04	4054.338
9	0.18	0.14	0.9	0.9	10	800	−0.2959	8349.38	9115.336
10	0.18	0.14	0.90909	0.9	10	800	−0.35504	17464.72	

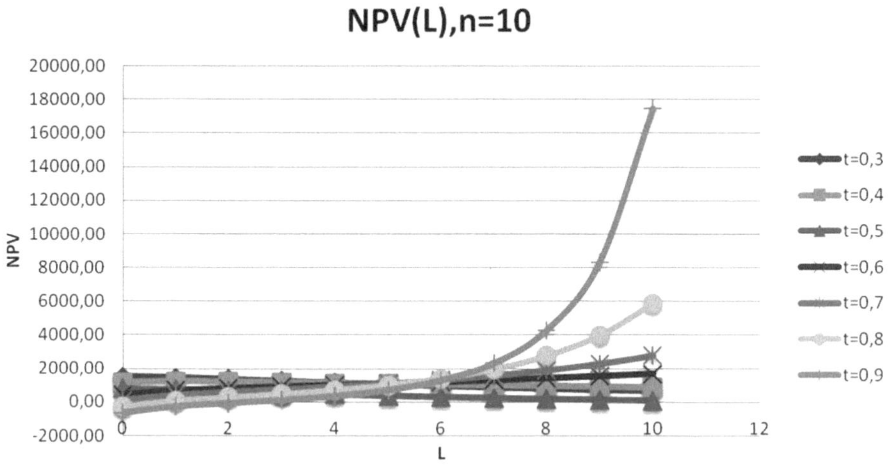

Fig. 28.3 Dependence of NPV on leverage level *L* at fixed levels of tax on profit rates *t* for 10-year project

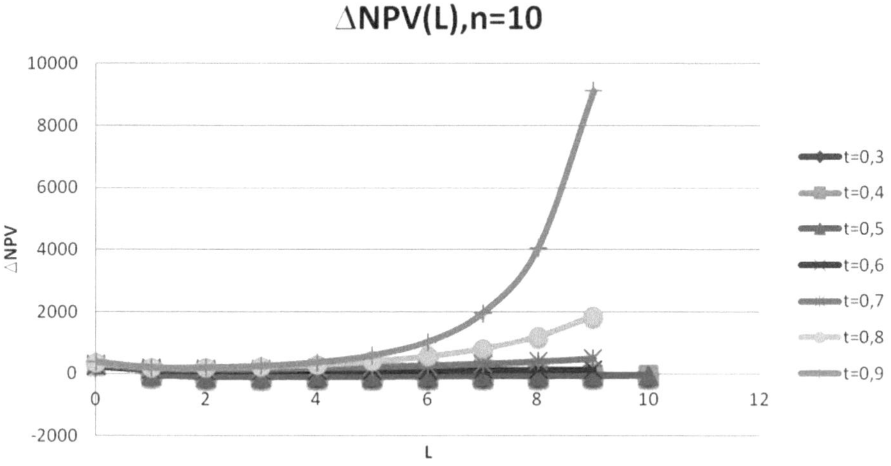

Fig. 28.4 Dependence of NPV on leverage level *L* at fixed levels of tax on profit rates *t* for 10-year project

28.2.1 *The Effectiveness of the Investment Project from the Perspective of the Equity Holders Only*

5-Year Project

For 5-year projects we get the following results (Tables 28.36, 28.37, 28.38, 28.39, 28.40, and 28.41, Figs. 28.11 and 28.12).

Table 28.15 Dependence of NPV and ΔNPV on leverage level L at fixed levels of tax on profit rates t for perpetuity project at $t = 0.3$

L	t	k_0	k_d	I	NOI	NPV	ΔNPV
0	0.3	0.18	0.14	1000	800	2111.111	−268.803
1	0.3	0.18	0.14	1000	800	1842.308	−269.426
2	0.3	0.18	0.14	1000	800	1572.881	−226.669
3	0.3	0.18	0.14	1000	800	1346.212	−188.404
4	0.3	0.18	0.14	1000	800	1157.808	−157.808
5	0.3	0.18	0.14	1000	800	1000	−133.662
6	0.3	0.18	0.14	1000	800	866.3383	−114.477
7	0.3	0.18	0.14	1000	800	751.8617	−99.0564
8	0.3	0.18	0.14	1000	800	652.8053	−86.509
9	0.3	0.18	0.14	1000	800	566.2963	−76.1777
10	0.3	0.18	0.14	1000	800	490.1186	

Table 28.16 Dependence of NPV and ΔNPV on leverage level L at fixed levels of tax on profit rates t for perpetuity project at $t = 0.4$

L	t	k_0	k_d	I	NOI	NPV	ΔNPV
0	0.4	0.18	0.14	1000	800	1666.667	−113.725
1	0.4	0.18	0.14	1000	800	1552.941	−181.011
2	0.4	0.18	0.14	1000	800	1371.93	−167.168
3	0.4	0.18	0.14	1000	800	1204.762	−145.631
4	0.4	0.18	0.14	1000	800	1059.13	−125.797
5	0.4	0.18	0.14	1000	800	933.3333	−108.995
6	0.4	0.18	0.14	1000	800	824.3386	−95.0283
7	0.4	0.18	0.14	1000	800	729.3103	−83.4322
8	0.4	0.18	0.14	1000	800	645.8781	−73.7569
9	0.4	0.18	0.14	1000	800	572.1212	−65.6277
10	0.4	0.18	0.14	1000	800	506.4935	

Table 28.17 Dependence of NPV and ΔNPV on leverage level L at fixed levels of tax on profit rates t for perpetuity project at $t = 0.5$

L	t	k_0	k_d	I	NOI	NPV	ΔNPV
0	0.5	0.18	0.14	1000	800	1222.222	27.77778
1	0.5	0.18	0.14	1000	800	1250	−98.4848
2	0.5	0.18	0.14	1000	800	1151.515	−109.848
3	0.5	0.18	0.14	1000	800	1041.667	−103.205
4	0.5	0.18	0.14	1000	800	938.4615	−93.2234
5	0.5	0.18	0.14	1000	800	845.2381	−83.3333
6	0.5	0.18	0.14	1000	800	761.9048	−74.4048
7	0.5	0.18	0.14	1000	800	687.5	−66.585
8	0.5	0.18	0.14	1000	800	620.915	−59.8039
9	0.5	0.18	0.14	1000	800	561.1111	−53.9341
10	0.5	0.18	0.14	1000	800	507.177	

Table 28.18 Dependence of NPV and ΔNPV on leverage level L at fixed levels of tax on profit rates t for perpetuity project at $t = 0.6$

L	t	k_0	k_d	I	NOI	NPV	ΔNPV
0	0.6	0.18	0.14	1000	800	777.7778	154.8753
1	0.6	0.18	0.14	1000	800	932.6531	−23.2191
2	0.6	0.18	0.14	1000	800	909.434	−55.9252
3	0.6	0.18	0.14	1000	800	853.5088	−62.0334
4	0.6	0.18	0.14	1000	800	791.4754	−60.7062
5	0.6	0.18	0.14	1000	800	730.7692	−57.0632
6	0.6	0.18	0.14	1000	800	673.706	−52.8156
7	0.6	0.18	0.14	1000	800	620.8904	−48.596
8	0.6	0.18	0.14	1000	800	572.2944	−44.6401
9	0.6	0.18	0.14	1000	800	527.6543	−41.0233
10	0.6	0.18	0.14	1000	800	486.631	

Table 28.19 Dependence of NPV and ΔNPV on leverage level L at fixed levels of tax on profit rates t for perpetuity project at $t = 0.7$

L	t	k_0	k_d	I	NOI	NPV	ΔNPV
0	0.7	0.18	0.14	1000	800	333.3333	266.6667
1	0.7	0.18	0.14	1000	800	600	43.13725
2	0.7	0.18	0.14	1000	800	643.1373	−7.02614
3	0.7	0.18	0.14	1000	800	636.1111	−23.4795
4	0.7	0.18	0.14	1000	800	612.6316	−29.2982
5	0.7	0.18	0.14	1000	800	583.3333	−30.9524
6	0.7	0.18	0.14	1000	800	552.381	−30.79
7	0.7	0.18	0.14	1000	800	521.5909	−29.8035
8	0.7	0.18	0.14	1000	800	491.7874	−28.4541
9	0.7	0.18	0.14	1000	800	463.3333	−26.9697
10	0.7	0.18	0.14	1000	800	436.3636	

Table 28.20 Dependence of NPV and ΔNPV on leverage level L at fixed levels of tax on profit rates t for perpetuity project at $t = 0.8$

L	t	k_0	k_d	I	NOI	NPV	ΔNPV
0	0.8	0.18	0.14	1000	800	−111.111	362.1749
1	0.8	0.18	0.14	1000	800	251.0638	98.59603
2	0.8	0.18	0.14	1000	800	349.6599	34.65386
3	0.8	0.18	0.14	1000	800	384.3137	10.40326
4	0.8	0.18	0.14	1000	800	394.717	−0.77759
5	0.8	0.18	0.14	1000	800	393.9394	−6.47072
6	0.8	0.18	0.14	1000	800	387.4687	−9.50257
7	0.8	0.18	0.14	1000	800	377.9661	−11.1173
8	0.8	0.18	0.14	1000	800	366.8488	−11.9282
9	0.8	0.18	0.14	1000	800	354.9206	−12.2633
10	0.8	0.18	0.14	1000	800	342.6573	

Table 28.21 Dependence of NPV and ΔNPV on leverage level L at fixed levels of tax on profit rates t for perpetuity project at $t = 0.9$

L	t	k_0	k_d	I	NOI	NPV	ΔNPV
0	0.9	0.18	0.14	1000	800	−555.556	440.3382
1	0.9	0.18	0.14	1000	800	−115.217	140.7493
2	0.9	0.18	0.14	1000	800	25.53191	66.13475
3	0.9	0.18	0.14	1000	800	91.66667	36.4966
4	0.9	0.18	0.14	1000	800	128.1633	21.83673
5	0.9	0.18	0.14	1000	800	150	13.58543
6	0.9	0.18	0.14	1000	800	163.5854	8.52995
7	0.9	0.18	0.14	1000	800	172.1154	5.243106
8	0.9	0.18	0.14	1000	800	177.3585	3.01188
9	0.9	0.18	0.14	1000	800	180.3704	1.447811
10	0.9	0.18	0.14	1000	800	181.8182	

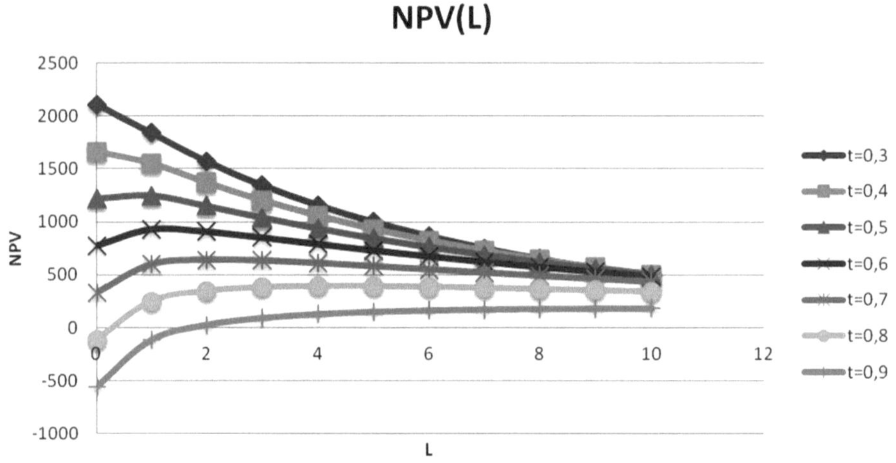

Fig. 28.5 Dependence of NPV on leverage level L at fixed levels of tax on profit rates t for perpetuity project

10-Year Project

For 10-year projects we get the following results (Tables 28.42, 28.43, 28.44, 28.45, 28.46, and 28.47, Figs. 28.13 and 28.14).

Perpetuity Limit

For perpetuity projects we get the following results (Tables 28.48 and 28.49, Fig. 28.15).

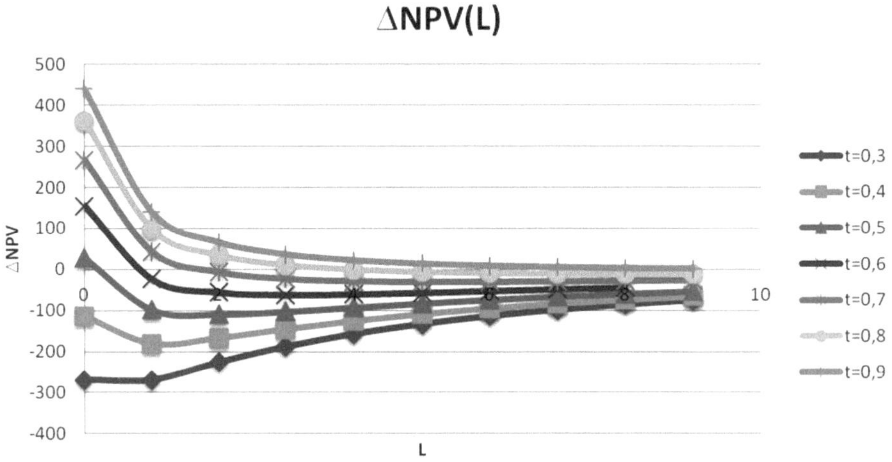

Fig. 28.6 Dependence of ΔNPV on leverage level *L* at fixed levels of tax on profit rates *t* for perpetuity project

28.2.2 *The Effectiveness of the Investment Project from the Perspective of the Equity and Debt Holders*

5-Year Project

For 5-year projects we get the following results (Tables 28.50, 28.51, 28.52, 28.53, 28.54, and 28.55, Figs. 28.16 and 28.17).

10-Year Projects

For 10-year projects we get the following results (Tables 28.56, 28.57, 28.58, 28.59, 28.60, and 28.61, Figs. 28.18, 28.19, 28.20, and 28.21).

Or more detailed.

It is seen from Fig. 28.19, that falling trend at $L = 0$, 2, and 4 alternates by growing trend at higher leverage levels $L = 6$, 8, and 10.

The observed at high leverage levels (starting from $L = 6$) increase of NPV with growth of the tax on profit rate *t* takes place at all values of *t*, that means that this is an entirely new effect in investments which can be applied in a real economic practice for optimizing of the management of investments.

Or more detailed.

Table 28.22 Dependence of NPV and ΔNPV on leverage level L at fixed levels of tax on profit rates t for 5-year project at $t = 0.3$

I	L	WACC	k_0	k_d	Wd	t	n	NOI	k_e	NPV	ΔNPV
1000	0	0.18	0.18	0.14	0	0.3	5	800	0.18	751.2158	4.922709
1000	1	0.14774	0.18	0.14	0.5	0.3	5	800	0.197488	756.1385	−36.8599
1000	2	0.13679	0.18	0.14	0.66667	0.3	5	800	0.214367	719.2786	−44.7663
1000	3	0.13127	0.18	0.14	0.75	0.3	5	800	0.231082	674.5122	−46.126
1000	4	0.12795	0.18	0.14	0.8	0.3	5	800	0.24773	628.3863	−45.4549
1000	5	0.12572	0.18	0.14	0.83333	0.3	5	800	0.264343	582.9313	−44.027
1000	6	0.12413	0.18	0.14	0.85714	0.3	5	800	0.280937	538.9044	−42.3084
1000	7	0.12294	0.18	0.14	0.875	0.3	5	800	0.297518	496.596	−40.4978
1000	8	0.12201	0.18	0.14	0.88889	0.3	5	800	0.314091	456.0982	−38.6879
1000	9	0.12127	0.18	0.14	0.9	0.3	5	800	0.330658	417.4103	−36.9239
1000	10	0.12066	0.18	0.14	0.90909	0.3	5	800	0.34722	380.4865	

Table 28.23 Dependence of NPV and ΔNPV on leverage level *L* at fixed levels of tax on profit rates *t* for 5-year project at $t = 0.4$

I	L	WACC	k_0	k_d	Wd	t	n	NOI	k_e	NPV	ΔNPV
1000	0	0.18	0.18	0.14	0	0.4	5	800	0.18	501.0421	64.13345
1000	1	0.13679	0.18	0.14	0.5	0.4	5	800	0.189578	565.1755	4.73089
1000	2	0.12201	0.18	0.14	0.66667	0.4	5	800	0.19803	569.9064	−9.5017
1000	3	0.11454	0.18	0.14	0.75	0.4	5	800	0.206172	560.4047	−14.7815
1000	4	0.11004	0.18	0.14	0.8	0.4	5	800	0.214184	545.6233	−17.1025
1000	5	0.10702	0.18	0.14	0.83333	0.4	5	800	0.22213	528.5208	−18.1709
1000	6	0.10486	0.18	0.14	0.85714	0.4	5	800	0.230037	510.3499	−18.6246
1000	7	0.10324	0.18	0.14	0.875	0.4	5	800	0.23792	491.7254	−18.7461
1000	8	0.10198	0.18	0.14	0.88889	0.4	5	800	0.245786	472.9793	−18.6762
1000	9	0.10096	0.18	0.14	0.9	0.4	5	800	0.253642	454.3031	−18.4911
1000	10	0.10014	0.18	0.14	0.90909	0.4	5	800	0.261488	435.812	

Table 28.24 Dependence of NPV and ΔNPV on leverage level L at fixed levels of tax on profit rates t for 5-year project at $t = 0.5$

I	L	WACC	k_0	k_d	Wd	t	n	NOI	k_e	NPV	ΔNPV
1000	0	0.18	0.18	0.14	0	0.5	5	800	0.18	250.8684	128.0669
1000	1	0.12572	0.18	0.14	0.5	0.5	5	800	0.181448	366.9353	41.1323
1000	2	0.10702	0.18	0.14	0.66667	0.5	5	800	0.181065	408.0676	22.57738
1000	3	0.09754	0.18	0.14	0.75	0.5	5	800	0.180162	430.645	15.19888
1000	4	0.09181	0.18	0.14	0.8	0.5	5	800	0.179041	445.8439	11.52994
1000	5	0.08797	0.18	0.14	0.83333	0.5	5	800	0.177806	457.3738	9.446706
1000	6	0.08522	0.18	0.14	0.85714	0.5	5	800	0.176505	466.8205	8.154973
1000	7	0.08315	0.18	0.14	0.875	0.5	5	800	0.175162	474.9755	7.302458
1000	8	0.08153	0.18	0.14	0.88889	0.5	5	800	0.173792	482.278	6.713275
1000	9	0.08024	0.18	0.14	0.9	0.5	5	800	0.172401	488.9912	6.291579
1000	10	0.07918	0.18	0.14	0.90909	0.5	5	800	0.170996	495.2828	

Table 28.25 Dependence of NPV and ΔNPV on leverage level L at fixed levels of tax on profit rates t for 5-year project at $t = 0.6$

I	L	WACC	k_0	k_d	Wd	t	n	NOI	k_e	NPV	ΔNPV
1000	0	0.18	0.18	0.14	0	0.6	5	800	0.18	0.694727	160.0655
1000	1	0.11454	0.18	0.14	0.5	0.6	5	800	0.173086	160.7602	70.95083
1000	2	0.09181	0.18	0.14	0.66667	0.6	5	800	0.163424	231.711	49.78654
1000	3	0.08024	0.18	0.14	0.75	0.6	5	800	0.15296	281.4976	42.12961
1000	4	0.07323	0.18	0.14	0.8	0.6	5	800	0.142153	323.6272	39.00243
1000	5	0.06853	0.18	0.14	0.83333	0.6	5	800	0.131168	362.6296	37.85743
1000	6	0.06515	0.18	0.14	0.85714	0.6	5	800	0.120078	400.487	37.74625
1000	7	0.06262	0.18	0.14	0.875	0.6	5	800	0.108922	438.2333	38.25211
1000	8	0.06064	0.18	0.14	0.88889	0.6	5	800	0.097721	476.4854	39.17065
1000	9	0.05905	0.18	0.14	0.9	0.6	5	800	0.086488	515.656	40.39427
1000	10	0.05775	0.18	0.14	0.90909	0.6	5	800	0.075231	556.0503	

Table 28.26 Dependence of NPV and ΔNPV on leverage level L at fixed levels of tax on profit rates t for 5-year project at $t = 0.7$

I	L	WACC	k_0	k_d	Wd	t	n	NOI	k_e	NPV	ΔNPV
1000	0	0.18	0.18	0.14	0	0.7	5	800	0.18	−249.479	195.3877
1000	1	0.10324	0.18	0.14	0.5	0.7	5	800	0.16448	−54.0913	92.40829
1000	2	0.07635	0.18	0.14	0.66667	0.7	5	800	0.145057	38.31702	69.68464
1000	3	0.06262	0.18	0.14	0.75	0.7	5	800	0.124461	108.0017	63.19622
1000	4	0.05427	0.18	0.14	0.8	0.7	5	800	0.103355	171.1979	62.43475
1000	5	0.04866	0.18	0.14	0.83333	0.7	5	800	0.081982	233.6326	64.46977
1000	6	0.04464	0.18	0.14	0.85714	0.7	5	800	0.060453	298.1024	68.26526
1000	7	0.0416	0.18	0.14	0.875	0.7	5	800	0.038822	366.3677	73.42597
1000	8	0.03924	0.18	0.14	0.88889	0.7	5	800	0.017124	439.7936	79.82524
1000	9	0.03734	0.18	0.14	0.9	0.7	5	800	−0.00462	519.6189	87.47314
1000	10	0.03578	0.18	0.14	0.90909	0.7	5	800	−0.0264	607.092	

Table 28.27 Dependence of NPV and ΔNPV on leverage level L at fixed levels of tax on profit rates t for 5-year project at $t = 0.8$

I	L	WACC	k_0	k_d	Wd	t	n	NOI	k_e	NPV	ΔNPV
1000	0	0.18	0.18	0.14	0	0.8	5	800	0.18	−499.653	221.1945
1000	1	0.09181	0.18	0.14	0.5	0.8	5	800	0.155616	−278.458	103.2179
1000	2	0.06064	0.18	0.14	0.66667	0.8	5	800	0.125907	−175.24	78.69376
1000	3	0.04464	0.18	0.14	0.75	0.8	5	800	0.094544	−96.5465	73.65304
1000	4	0.03489	0.18	0.14	0.8	0.8	5	800	0.062454	−22.8934	76.11182
1000	5	0.02833	0.18	0.14	0.83333	0.8	5	800	0.029979	53.21839	82.93806
1000	6	0.02361	0.18	0.14	0.85714	0.8	5	800	−0.00272	136.1565	93.28915
1000	7	0.02005	0.18	0.14	0.875	0.8	5	800	−0.03557	229.4456	107.1925
1000	8	0.01728	0.18	0.14	0.88889	0.8	5	800	−0.06852	336.6381	125.1636
1000	9	0.01505	0.18	0.14	0.9	0.8	5	800	−0.10154	461.8017	148.1245
1000	10	0.01322	0.18	0.14	0.90909	0.8	5	800	−0.13462	609.9262	

Table 28.28 Dependence of NPV and ΔNPV on leverage level L at fixed levels of tax on profit rates t for 5-year project at $t = 0.9$

I	L	WACC	k_0	k_d	Wd	t	n	NOI	k_e	NPV	ΔNPV
1000	0	0.18	0.18	0.14	0	0.9	5	800	0.18	−749.826	236.5329
1000	1	0.08024	0.18	0.14	0.5	0.9	5	800	0.14648	−513.293	100.4127
1000	2	0.04464	0.18	0.14	0.66667	0.9	5	800	0.105908	−412.881	71.49601
1000	3	0.02627	0.18	0.14	0.75	0.9	5	800	0.063074	−341.385	65.36206
1000	4	0.01505	0.18	0.14	0.8	0.9	5	800	0.019228	−276.023	68.55205
1000	5	0.00747	0.18	0.14	0.83333	0.9	5	800	−0.02516	−207.471	77.93497
1000	6	0.00202	0.18	0.14	0.85714	0.9	5	800	−0.06987	−129.536	93.29474
1000	7	−0.0021	0.18	0.14	0.875	0.9	5	800	−0.11479	−36.2409	115.8841
1000	8	−0.0053	0.18	0.14	0.88889	0.9	5	800	−0.15985	79.64322	148.3017
1000	9	−0.0079	0.18	0.14	0.9	0.9	5	800	−0.20501	227.945	194.9563
1000	10	−0.01	0.18	0.14	0.90909	0.9	5	800	−0.25024	422.9012	

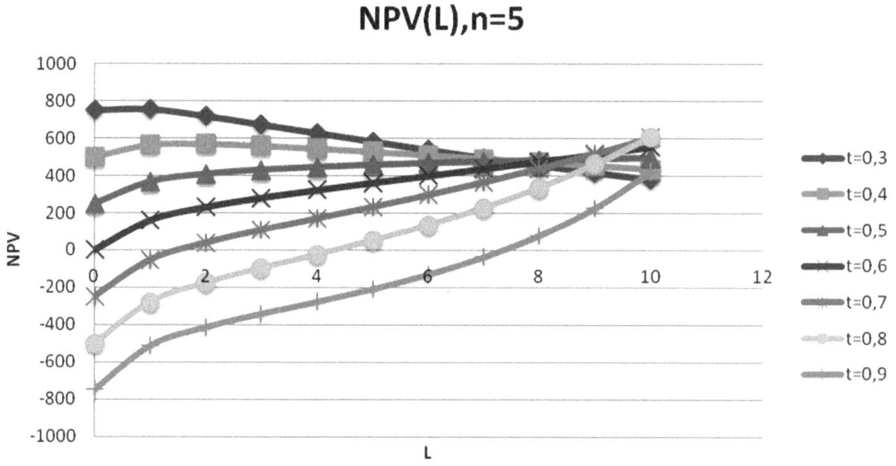

Fig. 28.7 Dependence of NPV on leverage level *L* at fixed levels of tax on profit rates *t* for 5-year project

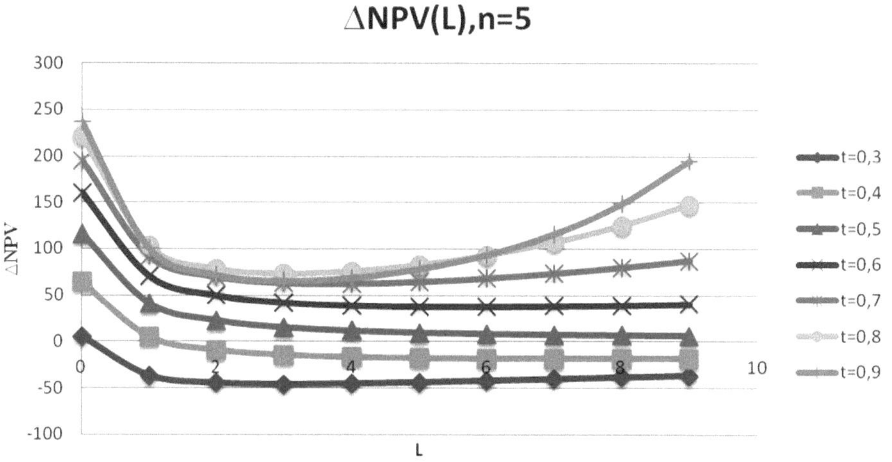

Fig. 28.8 Dependence of ΔNPV on leverage level *L* at fixed levels of tax on profit rates *t* for 5-year project

Table 28.29 Dependence of NPV and ΔNPV on leverage level L at fixed levels of tax on profit rates t for 10-year project at $t = 0.3$

I	L	WACC	k_0	k_d	Wd	t	n	NOI	k_e	NPV	ΔNPV
1000	0	0.18	0.18	0.14	0	0.3	10	800	0.18	1520.688	−51.935
1000	1	0.14854	0.18	0.14	0.5	0.3	10	800	0.19907	1464.753	−101.47
1000	2	0.13773	0.18	0.14	0.66667	0.3	10	800	0.217182	1363.283	−104.843
1000	3	0.13226	0.18	0.14	0.75	0.3	10	800	0.235022	1258.441	−100.263
1000	4	0.12895	0.18	0.14	0.8	0.3	10	800	0.252747	1158.178	−93.8156
1000	5	0.12674	0.18	0.14	0.83333	0.3	10	800	0.270413	1064.362	−87.1129
1000	6	0.12515	0.18	0.14	0.85714	0.3	10	800	0.288045	977.2493	−80.6768
1000	7	0.12396	0.18	0.14	0.875	0.3	10	800	0.305654	896.5725	−74.6812
1000	8	0.12303	0.18	0.14	0.88889	0.3	10	800	0.323249	821.8913	−69.1705
1000	9	0.12228	0.18	0.14	0.9	0.3	10	800	0.340834	752.7207	−64.1369
1000	10	0.12167	0.18	0.14	0.90909	0.3	10	800	0.358411	688.5839	

Table 28.30 Dependence of NPV and ΔNPV on leverage level L at fixed levels of tax on profit rates t for 10-year project at $t = 0.4$

I	L	WACC	k_0	k_d	Wd	t	n	NOI	k_e	NPV	ΔNPV
1000	0	0.18	0.18	0.14	0	0.4	10	800	0.18	1157.161	61.10029
1000	1	0.13773	0.18	0.14	0.5	0.4	10	800	0.191455	1218.262	−18.517
1000	2	0.12303	0.18	0.14	0.66667	0.4	10	800	0.201083	1199.745	−35.7995
1000	3	0.11554	0.18	0.14	0.75	0.4	10	800	0.210169	1163.945	−41.1321
1000	4	0.111	0.18	0.14	0.8	0.4	10	800	0.21902	1122.813	−42.6427
1000	5	0.10796	0.18	0.14	0.83333	0.4	10	800	0.22775	1080.17	−42.6081
1000	6	0.10577	0.18	0.14	0.85714	0.4	10	800	0.236408	1037.562	−41.8669
1000	7	0.10413	0.18	0.14	0.875	0.4	10	800	0.24502	995.6955	−40.7844
1000	8	0.10284	0.18	0.14	0.88889	0.4	10	800	0.253602	954.9111	−39.5386
1000	9	0.10182	0.18	0.14	0.9	0.4	10	800	0.262161	915.3725	−38.2228
1000	10	0.10097	0.18	0.14	0.90909	0.4	10	800	0.270705	877.1497	

Table 28.31 Dependence of NPV and ΔNPV on leverage level L at fixed levels of tax on profit rates t for 10-year project at $t = 0.5$

I	L	WACC	k_0	k_d	Wd	t	n	NOI	k_e	NPV	ΔNPV
1000	0	0.18	0.18	0.14	0	0.5	10	800	0.18	797.6345	−44.2013
1000	1	0.14484	0.20056	0.14	0.5	0.5	10	800	0.219687	753.4333	−104.707
1000	2	0.13178	0.20836	0.14	0.66667	0.5	10	800	0.255351	648.7259	−103.671
1000	3	0.125	0.21253	0.14	0.75	0.5	10	800	0.290005	545.0552	−93.6005
1000	4	0.12084	0.21513	0.14	0.8	0.5	10	800	0.32421	451.4547	−82.6797
1000	5	0.11803	0.2169	0.14	0.83333	0.5	10	800	0.358168	368.775	−72.7174
1000	6	0.116	0.21819	0.14	0.85714	0.5	10	800	0.391982	296.0576	−64.0269
1000	7	0.11446	0.21917	0.14	0.875	0.5	10	800	0.425702	232.0307	−56.5734
1000	8	0.11326	0.21994	0.14	0.88889	0.5	10	800	0.459362	175.4573	−50.1953
1000	9	0.1123	0.22055	0.14	0.9	0.5	10	800	0.492979	125.262	−44.7383
1000	10	0.11151	0.22106	0.14	0.90909	0.5	10	800	0.526564	80.52365	

Table 28.32 Dependence of NPV and ΔNPV on leverage level L at fixed levels of tax on profit rates t for 10-year project at $t = 0.6$

I	L	WACC	k_0	k_d	Wd	t	n	NOI	k_e	NPV	ΔNPV
1000	0	0.18	0.18	0.14	0	0.6	10	800	0.18	438.1076	244.5668
1000	1	0.11554	0.18	0.14	0.5	0.6	10	800	0.175084	682.6744	125.4304
1000	2	0.09247	0.18	0.14	0.66667	0.6	10	800	0.165407	808.1048	101.3662
1000	3	0.08054	0.18	0.14	0.75	0.6	10	800	0.154162	909.471	95.93652
1000	4	0.07324	0.18	0.14	0.8	0.6	10	800	0.142206	1005.408	97.02629
1000	5	0.06831	0.18	0.14	0.83333	0.6	10	800	0.129869	1102.434	101.3309
1000	6	0.06476	0.18	0.14	0.85714	0.6	10	800	0.117304	1203.765	107.7015
1000	7	0.06207	0.18	0.14	0.875	0.6	10	800	0.10459	1311.466	115.7055
1000	8	0.05998	0.18	0.14	0.88889	0.6	10	800	0.091775	1427.172	125.2046
1000	9	0.05829	0.18	0.14	0.9	0.6	10	800	0.078888	1552.376	136.2059
1000	10	0.0569	0.18	0.14	0.90909	0.6	10	800	0.065947	1688.582	

Table 28.33 Dependence of NPV and ΔNPV on leverage level L at fixed levels of tax on profit rates t for 10-year project at $t = 0.7$

I	L	WACC	k_0	k_d	Wd	t	n	NOI	k_e	NPV	ΔNPV
1000	0	0.18	0.18	0.14	0	0.7	10	800	0.18	78.58071	310.4744
1000	1	0.10413	0.18	0.14	0.5	0.7	10	800	0.166255	389.0551	177.1979
1000	2	0.0765	0.18	0.14	0.66667	0.7	10	800	0.145497	566.253	160.5402
1000	3	0.06207	0.18	0.14	0.75	0.7	10	800	0.122295	726.7932	169.9063
1000	4	0.05319	0.18	0.14	0.8	0.7	10	800	0.097953	896.6995	191.9196
1000	5	0.04716	0.18	0.14	0.83333	0.7	10	800	0.072986	1088.619	224.1724
1000	6	0.04281	0.18	0.14	0.85714	0.7	10	800	0.04764	1312.791	267.4019
1000	7	0.03951	0.18	0.14	0.875	0.7	10	800	0.022047	1580.193	323.9861
1000	8	0.03692	0.18	0.14	0.88889	0.7	10	800	−0.00372	1904.18	397.7601
1000	9	0.03484	0.18	0.14	0.9	0.7	10	800	−0.0296	2301.94	494.3094
1000	10	0.03313	0.18	0.14	0.90909	0.7	10	800	−0.05558	2796.249	

Table 28.34 Dependence of NPV and ΔNPV on leverage level L at fixed levels of tax on profit rates t for 10-year project at $t = 0.8$

I	L	WACC	k_0	k_d	Wd	t	n	NOI	k_e	NPV	ΔNPV
1000	0	0.18	0.18	0.14	0	0.8	10	800	0.18	−280.946	355.2633
1000	1	0.09247	0.18	0.14	0.5	0.8	10	800	0.156938	74.31716	204.8466
1000	2	0.05998	0.18	0.14	0.66667	0.8	10	800	0.123925	279.1638	198.4863
1000	3	0.04281	0.18	0.14	0.75	0.8	10	800	0.087223	477.6501	232.7806
1000	4	0.03215	0.18	0.14	0.8	0.8	10	800	0.048741	710.4306	297.8303
1000	5	0.02488	0.18	0.14	0.83333	0.8	10	800	0.009262	1008.261	400.9259
1000	6	0.0196	0.18	0.14	0.85714	0.8	10	800	−0.03083	1409.187	560.3743
1000	7	0.01558	0.18	0.14	0.875	0.8	10	800	−0.07133	1969.561	809.4944
1000	8	0.01243	0.18	0.14	0.88889	0.8	10	800	−0.11211	2779.056	1207.425
1000	9	0.00989	0.18	0.14	0.9	0.8	10	800	−0.1531	3986.481	1861.112
1000	10	0.0078	0.18	0.14	0.90909	0.8	10	800	−0.19424	5847.593	

Table 28.35 Dependence of NPV and ΔNPV on leverage level L at fixed levels of tax on profit rates t for 10-year project at $t = 0.9$

I	L	WACC	k_0	k_d	Wd	t	n	NOI	k_e	NPV	ΔNPV
1000	0	0.18	0.18	0.14	0	0.9	10	800	0.18	−640.473	375.0978
1000	1	0.08054	0.18	0.14	0.5	0.9	10	800	0.147081	−265.375	194.2145
1000	2	0.04281	0.18	0.14	0.66667	0.9	10	800	0.100417	−71.1609	186.9461
1000	3	0.02258	0.18	0.14	0.75	0.9	10	800	0.048303	115.7852	239.586
1000	4	0.00989	0.18	0.14	0.8	0.9	10	800	−0.00655	355.3712	356.4744
1000	5	0.00117	0.18	0.14	0.83333	0.9	10	800	−0.06297	711.8456	583.6021
1000	6	−0.0052	0.18	0.14	0.85714	0.9	10	800	−0.12039	1295.448	1032.383
1000	7	−0.0101	0.18	0.14	0.875	0.9	10	800	−0.17847	2327.831	1967.211
1000	8	−0.0139	0.18	0.14	0.88889	0.9	10	800	−0.23701	4295.041	4054.338
1000	9	−0.017	0.18	0.14	0.9	0.9	10	800	−0.2959	8349.379	9115.336
1000	10	−0.0195	0.18	0.14	0.90909	0.9	10	800	−0.35504	17464.72	

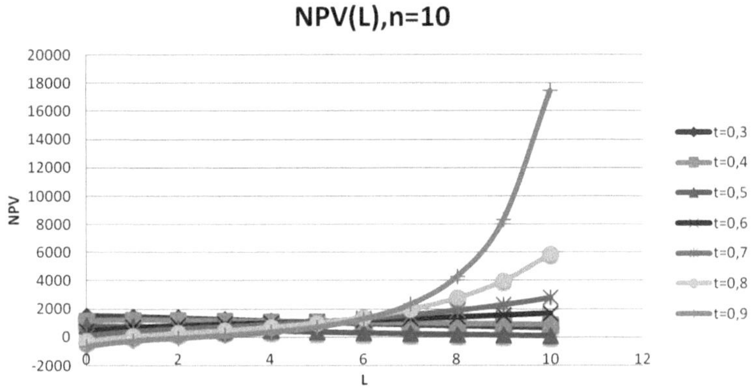

Fig. 28.9 Dependence of NPV on leverage level *L* at fixed levels of tax on profit rates *t* for 5-year project

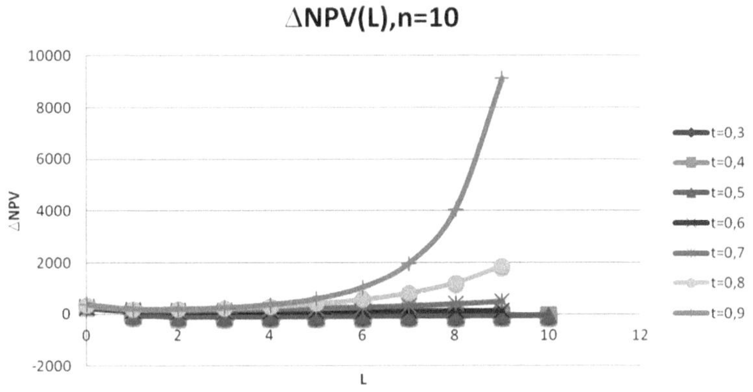

Fig. 28.10 Dependence of ΔNPV on leverage level *L* at fixed levels of tax on profit rates *t* for 5-year project

Table 28.36 Dependence of NPV and ΔNPV on tax on profit rate *t* at fixed leverage level *L* for 5-year project at $L = 0$

I	L	WACC	k_0	k_d	Wd	t	n	NOI	k_e	NPV	ΔNPV
1000	0	0.18	0.18	0.14	0	0	5	800	0.18	1501.74	−250.174
1000	0	0.18	0.18	0.14	0	0.1	5	800	0.18	1251.56	−250.174
1000	0	0.18	0.18	0.14	0	0.2	5	800	0.18	1001.39	−250.174
1000	0	0.18	0.18	0.14	0	0.3	5	800	0.18	751.22	−250.174
1000	0	0.18	0.18	0.14	0	0.4	5	800	0.18	501.04	−250.174
1000	0	0.18	0.18	0.14	0	0.5	5	800	0.18	250.87	−250.174
1000	0	0.18	0.18	0.14	0	0.6	5	800	0.18	0.69	−250.174
1000	0	0.18	0.18	0.14	0	0.7	5	800	0.18	−249.48	−250.174
1000	0	0.18	0.18	0.14	0	0.8	5	800	0.18	−499.65	−250.174
1000	0	0.18	0.18	0.14	0	0.9	5	800	0.18	−749.83	−250.174
1000	0	0.18	0.18	0.14	0	1	5	800	0.18	−1000.00	

Table 28.37 Dependence of NPV and ΔNPV on tax on profit rate t at fixed leverage level L for 5-year project at $L = 2$

I	L	WACC	k_0	k_d	Wd	t	n	NOI	k_e	NPV	ΔNPV
1000	2	0.18	0.18	0.14	0.66667	0	5	800	0.26	1108.06	−128.983
1000	2	0.16577	0.18	0.14	0.66667	0.1	5	800	0.245315	987.07	−129.204
1000	2	0.15137	0.18	0.14	0.66667	0.2	5	800	0.230116	857.87	−138.59
1000	2	0.13679	0.18	0.14	0.66667	0.3	5	800	0.214367	719.28	−149.372
1000	2	0.12201	0.18	0.14	0.66667	0.4	5	800	0.19803	569.91	−161.839
1000	2	0.10702	0.18	0.14	0.66667	0.5	5	800	0.181065	408.07	−176.357
1000	2	0.09181	0.18	0.14	0.66667	0.6	5	800	0.163424	231.71	−193.394
1000	2	0.07635	0.18	0.14	0.66667	0.7	5	800	0.145057	38.32	−213.557
1000	2	0.06064	0.18	0.14	0.66667	0.8	5	800	0.125907	−175.24	−237.64
1000	2	0.04464	0.18	0.14	0.66667	0.9	5	800	0.105908	−412.88	−266.698
1000	2	0.02833	0.18	0.14	0.66667	1	5	800	0.084989	−679.58	

Table 28.38 Dependence of NPV and ΔNPV on tax on profit rate t at fixed leverage level L for 5-year project at $L = 4$

I	L	WACC	k_0	k_d	Wd	t	n	NOI	k_e	NPV	ΔNPV
1000	4	0.18	0.18	0.14	0.8	0	5	800	0.34	808.33	−51.044
1000	4	0.16291	0.18	0.14	0.8	0.1	5	800	0.31053	757.29	−59.2842
1000	4	0.14556	0.18	0.14	0.8	0.2	5	800	0.279807	698.00	−69.6148
1000	4	0.12795	0.18	0.14	0.8	0.3	5	800	0.24773	628.39	−82.763
1000	4	0.11004	0.18	0.14	0.8	0.4	5	800	0.214184	545.62	−99.7794
1000	4	0.09181	0.18	0.14	0.8	0.5	5	800	0.179041	445.84	−122.217
1000	4	0.07323	0.18	0.14	0.8	0.6	5	800	0.142153	323.63	−152.429
1000	4	0.05427	0.18	0.14	0.8	0.7	5	800	0.103355	171.20	−194.091
1000	4	0.03489	0.18	0.14	0.8	0.8	5	800	0.062454	−22.89	−253.129
1000	4	0.01505	0.18	0.14	0.8	0.9	5	800	0.019228	−276.02	−339.472
1000	4	−0.0053	0.18	0.14	0.8	1	5	800	−0.02658	−615.49	

Table 28.39 Dependence of NPV and ΔNPV on tax on profit rate t at fixed leverage level L for 5-year project at $L = 6$

I	L	WACC	k_0	k_d	Wd	t	n	NOI	k_e	NPV	ΔNPV
1000	6	0.18	0.18	0.14	0.85714	0	5	800	0.42	574.85	−6.23744
1000	6	0.16168	0.18	0.14	0.85714	0.1	5	800	0.375729	568.61	−11.31
1000	6	0.14306	0.18	0.14	0.85714	0.2	5	800	0.329434	557.30	−18.3975
1000	6	0.12413	0.18	0.14	0.85714	0.3	5	800	0.280937	538.90	−28.5544
1000	6	0.10486	0.18	0.14	0.85714	0.4	5	800	0.230037	510.35	−43.5294
1000	6	0.08522	0.18	0.14	0.85714	0.5	5	800	0.176505	466.82	−66.3335
1000	6	0.06515	0.18	0.14	0.85714	0.6	5	800	0.120078	400.49	−102.385
1000	6	0.04464	0.18	0.14	0.85714	0.7	5	800	0.060453	298.10	−161.946
1000	6	0.02361	0.18	0.14	0.85714	0.8	5	800	−0.00272	136.16	−265.692
1000	6	0.00202	0.18	0.14	0.85714	0.9	5	800	−0.06987	−129.54	−458.495
1000	6	−0.0202	0.18	0.14	0.85714	1	5	800	−0.14147	−588.03	

Table 28.40 Dependence of NPV and ΔNPV on tax on profit rate t at fixed leverage level L for 5-year project at $L = 8$

I	L	WACC	k_0	k_d	Wd	t	n	NOI	k_e	NPV	ΔNPV
1000	8	0.18	0.18	0.14	0.88889	0	5	800	0.5	389.30	23.47489
1000	8	0.16099	0.18	0.14	0.88889	0.1	5	800	0.440923	412.78	22.6083
1000	8	0.14167	0.18	0.14	0.88889	0.2	5	800	0.379039	435.38	28.71462
1000	8	0.12201	0.18	0.14	0.88889	0.3	5	800	0.314091	456.10	28.88106
1000	8	0.10198	0.18	0.14	0.88889	0.4	5	800	0.245786	472.98	9.298693
1000	8	0.08153	0.18	0.14	0.88889	0.5	5	800	0.173792	482.28	−5.79258
1000	8	0.06064	0.18	0.14	0.88889	0.6	5	800	0.097721	476.49	−36.6918
1000	8	0.03924	0.18	0.14	0.88889	0.7	5	800	0.017124	439.79	−103.156
1000	8	0.01728	0.18	0.14	0.88889	0.8	5	800	−0.06852	336.64	−256.995
1000	8	−0.0053	0.18	0.14	0.88889	0.9	5	800	−0.15985	79.64	−652.415
1000	8	−0.0286	0.18	0.14	0.88889	1	5	800	−0.25759	−572.77	

Table 28.41 Dependence of NPV and ΔNPV on tax on profit rate t at fixed leverage level L for 5-year project at $L = 10$

I	L	WACC	k_0	k_d	Wd	t	n	NOI	k_e	NPV	ΔNPV
1000	10	0.18	0.18	0.14	0.90909	0	5	800	0.58	239.23	43.49836
1000	10	0.16056	0.18	0.14	0.90909	0.1	5	800	0.506115	282.73	46.87962
1000	10	0.14078	0.18	0.14	0.90909	0.2	5	800	0.428635	329.61	50.87829
1000	10	0.12066	0.18	0.14	0.90909	0.3	5	800	0.34722	380.49	55.32552
1000	10	0.10014	0.18	0.14	0.90909	0.4	5	800	0.261488	435.81	59.47085
1000	10	0.07918	0.18	0.14	0.90909	0.5	5	800	0.170996	495.28	60.76747
1000	10	0.05775	0.18	0.14	0.90909	0.6	5	800	0.075231	556.05	51.04171
1000	10	0.03578	0.18	0.14	0.90909	0.7	5	800	−0.0264	607.09	2.834166
1000	10	0.01322	0.18	0.14	0.90909	0.8	5	800	−0.13462	609.93	−187.024
1000	10	−0.01	0.18	0.14	0.90909	0.9	5	800	−0.25024	422.90	−985.964
1000	10	−0.034	0.18	0.14	0.90909	1	5	800	−0.37429	−563.06	

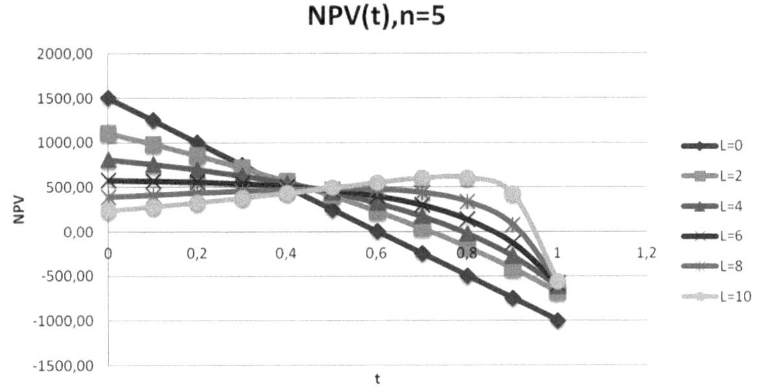

Fig. 28.11 Dependence of NPV on tax on profit rate *t* at fixed leverage level *L* for 5-year project

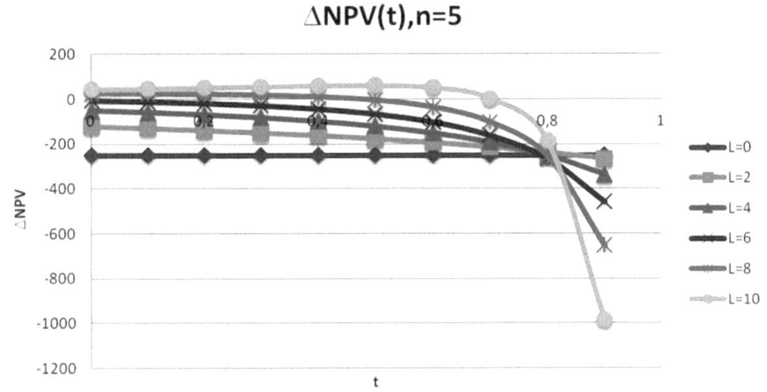

Fig. 28.12 Dependence of ΔNPV on tax on profit rate *t* at fixed leverage level *L* for 5-year project

Table 28.42 Dependence of NPV and ΔNPV on tax on profit rate *t* at fixed leverage level *L* for 10-year project at $L = 0$

I	L	WACC	k_0	k_d	Wd	t	n	NOI	k_e	NPV	ΔNPV
1000	0	0.18	0.18	0.14	0	0	10	800	0.18	2595.27	−359.527
1000	0	0.18	0.18	0.14	0	0.1	10	800	0.18	2235.74	−359.527
1000	0	0.18	0.18	0.14	0	0.2	10	800	0.18	1876.22	−359.527
1000	0	0.18	0.18	0.14	0	0.3	10	800	0.18	1520.69	−359.527
1000	0	0.18	0.18	0.14	0	0.4	10	800	0.18	1157.16	−359.527
1000	0	0.18	0.18	0.14	0	0.5	10	800	0.18	797.63	−359.527
1000	0	0.18	0.18	0.14	0	0.6	10	800	0.18	438.11	−359.527
1000	0	0.18	0.18	0.14	0	0.7	10	800	0.18	78.58	−359.527
1000	0	0.18	0.18	0.14	0	0.8	10	800	0.18	−280.95	−359.527
1000	0	0.18	0.18	0.14	0	0.9	10	800	0.18	−640.47	−359.527
1000	0	0.18	0.18	0.14	0	1	10	800	0.18	−1000.00	

Table 28.43 Dependence of NPV and ΔNPV on tax on profit rate t at fixed leverage level L for 10-year project at $L = 2$

I	L	WACC	k_0	k_d	Wd	t	n	NOI	k_e	NPV	ΔNPV
1000	2	0.18	0.18	0.14	0.66667	0	10	800	0.26	1771.84	−125.137
1000	2	0.16618	0.18	0.14	0.66667	0.1	10	800	0.24654	1646.71	−135.437
1000	2	0.1521	0.18	0.14	0.66667	0.2	10	800	0.232299	1511.27	−147.987
1000	2	0.13773	0.18	0.14	0.66667	0.3	10	800	0.217182	1363.28	−163.538
1000	2	0.12303	0.18	0.14	0.66667	0.4	10	800	0.201083	1199.74	−183.17
1000	2	0.10796	0.18	0.14	0.66667	0.5	10	800	0.183875	1028.58	−208.47
1000	2	0.09247	0.18	0.14	0.66667	0.6	10	800	0.165407	808.10	−241.852
1000	2	0.0765	0.18	0.14	0.66667	0.7	10	800	0.145497	566.25	−287.089
1000	2	0.05998	0.18	0.14	0.66667	0.8	10	800	0.123925	279.16	−350.325
1000	2	0.04281	0.18	0.14	0.66667	0.9	10	800	0.100417	−71.16	−442.002
1000	2	0.02488	0.18	0.14	0.66667	1	10	800	0.074631	−513.16	

Table 28.44 Dependence of NPV and ΔNPV on tax on profit rate t at fixed leverage level L for 10-year project at $L = 4$

I	L	WACC	k_0	k_d	Wd	t	n	NOI	k_e	NPV	ΔNPV
1000	4	0.18	0.18	0.14	0.8	0	10	800	0.34	1226.89	−475.625
1000	4	0.18316	0.20056	0.14	0.8	0.1	10	800	0.411801	751.26	−139.495
1000	4	0.17256	0.20836	0.14	0.8	0.2	10	800	0.414791	611.77	−74.4553
1000	4	0.15772	0.21253	0.14	0.8	0.3	10	800	0.396579	537.31	−54.8815
1000	4	0.14072	0.21513	0.14	0.8	0.4	10	800	0.367589	482.43	−52.0879
1000	4	0.12227	0.2169	0.14	0.8	0.5	10	800	0.331331	430.34	−60.1971
1000	4	0.10261	0.21819	0.14	0.8	0.6	10	800	0.289071	370.14	−80.4792
1000	4	0.08181	0.21917	0.14	0.8	0.7	10	800	0.241069	289.66	−128.753
1000	4	0.0598	0.21994	0.14	0.8	0.8	10	800	0.186991	168.91	−202.021
1000	4	0.0364	0.22055	0.14	0.8	0.9	10	800	0.126013	−33.11	−382.685
1000	4	0.01135	0.22106	0.14	0.8	1	10	800	0.056769	−415.80	

Table 28.45 Dependence of NPV and ΔNPV on tax on profit rate t at fixed leverage level L for 10-year project at $L = 6$

I	L	WACC	k_0	k_d	Wd	t	n	NOI	k_e	NPV	ΔNPV
1000	6	0.18	0.18	0.14	0.85714	0	10	800	0.42	847.62	37.06285
1000	6	0.16218	0.18	0.14	0.85714	0.1	10	800	0.379295	884.68	42.53552
1000	6	0.14392	0.18	0.14	0.85714	0.2	10	800	0.335473	927.22	50.03126
1000	6	0.12515	0.18	0.14	0.85714	0.3	10	800	0.288045	977.25	60.31305
1000	6	0.10577	0.18	0.14	0.85714	0.4	10	800	0.236408	1037.56	74.23617
1000	6	0.08569	0.18	0.14	0.85714	0.5	10	800	0.179813	1111.80	91.96624
1000	6	0.06476	0.18	0.14	0.85714	0.6	10	800	0.117304	1203.76	109.0267
1000	6	0.04281	0.18	0.14	0.85714	0.7	10	800	0.04764	1312.79	96.39537
1000	6	0.0196	0.18	0.14	0.85714	0.8	10	800	−0.03083	1409.19	−113.739
1000	6	−0.0052	0.18	0.14	0.85714	0.9	10	800	−0.12039	1295.45	−1669.51
1000	6	−0.032	0.18	0.14	0.85714	1	10	800	−0.22431	−374.07	

Table 28.46 Dependence of NPV and ΔNPV on tax on profit rate t at fixed leverage level L for 10-year project at $L = 8$

I	L	WACC	k_0	k_d	Wd	t	n	NOI	k_e	NPV	ΔNPV
1000	8	0.18	0.18	0.14	0.88889	0	10	800	0.5	572.25	67.74227
1000	8	0.16152	0.18	0.14	0.88889	0.1	10	800	0.445653	640.00	80.91054
1000	8	0.14255	0.18	0.14	0.88889	0.2	10	800	0.386973	728.91	100.9849
1000	8	0.12303	0.18	0.14	0.88889	0.3	10	800	0.323249	821.89	133.0198
1000	8	0.10284	0.18	0.14	0.88889	0.4	10	800	0.253602	954.91	187.1372
1000	8	0.08188	0.18	0.14	0.88889	0.5	10	800	0.17692	1142.05	285.1235
1000	8	0.05998	0.18	0.14	0.88889	0.6	10	800	0.091775	1427.17	477.0078
1000	8	0.03692	0.18	0.14	0.88889	0.7	10	800	−0.00372	1904.18	874.876
1000	8	0.01243	0.18	0.14	0.88889	0.8	10	800	−0.11211	2779.06	1515.986
1000	8	−0.0139	0.18	0.14	0.88889	0.9	10	800	−0.23701	4295.04	−4645.92
1000	8	−0.0426	0.18	0.14	0.88889	1	10	800	−0.38375	−350.88	

Table 28.47 Dependence of NPV and ΔNPV on tax on profit rate t at fixed leverage level L for 10-year project at $L = 10$

I	L	WACC	k_0	k_d	Wd	t	n	NOI	k_e	NPV	ΔNPV
1000	10	0.18	0.18	0.14	0.90909	0	10	800	0.58	365.08	85.01433
1000	10	0.16109	0.18	0.14	0.90909	0.1	10	800	0.512008	450.10	103.8568
1000	10	0.14168	0.18	0.14	0.90909	0.2	10	800	0.438456	553.96	134.6286
1000	10	0.12167	0.18	0.14	0.90909	0.3	10	800	0.358411	688.58	188.5659
1000	10	0.10097	0.18	0.14	0.90909	0.4	10	800	0.270705	877.15	292.3814
1000	10	0.07944	0.18	0.14	0.90909	0.5	10	800	0.173857	1169.53	519.0512
1000	10	0.0569	0.18	0.14	0.90909	0.6	10	800	0.065947	1688.58	1107.667
1000	10	0.03313	0.18	0.14	0.90909	0.7	10	800	−0.05558	2796.25	3051.344
1000	10	0.0078	0.18	0.14	0.90909	0.8	10	800	−0.19424	5847.59	11617.12
1000	10	−0.0195	0.18	0.14	0.90909	0.9	10	800	−0.35504	17464.71	−17800.8
1000	10	−0.0496	0.18	0.14	0.90909	1	10	800	−0.54557	−336.13	

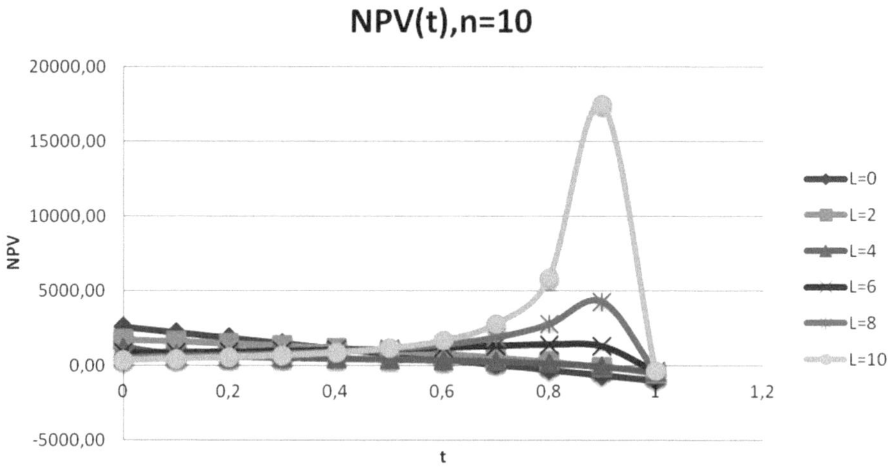

Fig. 28.13 Dependence of NPV on tax on profit rate *t* at fixed leverage level *L* for 10-year project

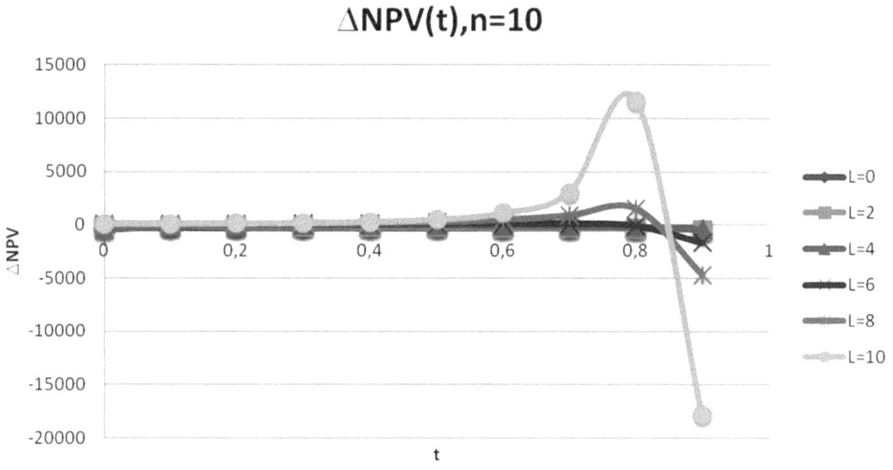

Fig. 28.14 Dependence of ΔNPV on tax on profit rate *t* at fixed leverage level *L* for 10-year project

Table 28.48 Dependence of NPV and ΔNPV on tax on profit rate t at fixed leverage level L for perpetuity project at $L = 8$

t	L	k_0	k_d	I	NOI	NPV	ΔNPV
0	8	0.18	0.14	1000	800	−32111.1	−5003.72
0.1	8	0.18	0.14	1000	800	−37114.8	−7713.12
0.2	8	0.18	0.14	1000	800	−44,828	−12659.2
0.3	8	0.18	0.14	1000	800	−57487.2	−23,580
0.4	8	0.18	0.14	1000	800	−81067.2	−57266.1
0.5	8	0.18	0.14	1000	800	−138,333	−326,333
0.6	8	0.18	0.14	1000	800	−464,667	816960.8
0.7	8	0.18	0.14	1000	800	352294.1	−222,466
0.8	8	0.18	0.14	1000	800	129828.3	−49,261
0.9	8	0.18	0.14	1000	800	80567.25	−21567.3
1	8	0.18	0.14	1000	800	59,000	

Table 28.49 Dependence of NPV and ΔNPV on tax on profit rate t at fixed leverage level L for perpetuity project at $L = 10$

t	L	k_0	k_d	I	NOI	NPV	ΔNPV
0	10	0.18	0.14	1000	800	−41,000	−9090.91
0.1	10	0.18	0.14	1000	800	−50090.9	−15909.1
0.2	10	0.18	0.14	1000	800	−66,000	−32692.3
0.3	10	0.18	0.14	1000	800	−98692.3	−99450.5
0.4	10	0.18	0.14	1000	800	−198,143	−8.6E+19
0.5	10	0.18	0.14	1000	800	−8.6E+19	8.65E+19
0.6	10	0.18	0.14	1000	800	202,750	−99632.4
0.7	10	0.18	0.14	1000	800	103117.6	−33006.5
0.8	10	0.18	0.14	1000	800	70111.11	−16374.3
0.9	10	0.18	0.14	1000	800	53736.84	−9736.84
1	10	0.18	0.14	1000	800	44,000	

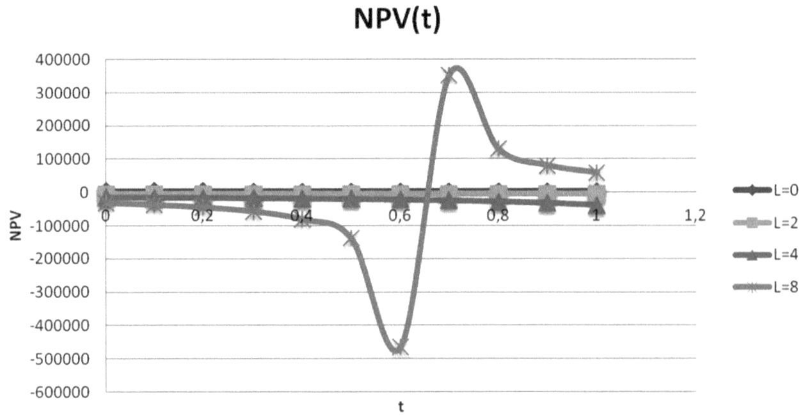

Fig. 28.15 Dependence of ΔNPV on tax on profit rate t at fixed leverage level L for perpetuity project

Table 28.50 Dependence of NPV and ΔNPV on tax on profit rate t at fixed leverage level L for 5-year project at $L = 0$

I	L	WACC	k_0	k_d	Wd	t	n	NOI	k_e	NPV	ΔNPV
1000	0	0.18	0.18	0.14	0	0	5	800	0.18	1501.737	−250.174
1000	0	0.18	0.18	0.14	0	0.1	5	800	0.18	1251.563	−250.174
1000	0	0.18	0.18	0.14	0	0.2	5	800	0.18	1001.389	−250.174
1000	0	0.18	0.18	0.14	0	0.3	5	800	0.18	751.2158	−250.174
1000	0	0.18	0.18	0.14	0	0.4	5	800	0.18	501.0421	−250.174
1000	0	0.18	0.18	0.14	0	0.5	5	800	0.18	250.8684	−250.174
1000	0	0.18	0.18	0.14	0	0.6	5	800	0.18	0.694727	−250.174
1000	0	0.18	0.18	0.14	0	0.7	5	800	0.18	−249.479	−250.174
1000	0	0.18	0.18	0.14	0	0.8	5	800	0.18	−499.653	−250.174
1000	0	0.18	0.18	0.14	0	0.9	5	800	0.18	−749.826	−250.174
1000	0	0.18	0.18	0.14	0	1	5	800	0.18	−1000	

28.3 Conclusions

Within modern theory of capital cost and capital structure by Brusov–Filatova–Orekhova (BFO theory) and created within this theory modern investment models influence of growth of tax on profit rate on the efficiency of the investment is investigated. It has been shown that for arbitrary duration projects as well as for perpetuity projects the growth of tax on profit rate change the nature of the NPV dependence on leverage at some value t^*: there is a transition from diminishing function NPV(L) when $t < t^*$ to growing function NPV(L). The t^* value depends on the duration of the project, cost of capital (equity and debt) values and other parameters of the project.

At high leverage levels this leads to qualitatively new effect in investments: growth of the efficiency of the investments with growth of tax on profit rate. Discovered effects take place under consideration from the point of view of owners of equity capital as well as from the point of view of owners of equity and debt capital.

The observed at high leverage levels (starting from $L = 6$) increase of NPV with growth of the tax on profit rate t takes place at all values of t, that means that this is an entirely new effect in investments which can be applied in a real economic practice for optimizing of the management of investments.

So, *two very important qualitatively new effects in investments has been discovered:*

1. *change the character of NPV dependence on leverage;*
2. *growth of the efficiency of the investments with growth of tax on profit rate.*

Both effects could be used in practice to optimize the investments.

Table 28.51 Dependence of NPV and ΔNPV on tax on profit rate t at fixed leverage level L for 5-year project at $L = 2$

I	L	WACC	k_0	k_d	Wd	t	n	NOI	k_e	NPV	ΔNPV
1000	2	0.18	0.18	0.14	0.66667	0	5	800	0.26	1108.057	−128.983
1000	2	0.16577	0.18	0.14	0.66667	0.1	5	800	0.245315	987.0733	−129.204
1000	2	0.15137	0.18	0.14	0.66667	0.2	5	800	0.230116	857.869	−138.59
1000	2	0.13679	0.18	0.14	0.66667	0.3	5	800	0.214367	719.2786	−149.372
1000	2	0.12201	0.18	0.14	0.66667	0.4	5	800	0.19803	569.9064	−161.839
1000	2	0.10702	0.18	0.14	0.66667	0.5	5	800	0.181065	408.0676	−176.357
1000	2	0.09181	0.18	0.14	0.66667	0.6	5	800	0.163424	231.711	−193.394
1000	2	0.07635	0.18	0.14	0.66667	0.7	5	800	0.145057	38.31702	−213.557
1000	2	0.06064	0.18	0.14	0.66667	0.8	5	800	0.125907	−175.24	−237.64
1000	2	0.04464	0.18	0.14	0.66667	0.9	5	800	0.105908	−412.881	−266.698
1000	2	0.02833	0.18	0.14	0.66667	1	5	800	0.084989	−679.579	

Table 28.52 Dependence of NPV and ΔNPV on tax on profit rate t at fixed leverage level L for 5-year project at $L = 4$

I	L	WACC	k_0	k_d	Wd	t	n	NOI	k_e	NPV	ΔNPV
1000	4	0.18	0.18	0.14	0.8	0	5	800	0.34	808.3293	−51.044
1000	4	0.16291	0.18	0.14	0.8	0.1	5	800	0.31053	757.2853	−59.2842
1000	4	0.14556	0.18	0.14	0.8	0.2	5	800	0.279807	698.0011	−69.6148
1000	4	0.12795	0.18	0.14	0.8	0.3	5	800	0.24773	628.3863	−82.763
1000	4	0.11004	0.18	0.14	0.8	0.4	5	800	0.214184	545.6233	−99.7794
1000	4	0.09181	0.18	0.14	0.8	0.5	5	800	0.179041	445.8439	−122.217
1000	4	0.07323	0.18	0.14	0.8	0.6	5	800	0.142153	323.6272	−152.429
1000	4	0.05427	0.18	0.14	0.8	0.7	5	800	0.103355	171.1979	−194.091
1000	4	0.03489	0.18	0.14	0.8	0.8	5	800	0.062454	−22.8934	−253.129
1000	4	0.01505	0.18	0.14	0.8	0.9	5	800	0.019228	−276.023	−339.472
1000	4	−0.0053	0.18	0.14	0.8	1	5	800	−0.02658	−615.495	

Table 28.53 Dependence of NPV and ΔNPV on tax on profit rate t at fixed leverage level L for 5-year project at $L = 6$

I	L	WACC	k_0	k_d	Wd	t	n	NOI	k_e	NPV	ΔNPV
1000	6	0.18	0.18	0.14	0.85714	0	5	800	0.42	574.8492	−6.23744
1000	6	0.16168	0.18	0.14	0.85714	0.1	5	800	0.375729	568.6118	−11.31
1000	6	0.14306	0.18	0.14	0.85714	0.2	5	800	0.329434	557.3019	−18.3975
1000	6	0.12413	0.18	0.14	0.85714	0.3	5	800	0.280937	538.9044	−28.5544
1000	6	0.10486	0.18	0.14	0.85714	0.4	5	800	0.230037	510.3499	−43.5294
1000	6	0.08522	0.18	0.14	0.85714	0.5	5	800	0.176505	466.8205	−66.3335
1000	6	0.06515	0.18	0.14	0.85714	0.6	5	800	0.120078	400.487	−102.385
1000	6	0.04464	0.18	0.14	0.85714	0.7	5	800	0.060453	298.1024	−161.946
1000	6	0.02361	0.18	0.14	0.85714	0.8	5	800	−0.00272	136.1565	−265.692
1000	6	0.00202	0.18	0.14	0.85714	0.9	5	800	−0.06987	−129.536	−458.495
1000	6	−0.0202	0.18	0.14	0.85714	1	5	800	−0.14147	−588.03	

Table 28.54 Dependence of NPV and ΔNPV on tax on profit rate t at fixed leverage level L for 5-year project at $L = 8$

I	L	WACC	k_0	k_d	Wd	t	n	NOI	k_e	NPV	ΔNPV
1000	8	0.18	0.18	0.14	0.88889	0	5	800	0.5	389.3004	23.47489
1000	8	0.16099	0.18	0.14	0.88889	0.1	5	800	0.440923	412.7753	22.6083
1000	8	0.14167	0.18	0.14	0.88889	0.2	5	800	0.379039	435.3836	20.71462
1000	8	0.12201	0.18	0.14	0.88889	0.3	5	800	0.314091	456.0982	20.88106
1000	8	0.10198	0.18	0.14	0.88889	0.4	5	800	0.245786	472.9793	9.298693
1000	8	0.08153	0.18	0.14	0.88889	0.5	5	800	0.173792	482.278	−5.79258
1000	8	0.06064	0.18	0.14	0.88889	0.6	5	800	0.097721	476.4854	−36.6918
1000	8	0.03924	0.18	0.14	0.88889	0.7	5	800	0.017124	439.7936	−103.156
1000	8	0.01728	0.18	0.14	0.88889	0.8	5	800	−0.06852	336.6381	−256.995
1000	8	−0.0053	0.18	0.14	0.88889	0.9	5	800	−0.15985	79.64322	−652.415
1000	8	−0.0286	0.18	0.14	0.88889	1	5	800	−0.25759	−572.772	

Table 28.55 Dependence of NPV and ΔNPV on tax on profit rate t at fixed leverage level L for 5-year project at $L = 10$

I	L	WACC	k_0	k_d	Wd	t	n	NOI	k_e	NPV	ΔNPV
1000	10	0.18	0.18	0.14	0.90909	0	5	800	0.58	239.2302	43.49836
1000	10	0.16056	0.18	0.14	0.90909	0.1	5	800	0.506115	282.7286	46.87962
1000	10	0.14078	0.18	0.14	0.90909	0.2	5	800	0.428635	329.6082	50.87829
1000	10	0.12066	0.18	0.14	0.90909	0.3	5	800	0.34722	380.4865	55.32552
1000	10	0.10014	0.18	0.14	0.90909	0.4	5	800	0.261488	435.812	59.47085
1000	10	0.07918	0.18	0.14	0.90909	0.5	5	800	0.170996	495.2828	60.76747
1000	10	0.05775	0.18	0.14	0.90909	0.6	5	800	0.075231	556.0503	51.04171
1000	10	0.03578	0.18	0.14	0.90909	0.7	5	800	−0.0264	607.092	2.834166
1000	10	0.01322	0.18	0.14	0.90909	0.8	5	800	−0.13462	609.9262	−187.024
1000	10	−0.01	0.18	0.14	0.90909	0.9	5	800	−0.25024	422.9017	−985.964
1000	10	−0.034	0.18	0.14	0.90909	1	5	800	−0.37429	−563.062	

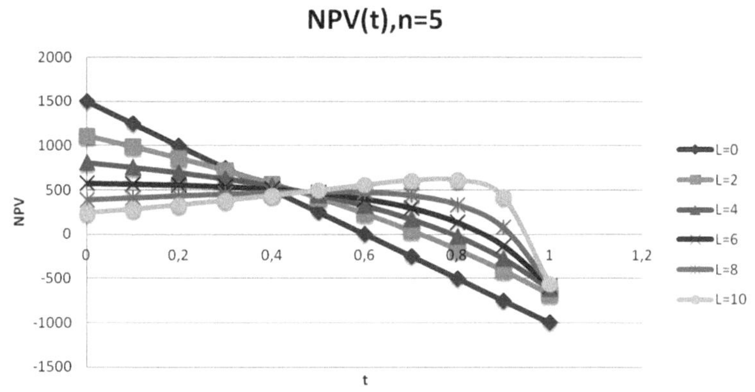

Fig. 28.16 Dependence of NPV on tax on profit rate t at fixed leverage level L for 5-year project

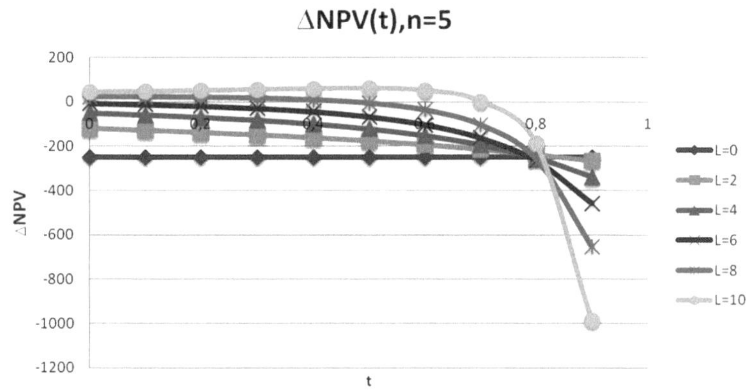

Fig. 28.17 Dependence of ΔNPV on tax on profit rate t at fixed leverage level L for 5-year project

Table 28.56 Dependence of NPV and ΔNPV on tax on profit rate t at fixed leverage level L for 10-year project at $L = 0$

I	L	WACC	k_0	k_d	Wd	t	n	NOI	k_e	NPV	ΔNPV
1000	0	0.18	0.18	0.14	0	0	10	800	0.18	2595.269	−359.527
1000	0	0.18	0.18	0.14	0	0.1	10	800	0.18	2235.742	−359.527
1000	0	0.18	0.18	0.14	0	0.2	10	800	0.18	1876.215	−359.527
1000	0	0.18	0.18	0.14	0	0.3	10	800	0.18	1520.688	−359.527
1000	0	0.18	0.18	0.14	0	0.4	10	800	0.18	1157.161	−359.527
1000	0	0.18	0.18	0.14	0	0.5	10	800	0.18	797.6345	−359.527
1000	0	0.18	0.18	0.14	0	0.6	10	800	0.18	438.1076	−359.527
1000	0	0.18	0.18	0.14	0	0.7	10	800	0.18	78.58071	−359.527
1000	0	0.18	0.18	0.14	0	0.8	10	800	0.18	−280.946	−359.527
1000	0	0.18	0.18	0.14	0	0.9	10	800	0.18	−640.473	−359.527
1000	0	0.18	0.18	0.14	0	1	10	800	0.18	−1000	

Table 28.57 Dependence of NPV and ΔNPV on tax on profit rate t at fixed leverage level L for 10-year project at $L = 2$

I	L	WACC	k_0	k_d	Wd	t	n	NOI	k_e	NPV	ΔNPV
1000	2	0.18	0.18	0.14	0.66667	0	10	800	0.26	1771.845	−125.137
1000	2	0.16618	0.18	0.14	0.66667	0.1	10	800	0.24654	1646.708	−135.437
1000	2	0.1521	0.18	0.14	0.66667	0.2	10	800	0.232299	1511.27	−147.987
1000	2	0.13773	0.18	0.14	0.66667	0.3	10	800	0.217182	1363.283	−163.538
1000	2	0.12303	0.18	0.14	0.66667	0.4	10	800	0.201083	1199.745	−183.17
1000	2	0.10796	0.18	0.14	0.66667	0.5	10	800	0.183875	1020.575	−208.47
1000	2	0.09247	0.18	0.14	0.66667	0.6	10	800	0.165407	808.1048	−241.852
1000	2	0.0765	0.18	0.14	0.66667	0.7	10	800	0.145497	566.253	−287.089
1000	2	0.05998	0.18	0.14	0.66667	0.8	10	800	0.123925	279.1638	−350.325
1000	2	0.04281	0.18	0.14	0.66667	0.9	10	800	0.100417	−71.1609	−442.002
1000	2	0.02488	0.18	0.14	0.66667	1	10	800	0.074631	−513.163	

Table 28.58 Dependence of NPV and ΔNPV on tax on profit rate t at fixed leverage level L for 10-year project at $L = 4$

I	L	WACC	k_0	k_d	Wd	t	n	NOI	k_e	NPV	ΔNPV
1000	4	0.18	0.18	0.14	0.8	0	10	800	0.34	1226.885	–475.625
1000	4	0.18316	0.20056	0.14	0.8	0.1	10	800	0.411801	751.2601	–139.495
1000	4	0.17256	0.20836	0.14	0.8	0.2	10	800	0.414791	611.7654	–74.4553
1000	4	0.15772	0.21253	0.14	0.8	0.3	10	800	0.396579	537.3101	–54.8815
1000	4	0.14072	0.21513	0.14	0.8	0.4	10	800	0.367589	482.4286	–52.0879
1000	4	0.12227	0.2169	0.14	0.8	0.5	10	800	0.331331	430.3407	–60.1971
1000	4	0.10261	0.21819	0.14	0.8	0.6	10	800	0.289071	370.1436	–80.4792
1000	4	0.08181	0.21917	0.14	0.8	0.7	10	800	0.241069	289.6644	–120.753
1000	4	0.0598	0.21994	0.14	0.8	0.8	10	800	0.186991	168.9112	–202.021
1000	4	0.0364	0.22055	0.14	0.8	0.9	10	800	0.126013	–33.1095	–382.685
1000	4	0.01135	0.22106	0.14	0.8	1	10	800	0.056769	–415.795	

Table 28.59 Dependence of NPV and ΔNPV on tax on profit rate t at fixed leverage level L for 10-year project at $L = 6$

I	L	WACC	k_0	k_d	Wd	t	n	NOI	k_e	NPV	ΔNPV
1000	6	0.18	0.18	0.14	0.85714	0	10	800	0.42	847.6197	37.06285
1000	6	0.16218	0.18	0.14	0.85714	0.1	10	800	0.379295	884.6825	42.53552
1000	6	0.14392	0.18	0.14	0.85714	0.2	10	800	0.335473	927.218	50.03126
1000	6	0.12515	0.18	0.14	0.85714	0.3	10	800	0.288045	977.2493	60.31305
1000	6	0.10577	0.18	0.14	0.85714	0.4	10	800	0.236408	1037.562	74.23617
1000	6	0.08569	0.18	0.14	0.85714	0.5	10	800	0.179813	1111.799	91.96624
1000	6	0.06476	0.18	0.14	0.85714	0.6	10	800	0.117304	1203.765	109.0267
1000	6	0.04281	0.18	0.14	0.85714	0.7	10	800	0.04764	1312.791	96.39537
1000	6	0.0196	0.18	0.14	0.85714	0.8	10	800	−0.03083	1409.187	−113.739
1000	6	−0.0052	0.18	0.14	0.85714	0.9	10	800	−0.12039	1295.448	−1669.51
1000	6	−0.032	0.18	0.14	0.85714	1	10	800	−0.22431	−374.066	

Table 28.60 Dependence of NPV and ΔNPV on tax on profit rate t at fixed leverage level L for 10-year project at $L = 8$

I	L	WACC	k_0	k_d	Wd	t	n	NOI	k_e	NPV	ΔNPV
1000	8	0.18	0.18	0.14	0.88889	0	10	800	0.5	572.2536	67.74227
1000	8	0.16152	0.18	0.14	0.88889	0.1	10	800	0.445653	639.9958	80.91054
1000	8	0.14255	0.18	0.14	0.88889	0.2	10	800	0.386973	720.9064	100.9849
1000	8	0.12303	0.18	0.14	0.88889	0.3	10	800	0.323249	821.8913	133.0198
1000	8	0.10284	0.18	0.14	0.88889	0.4	10	800	0.253602	954.9111	187.1372
1000	8	0.08188	0.18	0.14	0.88889	0.5	10	800	0.17692	1142.048	285.1235
1000	8	0.05998	0.18	0.14	0.88889	0.6	10	800	0.091775	1427.172	477.0078
1000	8	0.03692	0.18	0.14	0.88889	0.7	10	800	−0.00372	1904.18	874.876
1000	8	0.01243	0.18	0.14	0.88889	0.8	10	800	−0.11211	2779.056	1515.986
1000	8	−0.0139	0.18	0.14	0.88889	0.9	10	800	−0.23701	4295.041	−4645.92
1000	8	−0.0426	0.18	0.14	0.88889	1	10	800	−0.38375	−350.883	

Table 28.61 Dependence of NPV and ΔNPV on tax on profit rate t at fixed leverage level L for 10-year project at $L = 10$

I	L	WACC	k_0	k_d	Wd	t	n	NOI	k_e	NPV	ΔNPV
1000	10	0.18	0.18	0.14	0.90909	0	10	800	0.58	365.0841	85.01433
1000	10	0.16109	0.18	0.14	0.90909	0.1	10	800	0.512008	450.0984	103.8568
1000	10	0.14168	0.18	0.14	0.90909	0.2	10	800	0.438456	553.9552	134.6286
1000	10	0.12167	0.18	0.14	0.90909	0.3	10	800	0.358411	688.5839	188.5659
1000	10	0.10097	0.18	0.14	0.90909	0.4	10	800	0.270705	877.1497	292.3814
1000	10	0.07944	0.18	0.14	0.90909	0.5	10	800	0.173857	1169.531	519.0512
1000	10	0.0569	0.18	0.14	0.90909	0.6	10	800	0.065947	1688.582	1107.667
1000	10	0.03313	0.18	0.14	0.90909	0.7	10	800	−0.05558	2796.249	3051.344
1000	10	0.0078	0.18	0.14	0.90909	0.8	10	800	−0.19424	5847.593	11617.12
1000	10	−0.0195	0.18	0.14	0.90909	0.9	10	800	−0.35504	17464.71	−17800.8
1000	10	−0.0496	0.18	0.14	0.90909	1	10	800	−0.54557	−336.131	

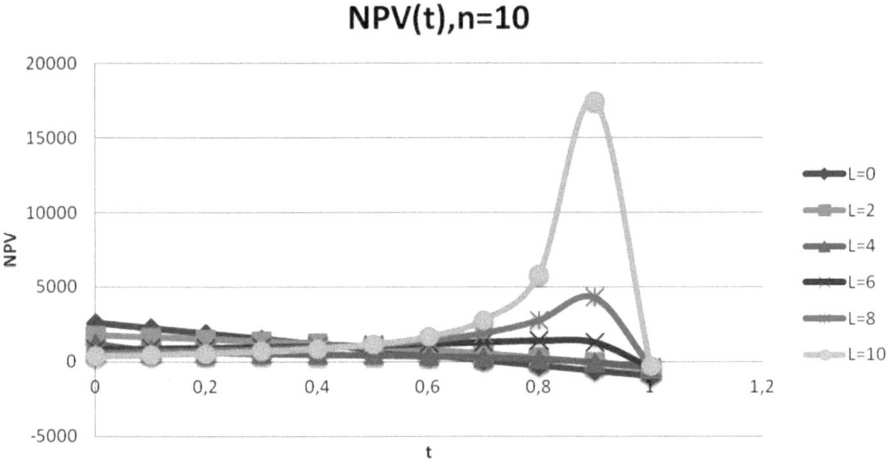

Fig. 28.18 Dependence of NPV on tax on profit rate *t* at fixed leverage level *L* for 10-year project

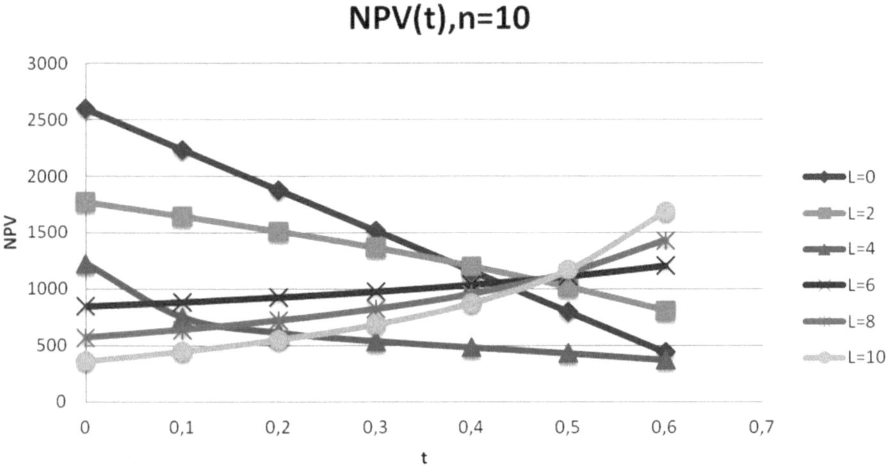

Fig. 28.19 Dependence of NPV on tax on profit rate *t* at fixed leverage level *L* for 10-year project (more detailed)

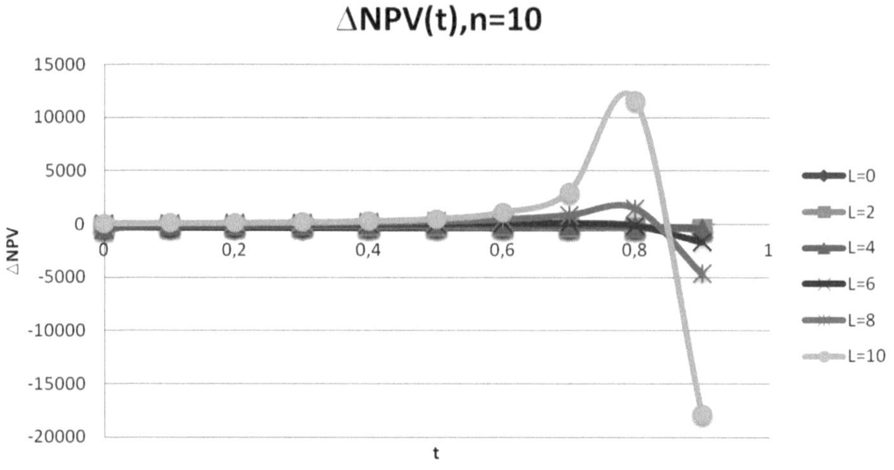

Fig. 28.20 Dependence of ΔNPV on tax on profit rate *t* at fixed leverage level *L* for 10-year project

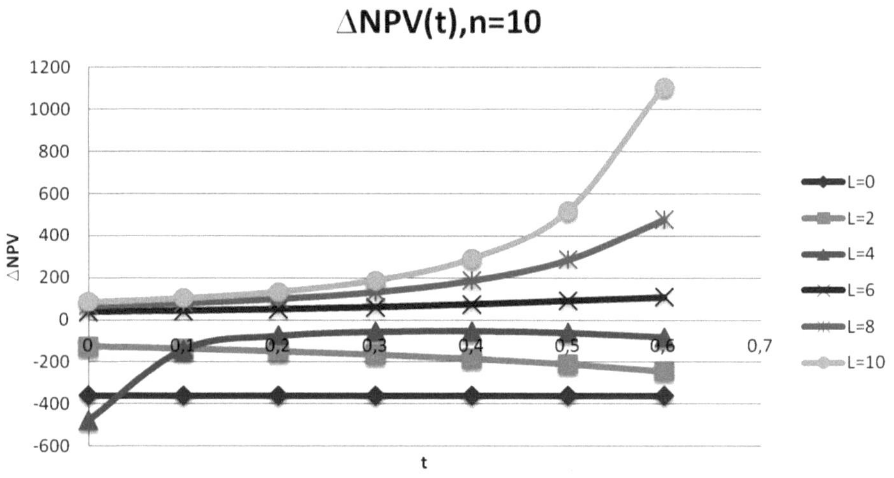

Fig. 28.21 Dependence of ΔNPV on tax on profit rate *t* at fixed leverage level *L* for 10-year project (more detailed)

References

Brusova, A. (2011). A comparison of the three methods of estimation of weighted average cost of capital and equity cost of company. *Financial Analysis: Problems and Solutions, 34*(76), 36–42.

Brusov, P. N., & Filatova, T. V. (2011). From Modigliani–Miller to general theory of capital cost and capital structure of the company. *Finance and Credit, 435*, 2–8.

Brusov, P., Filatova, T., Orehova, N., & Brusova, A. (2011a). Weighted average cost of capital in the theory of Modigliani–Miller, modified for a finite life–time company. *Applied Financial Economics, 21*(11), 815–824.

Brusov, P., Filatova, T., Orehova, N., et al. (2011b). From Modigliani–Miller to general theory of capital cost and capital structure of the company. *Research Journal of Economics, Business and ICT, 2*, 16–21.

Brusov, P., Filatova, T., Orehova, N., et al. (2011c). Influence of debt financing on the effectiveness of the investment project within the Modigliani–Miller theory. *Research Journal of Economics, Business and ICT (UK), 2*, 11–15.

Brusov, P., Filatova, T., Eskindarov, M., & Orehova, N. (2012a). Influence of debt financing on the effectiveness of the finite duration investment project. *Applied Financial Economics, 22*(13), 1043–1052.

Brusov, P., Filatova, T., Eskindarov, M., & Orehova, N. (2012b). Hidden global causes of the global financial crisis. *Journal of Reviews on Global Economics, 1*, 106–111.

Brusov, P., Filatova, P., & Orekhova, N. (2013a). Absence of an optimal capital structure in the famous tradeoff theory! *Journal of Reviews on Global Economics, 2*, 94–116.

Brusov, P., Filatova, T., & Orehova, N. (2013b). A qualitatively new effect in corporative finance: Abnormal dependence of cost of equity of company on leverage. *Journal of Reviews on Global Economics, 2*, 183–193.

Brusov, P., Filatova, P., & Orekhova, N. (2014a). Mechanism of formation of the company optimal capital structure, different from suggested by trade off theory. *Cogent Economics & Finance, 2*, 1–13. https://doi.org/10.1080/23322039.2014.946150

Brusov, P., Filatova, T., & Orehova, N. (2014b). Inflation in Brusov–Filatova–Orekhova theory and in its perpetuity limit – Modigliani – Miller theory. *Journal of Reviews on Global Economics, 3*, 175–185.

Filatova, T., Orehova, N., & Brusova, A. (2008). Weighted average cost of capital in the theory of Modigliani–Miller, modified for a finite life–time company. *Bulletin of the FU, 48*, 68–77.

Modigliani, F., & Miller, M. (1958). The cost of capital, corporate finance, and the theory of investment. *American Economic Review, 48*, 261–297.

Modigliani, F., & Miller, M. (1963). Corporate income taxes and the cost of capital: A correction. *American Economic Review, 53*, 147–175.

Modigliani, F., & Miller, M. (1966). Some estimates of the cost of capital to the electric utility industry 1954–1957. *American Economic Review, 56*, 333–391.

Part VIII
The Application of Modern Investment Models

Part VIII is devoted to the application of modern investment models created in this monograph in practice.

Chapter 29 conducts an analysis of the study of the effectiveness of investment projects of arbitrary duration (within the framework of the Brusov–Filatova–Orekhova theory). Chapter 30 studies the optimizing the investment structure of the telecommunication sector company. The role of the Central Bank and commercial banks in creating and maintaining a favorable investment climate in the country is examined in Chap. 31. Conclusions for the entire monograph are made in Chap. 32.

Chapter 29
The Analysis of the Exploration of Efficiency of Investment Projects of Arbitrary Duration (within Brusov–Filatova–Orekhova Theory)

Earlier (see in Brusov et al., 2022) we have conducted the analysis of effectiveness of investment projects within the perpetuity (Modigliani–Miller) approximation (Modigliani & Miller, 1958, 1963, 1966). In this Chapter the analysis of the obtained results on the exploration of efficiency of investment projects of arbitrary duration (within Brusov–Filatova–Orekhova theory (Brusov & Filatova, 2011; Brusov et al., 2011a, b, c, 2012a, b, 2013a, b, 2014a, b; Filatova et al., 2008; Brusova, 2011) is conducted.

29.1 The Effectiveness of the Investment Project from the Perspective of the Equity Holders Only

29.1.1 With the Division of Credit and Investment Flows

At a Constant Investment Value ($I = \text{const}$)

For NPV in this case the following expression has been obtained (Brusov & Filatova, 2011; Brusov et al., 2011a, b, c, 2012a, b, 2013a, b, 2014a, b)

$$NPV = -\frac{I}{1+L}\left[1+L\left((1-t)\left(1-\frac{1}{(1+k_d)^n}\right)+\frac{1}{(1+k_d)^n}\right)\right]+ \\ +\frac{NOI(1-t)}{k_e}\left(1-\frac{1}{(1+k_e)^n}\right). \tag{29.1}$$

Dependence of NPV on leverage level at fixed values of k_0 and k_d at a constant investment value ($I = \text{const}$) is shown in Table 29.1 and Figs. 29.1 and 29.2.

1. At the constant values of $\Delta k = k_0 - k_d$ NPV decreases with leverage at low values of k_0 (up to 20%) and grows at higher values of k_0 (from 25% to 30%). All curves

P. Brusov, T. Filatova, *Efficiency of Investment Projects*, Contributions to Finance and Accounting, https://doi.org/10.1007/978-3-032-01028-5_29

Table 29.1 n = 2, t = 0.2, NOI = 1200; I = 1000, $k_0 - k_d = const$

	k_0	$k_d \backslash L$	0.0	0.5	1.0	1.5	2.0	2.5	3.0	3.5	4.0	4.5
1	0.08	0.06	711.9	700.7	686.2	670.6	654.6	638.5	622.4	606.5	590.8	575.3
2	0.10	0.08	666.6	660.2	649.4	637.0	623.9	610.5	597.0	583.6	570.2	556.9
3	0.14	0.12	580.9	581.9	576.5	568.5	559.4	549.6	539.5	529.3	519.0	508.7
4	0.18	0.16	503.0	510.6	509.8	505.7	499.9	493.4	486.3	478.9	471.3	463.7
5	0.24	0.22	398.5	414.0	418.6	418.9	417.1	414.0	410.2	406.0	401.4	396.6
6	0.30	0.28	306.5	328.6	337.3	340.6	344.0	344.2	343.5	342.3	340.6	338.6
7	0.36	0.34	225.0	251.3	264.7	272.0	276.4	279.0	280.5	281.3	281.6	283.5
8	0.40	0.38	175.6	206.1	221.0	229.6	235.1	236.2	238.8	240.6	242.0	242.9
9	0.44	0.42	129.7	163.0	179.5	189.5	196.1	200.7	204.3	207.0	209.1	210.9
10	0.10	0.06	666.6	635.3	602.0	568.7	536.1	504.4	473.6	443.9	415.2	387.5
11	0.12	0.08	622.6	596.5	567.1	537.1	507.3	478.2	449.9	422.4	395.7	369.9
12	0.16	0.12	541.1	524.3	501.8	477.7	453.2	428.8	404.8	381.4	358.4	336.1
13	0.20	0.16	466.2	456.8	441.5	422.5	402.5	382.2	362.1	342.1	322.5	303.2
14	0.24	0.20	398.5	396.6	385.1	370.2	353.8	336.9	319.7	302.6	285.6	268.9
15	0.30	0.26	306.5	318.4	308.1	298.0	285.9	275.8	263.0	249.9	236.7	223.6
16	0.36	0.32	225.0	237.9	238.8	234.1	226.8	218.2	208.7	198.8	188.6	178.3
17	0.40	0.36	175.6	193.9	197.2	194.7	189.3	179.9	172.2	164.0	155.5	146.7
18	0.44	0.40	129.7	151.7	157.5	157.1	153.7	148.5	142.4	135.6	128.5	121.1

	k_0	$k_d \backslash L$	5.0	5.5	6.0	6.5	7.0	7.5	8.0	8.5	9.0	9.5	10.0
1	0.08	0.06	560.0	544.9	530.1	515.5	501.1	487.0	473.1	459.4	445.9	432.7	419.7
2	0.10	0.08	543.7	525.7	512.5	499.5	486.6	474.0	461.5	449.2	437.1	425.1	418.4
3	0.14	0.12	498.4	488.2	478.0	468.0	458.0	448.1	438.3	428.6	419.0	409.5	400.1
4	0.18	0.16	455.9	448.2	440.4	432.6	424.9	417.2	409.6	402.0	394.4	386.9	379.5
5	0.24	0.22	391.7	386.6	381.5	376.3	371.1	365.9	360.7	355.4	350.2	345.0	339.8
6	0.30	0.28	336.3	334.0	331.5	328.9	326.2	323.4	320.7	317.8	315.0	312.1	309.2

7	0.36	0.34	284.4	280.7	280.0	279.1	278.2	277.2	276.1	274.9	273.7	272.5	271.2
8	0.40	0.38	243.5	244.0	244.2	244.3	244.3	244.2	244.0	243.7	243.4	243.1	242.7
9	0.44	0.42	212.4	218.6	214.6	215.5	216.3	217.0	217.6	218.2	218.6	219.1	219.4
10	0.10	0.06	360.7	334.8	309.8	285.6	262.2	239.6	217.7	196.5	176.0	156.1	136.9
11	0.12	0.08	344.9	320.6	297.2	274.5	252.5	231.1	210.5	190.4	171.0	152.2	133.9
12	0.16	0.12	314.4	293.2	272.7	252.7	233.3	214.4	196.0	178.1	160.8	143.9	127.4
13	0.20	0.16	284.4	266.0	248.1	230.6	218.5	196.8	180.5	164.6	149.1	134.0	119.3
14	0.24	0.20	252.4	236.2	220.4	204.9	189.7	174.9	160.3	146.1	132.2	118.6	105.3
15	0.30	0.26	210.6	197.7	185.0	172.5	160.1	148.0	136.1	124.4	112.9	101.6	90.5
16	0.36	0.32	168.0	155.9	145.7	140.4	127.2	117.2	107.4	97.7	88.1	78.7	69.4
17	0.40	0.36	137.9	129.0	120.2	111.3	102.6	93.9	85.2	76.7	68.3	59.9	51.7
18	0.44	0.40	118.6	106.0	98.3	90.7	83.0	75.4	67.9	60.4	53.0	45.6	38.3

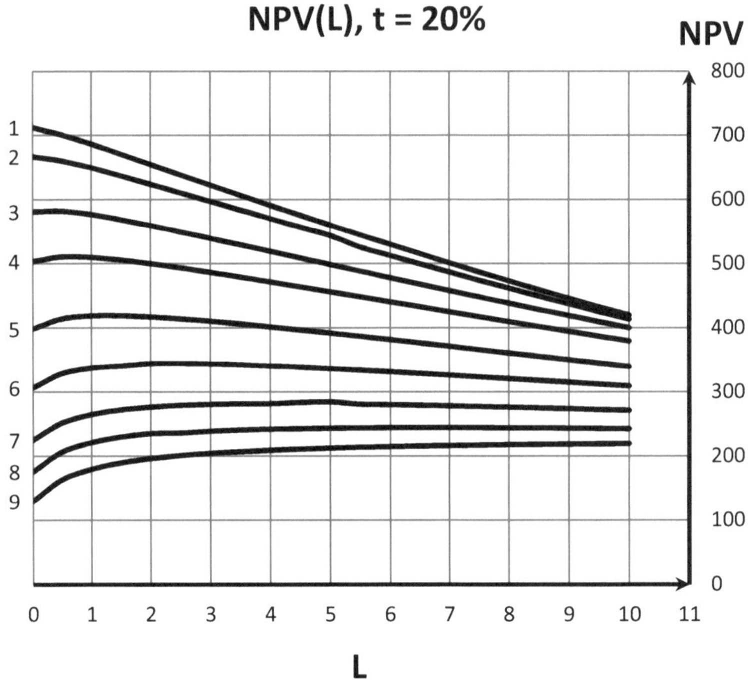

Fig. 29.1 Dependence of NPV on leverage level at fixed values of k_0 and k_d

NPV(L) are shifted down with growth of k_0. At small leverage levels L there is an optimum in the dependence of NPV(L).

2. At the constant values of k_0 NPV practically always decreases with leverage. Higher values of k_d (at the same value of k_0) correspond to higher lying curves NPV(L). At small leverage levels L at high value of k_0 (36–40%) there is an optimum in the dependence of NPV(L).
3. At the constant values of k_d NPV practically always decreases with leverage. Higher values of k_0 (at the same value of k_d) correspond to lower lying curves NPV(L). At small leverage levels L for some pairs of values k_0 and k_d (for example, k_0 (18%) and k_d (16%)) there is an optimum in the dependence of NPV(L).

At the Constant Equity Value (S = const)

For NPV in this case the following expression has been obtained (Brusov & Filatova, 2011; Brusov et al., 2011a, b, c, 2012a, b, 2013a, b, 2014a, b)

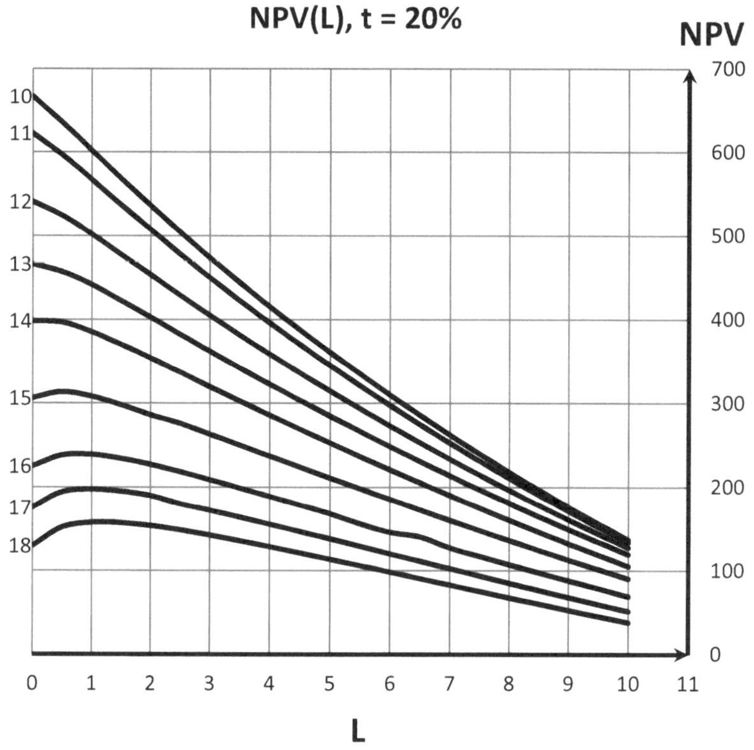

Fig. 29.2 Dependence of NPV on leverage level at fixed values of k_0 and k_d

$$NPV = -S\left[1 + L\left((1-t)\left(1 - \frac{1}{(1+k_d)^n}\right) + \frac{1}{(1+k_d)^n}\right)\right] + \\ + \frac{\beta S(1+L)(1-t)}{k_e}\left(1 - \frac{1}{(1+k_e)^n}\right). \tag{29.2}$$

Dependence of NPV on leverage level at fixed values of k_0 and k_d at a constant equity value ($S = \text{const}$) is shown in Table 29.2 and Figs. 29.3 and 29.4.

1. At the constant values of $\Delta k = k_0 - k_d$ NPV, as a rule, decreases with leverage. All curves NPV(L) are shifted down with growth of k_0 sometimes at small leverage level L values there is an optimum in the dependence of NPV(L). As it will be shown in Chap. 17 at the example with "Nastcom Plus" company, the dependence of NPV(L) strongly depends on the β parameter value and can have a marked optimum.
2. At the constant values of k_0 NPV practically always decreases with leverage. Higher values of k_d (lower values of $\Delta k = k_0 - k_d$) at the same value of k_0 correspond to higher lying curves NPV(L). Like previous paragraph, the

Table 29.2 n = 2, t = 0.2, S = 1000; β = 0.7, $k_O - k_d = const$

	k_O	k_d\L	0.0	0.5	1.0	1.5	2.0	2.5	3.0	3.5	4.0	4.5
1	0.08	0.06	−1.4	−7.3	−23.6	−50.0	−86.1	−131.8	−186.8	−250.8	−323.5	−404.7
2	0.10	0.08	−27.8	−41.4	−63.8	−94.9	−134.4	−182.1	−237.9	−301.5	−372.8	−451.5
3	0.14	0.12	−77.8	−107.4	−143.9	−187.3	−237.3	−294.0	−357.0	−426.4	−501.9	−583.5
4	0.18	0.16	−123.2	−167.5	−217.2	−272.2	−332.3	−397.5	−467.8	−542.9	−623.0	−707.8
5	0.24	0.22	−184.2	−249.1	−317.6	−389.7	−465.4	−544.7	−627.4	−718.6	−803.2	−896.1
6	0.30	0.28	−237.9	−321.3	−407.4	−496.3	−583.0	−674.4	−767.7	−862.9	−960.2	−1059.3
7	0.36	0.34	−285.4	−386.7	−487.6	−589.6	−692.5	−796.5	−901.4	−1007.3	−1114.3	−1216.1
8	0.40	0.38	−314.2	−424.8	−535.9	−647.5	−759.5	−877.3	−990.8	−1104.8	−1219.3	−1334.3
9	0.44	0.42	−341.0	−461.4	−581.9	−702.4	−822.9	−943.5	−1064.0	−1184.7	−1305.3	−1426.0
10	0.10	0.06	−27.8	−64.5	−121.8	−198.6	−293.5	−405.7	−534.0	−677.6	−835.6	−1007.3
11	0.12	0.08	−53.5	−97.1	−159.8	−240.6	−338.4	−452.2	−581.2	−724.6	−881.6	−1051.5
12	0.16	0.12	−101.1	−157.8	−231.0	−319.7	−423.1	−540.6	−671.4	−814.8	−970.3	−1137.3
13	0.20	0.16	−144.7	−214.6	−296.8	−393.4	−502.8	−624.4	−757.7	−902.1	−1057.2	−1222.4
14	0.24	0.20	−184.2	−265.3	−358.6	−463.6	−579.9	−706.9	−844.3	−991.6	−1148.5	−1314.5
15	0.30	0.26	−237.9	−335.4	−443.1	−560.7	−688.0	−818.1	−960.5	−1111.1	−1269.5	−1435.4
16	0.36	0.32	−285.4	−399.1	−519.3	−647.0	−782.0	−924.2	−1073.2	−1229.0	−1391.2	−1559.8
17	0.40	0.36	−314.2	−436.2	−565.0	−700.4	−842.1	−995.4	−1150.0	−1310.5	−1476.8	−1648.7
18	0.44	0.40	−341.0	−471.9	−608.7	−751.3	−899.4	−1053.0	−1212.0	−1376.1	−1545.3	−1719.4

	k_O	k_d\L	5.0	5.5	6.0	6.5	7.0	7.5	8.0	8.5	9.0	9.5	10.0
1	0.08	0.06	−494.2	−591.7	−697.1	−810.2	−930.6	−1058.3	−1193.0	−1334.6	−1482.9	−1637.7	−1798.8
2	0.10	0.08	−537.5	−649.7	−752.6	−862.5	−979.1	−1102.3	−1232.0	−1367.9	−1510.0	−1658.0	−1811.9
3	0.14	0.12	−671.0	−764.3	−863.3	−967.9	−1077.9	−1193.3	−1314.0	−1439.8	−1570.7	−1706.5	−1847.2
4	0.18	0.16	−797.3	−891.4	−990.0	−1093.1	−1200.7	−1312.5	−1428.6	−1548.9	−1673.4	−1801.8	−1934.3
5	0.24	0.22	−992.4	−1092.0	−1194.8	−1300.8	−1410.0	−1522.3	−1637.8	−1756.3	−1877.8	−2002.3	−2129.7
6	0.30	0.28	−1160.4	−1263.4	−1368.4	−1475.2	−1583.8	−1694.4	−1806.8	−1921.0	−2037.0	−2154.9	−2274.5

7	0.36	0.34	−1320.0	−1441.0	−1551.9	−1663.7	−1776.6	−1890.4	−2005.2	−2120.9	−2237.6	−2355.3	−2473.9
8	0.40	0.38	−1449.8	−1565.7	−1682.1	−1799.0	−1916.3	−2034.1	−2152.4	−2271.2	−2390.5	−2510.2	−2630.3
9	0.44	0.42	−1546.7	−1667.4	−1788.2	−1909.0	−2029.8	−2150.6	−2271.5	−2392.4	−2518.3	−2634.2	−2755.2
10	0.10	0.06	−1191.8	−1388.5	−1596.8	−1816.0	−2045.5	−2284.9	−2533.7	−2791.3	−3057.4	−3331.6	−3618.4
11	0.12	0.08	−1233.5	−1427.2	−1631.8	−1846.9	−2071.9	−2306.4	−2549.9	−2801.9	−3062.1	−3330.1	−3605.5
12	0.16	0.12	−1315.2	−1503.5	−1701.8	−1909.6	−2126.4	−2352.0	−2585.8	−2827.5	−3076.8	−3333.3	−3596.8
13	0.20	0.16	−1397.5	−1581.9	−1775.2	−1977.2	−2187.4	−2405.4	−2631.1	−2864.1	−3104.0	−3350.7	−3603.9
14	0.24	0.20	−1489.3	−1672.5	−1863.9	−2063.1	−2269.8	−2483.7	−2704.6	−2932.2	−3166.3	−3406.6	−3653.0
15	0.30	0.26	−1608.7	−1789.0	−1976.2	−2169.9	−2370.1	−2576.4	−2788.8	−3006.9	−3230.7	−3459.9	−3694.4
16	0.36	0.32	−1734.5	−1922.0	−2108.6	−2279.8	−2491.2	−2694.0	−2902.0	−3115.0	−3332.9	−3555.6	−3782.8
17	0.40	0.36	−1826.0	−2008.5	−2196.3	−2389.1	−2586.7	−2789.2	−2996.3	−3207.9	−3424.0	−3644.3	−3868.9
18	0.44	0.40	−1898.3	−2082.0	−2270.2	−2463.0	−2660.1	−2861.5	−3067.1	−3276.7	−3490.4	−3708.0	−3929.4

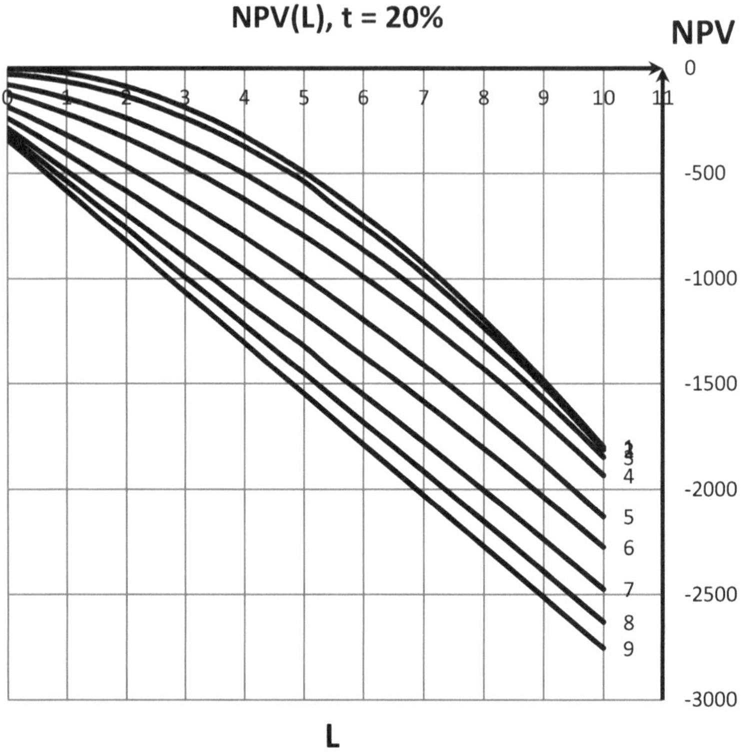

Fig. 29.3 Dependence of NPV on leverage level at fixed values of k_0 and k_d

dependence of NPV(L) strongly depends on the β parameter value and can have a marked optimum.

3. At the constant values of k_d NPV practically always decreases with leverage. Higher values of k_0 (higher values of $\Delta k = k_0 - k_d$) at the same value of k_d correspond to lower lying curves NPV(L). At small leverage levels L for projects with durations above 2 years for some pairs of values k_0 and k_d (for example, for 5-year and 7-year projects with k_0 (8%) and k_d (6%); k_0 (10%) and k_d (6%)) there is an optimum in the dependence of NPV(L). This optimum could be a marked one at other values of parameter β.

29.1.2 *Without Flows Separation*

At a Constant Investment Value (I = const)

For NPV in this case the following expression has been obtained (Brusov & Filatova, 2011; Brusov et al., 2011a, b, c, 2012a, b, 2013a, b, 2014a, b)

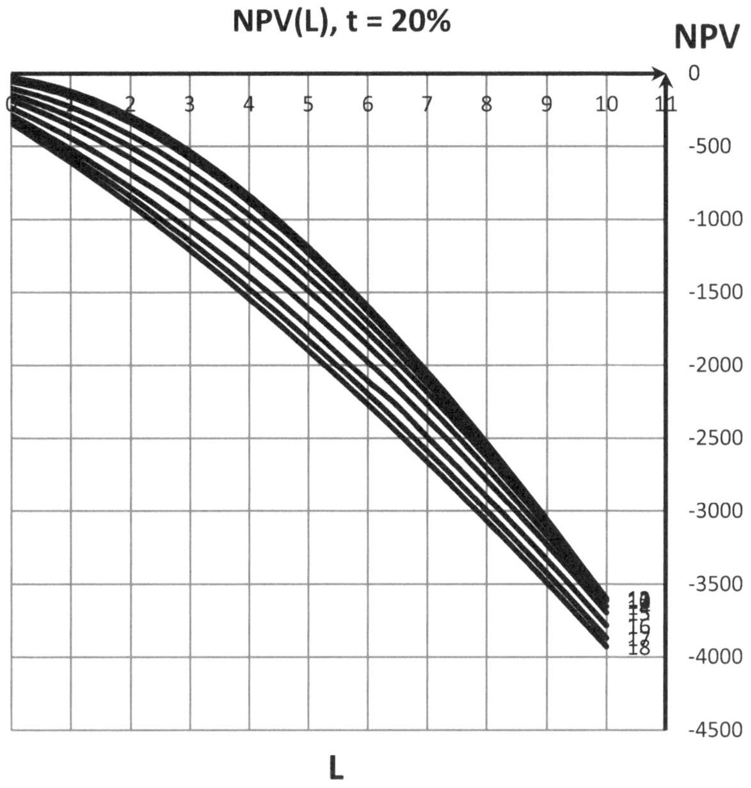

Fig. 29.4 Dependence of NPV on leverage level at fixed values of k_0 and k_d

$$NPV = -\frac{I}{1+L}\left[1 + L\frac{k_d(1-t)}{WACC}\left(1 - \frac{1}{(1+WACC)^n}\right) + \frac{L}{(1+WACC)^n}\right] + \\ + \frac{NOI(1-t)}{WACC}\left(1 - \frac{1}{(1+WACC)^n}\right). \tag{29.3}$$

Dependence of NPV on leverage level at fixed values of k_0 and k_d at a constant investment value ($I =$ const) is shown in Table 29.3 and Figs. 29.5 and 29.6.

1. At the constant values of $\Delta k = k_0 - k_d$ NPV demonstrates a limited growth with leverage with output into saturation regime. The main increase in NPV occurs at small values of leverage levels $L \leq 3$. With growth of k_0 (and k_d) the curves NPV (L) are lowered. Optimum in the dependence of NPV(L) is absent.
2. At the constant values of k_0 NPV demonstrates a limited growth with leverage with output into saturation regime. All curves NPV(L) at the constant values of k_0 and different values of k_d are started (at L = 0) from one point, the higher values of k_d (and, respectively, the lower values of $\Delta k = k_0 - k_d$) correspond to lower

Table 29.3 $n = 2$, $t = 0.2$, $NOI = 1200$; $I = 1000$, $k_0 - k_d = const$

	k_0	k_d\L	0.0	0.5	1.0	1.5	2.0	2.5	3.0	3.5	4.0	4.5
1	0.08	0.06	711.9	740.5	752.6	759.1	763.1	765.8	767.8	769.2	770.4	771.3
2	0.10	0.08	666.6	699.5	718.0	720.2	724.6	727.5	729.6	731.1	732.3	733.3
3	0.14	0.12	580.9	621.2	637.3	645.7	650.7	654.0	656.3	658.1	659.4	660.4
4	0.18	0.16	503.0	549.5	567.7	577.0	582.5	586.1	588.6	590.4	591.8	592.9
5	0.24	0.22	398.5	452.1	472.8	483.1	489.1	492.9	495.6	497.5	498.9	500.0
6	0.30	0.28	306.5	365.7	387.9	398.9	405.6	409.6	412.3	414.2	415.6	416.7
7	0.36	0.34	225.0	287.8	311.9	323.5	330.0	334.1	336.8	338.8	340.2	341.3
8	0.40	0.38	175.6	241.2	265.5	277.3	283.8	287.6	290.3	292.3	293.7	294.7
9	0.44	0.42	129.7	197.2	222.0	233.9	240.4	244.5	247.1	248.9	250.3	251.3
10	0.10	0.06	666.6	706.0	723.5	733.3	739.5	743.8	746.9	749.3	751.2	752.7
11	0.12	0.08	622.6	666.0	684.8	695.2	701.8	706.2	709.5	712.0	718.9	715.5
12	0.16	0.12	541.1	591.1	612.2	623.6	630.6	635.4	638.8	641.4	643.4	645.0
13	0.20	0.16	466.2	521.6	545.1	557.3	564.7	569.7	573.2	575.9	577.9	579.5
14	0.24	0.20	398.5	458.5	482.8	495.6	503.3	508.4	512.0	514.6	516.7	518.3
15	0.30	0.26	306.5	371.9	397.7	411.0	418.9	424.4	428.1	430.7	432.8	434.4
16	0.36	0.32	225.0	293.7	321.3	335.1	343.2	348.5	352.1	354.7	356.7	358.2
17	0.40	0.36	175.6	247.1	274.7	288.6	296.7	301.5	305.1	307.7	309.7	311.2
18	0.44	0.40	129.7	202.9	231.0	244.9	253.0	258.1	261.6	264.1	265.9	267.4

	k_0	k_d\L	5.0	5.5	6.0	6.5	7.0	7.5	8.0	8.5	9.0	9.5	10.0
1	0.08	0.06	772.1	772.7	773.2	773.7	774.1	774.4	774.7	775.0	775.2	775.4	775.6
2	0.10	0.08	734.1	734.4	734.9	735.4	735.8	736.1	736.5	736.7	737.0	737.2	737.4
3	0.14	0.12	661.3	661.9	662.5	663.0	663.4	663.8	664.1	664.4	664.7	664.9	665.1
4	0.18	0.16	593.7	594.4	595.0	595.5	596.0	596.3	596.7	597.0	597.2	597.4	597.6
5	0.24	0.22	500.9	501.6	502.2	502.7	503.2	503.5	503.8	504.1	504.4	504.6	504.8
6	0.30	0.28	417.6	418.3	418.9	419.4	419.8	420.1	420.4	420.7	420.9	421.1	421.3

7	0.36	0.34	342.2	342.7	343.3	343.7	344.1	344.4	344.7	344.9	345.1	345.3	345.4
8	0.40	0.38	295.5	296.2	296.7	297.1	297.5	297.8	298.0	298.2	298.4	298.6	298.7
9	0.44	0.42	252.0	252.6	253.1	253.5	253.8	254.1	254.3	254.5	254.7	254.8	254.9
10	0.10	0.06	754.0	755.1	755.9	756.7	757.4	758.0	758.5	759.0	759.4	759.7	760.1
11	0.12	0.08	716.8	717.8	718.7	719.5	720.2	720.8	721.3	721.8	722.2	722.6	722.9
12	0.16	0.12	646.3	647.4	648.4	649.2	649.8	650.4	651.0	651.4	651.8	652.2	652.6
13	0.20	0.16	580.9	582.0	582.9	583.7	584.3	584.9	585.5	585.9	586.3	586.7	587.0
14	0.24	0.20	519.6	520.7	521.6	522.4	523.1	523.6	524.1	524.6	525.0	525.4	525.7
15	0.30	0.26	435.6	436.7	437.6	438.3	438.9	439.5	440.0	440.4	440.8	441.1	441.4
16	0.36	0.32	359.4	360.3	361.2	362.1	362.6	363.1	363.5	363.9	364.2	364.5	364.8
17	0.40	0.36	312.4	318.4	314.2	314.9	315.4	315.9	316.4	316.7	317.1	317.3	317.6
18	0.44	0.40	268.5	269.4	270.2	270.8	271.3	271.8	272.2	272.5	272.8	273.1	273.3

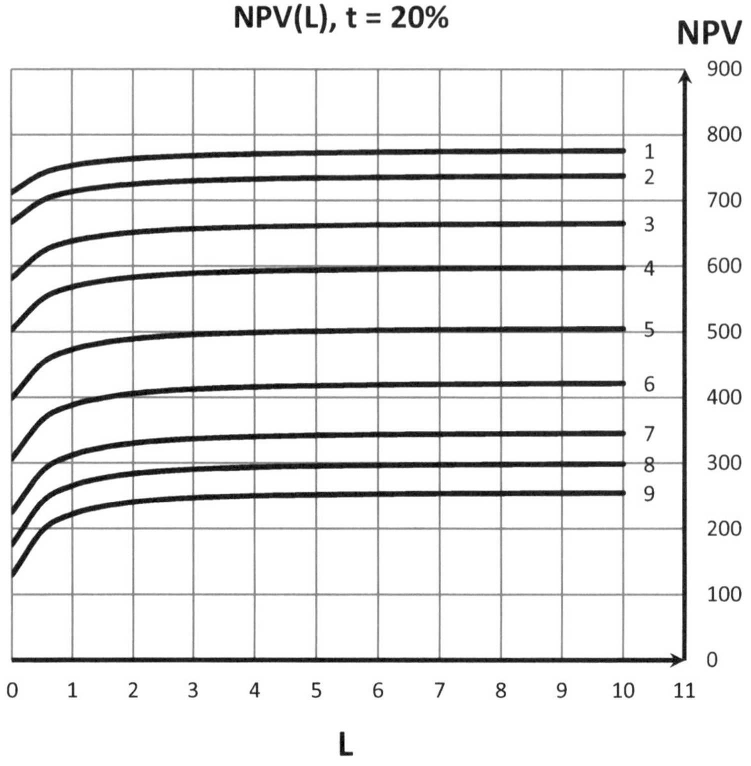

Fig. 29.5 Dependence of NPV on leverage level at fixed values of k_0 and k_d

lying curves NPV(L). Optimum in the dependence of NPV(L) is absent. With growth of NOI all curves NPV(L) are shifted practically parallel upwards.

Is of interest the crossing of individual curves NPV(L) at certain leverage levels. This means the equivalence of projects with different pairs of k_0 and k_d at this leverage level (see, for example, n = 7; L = 2.5; (k_0 and k_d) = (18;14) and (24;10).

3. At the constant values of k_d NPV shows a limited growth with leverage with output into saturation regime. All curves NPV(L) at the constant values of k_0 and different values of k_d are started (at L = 0) from one point, the higher values of k_0 (and, respectively, the higher values of $\Delta k = k_0 - k_d$) correspond to lower lying curves NPV(L). Optimum in the dependence of NPV(L) is absent. The crossing of individual curves NPV(L) at certain leverage levels (like point 2)) was not observed up to 10-year projects.

At a Constant Equity Value (S = const)

For NPV in this case the following expression has been obtained (Brusov & Filatova, 2011; Brusov et al., 2011a, b, c, 2012a, b, 2013a, b, 2014a, b)

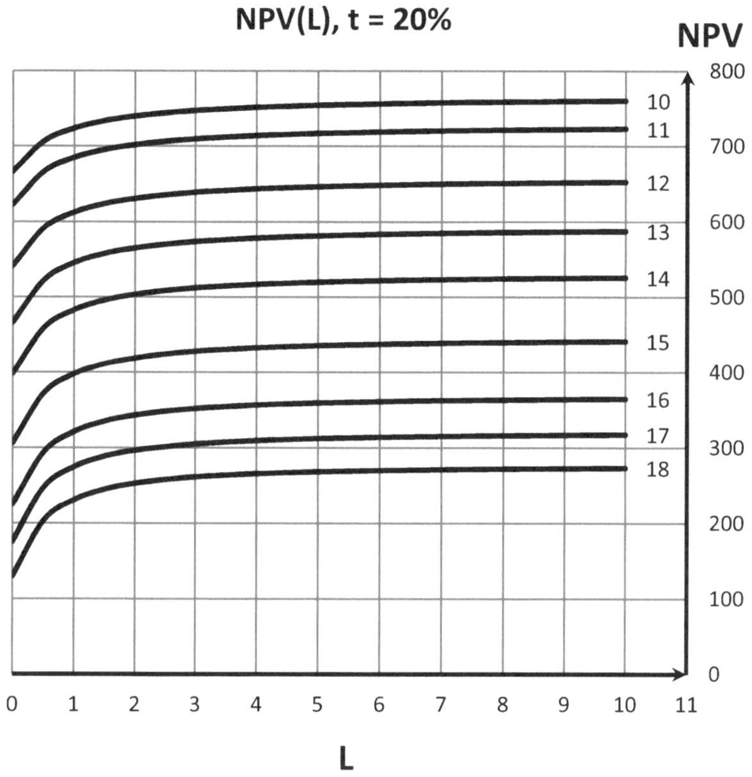

Fig. 29.6 Dependence of NPV on leverage level at fixed values of k_0 and k_d

$$NPV = -S\left[1 + \frac{Lk_d(1-t)}{WACC}\left(1 - \frac{1}{(1+WACC)^n}\right) + \frac{L}{(1+WACC)^n}\right] + \\ + \frac{\beta S(1+L)(1-t)}{WACC}\left(1 - \frac{1}{(1+WACC)^n}\right). \tag{29.4}$$

Dependence of NPV on leverage level at fixed values of k_0 and k_d at a constant equity value ($S = \text{const}$) is shown in Table 29.4 and Figs. 29.7 and 29.8.

1. At the constant values of $\Delta k = k_0 - k_d$ NPV shows as an unlimited growth with leverage and unlimited descending with leverage. It is interesting to note, that the credit rate value $k_d \approx 12\%$ turns out to be a boundary at all surveyed values of $\Delta k = k_0 - k_d$, equal to 2%, 4%, 6%, 10% (it separates the growth of NPV with leverage from descending of NPV with leverage) for 2-year projects. In other words, with growth of k_d the transition from the growth of NPV with leverage to its descending with leverage takes place, and at the credit rate $k_d \approx 12\%$ NPV does not depends on the leverage level at all surveyed values of k_0. For 5-years

Table 29.4 n = 2, t = 0.2, S = 1000; β = 0.7, $k_0 - k_d = const$

	k_0	k_d\L	0.0	0.5	1.0	1.5	2.0	2.5	3.0	3.5	4.0	4.5
1	0.08	0.06	−1.4	32.9	62.7	90.6	117.5	143.9	170.0	195.9	221.6	247.1
2	0.10	0.08	−27.8	−2.5	16.9	33.8	49.6	64.6	79.2	93.5	107.6	121.5
3	0.14	0.12	−77.8	−69.8	−70.1	−73.7	−79.1	−85.4	−92.4	−99.8	−107.4	−115.3
4	0.18	0.16	−123.2	−131.6	−150.2	−173.0	−198.0	−224.2	−251.2	−278.8	−306.7	−335.0
5	0.24	0.22	−184.2	−215.3	−259.2	−308.4	−360.4	−418.9	−468.4	−523.6	−579.3	−635.3
6	0.30	0.28	−237.9	−289.8	−356.6	−429.7	−505.9	−584.2	−663.6	−743.8	−824.6	−905.8
7	0.36	0.34	−285.4	−356.7	−444.0	−538.7	−637.0	−737.6	−839.5	−942.3	−1045.8	−1150.1
8	0.40	0.38	−314.2	−396.9	−497.4	−605.4	−717.3	−831.3	−946.9	−1063.6	−1180.9	−1298.7
9	0.44	0.42	−341.0	−434.8	−547.3	−667.9	−792.6	−919.7	−1048.4	−1178.1	−1308.6	−1439.6
10	0.10	0.06	−27.8	9.6	42.6	73.7	103.9	133.7	163.0	192.2	221.2	250.1
11	0.12	0.08	−53.5	−25.0	−2.2	18.2	37.4	56.0	74.1	92.0	109.6	127.1
12	0.16	0.12	−101.1	−89.8	−86.4	−86.4	−88.0	−90.6	−93.8	−97.3	−101.2	−105.3
13	0.20	0.16	−144.7	−149.8	−164.2	−183.1	−204.2	−226.5	−249.5	−273.1	−297.0	−321.2
14	0.24	0.20	−184.2	−204.4	−236.3	−273.0	−312.2	−352.8	−394.3	−436.4	−478.9	−521.8
15	0.30	0.26	−237.9	−279.4	−334.8	−396.0	−460.2	−526.1	−593.2	−661.0	−729.3	−798.1
16	0.36	0.32	−285.4	−346.8	−423.3	−506.6	−593.5	−682.4	−772.6	−863.6	−955.2	−1047.3
17	0.40	0.36	−314.2	−387.3	−477.3	−574.4	−675.2	−778.0	−882.2	−987.5	−1093.4	−1199.7
18	0.44	0.40	−341.0	−425.5	−527.9	−637.9	−751.8	−868.1	−985.8	−1104.5	−1223.9	−1343.8

	k_0	k_d\L	5.0	5.5	6.0	6.5	7.0	7.5	8.0	8.5	9.0	9.5	10.0
1	0.08	0.06	272.6	298.1	323.4	348.8	374.1	399.3	424.6	449.8	475.0	500.2	525.4
2	0.10	0.08	135.3	149.4	163.1	176.8	190.4	204.1	217.6	231.2	244.7	258.2	271.7
3	0.14	0.12	−123.4	−131.6	−139.8	−148.2	−156.6	−165.1	−173.6	−182.2	−190.8	−199.4	−208.0
4	0.18	0.16	−363.4	−391.9	−420.6	−449.4	−478.3	−507.2	−536.2	−565.3	−594.4	−623.5	−652.7
5	0.24	0.22	−691.5	−748.0	−804.6	−861.4	−918.2	−975.2	−1032.2	−1089.2	−1146.3	−1203.5	−1260.7
6	0.30	0.28	−987.4	−1069.1	−1151.1	−1233.2	−1315.5	−1397.8	−1480.2	−1562.7	−1645.3	−1727.9	−1810.6

7	0.36	0.34	−1254.7	−1358.6	−1463.4	−1568.4	−1673.5	−1778.7	−1884.1	−1989.5	−2094.9	−2200.5	−2306.1
8	0.40	0.38	−1417.0	−1535.5	−1654.3	−1773.3	−1892.5	−2011.7	−2131.1	−2250.6	−2370.1	−2489.7	−2609.4
9	0.44	0.42	−1571.0	−1702.7	−1834.7	−1966.9	−2099.3	−2231.8	−2364.4	−2497.1	−2629.8	−2762.7	−2895.6
10	0.10	0.06	278.9	307.6	336.3	365.0	393.6	422.1	450.7	479.2	507.8	536.3	564.8
11	0.12	0.08	144.5	161.8	179.1	196.3	218.4	230.5	247.6	264.6	281.7	298.7	315.7
12	0.16	0.12	−109.5	−118.9	−118.3	−122.9	−127.5	−132.1	−136.8	−141.5	−146.2	−151.0	−155.8
13	0.20	0.16	−345.6	−370.1	−394.7	−419.5	−444.3	−469.2	−494.2	−519.2	−544.2	−569.3	−594.4
14	0.24	0.20	−564.9	−608.1	−651.5	−695.1	−738.7	−782.4	−826.1	−869.9	−918.8	−957.7	−1001.6
15	0.30	0.26	−867.1	−936.3	−1005.7	−1075.3	−1145.0	−1214.7	−1284.6	−1354.5	−1424.5	−1494.5	−1564.6
16	0.36	0.32	−1139.8	−1232.0	−1324.9	−1419.3	−1511.6	−1604.9	−1698.3	−1791.8	−1885.3	−1978.9	−2072.5
17	0.40	0.36	−1306.5	−1418.5	−1520.8	−1628.3	−1735.9	−1843.6	−1951.4	−2059.3	−2167.3	−2275.3	−2383.4
18	0.44	0.40	−1464.2	−1584.8	−1705.6	−1826.7	−1947.9	−2069.2	−2190.7	−2312.2	−2433.8	−2555.5	−2677.2

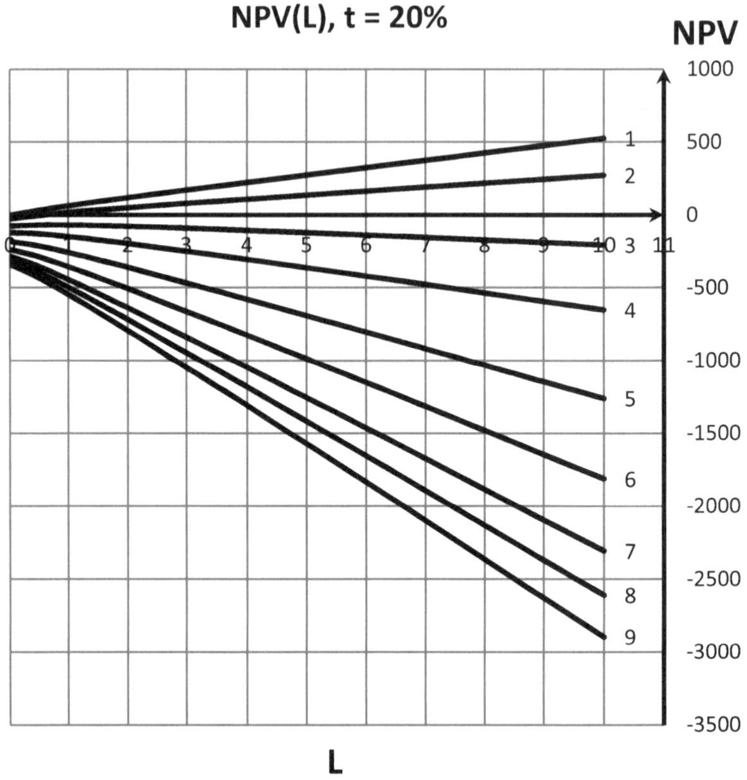

Fig. 29.7 Dependence of NPV on leverage level at fixed values of k_0 and k_d

projects this boundary credit rate is equal to 16–18%, for 7-year and 10-year projects it is equal to 12–15%.

Thus, we come to conclusion, that for arbitrary duration projects NPV grows with leverage at a credit rate $k_d < 12\ \% - 18\%$ and NPV decreases with leverage at a credit rate $k_d > 12\ \% - 18\%$ (project remains effective up to leverage levels $L = L_0$, $NPV(L_0) = 0$). Optimum in the dependence of NPV(L) is absent.

2. At the constant values of k_d NPV shows two types of behavior: (a) an unlimited growth with leverage, (b) NPV reaches maximum at relatively low leverage level (L < 1), then it unlimited descends with leverage.

NPV grows with leverage at a credit rate $k_d < 8\ \% - 10\%$ and NPV decreases with leverage at a credit rate $k_d < 8\ \% - 10\%$ (project remains effective up to leverage levels $L = L_0$, $NPV(L_0) = 0$). All curves NPV(L) at the constant values of k_0 and different values of k_d are started (at L = 0) from one point, the higher values of k_0 (and higher values of $\Delta k = k_0 - k_d$) correspond to more low lying curves NPV(L). Optimum in dependence of NPV(L) is absent. This is observed for projects of all analyzed duration-frames. The first type of dependence of NPV

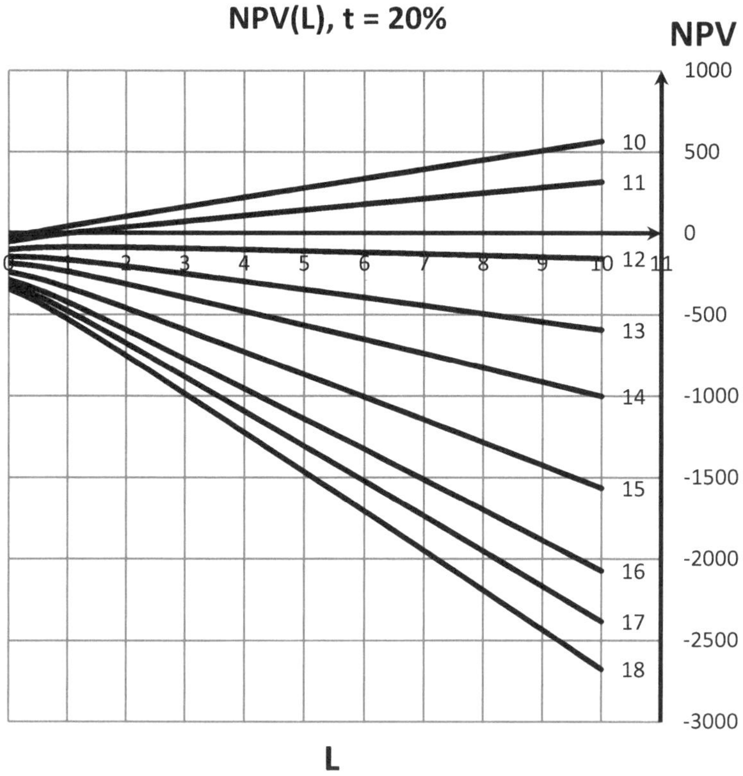

Fig. 29.8 Dependence of NPV on leverage level at fixed values of k_0 and k_d

(L) has a place mainly for pairs of values k_0 and k_d up to 16–18%, while the second type—for higher pairs of values k_0 and k_d and irrespective of the duration of the project.

Thus, for the projects of all analyzed durations the second type of dependence of NPV(L) has a place for $k_d = 16\ \%$; $k_0 = 18\ \% \div 24\ \%$; $k_d = 20\ \%$; $k_0 = 24\ \% \div 44\ \%$; $k_d = 24\ \%$; $k_0 = 30\ \% \div 44\ \%$; in the case 2-year project another pair ($k_d = 12\ \%$; $k_0 = 14\%$) is added yet.

3. At the constant values of k_0 NPV as well as in case of constant values of $\Delta k = k_0 - k_d$ shows as an unlimited growth with leverage and unlimited descending with leverage. An analysis of the data leads to the same conclusion, that and, in paragraph (1): at arbitrary duration of project NPV is growing with leverage at the credit rate $k_d < 18\%$ and NPV decreases with leverage at a credit rate $k_d < 18\%$ (project remains effective up to leverage levels $L = L_0$, $NPV(L_0) = 0$).

It should be noted that this pattern should be taken into account by the Regulator which should regulate normative base in such a way that credit rates, that are associated with central bank basic rate, not exceed, say, 18%.

All curves NPV(L) at the constant values of k_0 and different values of k_d are started (at L = 0) from one point, the higher values of k_d (and lower values of

$\Delta k = k_0 - k_d$) correspond to more low lying curves NPV(L). Optimum in dependence of NPV(L) is absent.

29.2 The Effectiveness of the Investment Project from the Perspective of the Owners of Equity and Debt

29.2.1 With the Division of Credit and Investment Flows

At a Constant Investment Value (I = const)

For NPV in this case the following expression has been obtained (Brusov & Filatova, 2011; Brusov et al., 2011a, b, c, 2012a, b, 2013a, b, 2014a, b)

$$NPV = -I\left[1 - \frac{Lt}{1+L}\left(1 - \frac{1}{(1+k_d)^n}\right)\right] + \frac{NOI(1-t)}{k_e}\left(1 - \frac{1}{(1+k_e)^n}\right) \quad (29.5)$$

Dependence of NPV on leverage level at fixed values of k_0 and k_d at a constant investment value (I = const) is shown in Table 29.5 and Figs. 29.9 and 29.10.

1. At the constant values of $\Delta k = k_0 - k_d$ NPV, as a rule, reaches an optimum at relatively low leverage level (L < 1), then decreases with leverage. All curves NPV(L) for the same values of k_0 are started at the same point, but with growth of $\Delta k = k_0 - k_d$ (respectively, decreasing of k_d) they are shifted into region of lower values of NPV, and descending speed decreases with leverage and for not too high values of k_0 and k_d curves NPV(L) practically output into saturation regime. For higher values of k_0 and k_d saturation regime does not occur and after optimum (sometimes, but more seldom, without optimum) falling trend is still present.
2. At the constant values of k_0 NPV practically always decreases with leverage, very rarely only (for individual values of k_0 and k_d) demonstrating the presence of optimum at low leverage levels (L < 1) (it should be noted that, with the increase of the duration of the project the number of curves NPV(L) having optimum is growing, while remaining to be not very large). All curves NPV(L) at the constant values of k_0 and different values of k_d are started (at L = 0) from one point, the higher values of k_d (and lower values of $\Delta k = k_0 - k_d$) correspond to higher lying curves NPV(L). Descending speed decreases with leverage.
3. At the constant values of k_d NPV practically always decreases with leverage, and the existence of optimum at low levels leverage (L < 1) is a rare exception. All curves NPV(L) at the constant values of k_0 and different values of k_d are started (at L = 0) from one point, the higher values of k_0 (and higher values of $\Delta k = k_0 - k_d$) correspond to lower lying curves NPV(L). Descending speed increases with leverage.

Table 29.5 $n = 2$, $t = 0.2$, $NOI = 1200$; $I = 1000$, $k_0 - k_d = const$

	k_0	k_d\L	0.0	0.5	1.0	1.5	2.0	2.5	3.0	3.5	4.0	4.5
1	0.08	0.06	711.9	700.7	686.2	670.6	654.6	638.5	622.4	606.5	590.8	575.3
2	0.10	0.08	666.6	660.2	649.4	637.0	623.9	610.5	597.0	583.6	570.2	556.9
3	0.14	0.12	580.9	581.9	576.5	568.5	559.4	549.6	539.5	529.3	519.0	508.7
4	0.18	0.16	503.0	510.6	509.8	505.7	499.9	493.4	486.3	478.9	471.3	463.7
5	0.24	0.22	398.5	414.0	418.6	418.9	417.1	414.0	410.2	406.0	401.4	396.6
6	0.30	0.28	306.5	328.6	337.3	340.6	344.0	344.2	343.5	342.3	340.6	338.6
7	0.36	0.34	225.0	251.3	264.7	272.0	276.4	279.0	280.5	281.3	281.6	283.5
8	0.40	0.38	175.6	206.1	221.0	229.6	235.1	236.2	238.8	240.6	242.0	242.9
9	0.44	0.42	129.7	163.0	179.5	189.5	196.1	200.7	204.3	207.0	209.1	210.9
10	0.10	0.06	666.6	635.3	602.0	568.7	536.1	504.4	473.6	443.9	415.2	387.5
11	0.12	0.08	622.6	596.5	567.1	537.1	507.3	478.2	449.9	422.4	395.7	369.9
12	0.16	0.12	541.1	524.3	501.8	477.7	453.2	428.8	404.8	381.4	358.4	336.1
13	0.20	0.16	466.2	456.8	441.5	422.5	402.5	382.2	362.1	342.1	322.5	303.2
14	0.24	0.20	398.5	396.6	385.1	370.2	353.8	336.9	319.7	302.6	285.6	268.9
15	0.30	0.26	306.5	318.4	308.1	298.0	285.9	275.8	263.0	249.9	236.7	223.6
16	0.36	0.32	225.0	237.9	238.8	234.1	226.8	218.2	208.7	198.8	188.6	178.3
17	0.40	0.36	175.6	193.9	197.2	194.7	189.3	179.9	172.2	164.0	155.5	146.7
18	0.44	0.40	129.7	151.7	157.5	157.1	153.7	148.5	142.4	135.6	128.5	121.1

	k_0	k_d\L	5.0	5.5	6.0	6.5	7.0	7.5	8.0	8.5	9.0	9.5	10.0
1	0.08	0.06	560.0	544.9	530.1	515.5	501.1	487.0	473.1	459.4	445.9	432.7	419.7
2	0.10	0.08	543.7	525.7	512.5	499.5	486.6	474.0	461.5	449.2	437.1	425.1	418.4
3	0.14	0.12	498.4	488.2	478.0	468.0	458.0	448.1	438.3	428.6	419.0	409.5	400.1
4	0.18	0.16	455.9	448.2	440.4	432.6	424.9	417.2	409.6	402.0	394.4	386.9	379.5
5	0.24	0.22	391.7	386.6	381.5	376.3	371.1	365.9	360.7	355.4	350.2	345.0	339.8
6	0.30	0.28	336.3	334.0	331.5	328.9	326.2	323.4	320.7	317.8	315.0	312.1	309.2

(continued)

Table 29.5 (continued)

	k_0	$k_d \backslash L$	5.0	5.5	6.0	6.5	7.0	7.5	8.0	8.5	9.0	9.5	10.0
7	0.36	0.34	284.4	280.7	280.0	279.1	278.2	277.2	276.1	274.9	273.7	272.5	271.2
8	0.40	0.38	243.5	244.0	244.2	244.3	244.3	244.2	244.0	243.7	243.4	243.1	242.7
9	0.44	0.42	212.4	218.6	214.6	215.5	216.3	217.0	217.6	218.2	218.6	219.1	219.4
10	0.10	0.06	360.7	334.8	309.8	285.6	262.2	239.6	217.7	196.5	176.0	156.1	136.9
11	0.12	0.08	344.9	320.6	297.2	274.5	252.5	231.1	210.5	190.4	171.0	152.2	133.9
12	0.16	0.12	314.4	293.2	272.7	252.7	233.3	214.4	196.0	178.1	160.8	143.9	127.4
13	0.20	0.16	284.4	266.0	248.1	230.6	218.5	196.8	180.5	164.6	149.1	134.0	119.3
14	0.24	0.20	252.4	236.2	220.4	204.9	189.7	174.9	160.3	146.1	132.2	118.6	105.3
15	0.30	0.26	210.6	197.7	185.0	172.5	160.1	148.0	136.1	124.4	112.9	101.6	90.5
16	0.36	0.32	168.0	155.9	145.7	140.4	127.2	117.2	107.4	97.7	88.1	78.7	69.4
17	0.40	0.36	137.9	129.0	120.2	111.3	102.6	93.9	85.2	76.7	68.3	59.9	51.7
18	0.44	0.40	118.6	106.0	98.3	90.7	83.0	75.4	67.9	60.4	53.0	45.6	38.3

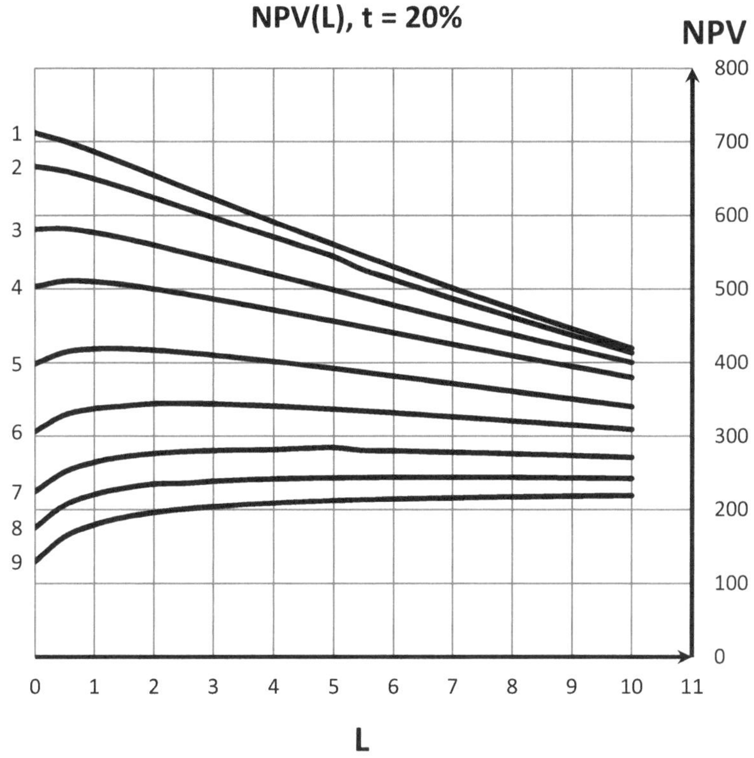

Fig. 29.9 Dependence of NPV on leverage level at fixed values of k_0 and k_d

At a Constant Equity Value (S = const)

For NPV in this case the following expression has been obtained (Brusov & Filatova, 2011; Brusov et al., 2011a, b, c, 2012a, b, 2013a, b, 2014a, b)

$$NPV = -S\left[1 + L - tL\left(1 - \frac{1}{(1 + k_d)^n}\right)\right] + \frac{\beta S(1 + L)(1 - t)}{k_e} \times \left(1 - \frac{1}{(1 + k_e)^n}\right) \quad (29.6)$$

Dependence of NPV on leverage level at fixed values of k_0 and k_d at a constant equity value (S = const) is shown in Table 29.6 and Figs. 29.11 and 29.12.

1. At the constant values of $\Delta k = k_0 - k_d$ NPV always decreases with leverage (existence of an optimum (at relatively low leverage level (L < 1)), practically has not been observed. All curves NPV(L) for the same values of k_0 are started at the same point, but with growth of k_0 (and, respectively, growth of k_d) they are shifted

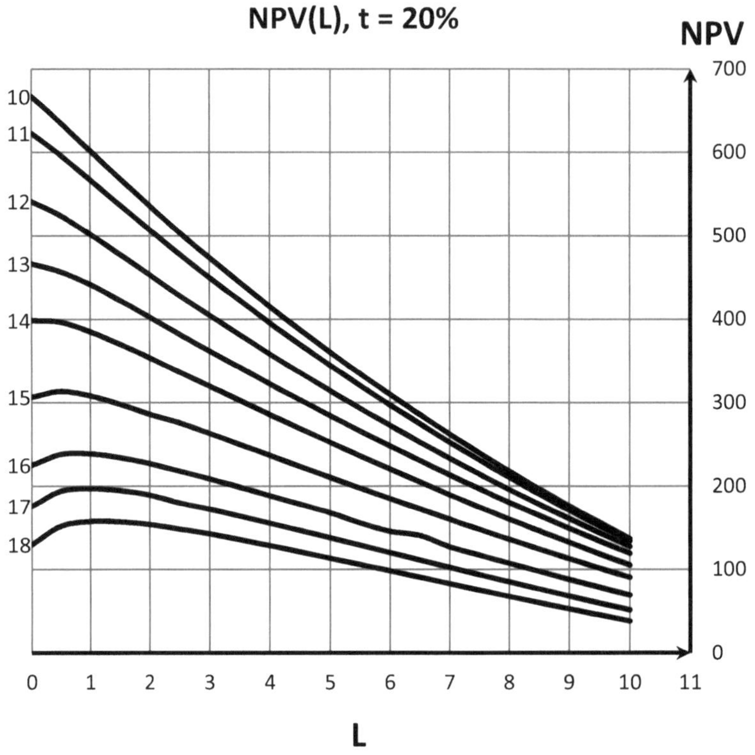

Fig. 29.10 Dependence of NPV on leverage level at fixed values of k_0 and k_d

into region of lower values of NPV, and descending speed increases with leverage.

The values of $\Delta k = k_0 - k_d$, equal to 2%; 4%; 6% and 10% have been used.

With growth of $\Delta k = k_0 - k_d$ a narrowing of NPV(L) curve cluster takes place (the width of cluster is decreased), the difference between curvers becomes less and less, and at $\Delta k = 10\%$ curve cluster is practically transformed into one wide line. The marked pattern has a place for projects of all examined duration projects (2, 5, 7, 10 years).

2. At the constant values of k_0 NPV always decreases with leverage. All curves NPV (L) for the same values of k_0 are started at the same point, but with growth of k_d (and, respectively, decrease of $\Delta k = k_0 - k_d$) they are shifted into region of higher values of NPV, and descending speed decreases with leverage. The width of the NPV(L) curves cluster is decreased with increase of the duration of the project.
3. At the constant values of k_0 NPV always decreases with leverage. All curves NPV (L) for the same values of k_0 are started at the same point, but with growth of k_0 (and, respectively, increase of $\Delta k = k_0 - k_d$) they are shifted into region of lower values of NPV, and descending speed increases with leverage. The width of the NPV(L) curves cluster is decreased with increase of the duration of the project.

Table 29.6 n = 2, t = 0.2, S = 1000; β = 0.7, $k_0 - k_d$ = const

	k_0	k_d\L	0.0	0.5	1.0	1.5	2.0	2.5	3.0	3.5	4.0	4.5
1	0.08	0.06	−1.4	−7.3	−23.6	−50.0	−86.1	−131.8	−186.8	−250.8	−323.5	−404.7
2	0.10	0.08	−27.8	−41.4	−63.8	−94.9	−134.4	−182.1	−237.9	−301.5	−372.8	−451.5
3	0.14	0.12	−77.8	−107.4	−143.9	−187.3	−237.3	−294.0	−357.0	−426.4	−501.9	−583.5
4	0.18	0.16	−123.2	−167.5	−217.2	−272.2	−332.3	−397.5	−467.8	−542.9	−623.0	−707.8
5	0.24	0.22	−184.2	−249.1	−317.6	−389.7	−465.4	−544.7	−627.4	−718.6	−803.2	−896.1
6	0.30	0.28	−237.9	−321.3	−407.4	−496.3	−583.0	−674.4	−767.7	−862.9	−960.2	−1059.3
7	0.36	0.34	−285.4	−386.7	−487.6	−589.6	−692.5	−796.5	−901.4	−1007.3	−1114.3	−1216.1
8	0.40	0.38	−314.2	−424.8	−535.9	−647.5	−759.5	−877.3	−990.8	−1104.8	−1219.3	−1334.3
9	0.44	0.42	−341.0	−461.4	−581.9	−702.4	−822.9	−943.5	−1064.0	−1184.7	−1305.3	−1426.0
10	0.10	0.06	−27.8	−64.5	−121.8	−198.6	−293.5	−405.7	−534.0	−677.6	−835.6	−1007.3
11	0.12	0.08	−53.5	−97.1	−159.8	−240.6	−338.4	−452.2	−581.2	−724.6	−881.6	−1051.5
12	0.16	0.12	−101.1	−157.8	−231.0	−319.7	−423.1	−540.6	−671.4	−814.8	−970.3	−1137.3
13	0.20	0.16	−144.7	−214.6	−296.8	−393.4	−502.8	−624.4	−757.7	−902.1	−1057.2	−1222.4
14	0.24	0.20	−184.2	−265.3	−358.6	−463.6	−579.9	−706.9	−844.3	−991.6	−1148.5	−1314.5
15	0.30	0.26	−237.9	−335.4	−443.1	−560.7	−688.0	−818.1	−960.5	−1111.1	−1269.5	−1435.4
16	0.36	0.32	−285.4	−399.1	−519.3	−647.0	−782.0	−924.2	−1073.2	−1229.0	−1391.2	−1559.8
17	0.40	0.36	−314.2	−436.2	−565.0	−700.4	−842.1	−995.4	−1150.0	−1310.5	−1476.8	−1648.7
18	0.44	0.40	−341.0	−471.9	−608.7	−751.3	−899.4	−1053.0	−1212.0	−1376.1	−1545.3	−1719.4

	k_0	k_d\L	5.0	5.5	6.0	6.5	7.0	7.5	8.0	8.5	9.0	9.5	10.0
1	0.08	0.06	−494.2	−591.7	−697.1	−810.2	−930.6	−1058.3	−1193.0	−1334.6	−1482.9	−1637.7	−1798.8
2	0.10	0.08	−537.5	−649.7	−752.6	−862.5	−979.1	−1102.3	−1232.0	−1367.9	−1510.0	−1658.0	−1811.9
3	0.14	0.12	−671.0	−764.3	−863.3	−967.9	−1077.9	−1193.3	−1314.0	−1439.8	−1570.7	−1706.5	−1847.2
4	0.18	0.16	−797.3	−891.4	−990.0	−1093.1	−1200.7	−1312.5	−1428.6	−1548.9	−1673.4	−1801.8	−1934.3
5	0.24	0.22	−992.4	−1092.0	−1194.8	−1300.8	−1410.0	−1522.3	−1637.8	−1756.3	−1877.8	−2002.3	−2129.7
6	0.30	0.28	−1160.4	−1263.4	−1368.4	−1475.2	−1583.8	−1694.4	−1806.8	−1921.0	−2037.0	−2154.9	−2274.5

(continued)

Table 29.6 (continued)

	k_0	k_d\L	5.0	5.5	6.0	6.5	7.0	7.5	8.0	8.5	9.0	9.5	10.0
7	0.36	0.34	−1320.0	−1441.0	−1551.9	−1663.7	−1776.6	−1890.4	−2005.2	−2120.9	−2237.6	−2355.3	−2473.9
8	0.40	0.38	−1449.8	−1565.7	−1682.1	−1799.0	−1916.3	−2034.1	−2152.4	−2271.2	−2390.5	−2510.2	−2630.3
9	0.44	0.42	−1546.7	−1667.4	−1788.2	−1909.0	−2029.8	−2150.6	−2271.5	−2392.4	−2518.3	−2634.2	−2755.2
10	0.10	0.06	−1191.8	−1388.5	−1596.8	−1816.0	−2045.5	−2284.9	−2533.7	−2791.3	−3057.4	−3331.6	−3618.4
11	0.12	0.08	−1233.5	−1427.2	−1631.8	−1846.9	−2071.9	−2306.4	−2549.9	−2801.9	−3062.1	−3330.1	−3605.5
12	0.16	0.12	−1315.2	−1503.5	−1701.8	−1909.6	−2126.4	−2352.0	−2585.8	−2827.5	−3076.8	−3333.3	−3596.8
13	0.20	0.16	−1397.5	−1581.9	−1775.2	−1977.2	−2187.4	−2405.4	−2631.1	−2864.1	−3104.0	−3350.7	−3603.9
14	0.24	0.20	−1489.3	−1672.5	−1863.9	−2063.1	−2269.8	−2483.7	−2704.6	−2932.2	−3166.3	−3406.6	−3653.0
15	0.30	0.26	−1608.7	−1789.0	−1976.2	−2169.9	−2370.1	−2576.4	−2788.8	−3006.9	−3230.7	−3459.9	−3694.4
16	0.36	0.32	−1734.5	−1922.0	−2108.6	−2279.8	−2491.2	−2694.0	−2902.0	−3115.0	−3332.9	−3555.6	−3782.8
17	0.40	0.36	−1826.0	−2008.5	−2196.3	−2389.1	−2586.7	−2789.2	−2996.3	−3207.9	−3424.0	−3644.3	−3868.9
18	0.44	0.40	−1898.3	−2082.0	−2270.2	−2463.0	−2660.1	−2861.5	−3067.1	−3276.7	−3490.4	−3708.0	−3929.4

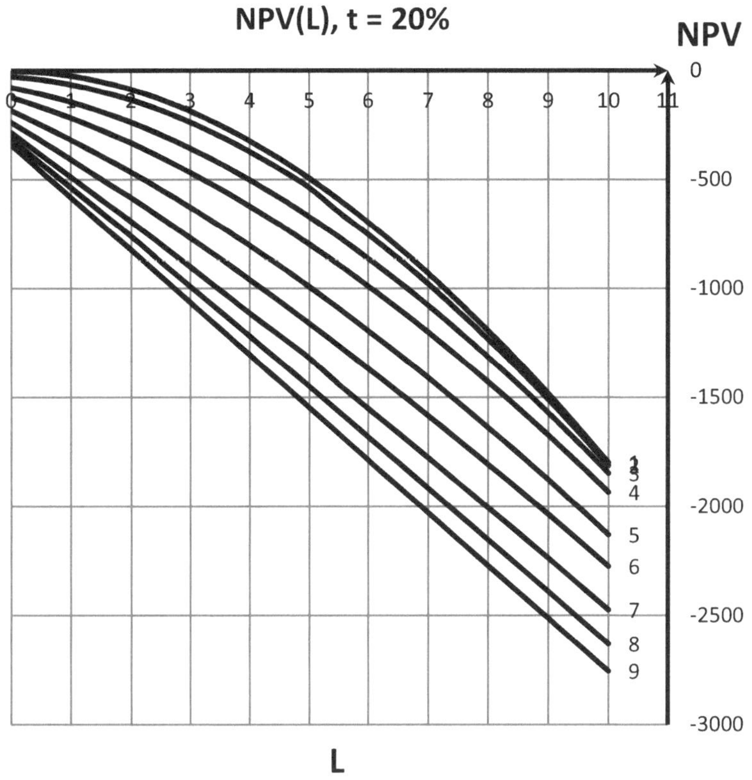

Fig. 29.11 Dependence of NPV on leverage level at fixed values of k_0 and k_d

29.2.2 *Without Flows Separation*

At a Constant Investment Value (*I* = const)

For NPV in this case the following expressions have been obtained (Brusov & Filatova, 2011; Brusov et al., 2011a, b, c, 2012a, b, 2013a, b, 2014a, b)

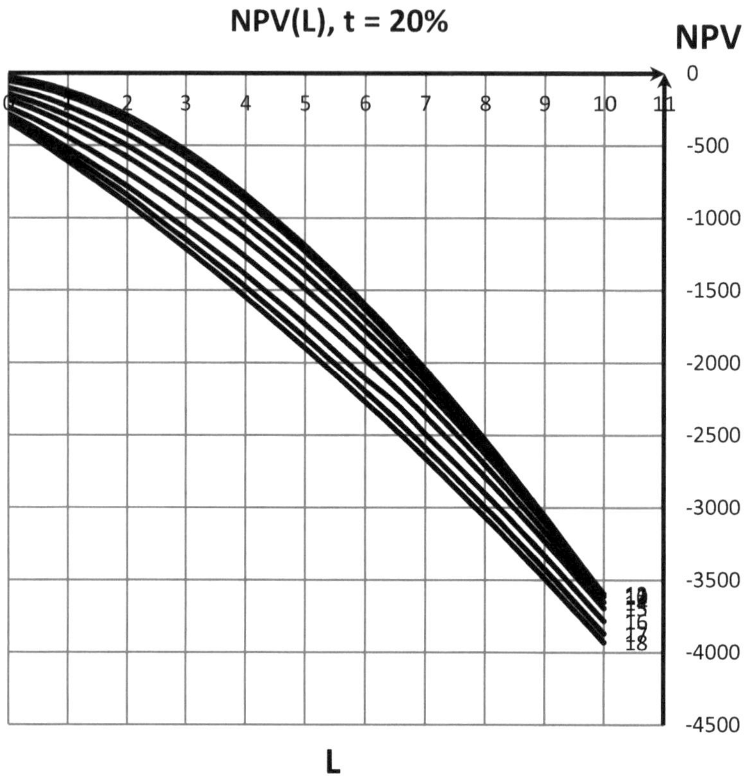

Fig. 29.12 Dependence of NPV on leverage level at fixed values of k_0 and k_d

$$NPV = -I + \frac{NOI(1-t) + k_d D t}{WACC}\left(1 - \frac{1}{(1+WACC)^n}\right)$$

$$NPV = -I + \frac{NOI(1-t) + k_d D t}{k_0\left(1 - \gamma\frac{L}{1+L}\right)}\left(1 - \frac{1}{\left(1 + k_0\left(1 - \gamma\frac{L}{1+L}\right)\right)^n}\right)$$

$$\begin{aligned} NPV = -I\left[1 - \frac{k_d t \dfrac{L}{1+L}}{k_0\left(1 - \gamma\dfrac{L}{1+L}t\right)}\left(1 - \frac{1}{\left(1 + k_0\left(1 - \gamma\frac{L}{1+L}t\right)\right)^n}\right)\right] + \\ + \frac{NOI(1-t)}{k_0\left(1 - \gamma\dfrac{L}{1+L}t\right)}\left(1 - \frac{1}{\left(1 + k_0\left(1 - \gamma\frac{L}{1+L}t\right)\right)^n}\right). \end{aligned} \quad (29.7)$$

Dependence of NPV on leverage level at fixed values of k_0 and k_d at a constant investment value ($I = \text{const}$) is shown in Table 29.7 and Figs. 29.13 and 29.14.

1. At the constant values of $\Delta k = k_0 - k_d$ NPV demonstrates a limited growth with leverage with output into saturation regime. The main increase in NPV occurs at

Table 29.7 $n = 2$, $t = 0.2$, $NOI = 1200$; $I = 1000$, $k_0 - k_d = const$

	k_0	k_d\L	0.0	0.5	1.0	1.5	2.0	2.5	3.0	3.5	4.0	4.5
1	0.08	0.06	711.9	731.8	741.8	747.8	751.9	754.8	757.0	758.6	760.0	761.1
2	0.10	0.08	666.6	692.1	705.0	712.9	718.1	721.9	724.7	726.9	728.7	730.2
3	0.14	0.12	580.9	615.9	633.9	644.7	652.0	657.3	661.2	664.3	666.7	668.8
4	0.18	0.16	503.0	546.2	568.4	581.9	590.9	597.5	602.4	606.2	609.3	611.8
5	0.24	0.22	398.5	451.5	478.8	495.5	506.8	514.9	521.0	525.7	529.5	532.7
6	0.30	0.28	306.5	367.3	398.7	417.7	431.7	441.1	448.3	453.9	458.4	462.0
7	0.36	0.34	225.0	291.2	326.7	348.5	363.2	373.8	381.8	388.1	393.1	397.6
8	0.40	0.38	175.6	246.0	283.0	305.8	321.2	331.5	340.0	346.7	352.1	356.6
9	0.44	0.42	129.7	203.1	241.7	265.6	281.7	293.4	302.3	309.2	314.8	319.4
10	0.10	0.06	666.6	686.0	695.8	701.8	705.7	708.6	710.7	712.4	718.7	714.8
11	0.12	0.08	622.6	647.4	660.0	667.6	672.7	676.3	679.1	681.2	682.9	684.3
12	0.16	0.12	541.1	575.2	592.7	603.2	610.3	615.4	619.3	622.3	624.7	626.6
13	0.20	0.16	466.2	507.9	530.4	543.6	552.4	558.8	563.6	567.3	570.3	572.8
14	0.24	0.20	398.5	447.3	472.4	487.7	498.0	505.5	511.1	515.4	518.9	521.8
15	0.30	0.26	306.5	363.6	393.1	411.0	423.0	432.8	439.5	444.7	448.9	452.3
16	0.36	0.32	225.0	287.9	321.6	342.3	356.3	366.3	373.9	379.9	384.6	388.5
17	0.40	0.36	175.6	243.0	278.3	300.1	314.8	324.5	332.7	339.1	344.2	348.4
18	0.44	0.40	129.7	200.3	237.4	260.2	275.8	287.0	295.5	302.1	307.4	311.8

	k_0	k_d\L	5.0	5.5	6.0	6.5	7.0	7.5	8.0	8.5	9.0	9.5	10.0
1	0.08	0.06	762.0	762.8	763.5	764.1	764.6	765.1	765.5	765.8	766.1	766.4	766.7
2	0.10	0.08	731.4	731.4	732.3	733.0	733.7	734.3	734.8	735.3	735.7	736.0	736.4
3	0.14	0.12	670.5	671.9	673.1	674.2	675.1	675.9	676.7	677.3	677.9	678.4	678.9
4	0.18	0.16	618.9	615.7	617.2	618.6	619.7	620.7	621.7	622.5	623.2	623.9	624.5
5	0.24	0.22	535.3	537.5	539.4	541.1	542.5	543.8	544.9	546.0	546.9	547.7	548.5
6	0.30	0.28	465.1	467.7	470.0	471.9	473.6	475.1	476.5	477.7	478.7	479.7	480.6

(continued)

Table 29.7 (continued)

	k_0	k_d\L	5.0	5.5	6.0	6.5	7.0	7.5	8.0	8.5	9.0	9.5	10.0
7	0.36	0.34	401.3	403.6	406.1	408.3	410.2	411.8	418.3	414.7	415.9	417.0	418.0
8	0.40	0.38	360.2	363.4	366.1	368.4	370.5	372.3	373.9	375.3	376.6	377.8	378.9
9	0.44	0.42	323.2	326.4	329.2	331.7	333.8	335.7	337.4	338.9	340.2	341.4	342.6
10	0.10	0.06	715.7	716.5	717.2	717.7	718.2	718.7	719.1	719.4	719.8	720.0	720.3
11	0.12	0.08	685.5	686.5	687.3	688.1	688.7	689.3	689.8	690.2	690.7	691.0	691.4
12	0.16	0.12	628.3	629.7	630.9	631.9	632.8	633.6	634.3	634.9	635.5	636.0	636.5
13	0.20	0.16	574.8	576.6	578.1	579.4	580.5	581.5	582.4	583.2	583.9	584.6	585.2
14	0.24	0.20	524.2	526.2	528.0	529.5	530.8	532.0	533.0	533.9	534.8	535.5	536.2
15	0.30	0.26	455.2	457.6	459.7	461.6	463.2	464.6	465.8	466.9	467.9	468.9	469.7
16	0.36	0.32	391.8	394.1	396.6	399.7	400.8	402.4	403.8	405.1	406.3	407.3	408.2
17	0.40	0.36	351.9	354.9	357.5	359.7	361.7	363.4	364.9	366.3	367.5	368.7	369.7
18	0.44	0.40	315.5	318.6	321.3	323.6	325.7	327.5	329.1	330.5	331.8	333.0	334.0

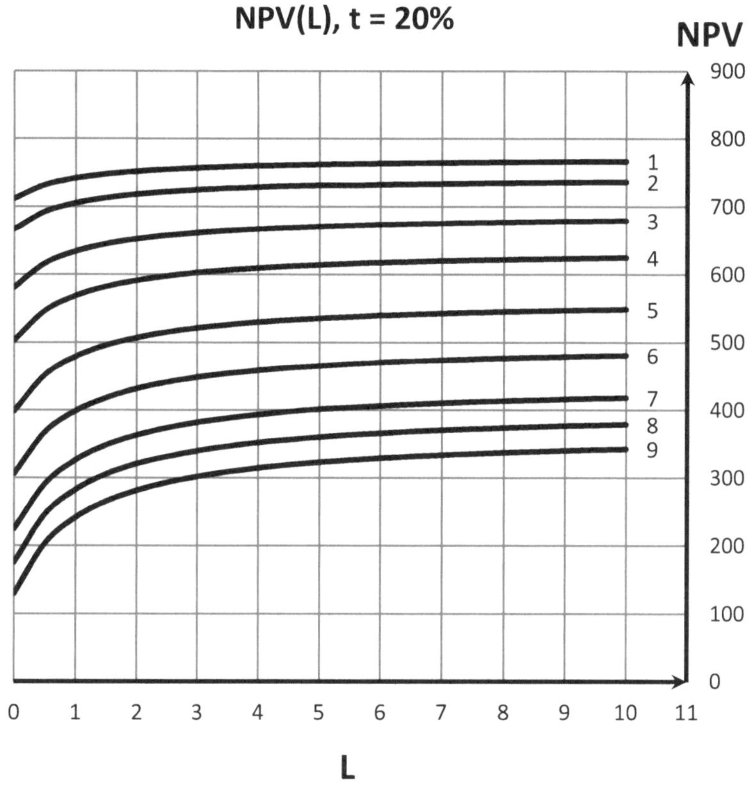

Fig. 29.13 Dependence of NPV on leverage level at fixed values of k_0 and k_d

small values of leverage levels $L \leq 3 \div 4$. With growth of k_0 (and k_d) the curves NPV(L) are lowered. Optimum in the dependence of NPV(L) is absent.

2. At the constant values of k_0 NPV demonstrates a limited growth with leverage with output into saturation regime. The main increase in NPV occurs at low values of leverage levels $L \leq 4 \div 5$. All curves NPV(L) at the constant values of k_0 and different values of k_d are started (at L = 0) from one point, the higher values of k_d (and, respectively, the lower values of $\Delta k = k_0 - k_d$ correspond to higher lying curves NPV(L). This distinguishes this case from consideration from the point of view of equity capital owners, where ordering of curves NPV(L) with growth of k_d is an opposite. Optimum in the dependence of NPV(L) is absent. With growth of project duration the distance between curves NPV(L), corresponding to different pairs of values k_0 and k_d, increases.
3. At the constant values of k_d NPV growth with leverage with output into saturation regime. All curves NPV(L) at the constant values of k_0 and different values of k_d are started (at L = 0) from one point, the higher values of k_0 (and, respectively, the higher values of $\Delta k = k_0 - k_d$ correspond to lower lying curves NPV(L). With

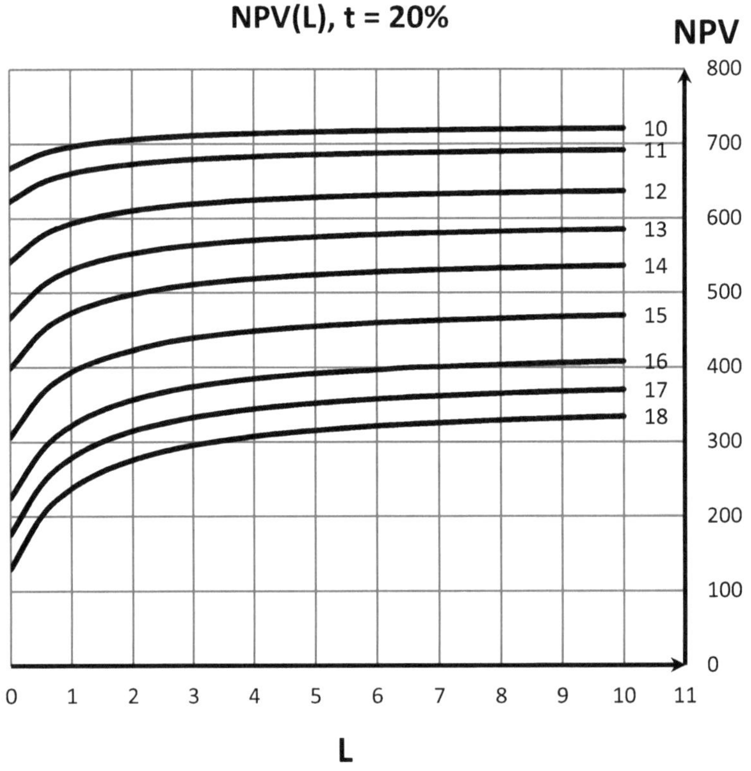

Fig. 29.14 Dependence of NPV on leverage level at fixed values of k_0 and k_d

growth of $\Delta k = k_0 - k_d$ the distance between curves NPV(L), corresponding to different pairs of values k_0 and k_d, decreases.

Optimum in the dependence of NPV(L) is absent.

At a Constant Equity Value (S = const)

For NPV in this case the following expression has been obtained (Brusov & Filatova, 2011; Brusov et al., 2011a, b, c, 2012a, b, 2013a, b, 2014a, b)

$$NPV = -S\left[1 + L - \frac{k_d L t}{WACC}\left(1 - \frac{1}{(1 + WACC)^n}\right)\right] + \\ + \frac{\beta S(1 + L)(1 - t)}{WACC}\left(1 - \frac{1}{(1 + WACC)^n}\right). \tag{29.8}$$

Dependence of NPV on leverage level at fixed values of k_0 and k_d at a constant equity value (S = const) is shown in Table 29.8 and Figs. 29.15 and 29.16.

1. At the constant values of $\Delta k = k_0 - k_d$ NPV shows as an unlimited growth with leverage and unlimited descending with leverage. There is a boundary credit rate

Table 29.8 $n = 2$, $t = 0.2$, $S = 1000$; $\beta = 0.7$, $k_0 - k_d = const$

	k_0	k_d\L	0.0	0.5	1.0	1.5	2.0	2.5	3.0	3.5	4.0	4.5
1	0.08	0.06	−1.4	19.8	41.1	62.5	83.9	105.3	126.8	148.2	169.7	191.1
2	0.10	0.08	−27.8	−18.6	1.0	15.6	30.3	45.1	59.9	74.7	89.5	104.3
3	0.14	0.12	−77.8	−77.7	−77.0	−76.1	−75.1	−74.0	−72.9	−71.8	−70.6	−69.4
4	0.18	0.16	−123.2	−136.4	−148.8	−160.8	−172.6	−184.3	−196.0	−207.6	−219.2	−230.7
5	0.24	0.22	−184.2	−216.3	−247.1	−277.3	−307.3	−337.1	−366.8	−396.5	−426.1	−455.7
6	0.30	0.28	−237.9	−287.3	−335.2	−382.5	−427.6	−473.6	−519.5	−565.3	−611.0	−656.7
7	0.36	0.34	−285.4	−351.6	−414.4	−476.2	−537.6	−598.7	−659.6	−720.4	−781.2	−840.5
8	0.40	0.38	−314.2	−389.7	−462.5	−534.1	−605.1	−677.6	−748.0	−818.3	−888.5	−958.6
9	0.44	0.42	−341.0	−426.0	−508.0	−588.6	−668.7	−748.3	−827.7	−906.9	−986.0	−1065.1
10	0.10	0.06	−27.8	−20.4	−12.7	−5.1	2.7	10.4	18.2	26.0	33.7	41.5
11	0.12	0.08	−53.5	−52.8	−51.9	−50.9	−49.9	−48.8	−47.6	−46.5	−45.4	−44.2
12	0.16	0.12	−101.1	−118.6	−125.5	−137.3	−148.9	−160.5	−172.0	−183.5	−195.0	−206.5
13	0.20	0.16	−144.7	−170.2	−193.6	−217.4	−241.0	−264.6	−288.1	−311.5	−334.9	−358.3
14	0.24	0.20	−184.2	−221.2	−257.1	−292.6	−327.8	−362.9	−397.9	−432.9	−467.8	−502.7
15	0.30	0.26	−237.9	−291.7	−344.2	−396.1	−447.8	−496.9	−547.6	−598.2	−648.7	−699.2
16	0.36	0.32	−285.4	−355.6	−422.6	−488.7	−554.4	−619.8	−685.1	−750.3	−815.5	−880.5
17	0.40	0.36	−314.2	−393.4	−470.1	−545.7	−620.8	−697.6	−772.2	−846.6	−920.9	−995.2
18	0.44	0.40	−341.0	−429.5	−515.1	−599.6	−683.5	−767.0	−850.3	−933.4	−1016.4	−1099.3

	k_0	k_d\L	5.0	5.5	6.0	6.5	7.0	7.5	8.0	8.5	9.0	9.5	10.0
1	0.08	0.06	212.6	234.0	255.5	277.0	298.5	319.9	341.4	362.9	384.3	405.8	427.3
2	0.10	0.08	119.1	130.1	144.6	159.1	173.6	188.1	202.6	217.1	231.6	246.1	260.6
3	0.14	0.12	−68.2	−67.0	−65.8	−64.6	−63.3	−62.1	−60.9	−59.6	−58.4	−57.2	−55.9
4	0.18	0.16	−242.3	−253.8	−265.3	−276.8	−288.3	−299.8	−311.3	−322.8	−334.2	−345.7	−357.2
5	0.24	0.22	−485.2	−514.8	−544.3	−573.8	−603.3	−632.8	−662.3	−691.8	−721.3	−750.7	−780.2
6	0.30	0.28	−702.3	−747.9	−793.5	−839.0	−884.6	−930.1	−975.6	−1021.1	−1066.6	−1112.1	−1157.6

(continued)

Table 29.8 (continued)

	k_0	k_d\L	5.0	5.5	6.0	6.5	7.0	7.5	8.0	8.5	9.0	9.5	10.0
7	0.36	0.34	−900.1	−963.2	−1023.8	−1084.3	−1144.9	−1205.5	−1266.0	−1326.5	−1387.0	−1447.5	−1508.1
8	0.40	0.38	−1028.7	−1098.7	−1168.7	−1238.6	−1308.6	−1378.5	−1448.4	−1518.3	−1588.2	−1658.1	−1727.9
9	0.44	0.42	−1144.0	−1222.9	−1301.8	−1380.6	−1459.4	−1538.2	−1617.0	−1695.7	−1774.4	−1853.2	−1931.9
10	0.10	0.06	49.3	57.1	64.9	72.7	80.5	88.2	96.0	103.8	111.6	119.4	127.2
11	0.12	0.08	−43.1	−41.9	−40.7	−39.6	−38.4	−37.2	−36.1	−34.9	−33.7	−32.6	−31.4
12	0.16	0.12	−218.0	−229.4	−240.9	−252.3	−263.8	−275.2	−286.6	−298.1	−309.5	−320.9	−332.4
13	0.20	0.16	−381.6	−405.0	−428.3	−451.7	−475.0	−498.3	−521.6	−544.9	−568.3	−591.6	−614.9
14	0.24	0.20	−537.5	−572.4	−607.2	−642.0	−676.8	−711.6	−746.5	−781.2	−816.0	−850.8	−885.6
15	0.30	0.26	−749.6	−800.1	−850.5	−900.9	−951.3	−1001.6	−1052.0	−1102.3	−1152.7	−1203.0	−1253.4
16	0.36	0.32	−945.6	−1012.2	−1077.2	−1137.5	−1205.5	−1270.4	−1335.4	−1400.3	−1465.2	−1530.1	−1595.0
17	0.40	0.36	−1069.4	−1143.6	−1217.8	−1291.9	−1366.0	−1440.1	−1514.2	−1588.3	−1662.4	−1736.4	−1810.5
18	0.44	0.40	−1182.2	−1265.0	−1347.8	−1430.6	−1518.3	−1596.0	−1678.7	−1761.4	−1844.1	−1926.7	−2009.4

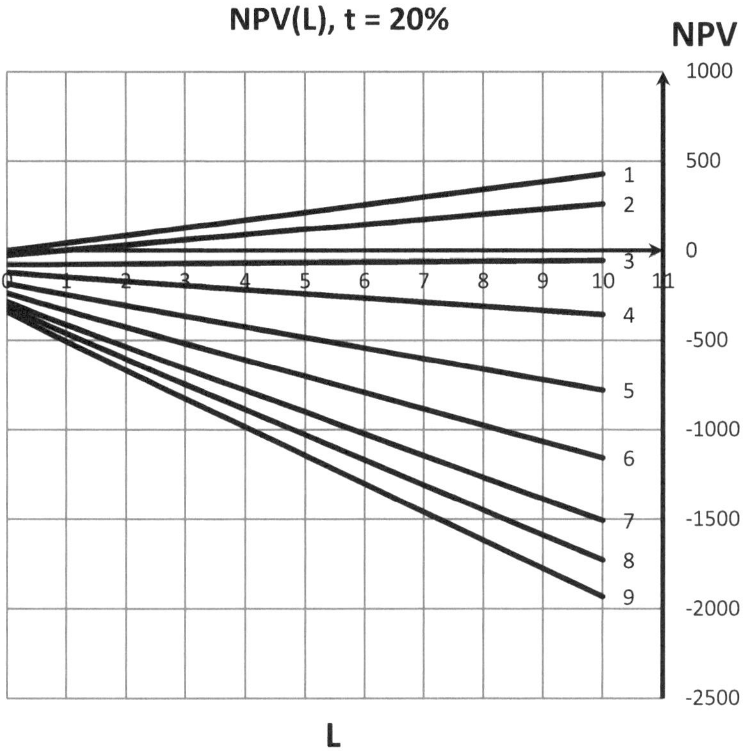

Fig. 29.15 Dependence of NPV on leverage level at fixed values of k_0 and k_d

value, which separates the growth of NPV with leverage from descending of NPV with leverage. This rate depends on the values $\Delta k = k_0 - k_d$, equal to 2%, 4%, 6%,10%, weakly depends on project duration and are within region 8–20%. Thus, we come to conclusion, that for arbitrary duration projects NPV grows with leverage at a credit rate $k_d < 8\ \% - 20\%$ and NPV decreases with leverage at a credit rate $k_d > 8\ \% - 20\%$ (project remains effective up to leverage levels $L = L_0$, $NPV(L_0) = 0$). Optimum in the dependence of NPV(L) is absent.

2. At the constant values of k_0 (similar to the case of $\Delta k = k_0 - k_d$) NPV shows as an unlimited growth with leverage and unlimited descending with leverage.

   All curves NPV(L) at the constant values of k_0 and different values of k_d are started (at L = 0) from one point, the higher values of k_d (and lower values of $\Delta k = k_0 - k_d$) correspond to higher lying curves NPV(L). Optimum in dependence of NPV(L) is absent.

3. At the constant values of k_d NPV as well as in case of constant values of $\Delta k = k_0 - k_d$ shows as an unlimited growth with leverage and unlimited descending with leverage. There is a boundary credit rate value, which separates the growth of NPV with leverage from descending of NPV with leverage. This rate depends on the values $\Delta k = k_0 - k_d$, equal to 2%, 4%, 6%, 10%, weakly depends on project

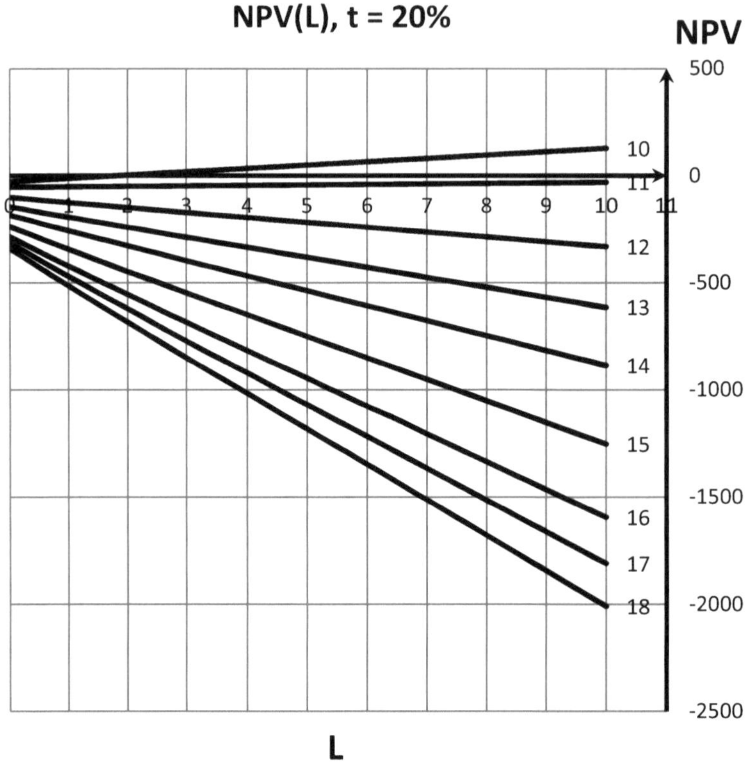

Fig. 29.16 Dependence of NPV on leverage level at fixed values of k_0 and k_d

duration and is within region 8–20%. Thus, we come to conclusion, that for arbitrary duration projects NPV grows with leverage at a credit rate $k_d < 8\ \% - 20\%$ and NPV decreases with leverage at a credit rate $k_d > 8\ \% - 20\%$ (project remains effective up to leverage levels $L = L_0$, $NPV(L_0) = 0$). Optimum in the dependence of NPV(L) is absent.

29.3 The Elaboration of Recommendations on the Capital Structure of Investment of Enterprises, Companies, Taking into Account All the Key Financial Parameters of Investment Project

29.3.1 General Conclusions and Recommendations on the Definition of Capital Structure of Investment of Enterprises

As indicates the dependence NPV(L) at different *I, NOI, S, β,* the changing of first two parameters, as a rule, leads to the shift of curves NPV(L) in the vertical direction only (parallel offset), without changing of characteristic points of these curves, of type L* (value of L, where NPV(L) reaches optimum (if available). Only the maximum permissible leverage level L_0 is changed (in case of descending of NPV with leverage).

This opens the way for tabulation of the results obtained in the case of constant value of investment. In other words this fact is the basis for the use of our tables and graphs to estimate the optimal debt level for the investor. Thus, obtained by us tables and graphs allows to determine value of L^*, knowing only k_0 and k_d for investment project. k_d is the credit rate, which is determined by creditor, but determination of parameter k_0 is always a complicate problem. This has been noted by several researchers, and we can also add that the parameter k_0 is one of the most important parameter in both used theories Modigliani–Miller (Modigliani & Miller, 1958, 1963, 1966) and Brusov-Filatova-Orekhova (Brusov & Filatova, 2011; Brusov et al., 2011a, b, c, 2012a, b, 2013a, b, 2014a, b; Filatova et al., 2008).

In contrast to the parameters I, NOI a change of parameters S, β, both individually and simultaneously, can significantly change the nature of the curves NPV(L), i.e., the dependence of NPV on the leverage level. Thus, with change of β NPV(L) can be changed from decreasing functions to function, having an optimum. Such studies have been conducted on the example of "Nastcom Plus" company. This means the inability of tabulation of the results obtained in the case of constant value of equity capital S: in this case, one will need to use the formulas we have received to determine the NPV at the existing level leverage, as well as to optimizing the existing investment structure.

In the case of a constant invested capital I with the division of flows (with using of two discount rates) in the case of finite duration projects the descending of NPV with leverage is possible, as well as the existence of optimum.. Without flows separation a moderate growth of NPV(L)—output to saturation has been observed. This indicates the limitations of the approach associated with the use of a single discount rate, which veils different variants of the NPV(L) dependence at different cost of equity and borrowed capital.

At constant equity value S with using of one discount rate (WACC approximation) one has either growth or decrease of NPV depending on credit rate k_d. We have found the boundary rates k_d, determining transition from growth to decrease of NPV. Because application of two discount rates (at flow separation) demonstrates the existence of optimum in this case, WACC approximation changes the type of dependence of NPV(L).

Thus, one can make the following general recommendations:

1. It is necessary to use an assessment of the efficiency of investment projects with flows separation;
2. In the case of a constant value of investment I a tabulation of the obtained results is possible, i.e. one can use obtained by us tables and graphs to estimate the optimal for the investor level of borrowing. Thus, obtained by us tables and graphs allows to determine value of L^*, knowing only k_0 and k_d for investment project.
3. At a constant equity value S as well as to determine the NPV at the existing level leverage, and to estimate the optimal for the investor level of borrowing received by us analytical expressions (formulas) should be used because the behavior of NPV(L) in this case depends strongly on S and β.

References

Brusova, A. (2011). A comparison of the three methods of estimation of weighted average cost of capital and equity cost of company. *Financial Analysis: Problems and Solutions, 34*(76), 36–42.

Brusov, P. N., & Filatova, T. V. (2011). From Modigliani–Miller to general theory of capital cost and capital structure of the company. *Finance and Credit, 435*, 2–8.

Brusov, P., Filatova, T., Orehova, N., & Brusova, A. (2011a). Weighted average cost of capital in the theory of Modigliani–Miller, modified for a finite life–time company. *Applied Financial Economics, 21*(11), 815–824.

Brusov, P., Filatova, T., Orehova, N., et al. (2011b). From Modigliani–Miller to general theory of capital cost and capital structure of the company. *Research Journal of Economics, Business and ICT, 2*, 16–21.

Brusov, P., Filatova, T., Orehova, N., et al. (2011c). Influence of debt financing on the effectiveness of the investment project within the Modigliani–Miller theory. *Research Journal of Economics, Business and ICT (UK), 2*, 11–15.

Brusov, P., Filatova, T., Eskindarov, M., & Orehova, N. (2012a). Influence of debt financing on the effectiveness of the finite duration investment project. *Applied Financial Economics, 22*(13), 1043–1052.

Brusov, P., Filatova, T., Eskindarov, M., & Orehova, N. (2012b). Hidden global causes of the global financial crisis. *Journal of Reviews on Global Economics, 1*, 106–111.

Brusov, P., Filatova, P., & Orekhova, N. (2013a). Absence of an optimal capital structure in the famous tradeoff theory! *Journal of Reviews on Global Economics, 2*, 94–116.

Brusov, P., Filatova, T., & Orehova, N. (2013b). A qualitatively new effect in corporative finance: Abnormal dependence of cost of equity of company on leverage. *Journal of Reviews on Global Economics, 2*, 183–193.

Brusov, P., Filatova, P., & Orekhova, N. (2014a). Mechanism of formation of the company optimal capital structure, different from suggested by trade off theory. *Cogent Economics & Finance, 2*, 1–13. https://doi.org/10.1080/23322039.2014.946150

Brusov, P., Filatova, T., & Orehova, N. (2014b). Inflation in Brusov–Filatova–Orekhova theory and in its perpetuity limit – Modigliani – Miller theory. *Journal of Reviews on Global Economics, 3*, 175–185.

Brusov, P., Filatova, T., & Orekhova, N. (2022). *Generalized Modigliani–Miller theory: Applications in corporate finance, investments, taxation and ratings* (p. 362). Springer Nature.

Filatova, T., Orehova, N., & Brusova, A. (2008). Weighted average cost of capital in the theory of Modigliani–Miller, modified for a finite life–time company. *Bulletin of the FU, 48*, 68–77.

Modigliani, F., & Miller, M. (1958). The cost of capital, corporate finance, and the theory of investment. *American Economic Review, 48*, 261–297.

Modigliani, F., & Miller, M. (1963). Corporate income taxes and the cost of capital: A correction. *American Economic Review, 53*, 147–175.

Modigliani, F., & Miller, M. (1966). Some estimates of the cost of capital to the electric utility industry 1954–1957. *American Economic Review, 56*, 333–391.

Chapter 30
Optimizing the Investment Structure of the Telecommunication Sector Company

30.1 Introduction

Investments in tangible and intangible assets play an important role in the activities of any company. They are a necessary condition for structural adjustment and economic growth and provide the creation of new and enhancement of existing basic funds and industries. The role of investment, which is always the one of the most important, is increased many times at the current stage.

For example, in Russia a priority of budget will be the reduced of dependence of the price of oil and gas. The main issue that helps at least to start the movement on this way is, of course, investments. In this way, the role of investment at the present stage is indeed increasing dramatically. In this regard, the role of the evaluation of the efficiency of investment projects, which in the context of scarcity and limited investment resources allows the realization of the most effective projects, increases. Since virtually all investment projects use debt financing, the purpose of the study of the impact of debt financing and capital structure on the efficiency of investment projects, determining the optimal level leverage, is especially actual at the present time.

The hope to determine the optimal capital structure, in which one or more parameters of the efficiency of the project (NPV, IRR, etc.) are maximum, more than half a century has encouraged researchers to deal with the issue. Some of the major problems in the assessment of the effectiveness of the projects are the following:

- What are financial flows and why are they necessary to take into account when calculating parameters of efficiency of the project (NPV, IRR, etc.)?
- How many discount rate should be used for financial flows associated with investments?
- How can these discount rates be accurately determined?

P. Brusov, T. Filatova, *Efficiency of Investment Projects*, Contributions to Finance and Accounting, https://doi.org/10.1007/978-3-032-01028-5_30

Discussion concerning the first two problems is ongoing. On the third issue, one needs to note that, in recent years, there has been a significant progress in the accurate determination of the cost of the equity capital of the company and its weighted average cost, which as time and are the discount rate when evaluating the NPV of the project. The progress is associated with work performed by Brusov, Filatova, and Orekhova (Brusov & Filatova, 2011; Brusov et al., 2011a, b, c, 2012a, b, 2013a, b, 2014a, b; Filatova et al., 2008; Brusova, 2011; Brusov et al., 2015, 2018a, b, c, d), in which the general theory of capital cost of the company (equity cost as well as weighted average cost) was established and its dependence on leverage and on lifetime (age) of company was found for the companies with arbitrary lifetime (age). The main difference between their theory and theory by Modigliani and Miller is that the former one waives from the perpetuity of the companies, which leads to significant differences of a new theory from theory of Nobel laureates Modigliani and Miller (1958, 1963, 1966).

The lack of modern methods of evaluation of effectiveness of investment projects with account of the debt financing, with the correct assessment of discount rate, used in investment models, has identified the need for research. The establishment of such modern models, considering problem from the point of view of equity capital owners as well as from the point of view of equity and debt capital owners, with the use of modern theory by Brusov, Filatova, and Orekhova that assesses the equity capital cost and weighted average cost of capital of the company (Modigliani & Miller, 1958, 1963, 1966), which play the role of discount rate in the investment models, can significantly contribute to the problem of the assessment of investment projects' effectiveness.

30.2 Investment Analysis and Recommendations for Telecommunication Company "Nastcom Plus"

Based on the method, developed by the authors (Brusov & Filatova, 2011; Brusov et al., 2011a, b, c, 2012a, b, 2013a, b, 2014a, b; Filatova et al., 2008), let us analyze the efficiency of investments of one of the leading companies in the telecommunication sector "Nastcom Plus" for 2010–2012 from the point of view of optimal structure of investment. The source data for the analysis are presented in Table 30.1.

Quantity k_0 is the equity cost of financially independent company (or equity cost at zero leverage) and for "Nastcom Plus" is equal to 23.67% (Brusova, 2011). Here are also calculated dependence of weighted average cost of capital WACC and the equity cost k_e on leverage (Fig. 30.1).

Table 30.1 Data of "Nastcom Plus" for 2010–2012

Indicator	2010	2011	2012
Investment I, million dollars	1.124	2.05	2.763
Revenue, million dollars	7204.335	8232.172	9418.773
Net operating income for the year before taxing, NOI, million dollars	2161.3	2469.174	2826
Equity cost at zero leverage, k_0, %	23.67	23.67	23.67
Debt cost $\overline{k}_d$, %	8.26	7.4	6.69
Return on investment for 1 year, $\beta = I/\text{NOI}$	1.92	1.204	1.02
Amount of debt financing, %	35	50	50
Amount of equity, %	65	50	50
Leverage level, L	0.54	1	1
Amount of equity capital S, million dollars	730.6	1025	1381.5

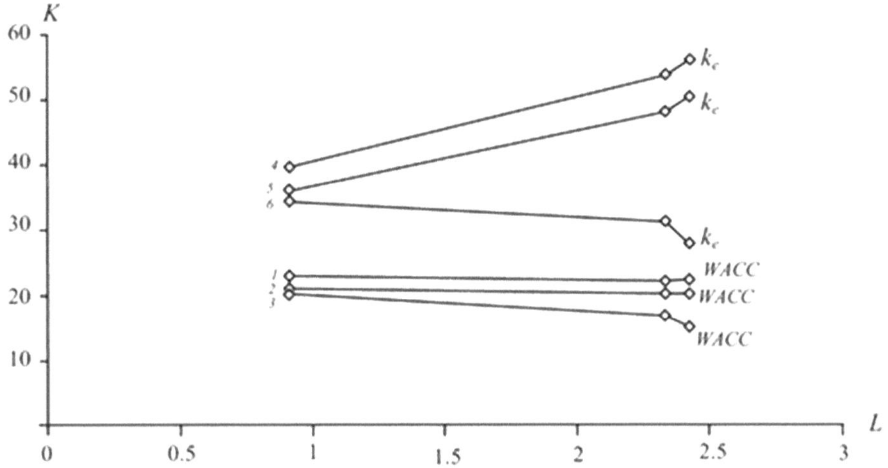

Fig. 30.1 Dependence of weighted average cost of capital WACC and the equity cost k_e on leverage: *1*, *4* within Brusov–Filatova–Orekhova theory; *2*, *5* within Modigliani–Miller theory; *3*, *6* within traditional approach

30.2.1 The Dependence of NPV on Investment Capital Structure

Analysis of investment will be continued with use of the formula provided in the works (Brusov & Filatova, 2011; Brusov et al., 2011a, b, c, 2012a, b, 2013a, b, 2014a, b; Filatova et al., 2008):

$$\begin{aligned} \mathrm{NPV} = & -S\left\{1 + L\left[(1-t)\left(1 - \frac{1}{(1+k_{\mathrm{d}})^n}\right) + \frac{1}{(1+k_{\mathrm{d}})^n}\right]\right\} \\ & + \frac{\beta S(1+L)(1-t)}{k_{\mathrm{e}}}\left[1 - \frac{1}{(1+k_{\mathrm{e}})^n}\right], \end{aligned} \tag{30.1}$$

Where:

NPV—net present value
S—equity capital amount
L—leverage level
t—tax on profit rate
k_{d}—debt cost
n—project duration
β—return on investment for 1 year
k_{e}—equity cost

Analysis of Investments in 2010 Using company's data (Table 30.1), we compute the WACC, k_{e}, and NPV (Tables 30.2, 30.3, and 30.4; Figs. 30.1 and 30.2).

In the company's investment in 2010, equity capital accounted for 65%, and debt 35%, i.e., the leverage level was equal to $L = 0.54$. The term for hardware depreciation, into establishment of which investment was done, was 5 years. Curve 2 (Fig. 30.2) corresponds to dependence of NPV on leverage level for the 5-year project. Optimum NPV is achieved when $L = 2$. At this leverage level, NPV = 3624.5 million dollars.

The level leverage with which "Nastcom Plus" has carried out its investment projects ($L = 0.54$) was lying far from optimum and provided NPV = 2979.2 million dollars, which is approximately 645 million dollars less than the optimal value of NPV.

Since the equipment can be operated, after depreciation, one can estimate the NPV for a more long-term perspective. For example, for the 7-year project, the optimal leverage value is $L = 2.0$, when the NPV = 4157.6 million dollars, which is 562.8 million dollars more than nonoptimal values of NPV = 3594.8 million dollars, obtained by the company. For the 10-year project, the optimal leverage level is $L = 1.5$, with NPV = 4509.1 million dollars, which is 422.9 million dollars more than nonoptimal values of NPV = 4086.2 million dollars, obtained by the company (Fig. 30.2; Tables 30.2, 30.3, and 30.4).

Analysis of Investments in 2011 Using company's data, we compute the WACC, k_{e}, and NPV (Tables 30.5, 30.6, and 30.7; Fig. 30.3).

In the company's investment in 2011, equity capital accounted for 50%, and debt capital for 50% as well, i.e., the leverage level was equal to $L = 1$. The term for hardware depreciation, into establishment of which investment was done, was 5 years. Curve 2 (Fig. 30.3) corresponds to dependence of NPV on leverage level for the 5-year project. Optimum NPV is achieved approximately at $L = 1$. More

Table 30.2 Weighted average cost of capital, WACC, in 2010

	Leverage level *L*										
Project duration *n*	0	0.5	1	1.5	2	2.5	3	3.5	4	4.5	5
2	0.2367	0.2284	0.2242	0.2217	0.2201	0.2189	0.2180	0.2173	0.2167	0.2163	0.2159
5	0.2367	0.2262	0.2209	0.2177	0.2156	0.2141	0.2130	0.2121	0.2114	0.2108	0.2103
7	0.2367	0.2256	0.2199	0.2166	0.2143	0.2127	0.2116	0.2106	0.2099	0.2093	0.2087
10	0.2366	0.2250	0.2190	0.2155	0.2131	0.2114	0.2101	0.2091	0.2083	0.2076	0.2071

Table 30.3 Equity cost, k_e, in 2010

Project duration n	Leverage level L										
	0	0.5	1	1.5	2	2.5	3	3.5	4	4.5	5
2	0.2367	0.3096	0.3824	0.4552	0.5280	0.6008	0.6736	0.7464	0.8192	0.8920	0.9649
5	0.2367	0.3063	0.3758	0.4452	0.5147	0.5841	0.6536	0.7230	0.7925	0.8619	0.9314
7	0.2367	0.3053	0.3738	0.4423	0.5107	0.5791	0.6480	0.7165	0.7850	0.8535	0.9220
10	0.2366	0.3044	0.3720	0.4396	0.5071	0.5746	0.6422	0.7097	0.7772	0.8447	0.9122

Table 30.4 Net present value, NPV, in 2010, million dollars

Project duration *n*	Leverage level *L*										
	0	0.5	1	1.5	2	2.5	3	3.5	4	4.5	5
2	910.5	1181.8	1358.3	1458.5	1496.3	1482.8	1426.6	1334.6	1212.4	1064.5	894.5
5	2371.6	2979.2	3347.7	3547.1	3624.5	3612.4	3533.6	3404.2	3236.1	3038.0	2816.2
7	2939.0	3594.8	3955.1	4121.6	4157.6	4103.6	3981.6	3821.6	3619.9	3397.1	3155.1
10	3444.5	4086.2	4396.9	4509.1	4497.4	4405.3	4258.9	4074.9	3863.8	3632.7	3386.5

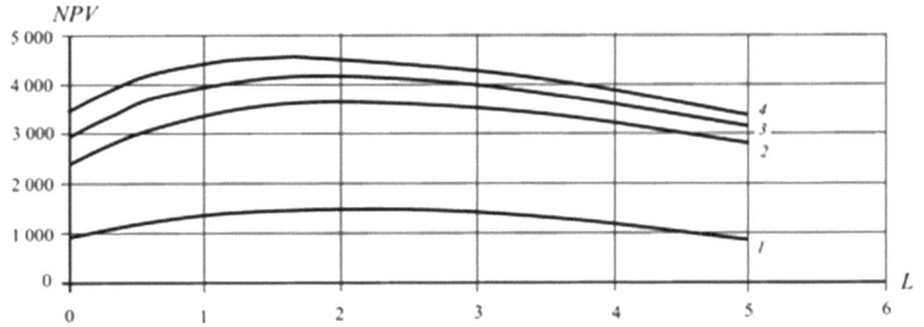

Fig. 30.2 Dependence of NPV on leverage L at $t = 20\%$ in 2010: *1* for 2-year income from investments; *2* for 5-year income from investments (the term for hardware depreciation); *3* for 7 years of investment income; *4* for 10 years of investment income

accurate calculations show that the optimal value of NPV = 2133.7 million dollars is achieved when $L = 1.1$.

The level leverage with which "Nastcom Plus" has carried out its investment projects ($L = 0.1$) was lying in the vicinity of optimum and provided NPV = 2131.8 million dollars, which is just two million dollars less than the optimal value of NPV. You can take it that, in 2011, investment in "Nastcom Plus" company has been carried out with almost optimal structure.

Analysis of Investments in 2012 Using company's data, we compute the WACC, k_e, and NPV (Tables 30.8, 30.9, and 30.10; Fig. 30.4).

The company's investment structure in 2012 was the same as in 2011: equity capital accounted for 50%, and debt capital for 50%, i.e., the leverage level was equal to $L = 1$. The term hardware for depreciation, into establishment of which investment was done, was 5 years. Curve 2 (Fig. 30.4) corresponds to dependence of NPV on leverage level for the 5-year project. Optimum NPV is achieved when $L = 0.5$. More accurate calculations show that the optimal value of NPV = 1987.7 million dollars is achieved when $L = 0.7$ (Table 30.11, Fig. 30.5).

The level leverage, with which "Nastcom Plus" has carried out its investment projects ($L = 0.1$), did not correspond to optimum value ($L = 0.7$) and provided NPV = 1954.6 million dollars, which is 31.7 million dollars less than the optimal value of NPV (Figs. 30.4 and 30.5; Tables 30.10 and 30.11).

Since the equipment can be operated after depreciation, one can estimate the NPV for a more long-term perspective. For example, for the 7-year project, the optimal leverage value is $L = 0.65$, when the NPV = 2580.6 million dollars, which is 51.7 million dollars more than nonoptimal values of NPV = 2528.9 million dollars, obtained by the company. For the 10-year project, the optimal leverage level $L = 0.55$, with NPV = 3043.5 million dollars, which is 98 million dollars more than nonoptimal values of NPV = 2945.5 million dollars, obtained by the company.

Table 30.5 Weighted average cost of capital, WACC, in 2011

Project duration n	Leverage level L										
	0	0.5	1	1.5	2	2.5	3	3.5	4	4.5	5
2	0.2367	0.3096	0.3824	0.4552	0.5280	0.6008	0.6736	0.7464	0.8192	0.8920	0.9649
5	0.2367	0.3063	0.3758	0.4452	0.5147	0.5841	0.6536	0.7230	0.7925	0.8619	0.9314
7	0.2367	0.3053	0.3738	0.4423	0.5107	0.5791	0.6480	0.7165	0.7850	0.8535	0.9220
10	0.2366	0.3044	0.3720	0.4396	0.5071	0.5746	0.6422	0.7097	0.7772	0.8447	0.9122

Table 30.6 Equity cost, k_e, in 2011

	Leverage level L										
Project duration n	0	0.5	1	1.5	2	2.5	3	3.5	4	4.5	5
2	0.2367	0.3142	0.3916	0.4690	0.5464	0.6239	0.7013	0.7787	0.8561	0.9335	1.0110
5	0.2367	0.3110	0.3853	0.4595	0.5338	0.6080	0.6822	0.7565	0.8307	0.9049	0.9791
7	0.2367	0.3100	0.3833	0.4565	0.5297	0.6029	0.6761	0.7493	0.8230	0.8963	0.9695
10	0.2366	0.3091	0.3813	0.4536	0.5258	0.5980	0.6702	0.7424	0.8146	0.8868	0.9589

Table 30.7 Net present value, NPV, of the company in 2011, million dollars

Project duration n	Leverage level L										
	0	0.5	1	1.5	2	2.5	3	3.5	4	4.5	5
2	418.8	460.5	415.8	302.3	133.3	−80.9	−332.5	−615.0	−923.5	−1254.0	−1603.1
5	1704.2	2025.4	2131.8	2089.9	1943.2	1721.0	1443.5	1124.7	774.8	400.9	8.5
7	2203.4	2558.0	2650.5	2576.1	2392.2	2134.4	1825.6	1480.7	1105.7	715.0	309.8
10	2648.2	2982.3	3028.1	2907.1	2685.2	2399.5	2071.6	1715.0	1338.0	946.0	542.9

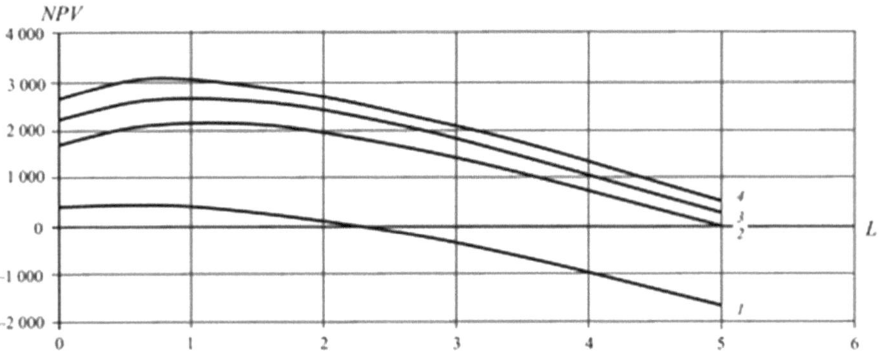

Fig. 30.3 Dependence of NPV on leverage L at t = 20% in 2011: *1* for 2-year income from investments; *2* for 5-year income from investments (the term for hardware depreciation); *3* for 7 years of investment income; *4* for 10 years of investment income

30.2.2 The Dependence of NPV on the Equity Capital Value and Coefficient β

Let us investigate the dependence of NPV on the equity capital value and coefficient β (Tables 30.12, 30.13, 30.14, and 30.15; Figs. 30.6, 30.7, 30.8, and 30.9).

With increase of the equity value, optimum is observed for all of the values S when leverage level is approximately equal to $L = 0.7$, and the optimum value as well as the NPV value is growing with increasing S, as long as the project remains effective (up to the leverage level approximately $L = 3.7$).

With the decrease of the return on investment ($\beta = 0.5$), the dependence of the NPV on leverage changes significantly: now NPV monotonically decreases with the leverage at all values of equity capital S (Fig. 30.7).

With the increase of the return on investment ($\beta = 1.5$), the NPV of the project has an optimum at all values of equity capital S at leverage level $L = 1.5$, and NPV (L) curve is going up with the increase in S until the project remains effective (up to the leverage level approximately $L = 7$). The optimum position (value L_0) almost does not depend on the equity value S ($L_0 = 1.5$). This means the possibility of a tabulation of obtained results by the known values k_0 and k_d for large values $\beta \geq 1$, when in dependence of NPV(L) there is an optimum.

With the further increase of the return on investment ($\beta = 2$), NPV of the project has an optimum for all values S at leverage level which is already approximately equal to $L = 1.8$, and the value of optimum as well as of NPV in general is growing with increasing S, until the project remains an effective (up to leverage level of order $L = 8.5$, see Fig. 30.9).

In this way, the analysis of the dependence of NPV on the equity value S and on the return on investment β allows us to conclude that in contrast to the parameters I and NOI, the change of parameters S and β, both individually and simultaneously, can significantly change the nature of the dependence of NPV on leverage level.

Table 30.8 Weighted average cost of capital, WACC, in 2012

	Leverage level L										
Project duration n	0	0.5	1	1.5	2	2.5	3	3.5	4	4.5	5
2	0.2367	0.2298	0.2264	0.2243	0.2229	0.2219	0.2212	0.2206	0.2202	0.2198	0.2195
5	0.2367	0.2278	0.2234	0.2207	0.2189	0.2176	0.2167	0.2159	0.2153	0.2148	0.2144
7	0.2367	0.2272	0.2224	0.2195	0.2176	0.2162	0.2152	0.2143	0.2137	0.2132	0.2127
10	0.2366	0.2265	0.2213	0.2183	0.2162	0.2147	0.2136	0.2127	0.2120	0.2115	0.2110

Table 30.9 Equity cost, k_e, in 2012

Project duration n	Leverage level L										
	0	0.5	1	1.5	2	2.5	3	3.5	4	4.5	5
2	0.2367	0.3180	0.3992	0.4805	0.5617	0.6430	0.7242	0.8055	0.8867	0.9680	1.0492
5	0.2367	0.3150	0.3933	0.4715	0.5497	0.6279	0.7062	0.7844	0.8626	0.9408	1.0190
7	0.2367	0.3140	0.3913	0.4685	0.5457	0.6229	0.7001	0.7772	0.8544	0.9316	1.0088
10	0.2366	0.3130	0.3892	0.4653	0.5415	0.6177	0.6938	0.7699	0.8461	0.9222	0.9983

Table 30.10 Net present value, NPV, in 2012, million dollars

Project duration n	Leverage level L										
	0	0.5	1	1.5	2	2.5	3	3.5	4	4.5	5
2	267.0	200.9	33.4	−214.0	−525.3	−888.3	−1293.4	−1733.4	−2202.4	−2695.8	−3209.9
5	1734.8	1968.9	1954.6	1771.8	1472.1	1089.6	647.1	160.6	−358.9	−903.4	−1467.2
7	2304.7	2566.9	2528.9	2304.4	1960.4	1537.2	1060.3	545.9	4.7	−556.0	−1131.2
10	2812.6	3041.6	2945.5	2667.4	2281.9	1829.9	1335.0	811.1	266.8	−292.2	−862.1

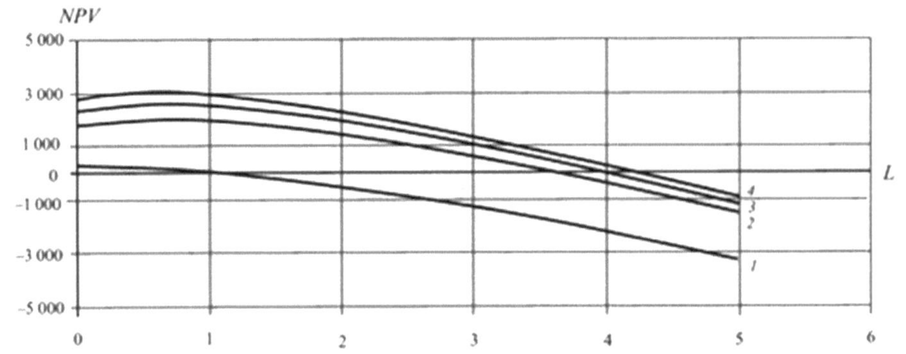

Fig. 30.4 Dependence of NPV on leverage L at $t = 20\%$ in 2012: *1* for 2-year income from investments; *2* for 5-year income from investments (the term for hardware depreciation); *3* for 7 years of investment income; *4* for 10 years of investment income

With the increase of return on investment (with the increased β), there is a transition from the monotonic decrease of NPV of the project with leverage (Fig. 30.7) to the existing of the optimum at all values of S (see Figs. 30.6 and 30.8). The growth of β leads to the growth of NPV as well as to the growth of the limit leverage value, up to which the project remains effective.

This means the inability of tabulation of the results, obtained in the general case of a constant value of equity capital; in this case, it is necessary to use the formulas obtained by authors in their works (Brusov & Filatova, 2011; Brusov et al., 2011a, b, c, 2012a, b, 2013a, b, 2014a, b; Filatova et al., 2008) to determine the NPV at the existing leverage level as well as for optimization of existing investment structure. Tabulation is possible only in the case of large β values ($\beta \geq 1$), when there is an optimum in the dependence of NPV on the leverage level.

30.3 Effects of Taxation on the Optimal Capital Structure of Companies in the Telecommunication Sector

We continue the analysis of activity of "Nastcom Plus." In this paragraph, we examine the effect of change of tax on profit rate, both in the case of its increase and decrease, on the optimal structure of investments at different project durations.

It is shown that the increase of tax on profit rate leads to the degradation (decreasing) of NPV, and the degradation is reduced with an increase of project duration. In particular, for the 5-year project (amortization period), when the tax on profit rate decreases on 1%, NPV decreases on 1.5–2.34% in different years.

The impact of change of tax on the profit rate on the optimum position while it exists (change optimum position) is changed for 2-year and 10-year projects in 2010 and for the 5-year project in the 2012 year [for 0.5–1 (in L units)]; nevertheless the optimum position turns out to be sufficiently stable (Figs. 30.10, 30.11, 30.12, 30.13,

Table 30.11 Net present value, NPV, at leverage level from 0.5 up to 1.05 in 2012, million dollars

Project duration n	Leverage level L											
	0.5	0.55	0.6	0.65	0.7	0.75	0.8	0.85	0.9	0.95	1	1.05
2	200.9	188.3	174.8	160.2	144.8	128.4	111.1	92.9	73.9	54.1	33.4	12.0
5	1968.9	1977.0	1982.8	1986.3	1987.7	1987.0	1984.2	1979.5	1973.0	1964.6	1954.6	1942.8
7	2566.9	2574.2	2578.8	2580.6	2580.0	2576.9	2571.5	2563.9	2554.2	2542.5	2528.9	2513.5
10	3041.6	3043.5	3042.4	3038.5	3032.0	3023.1	3011.7	2998.2	2982.6	2965.0	2945.5	2924.3

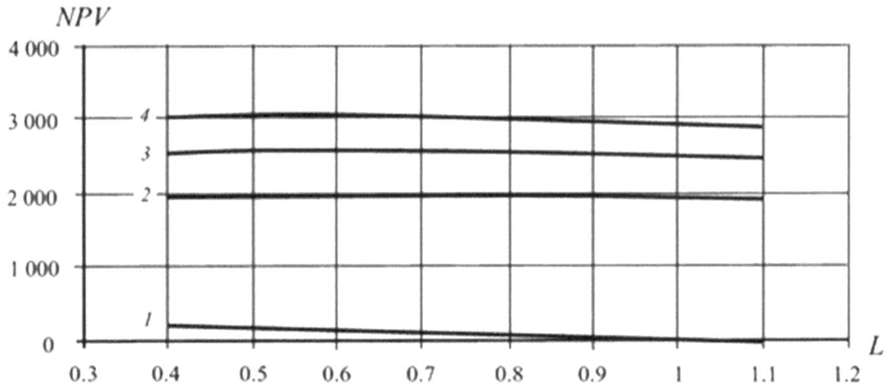

Fig. 30.5 Dependence of NPV on leverage L at $t = 20\%$ in 2012 in the vicinity of optimum: *1* for 2-year income from investments; *2* for 5-year income from investments (the term for hardware depreciation); *3* for 7 years of investment income; *4* for 10 years of investment income

30.14, 30.15, 30.16, 30.17, 30.18, 30.19, 30.20, and 30.21; Tables 30.16, 30.17, 30.18, 30.19, 30.20, and 30.21).

It is shown that the increase of tax on profit rate leads to the degradation (decreasing) of NPV, and the degradation is reduced with an increase of project duration. Note that for the 5-year project (amortization period), when the tax on profit rate decreases by 1%, NPV decreases by 2.34% in 2012, by 2.04% in 2011, and by 1.49% in 2010.

30.4 Conclusions

In 2010, the company "Nastcom Plus" worked at leverage level $L = 0.54$ instead of optimal value $L = 2.0$. The NPV loss amounted to 645 million dollars.

In 2012, the company worked at leverage level $L = 1.0$ instead of optimal value $L = 0.7$. The NPV loss amounted from 32 to 98 million dollars, depending on the term of operation of equipment.

The authors have evaluated effectiveness of investment at existing level of debt financing and have developed recommendations on the optimum level of leverage for the Russian company "Nastcom Plus" in 2010–2012.

The results indicate that if in 2011 the financial structure of the investment of "Nastcom Plus" was close to the optimal and NPV was only two million dollars less than the optimal value, in 2010, when the leverage level was $L = 0.54$, NPV was 645 million dollars less than optimal value (the optimal leverage level should be equal to $L = 2$).

In 2012 the leverage level, with which "Nastcom Plus" has carried out its investment projects ($L = 0.1$), did not correspond to optimum value ($L = 0.7$) and

Table 30.12 Net present value, NPV, at $\beta = 1.02$, million dollars

	Leverage level *L*										
Equity *S*	0	0.5	1	1.5	2	2.5	3	3.5	4	4.5	5
1382	1734.8	1968.9	1954.6	1771.8	1472.1	1089.6	647.1	160.6	−358.9	−903.4	−1467.2
1000	1255.7	1425.2	1414.8	1282.5	1065.6	788.7	468.4	116.2	−259.8	−653.9	−1062.0
1200	1506.8	1710.2	1697.8	1539.0	1278.7	946.4	562.1	139.5	−311.8	−784.7	−1274.4
1600	2009.1	2280.3	2263.7	2052.0	1705.0	1261.9	749.5	186.0	−415.7	−1046.3	−1699.2
1800	2260.3	2565.3	2546.7	2308.5	1918.1	1419.7	843.2	209.2	−467.6	−1177.1	−1911.6

Table 30.13 Net present value, NPV, at $\beta = 0.5$, million dollars

Equity S	Leverage level L										
	0	0.5	1	1.5	2	2.5	3	3.5	4	4.5	5
1382	146.1	−71.8	−411.5	−833.8	−1313.3	−1833.5	−2383.1	−2954.2	−3541.6	−4141.1	−4750.2
1000	105.7	−52.0	−297.9	−603.5	−950.7	−1327.2	−1725.0	−2138.4	−2563.6	−2997.6	−3438.4
1200	126.9	−62.4	−357.4	−724.2	−1140.8	−1592.6	−2070.0	−2566.1	−3076.3	−3597.1	−4126.1
1600	169.2	−83.2	−476.6	−965.6	−1521.0	−2123.5	−2760.0	−3421.5	−4101.7	−4796.1	−5501.4
1800	190.3	−93.6	−536.2	−1086.3	−1711.2	−2388.9	−3105.0	−3849.2	−4614.4	−5395.6	−6189.1

Table 30.14 Net present value, NPV, at $\beta = 1.5$, million dollars

Equity S	Leverage level L										
	0	0.5	1	1.5	2	2.5	3	3.5	4	4.5	5
1382	3201.2	3852.6	4138.6	4176.9	4043.3	3787.8	3444.2	3035.8	2578.9	2085.3	1563.3
1000	2330.2	2788.7	2995.8	3023.5	2926.8	2741.8	2493.1	2197.5	1866.7	1509.4	1131.6
1200	2780.7	3346.5	3594.9	3628.2	3512.1	3290.2	2991.7	2637.0	2240.1	1811.3	1357.9
1600	3707.5	4462.0	4793.2	4837.6	4682.8	4386.9	3989.0	3515.9	2986.8	2415.1	1810.6
1800	4171.0	5019.7	5392.4	5442.3	5268.2	4935.3	4487.6	3955.4	3360.1	2716.9	2036.9

Table 30.15 Net present value, NPV, at $\beta = 2.0$, million dollars

Equity S	Leverage level L										
	0	0.5	1	1.5	2	2.5	3	3.5	4	4.5	5
1382	4728.8	5814.9	6413.7	6682.3	6721.7	6598.5	6357.9	6030.8	5639.2	5198.5	4720.0
1000	3423.0	4209.1	4642.6	4837.0	4865.5	4776.3	4602.2	4365.4	4081.9	3762.9	3416.6
1200	4107.5	5050.9	5571.1	5804.4	5838.6	5731.6	5522.6	5238.5	4898.3	4515.5	4099.9
1600	5476.7	6734.5	7428.1	7739.2	7784.8	7642.1	7363.4	6984.7	6531.1	6020.6	5466.6
1800	6161.3	7576.4	8356.6	8706.6	8757.9	8597.4	8283.9	7857.8	7347.4	6773.2	6149.9

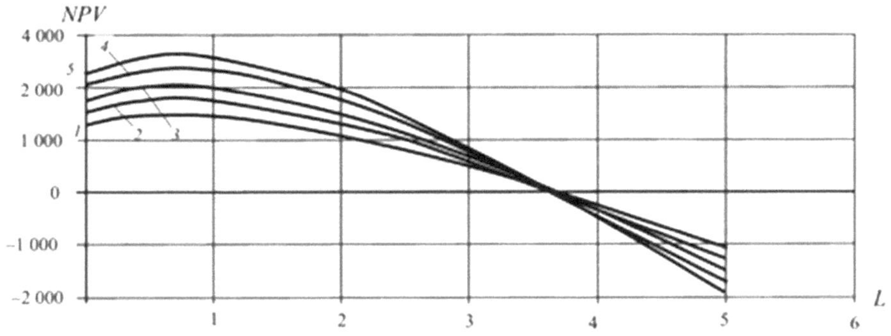

Fig. 30.6 Dependence of NPV on leverage at different values of equity cost S at $\beta = 1.02$, million dollars: *1* $S = 1382.2$ million dollars; *2* $S = 1000$ million dollars; *3* $S = 1200$ million dollars; *4* $S = 1600$ million dollars; *5* $S = 1800$ million dollars

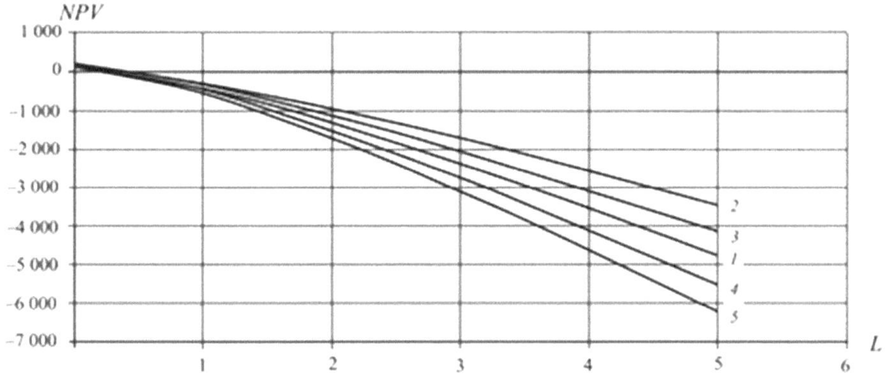

Fig. 30.7 Dependence of NPV on leverage at different values of equity cost S at $\beta = 0.5$, million dollars: *1* $S = 1382.2$ million dollars; *2* $S = 1000$ million dollars; *3* $S = 1200$ million dollars; *4* $S = 1600$ million dollars; *5* $S = 1800$ million dollars

provided NPV = 1954.6 million dollars, which is 31.7 million dollars less than the optimal value of NPV (Tables 30.10 and 30.11; Figs. 30.4 and 30.5).

Since the equipment can be operated after depreciation, one can estimate the NPV for a more long-term perspective. For example, for the 7-year project, the optimal leverage value is $L = 0.65$, when the NPV = 2580.6 million dollars, which is 51.7 million dollars more than nonoptimal values of NPV = 2528.9 million dollars, obtained by the company. For the 10-year project, the optimal leverage level $L = 0.55$, with NPV = 3043.5 million dollars, which is 98 million dollars more than nonoptimal values of NPV = 2945.5 million dollars, obtained by the company.

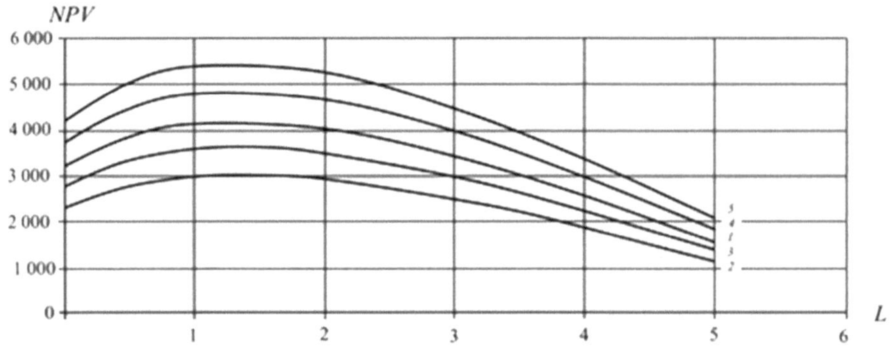

Fig. 30.8 Dependence of NPV on leverage at different values of equity cost *S* at $\beta = 1.5$, million dollars: *1* $S = 1382.2$ million dollars; *2* $S = 1000$ million dollars; *3* $S = 1200$ million dollars; *4* $S = 1600$ million dollars; *5* $S = 1800$ million dollars

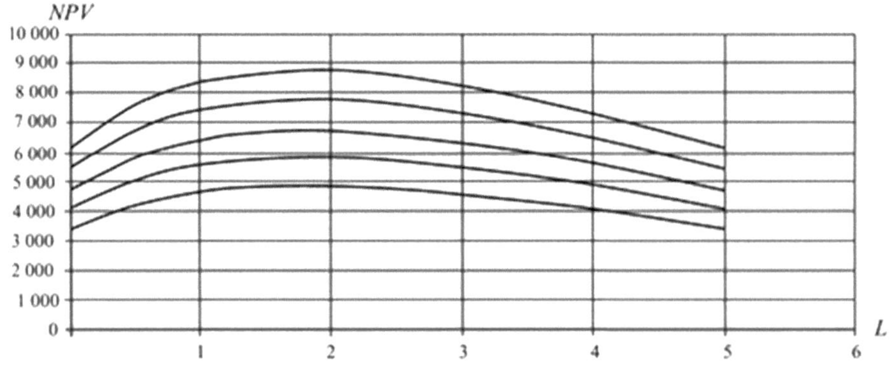

Fig. 30.9 Dependence of NPV on leverage at different values of equity cost *S* at $\beta = 2$, million dollars: *1* $S = 1382.2$ million dollars; *2* $S = 1000$ million dollars; *3* $S = 1200$ million dollars; *4* $S = 1600$ million dollars; *5* $S = 1800$ million dollars

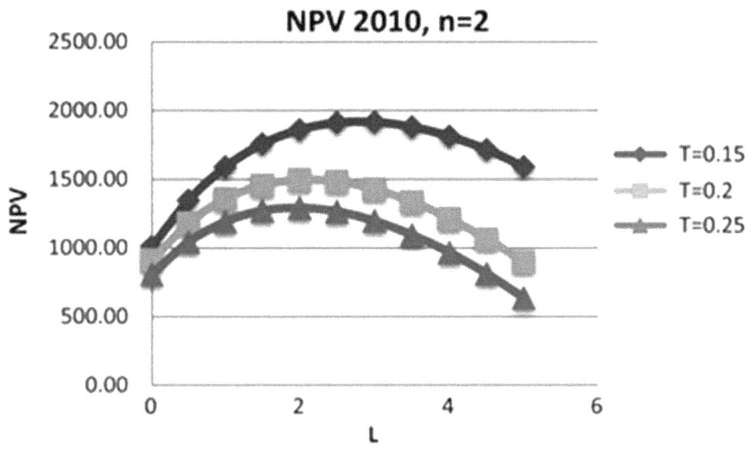

Fig. 30.10 Dependence of NPV on leverage level L for three values of tax on profit rate: $T = 15\%$, 20%, 25% for the 2-year project in 2010

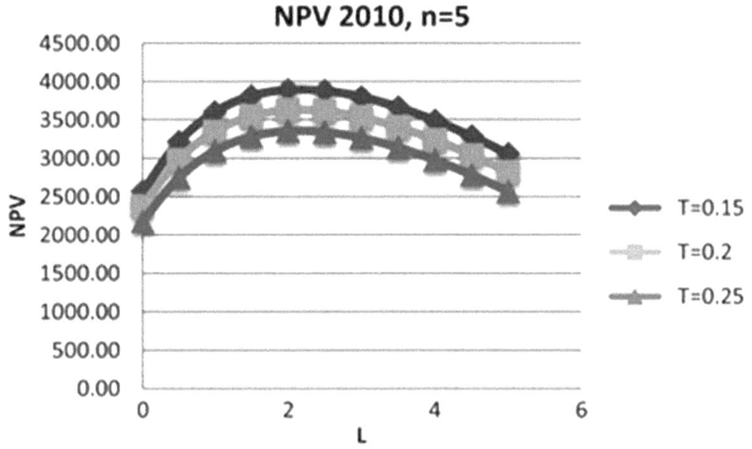

Fig. 30.11 Dependence of NPV on leverage level L for three values of tax on profit rate: $T = 15\%$, 20%, 25% for the 5-year project in 2010

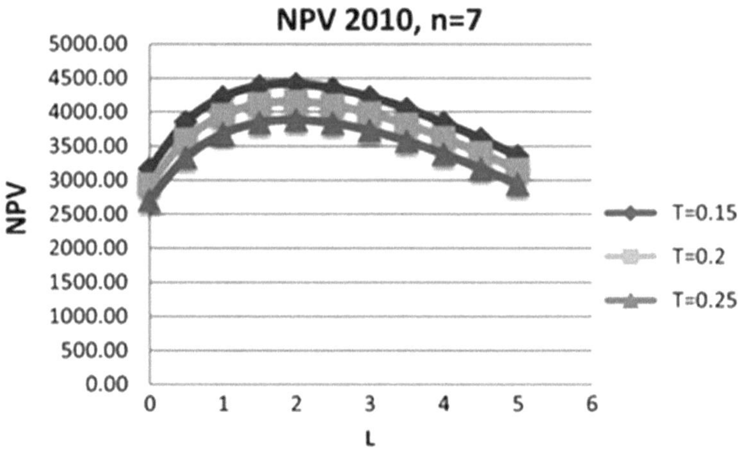

Fig. 30.12 Dependence of NPV on leverage level L for three values of tax on profit rate: $T = 15\%$, 20%, 25% for the 7-year project in 2010

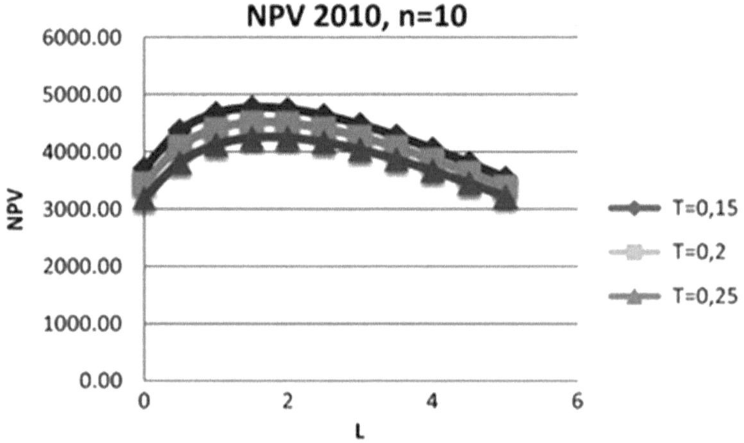

Fig. 30.13 Dependence of NPV on leverage level L for three values of tax on profit rate: $T = 15\%$, 20%, 25% for the 10-year project in 2010

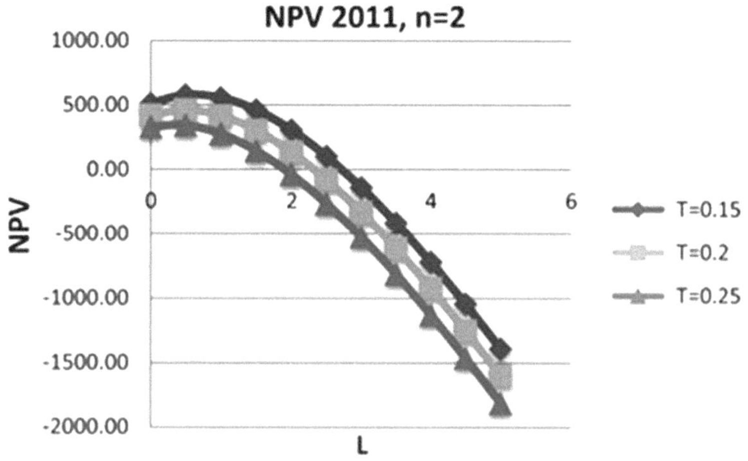

Fig. 30.14 Dependence of NPV on leverage level L for three values of tax on profit rate: $T = 15\%$, 20%, 25% for the 2-year project in 2011

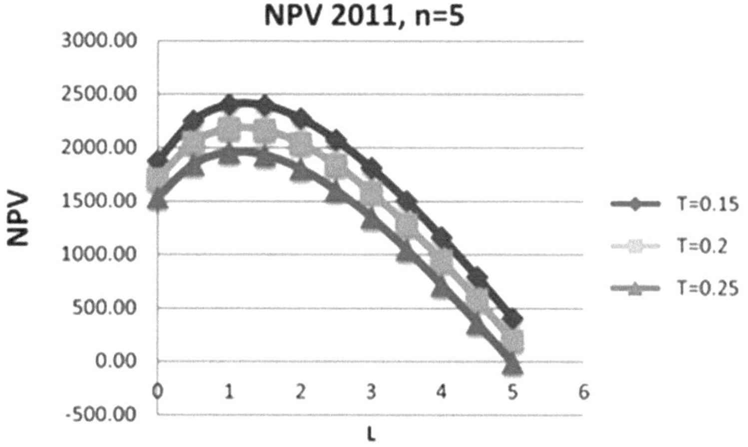

Fig. 30.15 Dependence of NPV on leverage level L for three values of tax on profit rate: $T = 15\%$, 20%, 25% for the 5-year project in 2011

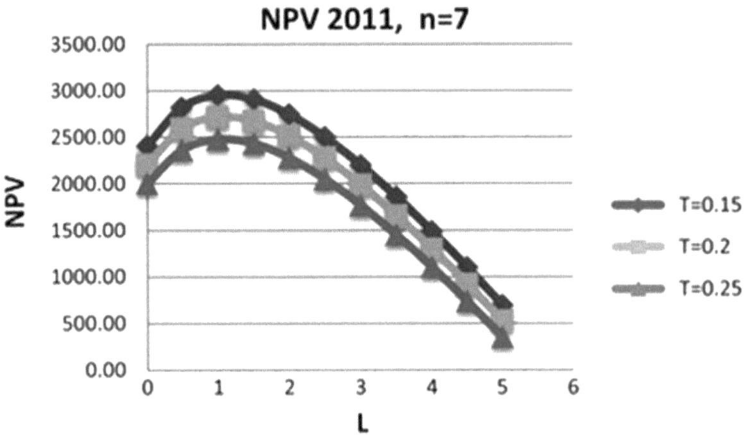

Fig. 30.16 Dependence of NPV on leverage level L for three values of tax on profit rate: $T = 15\%$, 20%, 25% for the 7-year project in 2011

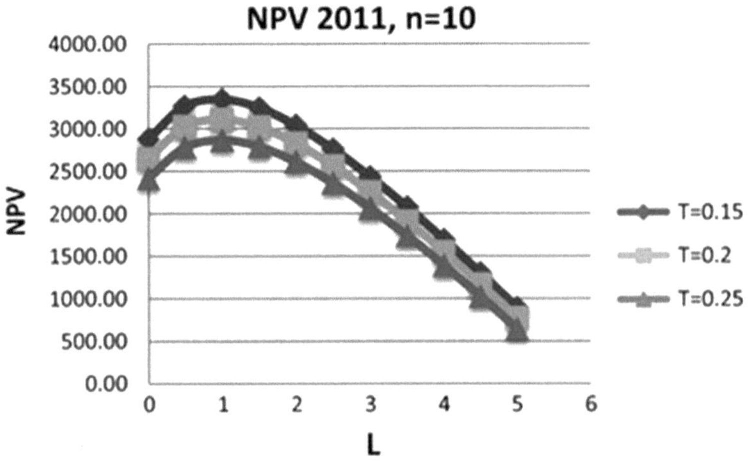

Fig. 30.17 Dependence of NPV on leverage level L for three values of tax on profit rate: $T = 15\%$, 20%, 25% for the 10-year project in 2011

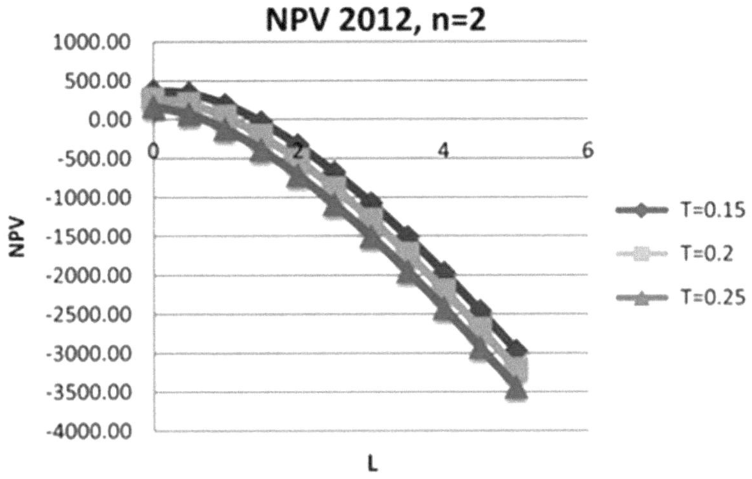

Fig. 30.18 Dependence of NPV on leverage level L for three values of tax on profit rate: $T = 15\%$, 20%, 25% for the 2-year project in 2012

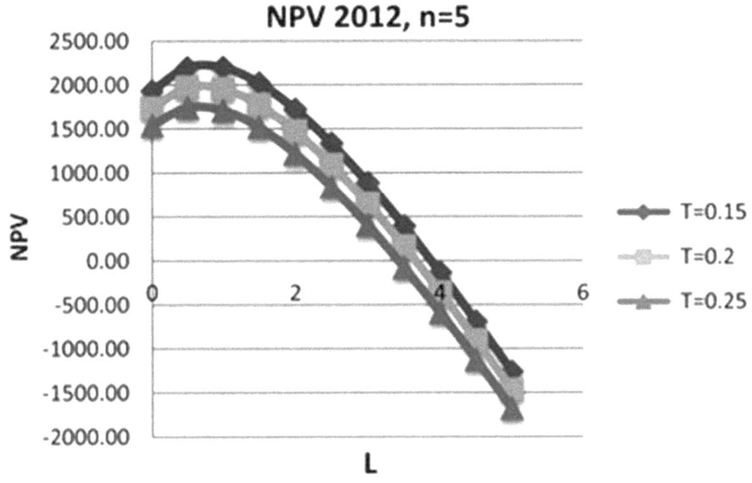

Fig. 30.19 Dependence of NPV on leverage level L for three values of tax on profit rate: $T = 15\%$, 20%, 25% for the 5-year project in 2012

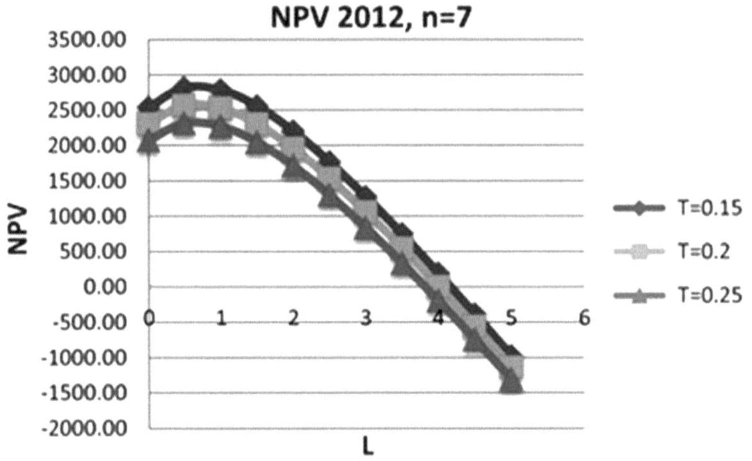

Fig. 30.20 Dependence of NPV on leverage level L for three values of tax on profit rate: $T = 15\%$, 20%, 25% for the 7-year project in 2012

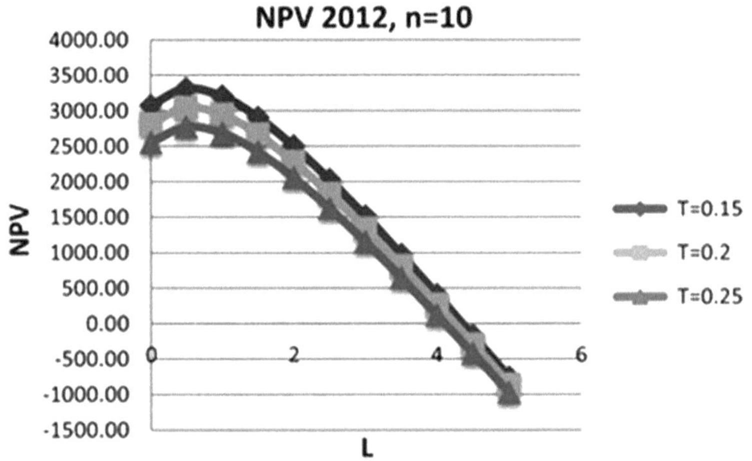

Fig. 30.21 Dependence of NPV on leverage level L for three values of tax on profit rate: $T = 15\%$, 20%, 25% for the 10-year project in 2012

Table 30.16 Dependence of optimum position L_0 of investment structure in 2010 on tax on profit rate and project duration

	n			
2010, *t*	2	5	7	10
15%	3	2	2	2
20%	2	2	2	1.5
25%	2	2	2	2

Table 30.17 Absolute and relative change of NPV with increase/decrease of tax on profit rate on 5% (1%) in 2010

2010	$t = 20–15\%$	$t = 25–20\%$	On 5%	On 1%
$n = 2$	418/487	204	34%	6.8%
5	266	273	7.45%	1.49%
7	263	271	6.88%	1.37%
10	245	256	5.56%	1.11%

Table 30.18 Dependence of optimum position L_0 of investment structure in 2011 on tax on profit rate and project duration

	n			
2011, t	2	5	7	10
15%	0.5	1	1	1
20%	0.5	1	1	1
25%	0.5	1	1	1

Table 30.19 Absolute and relative change of NPV with increase/decrease of tax on profit rate on 5% (1%) in 2011

2011	$t = 20–15\%$	$t = 25–20\%$	On 5%	On 1%
$n = 2$	119	118	25.8%	5.16%
5	222	223	10.2%	2.04%
7	236	239	8.75%	1.75%
10	238	242	7.7%	1.54%

Table 30.20 Dependence of optimum position L_0 of investment structure in 2010 on tax on profit rate and project duration

	n			
2012, t	2	5	7	10
15%	0	1	0.5	0.5
20%	0	0.5	0.5	0.5
25%	0	0.5	0.5	0.5

Table 30.21 Absolute and relative change of NPV with increase/decrease of tax on profit rate on 5% (1%) in 2012

2012	$t = 20–15\%$	$t = 25–20\%$	On 5%	On 1%
$n=2$	104	103	39%	7.8%
5	229	229	11.7%	2.34%
7	253	256	9.9%	1.98%
10	269	271	8.9%	1.74%

References

Brusov, P. N., & Filatova, T. V. (2011). From Modigliani–Miller to general theory of capital cost and capital structure of the company. *Finance and Credit, 435*, 2–8.

Brusov, P., Filatova, T., Orehova, N., & Brusova, A. (2011a). Weighted average cost of capital in the theory of Modigliani–Miller, modified for a finite life–time company. *Applied Financial Economics, 21*(11), 815–824.

Brusov, P., Filatova, T., Orehova, N., et al. (2011b). From Modigliani–Miller to general theory of capital cost and capital structure of the company. *Research Journal of Economics, Business and ICT, 2*, 16–21.

Brusov, P., Filatova, T., Orehova, N., et al. (2011c). Influence of debt financing on the effectiveness of the investment project within the Modigliani–Miller theory. *Research Journal of Economics, Business and ICT, 2*, 11–15.

Brusov, P., Filatova, T., Eskindarov, M., & Orehova, N. (2012a). Influence of debt financing on the effectiveness of the finite duration investment project. *Applied Financial Economics, 22*(13), 1043–1052.

Brusov, P., Filatova, T., Eskindarov, M., & Orehova, N. (2012b). Hidden global causes of the global financial crisis. *Journal of Reviews on Global Economics, 1*, 106–111.

Brusov, P., Filatova, P., & Orekhova, N. (2013a). Absence of an optimal capital structure in the famous tradeoff theory! *Journal of Reviews on Global Economics, 2*, 94–116.

Brusov, P., Filatova, T., & Orehova, N. (2013b). A qualitatively new effect in corporative finance: Abnormal dependence of cost of equity of company on leverage. *Journal of Reviews on Global Economics, 2*, 183–193.

Brusov, P., Filatova, P., & Orekhova, N. (2014a). Mechanism of formation of the company optimal capital structure, different from suggested by trade off theory. *Cogent Economics & Finance, 2*, 1–13. https://doi.org/10.1080/23322039.2014.946150

Brusov, P., Filatova, T., & Orehova, N. (2014b). Inflation in Brusov–Filatova–Orekhova theory and in its perpetuity limit – Modigliani – Miller theory. *Journal of Reviews on Global Economics, 3*, 175–185.

Brusov, P., Filatova, T., Orehova, N., & Eskindarov, M. (2015). *Modern corporate finance, investment and taxation* (1st ed., pp. 1–368). Springer International Publishing.

Brusov, P., Filatova, T., Orehova, N., & Kulk, W. I. (2018a). New meaningful effects in modern capital structure theory. *Journal of Reviews on Global Economics, 7*, 104–122.

Brusov, P., Filatova, T., Orehova, N., & Kulk, V. (2018b). A "golden age" of the companies: Conditions of its existence. *Journal of Reviews on Global Economics, 7*, 88–103.

Brusov, P., Filatova, T., Orehova, N., & Kulk, V. (2018c). Rating methodology: New look and new horizons. *Journal of Reviews on Global Economics, 7*, 63–87.

Brusov, P., Filatova, T., Orehova, N., & Kulk, V. (2018d). Rating: New approach. *Journal of Reviews on Global Economics, 7*, 37–62.

Brusova, A. (2011). A comparison of the three methods of estimation of weighted average cost of capital and equity cost of company. *Financial Analysis Problem Solution, 34*(76), 36–42.

Filatova, T., Orehova, N., & Brusova, A. (2008). Weighted average cost of capital in the theory of Modigliani–Miller, modified for a finite life–time company. *Bulletin of the FU, 48*, 68–77.

Modigliani, F., & Miller, M. (1958). The cost of capital, corporate finance, and the theory of investment. *The American Economic Review, 48*, 261–297.

Modigliani, F., & Miller, M. (1963). Corporate income taxes and the cost of capital: A correction. *The American Economic Review, 53*, 147–175.

Modigliani, F., & Miller, M. (1966). Some estimates of the cost of capital to the electric utility industry 1954–1957. *The American Economic Review, 56*, 333–391.

Chapter 31
The Role of the Central Bank and Commercial Banks in Creating and Maintaining of a Favorable Investment Climate in the Country

In this Chapter we study the role of the Central Bank and commercial banks in creating and maintaining of a favorable investment climate in the country. Within the framework of modern investment models created by the authors, the dependence of the efficiency of investments on the level of debt financing within a wide range of values of equity costs and debt capital costs under different project terms (long-term projects as well as projects of arbitrary duration) and different investment profitability coefficients β is investigated. The effectiveness of investments is determined by Net Present Value, NPV. The study is conducted within the framework of investment models with debt repayment at the end of the project term.

It is found that NPV depends practically linearly on leverage level L, increasing or decreasing depending on profitability coefficient β and credit rate values k_d. The cut off credit rate values k_d*, separating the range of increasing NPV(L) from range of decreasing NPV(L), are determined. The Central Bank should keep its key rate at the level which allow commercial banks keep their credit rates below the cut off credit rate k_d* values in order to create and maintain a favorable investment climate in the country.

31.1 Introduction

The investments play a very important role in an economy of each country. As a rule the the debt financing is always used in the investments. In current paper we determine the role of the Central Bank and commercial banks in creating and maintaining of a favorable investment climate in the country. Within the framework of modern investment models created by the authors, the dependence of the efficiency of investments on the level of debt financing within a wide range of values of equity capital costs and debt capital costs under different project terms (long-term projects as well as projects of arbitrary duration) and different investment profitability coefficients β is investigated. The effectiveness of investments is determined by

P. Brusov, T. Filatova, *Efficiency of Investment Projects*, Contributions to Finance and Accounting, https://doi.org/10.1007/978-3-032-01028-5_31

Net Present Value, NPV. The study is conducted within the framework of investment models with debt repayment at the end of the project term.

It is found that NPV depends practically linearly on leverage level L, increasing or decreasing depending on profitability coefficient β and credit rate values k_d. The cut off credit rate values k_d*, separating the range of increasing NPV(L) from range of decreasing NPV(L) are determined. The Central Bank should keep its key rate at the level which allow to commercial banks keep their credit rates below the cut off credit rate values k_d*in order to create and maintain a favorable investment climate in the country.

31.2 Investment Models with Debt Repayment at the End of the Project

The effectiveness of the investment project could be considered from two perspectives: the owners of equity and debt and the equity holders only. For each of these cases, NPV could be calculated in two ways: with the division of credit and investment flows (and thus discounting of the payments using two different rates) and without such a division (in this case, both flows are discounted using the same rate, as which can be, obviously, chosen WACC). For each of the four situations, two cases could be considered: (1) a constant value of equity S and (2) a constant value of the total invested capital $I = S + D$ (D is value of debt funds).

As it was stated above, the effectiveness of the investment project is considered from two perspectives: the owners of equity and debt and the equity holders only. In the first case, the interest and duty paid by owners of equity (negative flows) returned to the project because they are exactly equal to the flow (positive), obtained by owners of debt capital. The only effect of leverage in this case is the effect of the tax shield, generated from the tax relief: interest on the loan is entirely included into the cost and thus reduces the tax base. After-tax flow of capital for each period in this case is equal to

$$\mathrm{NOI}(1-t)+k_{\mathrm{d}}Dt \tag{31.1}$$

and the value of investments at the initial time moment $T = 0$ is equal to $-I = -S - D$.

Here *NOI* stands for net operating income (before taxes).

In the second case, investments at the initial time moment $T = 0$ are equal to $-S$ and the flow of capital for the period (in addition to the tax shields $k_d Dt$ it includes a payment of interest on a loan $-k_dD$):

$$(\mathrm{NOI}-k_{\mathrm{d}}D)(1-t). \tag{31.2}$$

Here, for simplicity, we suppose that interest on the loan will be paid in equal shares $k_d D$ during all periods. Note that principal repayment is made at the end of the last period.

We will consider the case of discounting, when operating and financial flows are not separated and both are discounted, using the general rate (as which, obviously, the weighted average cost of capital (WACC) can be selected). In this case for long-term (perpetuity) projects, the Modigliani–Miller formula (Modigliani & Miller, 1958, 1963, 1966) for WACC will be used and for projects of finite (arbitrary) duration Brusov–Filatova–Orekhova formula will be used (Brusov & Filatova, 2011; Brusov et al. 2011a, b, c, 2012a, b, 2013a, b, 2014a, b; Filatova et al., 2008; Brusova, 2011; Brusov et al. 2015, 2018a, b, c, d; 2012b, 2013b, 2014b).

Note that debt capital is the least risky, because interest on credit is paid after taxes in the first place. Therefore, the cost of credit will always be less than the equity cost, whether of ordinary or of preference shares $k_e > k_d$; $k_p > k_d$. Here k_e; k_p is the equity cost of ordinary or of preference shares consequently.

31.2.1 The Effectiveness of the Investment Project from the Perspective of the Equity Holders Only (Without Flows Separation)

In this case operating and financial flows are not separated and are discounted, using the general rate (as which, obviously, WACC can be selected).

The credit reimbursable at the end of the project (at the end of the period (n)) can be discounted either at the same rate WACC or at the debt cost rate k_d. Now we choose a uniform rate and the first option.

$$\begin{aligned}\mathrm{NPV} &= -S+\sum_{i=1}^{n}\frac{\mathrm{NOI}(1-t)-k_dD(1-t)}{(1+\mathrm{WACC})^i}-\frac{D}{(1+\mathrm{WACC})^n}\\ &= -S+\frac{\mathrm{NOI}(1-t)-k_dD(1-t)}{\mathrm{WACC}}\left(1-\frac{1}{(1+\mathrm{WACC})^n}\right)-\frac{D}{(1+\mathrm{WACC})^n}.\end{aligned} \tag{31.3}$$

At a Constant Value of Equity Capital (S = const)

Accounting that in the case S = const NOI is proportional to the invested capital, I, NOI $= \beta I = \beta S(1 + L)$, and substituting $D = LS$, we get

$$\mathrm{NPV} = -S + \frac{\mathrm{NOI}(1-t) - k_\mathrm{d}D(1-t)}{\mathrm{WACC}}\left(1 - \frac{1}{(1+\mathrm{WACC})^n}\right) - \frac{D}{(1+\mathrm{WACC})^n}, \tag{31.4}$$

$$\mathrm{NPV} = -S\left[1 + \frac{Lk_\mathrm{d}(1-t)}{\mathrm{WACC}}\left(1 - \frac{1}{(1+\mathrm{WACC})^n}\right) + \frac{L}{(1+\mathrm{WACC})^n}\right] + \frac{\beta S(1+L)(1-t)}{\mathrm{WACC}}\left(1 - \frac{1}{(1+\mathrm{WACC})^n}\right). \tag{31.5}$$

31.2.1.1 Modigliani–Miller Limit (Long-Term (Perpetuity) Projects)

In perpetuity limit ($n \to \infty$) (Modigliani–Miller limit) (turning to the limit $n \to \infty$ in the relevant equations), we have

$$\mathrm{NPV} = -S + \frac{\mathrm{NOI}(1-t) - k_\mathrm{d}D(1-t)}{\mathrm{WACC}}. \tag{31.6}$$

At a Constant Value of Equity Capital (S = const)

$$\mathrm{NPV} = -S + \frac{\mathrm{NOI}(1-t) - k_\mathrm{d}D(1-t)}{\mathrm{WACC}} \tag{31.7}$$

Substituting $D = LS$, we get

$$\begin{aligned}\mathrm{NPV} &= -S\left[1 + \frac{Lk_\mathrm{d}(1-t)}{\mathrm{WACC}}\right] + \frac{\mathrm{NOI}(1-t)}{\mathrm{WACC}} \\ &= -S\left[1 + \frac{Lk_\mathrm{d}(1-t)}{k_0(1-Lt/(1+L))}\right] + \frac{\beta S(1+L)(1-t)}{k_0(1-Lt/(1+L))}.\end{aligned} \tag{31.8}$$

In last equation we substituted the perpetuity (Modigliani–Miller) formula for WACC

$$\mathrm{WACC} = k0\left(1 - \frac{Lt}{1+L}\right). \tag{31.9}$$

Below we will investigate the dependence of the efficiency of investments on the level of debt financing within a wide range of values of equity costs k_0 and debt capital costs k_d under different project terms (long-term projects as well as projects of arbitrary duration) and different investment profitability coefficients β.

For long-term project calculations we use formulas (31.8) and (31.9), while for arbitrary duration project calculations we use formula (31.5) for NPV and BFO formula for WACC

$$\frac{1-(1+\mathrm{WACC})^{-n}}{\mathrm{WACC}}=\frac{1-(1+k_0)^{-n}}{k_0\left[1-\omega_{\mathrm{d}}T\left(1-(1+k_{\mathrm{d}})^{-n}\right)\right]}. \tag{31.10}$$

Here, S is the value of equity capital of the company, $w_{\mathrm{d}}=\frac{D}{D+S}$—the share of debt capital, $k_{\mathrm{e}}, w_{\mathrm{e}}=\frac{S}{D+S}$—the cost and the share of the equity capital of the company, and $L = D/S$—financial leverage.

31.3 Modigliani–Miller Limit (Long-Term (Perpetuity) Projects)

Let us start from the long-term projects. We will study the dependence of the efficiency of investments on the level of debt financing L for the values of equity costs k_0 from 6% up to 32% and for different debt capital costs and different investment profitability coefficient β values.

31.3.1 The Dependence of the Efficiency of Investments NPV on the Level of Debt Financing L *for the Values of Equity Costs* $k_0 = 0.2$

Below we represent the results of calculations for equity costs $k_0 = 0.2$; debt capital costs $k_{\mathrm{d}} = 0.18; 0.16; 0.14; 0.12; 0.10; 0.08; 0.06$; $S = 250$; tax on profit rate $t = 0.2$.

Results are shown in Tables 31.1, 31.2, and 31.3 and Figs. 31.1, 31.2, and 31.3.

1. β = 0.1

From Table 31.1 and Fig. 31.1 it is seen that NPV depends practically linearly on leverage level L, increasing or decreasing depending on credit rate value k_{d}. NPV(L) increases at credit rate $k_{\mathrm{d}} = 0.06$ and $k_{\mathrm{d}} = 0.08$. The cut off credit rate k_{d}* values, separating the range of increasing NPV(L) from range of decreasing NPV(L) for investment profitability coefficient β = 0.1, is equal to 0.1. At higher credit rate k_{d} values NPV(L) represents decreasing function.

2. β = 0.12 From Table 31.2 and Fig. 31.2 it is seen that NPV depends practically linearly on leverage level L, increasing or decreasing depending on credit rate value k_{d}. NPV(L) increases at credit rate $k_{\mathrm{d}} = 0.06$; $k_{\mathrm{d}} = 0.08$ and $k_{\mathrm{d}} = 0.1$. The cut off credit rate values k_{d}*, separating the range of increasing NPV(L) from range of decreasing NPV(L) for investment profitability coefficient β = 0.12, is equal to 0.12. At higher credit rate k_{d} values NPV(L) represents decreasing function.

Table 31.1 The dependence of NPV on the level of debt financing *L* for the values of equity costs $k_0 = 0.2$ and for different debt capital costs $k_d = 0.18; 0.16; 0.14; 0.12; 0.10; 0.08; 0.06$; $S = 250$; tax on profit rate $t = 0.2$ and investment profitability coefficient $\beta = 0.1$

L	NPV ($k_d = 0.18$)	NPV ($k_d = 0.16$)	NPV ($k_d = 0.14$)	NPV ($k_d = 0.12$)	NPV ($k_d = 0.1$)	NPV ($k_d = 0.08$)	NPV ($k_d = 0.06$)
0	−150.00	−150.00	−150.00	−150.00	−150.00	−150.00	−150.00
1	−227.78	−205.56	−183.33	−161.11	−138.89	−116.67	−94.44
2	−319.23	−273.08	−226.92	−180.77	−134.62	−88.46	−42.31
3	−414.71	−344.12	−273.53	−202.94	−132.35	−61.76	8.82
4	−511.90	−416.67	−321.43	−226.19	−130.95	−35.71	59.52
5	−610.00	−490.00	−370.00	−250.00	−130.00	−10.00	110.00
6	−708.62	−563.79	−418.97	−274.14	−129.31	15.52	160.34
7	−807.58	−637.88	−468.18	−298.48	−128.79	40.91	210.61
8	−906.76	−712.16	−517.57	−322.97	−128.38	66.22	260.81
9	−1006.10	−786.59	−567.07	−347.56	−128.05	91.46	310.98
10	−1105.56	−861.11	−616.67	−372.22	−127.78	116.67	361.11

Table 31.2 The dependence of NPV on the level of debt financing L for the values of equity costs $k_0 = 0.2$ and for different debt capital costs $k_d = 0.18; 0.16; 0.14; 0.12; 0.10; 0.08; 0.06$; $S = 250$; tax on profit rate $t = 0.2$ and investment profitability coefficient $\beta = 0.12$

L	NPV ($k_d = 0.18$)	NPV ($k_d = 0.16$)	NPV ($k_d = 0.14$)	NPV ($k_d = 0.12$)	NPV ($k_d = 0.1$)	NPV ($k_d = 0.08$)	NPV ($k_d = 0.06$)
0	−130.00	−130.00	−130.00	−130.00	−130.00	−130.00	−130.00
1	−183.33	−161.11	−138.89	−116.67	−94.44	−72.22	−50.00
2	−250.00	−203.85	−157.69	−111.54	−65.38	−19.23	26.92
3	−323.59	−250.00	−179.41	−108.82	−38.24	32.35	102.94
4	−392.86	−297.62	−202.38	−107.14	−11.90	83.33	178.57
5	−466.00	−346.00	−226.00	−106.00	14.00	134.00	254.00
6	−539.66	−394.83	−250.00	−105.17	39.66	184.48	329.31
7	−613.64	−443.94	−274.24	−104.55	65.15	234.85	404.55
8	−687.84	−493.24	−298.65	−104.05	90.54	285.14	479.73
9	−762.20	−542.68	−323.17	−103.66	115.85	335.37	554.88
10	−836.67	−592.22	−347.78	−103.33	141.11	385.56	630.00

Table 31.3 The dependence of NPV on the level of debt financing L for the values of equity costs $k_0 = 0.2$ and for different debt capital costs $k_d = 0.18; 0.16; 0.14; 0.12; 0.10; 0.08; 0.06$; $S = 250$; tax on profit rate $t = 0.2$ and investment profitability coefficient $\beta = 0.14$

L	NPV ($k_d = 0.18$)	NPV ($k_d = 0.16$)	NPV ($k_d = 0.14$)	NPV ($k_d = 0.12$)	NPV ($k_d = 0.1$)	NPV ($k_d = 0.08$)	NPV ($k_d = 0.06$)
0	–110.00	–110.00	–110.00	–110.00	–110.00	–110.00	–110.00
1	–138.89	–116.67	–94.44	–72.22	–50.00	–27.78	–5.56
2	–180.77	–134.62	–88.46	–42.31	3.85	50.00	96.15
3	–226.47	–155.88	–85.29	–14.71	55.88	126.47	197.06
4	–273.81	–178.57	–83.33	11.90	107.14	202.38	297.62
5	–322.00	–202.00	–82.00	38.00	158.00	278.00	398.00
6	–370.69	–225.86	–81.03	63.79	208.62	353.45	498.28
7	–419.70	–250.00	–80.30	89.39	259.09	428.79	598.48
8	–468.92	–274.32	–79.73	114.86	309.46	504.05	698.65
9	–518.29	–298.78	–79.27	140.24	359.76	579.27	798.78
10	–567.78	–323.33	–78.89	165.56	410.00	654.44	898.89

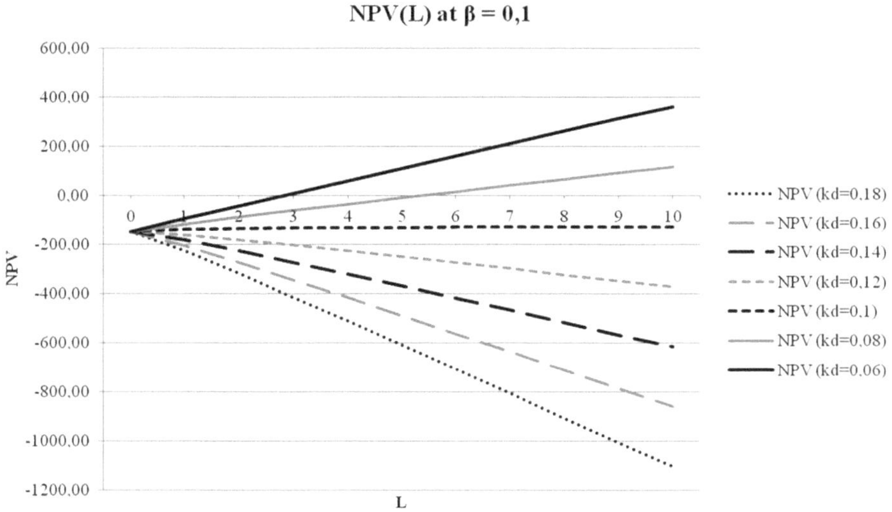

Fig. 31.1 The dependence of the Net Present Value, NPV on the leverage level L for the equity value $k_0 = 20\%$ and for different debt capital costs $k_d = 0.18$; 0.16; 0.14; 0.12; 0.10; 0.08; 0.06; $S = 250$; tax on profit rate $t = 0.2$ and investment profitability coefficient $\beta = 0.1$

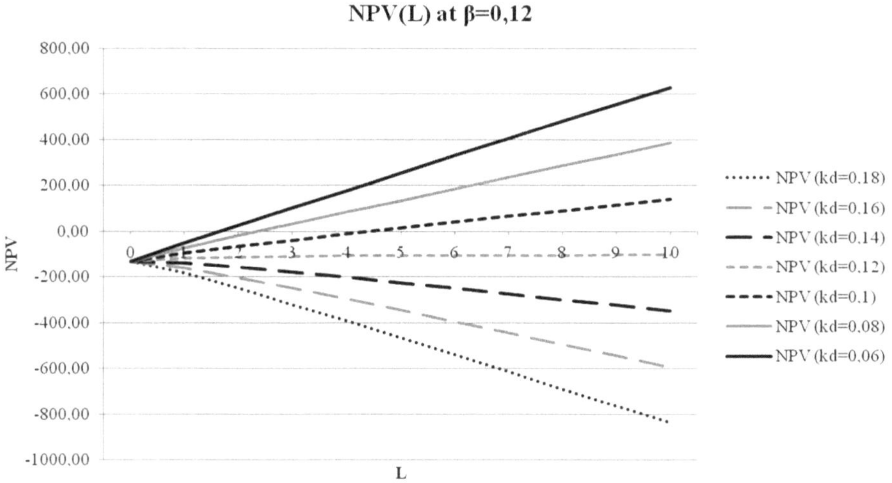

Fig. 31.2 The dependence of the Net Present Value, NPV, on the leverage level L for the equity value $k_0 = 20\%$ and for different debt capital costs $k_d = 0.18$; 0.16; 0.14; 0.12; 0.10; 0.08; 0.06; $S = 250$; tax on profit rate $t = 0.2$ and investment profitability coefficient $\beta = 0.12$

3. $\beta = 0.14$

From Table 31.3 and Fig. 31.3 it is seen that NPV depends practically linearly on leverage level L, increasing or decreasing depending on credit rate value k_d. NPV(L) increases at credit rate $k_d = 0.06$; $k_d = 0.08$; $k_d = 0.1$ and $k_d = 0.12$. The cut off credit

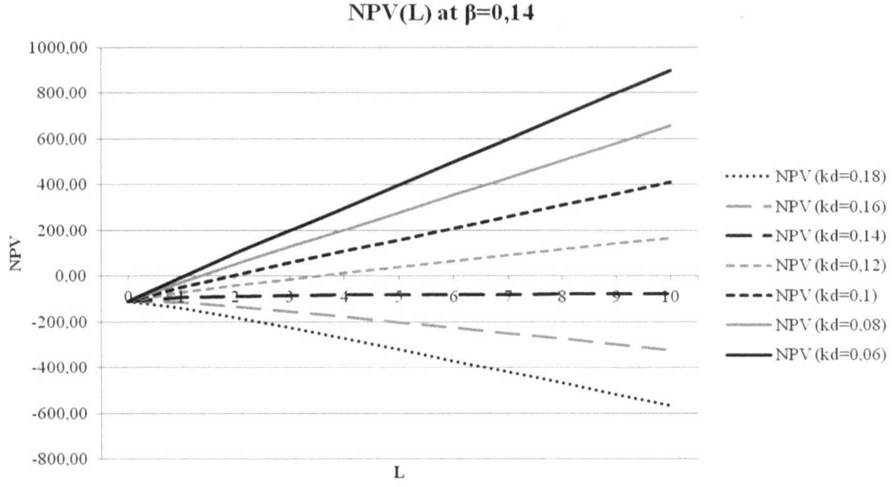

Fig. 31.3 The dependence of the Net Present Value, NPV, on the leverage level L for the equity value $k_0 = 20\%$ and for different debt capital costs $k_d = 0.18$; 0.16; 0.14; 0.12; 0.10; 0.08; 0.06; $S = 250$; tax on profit rate $t = 0.2$ and investment profitability coefficient $\beta = 0.14$

rate k_d* values separating the range of increasing NPV(L) from range of decreasing NPV(L) for investment profitability coefficient $\beta = 0.14$ is equal to 0.14. At higher credit rate k_d values NPV(L) represents decreasing function.

One can see that the cut off credit rate k_d* values separating the range of increasing NPV(L) from range of decreasing NPV(L) strongly correlates with investment profitability coefficient β and practically linearly depends on it.

For long-term projects (Modigliani–Miller limit) it was found that the cut off credit rate values k_d* are proportional to investment profitability coefficients β: it turns out that for equity capital cost $\boldsymbol{k_0} = \mathbf{0.2}$ the cut off credit rate values k_d* separating the range of increasing NPV(L) from range of decreasing NPV(L) are approximately equal to investment profitability coefficient β: for investment profitability coefficient $\beta = 0.1$ k_d* is equal to 0.1; for $\beta = 0.12$ k_d* is equal to 0.12; and for investment profitability coefficient $\beta = 0.14$ k_d* is equal to 0.14. The slope of the curve NPV(L) increases with investment profitability coefficient β for the same value of credit rate k_d.

31.3.2 The Dependence of the Efficiency of Investments NPV on the Level of Debt Financing L *for the Value of Equity Costs* $k_0 = 0.28$

Let us consider also the case of equity capital cost $k_0 = 0.28$ and debt capital cost $k_d = 6\%$; 8%; 10%; 12%; 14%; 16%; 18%; 20%; 22%; 24%.

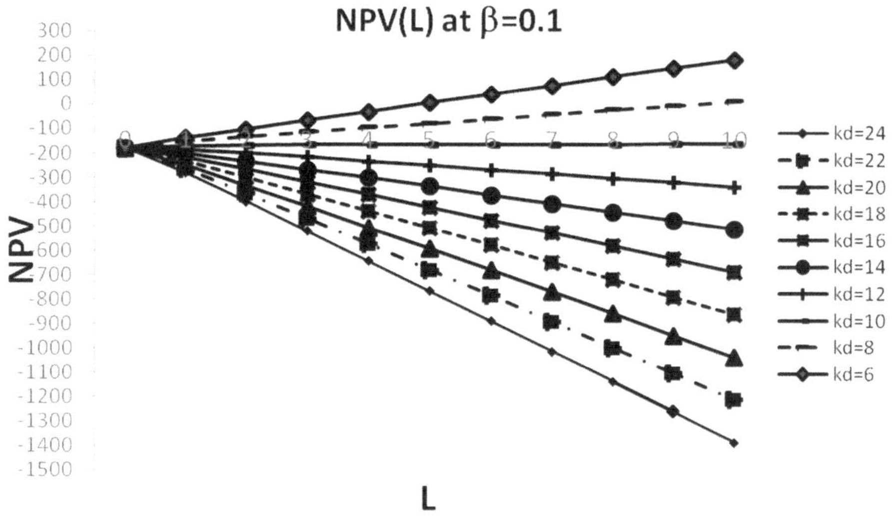

Fig. 31.4 The dependence of the Net Present Value, NPV on the leverage level L for the equity value $k_0 = 28\%$ and for different debt capital costs $k_d = 6\%$; 8%; 10%; 12%; 14%; 16%; 18%; 20%; 22%; 24%; $S = 250$; tax on profit rate $t = 0.2$ and investment profitability coefficient $\beta = 0.1$

It is seen from Fig. 31.4 that the cut off credit rate k_d* value separating the range of increasing NPV(L) from range of decreasing NPV(L) is equal to 10%.

It is seen from Fig. 31.5 that the cut off credit rate value k_d*, separating the range of increasing NPV(L) from range of decreasing NPV(L), is equal to 20%.

From Figs. 31.4 and 31.5 it follows that for $\beta = 0.1$ k_d* is equal to 0.1; and for investment profitability coefficient $\beta = 0.2$ k_d* is equal to 0.2.

We could come to conclusion that in perpetuity limit for both cases for the equity values $k_0 = 20\%$ and $k_0 = 28\%$ it turns out that the cut off credit rate k_d* values are equal to investment profitability coefficient β (and does not depend on the equity values k_0).

As we will see below this statement is not valid for the projects of arbitrary durations.

31.4 Projects of Finite (Arbitrary) Duration

Let us consider now the projects of arbitrary durations. We will study the dependence of the efficiency of investments on the level of debt financing L for the same values of equity costs k_0 from 6% up to 32%; for different debt capital costs k_d and different investment profitability coefficient β values as well as for different project durations.

For arbitrary duration project calculations we use formula (31.5) for NPV

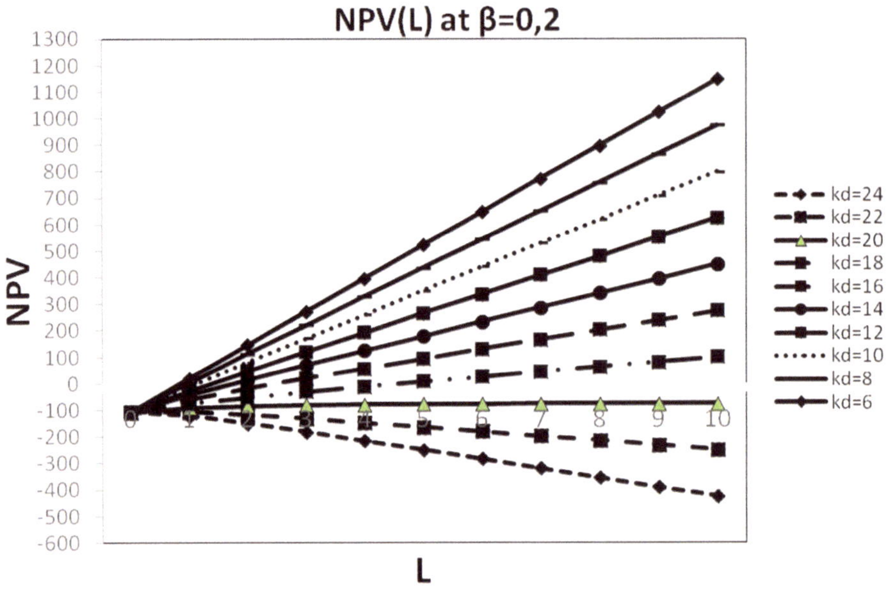

Fig. 31.5 The dependence of the Net Present Value, NPV on the leverage level L for the equity value $k_0 = 28\%$ and for different debt capital costs $k_d = 6\%; 8\%; 10\%; 12\%; 14\%; 16\%; 18\%; 20\%; 22\%; 24\%$; $S = 250$; tax on profit rate $t = 0.2$ and investment profitability coefficient $\beta = 0.2$

$$\mathrm{NPV} = -S\left[1 + \frac{Lk_d(1-t)}{\mathrm{WACC}}\left(1 - \frac{1}{(1+\mathrm{WACC})^n}\right) + \frac{L}{(1+\mathrm{WACC})^n}\right] + \frac{\beta S(1+L)(1-t)}{\mathrm{WACC}}\left(1 - \frac{1}{(1+\mathrm{WACC})^n}\right).$$

and BFO formula (31.10) for WACC

$$\frac{1-(1+\mathrm{WACC})^{-n}}{\mathrm{WACC}} = \frac{1-(1+k_0)^{-n}}{k_0\left[1-\omega_d T\left(1-(1+k_d)^{-n}\right)\right]}.$$

31.4.1 The Dependence of the Efficiency of Investments NPV on the Level of Debt Financing **L** *for the Values of Equity Costs* $\mathbf{k}_0 = 0.2$

Below we represent the results of calculations for equity costs $k_0 = 0.2$; debt capital costs $k_d = 0.18; 0.16; 0.14; 0.12; 0.10; 0.08; 0.06$; $S = 250$; tax on profit rate $t = 0.2$ and project duration n = 5. In next part we will compare the results for project durations $n = 5$ and $n = 3$.

Table 31.4 The dependence of NPV on the level of debt financing L for the values of equity costs $k_0 = 0.2$ and for different debt capital costs $k_d = 0.18; 0.16; 0.14; 0.12; 0.10; 0.08; 0.06$; $S = 250$; tax on profit rate $t = 0.2$; $\beta = 0.325$ and project duration $n = 5$

k_d	k_d	k_d	k_d	k_d	k_d	k_d	k_d
0	–55.62	–55.62	–55.62	–55.62	–55.62	–55.62	–55.62
1	–64.27	–51.88	–39.78	–27.84	–16.10	–4.63	6.60
2	–84.87	–58.97	–33.37	–8.12	16.70	41.06	64.88
3	–109.00	–69.04	–29.55	9.40	47.74	85.35	122.14
4	–134.55	–80.39	–26.81	26.01	78.04	129.06	178.95
5	–160.82	–92.33	–24.61	42.16	107.96	172.47	235.56
6	–187.68	–104.63	–22.73	58.04	137.67	215.71	292.05
7	–214.48	–117.19	–21.05	73.87	167.24	258.85	348.47
8	–241.83	–129.90	–19.51	89.50	196.72	301.91	404.83
9	–269.07	–142.72	–18.06	105.05	226.13	344.92	461.15
10	–296.16	–155.62	–16.68	131.54	255.50	387.89	517.45

For arbitrary duration project calculations we use formula (31.5) for NPV and BFO formula for WACC (31.10).

Results are shown in Tables 31.4, 31.5, and 31.6 and Figs. 31.4, 31.5, and 31.6.

1. β = 0.325

From Table 31.4 and Fig. 31.6 it is seen that NPV depends practically linearly on leverage level L, increasing or decreasing depending on credit rate value k_d. NPV(L) increases at credit rate $k_d = 0.06$; $k_d = 0.08$; $k_d = 0.1$ and $k_d = 0.12$. The cut off credit rate value k_d*, separating the range of increasing NPV(L) from range of decreasing NPV(L) for investment profitability coefficient $\beta = 0.325$, is equal to 0.14. At higher credit rates k_d values NPV(L) represents decreasing function.

2. β = 0.345

From Table 31.5 and Fig. 31.7 it is seen that NPV depends practically linearly on leverage level L, increasing or decreasing depending on credit rate value k_d. NPV(L) increases at credit rates $k_d = 0.06$; $k_d = 0.08$; $k_d = 0.1$; $k_d = 0.12$ and $k_d = 0.14$. The cut off credit rate value k_d*, separating the range of increasing NPV(L) from range of decreasing NPV(L) for investment profitability coefficient $\beta = 0.345$ is equal to 0.16. At higher credit rates k_d values NPV(L) represents decreasing function.

3. β = 0.365

From Table 31.6 and Fig. 31.8 it is seen that NPV depends practically linearly on leverage level L, increasing or decreasing depending on credit rate value k_d. NPV(L) increases at credit rate $k_d = 0.06$; $k_d = 0.08$; $k_d = 0.1$; $k_d = 0.12$; $k_d = 0.14$ and $k_d = 0.16$. The cut off credit rate value k_d*, separating the range of increasing NPV (L) from the range of decreasing NPV(L) for investment profitability coefficient $\beta = 0.365$, is equal to 0.18. At higher credit rates k_d values NPV(L) represents decreasing function.

One can see that the cut off credit rate values k_d*, separating the range of increasing NPV(L) from the range of decreasing NPV(L), strongly correlate with

Table 31.5 The dependence of NPV on the level of debt financing L for the values of equity costs $k_0 = 0.2$ and for different debt capital costs $k_d = 0.18$; 0.16; 0.14; 0.12; 0.10; 0.08; 0.06; $S = 250$; tax on profit rate $t = 0.2$; $\beta = 0.345$ and project duration $n = 5$

L	NPV ($k_d = 0.18$)	NPV ($k_d = 0.16$)	NPV ($k_d = 0.14$)	NPV ($k_d = 0.12$)	NPV ($k_d = 0.1$)	NPV ($k_d = 0.08$)	NPV ($k_d = 0.06$)
0	–43.66	–43.66	–43.66	–43.66	–43.66	–43.66	–43.66
1	–38.96	–26.63	–14.64	–2.83	8.77	23.08	31.15
2	–46.07	–23.39	4.97	29.97	54.50	78.55	102.02
3	–56.74	–17.11	22.02	60.58	98.47	135.61	171.89
4	–68.82	–15.10	37.99	90.29	141.71	192.09	241.28
5	–81.62	–13.68	53.42	119.54	184.58	248.28	310.49
6	–94.98	–12.63	68.53	148.51	227.22	304.31	379.58
7	–108.32	–11.83	83.45	177.41	269.74	360.22	448.59
8	–122.17	–11.18	98.22	206.13	312.16	416.07	517.55
9	–135.93	–10.64	112.91	234.77	354.52	471.86	586.48
10	–149.58	–10.18	127.53	263.36	396.83	527.61	655.38

Table 31.6 The dependence of NPV on the level of debt financing L for the values of equity costs $k_0 = 0.2$ and for different debt capital costs $k_d = 0.18; 0.16; 0.14; 0.12; 0.10; 0.08; 0.06$; $S = 250$; tax on profit rate $t = 0.2$; $\beta = 0.365$ and project duration $n = 5$

L	NPV ($k_d = 0.18$)	NPV ($k_d = 0.16$)	NPV ($k_d = 0.14$)	NPV ($k_d = 0.12$)	NPV ($k_d = 0.1$)	NPV ($k_d = 0.08$)	NPV ($k_d = 0.06$)
0	−31.70	−31.70	−31.70	−31.70	−31.70	−31.70	−31.70
1	−13.65	−1.39	10.49	22.18	33.65	44.80	55.69
2	−7.27	18.20	43.32	68.06	92.30	116.03	139.16
3	−4.48	34.82	73.59	111.76	149.21	185.87	221.64
4	−3.08	50.20	102.79	154.56	205.39	255.13	303.62
5	−2.41	64.96	131.45	196.91	261.19	324.10	385.43
6	−2.27	79.37	159.80	238.99	316.78	392.90	467.11
7	−2.17	93.53	187.94	280.94	372.24	461.60	548.72
8	−2.50	107.54	215.96	322.76	427.60	530.22	630.28
9	−2.79	121.44	243.87	364.50	482.91	598.80	711.81
10	−3.01	135.26	271.73	406.18	538.16	667.33	793.30

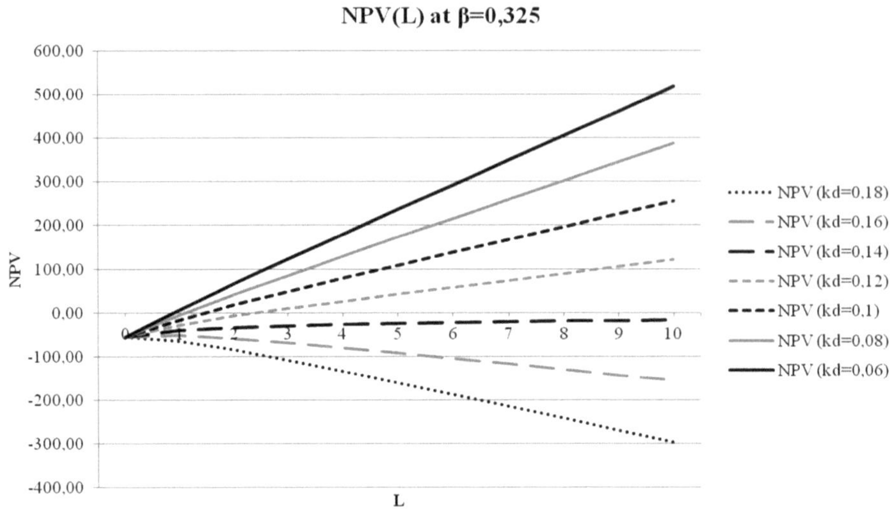

Fig. 31.6 The dependence of the Net Present Value, NPV, on the leverage level L for the equity value $k_0 = 20\%$ and for different debt capital costs $k_d = 0.18$; 0.16; 0.14; 0.12; 0.10; 0.08; 0.06; $S = 250$; tax on profit rate $t = 0{,}2$; investment profitability coefficient $\beta = 0.325$ and project duration $n = 5$

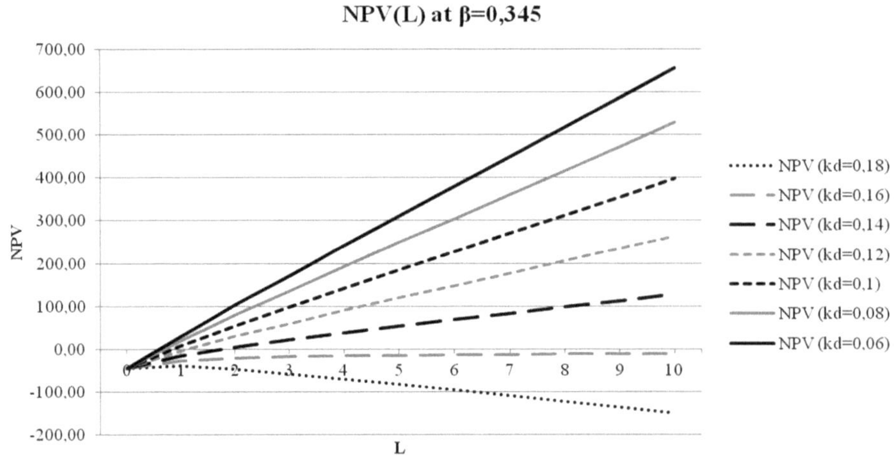

Fig. 31.7 The dependence of the Net Present Value, NPV, on the leverage level L for the equity value $k_0 = 20\%$ and for different debt capital costs $k_d = 0.18$; 0.16; 0.14; 0.12; 0.10; 0.08; 0.06; $S = 250$; tax on profit rate $t = 0.2$; investment profitability coefficient $\beta = 0.345$ and project duration $n = 5$

investment profitability coefficient β and practically linearly depends on it: k_d* linearly increases with profitability coefficient β.

For arbitrary duration projects results are as folloing. The efficiency of investments strongly depends on project duration and increases with duration. One can see,

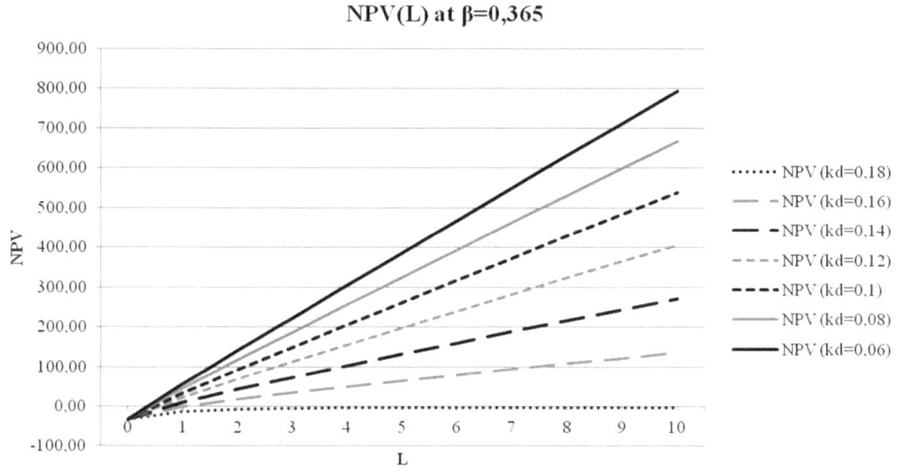

Fig. 31.8 The dependence of the Net Present Value, NPV, on the leverage level L for the equity value $k_0 = 20\%$ and for different debt capital costs $k_d = 0.18$; 0.16; 0.14; 0.12; 0.10; 0.08; 0.06; $S = 250$; tax on profit rate $t = 0.2$; investment profitability coefficient $\beta = 0.365$ and project duration $n = 5$

that the slope of the curve NPV(L) at project duration $n = 5$ is always higher, than for project duration $n = 3$. The efficiency of investments increases with project duration and is less than for long-term (perpetuity) projects. Transition to increasing NPV(L) behavior for finite duration projects requires much higher values of investment profitability coefficient β than in case of long-term (perpetuity) projects, where $k_d{}^*$ is approximately equal to β: for 5 years projects the cut off credit rate values $k_d{}^*$ for investment profitability coefficient $\beta = 0.325$ is equal to 0.14; for investment profitability coefficient $\beta = 0.345$ is equal to 0.16; for investment profitability coefficient $\beta = 0.365$ is equal to 0.18. Thus, for finite duration projects as well as for the long-term projects cut off credit rate values $k_d{}^*$ turn out to be proportional to investment profitability coefficients β, but investment profitability coefficients β, is approximately twice higher than $k_d{}^*$.

31.4.2 The Dependence of the Efficiency of Investments NPV on the Level of Debt Financing L for the Values of Equity Costs $k_0 = 0.28$

Below at Figs. 31.9, 31.10, 31.11, and 31.12 we present the results of calculations of the dependence of the Net Present Value, NPV, on the leverage level L for the equity value $k_0 = 28\%$ and for different debt capital costs $k_d = 0.24$; 0.22; 0.20; 0.18; 0.16; 0.14; 0.12; 0.10; 0.08; 0.06; $S = 250$; tax on profit rate $t = 0.2$; investment profitability coefficient $\beta = 0.1$ and project durations $n = 3$ and $n = 5$.

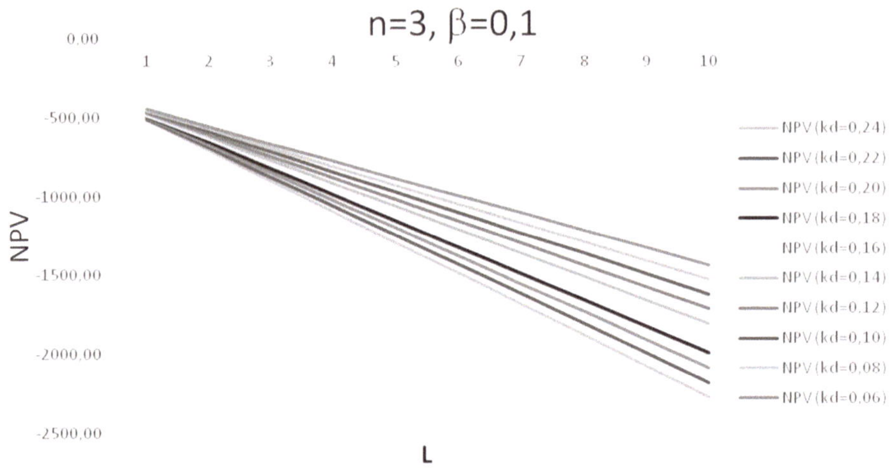

Fig. 31.9 The dependence of the Net Present Value, NPV on the leverage level L for the equity value $k_0 = 28\%$ and for different debt capital costs $k_d = 0.24; 0.22; 0.20; 0.18; 0.16; 0.14; 0.12; 0.10; 0.08; 0.06$; $S = 250$; tax on profit rate $t = 0.2$; investment profitability coefficient $\beta = 0.1$ and project duration $n = 3$

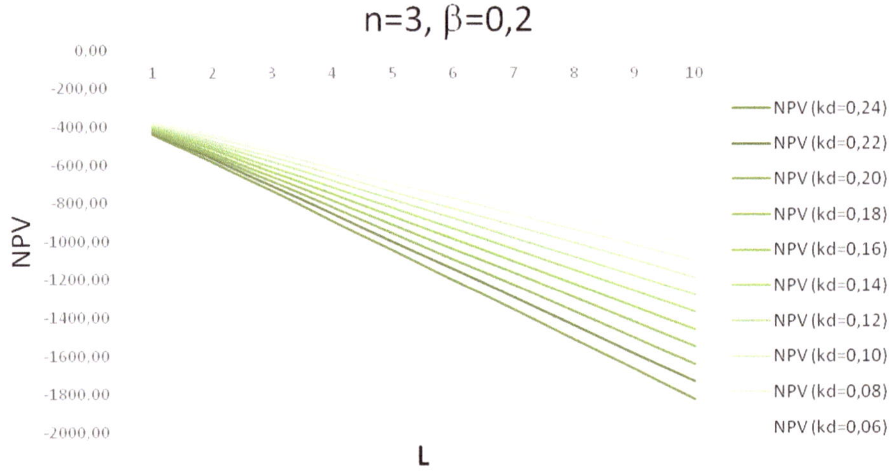

Fig. 31.10 The dependence of the Net Present Value, NPV on the leverage level L for the equity value $k_0 = 28\%$ and for different debt capital costs $k_d = 0.24; 0.22; 0.20; 0.18; 0.16; 0.14; 0.12; 0.10; 0.08; 0.06$; $S = 250$; tax on profit rate $t = 0.2$; investment profitability coefficient $\beta = 0.2$ and project duration $n = 3$

From the Figs. 31.9, 31.10, 31.11, and 31.12 one can make the following conclusions:

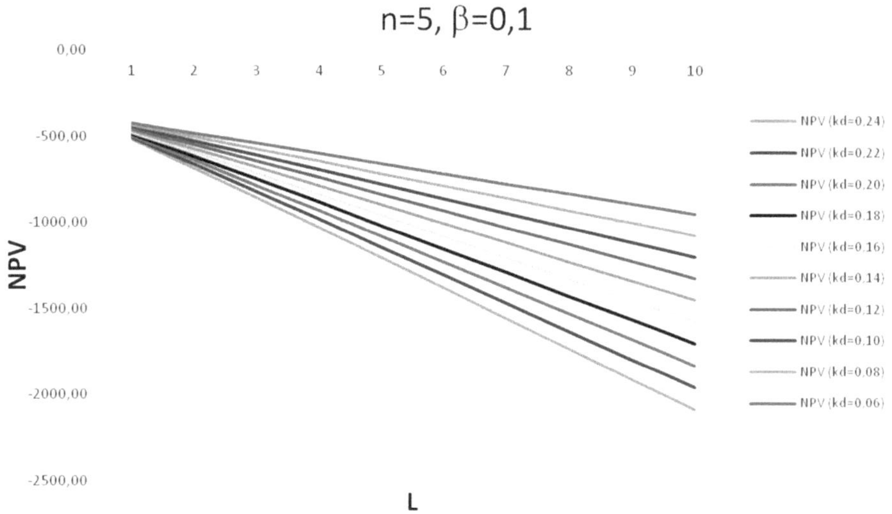

Fig. 31.11 The dependence of the Net Present Value, NPV on the leverage level L for the equity value $k_0 = 28\%$ and for different debt capital costs $k_d = 0.24$; 0.22; 0.20; 0.18; 0.16; 0.14; 0.12; 0.10; 0.08; 0.06; $S = 250$; tax on profit rate $t = 0.2$; investment profitability coefficient $\beta = 0.1$ and project duration $n = 5$

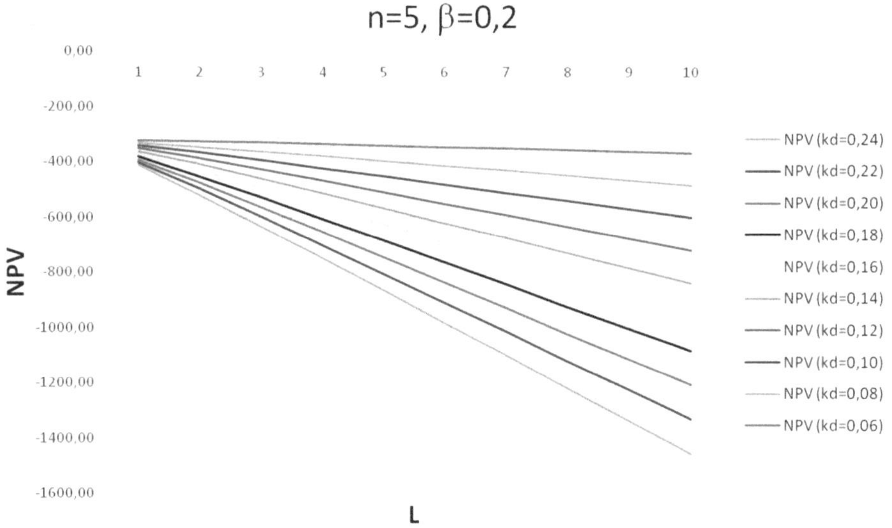

Fig. 31.12 The dependence of the Net Present Value, NPV, on the leverage level L for the equity value $k_0 = 28\%$ and for different debt capital costs $k_d = 0.24$; 0.22; 0.20; 0.18; 0.16; 0.14; 0.12; 0.10; 0.08; 0.06; $S = 250$; tax on profit rate $t = 0.2$; investment profitability coefficient $\beta = 0.2$ and project duration $n = 5$

1. NPV decreases with debt capital cost k_d.
2. NPV increases with investment profitability coefficient β as well as with project duration.
3. The cut off value k_d* has been reached at Figs. 31.9, 31.10, 31.11, and 31.12 only at profitability coefficient $\beta = 0.2$ for 5-year project and is equal to 6%; it will increases with investment profitability coefficient β. Bigger values of β, and/or longer durations n, and/or bigger values of equity capital S are required in order to demonstrate the presence of a cut off value k_d* for particular project.

31.5 The Dependence of the Net Present Value, NPV, on the Leverage Level L for Projects of Different Durations

We consider the case of equity cost (at $L = 0$) $k_0 = 14\%$ and fixed value of debt cost k_d = 0,04; 0,06; 0,08; 0,1; 0,12 and compare the results for projects of different duration: n = 3 years and n = 5 years (Tables 31.7, 31.8, 31.9, and 31.10; Figs. 31.13 and 31.14).

$k_d = 0.1$ (Tables 31.11, 31.12, 31.13, and 31.14; Figs. 31.15 and 31.16)
One can see, that at fixed credit rates k_0 NPV increases with project duration. The (negative) slope of NPV(L) curves decreases with project duration.

Table 31.7 The dependence of the Net Present Value, NPV, on the leverage level L for the equity cost value $k_0 = 14\%$ and for debt capital costs $k_d = 0.04$; $S = 250$; tax on profit rate $t = 0.2$; investment profitability coefficient $\beta = 0.1$ and for project durations $n = 3$

S	L	WACC	k_0	k_d	w_d	t	n	β	k_e	NPV
250	0	14.00%	0.14	0.04	0.00	0.2	3	0.1	0.14	–203.57
250	1	13.34%	0.14	0.04	0.50	0.2	3	0.1	0.23	–346.60
250	2	13.11%	0.14	0.04	0.67	0.2	3	0.1	0.33	–491.78
250	3	13.00%	0.14	0.04	0.75	0.2	3	0.1	0.42	–637.53
250	4	12.94%	0.14	0.04	0.80	0.2	3	0.1	0.52	–783.49
250	5	12.89%	0.14	0.04	0.83	0.2	3	0.1	0.61	–929.57
250	6	12.86%	0.14	0.04	0.86	0.2	3	0.1	0.71	–1075.71
250	7	12.84%	0.14	0.04	0.88	0.2	3	0.1	0.80	–1221.89
250	8	12.82%	0.14	0.04	0.89	0.2	3	0.1	0.90	–1368.10
250	9	12.80%	0.14	0.04	0.90	0.2	3	0.1	0.99	–1514.33
250	10	12.79%	0.14	0.04	0.91	0.2	3	0.1	1.09	–1660.57

Table 31.8 The dependence of the Net Present Value, NPV, on the leverage level L for the equity cost value $k_0 = 14\%$ and for debt capital costs $k_d = 0.04$; $S = 250$; tax on profit rate $t = 0.2$; investment profitability coefficient $\beta = 0.1$ and for project durations $n = 5$

S	L	WACC	k_0	k_d	w_d	t	n	β	k_e	NPV
250	0	14.00%	0.14	0.04	0.00	0.2	5	0.1	0.14	−181.34
250	1	13.26%	0.14	0.04	0.50	0.2	5	0.1	0.23	−272.31
250	2	13.01%	0.14	0.04	0.67	0.2	5	0.1	0.33	−366.58
250	3	12.88%	0.14	0.04	0.75	0.2	5	0.1	0.42	−461.70
250	4	12.80%	0.14	0.04	0.80	0.2	5	0.1	0.51	−557.17
250	5	12.75%	0.14	0.04	0.83	0.2	5	0.1	0.61	−652.81
250	6	12.72%	0.14	0.04	0.86	0.2	5	0.1	0.70	−748.55
250	7	12.69%	0.14	0.04	0.88	0.2	5	0.1	0.79	−844.36
250	8	12.67%	0.14	0.04	0.89	0.2	5	0.1	0.88	−940.21
250	9	12.65%	0.14	0.04	0.90	0.2	5	0.1	0.98	−1036.09
250	10	12.64%	0.14	0.04	0.91	0.2	5	0.1	1.07	−1131.99

Table 31.9 The dependence of the Net Present Value, NPV, on the leverage level L for the equity cost value $k_0 = 14\%$ and for debt capital costs $k_d = 0.04$; $S = 250$; tax on profit rate $t = 0.2$; investment profitability coefficient $\beta = 0.2$ and for project durations $n = 3$

S	L	WACC	k_0	k_d	w_d	t	n	β	k_e	NPV
250	0	14.00%	0.14	0.04	0.00	0.2	3	0.2	0.14	−157.14
250	1	13.34%	0.14	0.04	0.50	0.2	3	0.2	0.23	−252.69
250	2	13.11%	0.14	0.04	0.67	0.2	3	0.2	0.33	−350.39
250	3	13.00%	0.14	0.04	0.75	0.2	3	0.2	0.42	−448.64
250	4	12.94%	0.14	0.04	0.80	0.2	3	0.2	0.52	−547.12
250	5	12.89%	0.14	0.04	0.83	0.2	3	0.2	0.61	−645.71
250	6	12.86%	0.14	0.04	0.86	0.2	3	0.2	0.71	−744.36
250	7	12.84%	0.14	0.04	0.88	0.2	3	0.2	0.80	−843.05
250	8	12.82%	0.14	0.04	0.89	0.2	3	0.2	0.90	−941.77
250	9	12.80%	0.14	0.04	0.90	0.2	3	0.2	0.99	−1040.51
250	10	12.79%	0.14	0.04	0.91	0.2	3	0.2	1.09	−1139.26

31.6 Conclusions

We study the role of the Central Bank and commercial banks in creating and maintaining of a favorable investment climate in the country. Within the framework of modern investment models created by the authors, the dependence of the efficiency of investments on the level of debt financing within a wide range of values of equity capital costs and debt capital costs under different project terms (long-term projects as well as projects of arbitrary duration) and different investment profitability coefficients β is investigated. The effectiveness of investments is determined by net present value (Net Present Value, NPV). The study is conducted within the framework of investment models with debt repayment at the end of the project term.

Table 31.10 The dependence of the Net Present Value, NPV, on the leverage level L for the equity cost value $k_0 = 14\%$ and for debt capital costs $k_d = 0.04$; $S = 250$; tax on profit rate $t = 0.2$; investment profitability coefficient $\beta = 0.2$ and for project durations $n = 5$

S	L	WACC	k_0	k_d	w_d	t	n	β	k_e	NPV
250	0	14.00%	0.14	0.04	0.00	0.2	5	0.2	0.14	–112.68
250	1	13.26%	0.14	0.04	0.50	0.2	5	0.2	0.23	–132.50
250	2	13.01%	0.14	0.04	0.67	0.2	5	0.2	0.33	–155.57
250	3	12.88%	0.14	0.04	0.75	0.2	5	0.2	0.42	–179.49
250	4	12.80%	0.14	0.04	0.80	0.2	5	0.2	0.51	–203.76
250	5	12.75%	0.14	0.04	0.83	0.2	5	0.2	0.61	–228.20
250	6	12.72%	0.14	0.04	0.86	0.2	5	0.2	0.70	–252.73
250	7	12.69%	0.14	0.04	0.88	0.2	5	0.2	0.79	–277.33
250	8	12.67%	0.14	0.04	0.89	0.2	5	0.2	0.88	–301.97
250	9	12.65%	0.14	0.04	0.90	0.2	5	0.2	0.98	–326.65
250	10	12.64%	0.14	0.04	0.91	0.2	5	0.2	1.07	–351.34

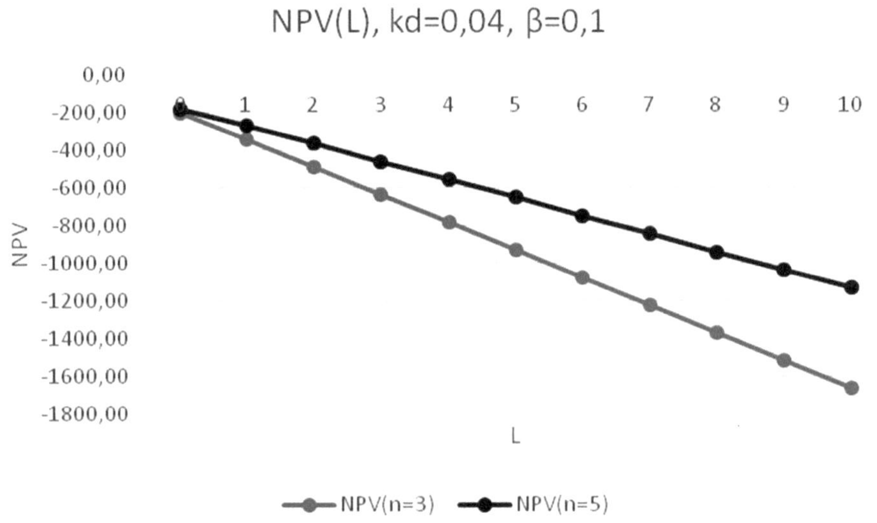

Fig. 31.13 The dependence of the Net Present Value, NPV, on the leverage level L for the equity cost value $k_0 = 14\%$ and for debt capital costs $k_d = 0.04$; $S = 250$; tax on profit rate $t = 0.2$; investment profitability coefficient $\beta = 0.1$ and for two project durations $n = 3$ and $n = 5$

It is found that NPV depends practically linearly on leverage level L, increasing or decreasing depending on profitability coefficient β and credit rate value k_d. The cut off credit rate values k_d* separating the range of increasing NPV(L) from range of decreasing NPV(L) are determined.

For long-term projects (Modigliani–Miller limit) it was found that the cut off credit rate values k_d* are proportional to investment profitability coefficients β: it turns out that for equity capital cost $k_0 = 0.2$ the cut off credit rate value k_d* separating the range of increasing NPV(L) from range of decreasing NPV(L) is

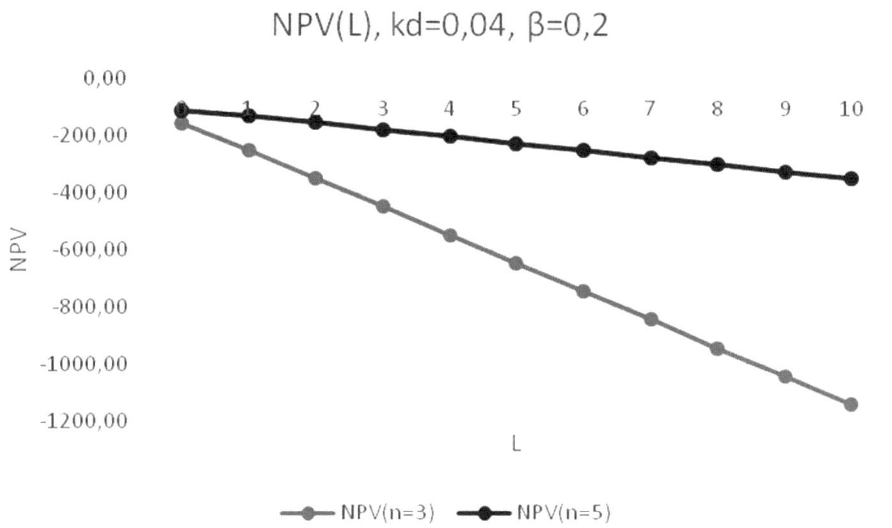

Fig. 31.14 The dependence of the Net Present Value, NPV, on the leverage level L for the equity cost value $k_0 = 14\%$ and for debt capital costs $k_d = 0.04$; $S = 250$; tax on profit rate $t = 0.2$; investment profitability coefficient $\beta = 0.2$ and for two project durations $n = 3$ and $n = 5$

Table 31.11 The dependence of the Net Present Value, NPV, on the leverage level L for the equity cost value $k_0 = 14\%$ and for debt capital costs $k_d = 0.1$; $S = 250$; tax on profit rate $t = 0.2$; investment profitability coefficient $\beta = 0.2$ and for project durations $n = 3$

S	L	WACC	k_0	k_d	w_d	t	n	β	k_e	NPV
250	0	14.00%	0.14	0.10	0.00	0.2	3	0.1	0.14	–203.57
250	1	12.51%	0.14	0.10	0.50	0.2	3	0.1	0.17	–377.91
250	2	12.01%	0.14	0.10	0.67	0.2	3	0.1	0.20	–557.74
250	3	11.76%	0.14	0.10	0.75	0.2	3	0.1	0.23	–739.00
250	4	11.61%	0.14	0.10	0.80	0.2	3	0.1	0.26	–923.85
250	5	11.51%	0.14	0.10	0.83	0.2	3	0.1	0.29	–1103.00
250	6	11.44%	0.14	0.10	0.86	0.2	3	0.1	0.32	–1285.31
250	7	11.39%	0.14	0.10	0.88	0.2	3	0.1	0.35	–1467.73
250	8	11.35%	0.14	0.10	0.89	0.2	3	0.1	0.38	–1650.23
250	9	11.31%	0.14	0.10	0.90	0.2	3	0.1	0.41	–1832.77
250	10	11.28%	0.14	0.10	0.91	0.2	3	0.1	0.44	–2015.35

approximately equal to investment profitability coefficient β: for investment profitability coefficient $\beta = 0.1$ k_d* is equal to 0.1; for $\beta = 0.12$ k_d* is equal to 0.12; and for investment profitability coefficient $\beta = 0.14$ k_d* is equal to 0.14. The slope of the curve NPV(L) increases with investment profitability coefficient β for the same value of credit rate k_d.

We come to conclusion that for long-term projects (in perpetuity limit) for both cases for the equity values $k_0 = 20\%$ and $k_0 = 28\%$ it turns out that the cut off credit rate values k_d* are equal to investment profitability coefficient β (and does not

Table 31.12 The dependence of the Net Present Value, NPV, on the leverage level L for the equity cost value $k_0 = 14\%$ and for debt capital costs $k_d = 0.1$; $S = 250$; tax on profit rate $t = 0.2$; investment profitability coefficient $\beta = 0.1$ and for project durations $n = 5$

S	L	WACC	k_0	k_d	w_d	t	n	β	k_e	NPV
250	0	14.00%	0.14	0.10	0.00	0.2	5	0.1	0.14	–181.34
250	1	12.41%	0.14	0.10	0.50	0.2	5	0.1	0.17	–317.91
250	2	11.88%	0.14	0.10	0.67	0.2	5	0.1	0.20	–462.97
250	3	11.61%	0.14	0.10	0.75	0.2	5	0.1	0.22	–610.30
250	4	11.45%	0.14	0.10	0.80	0.2	5	0.1	0.25	–758.57
250	5	11.34%	0.14	0.10	0.83	0.2	5	0.1	0.28	–907.31
250	6	11.26%	0.14	0.10	0.86	0.2	5	0.1	0.31	–1056.33
250	7	11.20%	0.14	0.10	0.88	0.2	5	0.1	0.34	–1205.52
250	8	11.16%	0.14	0.10	0.89	0.2	5	0.1	0.36	–1354.83
250	9	11.12%	0.14	0.10	0.90	0.2	5	0.1	0.39	–1504.22
250	10	11.09%	0.14	0.10	0.91	0.2	5	0.1	0.42	–1653.67

Table 31.13 The dependence of the Net Present Value, NPV, on the leverage level L for the equity cost value $k_0 = 14\%$ and for debt capital costs $k_d = 0.1$; $S = 250$; tax on profit rate $t = 0.2$; investment profitability coefficient $\beta = 0.2$ and for project durations $n = 3$

S	L	WACC	k_0	k_d	w_d	t	n	β	k_e	NPV
250	0	14.00%	0.14	0.10	0.00	0.2	3	0.2	0.14	–157.14
250	1	12.51%	0.14	0.10	0.50	0.2	3	0.2	0.17	–282.68
250	2	12.01%	0.14	0.10	0.67	0.2	3	0.2	0.20	–413.66
250	3	11.76%	0.14	0.10	0.75	0.2	3	0.2	0.23	–546.07
250	4	11.61%	0.14	0.10	0.80	0.2	3	0.2	0.26	–679.06
250	5	11.51%	0.14	0.10	0.83	0.2	3	0.2	0.29	–812.35
250	6	11.44%	0.14	0.10	0.86	0.2	3	0.2	0.32	–945.80
250	7	11.39%	0.14	0.10	0.88	0.2	3	0.2	0.35	–1079.36
250	8	11.35%	0.14	0.10	0.89	0.2	3	0.2	0.38	–1212.99
250	9	11.31%	0.14	0.10	0.90	0.2	3	0.2	0.41	–1346.67
250	10	11.28%	0.14	0.10	0.91	0.2	3	0.2	0.44	–1480.38

depend on the equity values k_0). This statement is not valid for the projects of arbitrary (finite) durations.

For arbitrary duration projects results are as folloing. The efficiency of investments strongly depends on project duration and increases with duration. One can see, that the slope of the curve NPV(L) at project duration $n = 5$ is always higher, than for project duration $n = 3$. The efficiency of investments increases with project duration and is less than for long-term (perpetuity) projects. Transition to increasing NPV(L) behavior for finite duration projects requires much higher values of investment profitability coefficient β than in case of long-term (perpetuity) projects, where k_d* is approximately equal to β: for example, for equity cost $k_0 = 0.20$ and 5 years projects the cut off credit rate values k_d* for investment profitability coefficient $\beta = 0.325$ is equal to 0.14; for investment profitability coefficient $\beta = 0.345$ is equal

Table 31.14 The dependence of the Net Present Value, NPV, on the leverage level L for the equity cost value $k_0 = 14\%$ and for debt capital costs $k_d = 0.1$; $S = 250$; tax on profit rate $t = 0.2$; investment profitability coefficient $\beta = 0.2$ and for project durations $n = 5$

S	L	WACC	k_0	k_d	w_d	t	n	β	k_e	NPV
250	0	14.00%	0.14	0.10	0.00	0.2	5	0.2	0.14	–112.68
250	1	12.41%	0.14	0.10	0.50	0.2	5	0.2	0.17	–175.18
250	2	11.88%	0.14	0.10	0.67	0.2	5	0.2	0.20	–246.02
250	3	11.61%	0.14	0.10	0.75	0.2	5	0.2	0.22	–319.09
250	4	11.45%	0.14	0.10	0.80	0.2	5	0.2	0.25	–393.09
250	5	11.34%	0.14	0.10	0.83	0.2	5	0.2	0.28	–467.55
250	6	11.26%	0.14	0.10	0.86	0.2	5	0.2	0.31	–542.29
250	7	11.20%	0.14	0.10	0.88	0.2	5	0.2	0.34	–617.20
250	8	11.16%	0.14	0.10	0.89	0.2	5	0.2	0.36	–692.22
250	9	11.12%	0.14	0.10	0.90	0.2	5	0.2	0.39	–767.32
250	10	11.09%	0.14	0.10	0.91	0.2	5	0.2	0.42	–842.47

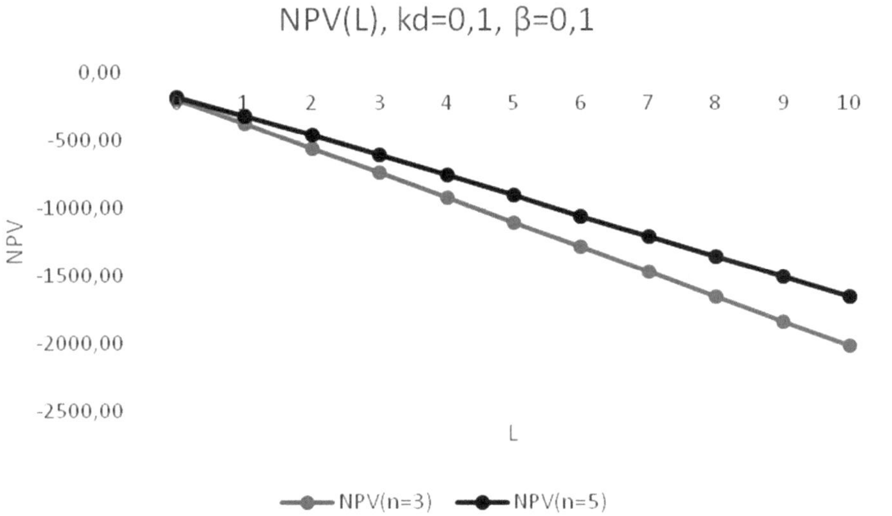

Fig. 31.15 The dependence of the Net Present Value, NPV on the leverage level L for the equity cost value $k_0 = 14\%$ and for debt capital costs $k_d = 0.1$; $S = 250$; tax on profit rate $t = 0.2$; investment profitability coefficient $\beta = 0.1$ and for two project durations $n = 3$ and $n = 5$

to 0.16; for investment profitability coefficient $\beta = 0.365$ is equal to 0.18. Thus, for finite duration projects as well as for the long-term projects cut off credit rate values k_d* turn out to be proportional to investment profitability coefficients β, but investment profitability coefficients β, is approximately twice higher than k_d*.

We develop a method of determination of the cut off credit rate values k_d*, separating the range of increasing NPV(L) from range of decreasing NPV(L). We have found the cut off credit rate k_d* values within a wide range of values of equity costs k_0 and debt capital costs k_d under different project terms (long-term projects as

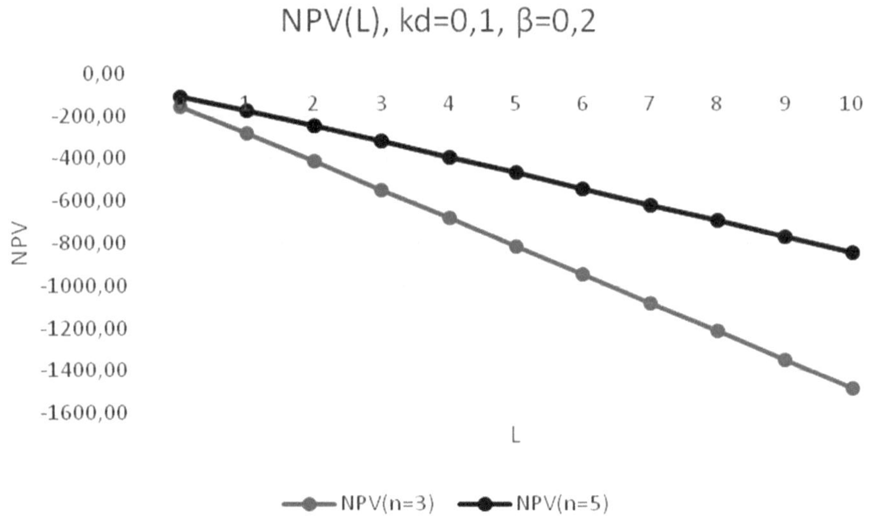

Fig. 31.16 The dependence of the Net Present Value, NPV, on the leverage level L for the equity cost value $k_0 = 14\%$ and for debt capital costs $k_d = 0.1$; $S = 250$; tax on profit rate $t = 0.2$; investment profitability coefficient $\beta = 0.2$ and for two project durations $n = 3$ and $n = 5$

well as projects of arbitrary duration) and different investment profitability coefficients β. Obtained results will help to the Central Bank keep its key rate at the level, which allow commercial banks keep their credit rates below the cut off credit rate values k_d* in order to create and maintain a favorable investment climate in the country.

References

Brusov, P. N., & Filatova, T. V. (2011). From Modigliani–Miller to general theory of capital cost and capital structure of the company. *Finance and Credit, 435*, 2–8.

Brusov, P., Filatova, T., Orehova, N., & Brusova, A. (2011a). Weighted average cost of capital in the theory of Modigliani–Miller, modified for a finite life–time company. *Applied Financial Economics, 21*(11), 815–824.

Brusov, P., Filatova, T., Orehova, N., et al. (2011b). From Modigliani–Miller to general theory of capital cost and capital structure of the company. *Research Journal of Economics, Business and ICT, 2*, 16–21.

Brusov, P., Filatova, T., Orehova, N., et al. (2011c). Influence of debt financing on the effectiveness of the investment project within the Modigliani–Miller theory. *Research Journal of Economics, Business and ICT, 2*, 11–15.

Brusov, P., Filatova, T., Eskindarov, M., & Orehova, N. (2012a). Influence of debt financing on the effectiveness of the finite duration investment project. *Applied Financial Economics, 22*(13), 1043–1052.

Brusov, P., Filatova, T., Eskindarov, M., & Orehova, N. (2012b). Hidden global causes of the global financial crisis. *Journal of Reviews on Global Economics, 1*, 106–111.

Brusov, P., Filatova, P., & Orekhova, N. (2013a). Absence of an optimal capital structure in the famous tradeoff theory! *Journal of Reviews on Global Economics, 2*, 94–116.
Brusov, P., Filatova, T., & Orehova, N. (2013b). A qualitatively new effect in corporative finance: Abnormal dependence of cost of equity of company on leverage. *Journal of Reviews on Global Economics, 2*, 183–193.
Brusov, P., Filatova, P., & Orekhova, N. (2014a). Mechanism of formation of the company optimal capital structure, different from suggested by trade off theory. *Cogent Economics & Finance, 2*, 1–13. https://doi.org/10.1080/23322039.20946150
Brusov, P., Filatova, T., & Orehova, N. (2014b). Inflation in Brusov–Filatova–Orekhova theory and in its perpetuity limit – Modigliani – Miller theory. *Journal of Reviews on Global Economics, 3*, 175–185.
Brusov, P., Filatova, T., Orehova, N., & Eskindarov, M. (2015). *Modern corporate finance, investments and taxation* (monograph) (1st ed., 368 pp). Springer Publishing.
Brusov, P., Filatova, T., Orehova, N., & Kulk, W. I. (2018a). New meaningful effects in modern capital structure theory. *Journal of Reviews on Global Economics, 7*, 104–122.
Brusov, P., Filatova, T., Orehova, N., & Kulk, V. (2018b). A "golden age" of the companies: Conditions of its existence. *Journal of Reviews on Global Economics, 7*, 88–103.
Brusov, P., Filatova, T., Orehova, N., & Kulk, V. (2018c). Rating methodology: New look and new horizons. *Journal of Reviews on Global Economics, 7*, 63–87.
Brusov, P., Filatova, T., Orehova, N., & Kulk, V. (2018d). Rating: New approach. *Journal of Reviews on Global Economics, 7*, 37–62.
Brusova, A. (2011). A comparison of the three methods of estimation of weighted average cost of capital and equity cost of company. *Financial Analysis Problem Solution, 34*(76), 36–42.
Filatova, T., Orehova, N., & Brusova, A. (2008). Weighted average cost of capital in the theory of Modigliani–Miller, modified for a finite life-time company. *Bulletin of the FU, 48*, 68–77.
Modigliani, F., & Miller, M. (1958). The cost of capital, corporate finance, and the theory of investment. *The American Economic Review, 48*, 261–297.
Modigliani, F., & Miller, M. (1963). Corporate income taxes and the cost of capital: A correction. *The American Economic Review, 53*, 147–175.
Modigliani, F., & Miller, M. (1966). Some estimates of the cost of capital to the electric utility industry 1954–1957. *The American Economic Review, 56*, 333–391.

Part IX
Conclusions

Chapter 32
Conclusions

Part I consists of two Chapters. The introduction (Chap. 1) discusses the shortcomings of existing methods for assessing the effectiveness of investment projects and describes innovative investment models created by the authors recently and well tested in practice. The structure of the monograph is described and the characteristics of the nine parts of which it consists are given. In Chap. 2 the main performance indicators of the investment project are described and compared. Among them: the payback period, discounted payback period, investment efficiency ratio, net present value, NPV, profitability index (profitability) of investments, internal rate of return (IRR), modified internal rate of return (MIRR). The concept of internal rate of return is considered in details.

One of the main parameters used in assessing the effectiveness of investment projects is the discount rate. **Part II** deals with the state of the theory of capital structure and cost of capital, which are used for valuation the discount rates. Two main theories of Modigliani–Miller (MM) and Brusov–Filatova–Orekhova (BFO) are discussed and analyzed, as well as their numerous modifications and generalizations (Brusov & Filatova, 2025, Brusov et al. 2023a, b, 2024a, b, c). Also discussed is the latest stage in the development of the theory of capital structure, which began a couple of years ago and is associated with the adaptation of two main theories of capital structure (Brusov–Filatova–Orekhova and Modigliani–Miller) to the established practice of the functioning of companies. This generalization takes into account the real working conditions of companies.

In **Part III** the modern investment models created by the authors and well tested in the real economy are described. These models are based on the first principles and are used in assessing the effectiveness of investment projects and to study various investment problems, such as the impact of debt financing, leverage, taxation, project duration, financing method and some other parameters on investment efficiency and other problems. But we use them in Chap. 18 to modify the methodology of project ratings. In this chapter we look at investment models with debt repayment at the end of the project and their application, and in Chap. 16 we look at investment models with equal debt repayment and their application.

P. Brusov, T. Filatova, *Efficiency of Investment Projects*, Contributions to Finance and Accounting, https://doi.org/10.1007/978-3-032-01028-5_32

Part IV describes innovative investment models with frequent payments of income tax and interest on debt. Chapter 17 describes investment models with frequent payments of income tax and interest on debt at the end of periods. Chapter 18 describes investment models with advance frequent payments of income tax and interest on debt. The results of two these models are completely different.

Part V discusses portfolio investments. A portfolio of securities is a collection of different financial assets owned by an investor, including stocks, bonds, mutual funds, and other investments. The purpose of a portfolio is to diversify investments, balance risk, and help achieve specific financial goals, such as capital growth or income. In Chap. 19 a portfolio of securities is considered. A detailed analysis of a portfolio of two and three securities is carried out. Markowitz portfolios are considered, as well as Tobin portfolios of n–securities. In Chap. 20 a portfolio of bonds is discussed in details.

Part VI is devoted to Ratings, Rating Methodologies and Ratings of Investment Projects.

Rating agencies play a very important role in economics. Their analysis of issuer's state, generated credit ratings of issuers help investors make reasonable investment decision, as well as help issuers with good enough ratings get credits on lower rates etc.

This Part suggests a new approach to rating methodology. Application of the Modigliani–Miller Theory in Rating Methodology is discussed in Chap. 21, while Chap. 22 is devoted to the application of the Modigliani–Miller Theory, modified for the case of advance payments of tax on profit, in rating methodologies.

In Chap. 23 a new approach to project ratings is described. The important features of current consideration as well as in previous studies are: (1) The adequate use of discounting of financial flows virtually not used in existing rating methodologies, (2) The incorporation of rating parameters (financial "ratios"), used in project rating, into considered modern investment models. This on the one hand allows use the powerful tools of this theories in the rating, and on the other hand it ensures the correct discount rates when discounting of financial flows. We discuss also the interplay between rating ratios and leverage level which can be quite important in rating. All these create a new base for rating methodologies. New approach to ratings and rating methodologies allows to issue more correct ratings of issuers, makes the rating methodologies more understandable and transparent.

In Chap. 24 further development of a new approach to project ratings has been done: it has been generalized for the general case of modern theory of capital structure (Brusov–Filatova–Orekhova (BFO) theory), which is valid for companies of arbitrary age.

Ratings of the investment projects of arbitrary durations is studied in Chap. 25.

In Chap. 26 ratings of investment projects of arbitrary duration with a uniform debt repayment is described.

Part VII is devoted to description of qualitatively new effects in investments, discovered by authors. In Chap. 27 a problem whether it is possible to increase taxing and conserve a good investment climate in the country is studied.

In Chap. 28 an abnormal influence of growth of tax on profit rate on the efficiency of the investment has been discovered and the answer on the question "Whether it is possible to increase of the investment efficiency, increasing tax on profit rate?" has been done.

Part VIII is devoted to the application of modern investment models created in this monograph in practice.

Chapter 29 conducts an analysis of the study of the effectiveness of investment projects of arbitrary duration (within the framework of the Brusov–Filatova–Orekhova theory). Chapter 30 studies the optimizing the investment structure of the telecommunication sector company. The role of the Central Bank and commercial banks in creating and maintaining a favorable investment climate in the country is examined in Chap. 31.

In **Part IX** Conclusions for the entire monograph are made in Chap. 32.

References

Brusov, P., & Filatova, T. (2025). *Business valuation and company value: New qualitative approaches considering business and financial risks* (850 pp). Springer Nature.

Brusov, P., Filatova, T., & Orekhova, N. (2023a). *The Brusov–Filatova–Orekhova theory of capital structure: Applications in corporate finance, investments, taxation and ratings* (769 pp). Springer Nature.

Brusov, P. N., Filatova, T. V., & Orekhova, N. P. (2023b). *Financial mathematics* (250 pp). Cambridge Scholars Publishing.

Brusov, P. N., Filatova, T. V., & Kulik, V. L. (2024a). Capital Asset Pricing Model (CAPM) 2.0: Account of business and financial risk. *Finance: Theory and Practice, 28*(2), 128–142. https://doi.org/10.26794/2587-5671-2024-28-2-128-142

Brusov, P. N., Filatova, T. V., & Kulik, V. L. (2024b). Application of the company's "Golden age" effect in the economic practice. *Finance: Theory and Practice, 28*(3), 61–83. https://doi.org/10.26794/2587-5671-2024-28-3-61-83

Brusov, P. N., Filatova, T. V., & Orekhova, N. P. (2024c). *Financial mathematics: Exercises and solutions* (275 pp). (Springer Texts in Business and Economics) Springer Nature.

MIX
Papier aus verantwortungsvollen Quellen
Paper from responsible sources
FSC® C105338

If you have any concerns about our products,
you can contact us on
ProductSafety@springernature.com

In case Publisher is established outside the EU,
the EU authorized representative is:
Springer Nature Customer Service Center GmbH
Europaplatz 3, 69115 Heidelberg, Germany

Printed by Libri Plureos GmbH
in Hamburg, Germany